POLLUTION ENGINEERING PRACTICE HANDBOOK

Paul N. Cheremisinoff, P.E.
Associate Professor of Environmental Engineering
New Jersey Institute of Technology
Newark, New Jersey

Richard A. Young
Editor, Pollution Engineering
Technical Publishing Company
Barrington, Illinois

Editors

ANN ARBOR SCIENCE
PUBLISHERS INC / THE BUTTERWORTH GROUP

Second Printing, 1976
Third Printing, 1981

Copyright © 1975 by Ann Arbor Science Publishers, Inc.
P. O. Box 1425, Ann Arbor, Michigan 48106

Library of Congress Catalog Card No. 74-14427
ISBN   0-250-40075-8

Manufactured in the United States of America
All Rights Reserved

PREFACE

Pollution in its many forms—air, water, noise and solid waste—can be controlled. There are, however, two major problems which continually confront the individuals responsible for energy/environmental control—**economics and information.** Each is directly proportional to the other.

No one wants to spend money for the sake of spending money. Most of the time, pollution control means spending and not receiving any direct economic replacement for company funds. Therefore, it is absolutely necessary to perform a total engineering analysis of any pollution problem to determine the single most effective and economic solution. This means pollution engineers and managers must have information to do a thorough job.

The methods and equipment to control pollution are largely old techniques—some technology dates back more than 100 years. What has really been lacking in the field of environmental control is the broad dissemination of available information on the tried and proven methods. When an engineer is required to find the most economical and effective solution to the emission of environmental pollutants, he really wants essentially one thing—practical information, and right now!

That is the type of information that the editors have attempted to provide in this handbook—data that discusses the "how to do it" methods of practical problem solving. Much of the information has been derived from articles previously published in *Pollution Engineering.* Many of these articles were written by the editors of this book in their capacities as editors of the magazine. Additional valuable information has been contributed by other authorities in the field.

The purpose of this handbook is to furnish engineers, managers and students with guidance and direction in this rapidly growing field. Technically and economically feasible techniques, methods, equipment and systems are assessed.

To become an effective pollution engineer, you must take the time to analyze the whole problem along with all its many interrelationships. Then, you must systematically select the commercially available equipment to abate the problem considering all the economics involved—immediate, short and long range.

Paul N. Cheremisinoff
Richard A. Young

INTRODUCTION

The United States, as a modern industrialized nation, is the product of constantly advancing technology. Our mobility, our affluence, and our high overall standard of living are manifestations of our technological progress. So is pollution.

Technology can and does create pollution. Fortunately, it can also be applied to control and abate the contamination of our environment. The scientists and engineers who once designed industrial processes that spewed out poisonous wastewater and smoke are now developing new processes that function just as efficiently without polluting our air and water. The costs of this technology are more than offset by the immediate benefits to public health and the general improvement in what we have come to call the "quality of life."

For example, our scientists and engineers have developed substantially improved auto emission controls and stack gas cleaning technologies to curb choking fumes and noxious air pollutants from motor vehicles and power plants. Technology also is being applied to clean up industrial and municipal wastewater discharges, to suppress noise and to recycle solid waste. Resource recovery systems, a number of which are being developed with EPA demonstration grants, are perhaps the ultimate example of effective pollution engineering since they take a problem and turn it into a profit. The metals, glass and energy (combustibles) that are recovered from our waste, along with the savings from reduced landfill operations, can actually enable these systems to pay for themselves.

In controlling pollution, whether by establishing discharge standards for new sources or compliance schedules for existing facilities, improvements in technology must and will be a driving force in achieving our environmental goals. Without a strong commitment from our scientific and engineering community, our capability to utilize "best practicable" and "best available" technology will turn out to be a retreat to the lowest common denominator. That would be inadequate, derelict and a sham.

EPA has a definite responsibility to stimulate and support the necessary scientific research and application of the research results to technology and thus provide a solid foundation for control standards and regulations that are economically and technically realistic. Beyond any doubt, poorly drawn standards and regulations are a serious disincentive for industry to accelerate research and development efforts in pollution control.

But "best practical" and "best available" are not absolute terms, nor does the setting of standards dictate the manner in which we should go about the task of meeting them. The degree to which we can improve the quality of our environment is directly proportional to the extent to which we find efficient and economical solutions to our environmental problems. A pollution control system with a low energy factor is probably going to be preferable to a system that imposes a heavy fuel penalty. A waste disposal system that recovers valuable resources at a competitive cost should be preferable to one that doesn't. The logic is as obvious and overwhelming today as it was in the folk adage about what will happen when a man builds a better mousetrap.

Finally, we should avoid becoming wedded to any particular system or device as the definitive achievement or form of pollution control. The most valuable component of our scientific and technological sector has been and still is its creativity. An environment that encourages imaginative and inventive approaches to problem solving is going to have the best chance to solve our environmental problems. And our ability to find efficient, economical and practicable solutions to our environmental problems will be determined in large part by how effectively we apply the methodology and resources that are available to us. There is no greater waste than the waste of knowledge.

Russell E. Train
Administrator
Environmental Protection Agency

TABLE OF CONTENTS

CHAPTER 1

ORGANIZATION AND ADMINISTRATION
OF POLLUTION ENGINEERING AND CONTROL

POLLUTION CONTROL INTERRELATIONSHIPS[3,1]

Current social concern for protecting and preserving the environment
is leading to increased regulation of production. Because of pollution standards and the prospect of waste handling surcharges, it is no longer possible
to consider a pollution abatement problem in isolation. It has become
necessary to determine the impact control of one problem will have on
other parts of the environment. Fundamentally, pollution falls into four
broad categories: air pollution, water pollution, solid waste and noise
pollution. To a greater or a lesser degree, these are all interrelated, the
exact nature of the interrelationship depending on the firm's business, the
degree of control needed, and the predominant pollution problem.

Many polluters are trying to meet just the minimum standards now
in force, but this is only a short-run solution to the problem and may be
suboptimum from a long-range viewpoint. The optimum solution, on the
other hand, is generally a long-range one, incorporating a total system of
control.

Wet collection equipment in air pollution control more often than
not results in a wastewater problem, heretofore widely ignored. Therefore,
forthcoming standards for wastewater discharges will place additional
technical and economic burdens on air pollution control. Liquid wastes
resulting from air pollution control devices can no longer be discharged to
surface waters or sewers without critical review. (Actually, it has never
been realistic to solve an air pollution problem by transferring the undesirable pollutants to a liquid and discharging them to streams or sewers.)
Table 1-1 shows interrelationships and the need for simultaneous consideration of air and water quality requirements.

1

Table 1-1. Simultaneous Air-Water Quality Requirements

Parameters	Recommended Concentrations for Effluents to Surface Streams	Threshold Limit Values of Airborne Contaminants	Primary Ambient Air Quality Standards 1975
pH	6.5-9.0	(HNO_3) 5-mg/M^3, (H_2SO_4) 1 mg/M^3	
Suspended Solids	20.0	(Graphite) 15 mg/M^3	(Particulates)
Cd	0.05	0.2	0.075 mg/M^3 *
Cr	0.5	0.5	
Pb	0.1	0 0.2	
Cu	0.5	0.1 (fume), 1.0 (dust)	
Fe	2.0	1 (sol. iron salts), 10 (FeO fumes)	
Hg	0.5 mg/M^3	0.05, (0.01-alkyl) mg/M^3	
NH_3	2.5	25	
Sulfate	250		(SO_2) 0.14 **
Chloride	150	(chlorine) 1.0	
Fluoride	1.5	2.5 mg/M^3, (fluorine) 0.1	
Nitrate	45.0	(NO_2) 5.0	(NO_x) 0.05 **
Oil	10.0	(mist) 5.0 mg/M^3	

Values in ppm unless otherwise stated
* Annual geometric mean
** Annual arithmetic mean

Wet Collectors

A large variety of wet collectors is used for cleaning, cooling, and deodorizing of gas, particulate, and vapor emissions to the atmosphere. All these can contribute to the total quantity of liquid wastes discharged to waterways. By way of example, scrubbing liquids may include alkali additives for sulfur dioxide and hydrogen sulfide acid mist control or acids for ammonia recovery. These variations can also pose related problems in the selection of construction materials and auxiliary treatment and/or disposal systems.

Some of the more common wet-type air control apparatus include packed bed absorption columns, high-energy venturi scrubbers, wet-type precipitators and filtration systems.

High-energy wet inertial scrubbers are used where the scrubbing liquid polluted with the material can be recirculated. Electrostatic precipitators and fabric filters are used for particle removal and where the availability of water and disposal of it may present problems.

In venturi scrubber usage, if the sludge is disposed of without clarification, the operating costs will be proportionately low. But, capital cost of the venturi system may exceed that of an electrostatic precipitator system if sludge clarification and collection are mandatory.

Particulate Control

Particulate matter has traditionally been thought of as the principal air pollutant. Industrial processes, including industrial fuel burning, are responsible for about 50 percent of the particulate discharge in the form of dust, fumes, smoke or mist. Chief among the industries contributing to this discharge are gray iron foundries, steel mills, cement plants, petroleum refineries and paper mills. Almost invariably, the control methods considered include wet collectors—packed absorption columns, venturi scrubbers, mist eliminators or some other device. Some typical industrial applications of wet scrubbers for particulate removal are shown in Table 1-2.

Table 1-2. Typical Industrial Applications of Control Apparatus

Type	Typical Application
Spray Tower	Blast Furnace Gas, Fume Control
Spray Chambers	Smoke Abatement, Dust Cleaning, Kraft Paper
Venturi Throat	Cupolas, Foundries, Flue Gas, Abrasives, Rotary Kilns
Multiple Jet Venturi	Flyash, Coke Oven Gas, Lime Kilns
Flooded Dish Venturi	Pulverized Coal
Mechanically Induced Spray	Cupolas, Smoke, Iron Foundry
Disintegrator	Blast Furnace Gas
Mist Eliminators	Coke Quenching
Fabric Filter	Oil Mists, Radioactive and Toxic Dusts
Wetted Filters	Light Dust Loadings

The Steel Industry

Steel plants emit flue gases containing dust that is primarily oxides of iron. Sources of this dust include basic oxygen furnaces, blast furnaces and sinter plant gas washers. The larger particles are removed in gravity separators, whereas most of the fines are removed by wet scrubbers and wet-type electrostatic precipitators.

Dust from basic oxygen furnaces forms a substantial fraction of the particulate discharge from steel mills. Wet precipitators or venturi scrubbers may be used to intercept iron oxide dust generated when oxygen is blown into molten iron. Gas wash water is then stripped of coarse particles by cyclone separators or dragout tanks. Gases from open hearth furnaces may also contain high dust concentrations. These are usually treated by scrubbing. Dust particles collected in scrubbing liquors are sufficiently fine to require chemical coagulation prior to clarification. Electric furnace and coke oven gas also produce high dust loadings. The scrubbing waters contain high suspended solids that also require clarification before reuse.

In the foundry shop, adequate ventiliation is required to hold down airborne dust. The air is usually scrubbed by water spray units which will generate large amounts of suspended solid matter in the scrubbing liquid.

Thus, water effluents from furnace cooling, wet-type precipitators and wet scrubbers are thoroughly polluted. Primary clarification and settling can remove the major portion of the particulate matter from these effluents. Subsequent thickening and dewatering is then often employed to ease the handling of sludge. If gas wash water is in a closed-loop system for recycle, wash water flow can be reduced substantially so that the treatment volume is cut in hydraulic loading and flow. Recycling, however, may require a cooling system and acid treatment to stabilize the pH of the wash water to 7.0, since alkalinity will be caused by the build-up. From this discussion it can be seen that scrubbing systems in steel plants should be tied into the water treatment system. Therefore, any economic comparison of wet scrubbers with dry-type collection equipment should include part of the cost of wastewater effluent treatment.

Foundries

Gray iron foundry flue gases contain metal fumes, grease, dust and iron oxide in the form of particulate matter. Cupolas and core making and shakeout systems are the primary sources of emission. When trying to control these atmospheric contaminants, water pollution problems emerge. Scrubbing systems on cupola stacks and dust collection systems can capture a considerable amount of airborne solids. These contaminants are physically discrete, and settling lagoons can be used for sedimentation of the solids from the scrubber water. When mechanical equipment is employed for coarse solids removal, it can be followed by coagulation, clarifying, sludge dewatering and sludge collection.

Calcium Carbide Plants

Dust control in these types of plants utilizes wet collection devices, and almost all calcining kiln dust is water scrubbed. Electric furnace fumes are drawn in through wet scrubbers where water sprays remove the particulate matter. Baghouses can be used to collect fine dust from carbide crushing operations. This dust is then water-mixed and requires processing before disposal. Clarification produces sludge that may be stored in lagoons. Clarifier overflow is mixed with cooling water to reduce the solids concentration before discharge to sewers.

Petroleum Refineries

Sludges from petroleum refineries are a mixture of petroleum residue and inorganics. These sludges are generally atomized, and the organic matter burned in an incinerator. The unburned organics escape in exhaust gases and are collected in high-energy scrubbers. Air blowing of asphalts generates oil and tar mists, which are usually scrubbed with sea water. These waters may be passed through separators to reclaim tar and oil.

Other Industries

In pulp mills some of the more important sources of particulate matter are lime kilns and smelt tanks. Stack dust from lime kilns is collected by 85 to 95 percent efficient venturi scrubbers. Water sprays of 20 to 30 percent efficiency and mesh demisters of 80 to 90 percent efficiency are used on smelt tanks.

Paint-spraying operations can produce particulate matter that can be removed in wet air scrubbers. The water used as the scrubbing medium may be treated in the waste treatment plant.

A considerable amount of dust is produced by polishing and grinding operations in the plating industry. Water sprays are normally used for dust removal. Wet-type centrifugal separators are also used. Both systems result in substantial water pollution problems.

Water disposal from wet scrubbers is dependent on many factors, such as the quantity of the waste, slurry particle distribution, solids loading, recovery value and the corrosiveness of the solution. Some typical methods are compared in Table 1-3 and illustrated in Figure 1-1.

Gaseous and Vapor Contaminant Control

Many processes include extensive washing of a wide variety of products. Such washes usually result in dissolved chemicals in the effluents. Pulp and paper washes may contain organics, pulp particulates and dissolved chemicals; plating washing may result in dissolved metals, cyanides and chromates.

Any of the following methods may be used for removal of such dissolved chemicals, the choice depending on the concentrations, the flow rates, the space requirements, etc.: chemical precipitation, adsorption, ion exchange, solvent extraction, chemical oxidation, reverse osmosis, crystallization and dialysis. The potential water pollution from such effluents, including that from air pollution control equipment, cannot be underestimated.

The Steel Industry

Coke oven gas that is washed with water to remove dust must also be treated with acid to remove ammonia. It is treated with oils to absorb C_6H_6, $C_6H_5CH_3$ and $C_6H_4(CH_3)_2$, which are more valuable as by-products than as fuel. Sulfur compounds are sometimes removed by sodium carbonate. Such gas-cleaning systems result in an ammoniacal liquor containing cyanates, phenols, spent acid, spent caustic and spent adsorber solution, such as carbonate.

Table 1-3. Comparison of Solids Removal Methods from Wastewater Effluents

Method	Application	Advantages	Disadvantages
Settling (Sedimentation)	Readily settling solid particulates, 1 micron size, easily flocculated matter, design scale-up possible from lab tests.	Wastewater chemically treated and re-used, low operating costs, abrasive solids handled.	Large areas, compared to volume of particulate, required; pond ground water contamination; natural evaporation.
Liquid Cyclones	Concentration of solids.	Low initial cost, low maintenance, ability to handle abrasive solids, low space requirements.	High solids filtrate on overflow, high power requirements.
Continuous Centrifuge	Submicron slurries.	High collection efficiencies, low space requirements, large design variety	High capital and operating costs, susceptible to abrasion and corrosion.
Continuous Filtration	Where solids have some recovery value, are porous or are incompressible	Dewatered waste product, moderate space requirements	High initial cost, high maintenance and operational costs.

Note: Chemical treatments for liquid wastes include treatments with lime, soda ash, coagulants, carbon dioxide and corrosion-inhibitors.

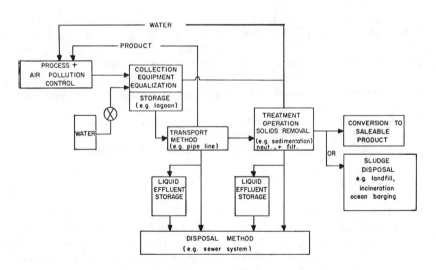

Figure 1-1. Generalized Flowsheet–Disposal of particulates
collected by air pollution control equipment.

Metal Fabricating

Surface-coating operations can produce organic solvent vapors. In paint-stripping operations, one of the more common chemical or solvent stripping methods is molten salt stripping, operated at 850 F. Gas-fired immersion burners are used for tank heating and immersion in the salt bath is followed by a hot water rinse. Mist and fumes generated are scrubbed by water. The effluent from the scrubbers has to be sent for treatment, even if it is only simple gravity skimming. Skimmer tanks that receive effluents at a uniform rate settle down the metal particles and remove free oils. If soluble oils are also present, further treatment—acid or alum addition—may be necessary. This treatment will adjust the pH, thus breaking down the emulsion. The oil is separated, and the remaining water is neutralized.

Plating

Fumes from nitric acid—NO and NO_2—sulfuric acid fumes, and gaseous hydrogen chloride are all given off during pickling, descaling and metal cleaning. Vapors of solvents from paint solvents and cleaners are also encountered in various electroplating operations. Fume washers and packed absorption columns using caustic will result in water pollution problems. However, since most plating operatins yield direct water wastes, contaminated water may be added to them for treatment prior to their discharge. Neutralization, separate chromium and cyanide waste treatment in plant process controls—such as solution regeneration and rinse water reconcentration—suspended solids separation and sludge handling constitute a procedure for plating wastewater treatment. Heavy metal removal by pH control, precipitation, or ion exchange may also be used.

Paper Mills

An example of treatment in a paper mill is an operation where organics are removed from the effluent, after evaporation in a fluid bed reactor. The combustion products are passed through a cyclone and a wet scrubber, so that an exhaust free of sulfur compounds is vented to the atmosphere. Weak liquor used in the wet scrubber as wash liquor goes to storage and is periodically dumped after dilution.

Sometimes water treatment may produce an air pollutant. For example, the treatment lagoon of a paperboard company gave off a terrible odor. A chemical oxidizer such as chlorine is turbulently mixed to eliminate the odor. Thus, an air pollution odor nuisance is controlled by an additional treatment step.

Electric Power Plants

Sulfur dioxide is removed from stacks at Bottusea and Bankside Stations in London. Alkaline river water is used for washing gas on grid scrubbers. The addition of manganese to scrubber effluent activates sulfite oxidation. Settling removes excess solids, and the final effluent with calcium sulfate in solution is diluted with cooling water and discharged to rivers. Such treatment would not, however, be acceptable for all discharges to fresh water bodies.

Incineration

Incineration is a principal control method used for solid and liquid wastes and atmospheric discharges. The wastewater from incinerator operations, however, requires greater attention than it has heretofore received, because of strict stream discharge standards and also the need to recycle waters. Recirculation will result in corrosion and erosion problems if the waters are not treated.

Even if there is no recirculation, stream and sewer regulations dictate that high thermal loads as well as suspended solids be reduced prior to discharge of the waters. For garbage-burning incinerators, water consumption is on the order of 0.25 to 5.0 gal/lb of refuse, the amount depending on the extent of recirculation. Air pollution control devices normally used are high-energy scrubbers, sprays and wetted baffle walls. A large variation of characteristics and quantity can be expected in the liquid waste effluent.

Combined Air-Water Pollution Control

Some practical examples involving a combined air/water pollution control approach follow.

Fat Hydrolysis

Entrained fatty acid droplets in exhaust flue gas cause an odor nuisance. A venturi scrubber is used to condense the droplets, which can be skimmed and recovered if a fat trap is used allowing the material to rise to the surface. At one plant, even small ppm concentrations in waters combined with sea water coalesced into clabber. An aeration-flotation unit was employed. Calcium hydroxide was reacted with fatty acids; air was pumped into the solution; and the mixture was sent under pressure to a settling tank through a distribution ring. Release of the pressure caused air to come out in fine bubbles, and insoluble heavy metal soaps were then skimmed off (see Figure 1-2).

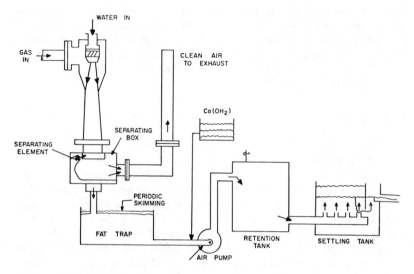

Figure 1-2. Hydrolysis of Fats—Collection of fat droplets and their disposal.

Ammonia Removal

How atmospheric, as well as steam contaminant, abatement could be brought to economic use is illustrated by the case of a Tennessee Valley Authority ammonia plant. To purify the nitrogen hydrogen mixture, ammoniacal copper solution was employed as a scrubbing agent. On heating for regeneration, ammonia and other nuisance gases were given off. The gases were combustible and could be used as fuel in the furnace, provided the ammonia was removed. The scrubbing operation resulted in a 5 percent $(NH_4)_2CO_3$ solution. Since it was not desirable to dump this solution into waterways, it was used to keep a condenser spray water pH at about 5.5 to inhibit corrosion. Thirty-six percent of the ammonia previously used as NH_4OH for this purpose was saved. The cleaned gases were sent to the furnace as a fuel. Thus combined, systems to clean air and water resulted in substantial economies.

Incinerator Process Waters

In attempting to characterize the incinerator process, a single-furnace unit operating at 300 tons/day with continuous feed, located at the Philadelphia Suburban Water Company, is a good example. After passing through a cooling water spray, the furnace gas enters a medium-energy scrubber, and cleaned flue gas is ejected from two stacks. The scrubber has a series of flooded, perforated plates separated by upward- and downward-directed sprays. Impingement plates help remove the particulate matter. Scrubber water is drained to the quench tank and then goes to a clarifier (110-min retention period). The supernatant flows to two lagoons (90-min retention period), passes through self-cleaning strainers, and then is recirculated to the scrubber and cooling water sprays. The scrubber uses about 4 gal of water per lb of refuse burned.

Acidic Exhausts

Many industries have processes that use acids in one form or another. Nitric, hydrochloric and sulfuric acids are widely used. HCl in small amounts is scrubbed, usually in packed bed columns that are hooked to the process source. The exhaust through the scrubber stack is then acid-free. However, spent water effluent from the scrubber is acidic and requires neutralization. This is normally provided by pumping the water to a sump tank, where an automatic alkali addition is made to adjust pH. The water leaving has a pH of 6.5 to 7.5 and can be safely discharged to sewers. In some instances, where water recirculation is desirable or practical, a possible acidic build-up in the scrubbing water is taken care of through automatic alkali adjustment. The water is, therefore, essentially neutral; and if discharge is necessary, it can be made without further neutralization. Figure 1-3 illustrates an HCl mist vapor removal system using a venturi scrubber with a neutralized recycle liquor system.

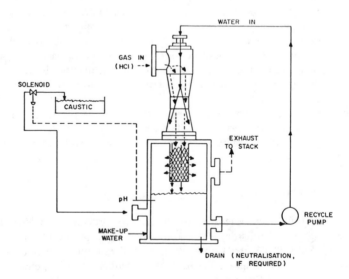

Figure 1-3. HCl mist-vapor removal—venturi scrubbing.

Fluorine Treatment

Of the varied problems peculiar to specific industries, the fluorine treatment at TVA facilities is a good example. Fluorine discharge without treatment could be disastrous. Superphosphate is made from phosphoric acid and pulverized phosphate; and during some of the production steps, 49 percent of the fluorine in the raw materials is given off as HF and SiF_2. A jet venturi scrubber removes 98 percent of the fluorine in a recirculation system. However, 100 gpm is replaced continuously as fresh water. The water content is 5000 ppm fluorine—equivalent to a ton per day. Phosphorous furnace slag, which is readily available and composed of calcium silicates and aluminates (93 percent smaller than 10 mesh), is utilized to react with the fluorine in wastewater and filter out suspended materials. All fluorine wastewater (average 700 ppm, 400 gpm) flows to a settling pond on the slag pile and then to holding (treatment) ponds, from which it percolates through the slag. Clear effluent eventually flows to the Tennessee River; 95 percent of the fluorine has been removed. Crust formation at the bottom of treatment ponds decreases the water flow through the slag pile and has to be removed periodically.

HOW TO MAKE POLLUTION CONTROL DECISIONS[62]

Analysis is an organized approach to decision making whereby alternate methods and courses of action are reviewed prior to decision making. Analysis of pollution potential from future facilities or equipment is normally investigated by the engineering department in an industrial organization. By using a flexible approach the pollution engineer can deal effectively with the complexities of environmental control. Then an analysis can be presented in an organized manner, examining the alternatives so that a decision can be made on the type of pollution control system.

A plan to build a new facility or modify an existing one constitutes need for action. This need is communicated to the pollution engineering department for analysis. The problem is then divided into several work areas.

In analyzing the problem, the pollution engineer should include the following three items in his work:

>*Acquisition of information*—site selection surveys, searches of existing and proposed regulations, analysis of processes, and study of governmental air pollution control techniques.
>*Processing of information*—data organized into a form that will enable management to make a decision.
>*Display of information*—organized presentation of alternatives for a pollution abatement program.

Information is acquired by thorough investigation of all sources which could in any way affect the use of the land, building, production processes or equipment. A pollution site survey should be made for a new plant and existing plants, prior to initiating a construction program. Sites should be evaluated in terms of:

1. *Meteorology*—wind speed and direction, temperature variations, stability conditions and precipitation,
2. *Air or water quality*—background levels,
3. *Effects*—observable adverse effects, proximity to sensitive receptors and
4. *Topography.*

A search of existing and proposed regulations should be made to find all pollution and land use restrictions which would affect operation.

Also, an analysis of the process should be made to determine number, location, and physical characteristics of discharge points, character and rate of emissions. Then processes should be analyzed in terms of the pollution control systems and equipment necessary to comply with regulations.

Information affecting pollution abatement plans should be identified and evaluated. All information must be considered in light of specific policy criteria established by the company. Included among these requirements are:

Goals of the organization—No new facility will be planned without pollution control.

Economics—Each pollution control device or abatement plan must be considered in relation to the profit making capacity of the plant, including tax relief and possible by-product recovery.

Personnel required—The number and technical training of staff required to operate the environmental control systems must be considered.

Abatement planning must be a balance of the equities between control, economics, and desired environmental quality. It is during this process that alternative solutions and possible outcomes must be determined by the pollution engineer.

Following the processing of information, data should be displayed to facilitate decision making by management. The key to such a display is a visual presentation of the maximum number of alternatives available, from which a final decision may be selected. By tabulating data in an orderly manner, obvious conclusions can be reached by simply comparing alternatives.

PR AND THE POLLUTION ENGINEER[103]

There is little doubt that the function of the pollution engineer is here to stay. However, since it is the management of a profit-threatening rather than a profit-building activity, the pollution engineer must be the devil's advocate to his company's management—an uncomfortable position at best.

Even more difficult, he must be prepared to face a public who may consider him more the devil himself than merely his advocate. Here, the name of the game is communications. A lot more is involved than the technical and economic problems of bringing pollution under control. The public must believe in the engineer's personal integrity before they can believe that his company is taking the necessary measures to control pollution—not just giving lip service to it.

Pollution Audit

A program is best started by a complete pollution audit. A thorough assessment should be made of the extent to which a company's product and processes may be contributing to environmental degradation. The audit should provide a detailed technical analysis of the company's performance against all applicable regulations and standards for pollution and waste disposal, and spell out what needs to be done to bring the company into compliance with the law and with standards of good practice.

The audit should not be entirely an engineering effort. If the company has public relations people, they should be involved. Engineers think primarily in terms of facts. Public relations people are more concerned with opinions. Where pollution is concerned, what people think you are doing becomes as important as what you are doing. People characteristically act on their opinions and use facts only if it suits their purposes. In other words, the audit should include all pollution sources—both real and imagined.

Position Paper

After the audit has been completed, the findings should be presented in a position paper for internal circulation to top management and other executives involved in settling corporate policy in pollution, authorizing expenditure of money for pollution control, or communicating on the subject internally or externally.

To meet this end the format for the position paper should include:

- An overall review of the major pollution problems in the area—air pollution, water pollution, solid waste and noise.
- A thorough analysis of the company's performance in each regard, both with respect to current legal standards and standards the company will have to meet in the long run.
- A step-by-step recapitulation of all the actions the company has taken to date to control pollution, plus a recommended program that the company initiate for future action, including a timetable for each phase. Any shortcomings in the long-range plan should not be overlooked.

- Explanations of the ways in which each employee can help eliminate pollution. Company policies and procedures are only as good as the people who carry them out.

Corporate Policy Statement

The position paper should be the foundation for a forthright corporate policy statement on pollution from the company's chief officer. The policy statement should make clear to all employees, the community, and the press that the company is dedicating itself to eliminating pollution to the full extent that technology makes possible. It should not seek to absolve the company from blame for past actions or to sugar-coat its performance. Nor should it overstate its future intentions.

That the company will do all it can is a key point. The pollution engineer can definitely influence this policy by the way he writes the position paper. He should not write the position paper as an exhortation to management to adopt a forthright policy, but write it on the assumption that management *wants* a forthright policy.

However, unless management thoroughly understands what it will take to bring pollution under control, both in dollars and time, the most forthright policy statement may later prove to be empty words.

Communications Program

Once top management has made its decision, the ball is back in the pollution engineer's hands to make sure that the decisions are carried out and that the program and its progress are communicated to all people concerned. This calls for a convincing communications program to earn employee and public confidence. The actual implementation of the communications program might well involve public relations people, but the technical guidance and ultimate responsibility for the general tone rest with the pollution engineer.

The tone must be one of mutual concern, avoiding at all costs the semblance of defensiveness or the unnecessary assumption of guilt. The company should present itself not just in terms of its own pollution control program, but as ready and willing to take an active role in any positive, community-wide pollution control program and to provide technical expertise and informational resources to the community.

A communications task force should be organized to speak publicly for the company on all matters pertaining to pollution. The task force should be made up of management, legal, industrial relations, operating, and pollution control people.

Don't select a team of people who are unreasonably pro-company and aching to take on those they regard as pollution nuts who are tearing at the fabric of our society. These spokesmen will only ensure that your company will become a prime target for civil action groups.

Employee Communications

A firm's employees are in daily contact with many people outside the company. Each employee should be regarded as a spokesman, and no effort spared to insure that he is well informed. Employees can be an unwitting source of misinformation and false rumors which are difficult to counter. Any bit of misinformation they carry has great weight with outsiders. After all, who is supposed to know better what is going on inside a company than its employees?

With this in mind, it is wise for the chief executive of the company to direct a personal letter to each employee at his home outlining the company's commitment to control pollution and seeking to enlist his support. Further, a series of staff meetings should be initiated, starting with the management group and fanning out to each department. The pollution program should be presented and discussed in open dialogue to make sure that any employee with questions or reservations has a chance to air them and get satisfactory answers.

Community Communications

The steps taken in community communications are very much the same as those taken for employee communications. First of all, news releases should be issued to the press, radio and TV as frequently as possible on each step the firm is taking to control pollution. Care should be taken that each release is definitely newsworthy and substantial in content, not just a whitewash job.

The communications task force, with a well prepared presentation, should be available to speak at PTA meetings, League of Women Voters meetings, schools, garden clubs, teacher groups, and the like. Much emphasis should be placed on talking with teachers, because they are usually very concerned and highly motivated but have little access to factual material on pollution in their locale except for news stories. They are extremely interested in local area ecology and the effects pollution has and will have on it, and they are frequently used as "technical experts" by activist groups.

Governmental Communications

It is just as important to communicate with governmental bodies, especially regulatory agencies, as with employees and the community. They must be able to understand the problems, and the company their's.

When working or communicating with any government agency, it helps to recognize it for what it really is—the institutionalization of public thought. To the unwary it may seem to be something of a plodding faceless body, immutable to change, and bound up in red tape and regulations.

This may be true until something stirs it—such as strong public opinion. It can then turn on and crush a polluter without batting an eye.

An industrial pollution engineer should keep a continuous dialogue going with both control agency administrators and engineers, and keep an ear to the ground. If problems occur with an agency, discuss them thoroughly in private before taking the case to the newspapers. When a polluter cries "harassment," it may well appear to the public that the agency is doing a good job.

Technical success is, of course, the final goal. But this will be more easily attained if the engineer makes a real effort to earn the support of his community and the public at large. That's what public relations is all about.

ESTABLISHING A CORPORATE DEPARTMENT
FOR POLLUTION CONTROL[24]

There are generally two categories of pollution control for a manufacturing enterprise: (1) those affecting the environment external to the plant; and (2) those affecting the working conditions within plant boundaries, the controls usually considered necessary for occupational safety and health.

These areas of pollution can also be divided into broad categories: (1) air pollution; (2) water pollution; (3) noise; and (4) solid wastes. The type of business will determine which of these subdivisions will be the dominant area of activity in pollution control. By way of example, Table 1-4 lists a number of industries and the dominant area(s) of pollution control in each.

Table 1-4. Predominant Areas of Pollution Control for Selected Industries

Type of Business	Predominant Areas of Pollution Control
Chemical Industry	Air - Water - Solid Wastes
Petroleum	Air - Water - Solid Wastes
Utilities	Air - Water - Solid Wastes
Machine Shop	Noise - Solid Wastes
Foundry	Air - Water - Solid Wastes - Noise
Forging	Noise
Fabricating	Noise - Solid Wastes
Plastics Manufacture	Noise - Solid Wastes

The formidable engineering problems associated with implementing a program to control the types of pollution listed above requires a special staff of experts. This is particularly true of larger industrial organizations and of those having more than one plant.

Defining the Problem

The necessary preliminary to any corporate activity is a clear, complete statement of the objective, formulated as a policy or a set of instructions. The resulting action must subordinate all secondary considerations to that stated goal. As an illustration of such a statement, the objective of a typical pollution control policy may be stated as follows: "It is the responsibility of the manufacturing enterprise to meet all governmental regulations, avoid the threat of shutdowns and fines, and improve public relations; it is equally important to avoid hasty decisions and to prevent disruption of normal plant operations."

Before any consideration is given to the specific controls required, clear-cut definitions of problems and group functions must be understood. These may be summarized as follows:

1. Reasons for pollution control. These may include nuisance value, public relations, odor, smoke, toxicity levels, recovery that may have some value and, of course, compliance with local, state and federal codes.

2. Identification of the problem. This will be through a process analysis and/or actual tests performed at the site of the existing facilities or new bench-scale or pilot plant work.

3. Establishment of quantitative parameters.

4. Determination of the regulations governing the geographical area. Obtain copies of all rules and regulations. Ascertain the degree of control required and make sure existing or projected operations are within the law. It is probably wise to overdesign, since laws are likely to be revised and stricter enforcement and more stringent requirements can be anticipated.

5. Determination of the degree of elimination or control required.

6. Decisions on engineering and economics.

7. Procurement of the permits needed to construct and operate the control facilities.

8. Specification and purchase of equipment.

9. Installation.

10. Start-up and operation.

Faced with these problems, the company must provide the services needed to handle all the requirements.

Organization Structure

The first step in developing any organization is the design of the structure. This requires determination of the duties to be performed and the selection of individuals to whom the duties will be assigned. Division of work on the basis of function is the foundation of this approach. This is, of course, necessary because:

1. The volume of work in an industrial organization of any size will be large.
2. Individuals differ in nature, capability and skill.
3. The range of knowledge required in an organization of any size is so vast that one person can have only a fraction thereof.

There is probably no single best way to provide pollution engineering service for a large multiplant corporation. Central pollution control corporate departments are as varied as the companies they serve. Facilities and service groups may be autonomous or function as an adjunct to a central engineering department.

Since the majority of pollution problems require engineering solutions, the principal members of any corporate environmental protection operation will be technical personnel: engineers, chemists, biologists, technicians. The one notable exception to this will be legal counsel, necessary because of the importance of compliance with public law and the rapidly changing legislative background. The interrelationships are shown in the accompanying organizational chart, Figure 1-4.

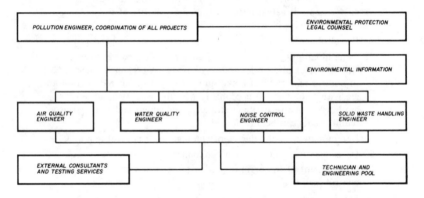

Figure 1-4. Organizational chart for central pollution engineering department.

Here a slight confusion about function may arise. Is the pollution control group functional or expert staff, or is it a line organization? The people who comprise the expert staff in an industrial organization normally have no line authority; their function is solely advisory.

In a modern industrial organization, the various experts may function at any one of the levels of authority. A simple illustration:

Line Executive	Staff Expert
President	Legal Counsel
Plant Manager	Management Consultant
Production Superintendent	Standards, Safety, Engineering
Foreman	Engineering

As a separate well-defined group, pollution control personnel can act as staff to all the line functions above as well as handle a line function of their own—the actual engineering and implementation of pollution control programs. The group will assume a completely staff function only if it becomes part of the engineering department itself.

The preceding consideration of organizational structure showed levels of authority and degrees of responsibility. These features are also found in the structure of the control organization. Here they are arranged in what is known as line of authority, line of command, or line of instruction running from a higher to a lower level of authority or the line of response, line of performance, or line of accountability, running from a lower to a higher level of authority. A typical line of authority is shown in Table 1-5, in which position titles are listed.

Table 1-5. Typical Lines of Authority and Response

Line of Authority		Line of Response
	President	
	V.P. Engineering–	
	Engineering Dept.	
	Environmental Control	
	Works or Plant Manager	
	Plant Superintendent	
	Maintenance or	
	Construction	
	Foreman	
	Worker	

The typical degrees of responsibility in the operating organization and the general duties of the incumbents of the positions are shown in Table 1-6. Environmental control, even though it is charged with implementation and engineering, has only staff functions in relation to the operating line.

Communication

Communication between the central pollution control staff and the plant operations at various locations is, of course, the heart of any successful program. The flow of information must be two-way, for there will be

Table 1-6. Typical Degrees of Responsibility and Corresponding Duties
As Related to Pollution Control and Line Personnel

Degree of Responsibility	1	2	3	4	5
Title	President	Works Manager	Superintendent	Foreman	Worker
Duties	Formulation of overall operating policies	Managerial policies managerial control	Operating control	Detailed control of operation	Performance of an assigned job

requests for assistance, policing of operations, checks on in-company compliance and enforcement.

From the standpoint of the central staff, the various plant operations must be informed of compliance requirements, codes and regulations and of the conditions that must be maintained. This necessitates dissemination of information by mailings and word of mouth; also required are on-the-spot inspections and review of operations to ensure satisfactory performance. It will be noted that environmental information is very prominent in the organization chart.

From the standpoint of local plant operations, the relationship can be that of consultant and client, or doctor and patient. Any problems imagined or **real**, suspected or actual, can be brought to the attention of the environmental control group for expert review and determination of action. The environmental control group should provide engineering advice and project management, as well as maintain contact with governmental enforcement agencies.

From the standpoint of other line or staff departments, the relationship of the environmental control group is that of staff.

Staffing the Group

The structure, or design of the organization, requires determination of the specific activities needed and arrangement of them in groups to be assigned on an as-needed basis to meet the local plant requirements. As shown in the chart, provision is made for air, water, noise and solid wastes control. The exact size of each section and the emphasis on each will be determined by the needs. Flexibility in engineering disciplines and reliance on outside consulting contractors will broaden the scope of any central corporate environmental group.

The typical central environmental protection team will be headed by a principal or chief pollution engineer charged with administering the pollution control policy. This team works with project, design, and plant engineers and operating personnel to implement policy; and it includes four environmental protection engineers chosen for their expertise in the handling of air, water, noise or solid wastes. Each of these engineers is charged with handling environmental protection problems in the specific discipline indicated. Naturally, because of the close interrelationships of the various problems, each engineer may call the others for aid on specific problems. Any new projects are reviewed for compliance with environmental policies or governmental codes, and recommendations are made. It is, therefore, the prime function of the central group constantly to review not only existing manufacturing activities, but any new or proposed work as well, to identify environmental problems and the economic burdens they may impose. For projects or problems outside the capabilities of the existing staff, consultants or special testing services may be employed.

Successful operation of a central environmental control operation can be based only on accurate information and wise decisions. Policies must be established to make the information and decisions effective. Technical competence determines the degree of success in implementing the program. No matter how experienced or skilled the staff is, however, in the final analysis it is guided by established corporate policies and by the criteria set for it by governmental enforcement agencies. Any policy and its resulting organization should be:

1. Flexible in application.
2. Subject to change and improvement.
3. Subject to enforcement to be effective.

ENVIRONMENTAL PROTECTION— A CORPORATE EXAMPLE[155]

The General Electric monogram appears on thousands of products, from aircraft engines, locomotives, and massive power-generating equipment to tiny batteries, toothbrushes, chemicals and plastics. This diversity in products has its counterpart in the plants that make these products. Distributed throughout 36 states, Puerto Rico, and more than a dozen foreign countries are more than 250 manufacturing plants, representing a tremendous range in size, age, complexity, type of location, and pollution problems.

Under General Electric policy each of these plants is expected to eliminate or limit "to lowest practicable levels all adverse environmental effects from its products, facilities and activities."

G.E. Organization

In July 1970 a vice-presidential task force completed a detailed examination of the environmental effects of General Electric plants and of the existing company programs for pollution abatement and control. In response to the task force recommendations, the objectives of these programs were given formal corporate recognition and definition in a specific policy statement, and a corporate-level Environmental Protection Operation was established to encourage the expansion, strengthening, and coordination of such programs by the company's operating components.

Under the overall direction of a Corporate Executive Office, the company is divided into ten major operating groups and two major corporate-level staff. The Corporate Executive Staff provides a continuing assessment of the long-range planning needs of the company. For the day-to-day operations, the Corporate Administrative Staff provides company-wide coordination in areas such as corporate finance, employee relations, legal operations, and public relations. One division of the CAS is Corporate Facilities Services, which includes the company's internal telephone and data processing operations, airplane fleet, and Real Estate and Construction Operation (RECO). RECO's primary functions are to manage the company's major construction projects, from initial conceptual study and site selection to plant start-up, and to provide services and engineering support to existing plants. The corporate Environmental Protection Operation (EPO) is a part of RECO.

The ten major operating groups are divided into a network of divisions and departments, which are components that provide the products and services the company sells to others. The policy directs that each of these operating components shall have procedures and clearly assigned responsibilities for environmental protection appropriate to its facilities and activities, and sets up guidelines for them to follow. It further assigns to the corporate Environmental Protection Operation responsibility for monitoring, counseling, appraising, and reporting on the environmental effects of these company-operated facilities and activities.

EPO performs its functions through a staff organized to provide technical, information, and legal services (Figures 1-5 and 1-6). Four engineers cover the major specialized areas of pollution control—air, water, noise and solid wastes—and also conduct annual on-site conformance appraisals of plant-level environmental protection programs. One engineer has responsibility, as environmental coordinator for new projects, for preventing environmental problems from arising in the new plants and facilities constructed each year by GE. An Environmental Information Center collects, evaluates, organizes and interprets environmental information relevant to the company's environmental programs and prepares packages of pertinent information in a form that is authoritative, timely, and useful to company management and operating personnel. An attorney is assigned full-time to EPO to provide its staff and company personnel with legal advice and interpretation concerning legislation, administrative regulations, and judicial decisions that affect the company's environmental protection programs. He also works with members of EPO's technical staff to prepare comments on proposed legislation and regulations for submission to government

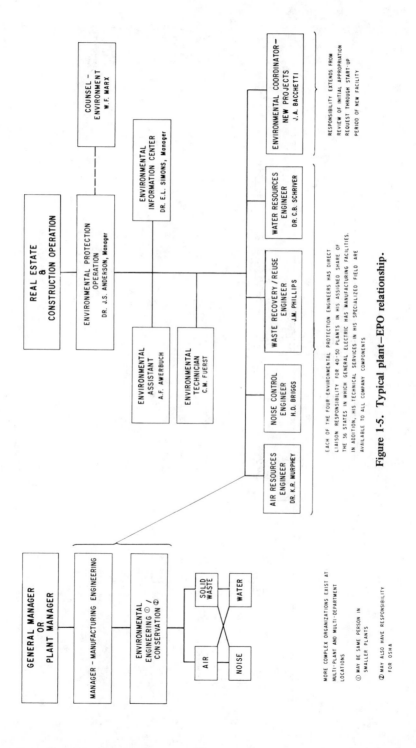

TYPICAL PLANT — EPO RELATIONSHIP

GENERAL MANAGER OR PLANT MANAGER

MANAGER – MANUFACTURING ENGINEERING

ENVIRONMENTAL ENGINEERING ① / CONSERVATION ②

SOLID WASTE
WATER
AIR
NOISE

REAL ESTATE & CONSTRUCTION OPERATION

ENVIRONMENTAL PROTECTION OPERATION
DR. J.S. ANDERSON, Manager

COUNSEL – ENVIRONMENT
W.F. MARX

ENVIRONMENTAL INFORMATION CENTER
DR. E.L. SIMONS, Manager

ENVIRONMENTAL ASSISTANT
A.F. AWERBUCH

ENVIRONMENTAL TECHNICIAN
C.M. FUERST

AIR RESOURCES ENGINEER
DR. K.R. MURPHEY

NOISE CONTROL ENGINEER
H.D. BRIGGS

WASTE RECOVERY / REUSE ENGINEER
J.M. PHILLIPS

WATER RESOURCES ENGINEER
DR. C.B. SCHRIVER

ENVIRONMENTAL COORDINATOR– NEW PROJECTS
J.A. BACCHETTI

RESPONSIBILITY EXTENDS FROM REVIEW OF INITIAL APPROPRIATION REQUEST THROUGH START-UP PERIOD OF NEW FACILITY

EACH OF THE FOUR ENVIRONMENTAL PROTECTION ENGINEERS HAS DIRECT LIAISON RESPONSIBILITY FOR 40-50 PLANTS IN HIS ASSIGNED SHARE OF THE 36 STATES IN WHICH GENERAL ELECTRIC HAS MANUFACTURING FACILITIES. IN ADDITION, HIS TECHNICAL SERVICES IN HIS SPECIALIZED FIELD ARE AVAILABLE TO ALL COMPANY COMPONENTS

MORE COMPLEX ORGANIZATIONS EXIST AT MULTI-PLANT AND MULTI-DEPARTMENT LOCATIONS

① MAY BE SAME PERSON IN SMALLER PLANTS

② MAY ALSO HAVE RESPONSIBILITY FOR OSHA

Figure 1-5. Typical plant–EPO relationship.

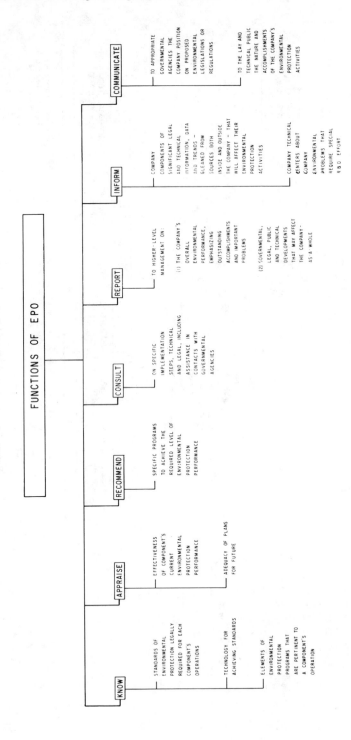

Figure 1-6. Functions of EPO.

agencies either directly by EPO, by company officers, or through trade associations and other appropriate groups.

Three major ingredients of a corporate environmental protection program, regardless of size, are management commitment, information and action.

Commitment

An essential starting point for an effective program is a firm and clear statement, originating from the company's top management, that it is committed to the goal of operating the company's plants and facilities in an environmentally acceptable manner. It is desirable to give this commitment formal expression in a written company policy. It is essential that this commitment be accompanied by the assignment of clearly delineated responsibility and authority to appropriate individuals for achieving this goal and for keeping management informed of significant developments that can affect the attainment of this goal.

Information

To meet this commitment a company must know the rules of the game. The kind of information you will need to keep up-to-date on the environmental laws and regulations with which your plants will be expected to comply comes from three major sources: state agencies, federal agencies, and private services.

State Agencies

The best starting points for building up the base of information you will need are the state agencies charged with responsibility for enforcing state environmental laws. Many states have consolidated all environmental activities within one department, while others maintain separate agencies for air and water pollution control. A list of the pertinent state agencies is available from the Office of Public Inquiries, Environmental Protection Agency, Waterside Mall, Room 206 W., 401 M Street SW, Washington, D.C. 20460. The Air Pollution Control Association (4400 Fifth Ave., Pittsburgh, Pa. 15213) publishes a directory of governmental air pollution agencies, and a similar listing of state and interstate water pollution control agencies is available from the Water Pollution Control Federation (3900 Wisconsin Ave., NW, Washington, D.C. 20016).

Federal Agencies

In more and more environmental areas the federal government is setting the guidelines and standards that the states must follow in controlling pollution within their borders. Furthermore, federal legislation (*e.g.*, the Clean Air Act Amendments of 1970) may require that the federal government (Environmental Protection Agency) must approve state pollution control plans and regulations before they can become effective and be enforced by the states. Where EPA finds a state's plans and regulations to be inadequate, it is empowered to establish and enforce its own substitutes.

All of these federal guidelines, standards, proposed regulations, notices of hearings, and final regulations are reported completely in the *Federal Register.* For an annual subscription fee of $25.00, payable to the Superintendent of Documents, U.S. Government Printing Office, Washington, D.C. 20402, you will receive five issues per week plus frequent indexes. Each issue is also indexed to allow easy scanning for items of environmental interest.

Private Services

Newspapers—You can follow significant environmental developments on a day-to-day basis by reading major metropolitan newspapers.

Newsletters—For more intensive, detailed coverage you should consult one or more of the weekly newsletters. The most comprehensive—and one of the most expensive—is the *Environment Reporter*, published by the Bureau of National Affairs, Washington, D.C.

Professional Societies—Professional societies today do not limit their activities solely to scientific and technical subjects, but in their publications and meetings also discuss legislative and regulatory developments pertinent to their areas of interest.

Trade Associations—Membership in an active and effective trade association is a good way to keep informed on developments affecting your particular business. Most multi-industry associations (*e.g.*, U.S. Chamber of Commerce, National Association of Manufacturers) have committees devoted to environmental affairs, but their coverage is devoted mainly to matters that have an impact on a large segment of industry. A directory of National Trade and Professional Associations is published annually by Columbia Books Inc., Washington, D.C.

Monthly Magazine—For a broader perspective you should read monthly environmental magazines like *Pollution Engineering.* Included among their technical articles are reviews of governmental, legal and regulatory developments.

Outside Counsel—To ensure that you will receive only information that is directly pertinent to your specific needs, you can retain the services of a law firm or lobbying agency to provide the necessary monitoring, screening and interpretation. Such services are most readily available from firms located in the national or state capitals.

Action

Commitment and information must be translated into action, which may be considered under two broad categories: (1) voluntary action to try to influence pending legislation or regulations, and (2) compulsory action to comply with existing legislation and regulations.

Influential Actions

As with any legislation, your views on pending environmental legislation can be communicated to the appropriate legislators through your own direct contacts (including presentation of oral testimony at committee hearings), through your own lobbyist, or through your trade association. Two government publications that present clear explanations of the federal legislative process are: "How Our Laws Are Made," by Joseph Fisher, Esq., for sale by the Superintendent of Documents, U.S. Government Printing Office, Washington, D.C. 20402, price 35 cents, and "Enactment of a Law" (Senate Document No. 35 - 90th Congress, 1st Session), by Floyd M. Riddick, for sale by the Superintendent of Documents, U.S. Government Printing Office, Washington, D.C. 20402, price 20 cents.

After pending legislation has become law, the agency charged with its enforcement develops proposed regulations for its implementation. Comments on these regulations may then be made, either as written (or written plus oral) statements to be presented at specifically scheduled public hearings or as written comments due at the issuing agency within some specified time (usually 30, 60 or 90 days) of the formal announcement (*e.g.,* publication in the *Federal Register.* Again, these comments can be made by individuals or by trade associations.

Compliance Actions

These fall into two broad categories: (1) those that involve submitting required information to government agencies, and (2) those that involve instituting specific pollution abatement procedures by certain dates. To be effective, both types of actions must be based upon a complete, honest, and technically reliable inventory of your environmental assets and liabilities. You should know the amount and composition of all materials that enter and leave the plant and the nature of the processes that transform intake into effluent. Traditionally, industry has concentrated its efforts on the amount and quality of the desired products, those that the company made for sale to its customers. Today a company must be equally knowledgeable about the amount and compositions of the waste products, which in the past may have been disposed of in the easiest and least costly way possible without regard to their effects on the environment outside the plant.

This initial inventory must be kept up-to-date by a systematic and accurate program of record-keeping. A file of understandable and dated records on important plant parameters is one important evidence of good faith when you arrive at the stage of reaching agreement with regulatory agencies on compliance schedules.

If you don't have the technical competence in-house to collect this information and to set up the record-keeping, retain the services of an outside consultant. Even the largest corporations, with extensive technical facilities of their own, recognize that for certain problems it is technically more effective and financially less expensive to use an independent specialist or technical organization than to build up the necessary in-house competence. Most technical journals in the environmental area print advertisements of professional consulting services, and the American Society for Testing and Materials (1916 Race Street, Philadelphia, Pa. 19103) publishes a Directory of Testing Laboratories - STP 333B ($3.50).

Abatement Compliance

The moment of truth arrives when you and a representative of a regulatory agency meet to discuss a specific monitoring and abatement schedule for a specific plant—the level of discharges that will be permitted, the testing that must be done to demonstrate compliance, and the dates by which certain agreed upon objectives must be reached.

You should approach such negotiations determined to demonstrate your knowledge of the problems involved, documentation of your actions to date, and an earnest willingness to comply with an abatement plan that is reasonable from a technical and economic point of view. Our experience with regulatory officials has been that most are reasonable men who are out to solve a problem rather than to penalize malefactors.

The regulatory agencies will discuss their requirements with you and will tell you the kinds of data upon which they will base their judgments of your operations. In some cases they may send their own inspectors to collect effluent samples and to make measurements at your plant site. After making tests of its own or examining your data, an agency will usually notify you of the specific parameters whose values it feels are in violation of its regulations and will then allow a specific period of time for you either to submit evidence that you have corrected the problem or to submit a proposed plan of action to reach a solution. You may propose end-of-the-line treatment to remove the offending pollutant from the effluent stream before it leaves your plant or you may find it more effective to modify your in-plant process to reuse or otherwise eliminate the actual production of the offending pollutant. The agency's principal interest is in the final result, provided that in solving one pollution problem you do not create another.

A note of caution: if you know that some of your emissions and discharges are in violation of existing regulations, don't delay your planning of remedial action until agencies are forced into taking action against you by outside complaints or by legal action directed against your operations. It is much more difficult to negotiate a mutually satisfactory abatement schedule when the government agency is under outside pressure to take enforcement action against you.

ENVIRONMENTAL FACTORS IN PLANT SITE SELECTION[61]

Five environmental factors merit close attention when selecting an industrial site: (1) location of emissions, (2) quality and quantity of pollutants, (3) control of pollutants, (4) impact on the environment and (5) compliance with regulations. Each of these should be analyzed when considering the effects of any of the four major environmental control categories.

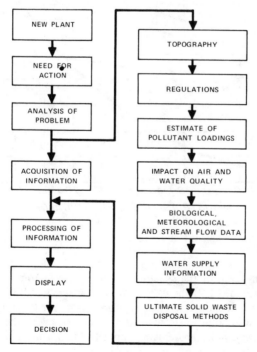

Figure 1-7. Site-selection decisions. Systematic approach to decision-making.

There is a relationship between industrial site-selection factors and the size of the industry locating in a specific area. Although a number of measures could be used for industrial size or the magnitude of production, land acreage is normally used as the size indicator.

Figure 1-8 illustrates the importance that is placed upon different site-selection factors with increase in size of the facility involved. Note that environmental considerations start at a high level and rapidly increase on the 10-point importance rating scale. Utilities rise only slightly, whereas public relations ascends from a rating of 0.5 to a rating of 10 as the industrial size expands to 100 acres. Transportation, which is the most important of the five factors considered in this figure, is given a high importance rating, regardless of the size of the industry.

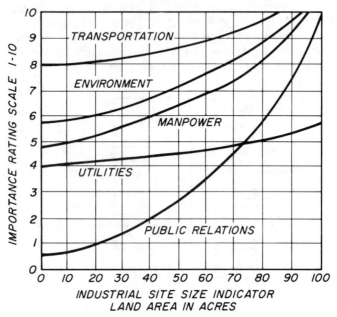

Figure 1-8. Dominant features of industrial site selection—"general."

A similar graph of the specific environmental factors involved in site selection (Figure 1-9) shows that air and water pollution control are the most important industrial site-selection factors for large industries. The importance of solid waste management declines with the increase of industrial size, and noise abatement, important to smaller plants, becomes less important as the size of the facility increases.

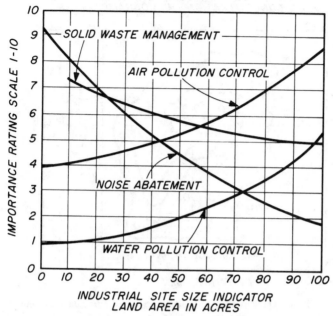

DOMINANT FEATURES OF INDUSTRIAL SITE SELECTION
"ENVIRONMENT"

Figure 1-9. Dominant features of industrial site selection—"environment."

Air pollution emission diffusion modeling is a good method for evaluating the new site to see if it meets proposed air quality standards and emission regulations. Selecting a new industrial site is a long and tedious process, the details of which a company usually keeps in strict confidence during the initial study period. To expedite the process and provide the industrial analyst and facility planner with proper data in their interpretation of air quality standards, an air quality information data bank should be established.

Much of the factual material required for diffusion modeling to make air quality estimates is simply background information on air quality in the area, meteorological statistics (wind speed, direction, stability), and emission levels. The data bank required for selection of the proper industrial site is vast.

At the present time, much background air quality information is being stored by the federal government in the Office of Air Programs of the U.S. Environmental Protection Agency in Durham, N.C. Meteorological data from most areas can be obtained from the National Weather Records Center in Asheville, N.C. Most of the emission facts needed for diffusion modeling decision-making can be obtained from the design of consulting engineers preparing plans for the new facility. Figure 1-10 illustrates concentrations of particulates in a general area.

AMBIENT AIR QUALITY LEVELS
EXPECTED PARTICULATE ARITHMATIC
MEAN CONCENTRATION - MICROGRAMS/METER^3

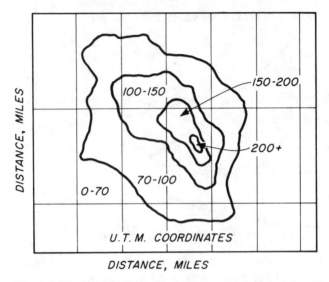

Figure 1-10. Ambient air quality levels, expected particulate arithmetic mean concentration—micrograms/meter^3.

State agencies have the responsibility for air pollution control implementation plans. These agencies prepare meteorological summaries, emission inventories, episode plans, air quality standards, emission standards and other specific components of the data bank necessary to prepare an implementation plan. Statistics gathered and assembled by the states, or by the consultants working for them, will provide a working tool for those considering selection of an industrial site.

As these facts become available, it will be possible, by utilizing the air quality display model developed by the Office of Air Programs of the Environmental Protection Agency, for the pollution engineer to judge the impact of his plan on any particular location. Similar models and techniques can also be used for water pollution control and solid waste management.

There are a number of parameters that can be used by engineers planning site selection. These include transportation, pollution, economic factors, etc. Regarding the environment, certain parameters must be taken into account such as fundamental acceptability, compliance with pollution control regulations and public reaction. These could be combined into some kind of a site-selection index, but, until this is done, it would be better for the engineer to realize that the only reason why pollution controls are being placed on industrial plants is regulations.

Even though an output measure of pollution levels provides a realistic guide, the most important influence is the specific requirements of the applicable pollution regulations. The relationship between the pollution that the plant puts out, the air quality standards for the area, and the background levels of pollution that now exist must be evaluated. This relationship can be expressed as an equation:

Air Quality Standard - Background Level = Maximum Plant Emission

Maximum plant emission is a value which can be compared with the actual output to determine if a proposed plant would be acceptable or not. This would certainly give the environmental engineer a guide with which to make a decision from a regional air quality standpoint.

Environmental planners and engineers would like to believe that the environment is the most important site-selection factor. However, the industrial segment of the community indicates that the profit motive is still paramount, even though the gap between a clean environment on the one hand and industrialization on the other is now closing.

It is interesting to note that air pollution control is now judged the most pressing problem of the environmental site-selection factors. The principal air pollution considerations in industrial site selection are: the area of emission, the quality of emissions, controlling the emissions, impact on the environment and complying with regulations.

The mathematical diffusion model technique is available to planners to test the influence of a new industry on the environment. Certain basic information is required in order to operate this model, such as: an emission inventory, meteorological information, and background air quality levels.

With the increase of available data, it is becoming easier for regional planners to make more accurate estimates of the effect a new plant in a specific location will have on the environment.

Table 1-7 provides some of the most critical site location or selection factors. Since no two industries are identical in their needs, each site location team will view this checklist in a different perspective.

Table 1-7. Environmental Site Selection Factors

A. Water Supply
- Water needs___process___cooling___potable___fire protection
- Water availability___public water supply___private water supply ___ground water___surface water
- Ground water geological potential (onsite)
- Water characteristics___chemical___bacteriological___corrosiveness
- Water distribution___amount available___pressure___variations ___proximity to site___size of lines
- Cost of water supply___extension of existing service___development of new supply___cost per 1000 gallons
- Water treatment requirements___process___cooling___boiler feed-water ___potable___other
- Special considerations___restriction on use___future supplies ___compatibility for use in process

B. Ecological Considerations Based on Anticipated Change in the Environment
- Discharges___gaseous___liquid___solid wastes; what are the ecological considerations?
- Existing area ecological relationships (use available background data and augment as necessary)
- Control measures to minimize ecological effects
- Wildlife propagation areas
- Terrestrial and aquatic areas
- Physical tolerance levels___ambient air quality standards___water quality standards___noise level standards___glare and/or lighting standards
- Nutrients
- Detrimental and beneficial development
- Buffer zones and green belts

C. Air Pollution Control Considerations
- Regional airshed standards
- State air pollution standards
- Local air pollution enforcement regulations and ordinances
- Meteorological conditions___wind direction and velocity variability, inversion frequency, intensity and height, and other microclimatology factors
- Proximity to population/employment centers
- Local topography
- Effect other area industrial emissions may have on the quality of new plant environment or on allowable emission rates

D. Wastewater Disposal
- Sewerage systems___stormwater___cooling water___process wastewater
- Anticipated mode of occurrence, flow and characteristics of plant wastewater discharges
- Proposed pollution loadings
- Toxic materials present
- Variations in flow and strength
- Variations in wastewater treatability
- Inplant control measures
- Onsite wastewater treatment and disposal possibilities
- Nearby water courses which may be considered for wastewater disposal
- Existing stream quality
- Water uses to be protected
- Stream quality standards
- Wastewater effluent standards
- State regulatory agencies concerned___permit requirements
- Stream flow characteristics, design critical flow
- Development of treatment design parameters
- Availability of a public sewerage system
- Pretreatment requirements if discharged to public sewers
- Sewer service charges and surcharges for industrial wastewaters

- Onsite underground disposal system ___ percolation rates
- Scavenger hauling of liquid wastes
- Emergency operation ___ electrical power dependability
- Performance reliability requirements

E. Solid Waste Disposal

- State and local regulatory agencies
- Disposal facilities available ___ incineration ___ sanitary landfill ___ other
- Local contract pickup and disposal ___ municipal control ___ competition between haulers
- Costs of solid waste disposal
- Dependability ___ lifetime of disposal facilities ___ probability of flooding
- Onsite disposal ___ incineration ___ landfill
- Responsibility ___ if public collector ___ if private collector ___ if other disposal
- Special handling and disposal practices required for industrial wastes

PREPARATION OF ENVIRONMENTAL IMPACT STATEMENTS[63]

An Environmental Impact Plan is a document presenting the results of a systematic study of all the potential effects of a proposed (or existing) facility or activity on its environment.

Philosophically, environmental impact statements are necessary to protect and to enhance the environment. They are particularly needed in industrialized areas to prevent any further degradation of the air, water, and land resources and to overcome undesirable conditions resulting from previously unplanned expansion.

Legally, the federal government requires the submission of an environmental impact plan in certain circumstances and undoubtedly will expand the application to other situations. So far, the enforcement has been felt primarily by operating units of the federal government and to a lesser extent by other governmental agencies or private companies whose activities are subject to federal regulation. The National Environmental Policy Act of 1969 directs all agencies of the federal government to "identify and develop methods and procedures which will insure that presently unquantified environmental amenities and values are given appropriate consideration in decision making along with economic and technical considerations." The Council on Environmental Quality, in furthering this Act, has set forth guidelines for preparation of the required environmental statements.

Practically, any industry contemplating new construction or major expansion or modification should seriously consider the advisability of preparing an environmental impact statement to meet applicable federal and, perhaps, state requirements.

All agencies of the federal government must submit environmental statements covering any federal action that affects the environment. Each federal agency has had different requirements, making uniformity in the preparation of statements difficult. Many states have now adopted legislation covering environmental impact statements at the local level.

According to the U.S. Environmental Protection Agency the following are screening criteria for determining whether proposals "significantly" affect the environment and require statements. Such proposals include (but are not limited to) those that:

1. Adversely affect the natural environment, disrupt ecosystems, imperil health, or offend the senses.
2. Represent small but cumulatively important projects, "foot-in-the-door" projects, *e.g.,* short segments of highway that initiate a commitment to a major route or circumferential highway system.
3. Present an individually minor but cumulatively major incursion, *e.g.,* a dredging permit for a small parcel of wetland which represents one more incursion on a scarce resource.
4. Set precedents.
5. Generate adverse secondary effects, *e.g.,* land development causing potential pollution problems.
6. Evoke concern by agencies and public interest groups; appear controversial.

Some of the areas in which statements may be required are:

1. Transportation: highways, airports, mass transportation.
2. Power generation: nuclear and fossil fuel power plants.
3. Water resources: reservoir and channelization projects, dredge and fill permits.
4. River basin and regional plans: water quality management plans.
5. Wastewater treatment facilities: EPA assisted facilities or those requiring federal discharge permits.
6. Urban-commercial-industrial development: HUD loans and mortgage guarantees for new towns, commercial-industrial complexes and multi-residential housing developments.
7. Economic Development Administration (EDA) loans and grants for commercial and industrial development.
8. Federal land management and special use permits: resort developments on national forest land, federal timber cutting and forest management practices.
9. Federal licenses and leases: offshore oil drilling.
10. Pesticide, herbicide, rodenticide and fungicide applications: federal policies and procedures, federal registration, federal cost-sharing practices or major technical assistance programs, individual applications involving federal financial or technical assistance that involve risks or persistence.

Another area that was added to the array of statement is that of power transmission lines. The 765-kw line going through Virginia came under attack by conservationists and alternate routes are being considered based upon environmental impact.

Perhaps of the most far-reaching concern to the industrialist and corporate management will be the statements that will have to be written for new building construction and plant site location. The federal review procedure is rather complicated as illustrated in Figure 1-11. The EPA has the responsibility to comment on the following: water quality and pollution control; air quality and pollution control; water supply and water hygiene; solid waste management; radiation problems; pesticide, herbicide, rodenticide and fungicide registration, use and control activities; hazardous and toxic materials; and noise.

Generally, an environmental impact statement contains the following:

1. Description of the proposed action,
2. Probable impact of the proposed action on the environment, including impact on ecological systems,
3. Probable adverse environmental effects which cannot be avoided,
4. Alternatives to the proposed action,
5. The relationship between local short-term uses of man's environment and the maintenance and enhancement of long-term productivity,
6. Any irreversible and irretrievable commitments of resources, and
7. Problems and objections raised by other federal, state, and local agencies and by private organizations and individuals, in the review process and the disposition of the issues involved.

Since an environmental impact plan generally deals with a planned or proposed facility or activity rather than an existing installation, the determination of the environmental impact requires significant departures from conventional investigations or studies: (1) identification and selection of data sources are more difficult; (2) new investigation procedures have to be developed to complete the synthesis of the estimated impact of the proposed facilities; (3) specific procedures will be required for evaluation of the impact on the various facets of the environment. Without the development of these tools specifically for the project at hand, it would be impossible to arrive at valid conclusions and recommendations.

PLANNING POLLUTION CONTROL PROJECTS BY PERT[4,5]

PERT (Program Evaluation and Review Technique) concepts may be applied easily to planning pollution control projects for existing plant equipment or processes. An objective project analysis utilizing such a technique and periodic reviews of the project status can increase the effectiveness of management supervision and control.

Application of the logic inherent in the formulation of a PERT network automatically yields several advantages:

1. The concept requires the identification of certain actions upon which subsequent actions depend.

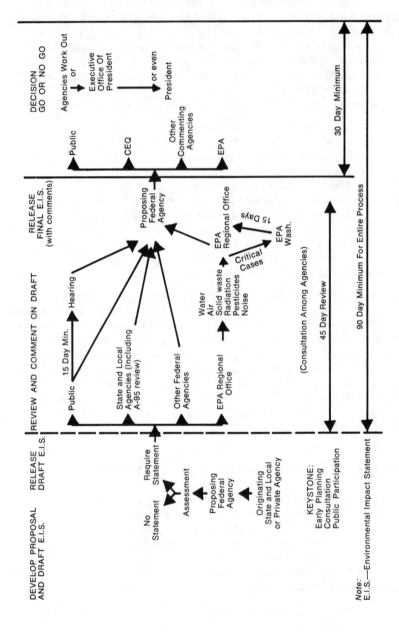

Figure 1-11. Flow of environmental impact statements.

Note:
E.I.S.—Environmental Impact Statement

2. Since there is distinct identification of durations for each activity, PERT encourages objective decision-making and identifies responsibilities.
3. Periodic analyses help evaluate the allocation of company resources for maximum efficiency.
4. Plans are more realistic since the estimates of activities have included allowances for the uncertainties of event occurrence.
5. Systematic scheduling shows a clear interrelationship with other activities.

The sequence of activities that comprise the project is established following some logical set of rules. The network shown in Figure 1-12, and outlined in Table 1-8, is composed of such a series of activities illustrating a project to provide air pollution control facilities. In some instances, activities may be accomplished either sequentially or simultaneously. Those activities that do not use up resources and do not accomplish a task are designated as dummy activities. They are shown simply to maintain continuity within the chart and to eliminate "dead ends." At the occurrence of event 2, the project begins, and with the occurrence of event 23 the project ends.

The first step toward the execution of the abatement project after it has been defined is the process analysis (2-3). This analysis might include the definition of possible pollutants, which can be determined from the material balances and/or the actual experimental tests conducted on stack/flue gases. It might also include determining the conditions of operations; for example, temperature, CFM, percent moisture, pressure, etc. Such data would enable evaluation of different methods available for air cleaning (3-4). Considerations may include economic and technical aspects as well as area and volume factors.

Availability of utilities (electricity, air, water supply, steam, etc.) and possible creation of water/solid waste disposal problems should also be evaluated. The most important factor is compliance with city, state and federal emission standards. The result of this analysis will be the selection of a particular kind of abatement method and equipment—for example, adsorption (fluidized or fixed bed), absorption (wet scrubbing, physical/chemical), electrostatic precipitators, cyclones or baghouses.

At this stage the decision as to whether the engineering design will be handled completely in-house should be made. Assuming that, apart from basic engineering requirements and final design approval, the vendor will provide the specified equipment, the next step would be to send out preliminary requests for quotes from several vendors (4-5).

At the same time, an in-plant analysis would indicate specifications for the provision of required services which would include electric power, water, pilings, etc. (4-6). The next sequential steps constitute approvals of quotes for equipment (5-7) and approvals for specifications on services (6-7).

Once the decision on the hardware has been made, the second phase of the project can be considered. This phase starts with the site selection (7-8). Space availability for the equipment and ductwork, proximity to the process, allowance for framework, and foundation questions have to be satisfied.

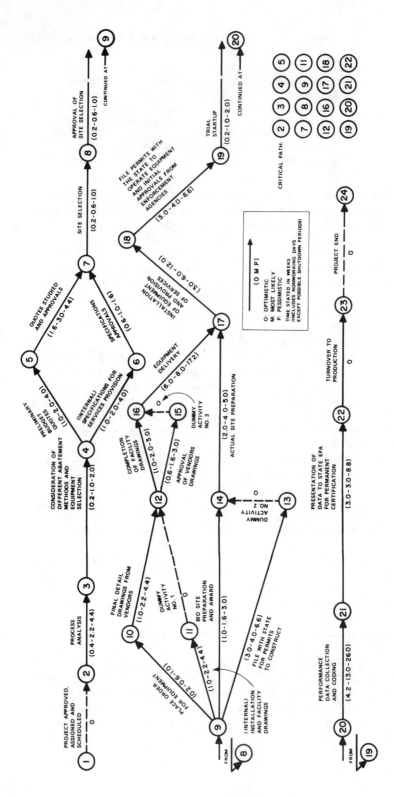

Figure 1-12. Preliminary network for air pollution abatement project for an existing process.

Table 1-8. Project Network Calculations

			Network Calculations			
Event	Preceding Event	Activity Description	Expected Duration (d_{ij}) (days)	Earliest Time (Forward Pass) $(T_E)_j$ (days)	Latest Time (Backward Pass) $(T_L)_j$ (days)	Total Float $(T_L)_j$-$(T_E)_j$ (days)
2	1	Project Start Dummy Activity	0	0	0	0
3	2	Process Analysis	15	15	15	0
4	3	Type of Method and Equipment Selection	7	22	22	0
5	4	Preliminary Budget Quotes	15	37	37	0
6	4	Specifications for Services Provision	15	37	52	15
7	5	Quotes Studied and Approvals	20	57	57	0
7	6	Specifications Approvals	5	42	57	15
8	7	Site Selection	3	60	60	0
9	8	Approval of Site Selection	3	63	63	0
10	9	Place Order for Equipment	3	66	68	2
11	9	Installation and Facility Drawings	16	79	79	0
12	10	Final Drawings from Vendors	11	77	79	2
12	11	Dummy Activity No. 1	0	79	79	0
13	9	File with State for Permits	26	89	133	44
14	9	Bid Site Preparation and Award	11	74	133	59
14	13	Dummy Activity No. 2	0	89	133	44
15	12	Approval of Vendors' Drawings	11	90	93	3
16	12	Completion of Facility Drawings	14	93	93	0
16	15	Dummy Activity No. 3	0	90	93	3
17	14	Actual Site Preparation	24	113	157	44
17	16	Equipment Delivery	64	157	157	0
18	17	Equipment Installation and Services Provision	45	202	202	0
19	18	File Permits to Operate Equipment	26	228	228	0
20	19	Trial Startup	7	235	235	0
21	20	Performance Data Collection and Coding	95	330	330	0
22	21	Data Presentation for Perm. State Certification	27	357	357	0
23	22	Turnover to Production	0	357	357	0
24	23	Project End Dummy Activity	0	357	357	0

An approval of site selection would thus mean that the equipment will have a site that is both safe and economical (8-9).

The next series of steps includes placement of the equipment purchase order with the vendor (9-10), solicitation of bids for site preparation (9-14), completion of construction drawings (9-11), and preparation of permit requests to be sent out to the state enforcement offices for approval to construct the pollution abatement system (9-13). Award of the site preparation contract is followed by actual preparation (14-17).

Once the order for equipment is placed, the vendor sends out the detail drawings (10-12) which are then studied and approved (12-15). At the same time, the facility drawings, based on the detail drawings from the vendor, are finalized (12-16). Approval of detailed equipment drawings and completion of drawings made in-house allows the vendor to begin fabricating the equipment for delivery (16-17).

Upon equipment delivery and completion of site preparation, the equipment may be installed and services provided as necessary (17-18). Since most state codes do not allow operation of a process without a permit, applications should be filed at this point (18-19). Initial permit approvals from the state agency lead to trial startup (19-20).

Data on the equipment operation and its performance are gathered and adequately coded (20-21). Such data may then be presented to the state EPA for issuance of the final certificate for the process and the equipment (21-22). This logically is the stage when pollution engineering hands the responsibility to production (22-23) and files the project as complete (23-24).

The estimates of duration shown are typical, based on several plants' experience with these activities over a long period of time. Three estimates are made to cover the range of different possibilities. The average weighted mean of expected duration is then calculated as:

$$d_{ij} = \frac{o + 4m + p}{6}$$

where:

d_{ij} = weighted mean of the expected duration events i and j.
o = most optimistic duration.
m = most likely duration.
p = most pessimistic duration.

Once the expected duration (d_{ij}) is known, the earliest occurrence of successive events $(T_E)_j$ may be calculated as:

$$(T_E)_j = \text{Maximum} \left[(T_E)_i + d_{ij} \right] \qquad \text{(Forward Pass)}$$

until the last event is reached. On the assumption that the last event has the same latest and earliest occurrence time, backward pass calculations yield the latest occurrence of each preceding event $(T_L)_i$ until the start event is reached:

$$(T_L)_i = \text{Minimum} \left[(T_L)_j + d_{ij} \right] \qquad \text{(Backward Pass)}$$

The difference between the earliest and latest times calculated gives the total "float" for each activity—a measure of the leeway within which the project can be completed by the scheduled date:

$$\text{Total Float} = (T_L)_j - (T_E)_j$$

Those activities that do not allow any leeway, *i.e.*, have zero float, are termed critical, and the path formed by such activities is the critical path. A delay on any of these activities would result in a delay of the scheduled completion date. In other words, the path along which the durations add up to give the longest schedule of time required for project completion is the critical path. The float times allow the estimation of the amount of resources to be allocated.

From the expected duration and the early and late start date calculations, early and late finish dates can be computed and the finish floats determined. The finish floats will be identical to start floats since the early and late finish directly depend on early and late starts.

In this work, no periodic analysis has been illustrated. However, in some cases, the network may be initially analyzed and tables prepared. As may be deemed convenient, the analysis may be reviewed periodically to update the scheduled dates and record any changes on the critical path and hence the completion date. If such a periodic analysis is undertaken, the change in path and float times would indicate a shift in the use of resources to achieve better efficiency.

It may be noticed that if there is an analysis conducted for update of schedule, some of the activities will have already started and, therefore, will have zero start float even if they are not on the critical path. Similarly, there may be some activities that have ended and will have zero finish float. A sample project, as described above, would not normally require computer calculations (this one didn't), and the analysis can be manually reviewed and updated. However, the choice will rest with the engineer.

Table 1-8 lists the activities in sequential order. Expected duration as calculated in Table 1-8 is listed next, followed by earliest and latest times which are calculated as shown previously. Table 1-9 lists these activities with their precedence requirements. Three duration estimates and the weighted mean are then listed. The T_E and T_L from Table 1-8 allow calculation of the actual start dates once a base date for the project start has been determined. Float and total duration values enable determination of the late start, early and late finish.

If a computer is used, the output report should be headed by an appropriate title and network identification. A run date in the top left corner will show the date of the computer analysis of the information prepared on the data date shown in the top right corner. Columns will comprise the event numbers, activity description, total duration for each activity, early and late starts followed by start float, and early and late finishes followed by finish float. The report should end with the line showing the scheduled start date of the project, its total duration, and

Table 1-9. Schedule—Design and Construction

Events i	Events j	Activity Identification	Activity Between the Two Events	Precedence Requirement (Activity Ident.)	Estimate of Duration (Weeks) Optimistic o	Most Likely m	Pessimistic p	Weighted Mean (o+4m+p)/6	Start Early	Start Late	Start Float	Finish Early	Finish Late	Finish Float (weeks)
1	2	1-2	Project Begun (Approved, Assigned & Scheduled)	None	–	–	–	0	4 Jan	4 Jan	0	4 Jan	4 Jan	0
2	3	2-3	Process Analysis (Experimental & Theoretical)	None	0.4	2.2	4.4	2.2*	4 Jan	4 Jan	0	19 Jan	19 Jan	0
3	4	3-4	Consideration of Diff. Methods & Selec. of Equip't	2-3	0.2	1.0	2.0	1.0*	20 Jan	20 Jan	0	26 Jan	26 Jan	0
4	5	4-5	Preliminary Budget Quotes	3-4	1.0	2.0	4.0	2.2*	27 Jan	27 Jan	0	10 Feb	10 Feb	0
4	6	4-6	Internal Specifications for Provn. of Services	3-4	1.0	2.0	4.0	2.2	27 Jan	10 Feb	2.2	10 Feb	23 Feb	2.2
5	7	5-7	Study & Approval of Quotes	4-5	1.6	3.0	4.4	3.0*	11 Feb	11 Feb	0	3 Mar	3 Mar	0
6	7	6-7	Final Approvals on Specifications	4-6	0.6	1.0	1.6	1.0	11 Feb	25 Feb	2.2	18 Feb	3 Mar	2.2
7	8	7-8	Selection of Site	5-7,6-7	0.2	0.6	1.0	0.6*	4 Mar	4 Mar	0	8 Mar	8 Mar	0
8	9	8-9	Approval on Selection of Site	7-8	0.2	0.6	1.0	0.6*	9 Mar	9 Mar	0	11 Mar	11 Mar	0
9	10	9-10	Orders Placed for Equipment	8-9	0.2	0.6	1.0	0.6	12 Mar	16 Mar	0.4	16 Mar	18 Mar	2.4
9	11	9-11	Internal Installation & Facility Drawings	8-9	1.0	2.2	4.4	2.4*	12 Mar	12 Mar	0	29 Mar	29 Mar	0
9	13	9-13	File for Permits to Construct	8-9	3.0	4.0	6.6	4.0	12 Mar	26 Apr	6.2	8 Apr	21 May	6.2
9	14	9-14	Bid & Award of Site Preparation	8-9	1.0	1.6	3.0	1.8	12 Mar	12 May	8.6	24 Mar	24 May	8.6
10	12	10-12	Detail Drawings from Vendors	9-10	1.0	2.2	4.4	1.8	17 Mar	19 Mar	0.4	29 Mar	31 Mar	0.4
11	12	11-12	Dummy	–	–	–	–	0	30 Mar	30 Mar	0	30 Mar	30 Mar	0
12	15	12-15	Approval on Vendors' Drawings	10-12	0.6	1.6	3.0	1.8	30 Mar	2 Apr	0.6	9 Apr	14 Apr	0.6
12	16	12-16	Completion of Installation & Facility Drawings	10-12	1.0	2.0	3.0	2.0*	30 Mar	30 Mar	0	12 Apr	12 Apr	0
13	14	13-14	Dummy	–	–	–	–	0	9 Apr	24 May	6.2	9 Apr	24 May	6.2
14	17	14-17	Site Preparation	9-14	2.0	4.0	5.0	3.6	12 Apr	25 May	6.2	5 May	17 Jun	6.2
15	16	15-16	Dummy	–	–	–	–	0	12 Apr	15 Apr	0.6	12 Apr	15 Apr	0.6
16	17	16-17	Delivery of Equipment	12-15	6.0	8.0	17.2	9.2*	15 Apr	15 Apr	–	17 Jun	17 Jun	0
17	18	17-18	Equipment Installation & Services Provision	14-17,16-17	3.0	6.0	12.0	6.6*	18 Jun	18 Jun	0	1 Aug	1 Aug	0
18	19	18-19	File Permits to Operate Equipment	17-18	3.0	4.0	6.6	4.0*	2 Aug	2 Aug	0	27 Aug	27 Aug	0
19	20	19-20	Trial Startup	18-19	0.2	1.0	2.0	1.0*	30 Aug	30 Aug	0	3 Sept	3 Sept	0
20	21	20-21	Data Collection & Coding	19-20	4.2	13.0	26.0	13.8*	6 Sept	6 Sept	0	9 Dec	9 Dec	0
21	22	21-22	Data Presentation to State for Perm. Certification	20-21	3.0	3.0	8.8	4.0*	10 Dec	10 Dec	0	6 Jan	6 Jan	0
22	23	22-23	Turnover to Production Certification	21-22	–	–	–	0*	7 Jan	7 Jan	0	7 Jan	7 Jan	0
23	24	23-24	Project End	22-23	–	–	–	0*	7 Jan	7 Jan	0	7 Jan	7 Jan	0

the calculated completion date. A re-analysis would result in a new report with updated dates and duration periods.

In the plan shown, cost estimates for each phase of the project have not been made. However, estimates and actual costs can be shown. Such a report would be similar to the one described above, but it would also show the percentage of actual completion and the actual cost incurred for each activity of the project. These actual data could be compared with estimated completion dates and costs, and an effective cost control could thus be maintained.

Some interesting points may be derived from the PERT network. For example, if all the activities on the critical path were completed in the optimistic duration, it would take less than 6 months for the whole project. On the other hand, if all the activities take the longest possible duration for their completion, it might be 22 months before the project is complete.

Using the weighted mean of duration, this project would take about 8 months from the conception of the project to the trial start-up. The calculated period of one year might be cut in half, depending on what priority it is given in the company.

Another factor to be noted is that the three activities directly relating to endorsement agencies can take close to three months for completion— the two of them falling on the critical path in this network require close to two months. These durations do not include the time during which data are compiled on the equipment performance for presentation to obtain permanent certification. On a pessimistic basis, these two activities might stretch to 3½ months.

AIR POLLUTION HEALTH EFFECTS[145]

Polluted air is not only an annoyance to our sense of smell, but it is a factor in the development of respiratory diseases which may cause disability and death. With an understanding of the health effects which air pollutants can create, a pollution engineer will appreciate the growing concern governmental agencies are showing toward this subject.

Polluted air may cause the eyes to burn and water, but the damaging effects take place in the respiratory passages. During a day of only moderate activity, over 300 cu ft of air is breathed into and out of an individual's lungs (see respiratory system diagram, Figure 1-13).

Air is inspired into the chest through the trachea, or windpipe. The trachea divides into two branches, the right and left main bronchi, which lead into the lungs. These bronchial tubes divide and redivide many times, finally ending in hundreds of tiny hollow twigs that carry the air into millions of minute air sacs called alveoli.

These tiny air cells give the normal lung its light, airy, spongy consistency. Walls of the air sacs are supplied with a network of blood vessels arranged so that only an extremely thin membrane separates the air in the alveoli from the blood. It is here, in the alveoli, that oxygen from the

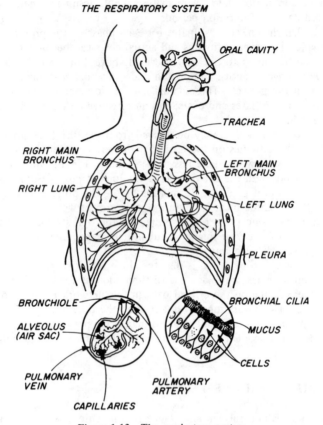

THE RESPIRATORY SYSTEM

Figure 1-13. The respiratory system.

inspired air passes into the blood and carbon dioxide passes from the blood to the air being expired.

The windpipe and smaller breathing tubes are lined with specialized cells from which sprout microscopic hairs called cilia. A thin layer of mucus, continuously secreted by other cells, coats the inner surface of the air passages. Under normal circumstances, the cilia beat rhythmically in such a way as to propel the mucus layer in an upward direction. As the mucus layer is swept toward the vocal cords, particulate matter which may have been breathed into the lungs and trapped on the sticky surface is carried up to the throat where, with a cough, it is cleared.

Current medical efforts are being directed toward earlier diagnosis and treatment of the lung diseases discussed below. These, coupled with efforts to control and eventually eliminate air contaminants, offer the best hope for prevention of health problems related to air pollution.

Occupational Diseases

There are several forms of chronic lung disease occurring in persons who are industrially exposed to air pollutants. One of these is silicosis, caused by the inhalation of air containing finely divided particles of silica. The end result is the formation of fibrous scar tissue in the lungs, greatly increasing shortness of breath. In another occupational disease, sensitive persons exposed to dusty, moldy vegetable materials or fertilizer may develop an inflammatory lung disease known as "farmer's lung."

Nonspecific Upper Respiratory Diseases

A well publicized study made in 1950 of a Maryland community compared two sizable groups of people who differed mainly in the degree of their exposure to air pollution. Those living in the area of higher exposure were found to have a higher incidence of common colds.

Other studies in industrial areas have shown increased frequency of absenteeism due to acute upper respiratory disease correlating with periods of high air pollution exposure.

Lung Cancer

A direct relationship has been established between cigarette smoking and the development of lung cancer. The situation is not so clear regarding the carcinogenic effect of industrial air contaminants, but facts highly suggestive of a positive relationship have emerged. Certain materials from polluted air, when collected and concentrated, can produce cancer when applied to the skin of rodents. Lung cancer has been increasing during the last 50 years, particularly in urban areas subjected to high levels of air pollution.

OSHA CHECKLIST[185]

There were particularly good reasons for the federal government to consider the new safety and health legislation which became the Williams-Steiger Occupational Safety and Health Act in 1970. Estimates indicate that annual on-the-job accidents kill 14,000 and injure 2.2 million employees.

It is important to appreciate that there are two aspects to OSHA—safety and health. OSHA requires that the history of health hazards at a plant be logged even as the company safety history is logged. Occupational illness is, according to OSHA, any abnormal condition or disorder (other than one resulting from an occupation injury) caused by exposure to environmental factors associated with his employment. It includes acute and chronic illnesses or diseases which may be caused by inhalation, absorption, ingestion, or direct contact, and which can be included in the categories listed below.

Occupational Skin Diseases or Disorders—Examples: contact dermatitis, eczema, or rash caused by primary irritants and sensitizers or poisonous plants; oil acne; chrome ulcers; chemical burns or inflammations.

Dust Diseases of the Lungs (Pneumoconioses)—Examples: silicosis, asbestosis, coal worker's pneumoconiosis, byssinosis, and other pneumoconioses.

Respiratory Conditions due to Toxic Agents—Examples: pneumonitis, pharyngitis, rhinitis or acute congestion due to chemicals, dusts, gases or fumes; farmer's lung.

Poisoning (Systemic effects of toxic materials)—Examples: poisoning by lead, mercury, cadmium, arsenic, or other metals; poisoning by carbon monixide, hydrogen sulfide or other gases; poisoning by benzol, carbon tetrachloride, or other organic solvents; poisoning by insecticide sprays such as parathion, lead arsenate; poisoning by other chemicals such as formaldehyde, plastics and resins.

Disorders due to Physical Agents (other than toxic materials)—Examples: heatstroke, sunstroke, heat exhaustion and other effects of environmental heat; freezing, frostbite, and effects of exposure to low tempeatures; caisson disease, effects of ionizing radiation (isotopes, X-rays, radium); effects of non-ionizing radiation (welding flash, ultraviolet rays, microwaves, sunburn).

Disorders due to Repeated Trauma—Examples: noise-induced hearing loss; synovitis and bursitis, Raynaud's phenomena; and other conditions due to repeated motion, vibration or pressure.

Other Occupational Illnesses—Examples: anthrax, brucellosis, infectious hepatitis, malignant and benign tumors, food poisoning, histoplasmosis, coccidioidomycosis.

Pollution engineering is deeply involved with controlling many on-the-job hazards that may contribute to health problems, outside and inside the plant. Since company problems vary even in the same industry, the best way to understand what OSHA is about is to review some basic check-points for industry.

Check for health hazards resulting from production—*i.e.*, toxic dusts, fumes, liquids, poisonous substances, radiation, noise, excessive heat. While most companies have an on-going program for the control of safety hazards, health problems have been overlooked in many cases. These problems should be evaluated and steps taken to correct conditions which are beyond acceptable limits, before an OSHA inspection takes place. Effective emergency procedures should be instituted. OSHA has put new effort into health-hazard research, which will tighten regulations as conclusive results are made known.

Ascertain whether your firm conforms to your industry standards. A firm can be in full compliance with all OSHA standards and still be cited under the law, if it is in violation of standards recognized within its own industry. It is virtually impossible to have OSHA standards for every circumstance in every industry. So, if there are recognized standards within an industry, even though the situation is not specifically covered by OSHA, these industry standards must be complied with.

Designate an individual or safety inspection team to check plant facilities to evaluate compliance with recognized safety and OSHA standards—before an OSHA inspection occurs.

Regulations cover the entire industrial process—for example, the following may be considered "serious" violations:

1. Access to portable extinguishers, fire hoses and sprinkler heads blocked.
2. Limited emergency egress—inadequate aisleways, stairways, fire escapes and exits.
3. Inadequate ventilation and air control—lack of make-up air.
4. Spray and dip painting not standard.
5. Excessive floor loadings in old wooden buildings.
6. Excessive noise.
7. Machinery without adequate guards.
8. Electrical tools, equipment receptacles without proper electrical grounds.
9. Lack of safety valves or blockage of valves on high-pressure steam lines.
10. Unprotected floor or wall openings.
11. Improper storage of flammable materials.
12. Lack of readily accessible emergency showers near hazardous acids and chemicals.
13. Lack of canopies on fork-lift trucks.
14. Inadequate fire protection or fire suppression systems.
15. Exposed energized electrical equipment.
16. Lack of identification on pipes, electrical switches, breakers.
17. Inaccessible shut-off valves.

Even so-called "non-serious" violations may bring penalties of up to $1000.

1. Unmarked traffic lanes or inside vehicle traffic.
2. Lack of proper guardrails at low level.
3. Lack of 3-wire electrical outlets.
4. Ladders not equipped with safety shoes.
5. Inadequate lighting for tasks and movement.

Make sure the supervisory staff understand requirements for reporting serious injuries. All fatal accidents and certain serious accidents resulting in the hospitalization of 5 or more must be reported to the OSHA regional office within 48 hours. This should be done by phone or telegraph.

Keep an injury log. This is another requirement, and one of the first things that a compliance officer would request to see. These must be kept current and available at each work location.

Inform employees of their rights. The law requires that employees be informed of their rights under the law. For example, they can request a physical examination under certain circumstances at your firm's expense, with results relayed to their personal physicians. Periodic training sessions would be the best means for informing employees. The use of the company newsletter and bulletins is also recommended. Even direct communication by memo could be helpful.

Document enforcement of OSHA standards. It may be necessary to prove that you really enforce safety standards within your company. Enforcement of standards is important in case of a serious accident to establish that the violation which resulted in the accident is not condoned by your company. Records, including evidence of disciplinary action taken against individuals, could be quite important.

Post the OSHA notice at each work location. This is a requirement and failure to post this notice could result in a citation.

Have copies of the standards available for employee use. At each location where employees work, keep a copy of the standards applicable to their work, accessible for them to read.

Prepare a plan of action in case of an OSHA inspection. It is best to determine in advance who is to accompany the inspector or the compliance officer as a representative of management and who is to represent the labor group. If more than one craft or trade is involved, a different labor representative may be needed for various sections of the inspection. However, these must be chosen by the employees. It would be advisable to do this in advance of any inspection.

It is important to understand that the written standards of OSHA, the Construction Safety Act, the Walsh-Healey Act, the NFPA, the National Electric Code and the ANSI Standards are all items with which a company can be in complete compliance, and yet still be cited for a violation of the OSHA law. The General Duty Clause of the Act is one under which most of the citations have been issued. These citations do not necessarily refer to any specific written standard. The General Duty Clause of the Act states that each employer "shall furnish to each of his employees employment and a place of employment which are free from recognized hazards that are causing or likely to cause death or serious physical harm to his employees; and (2) shall comply with the articles of Safety and Health Standards promulgated under this Act." It is impossible for the OSHA standards to encompass all of the recognized hazards in every type of industry. However, the safety rules of many companies or at least their industry safety rules often pinpoint recognized hazards within the industry. Even though standards are not mentioned in OSHA, the company can be cited for allowing its employees to violate the rules.

Measuring up to the varied and rigorous standards of OSHA is a challenge. With the federal government's intent to be effective in this area, and with organized labor in a watchdog role, all eyes will be checking for OSHA compliance.

GRAPHICAL DETERMINATION OF POLLUTION PROBLEMS[166]

The number and variety of air pollution surveillance locations grows larger each day. Beginning with manual intermittent sampling station techniques such as bubblers, dustfall buckets and sulfation plates, the degree of sophistication proceeds to continuously operating gas analyzers transmitting data by means of telemetry networks to real-time computer operated data acquisition systems. Once the data is recorded, however, be it handwritten on a piece of paper or recorded on a complex magnetic tape system, a common problem presents itself: what analysis should be made and how can it be made useful to the greatest number of people?

During study of air quality data for a substantial portion of the Metropolitan Washington Air Quality Control Region, an attempt was made to find more suitable graphic display of the data. A series of overlays were developed to present some of the relationships between such factors as pollutant source distribution, population densities, pollutant concentration levels, monitoring and site locations.

The overlays are designed to be placed over a base map, which in this case is a color-coded population density map prepared from government information. This type of display is more informative to a wider range of persons than statistical tables or tabulations of figures.

Preparing such maps is a fairly simple and straightforward process. Figure 1-14 is an overlay designed to show the relationship between where people live and where the major stationary sources of pollution are located. For simplicity, only two broad categories of sources are shown: those emitting over 100 tons per year, and those emitting over 1000 tons per year. Each source is shown with a plume extending downwind from the source location. This gives a graphic picture of where the sources are, where the pollution goes and whether or not the pollution produced affects greater or lesser population. The plumes may be reoriented to represent any desired wind direction.

The emission **plu**mes shown are for low wind velocities and have a horizontal dispersion angle of 12 degrees. They are shown approximately 5 miles in length and begin about a mile downwind from the source where the maximum ground level concentration occurs. Obviously, this technique can be refined to take into consideration stack height, time of day and alternate dispersion factors. In the more refined state we have a "poor man's diffusion model," which is simple to construct and gives a good first look at the characteristics of pollution on a large geographic scale.

Using this simple model it can be shown that there are specific wind directions which bring pollutants into contact with the largest number of people, where these people are located, and conditions under which sets of pollution sources overlap to cause pollution "hot spots." Using historical meteorological data, such a graphical display can be used to suggest pollution monitoring sites where public health or vegetation effects are predicted to be the greatest.

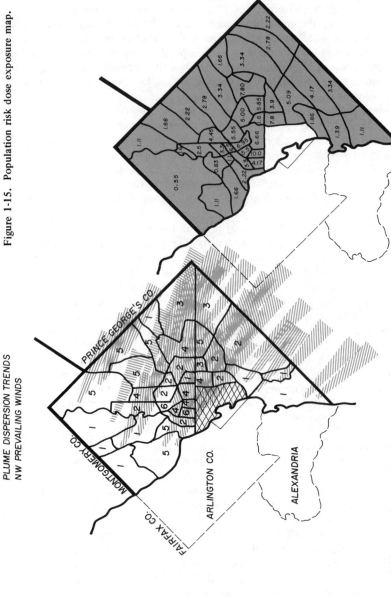

Figure 1-15. Population risk dose exposure map.

Figure 1-14. Air pollution dispersion over Washington, D.C. with NW winds.

Figure 1-15 is a combination of two overlays prepared to produce a population dose exposure map. It was prepared by multiplying the number of people per unit area by the average suspended particulate concentration for the same area. From an effects standpoint, this represents a risk map— the higher the number, the higher the risk. The numbers are chosen to represent a scale of 1 to 10, where 10 is simply the worst case for the data given. For simplicity, only six categories of population density were used.

A graphic presentation such as this can be useful in planning and evaluating long-range abatement strategy. This population dose map is for a single pollutant; however, the technique can also be used for multiple pollutants. Including vehicular traffic patterns and density adds another dimension. A more complete map of this type can be useful in land use planning, highway location, location of schools, hospitals and other munici-pal facilities which will either influence pollution levels or be adversely affected by increased pollution. Optimum monitoring sites are also suggested by more complete maps of this type.

MEASUREMENT OF POLLUTION IN AIR AND WATER[26]

Emphasis facing industry on eliminating or reducing pollution has made it increasingly important to be able to analyze qualitatively and quantitatively those substances that comprise air and water quality. Air pollution control activities in the United States, greatly expanded with the Clean Air Act of 1963 and its amendments through 1970, gave added impetus. Industrial water pollution control activities began to accelerate along parallel lines with clean water legislation and permit requirements for industrial wastewater discharges. To date, standards for maximum concentrations in the atmosphere and for flue gas emissions have been established for sulfur dioxide (SO_2), particulates, carbon monoxide (CO), photo-chemical oxidants, nitrous oxides (NO_x), and hydrocarbons (HC). Likewise, considerable interest has centered around the constituents of liquid waste discharges and their impact on water quality, the list of parameters often requiring extensive investigations.

The pollution engineer is faced with the problem of making these measurements for a number of reasons:

1. Such analytical data may be required for designing control equipment;
2. Instrumentation may be required for process control;
3. Data may be required for submission to government enforcement agencies to comply with source measurement requirements;
4. As legal evidence of compliance.

Sampling Policies

Pollution control measurements, be they in gaseous or liquid media, take two general forms: *sampling* and *continuous monitoring*. In both cases, a thorough knowledge of the process and the possible constituents in the resulting discharges is required for an intelligent program to be performed. A complete study is required before a sampling or continuous monitoring program is undertaken. The specific parameters to be measured of course may be dictated by process considerations or as required by governmental enforcement agencies. Measurement itself requires the following steps:

1. The source, sampling site selection and execution. This must be consistent with intended interpretation.
2. Sample transport—delivery of the sample to point of treatment, collection or the instrument.
3. Sample treatment—if necessary, preserve or convert to a form suitable for analyses and not affect the sample integrity.
4. Sample analyses by a suitably acceptable method. This generates the quantitative or qualitative data.
5. Information reduction and display.
6. Information codification and interpretation.

The state-of-the-art for measurement of pollution in air and water is still in the formative stage. Governmental regulations are subject to change and will undoubtedly be updated as monitoring devices become more developed. The following general outline is offered as prerequisites to any measurement program.

1. Knowledge of air and water regulations affecting the source or plant.
2. An intimate knowledge of processes that are the source of the discharge. This will initially identify what pollutants may be present and quite often emission rates may be calculated from available data.
3. An inventory of all air and water emission sources. This data should be quantitative with respect to flow volumes.
4. Process flow sheets may be useful showing raw materials, their quantities, process flow rates and fluid (gas or liquid) emission rates at critical process points.
5. Preliminary field measurements may be required to substantiate the previously referenced points.

Measurement of Air Pollution

Generally speaking, sampling techniques have been perfected for the major air pollutants: particulates, sulfur oxides, carbon monoxide, hydrocarbons and nitrous oxides. Other materials can be analyzed for, and methods and instrumentation may be available for specific requirements.

The following are the major sampling technique methods available and each has its advantages in relation to the problem at hand:

1. Manual sampling for off-site analysis. The principal method uses a previously evacuated vessel in which the sample is collected for laboratory analysis.

2. Sampling by pump and on-site analysis.
3. In-stack monitoring and analysis of the total stack effluent. This is usually an electro-optical measurement.
4. Remote sensing also restricted to electro-optical measurement or long path optical methods. This class of instruments includes atmospheric air monitoring systems.

Particulate Determination

Fundamental requirements for most stack measurements are the use of a pitot tube and thermocouples or thermometers to determine gas flow velocities and temperatures at the probe point in the stack. Aside from establishing gas volume during the test run this also insures that sampling is done at isokinetic rates.

Automatic sampling withdrawal is accomplished through a vacuum pump train as shown in Figure 1-16. This train contains a flow meter,

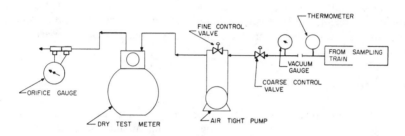

Figure 1-16. Vacuum pump train.

thermometers, and valves to adjust and check flow rates. Figure 1-17 illustrates the EPA sampling train. Probe and filter box are heated to maintain the sample above 250°F to prevent condensation. An ice bath contains four glass impingers, the first two being filled with distilled water, the third dry, and the fourth with silica gel to remove all moisture. As particulate matter is collected, pressure drop across the filter increases so that constant valve adjustment is required to maintain isokinetic conditions. On test completion, all loose particulate matter is collected from parts exposed to the stack gas for a total weight. Calculations are also required to prove isokinetic sampling and determine particulate matter and moisture per unit volume of flue gas.

Variations of these sampling trains are used. The ASME sampling train, for example, consists of a stainless steel nozzle and probe connected to a canister containing a filter to trap particulate matter. Both probe and canister are heated to prevent condensation. Adjustment valves and a vacuum pump complete the train. The Western Precipitation Method utilizes a ceramic thimble to filter out particulates.

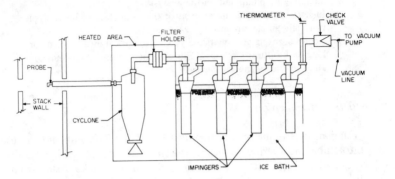

Figure 1-17. EPA sampling train.

Particle sizing, while not as yet a legal requirement, is often invaluable in control equipment design. Physical sizing may be achieved by wet or dry sieve screening where sufficient sample quantity is available. Microscopic analysis, light scattering techniques, and aerodynamic particle sizing are additional methods.

Smoke Monitoring

Density or opacity measurement of smoke from stacks has been a recognized analytical method for a long time. Maximillian Ringelmann developed his smoke charts in 1897 and Ringelmann numbers to compare smoke opacity have been widely used. The Ringelmann chart indicates intensity of a black plume and is relatively difficult to relate to red or yellow plumes. Measurements should be made by trained observers.

Instrumented smoke detectors and monitors have been available for quite some time. Most smoke monitors are transmission photometers, measuring light transmittance or the light intensity through the stack,

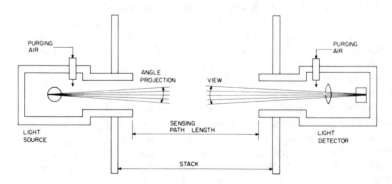

Figure 1-18. Straight through smoke monitor system.

having a light source on one side of the stack and a detector on the other side. Smoke detectors available commercially vary in price up to $5000.

Continuous Monitoring Devices
for Flue Gases

Continuous monitoring devices are commercially available for analyzing sulfur and nitrous oxides, carbon monoxide, hydrogen sulfide, particulate concentrations in either ambient air or stack flue gas emissions. These devices tend to be selective, analyzing only the portion of the gas intended. Systems include gas chromatography, photometry, chemiluminescence, fuel cell sensors, ultraviolet absorption, light scattering techniques, and beta-ray measuring of particulates. Most of these instruments and devices are in the $10,000 to $15,000 price range and have heretofore not achieved widespread use.

Measurement of Water Pollution

Improper identification of the nature or quantity of contaminant loading in wastewaters will invariably lead to the wrong control program. Wastewater sampling and flow data are the two most important factors of any water pollution control program. The analysis of samples identifies contaminants and their concentration, and flow determination totals the results.

As in the case of air pollution measurement, plant and process familiarity is the primary requisite. Plant drain layouts, sewers and surface discharges should be reviewed with actual examinations to determine direction and resultant outfall. In the absence of adequate drawings hidden or underground streams may be traced with dyes. Proper selection of sampling points assures samples that will yield representative and reliable flow and analytical data. A good place to begin sampling is at plant outfalls, since this data will characterize total plant effluent. Branch streams to main drains are also important and can be further studied to determine what plant or process operations contribute to the pollution load. The latter source is usually very specific in data contribution.

Once the drain, discharge and sample points have been established, it is necessary to determine the type of sample to be collected at each point. Samples are either grab or composite. A grab sample is a single sample, while the composite is one composed of a series of samples collected over a period of time and blended for analysis.

Composite Sampling

When flows are relatively uniform, samples can be manually composited by collecting a series of grab samples at given time intervals and blending. If flow rates vary, grab samples are proportioned to the flow. One major advantage to manual sampling is that visual observation of the streams is possible, particularly for changes such as color, solids, oil, etc.

Composite samples can also be collected automatically. A number of such automatic samplers are commercially available ranging in cost from $250 to $1500. Some of these devices can collect continuously in one container, while others will separate portions in a number of containers, thus providing a composite series of samples. Some automatic samplers can operate continuously or can be timed on an intermittent sequence.

Some automatic sampler problems include battery run-down, pump tubing and orifice plugging by suspended matter and problems of obtaining representative samples from streams that may be stratified or contain high quantities of suspended solids, oil or grease. Because these problems are sometimes difficult to overcome, allowances must be made when interpreting analytical results.

Continuous Monitors and Analyzers

The discussion on sampling has thus far assumed material collected will be analyzed in an off-site laboratory. Wastewater streams can be continuously and automatically monitored and analyzed by reliable on-stream effluent monitoring equipment currently on the market.

Many important parameters can be monitored and analyzed by such equipment. The limiting factor, however, is in the number of parameters that can be simultaneously monitored and analyzed. Many units can automatically analyze only one parameter, while more expensive units will analyze from 3 to 10 simultaneously.

In general, however, automatic on-stream analyzers are not utilized in initial or occasional plant wastewater sampling. Results of the survey from composite sampling may indicate whether automatic equipment can be profitably installed.

Regulatory agencies sometimes insist that certain parameters be monitored continuously to insure that they are within allowable discharge limits. This is particularly true of effluents from wastewater treatment facilities. Again, the sample point should be selected and equipment installed so that the sample collected is representative of the stream in question.

In addition to the selection of sample points, sampling methods and flow, other considerations are:

1. Volume of the sample required,
2. Analysis required, and
3. Proper handling and preservation of samples.

Sample volume depends upon the analyses required. Typical parameters are shown in Table 1-10.

Table 1-10. Typical Parameters that may be Required in
Wastewater Effluent Testing

pH	Nickel
Alkalinity or acidity	Lead
Suspended solids	Cadmium
Volatile suspended solids	Mercury
Settleable solids	Arsenic
Chloride	Zinc
Sulfate	Hexavalent chromium
Calcium	Total chromium
Magnesium	Total phosphate
Hardness	Total nitrogen
Aluminum	Ammonia nitrogen
Iron	Oil and grease
Copper	Total organic carbon (TOC)
Fluoride	Biochemical oxygen demand (BOD)
Manganese	Chemical oxygen demand (COD)
Cyanide	Phenol

After samples are collected, proper handling is important for valid test results. Many analyses require that samples be specially preserved immediately after collection to prevent characteristic changes of specific constituents.

Flow Measurement

Directly associated with sampling programs is the need to develop flow data for various effluent streams. Such data may be necessary to obtain in-plant water consumption figures and to quantify contaminants in the waste streams.

Floating objects, dyes and salt solutions are sometimes used to give the velocity of water effluents in sewers and open streams while other devices such as weirs give volumetric measurement and flow. If velocity measurements are made, the cross-sectional area of flow within the pipe that is occupied by the water must also be determined. Different types of devices can be used for flow measurement depending on channel features and flow volumes.

CHAPTER 2

PARTICULATE CONTROL OF AIR POLLUTION
DRY METHODS

ECONOMICS OF SELECTING
PARTICULATE CONTROL EQUIPMENT[164]

Solving one pollution problem often creates problems in another area. The control of particulate matter may affect process equipment. If the carrier is other than air, the chances are that a dual problem in air pollution control exists. When equipment for particulate control is to be purchased, the systems approach should be used. It is through this approach that the most in economics of control can be realized.

Particulate control economics are not different from the economics of control of water or other air pollutants. There are many variables that will affect the collection efficiency of particulates by a device; these will also affect the decision as to what particular brand of device to purchase. Economic factors to be considered include:

1. Optimum system for a particular application to collect particulates.
2. Operating costs over the expected life use of the system.
3. Maintenance costs over the expected life use of the system.
4. Conversion of the system to future applications.
5. Expansion or contraction of the system to future process capacities.
6. Optimum financial considerations to obtain use of the system.

Some of the factors to be considered in selecting an optimum particulate control system are characteristics of the process, carrier medium, the particulate and device limitations.

Process characteristics that affect collector system efficiency are carrier medium rate of flow in relation to time, particulate density, collector system efficiency requirements, pressure differential allowed, product quality requirements, process material rate of flow and quality variations in process material.

Carrier medium characteristics include many considerations of the particulate such as density, electrical conductivity, corrosiveness, toxicity, inflammability and explosiveness. Other characteristics include temperature, humidity, pressure dew points of condensable components and viscosity.

Particulate characteristics include (but are not limited to) shape, size and range of sizes in relation to time; density; and physicochemical properties such as tendencies to corrode, absorb moisture, agglomerate and to become toxic, sticky, inflammable, explosive or conduct electricity.

Device limitations are physical consisting of space and material considerations. These include floor area, height available, foundation requirements, specifications of the device to withstand temperature, pressure, corrosiveness and vibration.

Once these factors have been considered and a system chosen to effectively collect particulate matter, it is still not possible to economically justify the system. There is always more than one system that will accomplish the objective (Table 2-1). In economics, the prime goal is to

Table 2-1. Comparison of Particulate Control Systems

Method	Efficiency Percent by Weight	Installed cost per cfm
Dry Devices:		
Baghouse		
Med. temperature (250°F)	>99	0.75-$1.50
High temperature (500°F)	>99	1.50- 3.00
Gravity Chambers	35-93	0.10- 0.40
Cyclones		
General application	65-95	0.40- 0.52
Flyash	55-95	0.10- 0.20
Refractory kiln process	70-95	0.85- 1.75
Electrostatic precipitator		
Single-stage	75-99.9	1.25- 3.50
Two-stage	50-99.8	0.75- 2.50
Special applications	95-99+	3.50-15.00
	50-99 (All)	
Wet Devices:		
Spray chamber		0.25- 0.50
Cyclone or orifice scrubber		0.50- 1.50
Venturi scrubber (mild steel)		0.50- 2.00
Venturi scrubber (stainless steel)		1.00- 3.00

justify the selection of a system as the one that returns the most profit for the invested dollar. This is not an easy task. Often it is advantageous to consult equipment manufacturers to save time and costly errors. They may have resolved a similar problem before, or at least know what has or has not been tried (Table 2-2).

Operating cost is one of the other factors that must be considered, especially if power collection media or manpower are costly. Availability of water, electricity and trained manpower can and does effect equipment selection. Reclamation or disposal of particulate material can be decisive factors in operating costs. In any case a cost figure must be assigned to the operating factors for the life of the pollution control equipment.

Maintenance costs can seriously affect equipment selection. These are sometimes influenced by the state-of-the-art in solving such problems as bearing failure, abrasive wear, or filter life and efficiency. They are also influenced by availability of trained personnel and permitted shutdown time for breakdown and preventive maintenance. This in turn influences design reliability. Again a maintenance cost value must be designed to determine optimum system selection. Generalized equations for computing operating and maintenance costs are included in Table 2-3.

The life cycle use of the equipment must also be considered, as well as its adaptability to changing industrial processes. This includes the conversion of the equipment to either new processes or a change in a process such as production rate or process technique. In industries such as aerospace where production runs or process techniques may be short lived, this is a serious consideration. The pollution engineer must work extremely closely with the production and product assurance department to see that any process change is fully coordinated before being implemented. If this is not done, serious violations of the law, damage to pollution control equipment and variations in product quality can occur.

Once these considerations have been given a value, it would appear that the selection of a particulate collection system would be easy. This is not the case, for the terms of paying for the system may be a factor in the system selection. It is well to work closely with the corporate controller and be aware of the industry's financial situation. It may be more advantageous for the company to invest its money in some other venture that would return more profit than pollution control devices.

Not all particulate equipment manufacturers offer a lease-vs-buy option, though there are companies that will buy the equipment and lease it back. In a short-term use situation for the equipment, it is generally better to lease than to buy. Some manufacturers will sell the equipment with a credit guarantee toward exchange for other equipment of their manufacture at a later date. All other factors considered, the financial situation over the expected life use of the equipment can and does affect the ultimate selection of a particulate control system.

Another consideration in financing equipment is obtaining a federal grant to develop new technology for a particulate application. This is

Table 2-2. Industrial Process and Typical Particulate Control Method

Industrial Process	Particulates	Typical Control Method
Abrasive blasting	Dust	Baghouse or centrifugal scrubber
Aggregate plants	Dust	Spray and/or baghouse
Asphalt (hot) paving patch plants	Dust	Cyclone and centrifugal spray chamber or baffle spray tower
Carbon black production	Dust	Baghouse or cyclone and electrostatic precipitator
Cement batching plants	Dust	Baghouse, panel filter
Chemical milling	Mists	Wet collector, spray and baffle type
Coffee roasters	Dust, chaff, mists, smoke	Afterburners and/or cyclones
Dryers	Dust, smoke	Scrubber and/or baghouse and/or cyclones
Electroplating	Acid mists	Spray scrubber
Feed and grain mills	Dust	Cyclones or baghouse
Fish cannery	Dust, smoke	Cyclones and contact condenser scrubber
Food processing: 1. Deep fat frying	Smoke, food particles	Incineration and two-stage electrostatic precipitator
2. Smokehouses	Smoke, organic matter	Incineration or two-stage electrostatic precipitator with centrifugal scrubber
Frit smelters	Dust, fumes	Baghouse or venturi scrubber
Glass manufacturing: 1. Raw material	Dust	Baghouse or panel filter
2. Melting furnace	Various dusts	Cyclone scrubber or baghouse
3. Fanning machine	Smoke	Process control
Heat treating	Salt fumes	Baghouse
Insecticide manufacturing: 1. Dry	Toxic dusts	Baghouse
2. Liquid	Aerosols	Packed tower scrubber
Metallurgical processes: 1. Foundries a. Core ovens	Smoke, organic acid	Afterburning
b. Sand equipment	Dust, smoke, organic vapors	Baghouse or scrubber
2. Metal separation processes	Smoke, dust, fumes	Baghouse and/or afterburning
3. Lead refining	Smoke, fumes, oil vapor and dust	Afterburning and baghouse

Industrial Process	Particulates	Typical Control Method
4. Aluminum melting	Smoke, fumes	Horizontal wet cyclone or dynamic collector or packed column water scrubber and baghouse or ultrasonic agglomerator followed by multiple dry cyclone or electrostatic precipitator
5. Brass and bronze melting	Dusts, metallic fumes	Baghouse
6. Iron casting	Metallic fumes, smoke, oil vapor	Baghouse or electrostatic precipitator with auxiliary devices
7. Steel manufacturing	Fumes	Baghouse or electrostatic precipitator with condition gas stream
Mineral wood furnace	Fumes, Aerosols, oil vapors, wood fibers	Baghouse, afterburning
Paint baking ovens	Smoke, Aerosols	Process control
Paper mills, kraft	Chemical dusts	Electrostatic precipitator, venturi scrubbers
Perlite furnace	Dust	Baghouse
Petroleum equipment:		
1. Airblown asphalt	Aerosols	Scrubber and/or incineration
2. Catalyst regeneration	Dust, oil mists, aerosols	Cyclones, wet and/or dry, and electrostatic precipitators
3. Storage facilities	Aerosols	Scrubbers for water-soluble products
Phosphoric acid	Fumes, acid mist	Electrostatic precipitator or venturi scrubber and packed tower or packed tower and glass fiber packed filter
Pipe coating	Fumes	Baffle water spray chamber
Pneumatic conveyors	Dust	Baghouse
Portland cement	Alkali and dusts	Baghouse, electrostatic precipitator, mechanical collectors
Roofing felt saturators	Mists	Two-stage precipitators or baghouse or spray scrubbers
Rubber compounding equipment	Dusts, fumes, oil mists	Baghouse
Sulphuric acid	Acid mists	Electrostatic precipitators, (tube type) or packed bed separators or wire mesh eliminator or ceramic filters or sonic agglomeration and cyclones
Surface coating	Paint particles	Filter pads, water spray, baffle plates
Woodworking equipment	Dust, chips	Cyclone or baghouse
Zinc galvanizing	Oil mist, fumes	Baghouse, electrostatic precipitator

Table 2-3. Generalized Annual Operating and Maintenance Cost Equations For Particulate Control Equipment

Collector	Equation
Mechanical centrifugal	$G = S \dfrac{[0.7457PHK + M]}{6356E}$
Wet	$G = S \dfrac{[0.7457HK\ (Z + Qh) + WHL + M]}{1980}$
Electrostatic	$G = S\ [\ (JHK + M)\ ]$
Baghouse	$G = S \dfrac{[0.7457PHK + M]}{6356E}$

where:

G = annual costs in dollars for operating and maintenance
E = fan efficiency expressed as a decimal
H = hours of operation annually
h = elevation of pumping liquor in circulating system for collector in feet
J = kilowatts per acfm
K = cost of electricity in dollars per kilowatt-hour
L = cost of liquid in dollars per acfm
*M = maintenance cost per acfm in dollars per cfm
P = pressure drop in inches of water
Q = water circulation in gallons per acfm
S = design capacity of the unit in acfm
W = make-up liquid rate in gallons per hour per acfm
Z = total power input required for a specific scrubbing efficiency in horsepower per acfm

*Typical values for M are:

Collector	Dollars per acfm		
	Low	Medium	High
Baghouse	0.02	0.05	0.08
Electrostatic			
High-voltage	0.01	0.02	0.03
Low-voltage	0.005	0.014	0.02
Gravitational and dry centrifugal	0.005	0.015	0.025
Wet collectors	0.02	0.04	0.06

available if subsequent benefit to overall air pollution control is probable, and of course the information becomes part of the public domain. Many states allow a tax exemption, either sales or property or both, on air pollution control equipment. There is also a corporate tax consideration in lease-vs-buy, in that leased expense can be written off immediately while capital investment must be written off over a stipulated period of years. Which will be the most advantageous for a particular company depends on financial circumstances and business climate.

FABRIC FILTERS FOR DUST COLLECTION[38]

Use of fabric filters in the separation of particulate matter from gas streams may be, under certain conditions, the most efficient and economic of the principal methods employed.

In simple terms, a fabric filter is a device used for separating suspended impurities from process gases and liquids. Since the turn of the century, fabric filters, or baghouses, have been widely used in the mining industry. Application has spread to a wide variety of industries, making filters one of the most popular means to effective industrial air pollution control.

For years filtration has been considered more of an art than a science. Technical advances have improved filtration procedures, but there is no general method of application. Each problem still must be handled separately, taking into consideration its own special characteristics. Experience seems to be the key to successful application of filtration systems.

Dry dust filters are readily available in sizes from a few square feet up to several hundred thousand square feet of cloth. Gas flows that can be handled by individual units range from less than 100 to more than 1,000,000 cfm. A typical manufacturing plant may have many individual units in use, depending on applications, for everything from product recovery on a small process to air pollution control for an entire plant.

The total number of fabric filters in use in the United States has been estimated to exceed 100,000 units. Approximate unit sales run in excess of $25,000,000 for 8000 or more new installations per year, plus an additional 100,000,000 sq ft of filter fabrics sold at a value exceeding $15,000,000.

A fabric filter is composed of woven or felted textile material, through which the contaminated gas is passed to separate the particles. As the cake builds up on the cloth, resistance to gas flow increases. Deposits, therefore, must be removed periodically by vigorous cleaning of the cloth. This cleaning reduces the gas flow resistance and maintains the proper pressure drop across the filter.

Cloth filters are used in the control of dust sizes varying from submicron range fumes to powders 200 microns in diameter, and for filtering particle concentrations from 0.0001 grams per cubic meter (urban atmospheric dust) to 1000 grams per cubic meter (pneumatic conveying). Fabric filters are also finding use in providing a substrate for support of granular reactants or adsorbents, making gaseous components recovery possible.

The fabric filter's design is similar to that of a large vacuum cleaner. It consists of bags of various shapes made up of a porous fabric. The fabric must be able to withstand thermal, chemical, and mechanical rigors.

Filter bags come in two major designs—flat (envelope) bags, and round (tubular) bags (Figures 2-1 and 2-2).

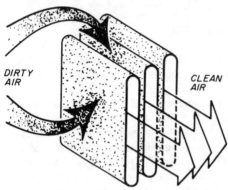

Figure 2-1. Envelope fabric filters can be mounted close together, offering large amount of filter area in relatively small space. Envelopes collect dust on outside surface and can be cleaned either by shaking or by reverse flow of gas or air.

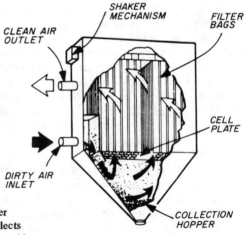

Figure 2-2. Typical tubular filter baghouse compartment collects dust inside tubes and is cleaned by mechanical shaker mechanism. Cleaning cycle is controlled automatically by timer or differential pressure-sensing device.

Envelope-type bags capture the dust from the airstream on the outside of the bag which is prevented from collapsing by internal frames. This design offers the greatest surface area contact between cloth and air (made possible by close spacing of the envelopes). Occasionally, the dust bridges or plugs the spacing between the envelopes. In such cases it becomes necessary to remove every other bag.

Tubular bags are open at one end and closed at the other, and come in a variety of forms. Sometimes the bags are sewed together in groups to form a multibag system. Disadvantages of this particular arrangement are that bag adjustment is limited and bag replacement may be a costly procedure.

In other tubular designs, gas enters either from the top or bottom. When the gas stream enters the system at the bottom, it passes through the hopper and upward into the filtering elements, allowing for partial separation by settling of the coarse particles in the hopper. Thus, the fabric handles only the lighter dust.

When the gas stream enters from the top, the dust load passes longitudinally through the entire tube length before reaching the hopper. Top entry baghouses generate a dead gas pocket in the hopper which can present a condensation problem when moisture-laden gas is being cleaned.

Gas flow direction in tubular bags can be either outside-in or inside-out. If outside-in, a frame is installed inside the bag to prevent collapsing. Arresting the dust on the outside of the bag requires periodic inspection of the unit on the dirty gas side. Shortened bag life may be experienced because of bag and frame contact.

Various theories have been devised to explain how air filtration works, of which four principally are drawn upon to describe the phenomena:

> *Inertial collection* occurs when, because of its inertia, a particle does not change direction with the gas and impinges on a fiber placed perpendicular to the gas flow direction. Inertial collection is the basic mechanism in woven media.
>
> *Interception* takes place when an inertialess particle (which does not cross the fluid streamlines) comes in contact with a fiber solely because of the fiber's size.
>
> *Brownian movement* diffuses submicron particles, thus increasing the chance of contact between the particles and collecting surfaces.
>
> *Electrostatic forces* may have a major effect, depending on the electrostatic charges of the particles and the filter media. In many cases, an electrostatic force may help the filter attract and capture particles.

Filter Cleaning Methods

There are many variations in the configurations of baghouse construction, but most can be placed into any one of three broad categories, according to the filter cleaning method employed. These categories are shaking, reverse flow and reverse jet.

Cleaning by Shaking

Cleaning by shaking may be accomplished manually, mechanically, intermittently, automatically, or continuously. Dust collection can be either inside or outside the filter tubes, and gas flow can be either up or down.

Intermittent mechanical shaking is the most common for general filtering jobs. It allows the gas flow to be interrupted at intervals to shake the excess dust off the tubes into the dust hopper below. Particular care must be taken in the design of the shaker mechanism, since it must suit the particular cloth being used. Intermittent shakers require little maintenance, and, because the tube vibrating mechanism driving gear is external, repairs and cleaning are easy. Shaking usually is controlled automatically either by a timer or by a pressure-sensing system.

Continuous operation, automatic shaking types are used when it is not permissible to shut down the system for tube shaking. The shaking apparatus is fully automatic. More than two compartments comprise each unit. Each compartment has its own damper in the inlet and outlet gas lines, allowing the compartment to be isolated automatically from the gas flow for a few minutes. During this period, the excess cake is shaken from the filter tubes. The remaining compartments handle the total gas volume while any one compartment is being cleaned.

Reverse Flow Cleaning

Using suction, pressure or both, the flow of clean gas (or air) through the filter is reversed, thus breaking the dust cake off the fabric. When the direction of flow through the filter fabric is changed, sides of the filter bag collapse and the dust falls into the hopper. Normally, baghouses of this type also are compartmentalized so that one compartment may be cleaned while the others handle the gas flow. The cleaning cycle can be controlled either by a programmer or by a monitoring system.

Reverse Jet Cleaning

This system employs compressed air to remove the filter cake from the fabric. There are two commonly used types: one has a pressure nozzle mounted at the top of each tubular bag; the other uses a traveling device to distribute the compressed air jet over the entire fabric area.

With the static nozzle system, dust is collected on the outside of the bag. During the cleaning cycle, high-pressure air is released into the bag, blowing the dust off the fabric.

Traveling devices are either rings around tubular bags or pipes that move across the surface of envelope bags. In both cases, the baghouse need not be compartmentalized because dust is collected on the inside of the bags, and cleaning is continuous.

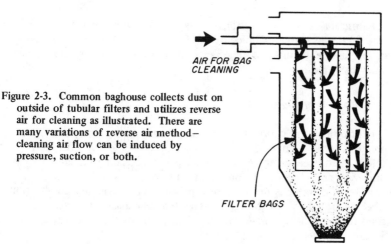

AIR FOR BAG
CLEANING

Figure 2-3. Common baghouse collects dust on outside of tubular filters and utilizes reverse air for cleaning as illustrated. There are many variations of reverse air method—cleaning air flow can be induced by pressure, suction, or both.

FILTER BAGS

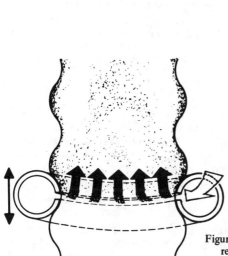

Figure 2-4. Popular application of reverse jet cleaning method uses ring with a slot on the inside circumference around the filter bag. Air is pumped through ring as it travels up and down length of bag.

Applications

Fabric dust filter applications are as varied as the types of systems and the fabrics available. Choice of the right design depends on the particular problem encountered. The range of possibilities is best illustrated by a few examples.

Fabric Filtration of Flue Gases

Fabric filtration is proving to be useful in cleaning flue gases from incinerators. One incinerator, for instance, comprises a multichamber, refractory-lined, steel-cased furnace and a dry gas cleaning system. Its gas cleaning system includes a cyclone, fabric filters, and a high efficiency filter. The incinerator is made up of a primary and secondary combustion chamber separated by two bridge walls with a downpass. This system is designed for great heat storage capacity. Any one of the filter designs previously discussed could be used here.

Type of waste encountered consists of paper, mops, rags, rubber, and plastics. The major problem with using cloth filters for this application is temperature. Hot flue gases from the furnace have to be suitably cooled so filtration can be carried out below the maximum temperature the fabric can withstand. This cooling is achieved by mixing the hot flue gas with atmospheric air.

Use of fabric filters on incinerators is still under study, but the possibilities are promising. Studies on special types of fire and chemical resistant fabrics are still incomplete and a great deal more research remains to be done.

Carbon Black Production Plants

Carbon black is generated by burning oil or tar in furnaces. The exhaust gases leave the furnace at a temperature of around 2000°F. The dust load is normally about 20 grains/cu ft. This is one application in which the fabric filter has proved to be highly successful. Particle size of carbon black is on the order of 15 to 100 millimicrons. The gases also contain corrosive acids. Thus, high operating temperatures are necessary, and corrosion is a major problem. Silicone-treated fiberglass filters are now being used in many installations.

A typical baghouse may collect around 35 tons of carbon black dust per day operating with continuous automatic cleaning by means of clean gas reverse flow. Such a system may consist of six compartments, each housing 400 fiberglass tubes. These plants operate at temperatures of about 400 to 500°F. Thus, the exhaust gases leaving the furnace must be cooled rapidly (from 200 to 600°F).

The cooling process is generally handled by the injection of high-pressure water into the gas stream. Such a system operates with a high degree of efficiency, and the fiberglass filter tubes usually last from one to two years under continuous use. However, graphite-coated fiberglass cloth may replace plain fiberglass filters in the near future because they can be used at higher gas operating temperatures.

Foundries

Generally, the filter unit employed by foundry operations is a continuous, automatic type with mechanical vibration tube cleaning. Captured dust in the storage hoppers is removed continuously by conveyors to one central disposal area. The dust is then subjected to wet mixers before being deposited into storage hoppers for disposal.

Metals Refinery Industry

The nonferrous smelting industry has found great use for fabric filters. Here, the dust is a valuable product. Several applications include:

Copper Roasters—Filter tubes usually are made of fiberglass which enables gases to be handled at temperatures of 400 to 450°F. Cleaning dusts from the tube in one such plant is done by a reverse flow shaking system. Tube life is approximately 12 years.

Zinc Ores—One filter system for a plant handling zinc ores cleans some 42,000 cfm of exhaust gases a day at temperature ranges of 400 to 500°F. The plant is arranged in four compartments on each side of a balloon flue where the heavy dust settles out, leaving only a fine dust which is drawn to the fabric filters. One thousand fiberglass tubes are housed within the system, and are cleaned by reverse flow and one cycle of mechanical vibration per day.

Other typical areas of nonferrous applications include beryllium plants, lead and zinc blast furnaces, ferro-silicon and silicon metal furnaces, and aluminum smelting. These applications are successful, even though difficult for fabric filters because of severe corrosion and temperatures. Care must be taken in the design and selection of the proper fabric.

Steel Industry

Oxygen injection into molten steel generates severe dust and fume problems. Control and cleaning of the fine red fumes discharged from an electric arc melting furnace, for example, can be handled by cloth filters. The fumes must be contained within a minimum volume and cleaned to less than 0.05 grains per cu ft at normal temperature and pressure.

Capturing process emissions can be solved by:

1. Installing exhaust hoods above the furnaces. However, this solution has the disadvantage of the extremely high gas volumes resulting from dilution.
2. Containing fumes by building fume hoods directly onto the furnace roof (or as near to the source of emission as possible).
3. Using a semi-direct hood similar to the above, but with an additional vent port in the furnace roof—the fumes being drawn through the port into the hood system.
4. Employing direct extraction. An elbow connection onto a vent hole in the furnace roof directs the exhaust fumes into a closed exhaust system. Air is mixed with the gases near the takeoff, and burners incinerate the gases, which are then spray-cooled before passing to the dust collector. Such a method results in small exhaust gas volumes and uses the smallest filter unit.

Iron oxide in the fumes is of a very small particle size, and a high-temperature fabric filter is necessary. Filter units operating at a filter ratio of 3:1 and a temperature of around 275 F using terylene filter tubes are popular for smaller furnaces. Bag life is usually around five years.

Larger furnaces require higher operating temperatures, and it is desirable to minimize cooling of the hot fumes. Continuous, automatic filters made of fiberglass filter tubes are used because their operating temperature is approximately 500 to 600 F. Fabric filters are proving to be an economical and efficient approach to handling these solid airborne discharges.

Boiler Plants

Currently, intensive research is being done in testing the applicability of cloth filters to boiler emissions. A recent pilot test indicated efficiencies by weight in excess of 99 percent of airborne particulates. Tests were run on a pulverized fuel-fired boiler with dust loads of 3.7 grains/scfm and with 50 percent of the particles less than 10 microns in diameter.

Chemical Industry

The chemical industry probably makes the greatest and most varied use of fabric filters. Exhaust systems on carbide furnaces are one example. In one such plant, a typical filter includes a six-compartment, automatic unit which houses 2160 filter tubes. The handling capacity is 30,000 cfm of hot gases from the furnace. Mechanical vibration at regular intervals cleans the bags. The entire process is controlled automatically and operated continuously.

Other applications include filters for coke and limestone dust from exhaust systems. Dryers, grinding mills and booster systems may all require filter bags.

Cement Product Plants

Fiberglass filter baghouses are being used to clean the high-temperature exhaust gases from cement kilns of wet, dry and semi-dry processes in many plants.

A disadvantage of the fabric filter, as applied here, is that it is too efficient. Filters can trap virtually all the dust, including very fine particles, which are primarily alkalis. This dust is not suitable for feeding back into the kiln, so the problem must be overcome by installing a multi-tubular mechanical collector upstream of the bag filter to remove the larger dust particles. The larger particulates are recycled to the kiln, while the fine dust is collected by the filter system.

The list of possibilities for industrial application of fabric filters is practically endless. See Table 2-4 for listing of additional areas in which dust collectors may be employed.

Table 2-4. Typical Fabric Filter Pollution Control Application

Industry	Fabric Filters Applicable To
Building Materials	rock wool chambers
Rock Products (Mining and Quarry)	dryers; crushing plants; material handling systems; packing systems
Food and Milling	(this includes grain elevators and processing equipment) process handling equipment; dryers; sawdust handling
Metalworking (Mining and Smelting)	grinding and polish dryers; kilns
Paper	process equipment; sawdust systems
Woodworking	sawdust and shaving systems; pattern shop

FABRIC FILTER TYPES[39]

Paramount to a good filtering system is the proper choice of fabric. Fabrics are the heart of fabric filter systems. They determine a collector's size and ultimate cost. For example, a filter system that accommodates 100 cfm of air at 2 cfm per sq ft of cloth will be almost twice the size of a 100 cfm unit designed for 4 cfm per sq ft of cloth. In addition, the fabric used can affect power requirements. Principal point of resistance in a control system is the filter fabric with its deposited dust cake. The cloth can also determine collector efficiency. Design of the fabric filters helps the filtrate to retain particles of submicron size and maintain efficient cleaning.

Natural fibers, such as cotton and wool and some manmade fibers, come in staple (spun) form. Spun form means that the individual fibers are limited to a few inches in length. Fiber length is normally random, but some manmade fibers are cut to a specific length. Staple fibers can produce fabric characteristics that filament (continuous) fibers cannot—yarns with greater bulk and thickness, higher weight, greater usable surface area, and a higher permeability to air flow.

Most manmade fibers are of the filament form. Multifilament fabrics are the most widely used and can produce fabric characteristics not always found in spun forms. These characteristics are light weight, high tensile strength, high dimensional stability, resistance to abrasion and easy cleaning.

Filter fabrics can be either of woven cloth or felted cloth. Woven fabrics are made to retain a residual cake of dust on the collection surface and to serve as a grid for a filter cake. They are generally associated with lower filter ratios (air-to-cloth ratios), depending upon the particular application (see Table 2-5). Another term denoting filter ratio is "filtration velocity." This velocity can vary from 1:5:1 to 4:1 fpm. With special treatment (such as napping or blending procedures), filtration ratios can be as high as 10:1.

Felted fabrics, on the other hand, are associated with high filter ratios (4:1 to 12:1 and, sometimes, higher). Felted bags are more expensive than woven. Also, the choice of a felted fiber is limited—not all fibers can be felted.

Felted baghouses are normally less efficient than systems using woven media. They do not collect extremely fine dusts or aerosols because embedding of the fine particles in the felt makes cloth cleaning difficult. High-pressure reverse jet or jet pulse at frequent cleaning intervals is the method most often employed for cleaning.

Fabric Selection

The proper fabric must fulfill two requirements. First, it must be able to collect fine particulate matter to a high degree of efficiency, and thus, clean the air. Second, and equally important, is that the fabric must have the ability to adequately clean itself of the collected contaminants in order to maintain an efficient system.

Numerous considerations involved in the choosing of the cloth media must be understood. These considerations are choice of yarn, fiber, design, finish, and fabrication. Other important factors are electrostatic characteristics of the dust, type of equipment used and method of operation, operating conditions (including temperatures encountered, moisture, corrosive properties of gases), capacity of unit, pressure drops, variations in operations and fabric cleaning method.

Fiber size is another important factor. In general, small-diameter filament offers several advantages. The smaller the fibers, the smaller the

Table 2-5. Air-to-Cloth Ratios for Various Common Baghouse Applications*

The following information is based on dry dust and air with only the moisture content of the surrounding atmospheric humidity. Some applications of these dusts will necessitate lower ratios and in some cases higher velocities in the branch lines. Excessive moisture and temperature can void all of these recommendations and in such cases each application should be treated with extreme care.

Alumina:	Use a ratio of 2.25:1 and a branch pipe velocity of 4500 fpm.
Aluminum Oxide:	Use a ratio of 2:1 and a branch pipe velocity of 4500 fpm.
Abrasives:	Use a ratio of 3.0:1 and a branch pipe velocity of 4500 fpm.
Asbestos:	Use a ratio of 2.75:1 and a branch pipe velocity of 3500 to 4000 fpm—a preliminary cyclone collector, and special hoppers with large gates or valves should be incorporated in the ventilation system and dust filter.
Buffing Wheels:	Use a ratio of 3 to 3.25:1 and a branch pipe velocity of 3500 to 4000 fpm—a preliminary cyclone collector, flame-retardant cloth and a spark arrester are recommended.
Bauxite:	Use a ratio of 2.5:1 and a branch pipe velocity of 4500 fpm.
Baking Powder:	Use a ratio of 2.25:1 to 2.5:1 and a branch pipe velocity of 4000 to 4500 fpm.
Bronze Powder	Use a ratio of 2.1 maximum and a branch pipe velocity of 5000 fpm.
Brunswick Clay:	Use a ratio of 2.25:1 and a branch pipe velocity of 4000 to 4500 fpm.
Carbon:	Use a maximum ratio of 2:1 and a branch pipe velocity of 4000 to 4500 fpm.
Coke:	Use a ratio of 2.25:1 and a branch pipe velocity of 4000-4500 fpm. Pressure relief provisions, grounded bags, and special electricals (II G) are recommended.
Charcoal:	Use a ratio of 2.25:1 and a branch pipe velocity of 4500 fpm. Pressure relief, grounded bags, and special electricals (II G) are recommended.
Cocoa:	Use a ratio of 2.25:1 and a branch pipe velocity of 4000 fpm. Grounded bags, pressure relief provisions, sprinkler fittings, special electricals (II G) are recommended.
Chocolate:	Use a ratio of 2.25:1 and a branch pipe velocity of 4000 fpm. Grounded bags, pressure relief provisions, sprinkler fittings, and special electricals (II G) are recommended.
Cork:	Use a ratio of 3:1 and a branch pipe velocity of 3000 to 3500 fpm. Special hoppers with large gates or valves, pressure relief provisions, and flame-retardant cloth, are recommended.
Ceramics:	Use a ratio of 2.5:1 and a branch pipe velocity of 4000 to 4500 fpm.
Clay:	Use a ratio of 2.25:1 and a branch pipe velocity of 4000 to 4500 fpm.
Chrome Ore:	Use a ratio of 2.5:1 and a branch pipe velocity of 5000 fpm.
Cotton:	Use a ratio of 3.5:1 and a branch pipe velocity of 3500 fpm. Preliminary centrifugal cyclone collector, flame-retardant cloth, special large hopper and valves, and pressure relief provisions are recommended.

Flour:	Use a ratio of 2.5:1 and a branch pipe velocity of 3500 fpm. Pressure relief provisions and special electricals (II G) are recommended.
Flint:	Use a ratio of 2.5:1 and a branch pipe velocity of 4500 fpm.
Glass:	Use a ratio of 2.5:1 and a branch pipe velocity of 4000 to 4500 fpm.
Granite:	Use a ratio of 2.5:1 and a branch pipe velocity of 4500 fpm.
Gypsum:	Use a ratio of 2.5:1 and a branch pipe velocity of 4000 fpm.
Graphite:	Use a maximum ratio of 2:1 and a branch pipe velocity of 4500 fpm.
Iron Ore:	Use a ratio of 2:1 and a branch pipe velocity of 4500 to 5000 fpm.
Iron Oxide:	Use a ratio of 2:1 and a branch pipe velocity of 4500 fpm.
Lampblack:	Use a maximum ratio of 2:1 and a branch pipe velocity of 4500 fpm.
Leather:	Use a ratio of 3.5:1 and a branch pipe velocity of 3500 fpm. A preliminary cyclone collector and special hopper valves or gates are recommended.
Cement— Crushing:	Use a ratio of 1.5:1 maximum and a branch pipe velocity of 4500 fpm—a preliminary classifier section is recommended. Insulate case of filter.
Grinding:	(Separators, cooling etc.) Use a ratio of 2.25:1 maximum and a branch pipe velocity of 4000 fpm.
Conveying:	Use a ratio of 2.5:1 and a branch velocity of 4000 fpm.
Packers:	Use a ratio of 2.75:1 and a branch velocity of 4000 fpm.
Batch Spouts:	Use a ratio of 3.1 and a branch velocity of 4000 fpm.
Cosmetics:	Use a ratio of 2:1 and a branch pipe velocity of 4000 fpm.
Cleanser:	Use a ratio of 2.25:1 and a branch pipe velocity of 4000 fpm. Pressure relief provisions, flame-retardant cloth, and grounded bags are recommended.
Feeds and Grain:	Use a ratio of 3.25:1 and a branch pipe velocity of 3500 fpm. Pressure relief provisions and special electricals (II G) are recommended.
Feldspar:	Use a ratio of 2.5:1 and a branch pipe velocity of 4000 to 4500 fpm.
Fertilizer:	(Bagging operations) Use a ratio of 2.4:1 and a branch pipe velocity of 4000 fpm.
Fertilizer:	(Cooler, dryer) Use a ratio of 2.0:1 and a branch velocity of 4500 fpm.
Limestone:	Use a ratio of 2.75:1 and a branch pipe velocity of 4500 fpm.
Lead Oxide:	Use a ratio of 2.25:1 and a branch pipe velocity of 4500 fpm.
Lime:	Use a ratio of 2:1 and a branch pipe velocity of 4000 fpm.
Manganese:	Use a ratio of 2.25:1 and a branch pipe velocity of 5000 fpm.
Marble:	Use a ratio of 3:1 and a branch pipe velocity of 4500 fpm.
Mica:	Use a ratio of 2.25:1 and a branch pipe velocity of 4000 fpm.
Oyster Shell:	Use a ratio of 3:1 and a branch pipe velocity of 4500 fpm.
Paint Pigments:	Use a maximum ratio of 2:1 and a branch pipe velocity of 4000 fpm.

Paper:	Use a ratio of 3.5:1 and a branch pipe velocity of 3500 fpm. Special hopper discharge valves are recommended.
Plastics:	Use a ratio of 2.5:1 and a branch pipe velocity of 4500 fpm. Explosion relief provisions are usually required.
Quartz:	Use a ratio of 2.75:1 and a branch pipe velocity of 4500 fpm.
Rock:	Use a ratio of 3.25:1 and a branch pipe velocity of 4500 fpm.
Sanders:	Use a ratio of 3.25:1 and a branch pipe velocity of 4500 fpm. Fla Flame-retardant cloth and a preliminary spark arrester section is recommended.
Silica:	Use a ratio of 2.75:1 and a branch pipe velocity of 4500 fpm.
Slate:	Use a ratio of 2.75:1 and a branch pipe velocity of 4500 fpm.
Soap:	Use a ratio of 2.25:1 and a branch pipe velocity of 3500 fpm. Flame-retardant bags and pressure relief provisions are recommended.
Starch:	Use a ratio of 2.25:1 and a branch pipe velocity of 3500 fpm. Flame-retardant bags and pressure relief provisions are recommended.
Sugar:	Use a ratio of 2.25:1 and a branch pipe velocity of 4000 fpm. Pressure relief provisions are recommended.
Soapstone:	Use a ratio of 2.25:1 and a branch pipe velocity of 4000 fpm.
Talc:	Use a ratio of 2.25:1 and a branch pipe velocity of 4000 fpm.
Tobacco:	Use a ratio of 3.5:1 and a branch pipe velocity of 3500 fpm. Pressure relief provisions, flame-retardant cloth and special hoppers and valves are recommended.
Wood:	Use a ratio of 3.5:1 and a branch pipe velocity of 3500 fpm. Special hoppers; valves, and pressure relief provisions are recommended.

*Although all ratios shown are below 3.5:1, cloth filter systems have been designed that operate well at ratios as high as 17:1. Governing factors are the filter material, operating conditions, and type of dust.

Table 2-6. Fiber Selection Chart[59]

Chemical	Nylon	Polyesters	Polyethylene	Acrylics	Saran	Modacrylics	PVC	Teflon	Polypropylene
Acetaldehyde	C	A	A	A	A	B	B	A	A
Acetanilide	A	A	A	A	A	A	A	A	A
Acetic Acid	A	A	A	A	A	A	A	A	A
Acetone	A	A	A	A	A	C	B	A	A
Acetyl Chloride	C	B	B	B	C	A	B	A	B
Acrolein	C	B	B	B	B	B	C	A	B
Acrylic Acid	C	A	A	B	A	A	A	A	A
Aluminum Chloride	C	A	A	A	A	A	A	A	A
Aluminum Hydroxide	A	A	A	C	C	A	A	A	A
Aluminum Sulfate	B	A	A	A	A	A	A	A	A
Ammonium Carbonate	A	A	A	B	A	A	A	A	A
Ammonium Chloride	C	B	A	B	A	A	A	A	A
Ammonium Hydroxide	A	A	A	C	C	A	A	A	A
Amylacetate	B	A	B	B	B	B	B	A	B
Amylalcohol	A	A	A	A	A	A	A	A	A
Aniline	B	A	B	C	B	C	C	A	B
Antimony Trichloride	B	A	A	A	A	A	A	A	A
Barium Carbonate	A	A	A	A	A	A	A	A	A
Barium Chloride	A	A	A	A	A	A	A	A	A
Benzaldehyde	A	A	A	A	B	B	B	A	A
Benzene	A	A	B	A	C	C	B	A	B
Benzene Sulfonic Acid	C	A	C	C	B	A	A	A	A
Benzoic Acid	B	B	A	A	A	A	A	A	A
Benzoyl Chloride	C	B	B	A	C	C	C	A	B
Benzyl Acetate	A	A	A	A	A	B	A	A	A
Benzyl Chloride	A	A	A	A	C	B	A	A	A
Bismuth Acetate	A	A	A	A	A	A	A	A	A
Bismuth Subcarbonate	A	A	A	A	A	A	A	A	A
Bromine	C	B	B	C	C	B	C	A	B
Bromoacetic Acid	C	A	B	A	A	A	C	A	B
Bromobenzene	A	A	B	A	B	A	B	A	B
Butyl Acetate	A	A	A	A	B	B	C	A	A
Butyric Acid	C	A	A	A	A	A	A	A	A
Cadmium Chloride	A	A	A	A	A	A	A	A	A
Calcium Acetate	B	A	A	A	A	A	A	A	A
Calcium Carbonate	A	A	A	A	A	A	A	A	A
Calcium Chloride	C	A	A	A	A	A	A	A	A
Calcium Hydroxide	A	A	A	B	A	A	A	A	A
Calcium Oxalate	C	A	A	A	A	A	A	A	A

A–Recommended; B–Conditional; C–Unsatisfactory

Table 2-6, continued

Chemical	Nylon	Polyesters	Polyethylene	Acrylics	Saran	Modacrylics	PVC	Teflon	Polypropylene
Caprylic Acid	A	A	B	A	B	B	B	A	B
Carbolic Acid	C	C	B	C	B	B	C	A	A
Carbon Tetrachloride	A	A	B	A	A	A	B	A	B
Catechol	C	B	A	A	A	A	A	A	A
Cetyl Alcohol	A	A	A	A	A	A	B	A	A
Chloroacetic Acid	C	B	A	A	B	A	C	A	A
Chlorobenzene	A	A	B	A	C	B	C	A	B
Chloroform	A	A	C	A	B	B	C	A	B
Chromic Chloride	C	B	A	A	A	A	A	A	A
Chromium Trioxide	C	C	C	C	C	C	C	A	B
Cinnamylic Acid	A	A	B	A	B	C	C	A	B
Citric Acid	A	A	A	A	A	A	A	A	A
Cresylic Acid	C	A	C	A	C	B	C	A	C
Cupric Carbonate	A	A	A	A	A	A	A	A	A
Cupric Chloride	C	A	A	A	A	A	A	A	A
Cupric Sulfate	B	A	A	A	A	A	A	A	A
Cyclohexanone	A	A	C	A	C	C	C	A	C
Cyclopentanone	A	A	C	A	C	C	C	A	C
Cymene	A	A	C	A	C	C	C	A	B
Dextrin	A	A	A	A	A	A	A	A	A
Diacetin	A	A	A	A	B	B	C	A	A
Diallyphthalate	A	A	C	A	A	C	C	A	C
Dibenzylamine	A	A	B	A	A	C	C	A	C
Dibenzyl Ketone	A	A	B	A	B	C	A	B	A
Dibromobenzene	A	A	B	A	B	B	C	A	B
Dibutylamine	A	A	B	A	B	B	B	A	B
Dichloroacetic Acid	C	B	B	B	A	A	B	A	B
Dichlorobenzene	A	A	B	A	B	B	B	A	B
Diethanolamine	A	A	A	A	B	A	B	A	A
Diethylene Glycol	A	A	A	A	A	A	B	A	A
Dimethyl Phthalate	A	A	B	A	B	B	B	A	B
Dinitrobenzene	A	A	C	A	C	C	C	A	B
Diphenyl	B	C	C	C	C	C	C	A	C
Diphenylacetic Acid	B	A	A	A	B	A	B	A	A
Diphenylamine	A	A	C	A	B	B	B	A	C
Epichlorohydrin	A	A	B	A	C	C	C	A	B
Ethyl Acetate	A	A	B	A	C	A	A	A	B
Ethyl Alcohol	A	A	A	A	A	A	A	A	A
Ethyl Benzene	A	A	C	A	C	C	C	A	C

A—Recommended; B—Conditional; C—Unsatisfactory

Table 2-6, continued

Chemical	Nylon	Polyesters	Polyethylene	Acrylics	Saran	Modacrylics	PVC	Teflon	Polypropylene
Ethyl Benzoate	B	A	A	A	A	A	A	A	A
Ethyl Carbamate	A	A	B	A	A	A	B	A	B
Ethyl Carbonate	A	A	A	A	A	A	A	A	A
Ethyl Chloride	A	A	A	A	A	A	A	A	A
Ethyl Chloroacetate	A	A	A	A	B	A	A	A	A
Ethylene Bromide	A	A	B	A	C	B	A	A	B
Ethyl Ether	A	A	B	A	C	A	C	A	B
Ethyl Phenylacetate	B	A	A	A	A	A	A	A	A
Ethyl Propionate	A	A	B	A	B	A	B	A	B
Ferric Ammonium Sulfate	C	A	A	A	A	A	A	A	A
Ferric Chloride	C	A	A	A	A	A	A	A	A
Ferric Sulfate	C	A	A	A	A	A	A	A	A
Ferrous Chloride	C	A	A	A	A	A	A	A	A
Formaldehyde	A	A	A	A	A	A	B	A	A
Formic Acid	C	A	A	B	A	A	A	A	A
Furfural	A	A	B	A	C	A	C	A	B
Gallic Acid	B	A	A	A	A	A	A	A	A
Gasoline	A	A	B	A	B	B	A	A	B
Glycerine	A	A	A	A	A	A	A	A	A
Hexachlorobenzene	C	A	B	A	B	C	C	A	B
Hexane	A	A	C	A	B	A	C	A	C
Hydrobromic Acid	C	A	A	B	C	A	A	A	A
Hydrochloric Acid	C	A	A	A	A	A	A	A	A
Hydrofluoric Acid	C	C	A	C	C	A	B	A	A
Lactic Acid	B	A	A	A	A	A	A	A	A
Lead Acetate	A	A	A	A	A	A	A	A	A
Lithium Chloride	A	A	A	A	A	A	A	A	A
Magnesium Chloride	A	A	A	A	A	A	A	A	A
Methyl Acetate	A	A	B	A	B	B	C	A	B
Methyl Alcohol	A	A	A	A	A	A	A	A	A
Methyl Bromoacetate	A	A	A	A	B	B	C	A	A
Methyl Ethyl Ketone	A	A	B	A	A	C	C	A	B
Methyl Urea	A	A	A	A	A	A	A	A	A
Monoacetin	B	A	B	A	B	A	B	A	B
Naphthylamine	A	A	C	A	C	C	C	A	C
Nickel Sulfate	A	A	A	A	A	A	A	A	A
Nitric Acid	C	B	A	B	A	A	A	A	A
Nitrobenzene	A	A	C	A	C	C	C	A	B
Nitrobenzoic Acid	C	A	B	A	C	C	C	A	B

A—Recommended; B—Conditional; C—Unsatisfactory

Table 2-6, continued

Chemical	Nylon	Polyesters	Polyethylene	Acrylics	Saran	Modacrylics	PVC	Teflon	Polypropylene
Nitrotoluene	A	A	C	A	C	C	C	A	A
Octyl Alcohol	A	A	A	A	B	A	B	A	A
Oleic Acid	A	A	C	A	A	A	C	A	C
Oxalic Acid	C	A	A	A	A	A	A	A	B
Paraldehyde	A	A	A	A	A	B	B	A	A
Pentaerythritol	A	A	A	A	A	A	A	A	A
Petroleum Ether	A	A	B	A	B	A	C	A	B
Phenoxyacetic Acid	B	A	A	A	A	A	A	A	A
Phenol Ether	A	A	C	A	C	C	C	A	C
Phosphoric Acid	B	A	A	A	A	A	A	A	A
Phosphorous Trichloride	C	A	A	A	A	A	A	A	A
Phthalic Acid	B	A	A	A	A	A	A	A	A
Potassium Bisulfate	B	A	A	A	A	A	A	A	A
Potassium Dichromate	C	A	B	A	A	A	A	A	B
Potassium Hydroxide	A	A	A	A	A	A	A	A	A
Potassium Permanganate	C	A	C	C	B	A	B	A	C
Propionic Acid	A	A	B	A	A	A	C	A	B
Pyridine	B	A	B	C	B	C	C	A	B
Resorcinol	B	A	B	C	A	B	C	A	B
Salicylic Acid	C	A	A	A	A	A	A	A	A
Sodium Bisulfate	B	A	A	A	A	A	A	A	A
Sodium Hydroxide	A	C	A	C	B	A	A	A	A
Sodium Hypochloride	C	B	B	B	C	A	A	A	C
Sodium Sulfate	A	A	A	A	A	A	A	A	A
Succinic Acid	B	A	B	A	B	A	B	A	B
Sulfamic Acid	C	C	A	C	C	A	A	A	A
Sulfuric Acid	C	B	A	B	C	A	A	A	A
Tetrachloroethylene	A	A	B	A	A	A	A	A	B
Thiocarbamide	A	A	B	A	C	A	B	A	B
Toluene	A	A	C	A	C	B	C	A	C
Tribromobenzene	B	A	B	A	C	C	C	A	B
Trichloroacetic Acid	C	A	A	A	C	A	A	A	A
Trichloroethylene	A	A	B	A	A	A	A	A	B
Valeraldehyde	C	A	C	A	C	C	C	A	C
Xylene	A	A	C	A	C	B	C	A	C
Zinc Chloride	C	B	A	B	A	A	A	A	A
Zinc Sulfate	B	A	A	B	A	A	A	A	A

A–Recommended; B–Conditional; C–Unsatisfactory

Table 2-7. Physical and Chemical Properties of Fibers

	Acetate	Acrylics	Modacrylics	Polyamides	"Nomex" Nylon	Polyesters	Polyethylene	Polypropylene	Rayon Viscose	Saran	Teflon	Glass	Paper	Cotton	Silk	Wool
Specific Gravity	1.33	1.18	1.30	1.14	1.38	1.38	.92	.90	1.52	1.70	2.10	2.54		1.50	1.25	1.32
Max. Recommended Operating Temperature, F	175	275	180	250	450	300	150	225	275	160	500	550	200	225	175	200
Filament																
Breaking Elongation, Percent (Dry)	23–44	13–23	33–39	16–40	18	10–25	10–80	15–30	9–30	15–30	15	3				
Breaking Elongation, Percent (Wet)	30–54	13–23	32–39	28–42	15	19–25	20–80	15–30	14–40	15–25	15	2.5				
Staple																
Breaking Elongation, Percent (Dry)	35–40	20–50		16–50	19	18–50		15–35	9–30	15–25	15			3–7		25–35
Breaking Elongation, Percent (Wet)	30–45	26–55		18–46	16	18–50		15–35	14–40	15–25	15			3–7		25–50
Resistance to:																
Abrasion	G	G	G	E	E	E	G	E	G	G	F	P	F	G	G	G
Dry Heat	F	G	F	G	E	G	F	G	G	F	E	E	F	G	F	F
Moist Heat	F	G	F	G	E	F	F	F	G	F	E	E	F	G	F	F
Mineral Acids	P	G	G	P	F	G	G	E	P	G	E	E	P	P	F	F
Organic Acids	P	G	G	F	E	G	G	E	G	G	E	E	G	G	F	F
Alkalies	P	F	G	G	G	G	P	G	F	F	E	P	G	G	P	P
Oxidizing Agents	F	G	G	F	G	G	G	G	G	F	E	E	F	F	P	P
Solvents	F	E	G	E	E	E	G	G	G	G	E	E	E	E	G	F

E—Excellent; G—Good; F—Fair; P—Poor

(Courtesy of J.P. Stevens and Company, Inc.)

particles collected. Also, smaller pressure drop is encountered (because of a higher surface area and greater flexibility). Fabrics of large-diameter filaments have greater pressure drops.

The fabric's weight per square yard is an additional point that must be taken into account. Most dust collectors utilize woven fabrics that range in weight from 4 to 10 oz/sq yd. Heavier fabrics are characterized by a low count of heavier bulk yarns. Conversely, lighter fabrics usually have a high count of finer yarns with a higher number of smaller openings. Yarn weight and strength vary with the yarn diameter. Felted fabrics are normally thicker and heavier than woven fabrics.

Permeability of a fabric—cu ft of air per min per sq ft of media at a draft loss of 0.5 in. water gage—is significant. In general, the most open fabric which will operate at the required level of collection efficiency is desired. Multifilament fabrics are specified over a range of 10 to 50 cfm per sq ft and spun fabrics over a range of 20 to 100 cfm per sq ft. Lighter fabrics must have low permeability. High permeability (over 50 cfm) is associated with fabrics of a minimum weight of 9 oz per sq yd.

Abrasion is a major cause of fabric failure. Filament fibers are often more abrasion-resistant than the staple form. Thus, when choosing the fabric, this factor should be weighed heavily.

Shrinkage and *elongation* of the fabric can be a problem. Both shrinking and stretching will change the porosity or permeability of the cloth; the bag may become too large or too tight for cleaning. Excessive stretching can cause contact of adjacent filter elements and affect service life. It is desirable to select a fabric with good dimensional stability.

All of the previous points should be considered when selecting the proper fabric. Now, the choice can be further reduced by evaluating the various commercial cloths available (Table 2-8). Cotton and wool are natural fibers available only in staple (spun) form. The individual fibers are limited to a few inches in length.

Cotton bags are used in most standard installations. They withstand a maximum temperature of 195 F, but the recommended maximum operating temperature should be held below 190 F.

Wool bags are commonly used when the dust arrester is operating in a system with contaminants of a combustible nature, or when the fumes are at elevated temperatures and acidic. Wool bags can operate at temperatures up to 210 F, but a maximum of 200 F is usually recommended.

Nylon has more tensile strength and greater resistance to abrasion than cotton or wool. Nylon bags can operate under the same temperatures as cotton, but with very heavy and more abrasive dust loads. Although cotton bags are less expensive, nylon bags are often more economical in the long run because they last longer.

Orlon (Registered TM, E.I. DuPont) bags have a maximum operating temperature of 275 F. The orlon fabric is preshrunk by a heatset process. Preshrinking controls the tendency of the fabric to shrink at elevated temperatures; however, the bags may still shrink or stretch slightly at

Table 2-8. Fabric Fiber Selection Guide

Fiber	Tensile Strength	Abrasion Resistance	Recommended Max. Operating Temp., Degrees F Long Exposure	Recommended Max. Operating Temp., Degrees F Short Exposure	Chemical Resistance Acids	Chemical Resistance Alkalies	Will Support Combustion	Special Properties	Chemical Classification
Dacron	A	B	275	325	G	F	Yes	More rapid degradation may occur in presence of heat and moisture. Holds crease.	Polyester
Glass	A	D	500	600	F	P	No	Limited by poor flex-abrasion qualities. Finishes limit maximum temperature range.	Glass
Microtain	B	C	260	300	G	F	Yes	Good all-around fiber up to recommended temperatures.	Acrylic
Duratain	B	C	260	300	G	F	Yes	Good all-around fiber up to recommended temperatures.	Acrylic
Nylon	A	A	200	250	P	E	Yes	Stays soft and pliable when exposed to heat.	Polyamide
Nomex	A	B	425	500	F	E	No	Fairly expensive. Subject to hydrolysis when exposed to certain conditions.	Polyamide
Orlon	B	C	240	275	G	F	Yes		Acrylic
Polypropylene	A	B	200	250	E	E	Yes	Affected by some organic solvents. Will stretch if under load for long time.	Polypropylene
Teflon	C	C	450	500	E	E	No	Excellent chemical resistance. Expensive	Polyfluoro-Ethylene
Wool	C	C	200	250	F	P	No	Can be felted	Protein
Cotton	B	C	180	225	P	E	Yes		Cellulose

E—Excellent; G—Good; F—Fair; P—Poor

lower temperatures. This property makes periodic inspection of the bags necessary after they have been used. For example, the bottom bag band may have to be adjusted to relieve the tension of the bag and the strain on the bagholder because of shrinkage, or the bag may need tightening because of stretching. Most orlon bags are of spun rather than continuous-filament material. These bags operate well under conditions similar to those described for wool bags.

Acrylic fibers were introduced in 1953. They have a high resistance to acids and perform well at high operating temperatures.

Polyester fibers came to the market in 1955. At present, they are the most widely used of all synthetics. Like acrylics, they endure high-operating temperatures and are abrasion-resistant. However, neither polyesters nor acrylics are as good as nylon bags for long life and efficiency.

Dacron (Registered TM, E.I. DuPont) bags operate well under conditions similar to those described for wool, but they have a maximum operating temperature of 275 F. Most dacron bags are of a continuous-filament nature. They are usually heatset and siliconized, giving them superior resistance to the adhesion of almost any wet or dry dust. Siliconizing, therefore, makes the bags much easier to clean.

Fiberglass bags are used in dust and fume-collecting systems which have operating temperatures ranging from 275 to 600 F. The glass fabric is usually siliconized, also. Shaking mechanisms are not recommended when using glass bags because of the bags' low abrasive strength.

Teflon (Registered TM, E.I. DuPont) was introduced in 1964. It has an excellent resistance to temperature (up to 550 F). It does, however, have poor dimensional stability and only fair resistance to abrasion. It is one of the more expensive materials, limiting its use to extreme chemical atmospheres in most cases. Chemical resistance is outstanding.

Nomex (Registered TM, E.I. DuPont) has high temperature capabilities and is abrasion-resistant. It is used quite extensively in the mining industry, but, like Teflon, it is high in cost.

Cotton fiber is the most popular fabric in terms of yardage sold each year. Flex abrasion is, however, a major problem in cotton filter fabrics. Synthetic materials have three to four times the abrasion resistance of cotton, but they are higher in cost. Synthetics also have a higher electrostatic charge. The choice of a fabric, therefore, is determined primarily by the operating conditions encountered and economic considerations.

Obviously, fabric selection has a tremendous effect on the efficiency of a filter system. Efficiency of a fabric filter is defined in terms of either particle size or percent collection. When a range of particle sizes is present, an overall efficiency of the filtering can be quoted by weight of dust collected or by numbers of particles. The difference between weight and number efficiencies is very significant for wide ranges of particle sizes.

In the laboratory filtering systems are studied for efficiency by preparing finely divided particles in the form of dusts or liquid aerosols. Detection devices based on light scattering from small particles allows a direct scale reading, indicating numbers of particles in the gas. By comparing the number of particles entering the filter with the number leaving, the efficiency of a particular unit can be computed.

Discoloration is a rough indication of how the dust is stopped by the fibers. The smallest particles usually show great cloth penetration which is, of course, undesirable in some cases. Examination of cloth media clearly reveals where penetration occurs, as indicated by the darkest areas on the cloth.

Efficiency of a dry filter system relies principally on the laws of probability and chance. The number of particles passing through a filter is governed by the number of intervening fiber barriers in relation to the number of entering airborne particles. Thus, the first aim in designing a high-efficiency system is to have the maximum number of intervening fibers in the path of the moving particulates.

Choice of a cloth filter, based on the size of particles being removed, is narrowed to three types:

1. Those capable of removing submicron particles;
2. Those suitable for particles with diameters in the 1 to 10 micron range, and
3. Those which remove coarse dust (above 10 microns).

Cloth filters suitable for submicron particles are capable of removing practically all suspended particulates. Both large and small particles are trapped. On a weight basis, efficiencies better than 99.95 percent have been obtained. High-efficiency systems generally operate at low concentrations.

Under normal operations, the filter efficiency increases with use because of changes in the filter geometry and electrostatic effects which may develop. However, condensation can greatly reduce efficiency.

Filters for 1 to 10 micron particles rely primarily on the mechanism of inertial impingement. Factors affecting inertial impingement (and, thus, efficiency) are air velocity, particle size, and particle density. Increasing these factors usually improves collection. However, in some cases, these improvements can be overbalanced by a tendency for particles to become reentrained—most likely to occur with large particles of high velocities. Excessive loading of these filters with dry dust can also lead to reentrainment. Efficiencies of 90 to 95 percent have been achieved with these types.

Efficiency of coarse dust filters varies from 50 percent (extremely coarse material) to 99.9 percent, depending on the dust being collected. The mechanism of collection in this case is predominantly inertial, although sieve action does play a role.

It should be noted that efficiency tends to increase during the first few hours of operation. A filter system will start off at a certain efficiency, but, after some time, the extremely minute interstices or defects become clogged or bridged over by the captured dust. This bridging increases the efficiency. This increase does not, however, occur with coarse-fibered, moderate-efficiency media. In any case, the increase will diminish and efficiency will eventually level off.

VARIABLES AFFECTING EFFICIENCY[40]

By altering fabric design, filtration is greatly affected. Variables affecting collection efficiency are: (1) thread count, (2) yarn size, (3) napping and (4) twist of the yarns.

If thread count is reduced (either warp or fill), the permeability is increased because of the increase in pore area. Collection efficiency is, therefore, reduced. On the other hand, if the thread count is increased in either direction, permeability is decreased. A balanced weave, which is equal in warp and fill count, is desirable. A low-count fabric has the most variable permeability.

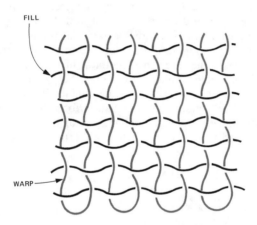

Figure 2-5. "Warp" and "fill" are names used to differentiate thread directions in cloth. Each cloth has thread count for both warp and fill.

Large yarn size results in a low total area and slow permeability. Conversely, yarn of small size is highly permeable, and efficiency, therefore, decreases.

Napping also affects cloth properties. Too much natural nap in staple fabrics causes beads or balls of dust to form on the fabric. Thus, napped, bulked, or filled yarns cannot be utilized in some applications. Multifill fabrics can be mechanically napped to produce a fabric of high efficiency under some conditions.

Generally, the higher the yarn twist, the higher the permeability and the lower the collection efficiency, because increasing the twist decreases the yarn diameter. When warp and fill yarns are of opposite twist, permeability and thickness increase. Fabric is more permeable and yarns are more compact when warp and fill yarns have the same kind of twist.

The particles to be removed play an important role in the selection of a fabric and filter efficiency. Specifically, the particulate density, concentration, velocity, and size are important. Each of these properties is interrelated with the pressure drop of the system, which is one of the most significant points affecting efficiency. Principal variables directly related to pressure drop are: (1) gas velocity, (2) cake resistance coefficient, (3) weight of cake per unit area and (4) air-to-cloth ratio.

The cake resistance coefficient is dependent on the particle size and shape, range of the particle sizes, and humidity. Weight of the cake per unit area is related to the concentration of particulates.

The design of fabric filters must satisfy two ideal criteria, namely, high efficiency and low pressure drop. Attempts have been made to correlate the filtration efficiency to the operating conditions of the filter. But pressure drop has emerged as the primary factor determining efficiency. Prediction of the pressure drop and knowledge of its dependence on operating conditions of the filter are necessary for sound design.

There are many conflicting procedures for predicting pressure drop across fibrous filters. The three chief methods of predicting pressure drop are based on the hydraulic radius theory, drag theory and dimensional analysis. Experimental investigation has determined dimensional analysis to be the most reliable and yields an equation for accurate pressure drop prediction. It is based on dimensional analysis of Darcy's law of flow through porous media and relates the pressure drop to the filter porosity. Darcy's original equation was:

$$\frac{PA}{hQ} = ku$$

where: A = cross sectional area of filter
P = pressure drop
h = filter thickness
Q = volumetric flow rate
k = permeability of the medium
u = fluid (gas) velocity

Darcy's empirical equations assumed the gas to behave ideally. Work done by Blake, Kozeny, and Carman resulted in the following equation:

$$\frac{PA}{hQ} = \frac{k_3 u S_0^2 (1-E)^2}{E^3}$$

where: k_3 = constant
S_0 = surface area/unit volume of solid material
E = bed porosity

By applying dimensional analysis, the following was devised by Davies:

$$k = \frac{PAd_E{}^2}{hQu} = 64(1\text{-}E)^{1.5}$$

$$(1 + 56\ (1\text{-}E)^3)$$

where: k = permeability coefficient
           d_E = effective fiber diameter

This formula has shown accurate results for fiber diameters ranging from 1.6 to 80 microns and filter porosities ranging from 0.700 to 0.994.

Still another formula has been found through dimensional analysis:

$$\frac{PAd^{-2}}{hQu} = k^{11}(1\text{-}E)^{1.5}$$

where: d^{-2} = mean square fiber diameter
           k^{11} = resistance coefficient

This expression predicts pressure drop for filters with porosities ranging from 0.88 to 0.96 and fiber diameters of about 0.1 to 3 microns.

A more useful formula not obtained through dimensional analysis is the following:

$$P = uV_s(k_o + k_1W)$$

again:
           P = pressure drop
           u = gas velocity
           V_s = air-cloth ratio (superficial velocity)
           k_o = weave resistance coefficient (values available in proper
                   literature)
           k_1 = cake resistance coefficient (dependent on shape,
                   concentration of particle, humidity)
           W = weight of cake per square foot of surface.

Use of these formulas (especially the last formula) makes possible accurate prediction of the pressure drop of a particular unit. Remember that low pressure drops are generally desirable for efficiency.

Under operating conditions, the pressure drop increases with time. This is due to the accumulation of collected material. In order for the gas to escape the baghouse, it must pass through a layer of collected dust, as well as through the cloth. Excessive pressure drop will eventually cause bag rupture or excessively low flow rates through the unit.

It has been observed that long cleaning cycles have higher average pressure drops than short cleaning cycles. This is significant because of the economical considerations in controlling pressure drop. The short cycle, however, tends to accelerate the rate of wear more so than longer cycles. Optimization between these two points has to be reached.

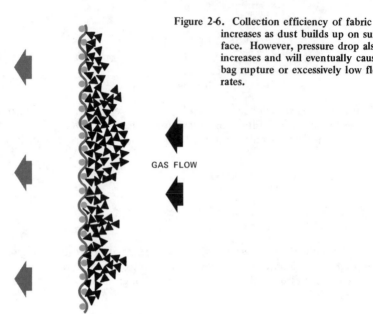

Figure 2-6. Collection efficiency of fabric increases as dust builds up on surface. However, pressure drop also increases and will eventually cause bag rupture or excessively low flow rates.

GAS FLOW

Many times moist flue gases must be handled. In such cases, liquid droplets captured by the filter can greatly alter efficiency by affecting the pressure drop. There is no simple method of predicting pressure drop when dealing with gases of this nature.

Pressure drop seems to be imposed not only by the basic resistance to gas-flow of the dry filter bed, but also by the liquor trapped in the bags. The liquid holdup is dependent on the gas velocity and the physical characteristics of the fibers. It contributes the major portion of the pressure drop. If liquid build-up is extreme, it will cause cloth-plugging or wet-caking of the collected dust on the cloth. In such cases, liquid repellent cloth is necessary for improved efficiency. However, the cloth's characteristics in relation to the particular liquor entrainment should be considered. Experimental investigation of synthetics indicates that silicone-treated material greatly improves dust collection performance in such cases.

Another factor to consider is the corrosive properties of the liquid. It has been found that a high resistance to chemical corrosion exists, when the cloth has been silicone-treated. For example, no noticeable deterioration in efficiency has been observed in filtering sulfuric acid mist in experiments of over 4600 hours. Under normal running conditions, bag life has been at least 5000 hours. The filter fabric can also be retreated with silicone for further use.

Prehumidification is another approach to preventing the deleterious effects of liquid mists. But it is still more convenient, and probably more economical, to use a filter medium possessing properties resistant to the particular mists.

Static electricity is generated when electrons are given up by one particle to another as they collide and separate. If the cloth is a poor conductor of static electricity, then the particles can charge the fabric. If particles and fabric are oppositely charged, attraction to the fabric takes place. If both fabric and particles have the same charge and sufficient difference in voltage exists, there is attraction. Static can improve or hinder the efficiency of the cloth–improve performance of a poor cloth or interfere with the cleaning of a fabric. It is advisable then to choose a fiber on the basis of the charge on the dust, as well as temperature and chemical abrasive properties, so that fabric and mist are compatible.

DESIGN AND OPERATION[41]

As with any gas cleaning equipment, proper fabric filter design features required for a specific application can be determined only by knowing the nature of the particulate collected, the properties of the gas stream, and the particular method of cleaning employed. Many times these factors are not clearly understood and investigated before purchasing the equipment. This situation leads to misapplications resulting in unsatisfactory dust collection. Proper design and application cannot be overemphasized.

Baghouse housings should not be taken for granted. Many fabric filters are purchased without a housing, because housings can be designed and installed to fit the particular operation and the space available. The housing can be of sheet metal, fabricated metal or other construction materials. In selecting the filter unit, take into account gages of metal, adequate stiffening and reinforcing, and accessibility–both externally and internally–to the cleaning unit for maintenance purposes. Use of structural baghouses with corrugated siding is becoming popular on large installations and should be considered, even though costs are higher.

Another important engineering parameter to consider is the amount of cloth needed. Cloth area needed depends on the quantity of gas handled, dust concentrations, and the specific flow-resistance property of the particulate deposit. Cloth area must be selected to operate at a particular pressure drop. The average filtration velocity on air-to-cloth ratio is often in the range of 1 to 15 cfm/sq ft. Gas flow rates higher than 50 ft/min can be achieved at moderate pressure drop for coarse dusts.

Other parameters of concern are the operating pressure drop, filtering velocity, and the time required between cleaning cycles (all of which affect operational and maintenance costs). In this regard, the following is a useful formula:

$$P = k_1 V$$

where: P = pressure drop across cloth, in. water
            k_1 = resistance of the cloth, in. water/ft/min
            V = gas flow velocity, ft/min

Time between cleaning cycles can be estimated from the pressure drop by using the following simplified equation:

$$\Delta P = \frac{k_2 V^2}{7000}(L)$$

where: ΔP = pressure drop as a function of time, in. water
            V = gas flow velocity, ft/min
            L = inlet dust concentration, grains/cu ft
            t = time, min
            k_2 = resistance coefficient of dust and cloth

Dust-fabric filter coefficients can be estimated from:

$$k_2 = \left(\frac{k}{g}\right)\left(\frac{U_f}{P_f}\right) s^2 \left(\frac{1-E}{E^3}\right)\left(\frac{1}{P}\right)$$

where: k = constant (dimensionless) obtained from the proper literature on particulate fabrics
            g = 32.2 ft/sec^2 (gravitational constant)
            U_f/P_f = gas viscosity/gas density, sq ft/sec
            P = density of particulate material, lb/cu ft
            S = specific surface area per unit volume of solids in the dust layer, sq ft/cu ft
            E = porosity (dimensionless) obtained from the proper literature

These equations are simplified versions of the complicated expressions used for finding exact values.

Experimental values of S (which can be approximated) and densities for specific dusts can be obtained from gas adsorption data available from the manufacturer of the baghouse unit. Values of E can also be obtained in this manner. It should be noted that the porosity E is affected by the drag caused by the gas flow through the layer and is a function of particle size and velocity.

The resistance coefficient k_2, then, is of primary importance. There are several fabric variables that influence this value. Bulkier yarns, napping, and/or felts tend to produce a more open deposit and lower values of k_2. Also, higher filtering velocities cause an increase in k_2 values. Thus, in calculating k_2, the areas to examine are the particle size, filtering velocity, and fabric parameters.

Another factor to consider is the discharge gas. The fabric filter's discharge can sometimes be recycled to the plant's interior, but it must be acceptable to health standards. Recycling offers the advantage of saving on heating or cooling of makeup air. However, special design considerations are involved.

Maintenance

After sizing the filter and choosing the fabric, method of fabric cleaning, and time between cleaning cycles, one should evaluate maintenance factors. Filter bag sizing and grouping can affect maintenance. Generally, a maximum length-over-diameter ratio of 30:1 provides a uniform filtration profile from the bottom to the top of the baghouse. It also contributes to a good entrance velocity at the filter bag inlet. A good design will have internal compartment walkways at the cell-plate level and at the bag-suspension level to provide for inspection and repairs.

Accessibility and location of the cleaning mechanism are very important. It is advisable, therefore, to have a cleaning mechanism with external drives and mechanical parts.

Other design features to note are placing, size, capacity, and accessibility of the dust hoppers. Dust-collecting hoppers usually come in inverted-pyramid or trough shapes. They should have adequate slope and should be positioned so as to provide a satisfactory means of removing the collected dust. The residue should not be allowed to remain in the hoppers for an extended period of time because fine particulates have a tendency to pack.

There are numerous variables affecting the sizing of a fabric collector. The major determining factor is the pressure drop. And, the filtering velocity (air-to-cloth ratio) varies for different applications depending on the filtration characteristics of the dust. There are formulas relating filtration velocity, grain loading, time, and pressure drop. Although these formulas give a great deal of information, personal experience and application knowledge are still the final criteria determining the design and sizing of the baghouse for a given problem.

It is advisable to examine known results in related applications and to allow for unknowns in the event that direct experience is lacking. If design parameters, factors determining efficiency, and the guidelines to proper fabric selection are followed, a highly efficient system at a modest cost can be installed. With proper design, operation and maintenance, fabric filter systems will collect more than 99.9 percent of the dust.

Costs

Optimization is the key to designing and operating efficiency of any piece of industrial equipment—including fabric filters. However, a favorable balance between efficiency and cost must be determined. It is useless to have a unit operating under total efficiency at tremendous costs. In judging baghouse systems, there are four primary cost areas: equipment, installation, operating and maintenance costs.

Equipment accounts for only a small fraction of the total cost of an installation. Electric motors, fans and conveyors for the collected dust vary in cost, depending upon the amount of gas handled and the size of the unit. Fan selection is particularly difficult. However, most manufacturers of fans do extensive testing on their equipment. Their product literature offers aid in determining size and cost.

Figure 2-7. Structural work can have a significant effect on cost of baghouse installation and must be considered with all other factors. This 8600 cfm system cost $150,000 when installed in 1970. It uses 28 x 36-in. steel ducts to carry exhaust fumes to the baghouse where particulate matter is collected in Dacron bags.

Housings also make up a part of equipment costs. Most housings are made of sheet metal, metal plate, or poured concrete. Sheet metal (iron, steel and aluminum are the most common types used) can be priced and identified by standard gage numbers (available in most engineering handbooks). In pricing these, take into account support beams for the baghouse.

Fabric is the final point to consider in the pricing of equipment. The range of costs varies from $1 to $100 per filter element (bag, tube or panel). Total cost for filter use in baghouses depends on the type of material, element size and fabrication costs. Some common causes of fabric failure are given in Table 2-9.

Table 2-9. Common Causes of Fabric Failure

Fabric failure from dust deposit interaction	Abrasion, wear
	Flexure wear failure
	Binding or plugging
	Burning due to sparks
	Humidity condensation
	Deposited dust hardens as result of inefficient cleaning, cake tears
Failure from mechanical deformities	Seams, sewing
	Tears at the top between cloth and support
Failure from poor design	Chafing against housing or other bags
	Tensioning caused by loose bag may cause tearing
	Seals around cloth-metal collars defective
	Cleaning carriage bag wear

Installation costs differ greatly from unit to unit. Erection costs can be divided into two broad categories: structural work and fabrication work. Structural work is priced on a weight basis. Structural mild-steel costs range from $155 to $176 per ton, and stainless-steel construction may run as high as $900 per ton. Fabrication costs also vary widely. A good guideline to bear in mind is that 20 to 60 sq ft of material may be fabricated by one man in an average 8-hour day.

A typical cost breakdown for a 10,000 cfm shaker-type baghouse installation is given in Table 2-10.

Maintenance and operating costs should also be evaluated. A rough estimate of annual operating and maintenance costs can be expressed in the following mathematical expression:

$$G = S \frac{0.7457}{6356E} \left[(PHK) + M \right]$$

where: G = annual cost, $
          S = design capacity of collector, cfm
          E = fan efficiency, generally assumed to be 60 percent
          M = annual maintenance costs, $/gal-hr
          P = pressure drop, in. water
          H = annual operating time, assumed 8760 hr
          K = power cost, $/kwh

It is apparent from the formula that the expenses involved in running a unit at a specified efficiency are dependent on:

1. Gas volume handled,
2. Pressure drop,
3. Operating time,
4. Cost of electricity, and
5. Mechanical efficiency of the motors, fans, pumps and cloth.

Table 2-10. Baghouse Installation Costs

Item	Cost ($/cfm)
1. Filter Fabric	$1.000
2. Fan and Motor	0.313
3. Duct Work	0.813
4. Conveying Equipment for Disposal	0.125
5. Foundation and Installation	0.350
6. Instrumentation	0.063
7. Engineering and Supervision	0.125
8. Freight	0.063
9. Startup Costs	0.125
TOTAL	$2.977

All these factors can affect the bag life and increase maintenance. The last formula roughly estimates annual costs. It is by no means an absolute method. Initial costs (equipment and installation) can be accurately determined and modified as required. Maintenance and operating costs, on the other hand, can only be estimated, and depend on good design.

ELECTROSTATIC PRECIPITATORS[36]

The efficient and economical collection of solid particulate from gas streams is one of the formidable problems facing industry today. Methods employed in particulate collection from exhaust gas streams may be broadly classified as mechanical or electrical.

The basic difference between the two general methods is reflected in the versatility in application of the electrical method. Here, the electrical or separation forces are directly applied to the particles themselves, whereas mechanical methods are applied to the entire gas stream. Because of the direct use of forces, electrical-separation systems, that is electrostatic precipitators, maintain only modest power requirements. This method is limited neither to large coarse particulate as are gravitational systems and cyclones, nor by resistance to the motion of the gas as are filters and scrubbers.

Even submicron size particles can be collected efficiently by electrostatic precipitation due to the relatively significant electrical forces acting on the particulate. In practice most units are capable of efficiencies of up to 99.9 percent. They display the broadest range of applications and most systems operate continuously at widely varying dust loads.

In general, precipitators have a high initial capital cost as compared to other types of collectors. However, one of the distinct advantages over other systems is the low operating pressure drop which results in power savings. Over the operating life of the unit, this reduces the overall cost differential.

Basic Design

There are two classes of electrostatic precipitators: one-stage and two-stage.

One-Stage Precipitators

Referred to as Cottrell precipitators, these are the most widely used in air pollution control. The basic elements are a source of unidirectional voltage, corona or discharge electrodes, collecting electrodes, and a means of removing the collected matter from the precipitator. One-stage systems combine both an ionizing and collecting step. There are two main types of one-stage electrostatic precipitators.

Plate Type—primarily used for dry dust removal. The grounded collecting electrodes are a series of parallel plates, which are encased in a shell. The discharge electrodes are suspended between the plates by means of insulators connected to a high voltage source.

Pipe Type—primarily used for the removal of liquid sludge particles and volatized fumes. A number of pipes are contained in a cylindrical shell under a header plate. The collecting electrodes are the pipes and the discharge electrodes are suspended within them.

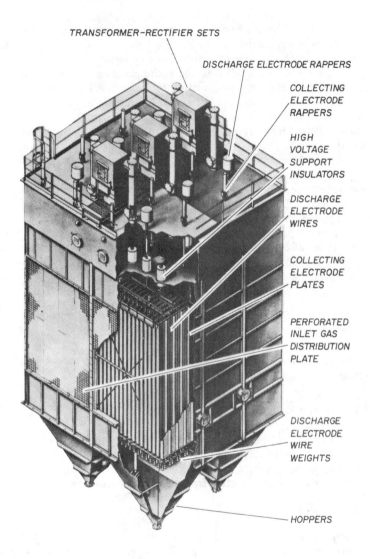

TRANSFORMER–RECTIFIER SETS

DISCHARGE ELECTRODE RAPPERS

COLLECTING
ELECTRODE
RAPPERS

HIGH
VOLTAGE
SUPPORT
INSULATORS

DISCHARGE
ELECTRODE
WIRES

COLLECTING
ELECTRODE
PLATES

PERFORATED
INLET GAS
DISTRIBUTION
PLATE

DISCHARGE
ELECTRODE
WIRE
WEIGHTS

HOPPERS

Figure 2-8. Elements of a three-section electrostatic precipitator.

Two-Stage Precipitators

Here, a pre-ionizing step is followed by collection. This type is primarily employed when particulate loadings are low and ozone generation must be minimized. The discharge plate is the positive electrode, which minimizes the generation of ozone. A schematic representation of each is shown in Figure 2-9.

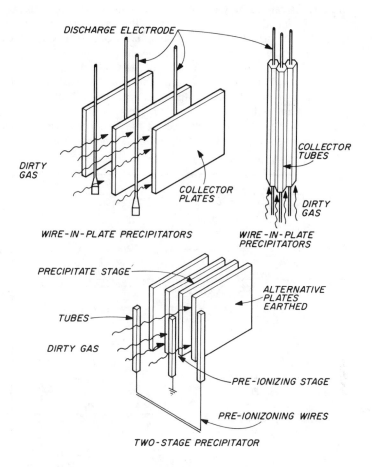

Figure 2-9. Three types of precipitators. (A) Wire-in-plate precipitator,
(B) Wire-in-tube precipitator; (C) Two-stage precipitator

In single-stage types a voltage differential is created between the discharge electrode and the collecting electrodes, thus generating a strong
electrical field between them. The contaminated gas is passed through
this field and a unipolar discharge of gas ions from the discharge electrode
attaches itself to the particles to be collected. This unipolar discharge of
gas ions (negative charge) is brought about at certain critical voltages
where air molecules become ionized. This ionization is evidenced by a
corona at the discharge electrode. The negative air ions move toward the
positive collecting electrode, while the positively charged particles migrate
toward the discharge surface. As the air ions move, they become attached

to neutral particles carried by the gas streams. This is the force causing
the suspended particulate to be transported to the appropriate electrode.
The particles that are attracted to the positive electrode, or discharge
electrode, dissipate their charge and become electrically neutral.

The particles must now be removed from the collecting electrode.
In a plate type precipitator, the solid particulate must be washed down,
usually with water or by rapping. In the pipe type, the collected material
is usually a liquid and will flow down off the electrode by gravity. The
collected matter is then placed in a hopper at the chamber bottom. Figure
2-10 shows the sequence of events which occur in a precipitator.

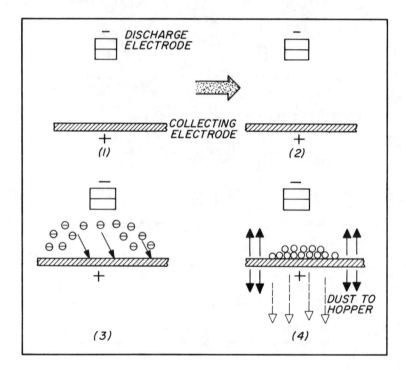

Figure 2-10. Sequence of events in an electrostatic precipitator. There are four steps
involved in electrostatic precipitation:
1. Generating a strong electrical field between electrodes.
2. Passing the suspended particles to be collected through the field whereupon
they are electrically charged by means of ionization.
3. The charged particles are then transported to a collecting surface, by means
of the force exerted on them by the electric field.
4. The electrically charged particles precipitated on the collecting surface are
neutralized and removed usually by rapping or shaking the collecting electrode.

Electrostatic Precipitator Components

Electrostatic precipitators consist of four main components:

1. Power supply unit, which provides high voltage unidirectional current,
2. Electrode system,
3. A means of removing the collected matter, and
4. A housing that provides an enclosed precipitator zone.

Power Supply

The power supply must deliver a unidirectional current to the electrodes at a potential very close to that which will generate arcing across the electrodes. This is important to ensure maximum operational efficiency. The power unit can be divided into four parts (Figure 2-11).

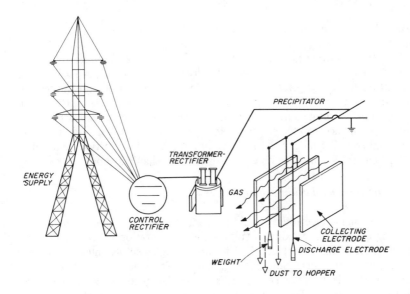

Figure 2-11. Schematic representation of the various components of an electrostatic precipitator.

Input Source of Power—Most installations are designed to require input voltage ranging from 110 volts to 440 volts. The power consumption generally ranges from 2.5 to 30 kVA.

High Voltage Rectifiers—There are a variety of designs on the market. The most widely used are the mechanical, electron tube and selenium rectifying systems.

Voltage Regulators—Maximum efficiency occurs at a voltage very near to that producing arcing between the electrodes. Because of this, it is necessary to include voltage regulation as an integral part of the power unit. This will prevent the establishment of a permanent power arc during periods of flashover.

Output Power Unit—An output power unit is also included. Systems can be obtained in a voltage range of 13,000 to 90,000 V and current of 20 to 400m a.

Electrode Unit

There are two types of electrodes. The discharge electrode is a high-voltage ionizer. These are generally negatively energized. The other type collecting electrodes are usually positive surfaces and must be insulated from the precipitator shell at all times.

Particle Removal from Electrodes

The electrode must be small or have sharp edges in order to facilitate the formation of a corona around the discharge electrode. A great deal of attention should be given to the collecting electrodes since the success of precipitation depends heavily on them. Collecting electrodes should display maximum collecting surface, have no tendency to buckle, and display corrosion resistance to the material being collected and its surrounding atmosphere. Rapping is the most common method of particle removal. Thus, electrodes must be able to withstand mechanical shock.

Housing of Precipitation Zones

Discharge electrodes should be accurately centered between the collecting electrodes. Maximum voltage and precipitation efficiency cannot be maintained if this is not done. The supporting structure must be stable to insure true centering. Sometimes weights are suspended on the bottom ends of the electrodes. This reduces the tendency to swing. Other designs have the electrodes fastened at the bottom to an insulated, stationary frame.

Properties of Dusts, Fumes and Mists[37]

In general, particle size and settling rate are the most important characteristic properties of suspended matter. Particles larger than 100μ may be excluded from the category of dispersions because they settle too rapidly. Particles of one micron or less in size will settle at slower rates, and are thus regarded as permanent suspensions. A thorough understanding and examination of the gas stream to be cleaned must be made.

Dusts

Dusts are formed by the pulverization of solid matter into small size particles. These may be the result of such processes as grinding, crushing, blasting and drilling. Particle sizes range from 1μ up to about $100\text{-}200\mu$. Dust particles are usually irregular in shape. Examples include flyash, rock dusts and flour. They are heterogeneous in size and structure.

Smoke

These include the derivatives from combustion of organic matter (wood, coal, tobacco), gaseous particle suspensions formed by chemical and photochemical reactions, condensations and volatilization. Smoke particles are extremely small, with sizes ranging from less than 0.01μ to 1μ. They are usually spherical in shape if of liquid or tarry composition. If they are of solid composition, an irregular shape is characteristic. *Smog* refers to a mixture of natural fog and industrial smoke.

Fumes

Fumes are formed by processes such as sublimation, condensation and combustion. Probably the most common type fumes are generated from the oxidation of metallic vapors or compounds. Smelter operations generate enormous volumes of metallurgical fumes of this type such as oxides of zinc, cadmium and beryllium. Particle sizes range from 0.1μ to 1μ.

Mists

Mists are formed by the condensation of vapors upon suitable nuclei. This gives rise to a suspension of small liquid droplets. Particle sizes of naturally occurring mists range between 5μ and 100μ. Submicron particles can be produced under special conditions. The terminology here is used rather loosely. For example, some industrial dispersoids such as sulfuric acid particulates are termed mists, even though they are one micron or less size particles, which would put them in the category of smokes or fumes.

Physical and chemical properties such as particle size, structure, rate of settling due to gravity, optical activity, affinity in absorbing electric charge, surface area to volume ratio, chemical catalytic activity, and reaction kinetics characterize particle suspensions. Most chemical and physical properties of particulates depend primarily on the amount of exposed surface area. For example, large surface areas of some dispersoids accelerate oxidation and increase solubility, evaporation, adsorption, catalysis and electrostatic adsorption.

Dusts and fumes absorb relatively large amounts of gas on their surfaces. Such a collection of particles, particularly when hot and very dry, takes on properties of liquids and gases. Under such conditions the fluid may be pumped through pipes, the same as a liquid. Dispersoids also display the ability to build up large electrical charges. They exhibit marked electrical activity such as building up high potentials on insulated electrodes when immersed in them. Electrostatic precipitators are largely successful due to the ability of dispersoids to adsorb large quantities of electrical charge.

The most important electrical properties of suspended particulates are their electric charge and conductivity. Most industrial dusts are electrically charged to a degree. The amounts of positively and negatively charged dust particles are approximately equal. Thus, the given suspension is electrically neutral on the whole.

In the case of dusts, the electrical charge is imparted by the mere separation of particles from the solid or bulk state. Particulates, which are initially in contact, have small contact potential differences at their surfaces. When they separate, the potential differences give rise to particle charges—one particle being negative, the other positive.

Smokes and fumes seem to have no charging mechanism. Particles are formed by condensation from the vapor stage. The source of ions is due to the flames that produce the vapor. Vapors formed at high temperatures in flames generate a charging effect. This is only apparent at very high temperatures; however, mists which are formed by condensation from vapor at low temperatures display no ion source. Such dispersoids are initially uncharged. They tend to absorb charge from the atmosphere, but this occurs over a long period of time.

Particle conductivity is a basic criteria involved in the collection, deposition, orientation and separation of particulates. Particles must display some conductivity in order for electrostatic precipitation to be effective. Nonconductive deposits or those with high resistivity on the collecting electrode will impede the flow of the corona ions. Excessive sparking and reduced precipitator current and voltage usually accompany this condition. All this leads to reduced precipitator efficiency. Careful examination must be made of the electrical properties of the matter to be collected, before installing a precipitator.

Performance

Efficiency can be defined as the percent or fraction of particulate entering the precipitator that is collected. Efficiency is primarily related to the total surface area of the collecting electrodes per unit volume of gas and is directly proportional to the particle rate velocity, often referred to as the precipitation rate parameter. The following expression shows the relationship:

$$\theta = 1 - e^{\left(-W\frac{A}{Q}\right)}$$

where: θ = efficiency
                 A = collecting electrodes surface area, sq ft
                 Q = volumetric gas flow rate, cfs
                 W = particle drift velocity, fps

The overall velocity of the gas stream is not a major parameter in the efficiency expression. Its importance is in optimizing precipitator performance. Note that it is not necessary to apply power to accelerate the gas to collect suspended particulates.

The particle drift velocity is obtained from the following expression:

$$W = \frac{E_o E_p{}^a}{\pi \eta} C$$

where: a   = particle radius, ft
                 E_o = charging field, volts/ft
                 E_p = collecting field, volts/ft
                 η = gas viscosity, lb_m ft^{-1} sec^{-1}
                 C = constant

In a single-stage precipitator E_o and E_p are approximately equal. Thus, the precipitation rate parameter is proportional to the square of the field strength. As shown in the above expression, it is directly proportional to the particle radius. Inertial systems, in contrast, depend much more strongly on particle size. Very high voltages are used in precipitators since field strength is of major importance. Generally, the voltage ranges from 60,000 to 100,000 volts (peak) in industrial designs.

Precipitator performance is sensitive to factors that affect the maximum operational voltage. These factors are primarily composition of the gas stream, density, and the electrical conductivity of the particles to be collected.

An important parameter to note is the sparkover voltage. Sparkover voltage is that voltage at which the gas becomes locally conductive, allowing a high current discharge. Gas composition and density largely determine the sparkover voltage. During each spark the electric fields diminish. This is a drastic waste of power. Gas density varies with temperature and pressure. However, this need not be a major factor affecting maximum allowable voltage when the operating conditions are allowed for in the initial precipitator design. Some precipitators have been designed to operate at temperatures as high as 1650 F and pressures as high as 800 psi.

Electrical resistivity can also greatly affect efficiency. Particles can lose their charge rapidly upon reaching collecting electrodes and can become reentrained in the passing gas stream if resistivity is too low. When resistivity is too high, the particles reaching the collecting electrodes cannot lose their charge due to the high resistivity already on the electrodes. A build-up of a voltage gradient across the layer of the collected material results. This build-up opposes the voltage that is used for collection. It can become high enough to cause a dielectric breakdown at the collecting electrode and thus significantly affects the sparkover voltage.

Proper sizing and selection of the operating conditions of the precipitators can control resistivity. It will vary with temperature rising to a peak with increasing temperature and then declining. Altering the temperature of the incoming gas streams can control most resistivity problems. Another solution might be to introduce a conditioning agent into the gas stream to control the resistivity of the particles. Water vapor is such an agent.

Gas velocity is another important property that can alter performance. It must not be so low as to prevent proper distribution through the cross section of the precipitator. Furthermore, it should not be so high that it carries collected matter off the collecting surfaces. Uniform inlet gas distribution must be achieved. Careful consideration should be given to duct and inlet area design to avoid eddies and jet flow of gas through the precipitator.

In obtaining the highest efficiency for a given dust load, precipitators are usually divided into electrically independent sections. These areas are powered and controlled separately from one another. Systems that are so designed can transmit voltage throughout the entire precipitator, and it is not limited by the least favorable transient conditions existing in any one component.

Likewise, the cycling of rappers for dislodging materials from the collecting electrodes is sectionalized. The inlet section of the precipitator encounters the heaviest particle load. It is obvious that the rapping load is the greatest at this point. Sectionalizing rapper control allows power input to this area to be tailored in handling this heavy load in the inlet section and reduced in areas encountering lesser loads. This aids in avoiding excessive rapping which tends to resuspend particles in the gas flow.

Electrostatic Precipitators in Industry

Some of the common operations and industries employing precipitators are briefly examined in this section. This is by no means a complete list, but it will indicate the wide range of uses of precipitators. Note that electrostatic precipitators have a dual role. They are not strictly used for air pollution control, but for product recovery as well. Still another use is in the cleaning of a gas stream of impurities, where the gas is the desired product.

Cement Industry

In cement plants, precipitators serve double duty: recovering process materials and reducing air pollution. Precipitators are used efficiently in both wet-process and dry-process cement kilns, shale and stone dryers, raw grinding mills and various other cement mill applications. Efficiencies as high as 99.9% are common in the cement industry.

Frequently, an integrated mechanical collector electrostatic precipitator is employed when large particles of low alkali content must be separated from very small particles of high alkali content. The mechanical collector removes the larger particles which are returned to the cement generating process. The precipitator collects the smaller particulate as refuse.

Pulp and Paper Industry

Precipitators are employed in handling air pollution problems generated in kraft and soda mills. These plants pulp hard and soft woods, in addition to processing bagasse and bamboo.

Precipitator designs include wet or dry bottom, depending on the particular process in this industry. In the wet-bottom type, captured salt cake dislodged from the collecting electrodes falls into a flat bottom that is flushed with black liquor. A rotating agitator is used to keep solids in suspension. In the dry-bottom design, trough hoppers or flat bottoms are used. Collected matter is removed by a drag scrapper.

Municipal Incinerators

Electrostatic precipitators have proven to be ideal in obtaining more efficient collection of particulates from incineration combustion gases. Efficiencies of 95% and better have been obtained in cleaning incinerator combustion gases, without causing any serious water pollution problems.

Precipitator designs for steam-generating or refactory-lined incinerator furnaces handle capacities of 25 tons per day and up.

Power Plants

Precipitators have been designed and are being employed with practically every type of boiler and for all types of coal and oil-fired boilers. Careful consideration must be given to operating parameters. Moisture content, sulfur content, ash content, and various other characteristics of coal and the ash that results from various combustions, all greatly influence precipitator design.

Iron and Steel Industry

Iron and steel production, more than any other industry, involves complex interdependent operations and constantly varying conditions. Control systems must be precisely matched to handle these. Furthermore, the systems must be designed to accommodate possible future alterations.

Specific areas in this industry where precipitators are employed include:

1. Basic oxygen furnaces—able to remove dust and fumes with efficiencies around 99.9%.
2. Open hearth furnaces—precipitators employed in both hot and cold metal shops.
3. Blast furnaces.

Drying Processes

Drying processes are employed in the production of such materials as cement, gypsum, bauxite and various ores. The object is to drive off the free moisture by heating moderately while the material is agitated slowly. A rotary kiln is commonly used. Generally, a large portion of the total feed is carried off as dust in the gas stream. This also includes products of combustion, excess air and moisture. The collected material is added to the main raw material stream. The cleaned gas stream is discharged to the atmosphere.

Nonferrous Metals

As in the iron and steel industry, nonferrous metals refining warrants proper pollution control techniques. Severe operating conditions are often encountered. Gas temperatures are usually high and acids are often employed. Thus proper steps in corrosion prevention are necessary. Care must be taken not to operate precipitators and mechanical collection systems below the dew point.

Nonferrous application primarily involves the treatment of off-gases from roasters, converters, blast furnaces, cupolas, and sinter machines. The main function is to reduce ores of copper, zinc, lead, tin, nickel, chromium, molybdenum and precious metals.

Acid Recovery Operations

Most acid precipitators collect either sulfuric or phosphoric acid. The stream usually passes through a scrubber or cooling tower, first, in order to saturate the gas and remove a portion of the acid. The gas is then sent into the precipitator where the acid is recovered as a fine mist. This collected material is added to the product stream. The discharge gases are either emitted to the atmosphere or to a secondary scrubber for the removal of any remaining SO_x.

Precipitators employed for this process are greatly limited. Material of construction is mainly lead or lead lined—limiting operating temperatures to 170 F. Phosphoric acid precipitators are constructed from various materials—carbon, stainless steel and rubber-lined vessels.

CYCLONES[35]

A working definition of a mechanical dust collector is a device which separates solid particulate from a dry gas or a liquid by employing inertial and/or gravitational forces. Cyclones have long been regarded as one of the simplest and most economical mechanical collectors. Chief advantages include high collection efficiency in certain applications, adaptability, and economy in power. However, their main disadvantage is their efficiency limitation to the collection of only large size particles. Cyclones are not usually suitable for dealing with dusts containing large amounts of particulates less than 10 microns in diameter. For such particulate characteristics and concentrations, more efficient collectors such as electrostatic precipitators, baghouses or wet scrubbers are employed.

There are many variations of the basic cyclone, and because of its simplicity and lack of moving parts, a wide variety of construction materials can be used to cover relatively high operating temperatures of up to 2000 F. Cyclones generally are efficient handling devices for a wide range of particulates, collecting particles ranging in size from 10 to above 2000 microns, and inlet loadings from less than one grain per cfm to greater than 100 grains per cfm.

Cyclones are employed in the following general applications:

1. Collection of coarse dust particles;
2. High solid gas concentrations particularly above 3 grains/scf;
3. Classification of particulate in sizes referenced;
4. Where extreme high collection efficiency is not critical; and
5. As precleaning units in line with higher efficiency collectors for fine particles.

Theory of Operation

A cyclone consists of a cylindrical shell fitted with: (1) a tangential inlet through which dusty gas enters, (2) an axial exit pipe for discharging the cleaned gas, and (3) a conical base with dust discharge facility (Figure 2-12).

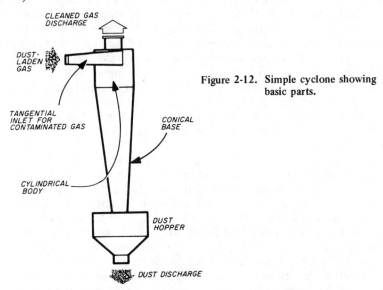

Figure 2-12. Simple cyclone showing basic parts.

Cyclones are commonly operated in a vertical position; however gravitational forces are not of major importance in particle collection. Thus, a system can also operate with its axis in a horizontal or inclined position. This can offer advantage in saving headroom; however, as will be shown later, certain operating problems do arise.

The flow pattern is complex in even the simplest cyclone. There are three main flow patterns employed which help to explain the method of operation.

1. *Descending spiral flow*—this pattern carries the separated dust down the walls of the cyclone to a dust hopper.
2. *Ascending spiral flow*—this rotates in the same direction as the descending spiral, but the cleaned gas is carried from the cyclone or the dust receptable to the gas outlet.
3. *Radially inward flow*—this feeds the gas from the descending to the ascending spiral.

The dust-collecting hopper is where the final separation occurs. In this component the total gas flow reverses direction and is fed to the ascending spiral flow.

The flow patterns are generated by the creation of a double vortex, which centrifuges the dust particles to the walls. The two distinct vortices present in a cyclone are: (1) large diameter descending helical current in the body and cone, and (2) ascending helix of smaller diameter extending up from the dust outlet section, through the gas outlet. At the walls they can be transported into the collecting hopper, which is isolated from the influence of the spinning gases. The gas spirals downward and upward at the inside. Upon entering the cyclone, the gas undergoes a redistribution of its velocity, so that the tangential component of velocity in a cyclone exceeds the value of the inlet-gas velocity by several times. It is important to note that as the gas spins in a vortex in the cyclone body, the tangential velocities increase as the axis of the cyclone is approached, at any horizontal plane. The tangential velocity at any radius appears to be relatively constant at all levels. However, tangential velocities at the extreme top of the cyclone are not included since the proximity of the cyclone cover slows the spin. When the downward air flow is smooth and unbroken, the dust particles flow spirally downward and pass out the bottom dust outlet without reentrainment.

As depicted in Figure 2-13, the dust-laden gas enters the tangential inlet and whirls through several revolutions in the body and cone, dropping its dust load. The clean gas is emitted through the axial cylindrical air outlet. The dust particles, which were uniformly dispersed in the entering gas stream, tend to concentrate in the layer of air next to the cyclone wall under the influence of centrifugal force. The helical motion of the main air stream down and the small quantity of air through the cyclone dust outlet project the separated dust solids into the receiving bin.

Air should be fed to cyclones in a radially thin layer, so that the radial distance through which a particle travels for separation from the air

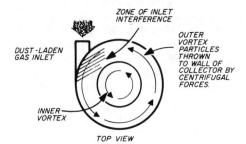

Figure 2-13. Cyclone schematic.

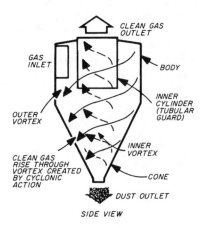

stream is at a minimum. It is important that the inlet be exactly tangent to the body, since separation is adversely affected by deviation. Flow through a cyclone outlet is vertical, being helically upward near the periphery of the outlet and downward near the center.

Collection Efficiency

Efficiency of a cyclone can be defined as the fractional weight of dust collected. It is a major parameter in the selection and design of a cyclone. Cyclones may be designed for any required efficiency; however, it is important to note that efficiency is a function of energy expended or space occupied. Thus, proper optimization to an acceptable efficiency at moderate pressure drop within reasonable space requirements is necessary. Efficiency can be markedly improved at the expense of pressure drop without tampering with space requirements.

The major parameter in the prediction of collection efficiency is particle size. For each particular cyclone design, there is a critical size of particle of a given density on which the centrifugal and inward viscous forces balance, so that the particle neither moves outward to the walls nor inward toward the cyclone axis. Particles larger than this critical diameter (called the "theoretical cut" and measured in microns) would be collected, while all smaller particulate would escape. Efficiency is enhanced as the axis of the cyclone is approached until the edge of the core is reached. This is so because the centrifugal force increases more rapidly than the inward drift up to the edge of the core. Therefore, particles rotate in orbits whose radii are dependent on the balance between the viscous forces due to the inward drift and outward centrifugal force. Particles will then be transferred to the outer orbits, where they pick up finer particles by collision or are captured by eddy currents. A knowledge of theoretical cut has very little direct significance in the prediction of collection efficiency. It does, however, give a rough indication. The most common method of estimating total collection efficiency is by plotting a fractional efficiency curve or grade-efficiency curve. A fractional efficiency curve is a plot of particle size versus percent collection. Figures 2-14 and 2-15 show the specific features of such a curve and a typical plot based on experimental data, respectively.

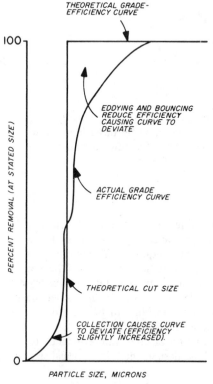

Figure 2-14. Grade efficiency curve. Note that the cyclone has 0% collection efficiency for all particles that are smaller than the theoretical cut size, and 100% efficiency for all larger sizes.

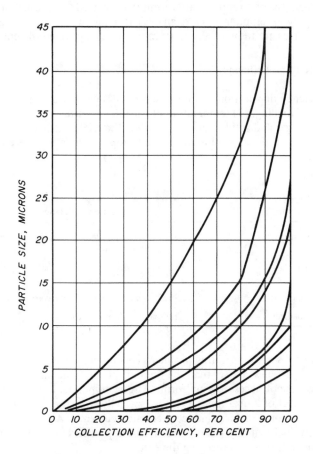

Figure 2-15. Collection efficiency, percent.

Figure 2-14 shows that the cyclone has zero percent efficiency for all particles smaller than the cut size and one hundred percent for all larger sizes. Under actual operating conditions, however, a considerable amount of particulates smaller than the cut size are separated with the coarser particles. This can be explained by collision between particulates or due to particle aggregation. Furthermore, a proportion of the particles larger than the cut escape collection (they are either carried into the inner vortex by eddies or by bouncing). Very little success has been achieved in predicting efficiency with particle sizes less than the cut. The degree of efficiency in this range largely depends on the properties of the dust.

It is apparent that the curve in Figure 2-15 may be used accurately in predicting the total collection efficiency of a cyclone, only if the particle size distribution of the dust is known. This must be determined in the laboratory by proper analysis of a representative sample for particle size distribution. Great accuracy in sampling and size analysis is required if the resulting grade efficiency curves are to be of any value in predicting cyclone efficiency.

Factors Affecting Efficiency

Efficiency depends on (1) pressure drop, (2) particle size distribution, (3) inlet dust loading, (4) temperature of the inlet gas stream and (5) specific gravity of the gas and solid. Table 2-11 shows the relationship of these factors to efficiency with the exception of pressure drop, which is considered separately.

Table 2-11. Factors on Efficiency

Condition	Effect on Efficiency
Temperature	Decreases as temperature increases due to gas viscosity changes
Velocity	Increases with velocity and falls off sharply below 25 ft/sec
Specific Gravity	Increases with higher specific gravity materials
Inlet Loading	Increases with inlet loading

There are several operating problems encountered that promote failure to collect all particles coarser than the theoretical cut. One major reason is that the drift is not necessarily uniform. At certain points, it may exceed the mean by a factor of 2 or 3 times. This is especially so at the ends of the cyclone, where the additional surfaces induce precession currents. A doubling or tripling of the drift velocity at any point would result in particles of 40 percent larger than the theoretical cut in reaching the exit.

Another effect causing poor separation of coarse particles from the gas is eddies in the vortex. Velocities of eddies normal to the main flow are generally 1/5 of the main flow. Thus, if the gases near the walls of the cyclone are spinning at about 50 ft/sec, which is general practice, the eddies may add 10 ft/sec or more to the inward drift velocity. The inward drift velocity should only be of the order of 1 ft/sec. The eddies will thus cause particles of the magnitude of three times or more the cut size to appear in the exit stream.

The generation of double eddies, superimposed in the vertical plane on the main flow, is also a problem. Figure 2-16 shows a schematic view of the eddy currents in a cyclone. These assist in the descending of the collected dust into the dust receptable but at the same time carry a portion of the collected dust back into the inner ascending vortex. Upswept dust has an opportunity to be reseparated while being transported to the clean gas exit. This is so because the gas spins very rapidly in the inner vortex. However, a large portion of dust does escape collection in this manner.

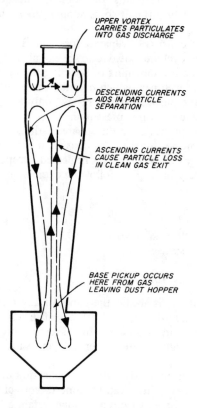

UPPER VORTEX
CARRIES PARTICULATES
INTO GAS DISCHARGE

DESCENDING CURRENTS
AIDS IN PARTICLE
SEPARATION

ASCENDING CURRENTS
CAUSE PARTICLE LOSS
IN CLEAN GAS EXIT

Figure 2-16. Eddying in a cyclone.

BASE PICKUP OCCURS
HERE FROM GAS
LEAVING DUST HOPPER

Base pick-up is a chief cause of excessive emissions. Large fitted dust hoppers function as a disengagement space for extracting dust from the return gas, but poorly designed cyclones can cause this problem. The effect is made worse by any additional vertical flow induced. The most common cause of this type of situation is a poorly scaled dust discharge.

Pick-up from the base along with re-entrainment from the walls can be arrested by irrigating cyclone walls. A 5 to 15 percent purge from the bottom of the cyclone also reduces base pick-up. However, the added expense usually outweighs any additional increase in efficiency.

Pressure Drop Through Cyclones

Pressure drop is one of the most important factors affecting efficiency and design, and thus deserves a separate discussion. The cyclone is usually the largest contributor to resistance of an entire pollution control system—very often contributing as much as 80 percent of the total resistance to the operating fan. Quite often, wide ranges of resistance are also encountered in units of identical dimensions due to adjustments for problems outside the cyclone itself. It is, therefore, extremely difficult to develop a general basis for cyclone resistance.

The total pressure in a cyclone consists of separate losses in (1) the inlet pipe, (2) the vortex and (3) the cyclone exit duct. An examination of these trouble-spots should be made before designing or selecting the cyclone. Mathematical expressions have been developed, which describe the pressure losses at these various points. For the purpose of this review, only a superficial examination of the overall pressure drop is presented.

The pressure drop across a cyclone varies as the square of the gas volume throughput, and is directly proportional to the density of the contaminated gas. Pressure's effect on gas density may be neglected except when dust concentrations are relatively high. The following expression shows this relationship:

$$P = K\rho v^2$$

where:
- P = cyclone pressure drop, in. of water
- K = pressure drop constant
- ρ = gas density, lb/cu ft
- v = entering gas velocity, ft/sec

Note that the pressure drop constant (**K**) is always the same for any one type of cyclone, regardless of height and diameter. Cyclones of different types quite often have a different pressure drop constant. This pressure drop constant can be obtained from manufacturers' literature and varies from 0.013 to 0.024.

Excessively large pressure drops can greatly harm the collection efficiency of the cyclone. Unfortunately, very little can be done to correct for pressure drop. The spinning gases, at the exit from a cyclone, retain a great deal of kinetic energy. Attempts have been made to arrest this by trying to recover all or part of this lost kinetic energy. A few methods employed include the following arrangements: (1) a tangential exit pipe, (2) a conical divergent pipe, (3) vanes in the exit pipe and (4) discs both inside and outside the pipe.

No noticeable recovery in the exit has been reported when the vortex-flow patterns and spinning speeds are kept constant. Some promising recovery has been reported through the use of vanes in and around the exit pipe. Some pressure recoveries have been reported but the spin of gases and thus the centrifugal effect in the cyclone has also been reduced. Thus, there is no real gain. A great deal more experimental work is needed.

Up to this point vortical flow through a cyclone, collection efficiency, secondary factors affecting separation, and pressure drop have been examined. The remainder of the discussion shall consider a few common designs and cyclones for special duties.

Large-Diameter Cyclones

Common cyclones, whose body diameters are 3 to 5 times the diameter of the inlet duct, are useful where large air handling capacity and moderate separating efficiency are required. The ratio of air volume per capital investment dollar is greater than for any other cleaning device. Applications of large-diameter cyclones include operations such as grinding, buffing, fibers and wood chip separation. Particle sizes are in a range of 50 microns or more for many of these functions. These cyclones are also suitable for fine powders such as ground pigments, flour, cosmetic dusts, talc, or finely divided particulate of this nature as well as some toxic dusts. High concentrations of dusts originating from foundries, tumbling, and sandblasting operations generally cannot be handled by single large bodied cyclones.

Small-Diameter Cyclones

Small-diameter cyclones and relatively great-length cyclone designs afford high separating efficiency for particles above 30 microns. The diameter in this type of design is rarely above four feet, and subsequently gas handling capacity is limited for single units. For greater air/gas loads than one unit can handle, multiple cyclones can be employed.

The theory in development of small-diameter, long-cone cyclones for collection of finely divided dusts is that the heavy particle entering the cyclone reaches the wall of the cyclone with comparatively small angular movement. The lighter particle entering at the same place must travel through a much greater angle to reach the cyclone wall. Hence, the smaller the particle size, the greater the angle of rotation. Also, the greater the number of convolutions of the separating vortex, the smaller the particles that can be separated.

High collection efficiencies are obtained from cyclones whose body diameters are less than one foot. For the highest separation, interiors must be made well polished and fitted with all eddy-forming projections eliminated. Efficiencies of more than 99% have been achieved for small cyclones on average particle sizes of 5 microns. The capacity of a 4-inch diameter cyclone of this type is 70 cfm with 2 inches of water pressure drop. The capacities for larger units of similar design are proportional to the square of the diameter. Again, because of limited capacity of a single unit, miniature cyclones can be installed in clusters and fed from the same duct. Greater pressure and more power is necessary to operate smaller combined units. Multitude cyclones consist of a number of small-diameter cyclones. The smaller diameters tend to increase the air stream's tangential velocity and centrifugal action. When the air flow is reversed, multiplying the effect producing a combination cyclone and skimmer. Note that since these are small diameters, the danger of chokage should not be overlooked.

Groups of Cyclones

Many times it is more advantageous to employ several cyclones in a parallel or series fashion. Figure 2-17 shows two cyclones in series. Groups of cyclones, under some conditions and applications, can handle larger quantities of gas and more efficiently.

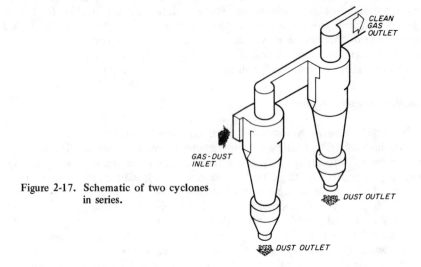

Figure 2-17. Schematic of two cyclones in series.

Parallel arrangements of cyclones can, however, present difficulties if proper design is not employed. Collection efficiency of a group in parallel could be less than one individual cyclone handling a comparable quantity of gas volume. This is brought about due to a difference in the pressure at the various cyclone dust outlets, causing a circulation of gas within the receptable from one cyclone to another.

The particular cyclone exhibiting a lower pressure at the outlet will discharge more gas from the hopper as it enters through the cyclone. The effect is that the ascending gas velocity increases without a corresponding increase in radial velocity, thus decreasing the overall efficiency. Fortunately, this can be overcome by minimizing the difference in pressure between the various dust discharges. This is done by making all the cyclones on one hopper dimensionally identical. Furthermore, conditions of each should be the same; *i.e.,* handle the same volume of gas with identical dust discharges and temperature.

Groups of cyclones in series are employed when it is desired to operate at a higher efficiency than is possible with one stage of collection alone or to handle higher gas volumes. A setup such as this may prove more economical than a single high-efficiency cylone that may have to handle a heavy concentration of abrasive dust. As shown from Figure 2-17, this involves a two stage process, a primary and secondary collector. A third may be necessary but is not commonly employed.

Water Cyclones

Water cyclones are used for the removal of solid particles from water and other liquids. The basic operating principles pertaining to the vorticies and flow patterns of gases outlined previously apply equally here. However, it is important to note that the density differences between particles and liquids are small. High spinning speeds cannot be employed. This is because of the very large pressure drops encountered. The density of water for example is some 800 times greater than air. Efficiency for this type system is not as high as with gas cyclones. Refer to Table 2-12 for a list of the various cyclone applications and efficiencies.

Table 2-12. **Typical Applications and Efficiencies of Single-Stage Cyclones**

Application	% < 10 microns	% Efficiency Range
Powerhouse flyash	20	90-95
	42	75-90
	65	55-65
Cement kilns	40	70-85
Asphalt	10	80-95
Aggregate kilns	30-40	80-90
Refractory kilns	40-50	70-80
Lime kilns	40-50	75-80
Fertilizer and chemical	40	80-85
Foundry	10-40	80-95
Chemical process drying	10-40	80-95
Incineration	20-40	65-75
Coal processing	10	90-97
Petroleum-cat cracking	0.6	99+
General	10-60	65-95

Limited Headroom

As was previously mentioned, gravity plays a very small role in particulate collection from the gas stream. The major force that impels the dust toward the apex of the cone is gas precession. Thus, horizontal and inclined cyclones theoretically should work as well as vertical designs. This is, of course, a tremendous space saver.

A horizontal cyclone offers the best advantage insofar as restricted spaces are concerned. However, in practice, severe difficulties arise during

operation. While relatively fine particles are transported into the dust recep-
tacle by the precession currents in the cone, coarser matter may fail to ascent
the lower edge of the cone. Accumulation may occur and eventually chokage
of the discharge. Figure 2-18 shows how this can happen. Inclined cyclones

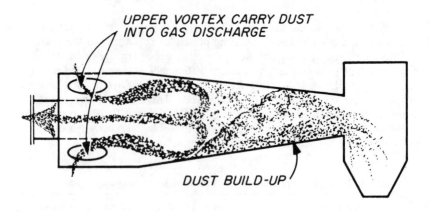

Figure 2-18. The horizontal cyclone offers low head-room; however,
dust tends to build up on the sloping wall which leads to chokage.

seem to obviate this problem. Their use can save a great deal of headroom—
especially when dealing with large cyclones. Figure 2-19 shows a typical
design.

Conclusion

Before actually designing or selecting a particular cyclone, careful analysis
must be made of the particulates to be collected along with an in-depth study
of operating parameters, *i.e.,* temperature, gas velocity, density, viscosity,
cyclone geometry and arrangement, pressure drop, and their relationship to
efficiency. The following is a general outline which may prove useful in
selecting the proper cyclone design.

1. Analyze a typical sample of the dust-laden gas to be cleaned making
   note of the following factors: temperature; corrosive properties, if
   any; concentration; particle size; and density.
2. Determine how much gas is to be handled.
3. Decide which type cyclone is necessary for the application. This can
   be determined by knowing the percent recovery desired and upon
   examining the proper literature for types and sizes of cyclones. Consult fractional efficiency curves available from equipment manufacturers.
4. Approach the vendor knowing the amount of gas to be handled, the
   type of cyclone and conditions listed above. The size (diameter, height,
   etc.) can then be determined intelligently.

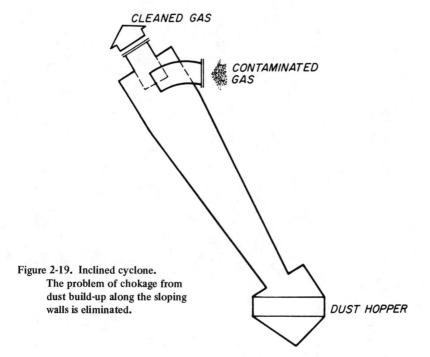

CLEANED GAS

CONTAMINATED GAS

DUST HOPPER

Figure 2-19. Inclined cyclone.
The problem of chokage from
dust build-up along the sloping
walls is eliminated.

CHAPTER 3

WET SCRUBBERS

APPLICATIONS AND LIMITATIONS[176]

Scrubbers permit contact between a scrubbing liquor and a gas stream containing solid or gaseous contaminants. This action allows the transfer of contaminants from the gaseous to a liquid phase.

Phase transfer for small particles requires high energy inputs, usually in the form of greater gas pressure drops across the scrubber. Low pressure drop scrubbers such as spray towers collect coarse dust in the range of 2 to 5 microns. High pressure drop venturi-type scrubbers are effective in removing 0.1 to 1.0 μ size particles.

The power requirement for a scrubber is often referred to as the contacting energy, and represents only that energy usefully consumed in gas-liquid contacting. Typically, a scrubber operating at 6 in. w.g. pressure drop should capture practically 100 percent of particles greater than 5 μ and about 90 percent of the 2 micron size particles. By increasing resistance to gas flow in the scrubber to about 10 in. w.g., it would be possible to collect about 90 percent of the particles in the 1 to 2 micron range. A typical efficiency curve for a wet scrubber operating at 6 in. w.g. pressure drop is shown in Figure 3-1.

Although collection efficiency is principally a function of particle size, the wettability of the particle also must be considered. Sand particles 0.5 to 1.0 micron in size are more easily collected than talc of the same size. In all cases, wet scrubber performance is unaffected by inlet dust loadings in the range of 1 to 10 grains per standard cubic foot.

For absorption of contaminated gases, a wet scrubber must provide gas-liquid turbulence and optimum contacting surfaces. In an absorption process such as sulfur dioxide scrubbing, the sulfur oxides must have a greater partial pressure in the gas stream than the vapor pressure existing above the scrubbing liquor. The difference between these pressures is the "driving force," which causes SO_2 to be transferred from the carrier gas stream into the scrubbing liquor.

125

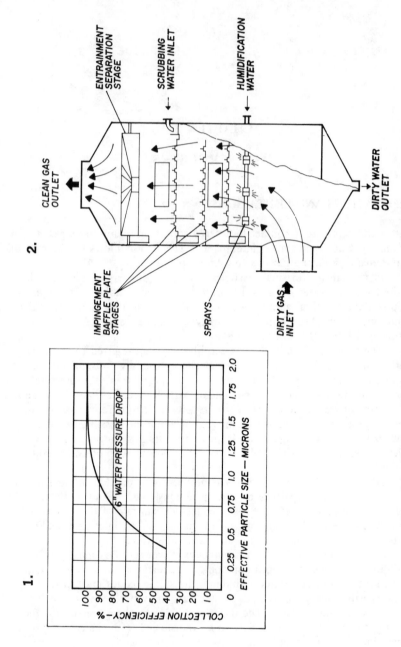

Figure 3-1. (1) Wet scrubber fractional efficiency curve; (2) Impingement scrubber.

Table 3-1. Scrubber Application–Quantities and Conversion Factors Most Often Used

Concentration	for gases = ppm
	for particulate = g/m^3; 1 gram/ft^3 = 2.29 g/m^3
Density	g/cm^3
Diffusivity	cm^2/sec
Dimensions	1 ft = 0.305 m
Flow Rate	for gases 1000 cfm = 28.3 m^3/min = 1700 m^3/hr
	for liquid rates = gpm
Liquid to Gas Rates	1 gal Mcf = 0.134 l/m^3 = 1.34 x 10^{-4} m^3/m^3
Power	1 hp = 0.746 kw
Power to Flow Rate	1 hp/Mcfm = 0.438 w/m^3/hr
Pressure Drop	1 in. w.c. = 2.54 cm w.c. = 2.489 mbar

To Convert From	To	Multiply By
centimeters of water	inches of water	0.394
cubic feet per minute	cubic meter per hour	1.70
cubic feet	cubic meters	0.0283
cubic meter per hour	cubic feet per minute	0.589
cubic meter per hour	gallons per minute	4.4
feet	meters	0.3048
gallons per minute	cubic meter per hour	0.227
grains per cubic foot	grams per cubic meter	2.29
grams per cubic meter	grains per cubic foot	0.437
inches of water	centimeters of water	2.54
meters	feet	3.28

   In the absorption of low concentrations of sulfur oxides from a coal-fired boiler flue gas, the partial pressure in the gas stream, at 0.13 vol-% SO_2, is about 1 mm Hg. To transfer the SO_2 to the liquor, its concentration in the liquid phase must be low enough to produce a vapor pressure above the liquid of practically zero. This liquor partial pressure is not only a function of SO_2 concentration, but also of temperature and pH. Therefore, the temperature of the flue gas, the volume of scrubbing liquor and the utilization of various alkaline liquors are the factors which determine the degree of absorption.

   Water is a relatively low-efficiency scrubbing solution because its initial pH is 7.0. This pH is reduced to 2 or 3, with attending high liquid vapor pressures, at the slightest absorption of SO_2. The use of an alkaline solution such as sodium carbonate effects good absorption. It produces a sodium monosulfate slurry having a pH in the range of 5 to 6.

In the control of coal-fired power plant emissions, SO_2 absorption and flyash removal can be simultaneously accomplished in a wet scrubber. This dual duty imposed on the wet scrubber is very common and, because of the different factors involved, it is important that enough data be provided for optimum design.

Although the prime purpose of any scrubber is maximum contact of the gas and liquor streams with the most effective use of the required pressure drop, a multitude of contact mechanisms are available. Some of these are spray contact, impingement, surface area, cyclone and venturi. The gas and liquor flows can be concurrent, counter-current, cross-flow and combinations of all three in the same scrubber.

Regardless of the contact mechanism used, all wet scrubbers are composed of three basic zones or operations: gas humidification, gas-liquor contacting and gas-liquor separation.

Impingement Scrubber

A typical liquor impingement design is shown in Figure 3-1. This type of scrubber contains one or more impingement plates. Influent gases, usually at some elevated temperature, enter the base of the scrubber and pass through the humidification stage where saturation occurs, temperature is reduced, and some of the coarse particles are removed.

Saturated gases continue to flow upward to the gas-liquor contacting zone where they pass through a series of perforated impingement plates. Each perforation is covered by a continuous baffle.. Each perforated plate, or stage, is flooded by the scrubbing liquor which flows across it normal to the direction of gas flow. As the gases pass through the perforations, they are impinged against the baffle above each perforation. This impingement action causes transfer of the dust particle from the gas stream to the liquor.

The gases flow upward through succeeding impingement plate stages and pass through the gas-liquor separation zone. Here, gas velocity is accelerated in an entrainment separator comprised of overlapping curved vanes. Reducing velocity in the open chamber above the vane section causes inertial separation of the liquid droplets. Liquid is returned through an overflow weir to the gas-liquor contacting zone.

Gas velocity through this type of scrubber is about 500 fpm at operating pressure drops in the order of 2 and 3 in. w.g. per stage. This scrubber can remove 97 percent by weight of particles above 1 micron.

Surface Area Scrubbers

These scrubbers utilize various packing materials and arrangements to break up the gas and liquor flow pattern (Figure 3-2). The packing is usually designed to expose maximum wetted surface area to the gas flow at minimum pressure drop. A number of packing materials are available including wood slats, random-sized stones, solid ceramic spheres and other ceramic and plastic shapes.

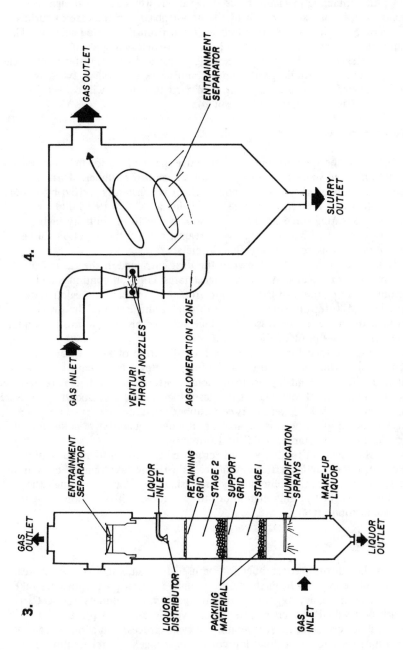

Figure 3-2. (3) Surface area scrubber; (4) Venturi scrubber.

All contacting mechanisms used in various wet scrubber designs are means of obtaining atomization of the liquor phase. In the case of surface area scrubbers, the gas stream atomizes liquid from the wetted surface. The liquor droplets serve as targets to which the particles cling.

A packed scrubber can produce high mass and heat transfer rates along with an ability to handle viscous liquids and heavy slurries. A two-stage scrubber, operating at a pressure drop of 8 to 10 in w.g., will collect 98 percent of particles greater than 1 micron.

Venturi Scrubbers

One of the simplest and most effective mechanisms for liquor droplet formation is the venturi. Figure 3-2 shows a venturi scrubber. Dust-laden gases flow through the venturi section, which is constantly wetted by liquor. At the throat of the venturi the gases, flowing at 12,000 to 18,000 fpm, produce a shearing force on the liquor stream due to the initially high velocity differential between the two streams. This shearing action causes the liquor to atomize into very fine droplets.

Impaction takes place between the dust entrained in the gas stream and the liquor droplets. As the gas decelerates, collision continues and agglomeration of the dust-laden liquor droplets takes place. Agglomerated particles and water droplets discharge through a diffuser into the lower chamber of the separator vessel. Impingement of this stream into the liquor reservoir removes most of the particulates.

Gases are then forced to reverse their direction of flow, through a section of packing where the remaining particulates can be removed by a minimal countercurrent liquor flow. Some venturi-type scrubbers are designed with an adjustable throat so that the gas velocities and corresponding pressure drops can be varied. A venturi-type scrubber operating in a pressure drop range of 30 to 40 in. w.g. is capable of an almost quantitative collection of particles in the size range of 0.2 to 1.0 micron.

Operating data for the venturi scrubber have proven that, for any particle size, the collection efficiency increases with increased energy consumption. Increased energy can be obtained by increasing gas velocities through the variable throat or by using high liquor-to-gas ratios. Either action increases the atomization effect.

Limitations

Although the wet scrubber can be used for handling hot gases, a steam plume will be emitted from the scrubber unless special precautions are taken.

When dry collection is required for recoverable products, the wet scrubber cannot be used. However, in cases where wet processing is practiced, the wet scrubber can return a recoverable slurry to process. In some areas of the country where water pollution control is stringent, water treatment of the scrubber liquor effluent must be considered or other types of air pollution control equipment must be used.

SCRUBBER SELECTION

Many different scrubbers are available from a broad range of companies. Additionally, numerous scrubber designs have been patented but are not currently manufactured. Because of the wide range in scrubber design, costs and performance, it is necessary to match a specific scrubber to an even more specific application. Generally speaking, however, it is possible to divide scrubbers by type and group based largely on operating principles.

Choice of the right scrubber for a particular application requires understanding and study of alternative methods. Some scrubbers are primarily designed for particulate collection, others for mass transfer. The prime requisite in both cases is good liquid-gas contact. The collection efficiency of a scrubber and the related cost determines the application of a particular scrubber to the job.

Deciding on the best method of controlling an air pollution problem requires that many alternatives be explored and compared. It is always useful to define the problem in terms of design criteria, whether one has to design, select, or up-grade performance. In all cases, configuration, size, liquid and gas flow rates, physical and chemical properties, pressure drop and economics are the input data required.

Emission concentration or rate limitations determine what must be achieved in control efficiency. Additionally, stack plume opacity, odor, and ground level concentrations tied to stack emissions by specific dispersion formulae may place additional burdens on design and selection. Particulate emissions are expressed in four ways: (1) maximum weight per hour emitted or allowable, lb/hr; (2) maximum concentration of particulate in the exhaust gas stream, g/m^3, grains/cu ft, or lb/1000 lb of gas; (3) maximum gas opacity emitted in terms of Ringelmann number, Ringelmann #1 = 20% opacity, #2 = 40%, #3 = 60%; (4) an emission rate related to an air quality standard by formula or chart based on atmospheric considerations. Allowable concentrations of gaseous pollutants are usually specified in terms of maximum fraction by volume of polluting gas in the discharged gas stream; the volume fraction is expressed as percent by volume, or as parts per volume (ppm) of the discharged gas.

In evaluation of alternate methods of air pollution control, economic considerations in addition to design criteria will often be the deciding factor. Capital investment, operating costs and other pollution problems (such as water contamination or noise) created in solving the original problem must be taken into account.

Spray Towers

Spray scrubbers collect particles or gases on liquid droplets and utilize spray nozzles for liquid atomization. The liquid droplets are formed by the liquid atomized in the nozzle. The spray is directed into a chamber suitably shaped to conduct the gas through the finely divided liquid. In a vertical tower the relative velocity between the droplets and the gas is eventually the terminal settling velocity of the droplets.

Spray towers are used for both particulate collection and mass transfer. They have low pressure drop, high scrubbing liquid rate, and are generally the least expensive scrubbing equipment. Spray chambers receive application in high solubility gas and liquid aerosol removal. The number of transfer units in this design is limited. Collection of particulate matter is by impaction and determined by the terminal settling velocity and diameter of the spray droplets. Relatively high collection efficiency is achieved for particulates larger than 5 microns in diameter. Since the pressure drop of this type of unit is low, usually 1 to 2 in. w.g., operating costs are also minimal. Spray towers are applicable where severe air cleaning conditions are not required.

Ejector Venturis

Ejector venturis use spray nozzles in which a high-pressure liquid spray is employed to collect particulate, or for mass transfer, and to move the gas. There is a high velocity between the liquid droplets and gas which affects particle separation. Collection efficiency is generally high for solid particles larger than 1 micron in diameter. Mass transfer is affected by the co-current flow of gas-liquid. The energy consumption per unit volume is relatively high in the form of scrubber liquid pumping costs.

Ejector venturis may stand by themselves or as the first stage of more complex air cleaning systems. The unit mechanism predominating in this type of scrubber is collection due to inertial impaction and is effected by liquid drops. Liquid drops behave very much as rigid spheres when collecting aerosol particles by inertial mechanism. Particles adhere upon striking the droplets depending upon the wettability of the particles. Particles uniformly distributed in the approaching gas stream are collected as the cross-sectional area is swept by the water spray.

Ejector venturis and spray scrubbers require droplets of uniformly small size to act effectively and avoid entrainment. Scrubbing liquid is usually recirculated to reduce liquid consumption so the nozzles must be capable of handling liquids of high solids concentration. Spray nozzles used may be classed under the following types: (1) pressure nozzles, (2) rotating nozzles, (3) gas atomizing nozzles and (4) sonic nozzles.

Venturi Type Scrubbers

The venturi scrubber and the flooded disc scrubber are control devices in which the gas and liquid are atomized in a moving gas stream. The relative velocity between gas and liquid aerosol droplets is high; gas velocities in the range of 200 to 400 fps are used. This high velocity promotes particle collection. Venturi type scrubbers use the converging-diverging sections as typically shown in Figure 3-3 with the liquid-gas contact at maximum in the venturi throat. Collection efficiency is proportional and increases with

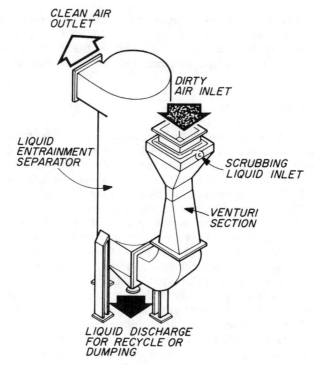

Figure 3-3.   Typical venturi scrubber.

pressure drop. High pressure drop venturi scrubbers, 40 to 60 in. w.g. and higher, efficiently collect submicron particles. Mass transfer is also effected by the co-current gas-liquid flow. A high degree of turbulence and liquid-gas contact is effected in the venturi throat.

By pressure drop adjustment and liquid flow rate or gas velocity variation, the venturi scrubber can achieve sufficiently high efficiencies to collect even submicron fumes. Scrubbers of this type are simple, due to the absence of moving parts, and generally offer relatively small capital investments. The high pressure drop, however, means high energy consumption as a liquid pumping cost.

In the flooded disc scrubber, liquid is introduced to a disc slightly upstream of the venturi throat. The liquid flows to the edge of the disc and is atomized. A typical arrangement is shown in Figure 3-3.

The venturi scrubber can be used in both particle collection and mass transfer. The liquid is introduced in the form of jets and enters the throat; in this region mass and particles may be transferred to the liquid jets. In the next region downstream, drops form and any further mass transport is to drops. Some liquid collects on the walls of the venturi and runs in rivulets and sheets. In the entrainment separator following the venturi, drops coalesce with one another, impinge on liquid surfaces and may exist for some period as quiescent or oscillating drops attached to liquid surfaces. There are additional zones of liquid jets and sheets. The velocity of the gas drops sufficiently in the separator section following the venturi, so that liquid droplets fall into the separator, since they are below their terminal settling velocity.

Packed Bed Scrubbers

Scrubbers, using mass packing, consist of a hollow tower containing packing elements such as rings, saddles or other manufactured elements. Scrubbing liquid is introduced at the top of the packing and trickles down through it. The packing breaks down the liquid flow into a high surface area film. The gas stream to be cleaned flows through the packing.

There are generally three types of packed towers: *counter-current, co-current,* and *cross-flow.* In the counter-current packed scrubber, as the name implies, liquid and gas travel in opposite directions. The gas flows upward while the liquid flows downward (Figure 3-4). Counter-current scrubbers have the greatest contact area and are therefore the most efficient and widely used of the three types. As the gas rises in a counter-current scrubber, some of its contaminant is lost. The clean liquid thus allows the remaining contaminant to be absorbed more readily.

In the co-current scrubber, the purest gas comes in contact with the most contaminated liquid at the bottom of the unit. The liquid with maximum contaminants flows with the gas containing minimum pollutants which could cause recontamination of the gas. The counter-current design is most efficient for scrubbing low or moderate solubility gases. Cross-flow units, however, can be designed to match the efficiency of the counter-current unit type.

The main disadvantage in the counter-current scrubbing systems is that the opposed streams may produce eddy currents which may significantly diminish either stream's velocity. This causes solids to deposit and build up at points of low velocity.

In the co-current flow scrubber both the gas and liquid move in the same direction. Usually this flow is downward because of gravitational considerations (Figure 3-5). Both the contaminated gas and scrubbing liquid enter at the top of the apparatus. Clean gas and contaminated liquid leave by different streams at the bottom. Exit stream gas is usually fed to an entrainment separator or demister. The co-current scrubber has greater handling capacity than the counter-current design because the liquid flows in the same direction. Higher liquid rates can be employed to wash solids through the packed bed.

The cross-flow scrubber (Figure 3-6), can be used for removal of coarse solids, liquid mists, as well as for mass transfer. Scrubbing liquid flows in from the top edge of the bed of packing, while the gas moves horizontally through it. This arrangement serves to wash any accumulated contaminants off the packing surface. Stability of the cross-flow unit is high.

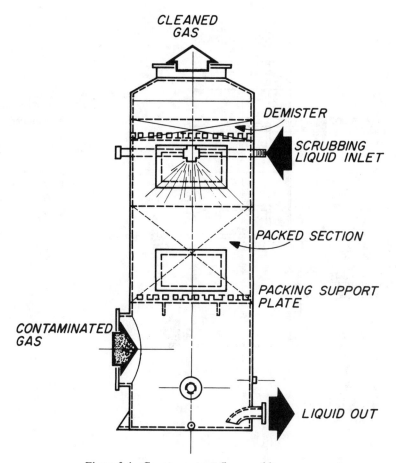

Figure 3-4. Counter-current flow scrubber.

Packed bed scrubbers have a long history in mass transfer operations. They can remove pollutant gases to any desired concentration, the limiting factor being economics. The higher the separation required, the greater will be the packing depth and pressure drop. Gases scrubbed in packed beds should not be heavily loaded with solid particulate as fouling of the packing material may result. Pressure drops through packed bed towers are usually low, typically on the order of 0.5 in. w.g./ft of packing.

Packed beds can also be used for mist elimination. For these cases, they are operated without any liquid rate. Collected mist drops through the packed bed and drains off the bottom. As in the case of mass transfer, collection efficiency can be raised indefinitely by increasing bed height.

The mechanisms involved in packed columns are mass transfer to sheets of liquid, transfer to oscillating drops and liquid separated on packing

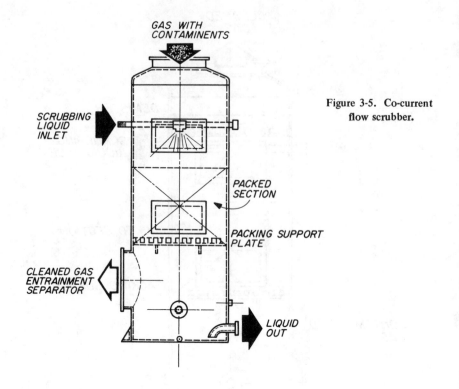

GAS WITH CONTAMINENTS

SCRUBBING LIQUID INLET

PACKED SECTION

PACKING SUPPORT PLATE

CLEANED GAS ENTRAINMENT SEPARATOR

LIQUID OUT

Figure 3-5. Co-current flow scrubber.

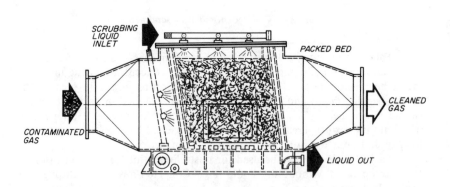

SCRUBBING LIQUID INLET

PACKED BED

CONTAMINATED GAS

CLEANED GAS

LIQUID OUT

Figure 3-6. Cross-flow packed scrubber.

elements, and transfer to curved surfaces. When operating flooded or partially flooded columns, additional mechanisms take place. Jets of gas, emerging from the openings of the packing into the bed of liquid froth layers, transfer bubbles of gas to liquid sheets formed along column walls and other surfaces. The liquid and gas mixing patterns, flow rates and equilibria data are essential to define the phase composition in contact at different points in the equipment.

Plate Scrubbers

Plate scrubbers consist of a vertical hollow tower with one or more trays or plates transversely mounted inside. Contaminated gas comes in at the bottom of the tower and passes through perforations, valves, slots, or other openings in each plate before exiting from the top. Scrubbing liquid is introduced at the top plate and flows successively across each plate as it moves downward to the liquid exit at the bottom. Gas passing through the openings of each plate contacts with the liquid flowing over it. The gas and liquid contact effects the mass transfer or particle removal for which this type of scrubber is designed. Plate scrubbers are named for the type of plates they contain. For example, a tower containing bubble caps will be called a bubble cap tower; a tower containing sieve plates is called a sieve plate tower. Plate scrubbers have been widely used for mass transfer operations and have the ability to remove gaseous pollutants to any desired concentration, if a sufficient number of plates are used.

Plate devices are also used for particle collection; sieve or perforated plates may be combined in a sieve plate tower. Some designs employ impingement baffles placed a short distance above each perforation on a sieve plate, forming an impingement plate. The impingement baffles are below the liquid level on the perforated plates and consequently are continuously washed clean of particles collected. Collection of particulate is further assisted by the atomization of the liquid flowing past the openings in the irrigated perforated plate and the subsequent venturi effect. Particulate collection efficiency is good for particle sizes greater than 1 micron in diameter. Pressure drop for this type of unit is generally low being on the order of 1 to 2 in. w.g./plate.

In the sieve plate scrubber, contact between gas and liquid phase is by means of jets and gas emerging from the perforation. These pass into the liquid and coalesce with other bubbles. The liquid phase is assumed to be thoroughly mixed when the column is small and the liquid drops through the perforation. If the plate is large, and if the liquid is introduced on one side of the plate, then the liquid is assumed to flow across the plate in plug-flow fashion. As in the design of all wet scrubbers, it is necessary to have equilibrium, kinetic, particle, temperature, and pressure data in order to define the transport phenomena involved.

Moving Bed Scrubbers

Moving bed scrubbers incorporate a zone of mobile packing, usually plastic, glass or marble spheres, where liquid and gas can intimately mix. A hollow shell contains a perforated plate on which the movable packing is placed. Contaminated gas passes upward through the packing while scrubbing liquid is sprayed up from the bottom through the perforated plate and down over the top of the moving bed. Gas velocity is sufficient to fluidize or move the packing material around during scrubber operation. This movement keeps the packing bed turbulent and the packing itself clean.

Moving bed scrubbers are used for particulate collection and mass transfer. Particle collection efficiencies are good on diameters over 1 micron. Several moving bed stages may be used in series. Pressure drops are typically 0.2 to 0.5 in. w.g./stage.

In the moving bed or turbulent contact absorber, there are shells or plastic balls of approximately the same density as water that move freely between upper and lower constraining plates or screens. The packing and liquid fed to the plate are violently agitated by the gas flow when in operation. The scrubbing mechanisms involved are transfer from gas emerging from the support plate, transfer to the liquid jets, and transfer to liquid sheets formed by the run-back of liquid along the column walls. Mixing of the gas phase in this apparatus is usually simple *plug-flow* type—where the liquid is perfectly mixed in the vertical direction but not in the horizontal. Liquid mixing is further complicated, depending upon the motion of the balls as well as the entrained liquid not attached to the packing.

Centrifugal Scrubbers

The characteristic of centrifugal scrubbers is that they impart a spinning motion to the gases passing through them. This spinning motion comes from the introduction of the gas to the scrubber tangentially or by directing the gas stream against stationary turning or swirl vanes. Spray can be directed through the rotating gas stream for particle capture by impaction upon the spray drops. Or a dry centrifugal collector, such as cyclone, can be wetted to bring reentrainment of particles collected. Sprays can be directioned from a central spray manifold or inward from collector walls. The latter design is more easily serviced since it is accessible from outside the scrubber.

Centrifugal scrubbers are generally used for particulate collection and operate at a pressure drop of 2 to 8 in. w.g. dependent upon design. This type of scrubber is usually most efficient on particle sizes of over 5 microns in diameter. Mass transfer applications are limited as in the case of simple spray towers.

The cyclonic spray scrubber uses a preformed spray from a manifold located either along the center of the axis of the cyclone or along the wall directed inward. The gas-liquid contact may be of cross-flow type, when the spray droplets are carried sufficiently by the gas flow part of the way before being thrown out to the wall by centrifugal force. The gas-liquid contact is a combination of cross-flow and co-current flow. The mass transfer mechanism in the cyclonic spray scrubber is primarily absorption by liquid droplets and by the liquid film at the wall of the scrubber.

An advantage of the cyclonic spray over the preformed spray is that the outward force on the droplets and their velocity relative to the gas is much higher than in free fall. The gas film coefficient is therefore higher than in the ordinary spray chamber.

Impingement Scrubbers

Impingement and entrainment scrubbers consist of a shell holding the scrubbing liquid. Gas entering the scrubber skims over the liquid surface to reach the gas exit duct. In skimming over the liquid, the gas atomizes some of the liquid into spray droplets. These droplets act in particle collection and as mass transfer surfaces. The gas exit duct is designed to turn the gas and droplet system flowing through it and to act as an entrainment separator. Impingement and entrainment scrubbers are generally used for particle collection. Their efficiencies are such that they are used only on particles larger than 10 microns in diameter. Pressure drops for this type system range from 4 to 20 in. w.g., although lower pressure drops are more common.

Baffle-Type Scrubbers

Baffle-type wet collectors are designed to effect changes in gas flow direction and velocity by means of solid surfaces or baffles. Flow direction may be changed and consequently flow patterns altered by louvers, zig-zag baffles, disc and donut baffles. Particle deposit occurs on the baffles due to impaction and interception. Liquid materials collected run down the baffle to a sump; solid particles can be washed intermittently from the baffle plates.

Baffle devices are generally used as entrainment separators collecting water droplets produced from upstream processes, such as scrubbers. Baffle collectors, capturing particles by changing the gas flow direction, generally have good efficiencies on particles greater than 20 microns in diameter. Baffle mist eliminators with large numbers of collecting surfaces effect good efficiency on particles or droplets down to 5 microns in diameter. Secondary flow collectors have been designed for efficiency of particles down to 1 or 2 microns in diameter. The pressure drop of these units is generally low, on the order of 1 to 3 in. w.g.

Fiber Packed Scrubbers

Like mass packing, fiber packing made from materials such as plastic, spun glass, fiberglass or steel is used to provide surfaces for particle collection and gas absorption. This type of packing generally has large void fractions with porosities ranging up to 99 percent. Packing fibers are usually small in diameter for efficient operation, yet strong enough to support collected particulates or liquid droplets without matting. Fiber packing is often used as a liquid drop entrainment separator following a scrubber, or to collect droplets from process sources. Fiber packings are often sprayed with a scrubbing liquid as an aid to gas-liquid contact for mass transfer. The wetted packing presents more liquid surface to the gas flowing through it than to the spray alone. When used in this fashion, it is analogous to the packed tower. Both mass transfer and particle collection capabilities are high. Low gas velocities, however, are necessary for efficient operations. Pressure drops are typically moderate in this type unit.

Fiber packings include materials ranging from fibers to vertical springs, diameters being less than 1 mm for the fiber and less than 10 percent volume solids. Compared to dumped or stacked packings, fibrous packings are usually installed in sections or even in complete column size units. Fibrous packing is made of inert fibers matted together. Knitted mesh packings are also available. They can consist of pads made by crimping knitted wire or plastic mesh in strips, then folding or coiling into a pad configuration. Woven fabric packings are made of fine wire screen in various shapes. Spring packings consist of many long vertical strings, which may be of pairs of helical wires.

In wettable fibrous packing, drops of liquid cover the fiber in a film. These drops eventually grow large enough to fall off the fiber. Some of the liquid thereby flows down the fiber to the next fiber as a film or drop, and some of the scrubbing liquid drops to the next fiber. Unit mechanisms involved are a mass transfer to liquid films, mass transfer to drops on the fibers, and mass transfer to falling drops. The amount of mass transfer in each of these mechanisms will vary considerably for different types of packing. The degree of wettability of the packing material will determine the degree a liquid film will drop on the packing surface.

Mechanically Aided Wet Scrubbers

Mechanically aided wet scrubbers are designs which incorporate a motor-driven device between the inlet and outlet of a scrubber. These devices can be fan blades used to move air through the scrubber. Solid particulate matter is collected by impaction on the fan or turbine blades as the gas moves through. Liquid is introduced at the hub of the rotating blades. The liquid runs over the blades, washing them of collected particles. Liquid caught by the fan housing drains into a sump and can be recirculated. Particle collection efficiency is good for particles larger than 5 microns. The advantages of these devices are low space and water requirements, with the principal disadvantage being a relatively high operating power cost.

Disintegrator scrubbers use a submerged motor-driven impeller to atomize scrubbing liquid into small droplets. The droplets fly from the impeller across the gas stream collecting particles on the way. Disintegrator scrubbers have good efficiency on submicron particles, but have a high operating cost due to the energy required to drive the impeller. Venturi scrubbers are often used since they can achieve the same efficiency as disintegrators, but at lower rates of energy consumption. Mechanically aided scrubbers are principally used for particulate collection. Mass transfer capabilities are low due to the small amount of liquid available as a contact surface. Pressure drops are generally low, on the order of 2 to 5 in. w.g.

Entrainment Separators

Where a gas or vapor passes through a liquid, complete separation of the phases is difficult. The gas flow will entrain and carry over with it varying sizes and quantities of liquid droplets. The size of the droplets determines the ease of separation. Droplets which are torn from liquid bulk are usually large—up to hundreds of microns in diameter. Droplets formed as a result of condensation or chemical reaction can be less than one micron in diameter. Particles such as these are removed from the gas stream by a large variety of devices. These include settling, impingement, diffusion and centrifugal separators. Droplets larger than 50 microns in diameter are easily settled by gravitational forces. In these cases, expanding the flow area and reducing the gas velocity is sufficient to separate such entrainment from the gas stream. Droplets larger than five microns can be collected by centrifugal and impingement collectors. Smaller droplets are either first agglomerated or collected by more efficient devices such as specially designed high-efficiency cylones or electrostatic precipitators.

Knitted wire or plastic mesh entrainment separators are used extensively, due to their low pressure drop and high collection efficiency. Four to six inch deep beds of 97 to 99 percent free film, knitted thin metal wire or plastic materials are used. In most cases the pressure drop is less than 1 in. w.g. Additionally, the wires of such separators furnish targets with which large particles must collide and be captured.

Devices using centrifugal force, in addition to impaction, are also used for entrainment separation. Cyclones are an example of centrifugal force entrainment separation and are used for removal of drops larger than five microns.

Entrainment spray removal is possible by a number of methods, which include:

1. Knitted wire mesh or plastic mesh demisters,
2. Swirl vanes or zig-zag vanes,
3. Cyclones,
4. Gravity settling chambers,
5. Knock-out pots in which the gas goes through a 180° reversal,
6. Packed section of column, and
7. Special designs where jets of gas impinge on a target plate.

Particle Collection Concepts

In the scrubbing of particulate matter from gases the principal concern is usually particles smaller than 10 microns in diameter. Particles larger than 10 microns are relatively easy to separate. The successful design and operation of wet scrubbers depends on knowing the size and properties of the small particles to be collected. It is important to know particle size, composition and formation method.

Among the particulate matter collected by wet scrubbers are dispersion aerosols. These result from processes such as grinding, solid and liquid atomization, and from the transport of powders in suspended state by air currents or vibration. Condensation aerosols are formed when supersaturated vapors condense or when gases react chemically forming a nonvolatile product. These latter aerosols are usually less than 1 micron in diameter as opposed to dispersion aerosols—which are in most cases considerably coarser and contain a wider range of particle size than condensation aerosols. In condensed aerosols, solid particles are often loose aggregates of a large number of primary particles of crystalline or spherical form. The dispersion aerosol usually consists of individual or slightly aggregated particles irregularly formed.

Dispersion aerosols with solid particles are called *dusts*. Condensation aerosols with a solid dispersed phase or a solid-liquid dispersion phase are classed as *smokes* or *fumes*. Condensation and dispersion aerosols with a liquid dispersion phase are called *mists*. This classification usually applies regardless of particle size, and distinction is sometimes difficult. In practice, a combination of dispersion and condensation aerosols is encountered. Different size particles behave differently—not only in regard to physical properties such as light scattering, evaporation rates, cooling and particle movement, but also the effect on mechanisms of particle removal from the gas phase. Particle sizes, volumes and weights may be obtained by microscopic sizing and density assumptions. Additionally, particle size and density may be quantified based on aerodynamic behavior of particles following Stokes' Law.

Mass Transfer Concepts

Mass transfer or diffusion is defined as gas or liquid components travelling in a solution or mixture from a region of high concentration to one of low concentration. Diffusion can take place from either stagnant conditions or where motion exists by eddy diffusion. Molecular diffusion is much slower than eddy diffusion, where whole masses of gas or liquid move. In either case, the absorbate will dissolve in the liquid and distribute itself uniformly throughout the scrubbing liquid until it is saturated. This latter solubility limit is an equilibrium condition, and no further dissolving and mass transfer can occur. Molecular diffusion, eddy diffusion, and finally equilibrium contact all take place in the gas phase, liquid phase or combined gas-liquid system. In scrubbing, one or more components is removed from the gas phase by absorption into the liquid phase. The interphase mass transfer rate is controlled by the diffusivities and the concentration differences between the phases.

In some cases the gaseous compounds react with the liquid, dissolving to form a new compound as with scrubbing acid fumes reacting with an alkali solution. In such cases the reaction rate constant and the reaction equilibrium constant are important. The absorption of gases is increased in transfer efficiency and capacity when the dissolved gas reacts in the liquid phase. The transfer efficiency increase depends on the order of the reaction. Additional capacity will depend upon the liquid phase reactant concentration and stoichometry of the reaction. Some very significant gas scrubbing operations involve mass transfer with a chemical reaction. Typical examples are the scrubbing of acid fumes and mists such as HCl, H_2SO_4, HNO_3, with alkali solutions of $NaOH$, Na_2CO_3; NH_4OH.

In the wet scrubbing gas phase, the equilibrium concentration of a soluble contaminant in liquid increases as the partial pressure of the soluble component in the gas phase increases. It is, therefore, usually more advantageous to scrub higher pressure gases, whereas the solubility may be only slight at lower partial pressures.

DESIGN AND COST OF HIGH-ENERGY SCRUBBERS[16]

Scrubber Operation

The basic high-energy scrubber configuration consists of a convergent-divergent vertical venturi section, an elbow plus connecting horizontal crossover section, and a vertical separator section (Figure 3-7). Waste gas from the process enters from the top of the converging section and flows downward. A concurrent flow of scrubber liquid is introduced in a manner that floods the surfaces of the converging venturi. The gas and liquid converge and are intimately mixed at the venturi throat where the scrubbing is accomplished. The diverging section assists energy recovery. The separator allows the falling liquid to escape through a port at the bottom, while the cleaned gas flows upward through a demister to remove entrained liquor and is exhausted out of the system by fan draft.

Sizing the Scrubber

In practice, the overall dimensions of the scrubber are sized to achieve the desired gas velocities within the various parts. Although there are variations among the scrubber manufacturers, the typical inlet gas velocity of high-energy orifice scrubbers ranges from 3300 to 3700 fpm, and the superficial inlet gas velocity to the separator is 600 fpm. The typical H/D (height to diameter) ratio of the separator is approximately 2:1.

Figure 3-8 shows the relationship of the various scrubber dimensions indicated in Figure 3-7 as a function of the inlet acfm. As an example, 95,000 acfm would necessitate a scrubber having the following dimensions:

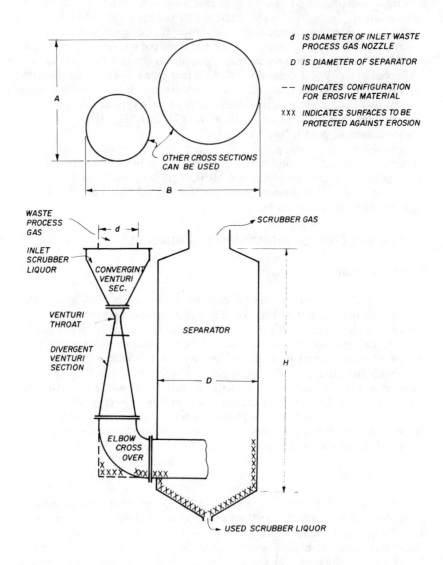

Figure 3-7. Basic high energy scrubber configuration.

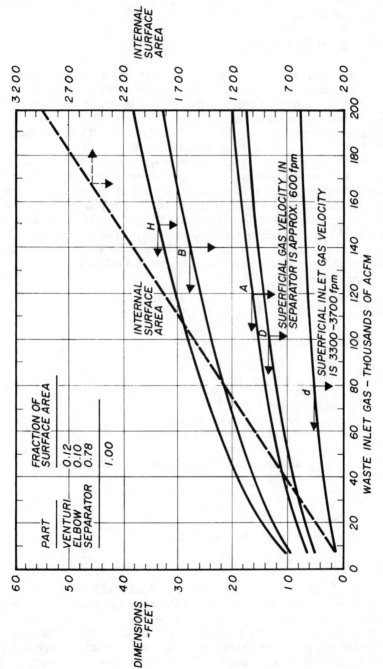

Figure 3-8. Relationship of scrubber dimensions to the amount of contaminated gas treated.

d = 6 ft, D = 13.2 ft, A = 15.2 ft, B = 23.2 ft, H = 27.2 ft, and an internal surface area of 1480 sq ft.

Determining Metal Thickness

The next step in predicting scrubber costs involves determining the required thickness of the metal. Additional ceramic liners are sometimes used to protect the convergent portion of the venturi section from the high inlet gas temperatures. However, for the sake of simplicity the same metal thickness is normally specified for the entire scrubber. This thickness value is fixed by the buckling load on the separator section.

In the design of the separator, operating experience has established that the used scrubber liquor outlet at the bottom of the separator should be made large enough to preclude liquid flooding of this section. In addition, the scrubber is normally installed so that it is not subjected to axial loading. Because of these factors, the sole design basis for establishing the metal thickness of the separator is the external pressure acting on the separator wall. This external pressure, which acts to buckle the separator, is related to the separator dimensions and material properties by the following equation:

$$W = \frac{Cbt}{SD/2} = \frac{55.416 \, bt}{SD}$$

where **b** is the critical buckling stress given by:

$$b = \frac{K\pi^2 E}{12(1-M^2)} \quad \frac{t}{H} = 26.21 \times 10^6 K\left(\frac{t}{H}\right)^2$$

and

C = Dimensional conversion factor
E = Modulus of elasticity, say 29×10^6
H = Separator height, in.
K = Buckling coefficient, defined in terms of Z[3] (Figure 3-9)
M = Poisson's ratio, say 0.3
S = Design safety factor, a minimum of 2
t = Separator wall thickness, in.
W = Net external pressure, in. of water
Z = Buckling correlation coefficient = $\dfrac{H^2 \; 1-M^2}{tD/2}$
π = 3.1416

The thickness as a function of **W**, the physical properties of the material, and the safety factor **S** can be calculated using the above relations. However, as a simplification, Figure 3-9 shows the required metal thickness plotted as a function of the acfm into the scrubber. The calculations used to develop this graph were based on the separator dimensions given in Figure 3-8, and on a design safety factor of 2. No corrosion or erosion allowance, however, is included.

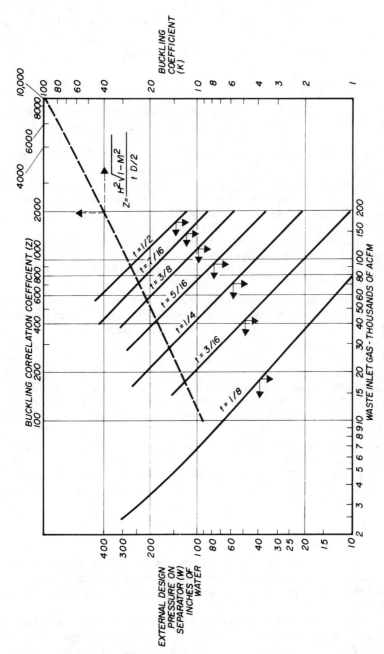

Figure 3-9. Metal thickness required in a scrubber as a function of the amount of gas treated.

Figure 3-9 indicates, for example, that when 95,000 acfm enters the scrubber, a 1/4-in.-thick wall is adequate (has a safety factor of 2) for a total pressure drop (scrubber plus demister plus internal cooler) of 44 in. of water; a 5/16-in. wall is sufficient for 78 in. of water, and a 3/8-in. wall is adequate for 120 in. of water.

Estimating Scrubber Cost

The cost of scrubbers varies greatly, as a function of the type of metal used for fabrication and the metal thickness. Figure 3-10 gives the f.o.b. cost of 1/8-in. wall carbon steel scrubber equipment, including venturi, elbow, separator sections, scrubber instrumentation and pumps. Figure 3-10 also gives the cost-multiplying factors for 316 stainless steel, 304 stainless steel, Cor Ten (Registered TM—U.S. Steel Co.), together with the additional cost of rubber and carbofax liners. For example, a 1/8-in. thick carbon steel scrubber handling 50,000 acfm costs approximately $12,200. The multiplier for 316 stainless steel is 2.63 for all metal thicknesses.

Figure 3-11 gives the cost thickness factors for carbon steel scrubbers. For example, the cost-thickness factor for 1/4 in. carbon steel handling 50,000 acfm is 1.61, giving a total cost for this type of material and thickness of $19,642.

In cases when it is desirable to construct the venturi, elbow and cross-over, and separator sections of different materials, a good estimate of the cost can be established by assuming that these sections contain approximately 12, 10, and 78 percent respectively of the total structural metal. Figures 3-10 and 3-11 are based on the middle of the range for bid requests received by Kaiser Engineers over a seven-year period. This cost data has been adjusted by cost index ratios to December 1970 prices. The bids covered an acfm range of 19,000 to 204,000 and a pressure drop range of 20 to 50 in. of water. Bid specifications covered equipment performance, and material thickness was not always specified. The bids included scrubbers whose venturi, crossover, and separator sections were made of different materials having various wall thicknesses.

In reapplying the cost data to these bids, in 90 percent of the cases, the ratio of mid-range bid price to estimated price using these curves equalled 1.08 ± 0.16.

Corrosion Protection

Corrosion protection can be provided by using liners, different material or construction, or a thicker material. Surprisingly, the cost of excess conservatism with respect to using thicker wall construction can be extremely high.

Table 3-2 was developed on the cost data presented for 50,000 acfm, and indicates the disproportionate costs of using the alternative thicker wall construction. For most purposes, therefore, the use of liners is the most practical alternative.

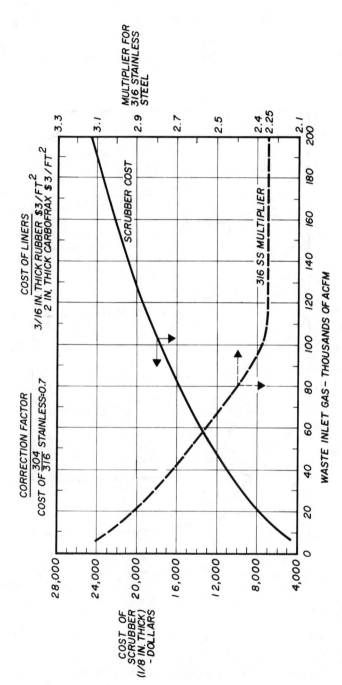

Figure 3-10. F.O.B. cost of a 1/8-in. wall carbon steel scrubber, including venturi, elbow, separator, pumps and instrumentation.

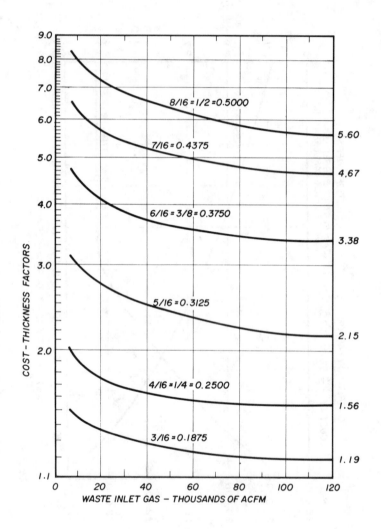

Figure 3-11.   Cost-thickness factors for carbon steel scrubbers.

Table 3-2. December 1970 Costs for 50,000 ACFM Scrubber

Scrubber	Scrubber Cost ($)
1/4-in.-thick carbon steel	19,600
5/16-in.-thick carbon steel	30,100
1/4-in.-thick 304 stainless steel	36,000
1/4-in.-thick 316 stainless steel	51,500
1/4-in.-thick carbon steel with 3/16-in.-thick rubber liner	22,200

For a given scrubber and constant acfm, the pressure drop can be changed using a variable venturi throat, which costs an additional $2000. The cost of adding power-assisted control to a variable venturi throat is approximately $1500. The power-assisted control feature is only needed when the overall design of the pollution control system calls for automatic adjustment of the variable venturi throat.

The cost figures for the variable venturi throat and power-assist controls are independent of the acfm into the scrubber and the scrubber pressure drop and are approximations that apply to all scrubbers.

Factor assembly of scrubber components is limited to weights and dimensions that can be shipped. Thus, the buyer need only erect small scrubbers, which are delivered ready for installation from the manufacturers. On larger scrubbers, however, the buyer must assemble and erect the components that are too large to ship preassembled.

The separator is the largest part of the scrubber and often causes shipping and handling problems. Separators with a diameter D of up to 8 ft (30,000 acfm) can usually be trucked to any location. Separators with D = 12 ft (80,000 acfm) can be routed by rail practically anywhere in the U.S. Separators having a diameter up to 15 ft (130,000 acfm) can be handled by rail to most locations with special routings. Units with dimension D greater than 15 ft will require field assembly.

Scrubber erection costs are subject to variation due to location. However, as a general estimate, the cost of scrubber erection, including foundations, equals one-half of the cost of an equivalent (same thickness) unlined carbon steel scrubber. Of course, this is only a very approximate method of estimating erection costs. A firmer estimate can be made after the site and installation arrangement have been established by using such information as scrubber weight and the installed height above grade.

The extra cost for field assembly F of the separator can be computed using the following formula:

$$F = 0.93 \, Nt + 3Y$$

where: $\quad$ N = number of inches to be welded
$\qquad\quad$ t = thickness of the plate in inches
$\qquad\quad$ Y = hourly cost of crane and crew

Sample Problem

Estimate the cost of a basic scrubber with the following input requirements: waste gas properties = air containing 51.5 percent water vapor,
inlet gas = 50,000 acfm at 600 F and 1 Atm (407.2 inches of water),
scrubber pressure drop = 50 in. of water, construction material = 304
stainless steel.

Figure 3-8 provides the basic dimensions of the scrubber. For a
50,000 acfm scrubber, the height is 21 ft and the diameter 10 ft. Assuming
a safety factor of 2, Figure 3-9 shows that for 50,000 acfm and 50 in. w.g.
pressure drop, the wall thickness should be 1/4 in.

Alternatively, Figure 3-9 and the equation on page 146 can be used
to find the buckling safety factor, based on the scrubber pressure drop and
the scrubber dimensions. For example, calculate the buckling safety factor
for a 1/4 in.-thick wall. In the equation, the term Z is defined as:

$$Z = \frac{H^2\sqrt{1\text{-}M^2}}{t\,D/2} = \frac{(250 \text{ inches})}{(0.25 \text{ inch})} \frac{2\sqrt{1\text{-}(0.3)^2}}{(120 \text{ inches})/2} = 4038$$

From Figure 3-9 the buckling coefficient **K** for Z is 70. From Equation 2,
the critical buckling stress is:

$$b = 26.21 \times 10^6 \ (70) \ \frac{(0.25)^2}{252} = 1806$$

Based on this value, the buckling safety factor, as determined from the
first equation is:

$$S = \frac{(54.416) \ (1806) \ (0.25)}{(120) \ (50)} = 4.1$$

This value represents the safety factor of the 1/4-in.-thick scrubber.
If the same calculations were carried out for the next lower wall thickness
value, *i.e.*, 3/16 in., the safety factor equals 1.9, thus dropping below the
accepted limit of 2. Therefore, the 1/4-in.-thick scrubber selected using
Figure 3-9 was accurate.

In calculating the total cost of the scrubber, Figure 3-10 shows that
a 1/8-in.-thick carbon-steel equivalent costs $12,200. Since the multiplier
for 316 stainless steel is 2.63 and the correction factor for 304 stainless
steel is 0.7, the cost of a comparable 1/8-in.-thick system in 304 stainless
is:

($12,200) x (2.63) x (0.7) = $22,460

Figure 3-11 shows that the thickness factor for a 1/4-in.-thick system
is 1.61. Therefore, the total cost of the scrubber as specified is:

($22,460) x (1.61) = $36,200

The next step involves the calculation of the erection cost, which are based on the cost data for carbon steel:

$$(12,200) \times (1.61) \times (0.5) = \$9,800$$

The combined scrubber and erection costs, therefore, amount to:

$$\begin{array}{r} \$36,200 \\ 9,800 \\ \hline \$46,000 \end{array}$$

The scrubber cost estimating procedures described are expected to predict the cost of a scrubber (for a given acfm, scrubber pressure drop, and material of construction) within less than ± 20 percent.

Ancillary Equipment—Fans, Fan Motor, Motor Starter and Fan Motor Coupling[17]

In the past, pollution engineers have had to estimate these costs from a myriad of different vendor catalogs and fan engineering handbooks, each using different terms and relationships that could not be easily interrelated. Now, by using the fact that the gases exiting the scrubber have only a restricted range of parameters, and by matching fan and fan motor data to these parameters, the pollution engineer can use simplified correlations for determining the cost of this equipment.

In practice, the fan is protected from high temperatures and particulate erosion by locating it downstream of the scrubber (Figure 3-12). The fan complex includes a single-inlet, single-width or a double-inlet, double-width fan, connected by a coupling to the fan motor and motor starter. [The double-inlet, double-width (DIDW) fan is essentially two fans with separate inlets in parallel on the same shaft and in the same housing.]

When an induced draft fan is used with a scrubber as shown, the scrubber system is below atmospheric pressure. The total gas pressure (TP) is the sum of the static pressure (SP) and the velocity pressure (VP), as follows:

$$TP = SP + VP$$

It is customary to keep the gas velocity (v) in the range of 3000 to 4500 fpm to prevent settling out of particulate matter. For systems whose gas densities are greater than that of standard dry air (0.075 lb/cu ft), the VP is calculated from the following equation to be approximately 0.6 in. w.g.:

$$VP = 0.83606 \times 10^6 dv^2$$

where **VP** is in in. w.g.; **d** is the gas density in lb/cu ft; and **v** is the gas velocity in fpm.

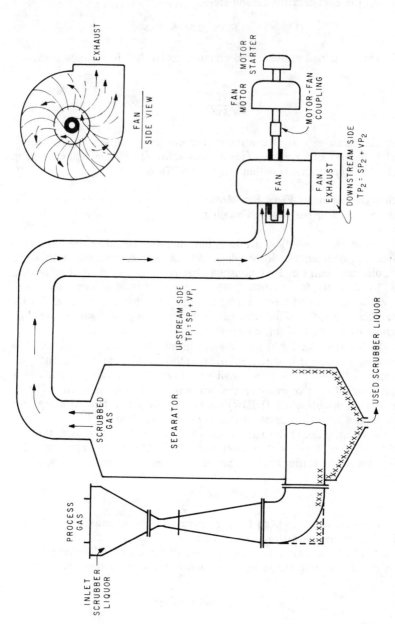

Figure 3-12. Scrubber system with connecting fan complex.

For high-energy orifice scrubbers, where most of the system pressure drop is in the scrubber, the scrubber pressure drop (10 to 80 in. w.g.) greatly exceeds the VP (0.6 in. w.g.), so that SP approaches TP throughout the system.

Fan manufacturers use a number of terms to describe fan performance, notably the fan static pressure (fST), fan total pressure (fTP), or fan equivalent static pressure (feSP). These terms are defined in terms of static and velocity pressures both upstream and downstream of the fan, as well as of the density factor (d_f), as follows:

$$d_f = d/0.075$$
$$fSP = SP_2\text{-}SP_1\text{-}VP_1$$
$$fTP = TP_2\text{-}TP_1$$
$$feSP = fSP/d_f$$

For cost estimating purposes, it is sufficient to describe fan performance in terms of the fTP and account for duct pressure losses by equating the fTP to 1.1 times the scrubber pressure drop.

Cost Data

Figure 3-13 shows the cost of an induced draft fan as a function of total acfm leaving the scrubber for a gas exit temperature of 180 F. Within the graph are separate curves for fTP of 10, 20, 40 and 60 in. w.g. These curves also indicate typical inlet fan arrangements (SISW or DIDW) and fan speeds.

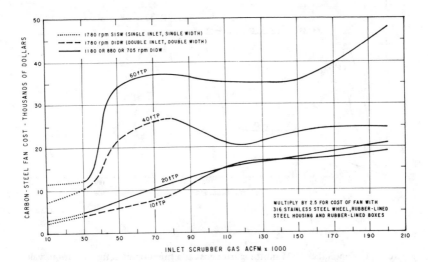

Figure 3-13. September 1971 fan cost as a function of acfm for various values of fTP at a gas temperature of 180 F. (Overall equipment index 324.1. Freight allowed to jobsite, Over 50,000 acfm, shipped knocked down.)

The fan design selected here for cost study is Arrangement No. 3, as designated by the Air Moving and Conditioning Association. This arrangement is considered superior for this application because the fan housing supports the fan bearings at both ends, thus offering less wear and promoting longer fan life.

Figures 3-14 through 3-16 show the uninstalled motor starter and fan motor costs as a function of total acfm leaving the scrubber. These curves were developed using estimated fan inertia data, together wtih the following general specifications:

1. Fan—continuous (24 hr per day) service, 40 F ambient.
2. Motor—NEMA B induction motor, horizontal operation, weather-protected, with roller-type anti-friction bearings and bearing seals at both sides of shaft. Load torque curve during starting must follow standard fan curve, terminating at 100 percent torque at full speed. For motors smaller than 250 electric hp, supply voltage is 460 v; for larger motors, supply voltage is 4000 v.
3. Coupling—Falk Steelflex, or equivalent.

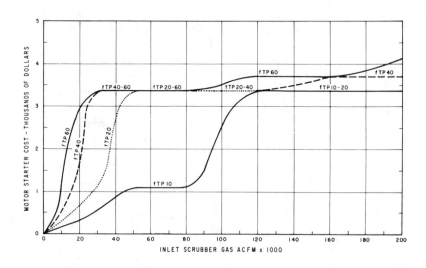

Figure 3-14. September 1971 motor starter cost as a function of acfm for various values of fTP. (Electrical power equipment index 310.0.)

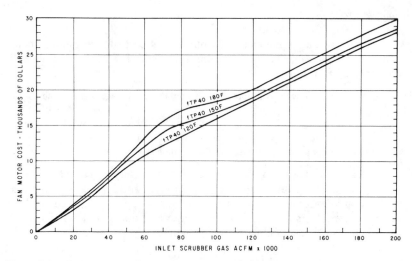

Figure 3-15. September 1971 fan motor cost for fTP value of 40 in. w.g. at saturated air temperatures of 120, 150, and 180 F. (Electrical power equipment index 310.0.)

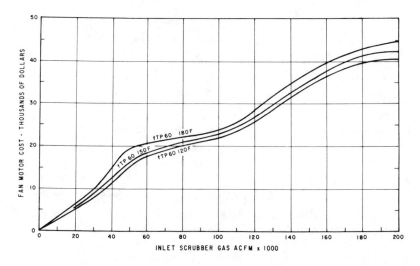

Figure 3-16. September 1971 fan motor cost for fTP value of 60 in. w.g. at saturated air temperatures of 120, 150, and 180 F. (Electrical power equipment index 310.0.)

Sample Problem

Estimate the September 1971 uninstalled cost of the scrubber fan complex for the same sample problem discussed. The input requirements are repeated as follows:

- Waste Inlet Gas Properties—Air containing 51.5 percent by volume water vapor
- Inlet Gas—50,000 acfm at 600 F and 1 Atmosphere (407.1 in. w.g.)
- Scrubber Pressure Drop—50 in. w.g.
- Material of Construction—304 stainless steel

Before turning to the cost curves, it is necessary to calculate the total acfm leaving the venturi portion of the scrubber and approaching the fan. At 70 F and 1 Atmosphere, 1 lb mole of gas occupies 386.94 cu ft. Based on the universal gas law equation (PV = nRT), at 600 F and 1 Atmosphere (407.2 in. w.g.), one lb mole of gas occupies:

$$386.94 \times \frac{460 + 600}{460 + 70} = 774 \text{ cu ft}$$

The 50,000 acfm of gas entering the scrubber is equivalent to 50,000/774 = 64.6 lb moles per min. The gas out of the scrubber is found by assuming that saturated gas leaves the scrubber, and that the enthalpy per lb of dry gas remains constant throughout the scrubber.

First, the enthalpy per pound of dry gas entering the scrubber is calculated. (The enthalpy of gas does not change appreciably with pressure changes of less than 1 Atmosphere.)

Component	Pound Moles (Total Pound Moles X Percent Vol)
Air	64.6(0.048) = 31.33
Water Vapor	64.6(0.515) = 33.27
	64.60

Component	Total Pounds (Pound Moles X Mol Wt)	Btu/lb	Total Btu
Air	(31.33) (28.97) = 907.63	139	126,161
Water Vapor	(33.37) (18.02) = 599.53	1335	800,373
			926,534

Then, the Btu per lb of dry gas is calculated to be 926,534/907.63 = 1020.8. The total pressure of the gas leaving the scrubber (based on the fTP being 110 percent of the scrubber pressure drop) is:

$$TP = 407.2 - 50(1.1) = 352.2 \text{ in. w.g.}$$

Now determine the total acfm leaving the scrubber. Using the enthalpy data, the scrubber exit gas temperature at 1020.8 Btu/lb dry gas can be extrapolated, as follows:

	357.2 in. w.g.	352.2 in. w.g.	347.2 in. w.g.
175 F	802	826	850
179 F ←		1020.8	
180 F	1020	1055	1090

With the scrubber exit gas temperature known to be 179 F, calculate the lb of water vapor per lb of dry gas at this temperature:

	357.2 in. w.g.	352.2 in. w.g.	347.2 in. w.g.
175 F	0.676	0.697	0.718
179 F		→ 0.858	
180 F	0.868	0.899	0.930

Therefore, there are 0.858 pounds of water per lb of dry gas, or 0.858(907.63) = 778 lb of water (43.23 lb moles of water). The total lb moles per min leaving the venturi is:

Component	Total Pounds	Pound Moles
Air	907.63	31.33
Water Vapor	778.75	43.23
	1686.38	74.56

At 179 F and 352.2 in. w.g., one lb mole occupies

$$386.94 \left[\frac{460 + 179}{460 + 70} \right] \left[\frac{407.2}{352.2} \right] = 539 \text{ cu ft}$$

and the total acfm leaving the venturi is calculated to be:

$$(74.56)(539) = 40,200 \text{ acfm}$$

Since the scrubbed gas leaving the venturi goes directly to the fan (in this case, no gas cooler), the fan parameters are:

$$fTP = 55 \text{ in. w.g.}$$
$$\text{Gas Rate} = 40,200 \text{ acfm at 179 F}$$

Now the cost curves of Figures 3-14 through 3-16 can be applied to determine the cost of the fan complex. Designing for 120 percent of fan capacity (1.2 x 55 in. w.g. = 66 in. w.g.), the cost of a *carbon-steel fan* can be seen from Figure 3-16 to be:

	fTP	$ Cost
	40	14,500
	60	22,500
(Extrapolated)	66	24,900 ←

The cost of the *fan motor* and *motor starter* can be based on a lesser margin, say 115 percent of fan motor requirements (or 1.15 x 55 in. = 63 in. w.g.). For 40,200 acfm at 179 F, the cost of the motor starter from Figure 3-14 is calculated to be $3,400. The cost of the fan motor (using Figures 3-15 and 3-16) is seen to be:

Figure	fTP	$ Cost
15	40	8,000
16	60	15,300
	63	→ 16,400 (Extrapolated)

The fan motor electric horsepower is given by the equation:

$$Hp = (fTP) (acfm) \qquad (1.573 \times 10^{-4})/E$$

where E (say, 0.65) is the efficiency of the fan multiplied by the efficiency of the motor. For $E = 0.85$, the fan motor electric horsepower is 613.

The cost of the *fan-motor coupling* is the last item to be determined. Since this can generally be considered as costing less than 1 percent of the combined unerected cost of a carbon-steel scrubber, fan, motor starter, and motor, this figure is seen to be 0.01 (24,900 + 3,400 + 16,400 + 20,400) = $650. (The actual cost of the fan-motor coupling in this application is approximately $250.)

Thus far, the total cost of the scrubber system can be summarized as follows:

	Component Cost (Sept. 1971)
304 SS scrubber for 50,000 acfm	$37,700
Carbon-steel fan	24,900
Motor starter	3,400
Fan motor	16,400
Fan-motor coupling	650
Total Cost	$83,050

Again, it is emphasized that these cost data represent September 1971 values, and chemical equipment cost index ratios must be used to bring the estimate fully up-to-date.

Table 3-3 summarizes the uninstalled costs of the scrubber system components for various values of fan inlet acfm and fTP. Quick cost checks can be made via this table. However, because of the many options available, it is recommended that the cost of a specific scrubber system, as well as system alternatives, be developed using the cost curves. Among the options available to the scrubber system designer are the use of multiple scrubbers instead of a single scrubber, selection of a buckling safety factor for the separator, selection of the material of construction, variations in duct work design, and the use of a gas cooler.

Table 3-3. Uninstalled Scrubber System Costs—September 1971—all carbon construction

Gas to Fan, acfm	10,0001		40,000		70,000		100,000		180,000	
fTP, in. w.g.	20	60	20	60	20	60	20	60	20	60
120 F Gas to Fan										
Percent of total cost for										
scrubber	49	21	56	36	50	39		42	62	62
fan	35	56	19	33	23	37	22	31	18	18
motor starter	6	7	9	5	7	3	5	3	3	2
motor	9	15	16	26	19	21	17	24	17	18
fan-motor coupling	1	1	<0.5	<0.5	1	<0.5	<0.5	<0.5	<0.5	<0.5
total uninstalled* cost, $	9,000	20,000	27,000	54,000	39,000	96,000	55,000	103,000	92,000	203,000
180 F Gas to Fan										
Percent of total cost for										
scrubber	46	20	53	31	48	37	54	39	60	59
fan	36	54	21	37	25	37	24	33	21	20
motor starter	6	7	8	5	7	3	5	3	3	2
motor	11	19	18	27	20	23	17	25	16	19
fan-motor coupling	1	<0.5	<0.5	<0.5	<0.5	<0.5	<0.5	<0.5	<0.5	<0.5
total uninstalled* cost, $	9,000	**21,000**	28,000	61,000	41,000	100,000	58,000	110,000	96,000	213,000

*for scrubber, fan, motor starter, motor, and fan-motor coupling

The Internal Gas Cooler[18]

The internal gas cooler installed within the housing of the separator consists of a water spray counter-current to the scrubbed gas flow, some form of extended gas-water contact surface, and a collection device so that the spray water used for cooling can be collected separately from the used scrubber liquor. In estimating the total cost of installing this type of gas cooler, since the exclusive use of spray trees is generally not acceptable, it is necessary to add the extra cost of some device, such as a bubble cap tray, which is needed for extended gas-liquid contact. Because this extended gas-liquid contact also promotes corrosion, the trays must be constructed of stainless steel.

As an estimate, the cost of adding a gas cooler can be approximated as the cost of 316 stainless steel bubble cap trays, costing $76 per sq ft gross surface area (December 1971, Marshall and Swift Chemical Industry Index 322.7). The $76 includes the cost of additional separator height needed to house the trays. Other devices used to ensure extended gas-liquid contact will have comparable prices.

Figure 3-17 shows the rate **r** of water removal in a gas cooler as a function of superficial gas velocity **G** and superficial liquid velocity **L** in the separator. This chart is used in calculating the number of bubble cap trays needed.

Example

Waste gas properties—air containing 51.5 percent by volume of water vapor; inlet gas—50,000 acfm at 600 F and 1 Atmosphere (407 in. w.g.); scrubber pressure drop—50 in. w.g.; material of construction—304 stainless steel. The scrubber separator is to be 10 ft in diameter and have a cross-sectional area of 78.5 sq ft. The gas coming out of the venturi portion of the scrubber should be 40,200 acfm at 179 F, 352.2 in. w.g., containing the following constituent breakdown:

Constituent	Total Pounds	Pound Moles	Percent by Volume
Air	908	31.3	42.0
Water Vapor	779	43.2	58.0
	1687	74.5	

The gas from the venturi section has an enthalpy of 1020.8 Btu/lb of dry gas, giving a total of (1020.8) (908) = 926,886 Btu.

Figure 3-18 is a schematic of the gas cooler operation for the sample problem. The procedure involves fixing three of the four cooler streams. Calculate the fourth stream by heat and mass balance laws, and then estimate the associated costs using the curves provided. The procedure is then repeated as required to evaluate the most economical system based on other combinations of the four cooler streams.

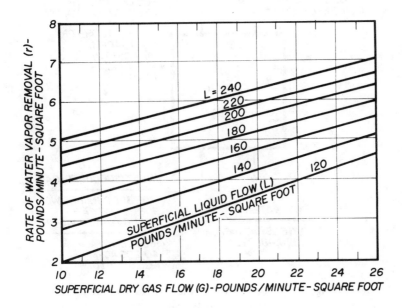

Figure 3-17. Rate of water removal as a function of gas (G) and liquid flow (L).

The inlet (scrubbed) gas stream is fixed by the gas flow into the scrubber and the operation of the venturi section of the scrubber. The inlet water
stream is based on water availability and on having a superficial liquid velocity
within the range shown in Figure 3-17. Using 10,000 lb water/min, **L** is
estimated to be 10,000/78.5 = 127 lb/sq ft.

As the outlet gas stream temperature approaches the inlet water stream
temperature, the size of the cooler increases. A temperature difference of
20 to 40 F between the inlet water and the outlet gas can be expected to
give a reasonable-sized cooler. If the exit gas temperature is assumed to be
110 F and the cooler pressure drop is neglected, the gas leaves the cooler
saturated with moisture at 110 F and 352.2 w.g. Using the moisture data
in the previous example, the gas contains 0.0693 lb water/lb dry air. The
exit gas volume is next calculated, based on the ideal gas laws.

$$386.94 \; \frac{460 + 110}{460 + 70} \; \frac{407.2}{352.2} = 481 \text{ cu ft/lb mole}$$

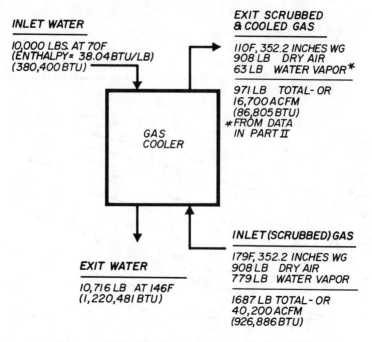

Figure 3-18. Schematic diagram of four cooler streams.

The pound-mole fractions of both air and water vapor under these new temperature and pressure conditions are calculated to be:

Constituent	Total lb/min	Total lb moles/min
Air	908	31.3
Water Vapor	33	3.5
		34.8 Total

Based on these calculations, the total acfm leaving the cooler is:

$$(34.8) (481) = 16,700 \text{ acfm}$$

The enthalpy of this outlet gas stream is found from the previous example to be 95.6 Btu/lb dry gas, giving a total of (908) (95.6) = 86,805 Btu.

Since 63 lb/min of water vapor still remain in the outlet gas stream (inlet water vapor equals 779 lb/min), the total water flow in the outlet water stream is calculated by mass balance to be

$$10,000 + 779 - 63 = 10,716 \text{ lb/min}$$

The enthalpy per pound of water is 1,220,481 Btu/10,716 lb = 113.89 Btu/lb which, using steam tables, corresponds to an outlet water stream temperature of 146 F.

With the four cooler streams established and the total acfm and temperature of the gas exiting the cooler known, it is now possible to calculate the reduced size of the separator needed and the cost of the integral cooler. Since the gas velocity through the separator can be increased from 660 to 780 fpm, the cross-sectional area of the cooler is 40,200 acfm/780 cu ft = 51.5 sq ft and the diameter is 8.1 ft.

The superficial dry gas flow is G = 908/51.5 = 17.6. The superficial liquid flow (L) can be calculated as the average water flow divided by the cross-sectional area, or

$$\frac{(10,000 + 10,716)}{2} \div 51.5 + 201$$

From Figure 3-17, the rate of water vapor removal (r) for G = 17.6 and L = 201 is 5.3. The number of equivalent separator trays is calculated to be:

$$\text{No. Trays} = \frac{\text{Water Removed from Gas}}{\text{r(Separator Cross-section)}}$$

$$= \frac{779 - 63}{(5.3)\,(51.5)} = 2.6 \text{ or} \sim 3$$

The cost of adding the gas cooler, therefore, is 3(51.5) ($76) = $11,700.

Because of the reduced acfm to the fan as compared to the same scrubber without the integral gas cooler, a savings is also achieved in the cost of the fan and fan accessories.

Interestingly, the capital cost reduction achieved by the addition of the cooler is not as important as the reduction in the horsepower requirements for the motor. This, in turn, decreases the operating cost of the system. Of course, the cost of the cooling water is not included and must be judged as a tradeoff, depending on the availability and cost of water.

SELECTING MATERIALS FOR WET SCRUBBING SYSTEMS[85]

One of the more difficult decisions facing the pollution engineer about to purchase a wet scrubber is selecting the correct materials. In wet scrubbing equipment, almost all types of corrosion, erosion, and temperature problems are encountered at one time or another. Often, several of these conditions may exist simultaneously. Therefore, careful selection of the scrubber material which will perform satisfactorily for a specific installation is extremely important. Moreover, the cost of scrubbing equipment must be held as low as possible since equipment may offer no economic return.

The following is a survey of materials used in the construction of commercially available wet scrubbers, materials most commonly recommended for certain scrubbing applications, and relative material costs.

Table 3-4. Typical Materials of Construction for Common Scrubbing Problems

Gas	Scrubbing Liquid	Scrubber Material
Ammonia	Water	Cast Iron, Steel, FRP, PVC, Ni-Resist
Chlorine	Water	Fiberglass, Haveg, PVC
Chlorine	Caustic	FRP, PVC, Kynar
Carbon Dioxide/Air	Caustic	Cast Iron, Steel, Ni-Resist
Hydrogen Chloride	Water or Caustic	FRP, PVC, Haveg Rubber Lined Steel
Hydrogen Fluoride	Water	FRP (with Dynel Shield), Rubber Lined Steel, Graphite lined, Kynar
Hydrogen Sulfide	Caustic	FRP, 316 SS, PVC
Hydrogen Sulfide	Sodium Hypochlorite	FRP, PVC, Kynar, Teflon
Nitric Acid	Water	FRP, 316 SS, 304 SS
Sulfur Dioxide	Caustic/or Lime Slurry	FRP, 316 SS (tends to pit)
Sulfuric Acid	Water	FRP, Alloy 20

Metallic Scrubber Materials

Carbon steel is probably the material most commonly specified for scrubber construction because it is easily fabricated and versatile. However, because of its limited corrosion resistance, which depends upon an oxide film, carbon steel is not recommended for scrubbing applications where dilute acids are present. It is recommended for dust collection systems or in scrubbers operating with basic solutions.

Cast iron, where available, is probably the most economical material for wet scrubbers. Units with few or no moving parts may be fabricated from it. Most cast iron scrubbers use stainless steel trim or other corrosion resistant alloys for spray nozzles and components where close dimensional tolerances must be maintained.

Cast iron may also be used in pilot plant scrubbing applications or where corrosion problems are not severe. Normal water supplies and air will not cause problems. However, SO_2 or similar gases in the fluid being scrubbed will cause the formation of dilute acids which will attack cast iron.

Ni Resist cast iron alloys containing 13.5 to 36 percent nickel are generally recommended for scrubbing applications where corrosion, erosion, high temperature or abrasion exist simultaneously. These alloys are particularly well suited to scrubbing of incinerator fumes or other fumes where high temperatures and flyash are present. Ni Resist alloys have much higher resistance to corrosion than plain cast iron or steel, but are not recommended where oxidizing agents are encountered.

Stainless steels most commonly used in wet scrubber fabrication include types 304 and 316, either in their standard or low carbon forms. Both of these materials form an oxide film and perform best under oxidizing conditions. They are not recommended, however, for use with hydrochloric acid or other reducing acid environments. Both have good temperature resistant properties and are often recommended for incinerator applications where relatively high temperatures are present. Extra low carbon types are recommended in both type 304 and 316 to prevent intergranular corrosion from scrubber fabrication.

Medium alloys such as Alloy 20 are used in many scrubber applications, particularly for sulfuric acid service. Materials such as Hastelloy C can also be used, but these alloys are not recommended unless absolutely necessary.

Plastic Scrubber Materials

Fiberglass-reinforced plastic (FRP) is used in wet scrubber construction because the material is economical, easily fabricated, lightweight, and has good resistance to both alkaline and acid environments. The particular FRP resin used for scrubber construction, however, should be carefully selected on the basis of its resistance to the environment. In general, FRP can be used for handling most strong acids or alkalis. Where necessary, an FRP scrubber can be equipped with a shield to prevent attack of the glass fibers in fluoride scrubbing. Limitations on the use of fiberglass include temperature (which generally must be below 220 F) and the presence of certain organic compounds which will attack the resin. FRP is used both as the basic material for scrubber fabrication, or as liner material for steel equipment which is subject to corrosion.

Polyvinyl chloride (PVC) has limited use as a material for wet scrubber construction because of its low temperature limitations (generally 140-160 F). However, this material can be used for low temperature applications involving acids, alkalis or dust. Most often, PVC is used for components such as spray nozzles or other standardized parts, as well as for packed tower grids and redistributors.

Furfuryl alcohol resin is particularly well suited to scrubber applications where the acid and alkali resistance of FRP is required, but where organic solvents such as Xyelene or Trichlorobenzene are present. Strength and temperature properties are similar to FRP which cannot be used with these organic compounds.

Haveg asbestos-filled resins of the phenolic type are normally recommended for scrubber applications where chlorine or hydrogen chloride are encountered.

Lining Materials

Generally speaking, linings can be used in many applications to hold down the cost of construction. While a rubber lining may be more expensive than FRP, in some applications it may be better suited. If the particular scrubbing problem involves high abrasion, for example, a rubber lining may be the ideal answer. The rubber tends to give and wear less rapidly than other lining materials.

Carbon graphite linings are often recommended for small-scale scrubbing operations because they are inert to most fluids except those containing strong oxidizing agents. Carbon graphite is not recommended for moving parts, for large units, or for use with designs involving complicated baffles.

Teflon (TM DuPont Co.) linings are suitable for very severe conditions involving HCl or similar reducing agents, bromine, or where temperature limitations will not allow the use of less expensive materials such as FRP. However, since Teflon cannot be bonded to a metal surface, a Teflon-lined steel unit requires multiple flanges to hold the Teflon sheet in place.

Kynar (TM), a material with properties similar to those of Teflon, is now available in sheet form bonded to a glass backing. This material can be used with FRP to produce a relatively inexpensive, inert material where FRP would be otherwise suitable except for the contaminants involved. Kynar-lined FRP units are recommended for scrubbing of chlorinated hydrocarbons and similar fluids.

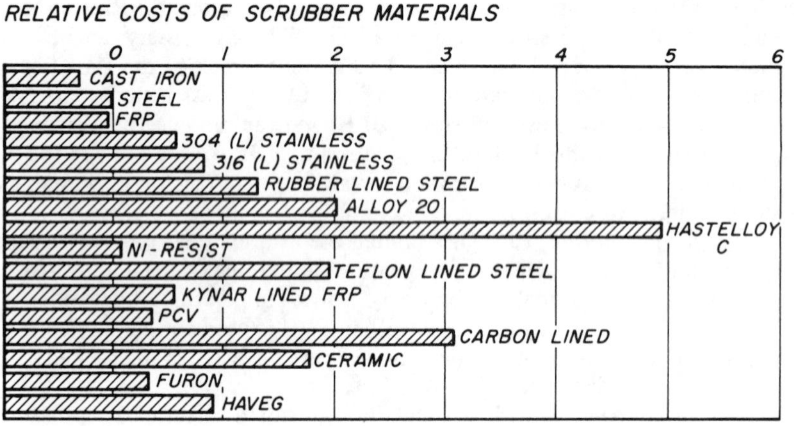

RELATIVE COSTS OF SCRUBBER MATERIALS

Figure 3-19. Relative costs of scrubber materials.

INCREASING VENTURI SCRUBBER EFFICIENCY WITH STEAM JET EJECTORS[84]

A typical venturi scrubber consists of a straight tube in which a static head is converted into velocity at the throat area. This high velocity through the throat permits the gas to disperse the collecting liquid into small droplets that collect the solid particles. Depending upon the particulate size and the particle density and concentration, the gas pressure drop often varies from 30 to 50 in. of water.

The conventional approach is to use a blower to induce flow through the unit. However, there are attendant operating problems.

Primary difficulty with the use of a high-static pressure blower stems from the nature of the scrubbing system. No scrubber is 100 percent efficient. A small percentage of the dust always passes through the unit and onto the fan blades. Normally, the characteristics of the dust at this point have been modified so that the smaller dust at the inlet to the scrubber has been agglomerated to form large particles. This material is wet and, depending upon its properties, may adhere to the blades of the blower, causing blower imbalance and leading to shutdowns.

In addition, if any gaseous contaminants are present at the inlet to the venturi scrubber, a small amount will also be present at the discharge. Further, if these gases are corrosive, the materials of construction for the fan must be selected accordingly—usually at increased expense.

An alternative to the use of high-static pressure blowers is the use of a *steam jet ejector* to induce the draft through the high-drop venturi unit.

A steam jet ejector is a device designed to convert the pressure energy of steam to velocity energy to induce flow. The ejector then recompresses the mixed gas flow by converting velocity energy back into pressure energy. Since a steam jet ejector has no moving parts, the presence of minimal amounts of dust from the discharge of the venturi scrubber will not seriously affect performance.

Ejectors of this type can be manufactured from most materials that can be fabricated. Prices typically run from $0.50 to $1.00 per cfm.

In existing systems, the steam jet may be located on the downstream side of the venturi unit in place of the blower. Figure 3-20 shows this type of arrangement. Note that the steam is shown as supplied by a waste heat

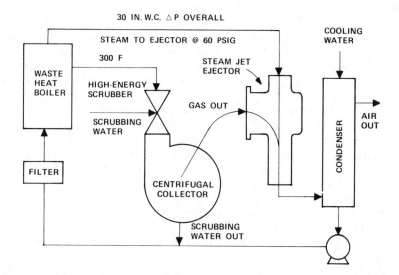

Figure 3-20.

boiler located upstream of the venturi unit, assuming the upstream temperature is sufficient to produce the required steam. In other cases, the steam
would have to be supplied by existing boilers or by additional equipment.
Table 3-5 shows the required steam consumption based on various pressure
drop venturi systems.

**Table 3-5. Steam Consumption of Steam Jet Ejectors
in Venturi Scrubber Systems**

		Consumption (lb Steam/lb Air)			
Steam Pressure		100 psig	80 psig	60 psig	40 psig
Pressure Drop Through Venturi	30	0.19	0.21	0.24	0.28
(in. of water)	40	0.22	0.25	0.28	0.33
	50	0.28	0.32	0.36	0.40

Some type of equipment is normally mounted on the discharge of an
ejector to limit noise and recover the steam as water. This device can be either
a silencer which simply cuts down noise or a condenser which not only reduces
noise, but recovers the heat value of the steam.

A steam jet ejector offers the advantages of low initial cost and low maintenance. Operating cost can also be kept to a minimum with use of a waste
heat boiler.

SUPPRESSING SCRUBBER STEAM PLUMES[140]

Water scrubber systems removing pollutants from combustion processes
can generate a supersaturated water vapor which becomes a visible white plume
as it leaves the stack. This plume may violate air pollution control ordinances
which restrict opacity emission of non-black plumes, or may be objectionable
only for aesthetic reasons.

There are several methods to avoid or eliminate a steam plume. An obvious method is to use air pollution control equipment which does not use
water in contact with the combustion gases. However, the gas temperature is
a determining factor in choosing a system of this type. If exhaust gases are in
the 500-600 F range or below, an electrostatic precipitator, baghouse, or cyclone
collector method may be used. If the effluent is above this temperature range,
it should be cooled. An electrostatic precipitator may be preceded by a waste
heat boiler for cooling the gases; however, a continuous "sink" for steam
generation is necessary.

Many plants with combustion processes have no way to use or dispose
of steam, and a medium-energy-type water scrubber is the only practical air
pollution control device. A scrubber can be designed to meet the most stringent
air pollution code limitations, but a steam plume will be formed. Use of a wet

scrubber or a spray chamber for cooling exhaust gases with water prior to entering an electrostatic precipitator will also result in a steam plume.

Steam Plume Formation

The degree of opacity of the fogged air or steam plume depends on a number of variables—the number or concentration of water vapor droplets and their size, the depth of field or plume thickness, and the lighting background of the sky. Droplet formation occurs at approximately the dewpoint temperature of the gas. Actual formation is influenced by dust nuclei.

Design conditions can be assumed (Figure 3-21) regarding formation of visible water droplets which produce fogged air.

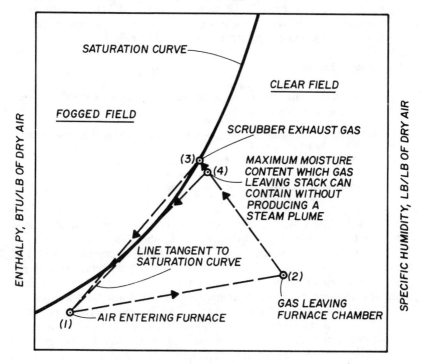

Figure 3-21. (1) Design conditions by which a steam plume may be formed can be plotted on a psychrometric chart. (2) The flow diagram shows the process equipment necessary to suppress a steam plume by transferring heat from the combustion gases and mixing with heated ambient air. (3) A steam plume may be suppressed by dehydrating the effluent and mixing it with heated ambient air. (4) A third method of steam suppression is to condense the moisture in the exhaust gases with cool water, then reheat the effluent before exhausting it.

For purposes of design, state points of mixtures on the chart to the right of the saturation curve are located in the clear field. State points of mixtures to the left of the saturation curve are located in the fogged field and will contain visible droplets of condensed water. (The degree to which these droplets will be visible can only be determined by field conditions.) By using the saturation curve on the psychrometric chart as the design condition at which the air-vapor mixture becomes visible, methods can be applied to process the combustion gases so that the state point of the resulting mixture with ambient air will remain in the clear field.

The maximum amount of moisture exhaust gases can contain (without resulting in a mixture with the ambient air in the fogged field of the chart) lies on a line tangent to the saturation curve as shown passing through point 4 in Figure 3-21. The maximum moisture content of stack gases that will not produce a steam plume is a function of ambient air temperature, relative humidity and temperature of the discharged gases.

Suppression Methods

Considering only thermodynamic means to suppress steam plumes, there are three possible methods:

System 1. Cooling of moist gases, followed by mixing with heated and relatively dry air.

System 2. Condensation of moisture by direct contact with water, then mixing with heated, ambient air.

System 3. Condensation of moisture by direct contact with water, then reheating scrubber exhaust gases.

In the first system, Figure 3-22, low-moisture-content ambient air may be reheated and mixed with the saturated scrubber exhaust. The state points of the process are shown on the system psychrometric chart. Exhaust combustion gases, point 2, are first cooled by mixing with ambient air. This tempering limits the temperature of the combustion gases entering the heat exchanger to prevent damage to the exchanger system metal.

Heat is then removed from the combustion gases between points 3 and 4. This heat is transferred to the ambient air, which is raised in temperature between points 1 and 1A when passing through the heat exchanger.

Combustion gases are next decreased in temperature along a constant wet-bulb between points 4 and 5 in the scrubber, leaving at a saturated condition, point 5. The maximum design moisture content of the mixture of combustion gases and heated ambient air leaving the system, point 6, is dependent upon the ambient air temperature at which the steam plume is to be suppressed. Since the moisture content of the ambient air is fixed for a given temperature condition, the wet-bulb temperature of the exhaust gases leaving the scrubber determines the moisture content of the gases and also the design temperature at which the steam plume will be suppressed.

In the second suppression method using a medium energy scrubber with an excess of cold water for cooling, the wet-bulb temperature of the

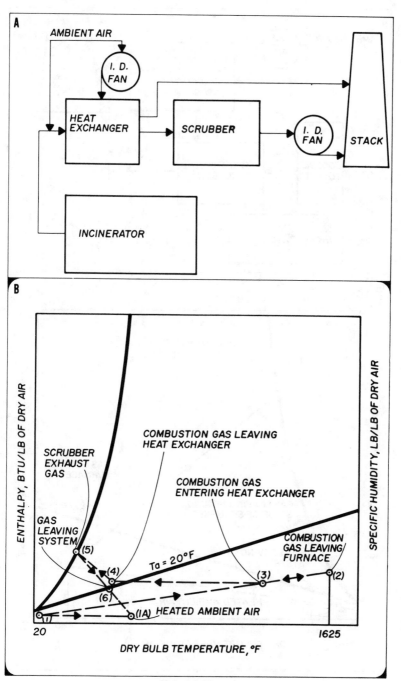

Figure 3-22.

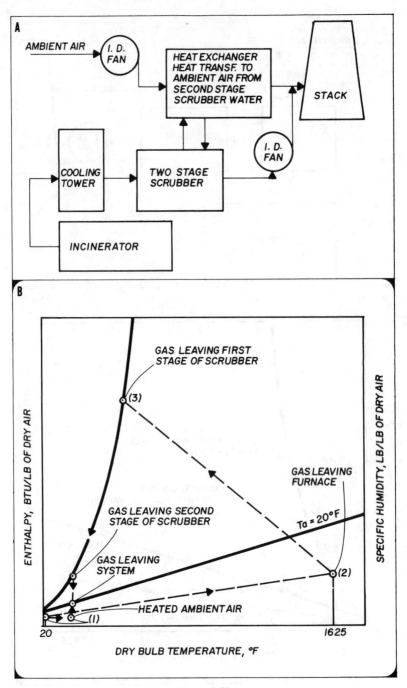

Figure 3-23.

scrubber exhaust gases can be reduced by direct contact with the water. Figure 3-23 shows a flow diagram in which the scrubber exhaust gases are mixed with heated, ambient air.

This system uses a separate supply fan for ambient air, permitting flexibility in the supply ratio of heated ambient air to combustion gas. The psychrometric chart shows the state points for this process.

The third suppression method is shown by the flow chart, Figure 3-24, in which moist furnace gases are dehydrated in the scrubber by direct contact with cooling water, followed by reheating to suppress the steam plume. Gases leaving the furnace, point 2, are cooled in a conditioning spray chamber to reduce and control the temperature of the gas entering the heat exchanger. The gases are then scrubbed and cooled in the two stages of the scrubber, points 4, 5, and 6. Leaving the scrubber, gases are reheated in the heat exchanger by heat transferred from the gases during cooling from points 3 and 4.

Economics

Although construction costs for any of these steam plume suppression systems are higher than for basic scrubber systems, operating costs are not proportionally higher. Operating costs of a scrubber include water which is lost by evaporation. The first two suppression systems discussed will not require significant amounts of makeup water even though they have larger power costs. The third requires large amounts of cooling water for the conditions shown. Normally, cost of this water would be prohibitive, unless it could be reprocessed in a cooling tower.

Steam plumes emitted from wet scrubbers controlling hot exhaust can and should be abated. In addition to saving water, abatement can avoid many roof maintenance and duct corrosion problems. Aesthetically, neighbors will feel that a plant has done its part to control air pollution when little or no emission is visibly exhausted.

VINYL PLASTIC MIST [15]

Vinyl polymers in general, and polyvinyl chloride in particular, have gained widespread market acceptance as versatile and economical materials for a wide variety of products. Processes creating finished vinyl products are equally varied. Basic vinyl processing operations have one characteristic in common—heat. High temperatures are required to polymerize plastisol and/or work the plasticized polymer. Accompanying these high processing temperatures is the discharge of plasticizer mists and vapors into the atmosphere.

There are several approaches to controlling the air pollution problems associated with plasticizer organic vapors in the exhaust gases of the various manufacturing processes. One approach is to revise raw materials or the basic process, reducing the pollutant concentrations. This approach is the logical first to try, but rarely is completely effective in solving the problems.

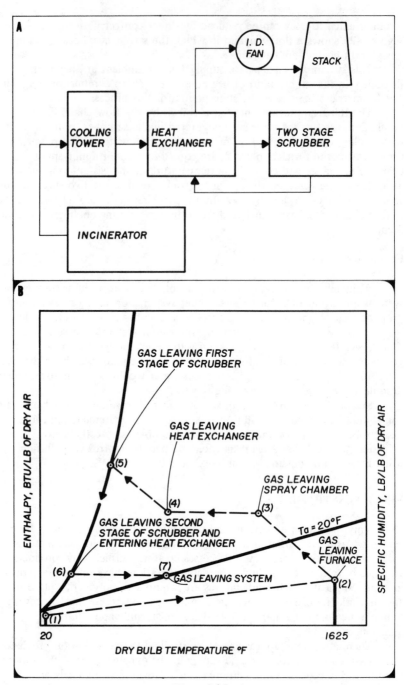

Figure 3-24.

Another control approach is incineration—heating the gases to high temperatures, often in excess of 1500 F, and reducing the organics to carbon dioxide and water. This approach often is technically feasible if the plasticizers are hydrocarbons. However, the plasticizers may also contain phosphorus, sulfur, chlorine, nitrogen, etc., in which cases the products of combustion can be much more obnoxious than the original organics. Also, new organic fluids are being used which are chemically designed not to burn or explode.

Fiber Mist Eliminators

A new systems approach to the recovery of plasticizers is the fiber bed mist eliminator. Simply stated, the unit is a fiber bed into which mist-laden gases enter, and out of which emerges a clean gas stream and a separated liquid stream. The fiber bed is designed and constructed to provide, where required, extremely high separation efficiencies.

A fiber bed is shown schematically in Figure 3-25. It consists of special fibers, vertically suspended, and enclosed in wire mesh screens. Mist-laden gases enter from the side of the bed and pass in a horizontal direction through the bed. Clean gases emerge from the bed and rise to exit from

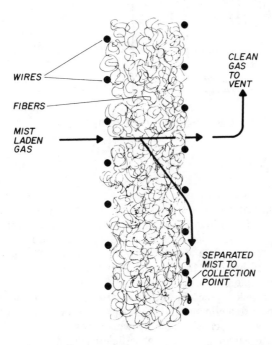

Figure 3-25. Schematic diagram of fiber bed mist remover.

the mist eliminator. Separated liquids flow downward and toward the outer screen, and ultimately drain down the outer edge of the bed. Figure 3-26 shows a typical fiber bed installation.

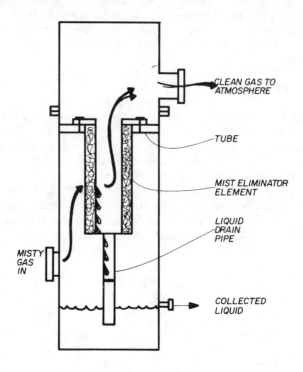

Figure 3-26. Fiber bed assembly.

Particle Separation

The three basic mechanisms for mist separation can best be described in the following manner. Consider a gas stream containing mist particles moving toward a fiber which is perpendicular to the direction of flow (Figure 3-27). The gas streamlines around the fiber. The momentum of larger particles (greater than about 1 micron) makes them deviate from the gas streamline and head for the fiber. Larger particles are thus separated through the principle of inertial impaction.

Smaller particles, generally smaller than one micron in diameter, tend to follow the gas streamline around obstacles. However, as is shown in Figure 3-27, they show considerable Brownian movement and diffuse from the gas to the surface of the fiber. A particle having a 0.1-micron diameter will have approximately 5 times the Brownian displacement of a 1.0-micron

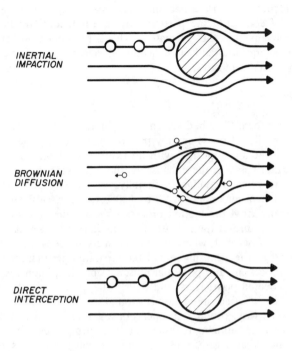

Figure 3-27. Mechanisms for fiber mist collection.

particle and about 15 times the Brownian displacement of a 5.0-micron particle. Through proper fiber bed design, submicron particles can thus be effectively collected. Utilizing Brownian movement, the collection efficiency increases as the particle size decreases because the Brownian displacement actually increases as the particle size decreases.

Particles may be collected also by direct interception. The particle may follow a gas streamline and be collected without inertial impaction or Brownian diffusion, if the streamline is relatively close to the fiber. If a particle with a diameter of 1 micron follows a gas streamline which passes within 0.5 microns of the fiber, the particle will touch the fiber and be collected.

These three mechanisms make this type of mist eliminator a highly efficient mechanical-type liquid entrainment separator. While the design's primary purpose is to remove low-micron and submicron liquid particles, it will also handle large particles at higher efficiencies. Essentially 100 percent collection efficiencies can be achieved with fiber bed mist eliminators for particles larger than 3 microns.

An interesting feature of fiber bed systems is that, with submicron particles, the collection efficiencies are actually increased slightly as the gas flow rate through the bed is reduced. At reduced flow rates, there is a greater residence time in the bed and thus greater exposure time to the fibers. Probability of contact of the particles with the individual fibers through Brownian movement is thus increased. This is in direct contrast to the typical high-energy inertial-type collection devices, where efficiencies fall off sharply at reduced rates.

Vinyl Mists

In order for a fiber bed system to separate pollutants, the contaminants must be in the liquid phase as they pass through the fiber bed. In many cases, this can be accomplished with no external cooling. In others, an inexpensive water spray in the exhaust ducting or a water-cooled heat exchanger is necessary to provide cooling. Generally, gas inlet temperatures to the element must be a maximum of 90 to 120 F to ensure efficient gas cleanup.

For many vinyl production processes the exhaust gas is at or close to ambient temperature in the exhaust system. Exhaust gases can be delivered directly to the fiber bed, where plasticizer mists are separated. Cleaned gases are exhausted to the atmosphere and the separated plasticizers are collected for further processing and potential reuse. Depending upon the cost of purifying the recovered plasticizers, and their value, part or all of the air pollution system may be justified on economic grounds.

For some vinyl production operations, gas cooling is necessary—such as in plastisol units utilizing high-temperature curing ovens. With this type operation, the exhaust gases must be cooled through a water spray chamber or other cooling device to condense the vapors prior to entrance into the fiber bed. The water-plasticizer effluent from the fiber bed drains to a settling tank for further separation of the water and oil phases. The recovered plasticizer is drained off the top of the settling tank and recovered for reuse.

SELECTING PUMPS FOR MINIMUM SCRUBBER MAINTENANCE[174]

Pumps are important components in any wet scrubber system. Pumps and attendant piping act as the important cardiovascular system of the scrubber installation. They move clean scrubbing liquid into the system, and sludge and waste matter out of the system. They also perform all liquid-handling jobs in between.

Two important facts must be considered when selecting pumps for service other than moving clear liquid:

1. Chemical nature of the slurry
2. Abrasive solids content of the slurry

Whenever possible, the pollution engineer should make a qualitative and quantitative analysis of the material to be pumped. Pump manufacturers will not accept responsibility for the selection of the pump "wetted" materials. They will make recommendations based only on the information provided by the buyer. The pollution engineer, therefore, must accept this responsibility, and for this reason he should know some of the pitfalls.

Even traces of an element appearing in a slurry should be noted, as they can have a definite bearing on the useful life of the pump. For example, a typical specification might include these facts about a pumped liquid:

1. pH value is 3 to 4, due to sulfuric acid content
2. Includes maximum concentration of 5 percent by weight of flyash solids, and
3. Contains traces of phosphorus pentoxide.

This kind of information will help the pump manufacturer and his metal specialist make the correct recommendation for materials.

Where the application includes both severe corrosion and abrasion probabilities, the choice is more difficult. There is a limited number of materials offering optimum resistance to both of these conditions. It may be necessary to sacrifice one quality, *i.e.,* corrosion resistance, to get the abrasion resistance necessary for longer pump life.

Impellers

Another important point to consider in pump selection is the type of impeller to be furnished, and impeller clearance adjustment provisions. An enclosed-type impeller with axial adjustable clearance is best for liquids with abrasive solids in suspension.

This clearance (often defined as axial clearance or perpendicular to fluid flow) with an adjustment provision allows the volumetric efficiency to be maintained at or near the original efficiency even though wear occurs. Generally this adjustment is accomplished from the exterior of the pump. The provision for clearance adjustment may make use of shims, an adjustable bearing cartridge with elements to move the entire pump rotating element, a front sideplate (wearplate) which may have integral adjustment provisions, axial adjustment wearing rings, or combinations of several of these design features. A pump with radial wearing rings and close radial clearance should never be used in abrasive service. Units of this design are specifically for clear liquids.

The fully open-type impeller may be selected for those services where corrosion is the most important consideration. Again, this should be furnished with an axial adjustable clearance. Open impellers are more popular in corrosive services because they are simpler to cast and machine and can be made in a variety of alloys.

Sealing

The pump stuffing box is still another important area of consideration in selecting a pump for severe service. Packing is almost universally used since it is still the most reliable method to avoid a rapid, complete pump breakdown. With a packed stuffing box, the pump may limp along even with severely worn parts until corrective maintenance can be scheduled. When a mechanical seal fails, however, immediate replacement is required.

Further, a packed pump is less expensive than one equipped with a mechanical seal. Seal water dilution is not generally a problem on scrubber applications. Therefore, the only consideration is the cost of the seal water.

Adjustable impeller clearance is designed for packed stuffing boxes as standard.

Mechanical seals should also be considered, but maintenance features should be thoroughly investigated. A definite need exists for an economical mechanical seal designed for low-suction pressures which allows slight liquid leakage. Most seals are generally designed for high operating pressures and near leak-free service. They can require elaborate installation and maintenance.

Any packed pump used for solids-handling requires a renewable, hard-alloy shaft sleeve. The abrasive qualities of the pumped product, plus frequent tightening and adjustment of the packing, will score any sleeve. So, it must be replaced when severe wear occurs. For longest life, shaft sleeves should have resistance to both abrasion and corrosion.

All solids-handling pumps should ideally be designed internally for low stuffing box pressures. This results in less sealing water pressure and reduced sealing water flow, which means less dilution and a smaller water bill. It also means that packing and shaft sleeve life will be extended and maintenance will be significantly lower.

Bearings

Protective features for bearings are another important consideration. Bearings that are effectively sealed off from contaminants will have a substantially longer life. One method of accomplishing this is to use auxiliary seals located outside the bearings. One device (Figure 3-28) is frequently referred to as a "taconite seal" because it first came into general use in the iron ore mining ranges.

This lip-style seal placed on the outside of the bearing provides a barrier of grease on the running surface of the shaft. Grease is allowed to bleed through close running clearances to the outside. This design completely prevents the entrance of contaminants into the bearing compartment. A vent hole is provided for the relief of excess grease, or overgreasing.

Taconite seals, used in conjunction with oil-lubricated bearings, result in an optimum lubrication/protection system. Oil is a preferred lubrication medium. A properly designed oil-lubricated bearing system is difficult to over-lubricate. For this reason, bearings will run cooler and last longer.

Drive Systems

Engineers designing air pollution control systems should not overlook the benefits of V-belt drive rather than direct-drive pumps. Belt-driven pumps

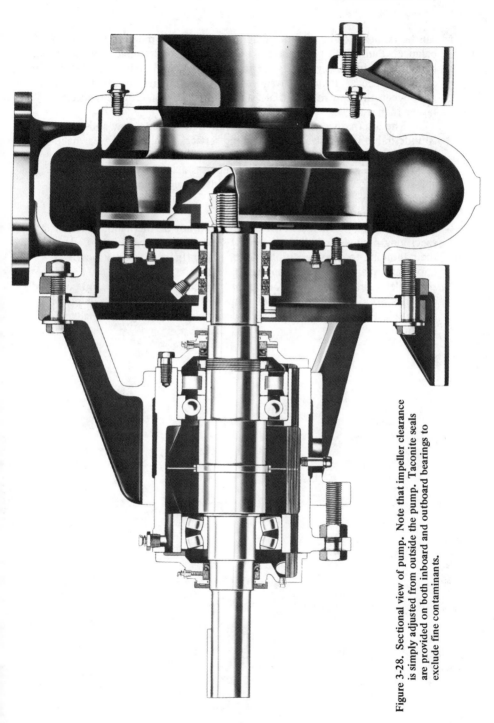

Figure 3-28. Sectional view of pump. Note that impeller clearance is simply adjusted from outside the pump. Taconite seals are provided on both inboard and outboard bearings to exclude fine contaminants.

can be tailored to the system by adjusting speeds as required for system balance. Pumps can be operated at a slower, more desirable speed when belt driven. In abrasive slurry services, all "wet-end" pump parts last longer at slower speeds. Pump wear and overall operating life are needlessly sacrificed when the unit must run at a faster direct-driven 1750-rpm speed rather than a slower V-belt 1280-rpm speed.

The alternative method of using valves to regulate flow and pressure is undesirable in abrasive solids-handling services. Valves will wear needlessly if operated at other than their optimum efficiency position (fully open) on fine abrasives. If the pump requires a special alloy to resist the abrasive/corrosive media, the valves should be made of similar metals. Special alloy valves are, however, extremely expensive, particularly in the 3 to 12-in. sizes.

Operating mechanical efficiency should not be the major criterion in specifying pumps for pollution control services. Engineers should adjust their thinking to consider utility rather than efficiency, as the most important quality they want in their pump.

Wet scrubber air pollution control systems are expensive and complex, and they are only as good as the pumps in their system. Without proper pump selection, these systems will become ineffective and eventually inoperable.

CHAPTER 4

FANS, BLOWERS AND VENTILATION

FANS AND BLOWERS[48]

The most common method for moving gases under moderate pressures is by means of some type of fan. The fan is significant and important in all industrial plants. It is the heart of any system that demands that air be supplied, circulated and removed in a way that provides a safe and comfortable environment. For industrial plants the needs of heating, ventilating, air conditioning and pollution control are fulfilled by this equipment.

There are two general classes of fans: *axial* and *centrifugal*. Axial fans employ propellers and are classed into three sub-types—*propeller, tube-axial* and *vane-axial*. Centrifugal fan flow is principally radial rather than axial. Centrifugal fans are also divided into three groups: *forward, backward* and *radial*. A distinction is made in engineering practice between fans for low pressure and centrifugal compressors for high pressure. A boundary separating the two classes of equipment is set at 7 percent increase in density of air from the inlet to the outlet. Fan action is below this density increase and the incompressibility of gas moved is assumed.

Choice of fan depends on flow volume required, static pressure, condition of air handled, available space, noise, operating temperature, efficiency and cost. Consideration should also be given to the drive system, whether it should be direct or belt-driven.

Axial Fans

The axial flow fan is used in systems which have low resistance levels. This type fan moves the air or gases parallel to the fan's axis of rotation. Axial flow fans use the screw-like action of their propellers to move the air in a straight-through parallel path. This screw-like action of the propeller causes a helical type flow pattern. *Propeller type* axial flow fans move air at pressures from 0 to 1 inch of water. Additional variations of the axial flow fan can move air at somewhat higher pressures.

A variation of the axial flow fan is the *tube-axial fan*—the basic axial flow fan encased in a cylinder. The fan's propeller in the cylinder helps to collect and direct the air flow. The tube-axial fan can move air or gas at pressures between 1/4 and 2-1/2 inches of water.

A second variation of the axial fan is known as the *vane-axial fan.* This is an adaptation of the tube-axial fan using air guide vanes mounted in the cylinder either on the entry or discharge side of the propeller. These vanes further increase the fan's efficiency and working pressures from 1/2 to 10 inches of water by straightening out the air or discharge flow.

Principal advantages of axial fans are their economy, installation simplicity and small space requirements. The principal disadvantage, aside from operating pressure limitations, is noise. This latter problem is usually apparent at maximum pressure levels. These fans are seldom used in duct systems because of the relatively low pressures developed. They are well adapted for moving large quantities of air against low pressures with free exhaust, as from a room to the outside.

Centrifugal Fans

Centrifugal fans or blowers move the air or gas perpendicular to the fan's axis of rotation. Air is drawn into the center of the revolving wheel, which is on a shaft containing the fan's blades. The gas stream then enters the spaces between the wheel's blades and is thrown out peripherally at high velocity and static pressure. As this occurs, additional air is drawn into the eye of the wheel. This type blower is used where the frictional resistance of the system is relatively high.

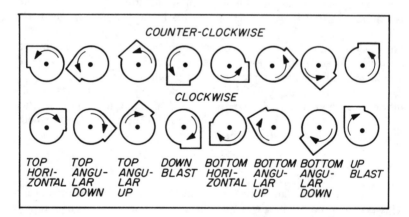

Figure 4-1. Centrifugal fan rotation and discharge. Two directions of discharge and sixteen discharge positions are possible with centrifugal fans. Rotation direction will be determined by the fan function and is specified according to the view from drive side.

There are various adaptations of the centrifugal fan, which are distinguished by the type of blade used. Blade types depend on space limitations, efficiency demanded by the system for particular load conditions, and allowable noise levels. There are three general types of blades that are used in blowers: forward-curved, backward-curved, and straight or radial type.

In the *forward-curved centrifugal fan*, the blade is inclined at the tip toward the direction of rotation. This is the most widely used centrifugal fan for general ventilation purposes. It operates at relatively low speeds and is generally used for producing high volume air flow and low static pressure. This type of fan is quiet, economical, space-efficient and lightweight. Because of the inherent design of its blade configuration and low operating speeds, the forward-curved fan cannot develop high static pressures.

Backward-curved fans are more suitable for higher static pressure operation. They operate at about twice the speed of forward-curved centrifugal fans, and have higher efficiency and a non-overloading horsepower curve. The higher operating speeds, however, require larger shaft and bearing size. Greater care must therefore be taken in system balance.

The *radial fan* has a blade curvature tangent to the radius at its outer tip. The radial type centrifugal fan is generally designed for handling low air volumes at relatively high static pressures. It is also suitable for handling high dust concentration air because of its wheel design.

Table 4-1. Relative Characteristics of Centrifugal Fans

	Forward Curved Blade	Backward Curved Blade	Radial Blade
First Cost	Low	High	Medium
Efficiency	Low	High	Medium
Operational Stability	Poor	Medium	Medium
Tip Speed	Low	High	Medium
Abrasion Resistance	Poor	Medium	Good
Sticky Material Handling	Poor	Medium	Good

Fan Selection

When selecting a fan, one must consider which fan will fit the purpose, while being the most economical to operate. Cost considerations before purchase include operating, maintenance and equipment costs. It is not necessary to design a new fan for each new application. Choice of a fan that fits the needs usually consists of selecting one commercially available from suppliers.

A fan's capacity is measured in cubic feet per minute. This is equivalent to the number of pounds of air or gas flowing divided by the number of pounds of gas or air per cubic foot at the system's inlet. In order to meet the fan's capacity, the right horsepower motor must be used to drive the fan. Belt-driven fans are used for motor requirements generally between 1 and 200 horsepower. Fans of this type are available in a larger number of standard sizes. Economical motor selection can be made, even when its speed is different from that of the fan, by selecting the proper belt-drive ratio. For fans requiring drive motors larger than 200 hp, direct-drive motors are generally used. Direct-drive fans are limited to the fan's motor speed. Principal advantages of this latter type drive are less required maintenance and less power transmission loss than from belt-driven fans.

When ordering a fan, the following data are required:

1. *Flow volume*—the volume of air the fan will handle at the actual temperature conditions that will prevail.
2. *Composition of the gas handled*—moisture, dust load, corrosives present, etc.
3. *Static pressure*—the resistance the fan will have to overcome to deliver the required volume of air. This includes the resistance or pressure drop in the total system the air flows through from process intake to exhaust stack exit.
4. *Operating temperature*—this parameter is important not only from the standpoint of affecting the volume of air handled, but it will determine in many cases the materials of construction.
5. *Efficiency*—the ability of the fan to handle the required volume and pressure with a minimum horsepower motor and expenditure of electrical energy. This parameter will determine the operating costs of the unit.
6. *Noise*—the best guide to the selection of a suitably quiet fan is successful previous performance.
7. *Space requirements and equipment layout*—this will include orientation of fan inlet and outlet since many options are available.
8. *Initial cost* of the equipment.

Space requirements and initial cost are usually secondary considerations. Two important factors in fan selection for ventilation, for example, are efficiency—which affects operating costs—and noise.

When ordering a fan, the customer must include information concerning the applicable size of duct work and the system for which the fan is to be used. From this information the supplier can make sure the fan will meet pressure requirements. To fulfill the requirements, the fan must be able to accelerate the air or gas from the velocity at the system's entrance to that of its exit. It must also be able to overcome any pressure differences within the system. Finally, it must be able to overcome frictional and shock losses encountered in the system. Additionally, fans achieving their final exhaust through stacks must maintain a minimum exit velocity (usually 60 fps minimum) to ensure that the exhaust stack gas will escape the turbulent wake of the stack. In many cases, it is desirable to have the gas exit velocity on the order of 90 or 100 fps. Another piece of information that the supplier should have is whether the fan will be subject to any unusual conditions. This is particularly important for fans and blowers that are used with air pollution control devices.

When the system requirements are known, the main points to be considered in fan selection are: efficiency, reliability of operation, size and weight, speed, noise and cost.

To assist customers in choosing a fan, manufacturers supply tables or curves that show the following factors for each fan size, operating against a wide range of static pressures:

1. Air volume handled, in cubic feet per minute at standard conditions (68 F, 50% R.H., weighing 0.07496 lb/cu ft).
2. Air velocity at the outlet.
3. Fan speed, rpm.
4. Brake horsepower.
5. Peripheral speed, or blade tip speed, fpm.
6. Static pressure, inches of water.

Tables listing fan capacities indicate the most efficient operating point by printing values in bold face, italics, or by some other designation.

Corrosion Resistance

Two major problems which fan and blower designs must overcome are excessive temperatures and corrosive atmospheres. Mild steel is good for fan construction in dry applications up to temperatures of 900 F. Temperatures exceeding this cause scaling. In such cases steel may be coated with a protective alloy.

Structural and corrosive problems arise at excessive high gas stream temperatures and/or corrosive atmospheres. High temperatures cause many materials to lose their strength and promote chemical reaction in the metal itself, such as scaling. Some methods used to solve these problems include lowering the gas temperature and controlling the concentration of corrosives in the exhaust gas. With lower temperatures, the fan can be coated with a layer of lead, vulcanized rubber, or plastic for corrosion protection. Fans fabricated of higher resistance metals such as stainless steel and monel can be used with excellent results if it is impractical to lower the temperature of a corrosive atmosphere. These latter systems, however, are substantially more expensive.

Fans fabricated from fiberglass-reinforced plastics are also used under corrosive conditions. Fiberglass plastics are strong, lightweight and economical, as well as corrosion resistant. Fans can also be coated with fiberglass plastics for protection. The maximum temperature at which fiberglass can be used is 200 F. Aluminum and aluminum alloys also have corrosion resistant properties and can be used for applications with a maximum operating temperature of 300 F.

Fan Noise

Fan noise is a complex mixture of sounds of various frequencies and intensities. The total pressure rise produced by a fan and the air volume delivered can be measured exactly. These quantities can be rated under pressure and volume of the fan. Sound energy for an absolute noise rating cannot be measured and is limited to comparative intensities of noise produced at some given point. Noise rating of a fan must specify the measurement positions or points. Size of the room, the form and material of the bounding surfaces will also have an effect on the noise intensity at a given point. It is therefore important that measurements be compared on a common basis such as the same room, at the same location, with a satisfactory noise level measuring instrument. These limitations should be recognized and noise level values from manufacturers be used as guides. The best guide to the selection of a suitably quiet fan is successful previous performance on a job similar to the one under consideration. For the reasons indicated, there is no such quantity as an absolute decibel rating of a fan.

Noise may be caused by factors other than the fan itself. For example, too high velocity of air in the duct work and improper construction of ducts and air passages, as well as unstable housings, walls, floors and foundations can cause noise. The importance of selecting a fan to suit the characteristics of the duct system accurately cannot be over-emphasized.

Where noise responsibility can be attributed to the fan itself, the cause may be improper selection of type or excessive speed for the size. The tip speed required for a specific capacity and pressure varies with the type of blade. An excessive tip speed for forward-curved blades may not be required for a backward-curved type. A fan operating considerably above its maximum efficiency is usually noisy.

Fan Laws

When a given fan is used for a specific system, the following fan laws apply:

1. The air capacity (cfm) varies directly as the fan speed.
2. The pressure (static, velocity or total) varies as the square of the fan speed.
3. The horsepower required varies as the cube of either the fan speed or capacity.
4. At constant speed and capacity, the pressure and horsepower vary directly as the density of the air.
5. At constant pressure, the speed, capacity and horsepower vary inversely as the square root of the density.
6. At constant weight delivered, the capacity, speed and pressure vary inversely as the density, and the horsepower varies inversely as the square of the density.

For conditions of constant static pressure at the fan outlet or fans of different sizes but same blade tip speed, πDR = constant:

7. The capacity and horsepower vary as the square of the wheel diameter ratio.
8. The speed varies inversely as the wheel diameter.

9. With constant static pressure, the speed, capacity and power vary inversely as the square root of the air density.

10. At constant capacity and speed, the horsepower and static pressure vary directly as the air density ratio.

At constant weight delivered:

11. The capacity, speed and pressure are inversely proportional to the density. Horsepower is inversely proportional to the square of the density.

These laws can be expressed mathematically, singly or in combination, as follows:

$$Q = A\,RD^3 \qquad H = B\,R^2D^2d \qquad P = C\,R^3D^5d$$

where:

Q = capacity, cfm
D = wheel diameter, ft
H = static pressure head, ft fluid flowing
P = horsepower, hp
R = speed, rpm
d = density or specific weight of air or gas, lb/cu ft
A,B,C = constants

If, when considering two fans, $A = A_1$, then $B = B_1$ and $C = C_1$. The fans are said to be operating at the same equivalent orifice, ratio of opening, point of operation, corresponding points or point of rating. This means the two fans are proportional and the above three equations are applicable, and the fans have identical efficiencies.

Example

A fan is rated to deliver 20,500 cfm at a static pressure of 2 in. of water (w.g.) when running at 356 rpm and requiring 5.4 hp. If the fan speed is changed to 400 rpm, what is the resulting cfm, static pressure and hp required at standard air conditions?

Solution

By fan laws, 1, 2 and 3

$$\text{Capacity} = 20,500 \left(\frac{400}{356}\right) = 23,042 \text{ cfm}$$

$$\text{Static pressure} = 2 \left(\frac{400}{356}\right)^2 = 2.53 \text{ in. of water}$$

$$\text{hp} = 5.4 \left(\frac{400}{356}\right)^3 = 7.67 \text{ hp}$$

Note: Standard air in fan tabulations is usually taken as air at 68 F, at 29.92 in. Hg and 50% relative humidity, weighing 0.07496 lb/cu ft (0.075 is most often used for approximate calculations).

Example

If in the previous example, in addition to speed change, the air handled was at 150 F, instead of standard 68 F, what capacity, static pressure and horsepower would be required?

Solution

Air density at 68 F and 29.52 in. Hg is 0.075 lb/cu ft

$$0.075 \left(\frac{460 + 68}{460 + 150}\right) \left(\frac{29.92}{29.92}\right) = 0.065 \text{ lb/cu ft.}$$

Density at 150 F and same barometric pressure is obtained by multiplying by absolute temperature and pressure ratio.

By fan law 4

Capacity	$= 23{,}042$ cfm at 150 F	
Static pressure	$= 2.53 \left(\dfrac{0.065}{0.075}\right)$	$= 2.19$ in. of water
hp	$= 7.67 \left(\dfrac{0.065}{0.075}\right)$	$= 6.65$ hp

Fundamental Formulas

Pressure in fan engineering is called *static pressure*. The pressure resulting from velocity impingement is called *velocity pressure*. The sum of static pressure and velocity pressure is the *total pressure*. Fan pressures are determined from duct pressure readings. The total pressure of a fan is the increase in total pressure through the fan as indicated by a differential reading between the fan inlet and outlet of two impact tubes facing the air current.

Static pressure (p_s) is the total pressure rise **p** less the velocity pressure in the fan inlet.

Velocity pressure (p_v) is the velocity pressure in the fan outlet, expressed in inches of water.

Velocity can be expressed in terms of velocity pressure as follows:

$$V = 18.3\sqrt{p_v/d} \quad \text{fps} \quad = 1{,}906\sqrt{p_v/d} \quad \text{fpm}$$

where d = density of gas in lb/cu ft.

Air horsepower or power-output of the fan

$$\text{Air hp} = \frac{62.3pQ}{12(33,000)} = 0.0001575\ pQ$$

where: Q = volume of air, cfm
p = pressure rise in inches of water

Efficiency of a fan is the ratio between output horsepower (air hp) and the input horsepower (bhp)

$$\text{efficiency} = \text{air hp/bhp}$$

Static efficiency of a fan is the ratio of static pressure power and the input horsepower.

Standard air density is 0.075 lb/cu ft. Fan pressures and horsepowers vary directly as air density.

Fan Characteristics

Fan performance can be best presented graphically. A chart usually plots volumes against pressures, horsepower inputs and efficiencies. The forms of the pressure and horsepower curves depend on blade type. Figure 4-2 shows a typical plot of fan performance, volume cfm, against total pressure, static pressure, horsepower and efficiencies. It is drawn for a given size fan at a given speed. Plots of more general application are also used,

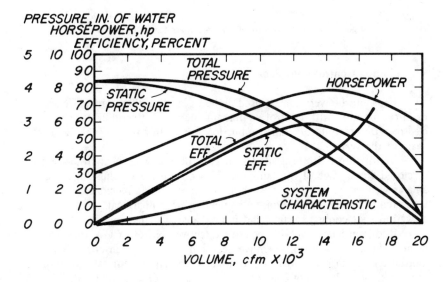

Figure 4-2. Typical fan characteristic curves.

since fans function closely to dimensional theory. Dimensionless plotting of fan curves is accepted practice. A dimensionless plot, Figure 4-3, shows percent of wide-open volume vs percent pressure, horsepower and efficiencies. These typical performance curves show how efficiency, pressure and power input vary with changing flow volume. Plots are based on fans operating at constant speed and standard air density.

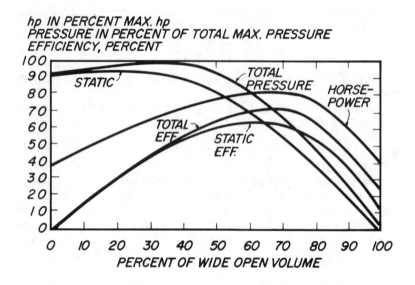

Figure 4-3. Typical plot of dimensionless fan characteristics.

Air Pollution Control with Fans and Blowers

Fans and blowers can be used alone as air pollution control devices, or in conjunction with control equipment such as wet scrubbers, baghouses, electrostatic precipitators and combustion units. In any case, the fan is the heart of the system.

Ventilation fans are used in heat control, removing heat from rooms, or closed areas. Size of the fan depends upon the work being done in the area as well as the equipment, such as furnaces, milling machines, etc. In industrial heat relief, insulation and shielding from high heat sources are used as well as spot cooling by fans and fan exhaust systems employing hoods.

Roof ventilators provide positive effective control of in-plant environment. These compact units remove heat and contamination efficiently at modest cost from work areas. Additionally, the equipment can incorporate split or combined heating control and room air can be recirculated. Mechanical ventilators have other advantages. Unit efficiency can be maintained regardless of weather conditions and often equipment can be located in otherwise wasted space.

Fans and blowers have innumerable applications as ventilating devices aside from their industrial use. Systems of fans and blowers are used in large traffic tunnels to reduce carbon monoxide concentrations from automotive exhaust to 2½ parts per 10,000 parts of air. Clean air is fed by blowers into the tunnel through a system of ducts located under the roadway. Exhaust air is removed by fans through ducts in the ceiling of the tunnel. Most traffic tunnels have two large fan rooms located at either end of the tunnel. Large garages also use ventilation systems of fans and blowers to remove carbon monoxide. Fan size units depend upon the total number of moving and idling cars in the garage. Most ventilating systems of this type remove between 2800 and 5600 cfm per car to reduce the level of carbon monoxide below 1 part per 10,000.

Underground mining operations also rely on fans and blowers for ventilation. Mine ventilation is a complex problem, requiring units that must supply a continuous flow of fresh air to mine shafts and tunnels, as well as remove dusts and fumes caused by the mining operation. Ventilators used in mines usually are capable of reversing their flow to prevent spread of dusts and fumes in case of fire.

Ventilation is important in the removal of odor and moisture in barns and animal shelters. In most barns exhaust fans are usually located 18 inches above the floor.

In-plant odor control involves the use of fans and blowers to force or induce contaminated air through various control devices. Industrial toxicants and odiferous materials include substances such as ammonia, carbon tetrachloride, phenol, ozone and hydrogen sulfide. Manufacturing processes generate contaminants including irritants, toxic dusts, fibrosis-producing-inert-allergy producing dusts, asphyxiants, inorganic and organic gases.

Activated carbon filters are used in conjunction with fans to control odors and contaminants consisting of organic substances. Fans are used to draw the contaminated air through a bed of activated carbon which absorbs the odors. All of the air may be passed through the carbon bed (a continuous bed system), or some may be diverted around the bed, making it a discontinuous bed. Continuous carbon beds are made of porous tubes filled with charcoal, or which have flat strips with charcoal granules glued to them. Most applications use continuous beds made of pleated or flat cells of charcoal or hollow cylinder canisters filled with charcoal. These absorb most odors in a single pass at air velocities between 50 to 120 fpm. Maximum recommended velocity for continuous bed absorbers is 250 fpm. Continuous bed absorbers are 95% efficient using from 5 to 50 lb of charcoal/1,000 cfm capacity, depending upon the required application.

Air washers are used to remove water-soluble vapors, dusts, gases and fumes resulting from plant processes. Air washers exhibit good efficiency on particles over 5 microns in size. The polluted air is drawn into the washer by the fan, and water is sprayed into the air perpendicular to the flow. Water and particulates land on a filter. The water trickles through and the particulate remains on the filter. Units such as these can also be adapted as humidifiers and dehumidifiers. Air velocity of such units ranges between 200 and 500 fpm, efficiency increasing with lower velocities. Between 2 and 5 gallons of water per minute are used per 1000 cfm for washing, depending on the application.

Dry filters are also widely employed. They consist of a bed or mat of fiberglass or fine synthetic fibers. This type filter actually increases in efficiency as a dust layer builds up acting as an additional filter surface. Low air velocities between 300 to 500 fpm also increase efficiency. When filters become dirty they can be washed and reused, or disposable filters may be thrown away and replaced.

DUCTING DESIGN[141]

The time to avert air flow problems in a plant's heating, ventilating and air conditioning ductwork is when the system is being designed. Once the ducts are in place, revisions can be extremely expensive not only from the standpoint of modification costs but also because of the disruption created.

There are two occasions when the engineer is presented with an opportunity to forestall ductwork problems. One occurs when a new plant is being designed, the other when an existing facility is being revised. In both instances, an understanding of the basic principles of practical duct design is essential.

General Principles

Pressure losses in ducting systems are caused by skin friction, flow separation, and changes in flow direction produced by bends, splits and takeoffs. Good duct design requires that such pressure losses be minimized so that the required pumping power can be kept as low as possible. Except for laminar or low velocity streamline flow, most pressure losses can be considered approximately proportional to the dynamic velocity head which is a function of the square of the duct velocity. Accordingly, the first basic principle of duct design is:

1. Maintain air flow at the lowest practical velocity by using adequately sized ducts.

When flow in a duct separates from the wall, as in a sudden expansion, localized flow reversals and high turbulence occur in the separated region. This condition causes high duct pressure losses. Consequently, the second general design principle is:

2. Maintain gradual deceleration of the air flow through good diffuser design. (A seven-degree diffuser half angle usually results in a good compromise.)

Rapid changes in flow direction, such as those created by sharp bends, can also result in flow separation and, consequently, large duct pressure drops. Therefore:

> *3. Use a generous turning radius wherever possible. A good rule of thumb to follow here is: turning radius should be 1.5 times the duct diameter.*

Another contributing factor in pressure losses is duct surface roughness which creates flow disturbance. Such disturbances, which are the result of protrusions into the fluid stream, cause form drag, local flow separation and increased pressure drop. Thus:

> *4. Keep the surface of the duct as smooth as possible. Preferably, keep the ratio of roughness protrusion height to duct diameter less than 0.0001.*

In some instances, duct resistances can be used to advantage. Screens, grills and other resistance elements in a duct can act to stabilize and strengthen air flow, reducing the possibility of flow separation.

Definitions and Terminology

Basic equations governing fluid flow through a ducting system are developed on the premises that mass and energy are conserved and that Newton's second law of motion is followed.

A segment of a ducting system in which the cross sectional area changes as the flow moves from one section to another is shown in Figure 4-4. The continuity equation requires that the mass of fluid per unit time entering section 1 must equal the mass of fluid per unit time leaving section 2. For

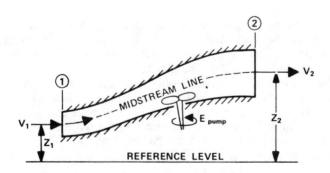

Figure 4-4. Typical duct segment with varying cross sectional area.

a compressible fluid, then,

$$\rho_1 A_1 V_1 = \rho_2 A_2 V_2$$

where symbols are defined in *Nomenclature*, p 201. For an incompressible fluid, where mass density ρ is constant, this equation becomes

$$A_1 V_1 = A_2 V_2$$

This expression of constant volume flow per unit time is valid for liquids or gases in motion where only small variations in density occur. In most ducting systems, air pressure and, hence, density do not vary substantially from atmospheric conditions. Therefore, assumption of incompressibility is acceptable for purposes of calculation. For example, a relatively high duct pressure level of 10 in. w.g. (referred to atmospheric pressure) is only 2.46 percent of standard atmospheric pressure (407 in. of water).

Bernoulli's classical equation for steady frictionless flow of an incompressible fluid along a streamline is:

$$Z + \frac{P_s}{\rho} + \frac{V^2}{2g_c} = \text{Constant}$$

where each of the terms may be interpreted as a form of energy:

Z = potential energy per pound of fluid based on an arbitrary reference level.

P_s/ρ = measure of the work the fluid can do by virtue of its sustained pressure (sometimes called the pressure energy).

$V^2/2g_c$ = kinetic energy per pound of fluid.

Application of this equation to the duct situation of Figure 4-4 gives

$$Z_1 - Z_2 + \frac{P_{s1} - P_{s2}}{\rho} + \frac{V_1{}^2 - V_2{}^2}{2g_c} = 0$$

This expression states that the difference in potential, pressure, and kinetic energies between sections 1 and 2 must be zero. Of course, this relationship is true only if the flow is frictionless. In reality, all fluids have viscosity and, therefore, offer resistance to deformation.

During flow, this resistance creates shear stresses which result in the conversion of mechanical energy to thermal energy, or heat. This thermal energy usually cannot be converted back to mechanical energy, resulting in a loss (E_{loss}) to the system. As compensation for this loss, a pump or fan is used to add energy (E_{pump}) to the flowing fluid.

Bernoulli's equation can now be rewritten to account for energy losses and additions between sections 1 and 2:

$$Z_1 + \frac{P_{s1}}{\rho} + \frac{V_1{}^2}{2g_c} + E_{pump} = Z_2 + \frac{P_{s2}}{\rho} + \frac{V_2{}^2}{2g_c} + E_{loss}$$

This expression can be simplified by introducing the concept of total pressure (P_T) and its components, static pressure (P_s) and velocity pressure or head. The relationship of these quantities is illustrated in Figure 4-5, which shows a section of ducting where manometers are being used to measure local pressure levels.

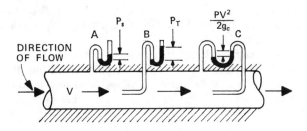

Manometer A (Top flush with wall of duct.)
    Measures the static pressure (P_s) which is a compressive unit force existing in the duct, and does not depend on the direction or magnitude of the fluid velocity.

Manometer B (Probe faces directly into direction of flow.)
    Measures the total pressure (P_T) at a given point in the system. Total pressure is the sum of the static pressure and the velocity head $(P_T = P_s + \rho V^2/2g_c)$.

Manometer C (Combination of manometers A and B.)
    Measures the velocity head or velocity pressure $\rho V^2/2g_c$ which is directly related to duct velocity and represents kinetic energy.

Figure 4-5. Duct pressure relationships.

In an actual ducting system, total pressure always decreases in the direction of flow because of mechanical energy losses. Static pressure and velocity head are mutually convertible; the magnitude of each is dependent on local duct cross sectional area which determines the flow velocity. Total pressure, which is the sum of static pressure and velocity head, is defined by

$$P_T = P_s + \frac{\rho V^2}{2g_c}$$

Incorporating this total pressure definition into the previous equation and simplifying, gives

$$Z_1 + \frac{P_{T1}}{\rho} + E_{pump} = Z_2 + \frac{P_{T2}}{\rho} + E_{loss}$$

This equation can be further simplified for most gases (including air) flowing in a duct since the potential energy term Z (height of fluid above a datum line) effectively can be neglected. Then,

$$\frac{P_{T_1} - P_{T_2}}{\rho} = E_{loss} - E_{pump}$$

This expression means, simply, that since the entrance and exit of a fluid ducting system are at atmospheric pressure, the loss (E_{loss}) in mechanical energy per pound of fluid flow must be balanced by pumping work (E_{pump}) on the system.

In fact, what happens is that the fluid flow through the duct adjusts itself until this condition is satisfied. For the purposes of this discussion the duct fluid will be treated as incompressible. This assumption considerably simplifies the equations and is sufficient for liquids and most gases at low flow velocities.

Pressure-Flow Matching

When a ducting system is selected, the total pressure drop needs to be matched to the output of the pumping device. Since duct resistance—and its associated pressure drop—is a function of cross sectional area, length, surface roughness, turning radius, etc. it can be represented as:

$$\Delta P_T = C_T Q^2$$

The application of this equation for matching three ducting system resistances to the output of an air moving device is shown in Figure 4-6.

The procedure used to accomplish the match is actually quite simple. A flow Q_{guess} (usually equal to the desired flow) is assumed, and the corresponding duct system pressure drop $\Delta P_{T_{guess}}$ is calculated. This determination allows the constant, C_T, in the previous equation to be evaluated and, then, the actual system pressure drop characteristic can be obtained (Figure 4-6). Superimposing the characteristic curve of the air-moving device on the system resistance plot locates the operating point of the intersection of the two curves.

At this point, identified as Q_A in Figure 4-6, the system requirements are exactly matched by the pump or fan output. If the fan is oversize and produces too much flow, the duct system resistance should be increased to balance the flow. For instance, if the resistance is increased so that system curve B or C is obtained, a reduced flow equal to Q_{guess} (also Q_B) or Q_C, respectively, results.

Calculating Duct Losses[142]

Calculating pressure losses in a plant's heating, ventilating, and air conditioning ductwork involves analyses of several variables.

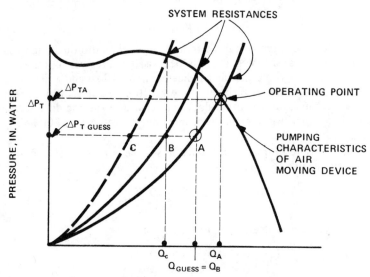

SYSTEM RESISTANCES

OPERATING POINT

PUMPING
CHARACTERISTICS
OF AIR
MOVING DEVICE

ΔP_{TA}

ΔP_T

$\Delta P_{T\ GUESS}$

C B A

PRESSURE, IN. WATER

Q_c Q_A

$Q_{GUESS} = Q_B$

VOLUME FLOW RATE, Q, CUBIC FEET PER MINUTE

Nomenclature

A	=	Area	P_T =	Total pressure
C_T	=	Overall system loss constant	Q =	Volumetric flow rate
E_{loss}	=	Energy loss to system	V =	Velocity
E_{pump}	=	Input pumping power	Z =	Height of fluid above datum
g_c	=	Gravitational constant		line
P	=	Pressure	ΔP =	Difference in pressure
P_s	=	Static pressure	ρ =	Density

Figure 4-6. Combination plot of system resistance and fan characteristics for analysis of pressure-flow matching.

Duct Losses from Friction

When long ducts are used, the effect of friction on pressure drop can be considerable. Frictional losses are a function of the duct surface condition and the type of fluid motion.

One type of duct flow is called laminar, because the fluid particles move essentially along a streamline or laminae (thin layers) in the direction of flow. A second type of flow, called turbulent, is characterized by fluid particles moving in a random or eddying motion, while, on an average, still moving in the direction of flow. The type of motion which predominates in a duct is measured by Reynolds number (N_{re}) which is defined as:

$$N_{re} = \rho DV/\mu$$

The Reynolds number can be thought of as the ratio of the local inertial force per unit area ($\rho V^3 g_c$) to the local viscous force per unit area ($\mu V/gD$). A low Reynolds number indicates laminar flow and a higher number is characteristic of turbulent flow situations. The transition from one flow mechanism to the other does not occur at a specific duct Reynolds number, but rather over a range. For most ducts this range is $2000 \leqslant N_{re\ transition} \leqslant 3500$.

Duct pressure drop from frictional loss alone is given as a function of the velocity head ($\rho V^2/2g_c$) by:

$$\Delta P_{friction} = f \frac{L}{D_H} \left(\frac{\rho V^2}{2g_c} \right)$$

where the hydraulic diameter (D_H) is defined as:

$$D_H = \frac{4(\text{duct cross sectional area})}{(\text{wetted duct perimeter})}$$

and the friction factor (f) is essentially a function of the Reynolds number and duct roughness (Figure 4-7). The straight line curves to the left are the laminar friction factors which depend on the duct cross sectional ratio A/B as well as the Reynolds number. For turbulent flows, the duct hydraulic diameter is used to obtain the Reynolds number from the earlier equation $N_{re} = \rho DV/\mu$ and the roughness to diameter ratio ϵ/D is estimated. Figure 4-7 is then used to estimate the friction factor.

Duct Dynamic Losses

Eddying motions, brought about by sudden changes in the direction and magnitude of the duct velocity, cause significant flow losses. These dynamic losses are a function of the local velocity head and can be determined from

$$\Delta P_{dynamic\ loss} = K_T \left(\frac{\rho V^2}{2g_c} \right)$$

The constant K_T, termed the dynamic loss coefficient, usually is experimentally determined. An enormous quantity of experimental data exists on the magnitude of this loss under a multitude of conditions—inlets, expansions, contractions, turning losses and the like. In this article, discussion will be limited to only a few important cases which demonstrate the technique.

Dynamic Losses from Area Changes—Perhaps the most important duct area change is at the inlet point. Various internal-duct-inlet designs are shown in Figure 4-8. The importance of eliminating sharp entrance corners to reduce the loss coefficient K_T is clearly indicated.

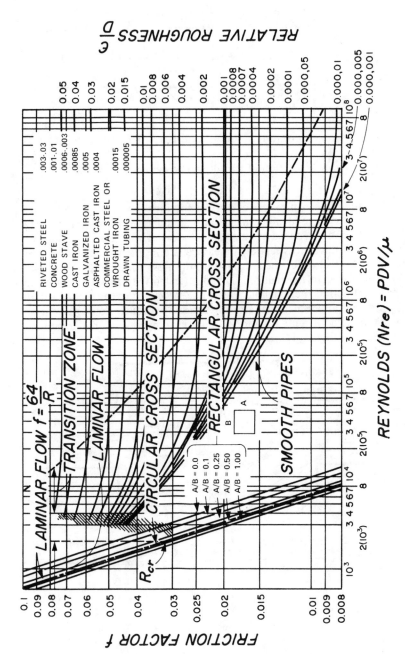

Figure 4-7. Plot for determining Reynolds number (N_{re}) as a function of friction factor and duct roughness.

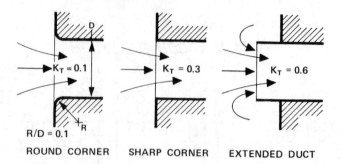

Figure 4-8. Effect of inlet designs on dynamic loss coefficient.

Dynamic losses in excess of normal frictional losses also result when a fast-moving stream suddenly expands into an enlarged cross sectional area or contracts into a reduced cross sectional area. The effect of a sudden contraction in duct cross sectional area is less important than that of an expansion. However, after the contraction, the flow continues to converge to an area smaller than the reduced duct size (called the vena contracta). In a sudden contraction the dynamic loss is due largely to expansion of the flow filling the duct cross sectional area after passing through the vena contracta. Dynamic loss coefficients for sudden changes in cross sectional area are summarized in Figure 4-9.

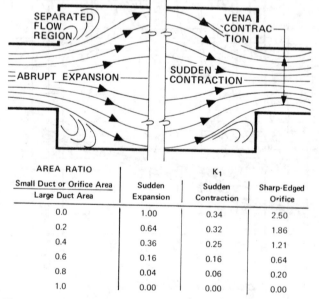

AREA RATIO	K_1		
Small Duct or Orifice Area / Large Duct Area	Sudden Expansion	Sudden Contraction	Sharp-Edged Orifice
0.0	1.00	0.34	2.50
0.2	0.64	0.32	1.86
0.4	0.36	0.25	1.21
0.6	0.16	0.16	0.64
0.8	0.04	0.06	0.20
1.0	0.00	0.00	0.00

Figure 4-9. Effect of sudden changes in cross sectional area and corresponding dynamic loss coefficients for duct.

Dynamic losses can be reduced by using transition pieces between ducts of different size. The effect a transition piece has on dynamic losses is illustrated in Figure 4-10.

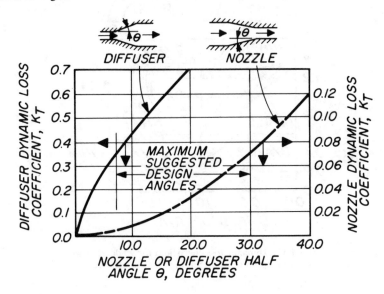

Figure 4-10. Dynamic loss coefficients for gradual changes in duct cross section.

D	=	Characteristic length of system, ft (often the duct diameter)	
f	=	Friction factor	
g_c	=	Gravitational constant	

K_T	=	Dynamic loss coefficient
L	=	Length of duct
N_{re}	=	Reynolds number
Q	=	Volumetric flow rate

Dynamic Losses from Changes in Flow Direction–Dynamic losses, caused by changes in flow direction can be significant. If the flow cannot adjust quickly enough to follow a sharp duct turn smoothly, separation and turbulence result, and an additional pressure drop occurs.

Figure 4-11 summarizes the dynamic losses of circular and rectangular ducts for 90-degree smooth turns. A minimum value of R/D or R/A = 1.5 is suggested. When the ratio of mean turn radius to duct diameter (or area) is smaller than this value, the losses increase dramatically. Anything above this value improves air flow. If it is impossible to increase R/A to above 1.5, a flow splitter should be used. This device will divide the flow, reducing the effective width A and, consequently, increasing R/A. When using splitters, it is a good rule to make the R/A for all of the flow paths equal.

In some instances, the duct elbow is constructed of separate pieces which are joined with mitered corners. The greater the number of transition pieces, the smoother the air flow. For example, Figure 4-12 shows that

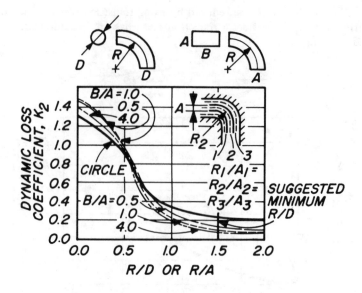

Figure 4-11. Dynamic loss coefficient for 90 degree turns.
Inset shows method of using turning vanes to balance airflow.

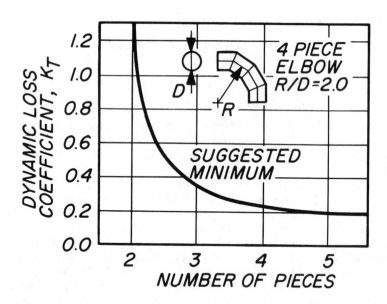

Figure 4-12. Dynamic losses in a fabricated elbow.

when an additional transition piece is inserted into the three-piece mitered corner section the dynamic loss coefficient K_T is reduced from 1.3 to 0.33. Thus, the value of the added piece is clear.

*Predicting Flow and Pressure Drop
in Multiple Branches*

A ducting system with multiple flow paths can be easily evaluated when each branch is treated as a section of unbranched ducting. In this manner, the previous techniques for predicting frictional and dynamic losses are applicable. The procedure is illustrated by the example diagrammed in Figure 4-13. Here, a single fan supplies the flow for three individual ducts.

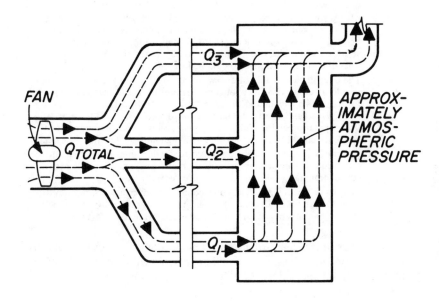

Figure 4-13. Flow in a multiple branch duct.

V	=	Fluid velocity, ft/sec	ρ =	Fluid density, lb-mass/ft^3
ΔP	=	Difference in pressure	D_H =	Hydraulic diameter
ϵ/D	=	Roughness to diameter ratio	Q_T =	Total conditions
μ	=	Absolute viscosity, lb-mass/ft-sec		

The pressure drop in each of the flow branches can be closely represented by

$$\Delta P = K_{Tn}Q_n^{\ 2}$$

where n represents branch 1, 2 or 3.

For each assumed value of ΔP, individual flows Q_1, Q_2 and Q_3 can be obtained. Total flow, $Q_{total} = Q_1 + Q_2 + Q_3$, at this assumed ΔP can then be found. From these results, the total system characteristic of ΔP versus Q_T can be cross-plotted with a fan output curve. Once the operating pressure level is evaluated, the individual path flows can be determined from the foregoing equation.

Summary

These general rules should be followed in designing a ducting system:

1. The flow medium should be conveyed as directly as possible at a velocity consistent with cost limitations imposed by materials, space and power.
2. Changes in flow direction should be minimized. When bends are required, a turning radius to duct diameter ratio not less than 1.5 should be used. If this requirement cannot be met, turning vanes or flow splitters should be used. If an elbow is of mitered construction, at least one transitional piece should be inserted.
3. The duct surface should be as smooth as possible, and steel or aluminum should be used. If surface roughness cannot be avoided, an allowance for it must be included in the estimated friction factor.
4. Abrupt increases in area should be avoided since they tend to cause flow separation and turbulence. When possible, expanding transitional segments should be utilized with a half angle not greater than 7 degrees.
5. Because acceleration tends to prevent separation, abrupt decreases in cross sectional area are not as important as rapid expansions. However, contraction half angles should not exceed 30 degrees.
6. The fan or air moving device selected must produce a pressure rise sufficient to match the total duct loss plus the losses caused by any other system components (filters, heat exchangers, washers, spray chambers, etc.

In practice, it is often difficult to construct the ducting system exactly as designed. For this reason, fans should be selected with a factor of safety. A fan pressure level approximately 15 percent above the design prediction will usually suffice.

Simplified Duct Sizing[182]

When designing a dust collection system, it is necessary to determine the size of ductwork needed and find the friction loss encountered. Using a nomograph will simplify and speed up these calculations.

On the accompanying nomograph, Figure 4-14, draw a straight line from scale **A** (air flow, cu ft/min) to scale **D** (velocity, ft/min). At the point of intersection with scale **B** you will find the minimum duct size needed to handle the air.

The nomograph also permits calculation of velocity pressure. Velocity pressure changes are proportional to changes of the velocity and can be read directly from scale **D**.

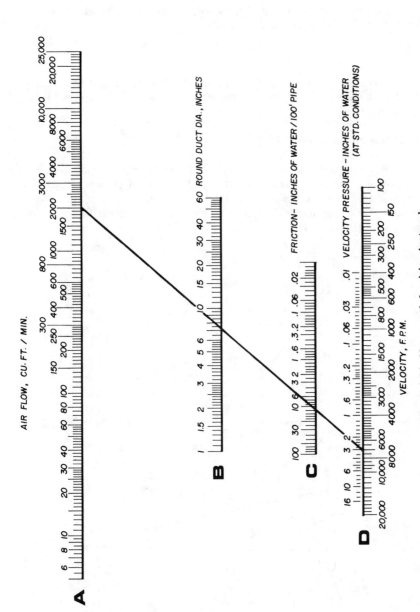

Figure 4-14. Nomograph for sizing ductwork.

Problem

Convey 2000 cfm of air at a velocity pressure of 3 in. of water. Determine the size of round dust required.

Answer

Connect 2000 on scale **A** with 3 in. **D**. Read the round duct size on scale **B** as 7¼ in. The velocity of the air in the duct will be approximately 7000 fpm.

Duct Weight Calculation[120]

When designing a ventilation system for makeup air or pollution exhaust, it is usually necessary to calculate the total number of pounds of material required to determine the cost for the system. The nomograph provides a rapid and easy method for determining the weight per linear foot of galvanized sheet steel (Figure 4-15).

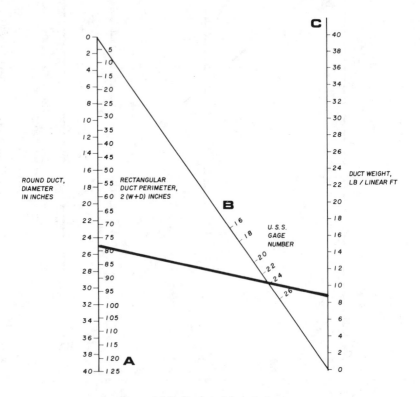

Figure 4-15. Duct weight calculator.

To calculate weight per linear foot, select either the diameter for round duct or the perimeter of rectangular duct on the **A** scale. On scale **B** select the required gage for the sheet steel of the duct. Draw a line from **A** to **B** extending it to scale **C**. Read the duct weight in pounds per linear foot on **C**.

Example

Find the total weight of a 25 in. round duct 35 ft long. Duct material is 24 gage sheet steel.

Solution

Align 25 on the round duct portion of Scale **A** with 24 on scale **B**, and read 8.7 lb per lin ft on scale **C**. Multiply 8.7 by 35, and find the weight of the duct to be 304.5 lb.

Sizing Roof Ventilators[34]

Roof ventilators can provide positive, effective control of in-plant environment. These compact units remove heat and contamination efficiently at modest cost. In addition, the equipment can incorporate split or combined heating control, and room air can be recirculated.

Mechanical ventilation has other advantages. The unit's efficiency can be maintained regardless of weather conditions. Also, the equipment can be located in otherwise wasted space.

For best results, care should be taken to select equipment of the proper type and size, and with sufficient handling capacity. Several factors should be examined:

- Size of room or building that the system is to service,
- Number of occupants and their jobs,
- Heat gains from equipment and solar radiation and
- Outside air temperature.

Each of these factors provides basic data needed in the calculations for sizing and unit selection.

The amount of ventilation needed for a specified building or work area can be determined by one of two methods: Rate of Air Change or Heat Load.

Both procedures are widely used, but the Rate of Air Change method is the simplest. It requires a close examination of manufacturers' literature and measurement of the area to be ventilated.

The Heat Load method is based on the quantity of air needed to remove heat generated in the work area.

Although there are no set rules for selecting the **proper** roof ventilation, two basic factors serve as guides. First, no matter what means is utilized to exhaust air from a space, equal quantities of make-up air must be provided. Second, the chosen system should be capable of handling the worst possible conditions for an indefinite length of time.

Rate of Air Change Method

The first step of the Rate of Air Change method is to determine the total volume of air contained in the area to be ventilated. This is easily obtained by multiplying the dimensions of its boundaries: length x width x height.

The second step is to find the rate of recommended air changes. Table 4-2 shows the estimated number of air changes per hour for various types of structures in temperature climate zones. In a hotter climate, at least a doubling of the air changes per hour is necessary.

Table 4-2. **Air Changes per Hour Recommended**

Situation	Air Changes per Hour
Boiler Rooms	30
Engine Rooms	45
Industrial Plant Buildings:	
General	20
Fumes and Moisture	30
Foundries	45
Forge Shop	45
Garages	15
Laboratories	15
Machine Shop	20
Mills (Dye House	30
(Paper)	20
(Textile)	15
Offices	10
Shops: General	10
Paint	30
Waiting Rooms	15
Warehouses	6-20

Another factor affecting these values is the presence of contaminated air. For example, the number of changes should be increased if the plant has a large amount of smoke from any source, making sure that the air contamination is brought down to a safe level.

Multiplying the number of air changes required per hour by the volume of the building results in the cubic feet of air that must be withdrawn each hour. Further examination of the building—its size, shape, roof area—and reference to manufacturers' literature will indicate the correct number of ventilating units. A rule-of-thumb in finding the number of ventilators needed is to provide one unit for every 15 or 20-ft bay.

After determining how many units are required, all that remains is to find the handling capacity and size required for each unit. Capacity is obtained by dividing the total volume of air handled per hour by the number of systems required. For convenience and in general practice, the amount of air per hour is converted to cubic feet per minute by dividing by 60. With this information as a guide, examine manufacturers' capacity tables to select the size and speed of a system to meet specifications and conditions.

Example—A plant building, where machinery generates irritable dust, is 45 ft x 100 ft x 35 ft. Determine the amount of ideal ventilation required by the Rate of Air Change method.

Solution—1. Volume determination: Volume = 45 ft x 100 ft x 35 ft = 157,500 cu ft.
2. From Table 4-2 (industrial plant buildings, fumes and moisture) the required air changes per hour are 30.
3. The amount of ideal ventilation is calculated as follows: Volume (157,500) x Air changes/hr (30) ÷ 60 min = 78,750 cfm
4. The plant has five 20-ft bays; an installation of five roof ventilators would give ideal distribution.
Now the handling capacity of each ventilator can be determined by dividing the total ventilation by the number of units required: 78,750 cfm ÷ 5 units = 15,750 cfm per ventilator
5. Refer to the manufacturers' capacity tables for selection of the unit.

Heat Loss Method

In this technique the amount of heat generated in the plant is the chief factor to consider in determining the ventilation required.
The heat in an industrial plant comes from three primary sources:

- people (body heat)
- equipment and heat-generating processes
- sun rays

The first of these, body heat, is generated proportional to the body structure, size of individual, physical activity, age, sex, health, nutrition, and working environment. Additional energy is expended as useful work. A person seated at rest generates 400 Btu per hour; during strenuous work, as much as 1400 Btu per hour per person is produced.
It is important, then, to know the number of people working in a specific area and the individual duties of each. Published guides listing heat dissipation rates of individuals for different occupations can help in determining heat loss to the environment.
Besides electric lights, which generate 3.4 Btu/hr/watt, electrical motors, pumps, etc., give off sizable amounts of heat (see Table 4-3). Some plant processes may generate latent heat in the form of moisture. Total energy liberated as heat from any chemical or mechanical process may be determined from a knowledge of fuel consumption, efficiencies, and other data.
The final factor contributing to heat gain in the working environment is sun rays striking walls and roof. Solar heat gains in a room or building

Table 4-3. Heat Generated by Electric Motors

Motor Size	
¼ hp	generates 4000 Btu/hr per hp
1 hp	generates 3400 Btu/hr per hp
5 hp	generates 3100 Btu/hr per hp
25 hp	generates 2900 Btu/hr per hp
100 hp	generates 2800 Btu/hr per hp

depend on several outdoor factors: atmospheric clarity, shading by trees or other structures, wind velocity, and location. Finding such heat gains requires knowledge of solar heat transmitting values of building materials, along with corresponding heat transmission per square foot per hour for these materials. Values for several materials are given in Tables 4-4 and 4-5.

Table 4-4. Solar Heat Gains Through Walls

Wall Construction	Wall Thickness Inches	Solar Heat Gain in Btu Per Sq Ft Per Hr Lat. 40 Deg North or South		
		East and West Walls	Southeast and Southwest Walls	South Wall
Brick—Solid Unplastered	4½	23	16.6	8.8
	9	16	12.4	6.4
	13½	13	9.7	5.2
Brick—Solid Plastered ½ in.	4½	19	13.8	7.6
	9	15	11.0	6.0
	13½	12	8.3	4.8
	18	10	6.9	4.0
	22½	8	5.5	3.2
Brick—Hollow Plastered ½ in.	11	10	6.9	4.4
	15½	8	6.9	3.6
	20	7	4.1	3.2
Concrete	6	22	17.9	8.8
	8	19	13.8	7.2
	10	16	12.4	6.4
	16	14	11.0	6.0
Stone	12	17	12.4	6.8
	18	14	9.7	5.6
	24	11	8.3	4.8
Wood—Tongued and Grooved	1	17	12.4	6.8
	1½	14	9.7	5.6
Sheets:				
Asbestos (Flat)	1/4	31	22	11.6
Corrugated Asbestos	–	41	27	14.4
Corrugated Iron	1/16	43	29	14.8
Corrugated Iron on 1-in. boards	–	14	11	5.6
Glass				
Bare Window Glass	–	185	151.8	156.0
Windows with Canvas Awning	–	52	42.8	12.0

Table 4-5. Solar Heat Gains Through Roofs—Maximum Heat Transmission

Roof Construction	Btu/Sq ft/Hr Lat. 40 deg N/S
FLAT	
Asphalt on 6-in. concrete	26.7
Asphalt on 6-in. concrete with 1-in. cork	9.2
Asphalt on 6-in. concrete with 2-in. cork	5.5
Asphalt on 6-in. hollow tile	22.1
Asphalt on 6-in. hollow tile with 1-in. cork	9.2
Asphalt on 6-in. hollow tile with 2-in. cork	5.5
Asphalt, 1-in. cork, 1¼-in. boards, joist and plaster ceiling	6.4
PITCHED	
Corrugated asbestos	68.1
Corrugated asbestos lined ½-in. boards	23.9

Not all the walls of a building are affected by solar radiation at the same time. For a rectangular structure only one or two walls at a time are exposed fully to the sun's rays. If a room or building has two sun exposures, only the side with the largest exposure should be considered; if three sides undergo exposure, then only the side with the greatest exposure is considered.

Calculations are based not only on the one wall receiving the greatest solar energy, but also on the hottest time of the day. The drawing gives a rough estimate of the percent of surface area that should be included in calculations for solar effects. These percentages apply only to northern latitudes; in southern latitudes, maximum radiation falls on the north wall from 10 a.m. to 2 p.m.

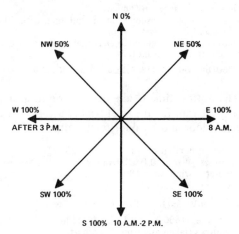

Figure 4-16. Arrows indicate solar intensity at various times of the day in different directions. The percentages shown can be used to calculate solar heat transmitted through wall areas.

After heat generated by the working environment and surroundings has been determined, the proper fan can be selected. The amount of ventilation needed can be calculated by tabulating the total heat gained through the structure's walls and roof along with heat generated by employees and machinery, and applying this figure to the following:

$$\text{Air Flow (cfm)} = \frac{\text{Total Energy or Btu/sq ft/hr}}{\text{Temperature Difference} \times 1.08}$$

The temperature gradient in the above expression is the difference between actual indoor and outdoor temperatures (which must be a minimum for the calculations).

In general, the rate of ventilation required increases as the temperature difference decreases. It is thus impossible to cool a building below the outside temperature by ventilation alone. By examining the structure's dimensions and noting the number of bays, the engineer can estimate how many units are required. The capacity of each discharge roof ventilator is determined by dividing the total ventilation amount (cfm) by the number of units required. Then, by referring to capacity tables, ventilators meeting the desired specifications can be chosen.

Example—A building has these specifications:

- Dimensions are 60 ft x 120 ft x 30 ft high.
- Walls are 8-in.-thick concrete.
- The longer walls have four windows 8 ft x 3 ft, and shorter walls have two windows 4 ft x 10 ft with corrugated asbestos awning covers and heavy drapes or shades.
- The roof is made of asphalt on 6-in. concrete.
- The shorter walls face east and west.
- There are 75 people employed in the building, with 50 engaged in light work and 25 in heavy work.
- The following sources generate electrical heat: twelve 1-hp electric motors and thirty 5-hp motors.

Determine ventilation by Heat Load method.

Solution—The hottest time of the day will occur around 4 p.m.; thus, calculate the solar heat gain for the west wall and roof. The average difference between indoor and outdoor temperatures is 7 F.

Area of concrete portion of wall: 60 ft x 30 ft	=	1800 sq ft
Area of windows: 4 ft x 10 ft x 2 windows	=	80 sq ft
Area of concrete - windows = 1800 - 80	=	1720 sq ft
Area of roof = 60 ft x 120 ft	=	7200 sq ft

Heat gains (Tables 4-4 and 4-5):

Through concrete: 1800 sq ft x 19 Btu/sq ft/hr	=	34,200 Btu/hr
Through windows (taken as corrugated asbestos rather than glass): 80 sq ft x 41	=	3,280 Btu/hr
Through roof: 7200 sq ft x 16.7 Btu/sq ft/hr	=	192,240 Btu/hr
From occupants:		
Light work: 50 people x 600 Btu/hr/person	=	30,000 Btu/hr
Heavy work: 25 people x 1400 Btu/hr/person	=	35,000 Btu/hr

From electric motors (Table 4-3)

12 x 1 hp x 3400 Btu/hr/hp	= 40,800 Btu/hr
30 x 5 hp x 3100 Btu/hr/hp	= 465,000 Btu/hr
Total heat gain	800,520 Btu/hr

Assume that inside temperature can exceed outside temperature by 7 F. Ventilation required will be:

$$\frac{\text{Total Heat Gain in Btu/sq ft/hr}}{\text{Temperature Difference} \times 1.08} = \frac{800,520}{7 \times 1.08} = 105,889 \text{ cfm}$$

Building has six 20 ft bays   6 units needed.

Therefore, 105,889 cfm ÷ 6 units = 17,648 cfm per ventilator.

Manufacturers' capacity tables will indicate the proper unit, in this case an 18,000 cfm unit. However it may be more desirable to use two 9000 cfm units per bay instead to get a more uniform air movement. In that case, twelve 9000 cfm units are needed.

The Psychrometric Chart [136]

Most pollution engineers sooner or later will be exposed to a project that requires a knowledge of terms and calculations common to industrial air pollution control. With the exception of a pencil and slide rule, the handiest tool for simplifying these calculations is a psychrometric chart. Psychrometrics, in the modern sense, means the evaluation of air properties and the processes which alter these properties. The field of psychrometrics has a special vocabulary often used in air pollution control work.

Absolute humidity or *humidity ratio* is the mass of water vapor per unit mass of dry air in a mixture of air and water vapor. This mixture is commonly called *gas.*

Relative humidity is the ratio of the partial pressure of the water vapor in a mixture to the saturation pressure of pure water at the same temperature.

Dry, DA, is the mixture of all the normal components of atmospheric air except water vapor.

Saturated air is a mixture of dry air and saturated water vapor or, alternatively, a mixture having relative humidity equal to 1.00.

Dewpoint temperature is the temperature at which the mixture, or gas, becomes saturated (or condensation begins) when a mixture of air and water vapor is cooled at constant pressure from an unsaturated state.

Dry bulb temperature, DB, is the actual temperature of the gas.

Wet bulb temperature is the temperature indicated by a thermometer having its bulb covered by a film of water, when the thermometer is exposed to an air-vapor mixture in turbulent flow.

Adiabatic saturation temperature is the temperature reached by an air stream after it has been saturated with water vapor with no sensible heat transfer. The wet bulb temperature and the adiabatic saturation temperature are numerically very close for air-vapor mixtures only. It is largely this fact that makes the wet bulb temperature useful. Adiabatic saturation occurs at constant enthalpy.

Humid volume is the volume occupied by one pound of dry air, with its water vapor, in a mixture.

A psychrometric chart graphically displays the above properties over a range of temperatures and humidities. Its great usefulness is in the fact that processes can be easily traced and computations are greatly simplified. A typical high temperature psychrometric chart is shown in Figure 4-17. The following examples explain the use of the psychrometric chart.

Example One

Convert 33,000 acfm at 300 F dry bulb and 0.10 lb of H_2O per pound of dry air to standard conditions (70 F and dry).

Solution—This problem asks, in effect, to determine the volume flow at standard conditions that will yield the same mass flow as the given conditions. This can be done by multiplying 33,000 acfm by the ratio of standard density to actual density. The density of air at standard conditions is 0.075 lb/cu ft. By referring to the psychrometric chart and schematic chart, Figure 4-18, it may be seen that, at 300 F dry bulb and 0.10 lb of H_2O per lb of dry air, the humid volume is about 21.5 ft^3/lb DA, therefore, the actual gas density is:

$$\frac{1 \text{ lb DA} + 0.10 \text{ lb } H_2O}{2.15 \text{ ft}^3/\text{lb DA}} = 0.0512 \text{ lb/ft}^3$$

$$\text{standard flow rate} = 33,000 \times \frac{0.0512}{0.075} = 22,528 \text{ acfm}$$

Example Two

Suppose the gas in Example One is sensibly cooled to 200 F DB. What is the new volume?

Solution—Since the gas is cooled without adding or condensing moisture, it must occur along a constant absolute humidity line, Figure 4-19. The new volume will be the initial volume multiplied by the ratio of final to initial humid volume:

$$\text{new volume} = 33,000 \text{ acfm} \times \frac{19 \text{ ft}^3/\text{lb DA}}{21.5 \text{ ft}^3/\text{lb DA}} = 29,163 \text{ acfm}$$

Example Three

How much water is required to cool the gas in Example One by adiabatic saturation, and what is the resultant gas volume?

Solution—For purposes of calculation, adiabatic saturation can be assumed to take place along a constant wet bulb line, Figure 4-20. The gas initially contains 0.10 lb, H_2O/lb DA, and after cooling contains 0.126

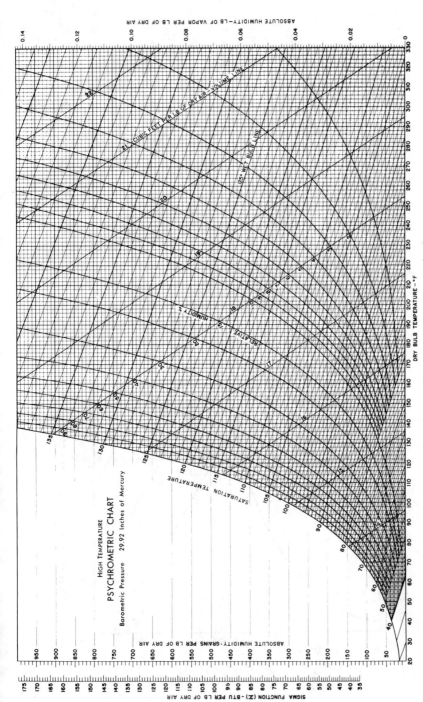

Figure 4-17. Typical high temperature psychrometric chart.

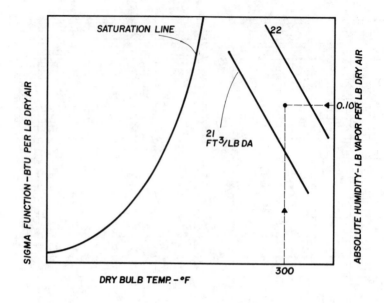

Figure 4-18

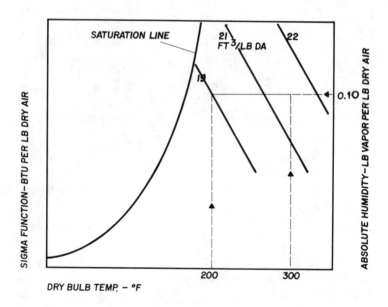

Figure 4-19

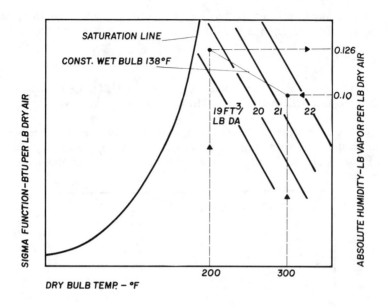

Figure 4-20.

0.126 lb H_2O/lb DA. Therefore, the added water is:

0.126 lb H_2O/lb DA - 0.10 lb H_2O/lb DA = 0.026 lb H_2O/lb DA

The mass flow rate is:

$$\frac{33,000 \text{ acfm}}{21.5 \text{ ft}^3/\text{lb DA}} = 1535 \text{ lb DA/min}$$

$$\text{water requirement} = \frac{1535 \text{ lb DA}}{\text{min}} \times \frac{0.026 \text{ lb } H_2O}{\text{lb DA}} \times \frac{\text{gal}}{8.3 \text{ lb } H_2O}$$

$$= 4.8 \text{ gpm}$$

$$\text{new volume} = \frac{33,000 \text{ acfm} \times 19.7 \text{ ft}^3/\text{lb DA}}{21.5 \text{ ft}^3/\text{lb DA}} = 30,237 \text{ acfm}$$

Example Four

What is the mass and volume flow rate of the water component of the gas mixture in Example One?

Solution—By referring to Example One, it is seen that each pound of dry air contains 0.10 lb of H_2O vapor. From Example Three, it was found that the mass flow rate of dry air was 1540 lb DA/min. Therefore:

$$\text{mass flow of vapor} = \frac{1535 \text{ lb DA}}{\text{min}} \times \frac{0.10 \text{ lb } H_2O}{\text{lb DA}} = 154 \text{ lb } H_2O/\text{min}$$

Since all components in a gas mixture occupy the same volume, the volume flow rate of water vapor is 33,000 acfm.

Example Five

How much heat energy must be added to 150,000 acfm of gas at 150 F DB and 0.02 lb H_2O/lb DA to raise the temperature to 300 F DB?

Solution—If the products of combustion are assumed to contribute little to the total mass, then the initial state can be represented by point A on Figure 4-21 and the final state is point B. Point B is determined by the

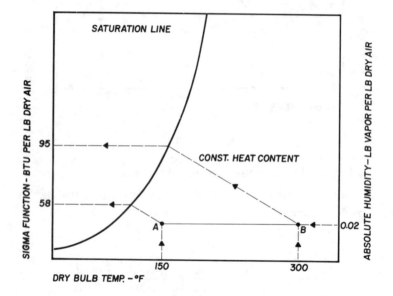

Figure 4-21

intersection of the required dry bulb temperature and a constant absolute humidity line. By referring to Figure 4-17, it can be seen that the heat energy content of the various combinations of gas mixtures is given by sigma function expressed in Btu/lb DA. The sigma function is analogous to enthalpy and differs from enthalpy by an amount equal to the enthalpy of liquid water at the adiabatic saturation temperature. In practice, either sigma function or enthalpy can be used for computation. Constant heat energy lines are approximately parallel to wet bulb lines.

heat energy at A = 58 Btu/lb DA
heat energy at B = 95 Btu/lb DA
heat energy added = (95 - 58) Btu/lb DA = 37 Btu/lb DA

$$\text{lb DA} = \frac{150{,}000 \text{ acfm}}{19.7 \text{ ft}^3/\text{lb DA}} = 7614 \text{ lb DA/min}$$

total heat added = 7614 lb DA/min x 37 Btu lb DA = 282,000 Btu/min

The above examples are only a sample of the many calculations needed by a pollution engineer. It should be apparent, however, that use of the psychrometric chart greatly facilitates these and many more calculations.

CHAPTER 5

AIR POLLUTION STACK TESTING
AND INSTRUMENTATION

INSTRUMENTATION FOR STACK MONITORING[21]

The Environmental Protection Agency has promulgated standards of performance for five categories of stationary sources, limiting gaseous and particulate emissions. So far, only new plants come under control, but it is expected that existing plants will soon receive the same attention and that other types of plants will be covered as soon as the necessary background information is developed.

In three of the original five plant categories, instrumentation for continuous monitoring is required. Operating experience with such monitors is limited and there is a bewildering array to choose from.

Affected Plants

Standards of Performance for new stationary sources were published by EPA in the *Federal Register*, Volume 36, Number 247, December 23, 1971. The plant categories affected by this law are:

1. Fossil-fuel-fired steam generating units of more than 250 million Btu/hr per input.
2. Nitric acid plants.
3. Sulfuric acid plants.
4. Incinerators of more than 50 tons/day charging rate.
5. Portland cement plants.

All five categories are required to undergo qualifying emission tests within 60 days of achieving normal production rate and no more than 180 days from initial start-up. These tests are performed by manual methods thoroughly detailed in the *Federal Register*.

In addition to the qualifying tests, the first three categories are required to have continuous emission monitoring instruments for visible solids and/or

225

sulfur dioxide and oxides of nitrogen. Specifically, they are:

Plant	Monitor
Steam generator (other than gas-fired	Smoke
	Sulfur dioxide (except where low sulfur fuel is used)
	Nitrogen oxides
Nitric acid plants	Nitrogen oxides
Sulfuric acid plants	Sulfur dioxide

Monitoring Instrumentation

The *Federal Register* does not specify the equipment to be used for emission monitoring. This leaves the pollution engineer in a quandary since he must choose from the rapidly increasing array of instrumental approaches and instrument types that are hitting the market.

There are three basic approaches to source monitoring. In the first, which may be called *extractive*, a continuous sample stream is drawn from the stack and transported to the analyzer, which can be mounted in any convenient location. This requires a probe mounted in the stack or duct, and some form of interface system to provide the analyzer with a sample that is in an appropriate state of cleanliness, temperature, pressure, and moisture content. This approach is the oldest and has provided the most experience to date.

The second approach may be called *in-situ* monitoring. The instrument is mounted either inside the plenum or just outside the stack. In the case of optical instruments, the source may be mounted on one side and the detector on the other, so that the instrument scans the full width of the stack. This method is the most common used for visible particulates or "smoke." A combination of the first and second methods is to mount the analyzer directly on the stack and draw the sample through it with little or no preconditioning. Obviously, this requires an instrument that can accept the sample in its natural state.

The third approach is to monitor the plume above the stack with a remote optical instrument. So far, this approach is in the research stage, whereas the other two methods have been reduced to a more practical state.

The in-situ across-the-stack approach and the remote method are claimed to offer an advantage over the extractive approach in that they provide an average reading rather than a point reading. However, it is theoretically feasible to use multiple extractive probes and obtain an integrated sample that is representative of the complete cross section.

In-situ monitoring eliminates the need for an interface sampling system. However, it has several inherent disadvantages. In-situ optical instruments involve the problem of keeping windows clean. This cleaning problem is minimized in dual wavelength instruments.

Table 5-1. Comparison of In-Situ and Extractive Type Air Pollution Monitors

Basic Approach	Analytical Principle	Number of Components per Analyzer	Measurable Components	Analyzer Price Range	Sample Handling Price Range
In-Situ	Optical	Single or Multiple	SO_2,NO_X	$10,000-$40,000	Not Required
In-Situ	Optical	Single	Smoke	$ 1,100-$ 5,000	Not Required
Extractive	Optical	Single or Multiple	SO_2,NO_X	$ 2,800-$ 6,000	$5,000-$8,000
Extractive	Wet Chemical	Single	SO_2,NO_X	$ 2,000-$ 6,000	$5,000-$8,000
Extractive	Electrochemical	Single or Multiple	SO_2,NO_X	$ 1,500-$ 3,500	$1,800-$3,000
Extractive	Chemi-luminescent	Single	NO_X	$ 5,000-$10,000	$3,000-$8,000
Extractive	Flame Photometric	Single	SO_2	$ 3,000-$ 6,000	$3,000-$8,000

Equipment Selection

When selecting monitoring equipment, it is wise to consider it as a complete system. Most instruments, however, are not offered this way. Performance specifications and prices are generally quoted on the basis of the analytical device alone. But what the pollution engineer needs is a complete system. Performance specifications and costs should be compared on this basis.

Site preparation and mounting provisions must be included, since they may be more involved for in-situ instruments than for the extractive type. Sampling systems for the extractive methods are normally part of the total expense and may affect total system performance. In-situ instruments must be more rugged to withstand the rigors of exposure. An inexpensive but frail instrument can turn out to be very expensive when mounted high on a stack. Above all, maintenance of the instrumentation system must be thoroughly considered.

STACK TESTING AND MONITORING[115]

Stack testing implies the determination of emission characteristics and quantities at one particular time. It is usually done with portable equipment and under close supervision to assure compliance with established procedures. For any given source, tests should be made at least twice: (1) to establish conditions *before* abatement equipment is installed, and (2) to determine the effectiveness of treatment *after* equipment has been installed.

Stack monitoring implies continuous determination of emission characteristics and quantities and usually requires permanent instrumentation and sampling systems. Its accuracy depends upon instrument maintenance and calibration and proper sample collection. Monitoring of emissions before installation of pollution control equipment is desirable, but may not be economically feasible. The equipment is relatively expensive and may not have a great enough range of reliable measurement capability for the emissions both before and after the abatement equipment has been installed.

Many states now require monitoring of emissions on a continuous or regular basis, with records kept and submitted to the controlling agency. Any periods of time that the emissions exceeded the allowable limits may have to be reported separately. The essential elements of source testing at a point are:

1. Measuring the total gas flow (volume).
2. Measuring the amount of sample withdrawn (volume).
3. Determining the amounts of specific contaminant materials collected in the gas sample.

The objective in making a test is to get an accurate, precise, and reliable determination of materials being released. A successful test program involves six basic steps—planning, preliminary evaluation, stack sampling, sample analysis, calculations, and report preparation.

Planning

Preliminary planning is essential to a successful source testing program. A checklist should be used to identify and record all pertinent information, including location, date, process, anticipated emissions, process operating and control equipment, types of tests, test conditions and equipment, and notes on discussions and observations of the site.

A stack data table should be filled out to describe the sampling locations. It may also be necessary to take samples in a duct leading to the stack or preceding a precipitator.

Preliminary Survey

The purpose of the preliminary survey is to prepare the way for the full crew required for the stack sampling tests. Typically, one or two men can make the preliminary survey so that the full crew of perhaps 5 to 10 men will have the right test equipment and be able to work at maximum efficiency. The preliminary survey team should get data on gas flow rate, moisture content, temperature, gas density, and some indication of emission level (order of magnitude). Instruments may be needed for these preliminary measurements (a pitot tube for gas flow, a condenser and calibrated flask for moisture content, a thermometer or thermocouple for temperatures, and an Orsat analyzer for gas density).

The survey team should also select the locations for the test ports and make sure that the test ports are prepared and accessible for the full crew when it arrives. Scaffolding may be required for access. There must be an agreement with the plant management as to who is responsible for providing access to the test points. Well-organized preliminary arrangements will go a long way toward assuring good performance of the test crew.

To begin a source sampling project, start with the selection of a sampling site in the stack or duct. Try to select a sampling site at least two stack diameters upstream from any flow disturbance such as a bend or expansion point. The site should also be at least eight diameters downstream from such a disturbance. If these conditions can be met conveniently, then sample the site at six points on each of two perpendicular diameters, as shown in Figure 5-1 (12 points total). This situation represents the minimum (or ideal) condition of stack sampling. The locations of the sampling (traverse)

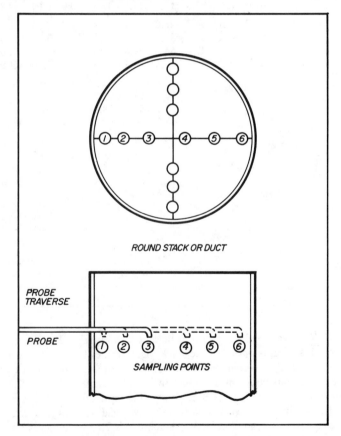

Figure 5-1. Ideal traverse sampling points in a round duct. It is necessary to make two traverses at right angles.

points are given in percentages of stack diameter from the inside wall. These locations are shown in Table 5-2 under the first column headed "6." The other columns give the locations of traverse points per diameter when the "ideal" conditions cannot be met. Additional points along the diameter must be sampled due to the flow disturbances closer than the desired limits.

Table 5-2. Location of Traverse Sampling Points in Circular Stacks (Percent of Stack Diameter Inside Wall to Traverse Point)

Point Number	Required Number of Traverse Points (Each Diameter) (Percent of Stack Diameter)									
	6 (Min)	8	10	12	14	16	18	20	22	24
1	4.4	3.3	2.5	2.1	1.8	1.6	1.4	1.3		1.1
2	14.7	10.5	8.2	6.7	5.7	4.9	4.4	3.9		3.2
3	29.5	19.4	14.6	11.8	9.9	8.5	7.5	6.7	6.9	5.5
4	70.5	32.3	22.6	17.7	14.6	12.5	10.9	9.7	8.7	7.9
5	85.3	67.7	34.2	25.0	20.1	16.9	14.6	12.9	11.6	10.5
6	95.6	80.6	65.8	35.5	26.9	22.0	18.8	16.5	14.6	13.2
7		89.5	77.4	64.5	36.6	28.3	23.6	20.4	18.0	16.1
8		96.7	85.4	75.0	63.4	37.5	29.6	25.0	21.8	19.4
9			91.8	82.3	73.1	62.5	38.2	30.6	26.1	23.0
10			97.5	88.2	79.9	71.7	61.8	38.8	31.5	27.2
11				93.3	85.4	78.0	70.4	61.2	39.3	32.3
12				97.9	90.1	83.1	76.4	69.4	60.7	39.8
13					94.3	87.5	81.2	75.0	68.5	60.2
14					98.2	91.5	85.4	79.6	73.9	67.7
15						95.1	89.1	83.5	78.2	72.8
16						98.4	92.5	87.1	82.0	77.0
17							95.6	90.3	85.4	80.6
18							98.6	93.3	88.4	83.9
19								96.1	91.3	86.8
20								98.7	94.0	89.5
21									96.5	92.1
22									98.9	94.5
23										96.8
24										98.9

When ideal conditions cannot be met, use Figure 5-2 to determine the number of sampling points in the stack. Determine the distance **A** upstream from the nearest disturbance in terms of duct diameters and drop a perpendicular from that point on the upper scale to the slanting line on the graph. Do the same for the distance **B**, running a line upward from the lower scale. Read the number of traverse points on the vertical scale and use whichever value is greater. In the sample, **A** is 1.5 and **B** is 7; **A** value gives about 24 points, while the **B** value gives only 18. The **A** value, being greater, will control the test (always use a multiple of four). With the number of traverse

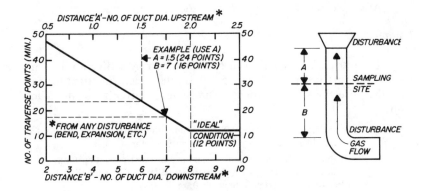

Figure 5-2. Determining the number of traverse sampling points
in a stack when ideal conditions do not exist.

points determined, use Table 5-2 to determine the locations of the points
along the perpendicular diameters.

If the duct is rectangular, first calculate the equivalent diameter from
the following equation:

$$\text{Equivalent diameter} = 2 \, \frac{\text{length x width}}{\text{length + width}}$$

Next, determine the number of traverse points for the sampling site
using the equivalent diameter and Figure 5-2, the same as for a circular stack.
Then, divide the rectangular cross section into as many equal areas as the
required number of traverse points, such that the ratio of the length to the
width of the elemental areas is between one and two. This is shown in
Figure 5-3 for the minimum case of 12 traverse points. The sampling location is at the centroid of each elemental area.

Figure 5-3. Divide a rectangular duct into
equal areas (same number of traverse
points); locate traverse points at the
centroid of each elemental area.

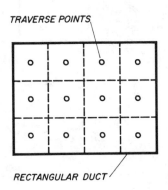

After selecting the sampling site, proceed as follows.

Equipment

The test crew begins the stack sampling test by setting up the sampling train at the first test port. Figure 5-4 illustrates a sampling train for particulates that meet EPA specifications. The equipment consists of the following items:

1. Probe, including nozzle, 1/8 to 1/2 in. (intercepts stack gases), heated sample line, stainless steel, with or without a glass liner.

2. Heated sample box, including glass cyclone and collection flask (collects particles down to 5 microns), filter holder with fiberglass filter (collects particles down to 0.3 microns).

3. Impinger box, with four impingers; No. 1 is a modified Smith-Greenberg unit containing 250 ml distilled water; No. 2 is a standard S-G unit containing 150 ml distilled water; No. 3 is a modified S-G empty; and No. 4 is a modified S-G containing 175 grams of silica gel; all connectors are glass. Impinger box is cooled by refrigeration unit or ice.

4. Separate parts—impinger thermometer (to insure that the gas sample is kept below 70 F in the impinger section), check valve, umbilical cord (connects sample box to metering box and can be 100 ft long).

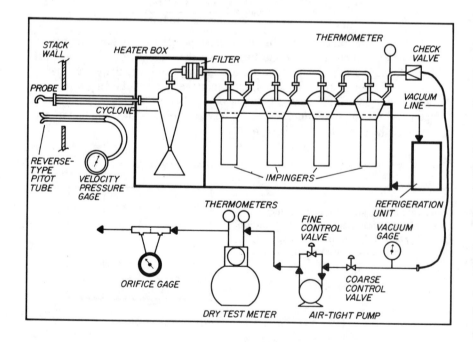

Figure 5-4. EPA-type sampling train.

5. Metering box, including vacuum gauge, main valve for coarse adjustment, vacuum pump, by-pass valve for fine adjustment, dry gas meter, orifice (exhausts to atmosphere) draft gauge (dry-type or manometer, to measure velocity pressure of the stack gas flow), thermometers, and draft gauge [dry-type or manometer, to measure pressure drop across orifice (measures sample flow)].

Procedure

In general, the sampling train is used only for collection of particular matter in the exhaust gases being tested. Gases such as CO_2, SO_2, H_2S, and nitrogen oxides consist of molecular size particles and do not condense in the equipment. Thus, they pass on through and are merely measured as a gas flow.

The sampling train is designed to intercept and collect particles down to at least 0.3 micron in size plus molecular constituents that will condense at normal atmospheric temperatures, 70 F. Thus, the sampling train consists of several components in series, each with a function to perform in the overall operation. These functions are: (1) intercept the stack gases in the flue; (2) collect the particulate matter by filters, cyclones, etc.; (3) condense and collect the condensibles; and (4) measure the flow of the residual dry gas.

In collecting particulate material, the ultimate objective is to get a measurement of the stack emissions to the atmosphere (pollutant mass rate) in units prescribed by the applicable codes (lb/hr, for example). This is done essentially by weighing the collected matter, measuring the total sample gas flow during the test, and dividing to get grains per cubic foot. This figure is then converted to pounds per hour by multiplying by the measured stack gas velocity.

Particulate sampling is done on sources where solid and liquid particles are being emitted. Particulate sampling involves the added complication of continuous stack gas velocity determination because of inertial effects of many particles. It is necessary to draw the sample gas into the sample probe at the same velocity as the average velocity of the flue gas stream at that point in order to obtain a representative sample. This condition is known as isokinetic sampling, and must be maintained throughout particulate sampling. At each point in the traverse, the sample flow rate must be adjusted to attain isokinetic conditions at the sample nozzle. Nomographs or other calculation aids are available to enable rapid determination of the required sample flow.

Gaseous sampling in flue gases is simpler than particulate sampling, principally because the gas molecules are small enough to be governed by the random nature of Brownian motion; inertial effects become insignificant. In continuous sampling it is necessary only to withdraw a sample from the flue at a known rate. The task of obtaining a representative sample, therefore, is considerably easier because the sampling rate can be independent of the velocity in the duct. This is true, of course, only for those pollutants that exist exclusively as gases. For pollutants such as mercury and fluorides which exist as both particulates and gases, an isokinetic sampling procedure must be followed.

Several methods are available for collection of gaseous constituents from flue gas streams. These include absorption into a liquid phase, collection in an evacuated container, collection in a flexible fabric bag, absorption on a solid material, and freezeout techniques. Each method is useful for particular application, depending on the temperature and moisture content of the flue gas, the material being analyzed, and the method of analysis used.

Monitoring

Since monitoring is optimally done on a continuous basis, efforts should be made to obtain representative samples of emissions. Static samplers draw from only one point in the stack or duct, so the optimum arrangement would be a traversing sampler.

Condensation of vapors presents a mechanical problem in sampling trains. Provisions must be made either to harmlessly condense the vapors but still account for them, or to prevent them from condensing until sample analysis is complete.

Monitoring devices for continuous measurement of particulate stack emissions fall into two categories: (1) those that sample the gas in the stack and measure the particulate concentration in the sample by such methods as beta radiation adsorption, frequency shift of a piezoelectric oscillator, and disturbance of an electrostatic field and (2) those that project a light beam of visible, infrared, or ultraviolet rays through the gas in a duct or stack and measure the attenuation or reflection of the beam by the particles in the gas.

Both types require frequent calibration and maintenance to insure reliable operation. For example, the lenses of the light beam attenuation types must be cleaned often so that dust settling on the lenses does not appreciably affect the instrument's operations.

Some of the commercial units available are:

> Beta radiation absorption: Research Applicance Co., Allison Park, Pa.
> Oscillator frequency alteration: Thermal Systems, Inc., St. Paul, Minn.
> Electrostatic field disturbance: Ikor, Inc., Burlington, Mass.
> Visible light attenuation: Lear Siegler, Inc., Englewood, Colorado
> Partial infrared spectrum: Bailey Meter Company, Wickliffe, Ohio

Measurement

Continuous measurement of gases, as opposed to analyzing grab samples, is not as far advanced. It is only in recent years that the manufacturers of these first generation instruments actually realized that the industrial environment is far more hostile and unpredictable than the laboratory one. Earlier experience has stimulated manufacturers to develop new techniques and methods to circumvent earlier application errors.

Over $500 million will be spent in this decade for measurement of air emissions. About 57 percent of the total instruments will be used for measuring stationary source emissions. About 13 percent will be for ambient air monitoring systems, and about 30 percent will be for instruments to check auto exhaust emissions. Expectations of these future needs have caused a rapid growth in both the number of new manufacturers and the techniques available to make a continuous measurement. Although the methods are continually changing, the basic principles of classification remain the same.

Classification and Selection

Classifying gas analyzers is usually done several ways: (1) selective or non-selective, (2) continuous or semi-continuous, (3) by operating principles.

An analyzer will be selective in analyzing a gas stream if it responds to only one specific component of the gas stream rather than detecting a physical property of the stream. An infrared analyzer is usually applied to monitor a single component in a process stream and within the restraints placed on its application, and should do so regardless of variations of other components in the same gas stream. However, a thermal conductivity analyzer measures the thermal conductivity of the total gas stream and is therefore non-selective since the thermal conductivity will vary as components in the gas stream vary.

Many analyzers give a continuously recorded or indicated picture of the pollutant concentration with time while others are time-dependent for analysis. The gas chromatograph is an example of a semi-continuous analyzer where column elution time of less than a minute to ten minutes is required before a recorded peak is complete.

A paper-tape monitor for hydrogen sulfide is another example of a semi-continuous analyzer. Here a time of 1 to 4 hours is required to determine the concentration of the sample. Other analyzers using thermal conductivity or non-dispersed infrared detectors are practically instantaneous on readout and are therefore classified as continuous.

The most common and probably preferred way of classifying process analyzers is on the basis of their operating principles. These involve the following properties:

1. Electromagnetic radiation
2. Chemical affinity or reactivity
3. Electrical or magnetic fields
4. Thermal or mechanical energy
5. Combinations or variations of these

In application, the final selection is based on several factors. Initial cost is always important. Plant familiarity with one type may influence the selection (favorably or unfavorably). Accuracy, reproducibility, and speed of response are as important as cost, particularly in control installations. Proper selection of the best analyzer for a given measurement requires a complete knowledge of the process variables such as compositions, temperatures, and pressures.

Selection is then based on sound understanding of the principles of operation of alternative instrument methods, and an equally sound understanding of the chemistry and operation of the processes.

Thermal Conductivity

This is one of the simplest methods for measuring the higher percentage concentrations of binary gases. Its principle of operation is based on the fact that each elemental gas has a characteristic ability to conduct heat at a different rate. A reference gas, usually the major constituent of the binary mixture, envelops one leg of a reference filament of a Wheatstone bridge. Another filament of the bridge circuitry detects the change in heat rate of the sample gas as a function of the bridge resistance.

The degree of unbalance between reference and sampling filament is converted to a quantitative percentage of the gas concentration. Common applications include: CO_2 in air, SO_2 in air, H_2 in air, H_2 in N_2, H_2 in O_2, O_2 in H_2, H_2 in Cl_2 and Cl_2 in air. Standard measurement ranges vary between 0 to 5 percent (V/V), up to 100 percent (V/V). Major manufacturers include: Beckman, Leeds & Northrup, Mine Safety Appliances, and Teledyne.

Infrared

Next to the gas chromatograph this is perhaps the most versatile and popular type of analyzer. Its operating principle is based on the unique absorption of electromagnetic radiation (infrared spectrum) for specific gases. The amount of absorption is proportional to the quantitative concentration of the gas. Although the electronic circuitry and detection components are somewhat complex, basically the reference gas is sealed in a membrane; the absorption of radiation by the reference gas is compared to the sampling gas stream. Common applications include: CO, CO_2, SO_2, CH_4, NH_3, and some hydrocarbons. Exceptions include O_2, H_2, N_2, and most diatomic gases, and the rare gases. Ranges vary from 0-10 ppm (V/V) up to 1000 ppm. Major suppliers include: Beckman, Leeds & Northrup, Mine Safety Appliances, and Teledyne.

Ultraviolet

Another widely used radiant energy analyzer is the ultraviolet unit that operates in the electromagnetic spectrum from 100 to 400 millimicrons. Its operating principle is similar to that of the infrared analyzers. Common applications include: NO_2, NO_x. Ranges vary from 0-10 ppm (V/V) up to 500 ppm. Manufacturers are: Beckman, DuPont, and Teledyne.

Colorimetric

This wet chemical method is a reference for SO_2 analysis. In this method a specific absorbent is used to form a stable nonvolatile complex with the gas to be analyzed and a subsequent reaction with an acid bleached compound gives a sensitive, specific, and temperature independent color reaction. The color reaction is identified by a dual beam visible light source with separate photocells. Other gas interferences such as SO_3, NH_3, CO are eliminated by the addition of a specific acid. Several applications include: SO_2, NO_2, H_2S, NO_x, HCHO, and total oxidants. Ranges vary from 0-1 ppm up to 0-10 ppm (V/V) and higher. There are numerous manufacturers; some are: Technicon, Beckman, DuPont, and Scientific Industries.

Chromatograph

Probably the most popular and widely used process analyzer today is the gas chromatograph. It is highly sensitive, extremely flexible to applications, and quite reliable with recent modifications to include solid state components. The gas chromatograph consists essentially of a six-part system: (1) carrier gas, (2) sample injection, (3) chromatograph column, (4) detector, (5) electronics, (6) recorder. A means of injecting a fixed-volume of sample into the flowing carrier is usually done by using linear or rotary valves that trap the sample in a void or tubing length and place it in the flowing carrier stream.

The heat of the chromatograph is the chromatographic column. Its function is to separate the sample components and thus permit the carrier gas to elute the component of interest to the detector as a binary (mixed with the carrier gas).

The detector monitors the gas discharged from the column. The most common detector system is of the thermal conductivity type. Flame-ionization detectors find increasing use in chromatographs because of the extreme sensitivity and some degree of selectivity. Thermal conductivity and flame ionization detectors are probably used in 96 to 98 percent of the process units in service today.

Common gas analysis applications include those mentioned previously plus many more. Ranges vary from a few ppb up to 5000 ppm (V/V). Some manufacturers are: Beckman, Phillips Electronics, and Honeywell.

Chemiluminescent

A chemiluminescent reaction of NO with ozone and subsequent optical detection has added new expectations for development. No reactant gas or wet chemicals are required, yet sensitivity in the ppb range makes this a promixing instrument for stationary-source monitoring. Common gas analysis: NO, NO_x. Ranges: 0-5 ppm (V/V) up to 1000 ppm (V/V). Manufacturers: Aerochem, and Combustion.

Table 5-3. Operating Principles and Costs of Gas Analyzers

Analyzer	Action	Operation	Operating Principle	Advantages/Disadvantages
Thermal Conductivity	Nonselective	Continuous	Measurement of thermal conductivity of gas stream	Inexpensive/Limited to binary streams
Nondispersive Infrared	Selective	Continuous	Infrared energy absorbed by component of interest	Versatile, fairly well tested, sensitive, responsive/Frequent zeroing and calibration, sensitive to water vapor
Nondispersive Ultraviolet	Selective	Continuous	Ultraviolet energy absorbed by component of interest	Sensitive/Limited use
Colorimetric	Selective	Continuous	Visible energy absorbed by material of interest	Simple/Calibration standard problem
Chromatograph	Selective	Semicontinuous	Separation of components by chromatographic column	Versatile, sensitive, very selective, multicomponents, thoroughly tested/Cyclic, difficult readout
Chemiluminescent	Selective	Semicontinuous	Optical detection	Very sensitive, no reaction or wet chemicals/Simple hardware wise, specific applications, very new, untested
Coulometric	Selective	Continuous	Titration of selected titrant to generate ionic current	Sensitive/Slow response, reagent inventory
Flame Ionization	Selective	Continuous	Ionization potential change when burning hydrogen in air	Sensitive, maintenance, reliability good in ambient conditions/Limited range, requires hydrogen generator
Spectrophotometer Visible, IR, UV	Selective	Semicontinuous	Optical comparison of radiation absorption	Versatile, sensitive/Primarily lab instrument, difficult readout, expensive
Paramagnetic Properties	Selective	Continuous	Attraction to magnetic field	Sensitive/Gases only, O_2 primarily
Selective Membrane	Selective	Continuous	Gas molecular bond rupture in proportion to concentration	Inexpensive, good selectivity/Not well tested
Electrochemical	Selective	Continuous	Electroxidation or reduction in a sealed unit	Good selectivity, sensitive/Slow response
Paper-Tape	Selective	Semicontinuous	Color change on a tape impregnated with lead acetate	No wet chemicals/Slow response

Coulometric

Coulometry is based on the principle of electrically generating a selected ion in a titration cell. The amount of current required to generate sufficient ions to maintain a zero reference value is directly proportional to the reduction of ions caused by the reactable gas (SO_2, H_2S, etc.). Applications include: SO_2, H_2S, CH_3SH, $(CH_3)S$, $(CH_3)_2S_2$. Ranges vary from a few ppm to 1000 ppm.

Flame Ionization

This method is primarily used to determine the concentrations of hydrocarbons present in a gas stream or ambient air. The flame formed when hydrogen burns in air contains a negligible number of ions. The introduction of traces of hydrocarbons into the flame results in a complex ionization, producing a large number of ions. A polarizing voltage applied between the burner jet and the collector produces an electrostatic field in the vicinity of the flame. This field is measurable with an amplifier circuit and is directly proportional to the hydrocarbon concentration. Major applications are hydrocarbons, inert gases, CH_3, primarily compounds with C-H bonds. Rangeability varies between 0-5 ppm and up to 0-25 ppm. Some suppliers are: Beckman and MSA.

Spectrophotometer

A complex series of lenses and filters transmit radiant energy of specific wavelength (visible, ultraviolet, infrared) to a gas sample where the absorption of radiant energy causes an imbalance between sample and reference beams which gives rise to an a-c electrical signal from the detector. This signal is amplified and powers a sensitive servomotor which drives a precision optical attenuator into the reference beam until the beams are of equal intensity. The recorder pen is directly coupled to the attenuator and thus directly records the sample absorption on a linear transmittance scale. Applications include SO_2, NH_3, and many derivates of phenols, alkenes, ketones. Rangeability from 1 ppm scale up to a factor of 10 ppm (V/V). Manufacturers are: Beckman and Phillips Electronics.

Paramagnetic Properties

Oxygen is paramagnetic; that is, it is attracted by a magnetic field. This paramagnetic property of oxygen, caused by its atomic and molecular structure, is inversely proportional to its absolute temperature. When oxygen is heated, it loses its paramagnetic property and becomes diamagnetic (repelled by a magnetic field). Successive heating and then cooling of a precision resistor of a Wheatstone bridge provides a current signal proportional to oxygen content of the gas. Prime use is for O_2; range 0–0.5 percent up to 100 percent. Manufacturers include: Bailey Meter, Beckman, L & N, and MSA.

Electrochemical Membrane–Type Polarographic Sensor

These small, inexpensive cartridge sensors have an excellent potential. The operating principle is a liquid-state nonohmic variable resistor in which a pollutant-selective activating surface ruptures the gas molecular bonds, releasing energy as a voltage signal proportional to the pollutant concentration. With sensitivities of 10-15 ppb, full scale ranges from 0.1 ppm to 0-10,000 ppm are available. Major applications include: NO_x, NO_2, SO_2, H_2S, and total sulfur. The manufacturer is Envirometrics.

Electrochemical Transducer

The sensor consists of an electrochemical cell which is covered with a membrane having a high permeability for SO_2. As SO_2 diffuses through the cell, an electrochemical reaction takes place which produces an electric current directly proportional to its concentration. The cell is impermeable for ions and large molecules. The cell does not respond to gases such as O_2, CO_2, CO, NO_2, O_3, or Cl_2. Recent applications are: SO_2, NO_x, NO_2, Cl_2. Range varies from 0-5 ppm (V/V) up to 5000 ppm. Major suppliers are: Dynasciences, and Theta-sensors.

Paper-Tape

When lead acetate-impregnated tape is exposed to H_2S, a reaction occurs causing the tape to turn black. The density of the black spot is measured in a transmission photometer. This measurement gives the average H_2S concentration for the sample period. Major applications include: H_2S, fluorides. Range varies between 0-1 ppm up to 0-20 ppm (V/V). Major suppliers include: Research Appliance Company.

SAMPLING TECHNIQUES[181]

To prove something exists, it should be capable of being measured. The biggest problems in existing environmental study are how to measure the amount of pollutants and the effectiveness of reliable and economical control. It can be frustrating to find that both money and effort were wasted because the initial parameters and results were wrongly measured.

The principal mistakes made in measuring are in using:

1. Improper measuring devices or apparatus,
2. Faulty equipment,
3. Wrong measurement techniques and methods,
4. Unskilled personnel, and
5. Improperly reported test data.

Problems and Errors

Results from wrong measuring and handling of equipment can be extremely misleading. For example, when an aspirator air or vacuum pump valve is closed before the ball valve (the valve located before the orifice, or shortly after the thimble holder), the negative pressure will draw air backwards through the sampler and draw dust out of the filter. Also, when a test probe is removed from a duct with the filter case tipped downward, dust falls off the filter and out the nozzle, giving an incorrect result.

When a probe is inserted into a test port, excess dust can enter the nozzle before the vacuum pump starts to pull dust-laden gases if the nozzle touches the duct wall. Results therefore can show more dust than is really present in the gas stream.

When gas is sampled from the top of a duct, the sampler nozzle can touch or go too near the dust layer at the bottom of the duct. This will cause the nozzle to draw extra dust into the sampler. Dust in the bottom of a duct can often be several inches deep.

Filters and thimbles may often have dust remaining on them from previous tests, thus causing errors. It is also important to remember that during a test, especially a short run, a large part of the dust may settle on the walls of the nozzle tube and thimble holder, and not be measured as part of the dust load.

There are only a few examples of what may happen if equipment is not properly handled. But even if every safety precaution is taken, results of measurement can be faulty and misleading. One of the most popular dust load measurements used is placing a nozzle in the dust-laden gas stream and drawing a known volume of gas through the nozzle into the sampling thimble. This dust is then measured and compared to the total dust-laden gas system.

Depending upon the gas velocity difference between the nozzle and duct (stack), more than a 100 percent error can be made. To eliminate this type of mistake, the "null" nozzle can be used. However, this nozzle may also give sampling errors if the velocity difference is larger than 1-2 in. water and the duct velocity is lower than 3000 fpm (Figure 5-5). Assume a test is made with known dust load, gas volume, and nozzle and stack gas velocities.

$$W_R = \text{Weight ratio (Error ratio)} = \frac{M_D}{C_D}, \text{where}$$

M_D = Measured dust sample, grains

C_D = Calculated dust, grains (how much it should be according to the sample volume if total dust load is known, where
$C_D = D_L \times M_V$

D_L = Known or total dust load, gr/cu ft

M_V = Measured volume, cu ft, where
$M_V = V_X \times A_N \times t$

V_X = V_N or V_S depending on load calculation based on nozzle or stack (duct) gas velocity, fpm

t   = Sampling time, min

A_N = Area of nozzle, sq ft

EXAMPLE: (V_N = 2000 fpm   t = 1 min

Assume   (V_S = 3000 fpm D_L = 6 gr/cu ft
(A_N = 0.001 sq ft M_D = 19 gr

M_{VN} = 2000 x 0.001 x 1 = 2 cu ft
M_{VS} = 3000 x 0.001 x 1 = 3 cu ft
C_{DN} = 2 x 6 = 12 gr
C_{DS} = 3 x 6 = 18 gr

$$V_R = \frac{V_N}{V_S} = \frac{2000}{3000} = 0.666 = \text{Velocity ratio}$$

$$W_{RN} = \frac{M_D}{C_{DN}} = \frac{19}{12} = 1.584 \text{ mistake = plus 58.4 percent}$$

$$W_{RS} = \frac{M_D}{C_{DS}} = \frac{19}{18} = 1.055 \text{ mistake = plus 5.5 percent}$$

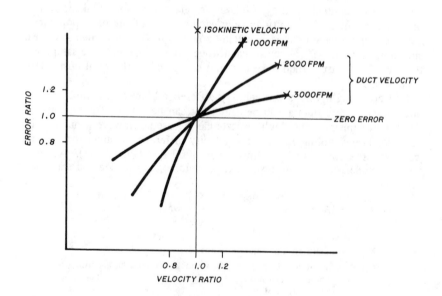

Figure 5-5. Sampling error ratios possible with low duct velocity using null nozzle.

The best method of measurement, of course, is when the sample is taken at the isokinetic velocity. However, in most cases this is very difficult. Error is also dependent on particle size and nozzle diameter.

$$\text{Ratio} = \frac{\text{Measured Dust}}{\text{Calculated Dust}}$$

Nozzle Inside Dia, In.	Particle Size, Micron	
	5-25	420-500
1/8	1.0-1.04	0.78-0.80
1/4	1.0-1.03	0.84-0.90
3/8	1	1

Therefore, it is to be remembered when sampling a gas stream that the accuracy of particle or gas sampling depends upon: (1) difference of stack and nozzle gas velocities, (2) particle size, (3) nozzle diameter, (4) number and length of sampling (5) division of gas flow area, (6) handling of equipment and (7) accuracy of equipment and report of test.

$$\text{Smaller nozzle velocity } (V_R = \frac{V_N}{V_S} = 1)$$

produces larger errors than a bigger nozzle (V_N) gas velocity, compared to stack (duct) gas velocity. Larger duct or stack gas velocity (3000 fpm or over) produces smaller error. Finer dust (5-25 micron), combined with high velocity, is more sensitive to deviation from isokinetic velocity. If particle size is very fine, 5 microns, sampling error is smaller if volume calculation is based on nozzle velocity instead of stack gas.

TAPE SAMPLERS[83]

Automatic instruments using filter paper tapes to sample particulate matter in ambient air have been in use for 20 years. Present state-of-the-art in sophisticated tape instruments makes them some of the most adaptable air sampling devices available today. Modified tape units can sample certain toxic gases in addition to particles as small as 1 micron in diameter. These instruments combined flexible design, modular subsystems, and portability to provide optimum versatility for a broad range of air sampling applications.

Automated filter tape air samplers are sensitive, timer-controlled, portable instruments that operate with a very high degree of repeatability. They can sample or monitor all types of particulate matter and certain gaseous pollutants in ambient air. Designed to collect cyclical samples of polluted atmosphere, and to operate unattended for extended time intervals, these tape instruments perform widely varied indoor and outdoor air sampling applications. When used outdoors, the instruments should be suitably enclosed for weather protection, and to discourage tampering.

Developed early in the 1950s, the tape sampler concept has evolved from a rather primitive device to a fully automated air sampling system. A tape instrument's collecting efficiency, including the particle sizes retained, is controlled by the type of filter medium used. The most commonly used filter paper tapes have a 3.4-micron pore size, which will retain most particles down to 1 micron or less, at the onset of sampling. Efficiency in small particle retention increases with continued filter loading.

Tape instruments have been adapted to sample hydrogen sulfide (H_2S) and fluoride (F) gases in the parts per billion (ppb) range. This adaptability for sampling gaseous pollutants is a very significant advantage. Research efforts currently are focused on developing new types of special filter tapes to expand the range of toxic gases that can be evaluated by means of tapes.

The most common applications for filter tape units involve continuous monitoring of atmospheric particles on an hourly or other cyclical basis. They also are widely used with an integral alert-alarm system when monitoring particulate matter or hydrogen sulfide in the air. These go into the alarm mode immediately if a preset, adjustable, ambient level of particles or H_2S is reached during any sampling sequence. Instruments with this alarm system also can send an automatic signal to activate remote audio-visual warning devices or process controls for remedial action.

Special applications for tape units are numerous. Tape instruments have visually demonstrated the performance of a dust collecting system by taking simultaneous air samples on both sides of the collector unit. They also have been used to demonstrate the overloading of air conditioning units in smoke-filled rooms, to plot automotive traffic patterns, to verify the correlation between wind direction and ambient pollutant levels, and to make other specialized ambient air evaluations.

Filter tape sampling instruments actually perform two distinctly different functions. In the simplest mode, these automatic instruments only sample ambient air. In this context, sample means that the instrument collects successive air samples during its preset cycle, and retains them passively. The tape must be removed and analyzed by secondary equipment or laboratory techniques to ascertain its pollutant content.

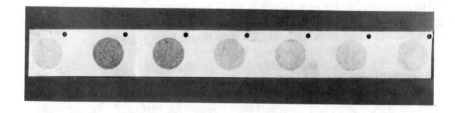

Figure 5-6. Filter tape section showing varying density of ambient particulate matter during portion of overall sampling cycle.

When a tape instrument operates as a monitor, in comparison, it not only takes cyclical air samples but also uses a built-in densitometer to evaluate each sample spot, and an integral output signal to telemeter quantitative measurements to a remote location. Typical remote pickup systems include indicating-recording instruments, central data collection stations, computer interfacing, alarm systems, or corrective control systems. A monitor's output signal usually can be adapted to any suitable secondary function that the user desires.

Cycles for sampling with filter tape instruments are regulated by a timer mechanism, which is determined by the unit's intended application. On units designed specifically as H_2S alarm monitors, the standard cycle is 60 minutes, with sampling sequences adjustable in one-minute increments. The timer mechanism furnished with standard sampler units provides ½, 1, 2, and 4-hr cycles. Another standard timer, supplied with the various monitor configurations, permits sampling cycles from 10 min to 3½ hr, in 10-min increments. Other optional timer systems can be adapted.

Regardless of their function, all filter tape instruments operate in the same manner. They use an electrically driven continuous-duty vacuum pump to obtain a measured volume of ambient air at timer-controlled intervals. This air sample then is drawn through a precision nozzle that clamps and shrouds the filter paper tape. As the air sample passes through the tape, entrained particulate matter or gases are deposited in a circular spot one inch in diameter.

Immediately after each sample is taken, the tape is automatically indexed to put a fresh section in the nozzle for the next sequence. Punched tapes are used to assure positive indexing throughout the fully automated cycle.

To prevent contamination of fresh tapes by ambient air, tape sampling instruments use clean, filtered air from the vacuum pump to maintain a slight positive pressure in the enclosed front compartment, which contains the tape spools, nozzle, densitometer (monitors only), adjustable timer, and other primary controls.

Pollutant Evaluation

Particle samples collected on tapes may be evaluated by two different techniques. One uses light transmitted through the sample spot, the other light reflected from the spot. In both techniques, the degree of soiling by particulate matter is determined by comparing the light reading of a sample spot (from 0 to 100 percent light transmission or reflectance) with a reading of 100 percent transmission or reflectance taken from the clean tape sections adjacent to the sample spot.

The light transmission technique provides readings based on the Coefficient of Haze (COH) values. The percent of light transmitted through a sample spot can be adapted to COH units by calculations or by means of conversion charts. This technique is widely used, and is becoming the

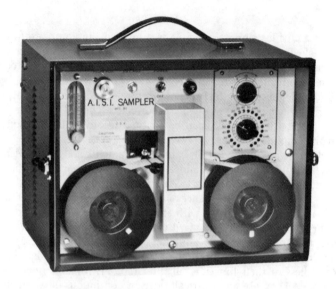

This automatic sampler collects ambient particulate samples on a 100-ft roll of filter paper tape. Its timer mechanism provides sampling cycles ranging from 10 min to 3½ hr, in 10-min increments. Unit has a push-to-test flowmeter that permits quick spot-checks of flow rates and features a special bypass design that prevents particle accumulation in the rotameter.

This sampling instrument has an integral densitometer that evaluates percent of light transmitted through the filter tape after each cyclical air sample is taken, and an inkless strip-chart recorder to log densitometer readings (from 0 to 100 percent transmission). Nozzle design positions both densitometer lamp and photocell outside of air sample stream, preventing their contamination by entrained particles. For evaluation, recorded densitometer measurements are converted to COH values by calculation or use of conversion charts.

Figure 5-7. Particulate samplers.

Figure 5-8. Reflectance-Transmission Particulate Monitor. Equipped with 600-ft rolls of filter tape, this automated instrument operates unattended for long periods of time, and takes dual readings on particulate samples. Its design combines dual light-sensing heads, providing both light transmission (COH) value) and light reflectance (RUD value) measurements on each sample spot. It also has two separate 0-10 mV output signals for telemetering, one for each type of sample measurement.

Figure 5-9. H_2S Alert-Alarm Monitor. Designed primarily for personnel protection in hazardous work areas, this totally enclosed instrument (purged with an inert gas) is used in explosive or flammable environments. This unit automatically goes into an alarm mode if ambient H_2S exceeds an adjustable, preset level during any sampling sequence.

Figure 5-10. Fluoride Sampler. This dual-tape instrument takes simultaneous samples of particulate and gaseous fluorides. Equipped with two different impregnated filter tapes (100-ft lengths), the unit collects particle samples on the upper tape, gaseous samples on the lower. After sampling, the tapes must be chemically processed to develop the sample spots, which then can be evaluated by spot evaluator instruments using the light transmission technique to obtain readouts in COH values.

preferred method since COH units now are a recognized measurement. In some instances, COH values are written into air pollution codes as an index of contaminant concentrations.

The reflected light technique provides readings based on Reflectance Unit Density (or dirt shade) values known as RUD units. If sampling times and rates remain constant, tape instruments can be equipped with meters that read directly in RUD units. Otherwise, calculations or conversion charts are used to adapt sample light reflectance readings, ranging from 0 to 100 percent, to RUD units. Tapes containing samples of particulate matter also can be evaluated by spectrographic analyses for a wide range of elements.

The filter tape monitor principle has now been combined with a beta radiation gage to develop an automatic AISI stack monitor. This advanced system automatically and continuously samples, measures, and records mass particulate emissions in stacks. Samples of particulate matter are collected on a filter tape and "read" by a beta gage. The quantitative evaluations then are transmitted to a recorder in the system's remote master control console.

Tapes presently used to sample gaseous pollutants, with one exception, must be processed before they can be evaluated. Chemical techniques are employed to develop the sample spots deposited on the tapes, which then can be evaluated by either light transmission (COH) or light reflectance (RUD) methods.

A lead acetate impregnated filter tape is used to sample hydrogen sulfide. Since the H_2S reacts with the lead acetate, it produces stains on the tape comparable to the spots obtained with particulate matter. The stain is evaluated by light transmission and compared to a curve to determine the H_2S concentration.

A mercuric chloride tape that also is used to sample H_2S must be exposed to ammonium hydroxide fumes to develop its sample spots. Wet-chemical methods are required to develop sample spots on a sodium hydroxide impregnated tape used to sample gaseous fluorides, and also on citric acid impregnated tapes that sample particulate fluorides.

Fluctuations in pollutant concentrations that occur during sampling cycles, and the rates of change, are determined by the location of consecutive sample spots on the automatically cycled tape.

Figure 5-11. Manual Spot Evaluator (COH). This manually operated spot evaluator uses light transmission to measure particulate or gas sample spots on filter tapes. If a 2-hr sampling cycle has been used, sample spots can be read directly in COH values. For other length sampling cycles, calculations or conversion tables are required. Tapes can be read in either direction, at operator's preference.

Limitations

Although tape instruments consistently collect samples of ambient pollutants, they are not suitable for air evaluations based on volumetric or mass measurements, such as the measurement of a certain weight of particulate matter in a specific volume of air. The same limitation applies, with occasional exceptions, when very high concentrations of particles or H_2S are present in process emissions. For these types of applications, more specialized air sampling instruments (like the Hi-Volume and Total Particulate Membrane Samplers) must be utilized.

These inherent limitations do not seriously restrict the versatility of filter tape air samplers and monitors. No instrument conceived to date can perform all possible air sampling functions. Virtually no other design offers adaptability comparable to that of the tape instruments, which can provide a number of performance characteristics, in addition to their overall sampling capabilities, through the use of modular subsystems and accessories.

CHAPTER 6

FUME COMBUSTION AND INCINERATION

FUME INCINERATION[128]

Objectionable odors are discharged to the atmosphere from commercial and industrial processes in the form of gases, mists or solids. Hydrogen sulfide, carbon disulfide, mercaptans, products from the decomposition of certain proteins, and petroleum hydrocarbons are common malodors. Besides these, there are a great many different kinds of odors that are potential nuisances.

A number of exhaust control methods have been successfully developed for applications where the pollutant is organic and in the form of a fume or gas. Organic particulates can also be handled with special consideration, depending on the application.

Incineration is an air pollution control process in which objectionable organic vapors or organic particulates are converted to harmless carbon dioxide and water vapor. Organic emissions are destroyed by exposure to high temperatures. Methods of incineration include: direct-flame, catalyst and direct-combustion. In direct-flame and catalyst incineration the concentration of contaminants must be well below the lower explosive limit. The concentration of organics is not usually permitted to exceed 25 percent of the lower explosive limit for the material. For toluene, used in metal finishing, concentrations should not exceed about 3000 ppm by volume. If the contaminant air mixture is in the flammable range, direct-combustion can be used such as in a flare.

Direct-Flame Fume Incineration

In direct-flame incineration the organic emissions in concentration well below the lower explosive limit are destroyed by exposure, under the proper conditions, to temperatures of 900 to 1400 F, in the presence of a flame. The actual temperature required to do an effective job depends on the specific pollutants involved and the design of the combustion chamber.

Direct-flame incineration is also referred to as afterburning, direct-flame oxidation, thermal oxidation, or thermal incineration.

The presence of a flame is important for contaminant removal. Evidence indicates that when using electric heat energy, much higher temperatures are required—1500 to 1800 F—to obtain the same efficiency achieved with a direct-flame system at 1000 to 1400 F. If satisfactory incineration is to be achieved at the lowest possible temperature, the type of flame and the design of the combustor are also important factors to be considered.

Because direct-flame incineration involves heating exhaust gases to high temperatures, heat recovery equipment to reduce fuel costs can usually be justified. The level of heat recovery selected will depend upon acceptable operating costs in each case. Once it has been established that direct-flame incineration is required for pollution control, heat recovery equipment makes operating costs reasonable and capital investment can be paid out in a relatively short time. In some cases this helps to justify direct-flame incineration as a practical solution to the problem, since average installed system payouts of one to three years can be obtained.

Direct-flame incineration can be highly effective. Experience has shown that direct-flame incineration systems can operate continuously at efficiencies of 90 to 99+ percent. Such systems can be readily adapted for automatic temperature control. This is an important consideration in processes where 100 percent continuous pollution abatement is required in spite of changing conditions.

The basic variables affecting the design of a direct-flame incinerator are:

1. Incineration temperature
2. Length of time the contaminant air is in contact with the flame and held at a specified temperature,
3. Amount of turbulence or mixing designed into the combustor.

Temperatures of 850 to 1500 F, velocities of 15-25 ft/sec and residence times of 0.30 to 0.50 seconds in general have been found to give satisfactory cleanup for most installations.

In order to properly design an effective direct-flame fume incineration system, the following information is required: flow to be handled—scfm; temperature and pressure of gases to be handled; list of contaminants involved—type and concentration; deposit problem, if any; fuel available— natural gas or oil; cost of fuel; number of hours of plant operations; and an indication if heat energy can be used elsewhere in the plant.

In most applications a simple flow diagram helps to ensure that the best and most practical fume control system is being chosen. Heat recovery equipment can usually be applied to direct-flame incineration systems in several ways. Equipment can be used to reduce the fuel input required to the combustor by preheating exhaust fumes entering it (Figure 6-1). In this system, the heat exchanger preheats contaminated waste gases before they enter the combustor and residence chamber where the pollutants are

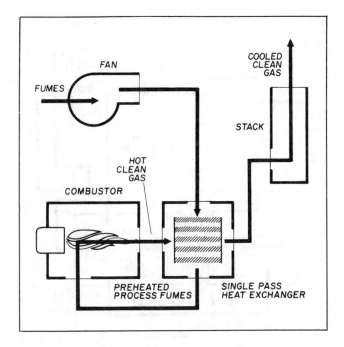

Figure 6-1. Forced draft direct-flame fume incineration system with a single pass primary heat exchanger

destroyed. The clean high-temperature gases discharging from the combustor are directed back through a heat exchanger to preheat additional contaminated gases before they enter the combustor. Heat recovery equipment can also be used to reduce process fuel requirements such as providing preheated air for a dryer.

Catalytic Fume Incineration

In catalyst incineration the presence of a catalyst allows the direct-flame oxidation process to proceed at a lower temperature and in the absence of a flame.

Incineration temperature for a catalyst system is the temperature out of the catalyst bed and ranges from 600 to 1000 F. If a catalyst system operates satisfactorily at about 600 F, it would probably be difficult to justify using heat recovery equipment. If incineration temperatures of 800 to 1000 F are required, the application of heat recovery equipment should be considered.

A typical catalytic-type incinerator with heat recovery is shown in Figure 6-2. The unit shown is operating with a preheat temperature of 900 F and an incineration temperature of 1000 F. The 100-degree rise in the catalyst bed is a result of the energy released from the contaminant being oxidized.

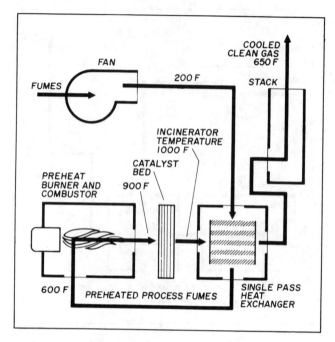

Figure 6-2. Catalytic-type fume incineration system with heat recovery.

Catalytic systems should not be used where poisons, suppressants, or fouling agents are present in the exhaust stream. A list of typical contaminants for the platinum family catalysts is shown in Table 6-1. In addition, no significant amounts of solids should be present. Direct-flame incineration should be used when solids or catalyst poisons are present.

Table 6-1. Poisons, Suppressants and Fouling Agents
for Platinum-Family Catalysts

Poisons	Heavy metals
	Phosphates
	Arsenic
Suppressants	Haolgens (both as elements and in compounds)
	Sulfur compounds
Fouling Agents	Inorganic particulate
	Alumina and silica dusts
	Iron oxides
	Silicones

In catalyst systems preheat temperature and space velocity through the bed of the catalyst are important variables affecting efficiency. Efficiency capabilities of 85 to 92 percent have been reported for properly maintained catalyst systems.

The Los Angeles County Air Pollution Control District has probably had more experience with incineration than any other district or institution in the country. When Rule 66 was passed to control organic solvent emissions it stated that, if incineration was to be used, the control system must have an efficiency of not less than 90 percent. In response to questions concerning Rule 66, the Los Angeles County Air Pollution Control District published a series of questions and answers. Question 45 read as follows: "Is catalytic incineration a satisfactory method of complying with this rule?" *Answer*: "Any air pollution control equipment capable of reducing the organic materials to the required quantities is acceptable. The catalytic incineration devices that have been tested so far by the Air Pollution Control District are judged to be incapable of meeting the requirements of Rule 66." As a result of this experience, fume incineration equipment installed in Los Angeles to comply with Rule 66 have all been direct-flame systems.

Direct Combustion (Flare)

In direct-combustion the organic emissions in concentrations in the flammable range are destroyed by burning as in a flare. This is the least costly form of incineration since the contaminant organic being emitted is used as the fuel. If process conditions fluctuate, an auxiliary fuel is usually made available to maintain a flammable mixture in the event the contaminant concentration drops below the lower explosive limit.

Effectiveness

After equipment is installed, its effectiveness can be measured in two ways—analytically by chemical or instrument analysis, or subjectively by use of an odor panel.

An acceptable method of analysis which measures specific contaminant concentration in and out is satisfactory; however, one must take into account any products that could be formed from combustion reactions such as aldehydes and other oxygenated derivatives or hydrocarbons which can be objectionable. The Los Angeles County APCD analysis method accounts for every carbon atom in the effluent but does not attempt to identify them. The sampling equipment required is relatively easy to use, and the cycle times required for complete analysis are reasonable—about two hours.

For an odor problem such as in a wire enameling plant, the odor panel method of quantitatively measuring odor is useful in determining the effectiveness of fume control equipment. As a guideline, if the odor strength of the effluent in a stack is reduced to less than 150 odor units per cubic foot, preferably in the range of 25 to 50 units, odor nuisances in a

community can be prevented. Odor units can be defined as the number of
dilutions with clean air required for a sample so that 50 percent of the odor
panel members do not detect any odor in a diluted sample.

The guideline of 25 to 50 odor units/scf can be refined by converting
the odor strength measurement (in odor units/scf) to an odor emission rate
(in odor units/min) taking into account the volume rate of stack exhaust
(in scfm). Using this odor emission rate as a guideline, it appears that up
to about 1 million units per minute is acceptable to avoid odor complaints
from a single stack or from the combined effluent from a group of stacks.
In some cases, however, other factors might have to be considered such as
stack height, distance to residences, topography of the area, discharge
velocity and meteorological conditions.

Rich Fume Incineration[89]

Since the nature of off-gases varies from one industry to another, one
single method of fume incineration cannot be equally efficient in all cases.
At the same time, one incineration process cannot be developed which is
tailormade to fit all different types of industrial off-gases.

For this reason, industrial fumes containing common air pollutants
can be divided into two major groups. The first type of fumes, termed *lean,*
contain relatively insignificant quantities of combustible pollutants and have
an ample supply of oxygen present for their combustion.

The other type of fumes, termed *rich*, contain a nonexplosive mixture
of gases with varying amounts of combustibles, inert gases and a relatively
insignificant concentration of oxygen. A few of the many processes where
lean fumes are generated include: paint drying, rubber compounding, solvent applications and oil quenching. Rich fumes are commonly encountered
in industries like carbon baking, wax melting, organic resin coating and
pyrolysis of carbonaceous material.

The following major steps are necessary for efficient incineration of
rich fumes: (1) handling of rich fumes; (2) addition of necessary air to
provide enough oxygen; (3) mixing of air and fumes; (4) raising the mixture temperature to accelerate the reactions; and (5) providing the necessary
residence time to complete the reactions.

Rich fumes normally encountered in industry are at a relatively high
temperature and contain some hydrocarbons and other types of combustibles
which can condense at low temperatures. The use of fans and other conventional air handling systems can create maintenance problems; hence, it is
necessary to provide some method of fume handling which has a minimum
of cold moving parts.

The rich fumes do not contain enough oxygen to complete the oxidation of the combustibles in the fumes. It is, therefore, necessary to introduce
the proper amount of air during the incineration process. This air should be
intimately mixed with the fumes. Mixing should be completed within a
short distance from the point of air introduction. Since the air is normally

taken from the atmosphere, it is relatively cold. Its mixing with hot fumes lowers the mixture temperature. Since the oxidation reaction is greatly dependent on temperature, the mixture temperature should be raised to a level where the reactions can be completed within a short time. An improper temperature may require a long residence time or result in smoke or other types of pollutants being formed.

In case of rich fumes containing a high percentage of combustibles, it is necessary to supply either additional air or another type of heat sink to reduce the incinerator temperature. The amount of excess air, of course, depends on the selected incineration temperature and the quality of the fumes.

In practical application of a rich fume incinerator, especially for batch processes, the fume volumes, heating value and temperature vary over a wide range. A good design should perform the incineration of fumes under all conditions without using an excessive amount of auxiliary fuel or air.

Background Research

In early 1967 a joint venture research program was established to perform the necessary research work concerning fume incineration. The studies involved work in the mixing of fumes with air in a variety of aerodynamic mixing systems. From these studies, it was concluded that a jet pump-type mixing device, in which two fluids (fumes and air or fuel) are introduced from two concentric openings, is one of the best practical mixing systems. The velocity profile and tracer concentration studies for two concentric jets in a confined space showed that the velocity and concentration of the flow stream becomes uniform within 4 to 6 diameters from the inlet. It was found that the same device can be used to provide pumping action for one of the fluids.

In the initial stages of this program, very little information was available concerning the incineration rates of fumes containing common pollutants. The range of recommended incineration temperatures and residence times varied considerably. There was a widespread belief that the presence of flame and free radicals associated with a flame were necessary for the incineration of hydrocarbon fumes.

During the course of the fume incineration research work, an extensive experimental program was carried out to measure the reaction rates of a broad class of industrial fumes. Based on this work, it was concluded that for efficient incineration of industrial fumes the air-fume mixture should be raised to about 1500 F or more. At these temperatures, the required residence time is less than 0.25 seconds. The presence of small smoke (carbon) particles and excessive amounts of inert gases will, of course, require slightly higher temperature and additional residence time.

Incinerator Design

Figure 6-3 shows the pilot-scale, rich fume incinerator installed in the research and development laboratory. It can be used as an integral part of

Figure 6-3. Pilot-model rich fume incinerator.

the fume generating equipment or as a central unit to serve a number of processes simultaneously.

Fumes from the source are transported to the incinerator by an annular jet pump. Ambient air at 10 in. w.g. pressure is used to provide the pumping action. The jet pump eliminates the need of a mechanical device like a fan for the fume transportation, and the problems associated with hydrocarbon condensation, soot deposition and corrosion of fan blades and other components. The cost of the high temperature fan with special alloy parts to resist the attack by various hydrocarbons can be considerable.

The fume pipe is located in the center of the incinerator and is insulated to prevent heat losses which can cause a drop in fume temperature and condensation of some of the hydrocarbons. The pumped air serves another purpose of supplying air for the combustion of volatiles and hydrocarbons in the fumes, and for providing the excess air required for the control of the incineration temperature.

Two or more burners, fired tangentially, provide the necessary heat to raise the mixture temperature in the case where the fumes do not contain enough combustibles. When the fumes contain enough combustibles, the burner firing is reduced considerably to save on fuel consumption.

For batch-type fume generating equipment, the amount of combustibles in the fumes increases from zero to maximum and then drops to zero. The burners used in this rich fume incinerator have a 10 to 1 turndown on ratio to obtain the highest possible available heat at the incineration temperature. Burners can be operated on oil or gas, direct-spark ignited, and used with a U.S. flame detection unit or flame rods.

The step in the incinerator promotes recirculation of hot gases from downstream and mixes them with the fresh mixture from the jet pump system. When the fumes do not need auxiliary heat from the burners, recirculation mixing increases the mixture temperature until the exothermic reactions between the combustibles and the oxygen in the mixture become significantly fast and supply additional heat.

The incinerator temperature is controlled by a proportional controller which regulates the air supply to the burners and the jet pump. When the incinerator temperature increases above the set point, the controller acts to close the air supply to the burner. This, in turn, reduces the fuel supply to the burner through a backloaded zero regulator. At the same time, the air supply to the jet pump is increased to provide excess air cooling effect. In case of lower than set point temperature, the controller action increases the air and fuel supply to the burner but cuts down the jet pump air. A certain minimum air and fuel supply to the burners and jet pump is always maintained. Figure 6-4 shows a typical temperature, burner air-fuel and jet pump air flow relation for the rich fume incinerator.

Economics of Rich Fume Incineration

The rich fume incinerator design and method of operation is aimed at the minimum use of auxiliary fuel under the practical conditions of fume

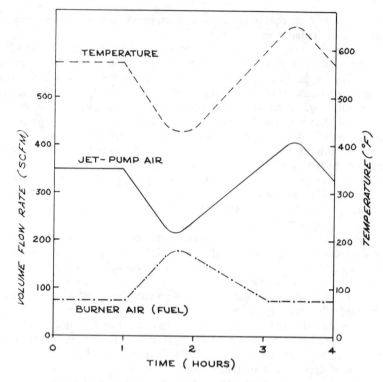

Figure 6-4. Fume temperature, jet-pump air, fuel input
relationship for a rich fume incinerator.

generation. The economics of the process are illustrated by the following
example:

Figure 6-5 shows the variation in fume volume, temperature, and the
volatile contents of the fumes for a typical fume generation process. The
heating value of the fume is directly related to the amount of volatiles.
For efficient incineration of these fumes, the fume temperature is raised to
1450 F. During the volatile evolution period of the process, the overall
heat input into the incinerator increases. This increase in the heat input
raises the temperature of the incinerated gases which is controlled by addition of extra air. When the gases do not contain a substantial amount of
volatiles, they are brought to 1450 F by using auxiliary fuel.

Figure 6-6 shows the theoretical and designed auxiliary fuel requirements for proper incineration. When the fumes are very rich (have a very
high heating value), the theoretical fuel demand is zero. But for safe operation of the unit and practical characteristics of various components of the
incinerator, the fuel consumption is at the rate shown by the dotted line.
The difference between the theoretical and actual demand of the auxiliary
fuel is shown by the shaded area. This difference will vary from one process

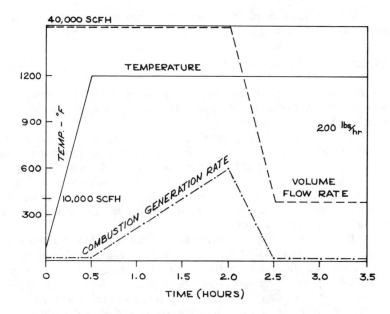

Figure 6-5. Variation in volume, temperature and volatile contents of fumes for typical generation process.

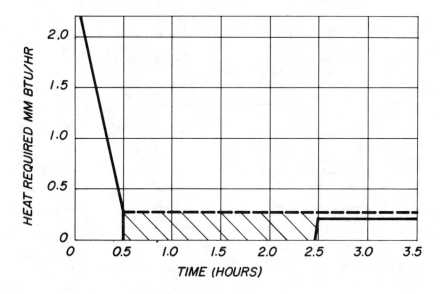

Figure 6-6. Theoretical and designed auxiliary fuel requirements for proper incineration.

to another, but this example shows that the design uses just enough additional fuel to make the system safe.

CATALYTIC INCINERATION[171]

Thermal and catalytic incineration are two methods of achieving the chemical union of certain organic hydrocarbons and oxygen. The end result should be carbon dioxide and water.

Thermal incineration requires that the temperatures of the process effluents be raised above the auto-ignition temperature of the organic pollutant and held there for a period of time. A reaction will occur between organic combustibles and oxygen, forming carbon dioxide and water. The reaction rate is affected by several variables, one of which is the excess of preheat temperature. The greater this excess above the auto-ignition temperature, the faster the reaction.

Catalysis is a reaction that depends upon the diffusion of a combustible vapor into the porous catalyst surface where it is adsorbed. Oxygen in a highly activated state is also adsorbed on the same surface. The oxidation reaction takes place at the porous surface. Following the reaction, combustion products are desorbed. A catalytic agent is a substance which changes the speed of a reaction, and may be recovered unaltered at the end of the reaction.

Table 6-2. Comparison of Thermal and Catalytic Incinerators

Assume: (1) Process gas volume is 6000 scfm at 300 F
(2) Standard solvents for the metal decorating industry: toluene, MEK, xylene, etc.
(3) Design for 90 percent conversion of hydrocarbon.
(4) Fuel cost is $0.75 per MMBtu.

Item	Thermal	Catalytic
Inlet volume, scfm	6000	6000
	8600	8600
Solvent at 25 percent LEL, gph	36	36
Operating Temp, F	1350	900
Preheat at zero solvent, MMBtu/hr	7.3	4.2
Preheat at full solvent, MMBtu/hr	3.0	zero
Inside cross sectional area, sq ft	11.1	6.0
Refractory thickness, in.	9	5
Approximate weight, lb	18,600	12,000
Cost of capital equipment	$18,400	$19,000
Fuel saving, MMBtu/hr	–	3.1
$/hr	–	2.33
$/5000 hr	–	11,650
Catalyst replacement	–	$7750
Payout of catalyst bed, hr	–	3300
Installation ratio	1/1	.75/1
Shipping cost ratio	1	.67/1
Maintenance factor	refractory stack burner	refractory catalyst burner

Thermal Afterburners

To meet many of the air pollution codes in the United States, a thermal afterburner must have the following requirements:

Temperature—capability of supplying preheat to raise process effluents to 1500 F.

Time—physical length and cross-sectional dimensions of unit to permit a minimum of 0.6 second total residence time in the unit with at least 0.2 second retention time in the secondary combustion chamber of the afterburner. Time required for the oxidation reaction starts after the process gases are raised to the thermal ignition temperature of the organic pollutant.

Turbulence—required to ensure complete mixing between the products of combustion of the preheat burner and the process effluents. This provides a homogeneous mixture at a uniform temperature profile. Turbulence is caused by a combination of high velocity gas travel and internal baffling. Velocities of 2100 fpm are considered nominal.

Catalytic Afterburners

The same three factors are required of catalytic afterburners as of thermal incinerators, but to different degrees. The ignition temperature of a combustible vapor in the presence of a catalyst is always lower than the auto-ignition temperatures required of thermal oxidation. Table 6-3 provides comparison of various materials for 90 percent conversion to carbon dioxide and water.

Table 6-3. Comparison of Temperatures Required to Convert Combustibles to CO_2 and H_2O

Combustible	Ignition Temp. F		Difference F	Catalyst/ Thermal %
	Thermal	Catalytic		
Benzene	1076	575	501	53.4
Toluene	1026	575	451	56.0
Xylene	925	575	350	62.2
Ethanol	738	575	163	77.9
MIBK	858	660	198	76.9
MED	960	660	300	68.8
Methane	1170	932	238	79.9
Carbon Monoxide	1128	500	628	44.3
Hydrogen	1065	250	815	23.5
Propane	898	500	398	55.7

Time for the oxidation reaction to start and proceed to completion is reduced by at least 1/5 to 1/10 of the time required for the thermal method. This reduction in residence time is important, since catalyst beds are generally from 3 to 15 in. thick in the direction of the air flow.

Turbulence is easily achieved in a catalytic system, since all catalyst beds inherently have both pressure drop and large surface areas. Air flow through the bed is purposely discontinuous to increase the probability of organic pollutant molecules striking an activated surface.

Catalyst metals generally used are platinum and palladium. Palladium is cheaper, but its activity is lower except for certain substances and classes of compounds. The high cost of these metals makes recovery from scrap catalyst economically feasible.

Advantages of Catalytic Afterburners

The combination of lower ignition temperatures, faster reaction time and inherent turbulence gives catalytic afterburners some distinct advantages:

1. Smaller units in both cross-sectional area, linear length and weight;
2. Lower operating temperature;
3. Faster startup and shutdown procedures permissible;
4. Less costly to install because of smaller weight and size;
5. Lower operating costs with regard to preheat fuel requirements;
6. Less maintenance on the construction materials and insulation;
7. Less need for expensive recuperative heat exchangers to reduce operating costs;
8. If recuperative heat exchangers are economically possible, a catalyst unit needs only standard materials of construction, while thermal units could require high nickel and stainless steel metals;
9. Needs less oxygen in the effluent stream, since it is more likely to operate closer to the theoretical oxygen acquired; and
10. Less smog-producing oxides of nitrogen are produced.

Disadvantages

1. Higher initial capital equipment cost;
2. Replacement of catalyst bed required on a periodic basis; and
3. Exhaust may contain substances that could act as a suppressant or even a poison to the catalyst.

Process parameters should be thoroughly reviewed before deciding upon a catalyst unit. A general rule says that a one-year catalyst life is necessary to justify the system on an economic basis.

Catalyst suppressants inhibit activity by masking the surface. They prevent combustible vapor from physically contacting the activated surface and prevent the union of the organic molecule and oxygen. Catalyst activity can be regenerated by removing the masking material. A condensable but combustible organic can be removed by raising the preheat temperature to 1000 to 1100 F for 2 to 3 hours. Or, by removing the bed, a solvent wash can dissolve the condensed organic.

Inorganic oxides such as iron oxide are removable by a mild organic acid wash. A 5 percent oxalic acid solution at 200 F and 2 to 3 hours soaking followed by a clear water rinse is usually enough to regenerate to 90 percent of the original activity of the catalyst. Water-soluble materials are removed with a mild detergent soaking and a clear water rinse.

COST COMPARISON FOR
BURNING FUMES AND ODORS[122]

Total costs for fume and odor control are difficult for the engineer to determine. He is usually besieged by equipment technical facts, equipment quality and company history. However, when it comes to a determination of the total cost of owning and operating the equipment, the purchaser receives little or no help. Unfortunately, this is because the supplier representative normally does not know the total cost, since his orientation is toward the selling price of his equipment.

In purchasing control equipment, the pollution engineer should make an in-depth study of all costs relating to the purchase and operation of the equipment. This caution is justified since the operating cost of fume and odor control equipment is usually the most, if not the only, important factor.

Selection of pollution control equipment is clearly a decision of two parts:

1. Equipment must operate to meet pollution control regulations, and
2. Equipment must be economically practical to install, maintain and operate.

Today's regulations are such that (with few exceptions)a recognized approach is to thermally destroy the organic fumes and odors by raising the process exhaust to 1400 F for 0.5 second. Part two of equipment selection process, however, offers some unexpected surprises.

Equipment

For all practical purposes, there are only three types of equipment available for installation on a process exhaust which will meet the 1400 F for 0.5 second requirement:

Afterburner

An afterburner as used in fume and odor control is a simple cylindrical steel shield, refractory-lined. It is normally mounted vertically with the process gases entering at the bottom where the burner (and combustion blower, if used) is located. Process gases combine with the combustion products raising the mixture to the required incineration temperature. The hot mixture then passes through the afterburner and is exhausted to atmosphere. Length of the afterburner determines the retention time of the hot mixture at the purification temperature.

Afterburner with Energy Recovery

This unit is a basic afterburner with the addition of a heat exchanger. Hot gases leave the afterburner and pass through a heat exchange unit having thin-wall metal tubes. Incoming process gases to the afterburner pass on the other side of the metal tubes—thus a heat transfer through the tubes. The thermal recovery efficiency of a heat exchanger is a measure of its ability to recover the heat used to raise the polluted process stream gases to the required temperature. Efficiency of an afterburner with heat recovery is normally about 40 percent. This, in effect, says that a 40 percent efficient system having 200 F process stream gases and a 1400 F purification temperature (*i.e.,* 1200 F temperature rise) will have a 920 F system exhaust temperature (1200 F x .40 − 480 F recovery). Efficiency should remain relatively constant providing that there is no build-up of oils or other materials that would reduce the heat transfer efficiency.

Thermal Regenerative System

This is an incineration system (Figure 6-7) that relies on the heat radiation capability of an inert ceramic. This system slowly passes hot gases through the heat retention material and, after a few minutes, the gas flow reverses to use this stored energy to preheat the incoming gases. Thermal recovery efficiency is primarily related to the surface area of the heat transfer media, although the velocity of the gases and the makeup of the process gas stream have an influence on the ultimate efficiency.

As they pass through the equipment, preheated gases are raised to the purification temperature. The time that gases are held at that temperature has a direct relationship to the volume of the process gases and the equipment dimensions. The normal condition, as with other types of equipment, is 0.5 second retention at the purification temperature.

Each of the three systems performs essentially the same fume and odor control function. Each system claims particular advantages over the others and, in the final analysis, one of these claims may be the final factor for determining which system to buy.

Oxides of Nitrogen

Another item that should be reviewed is the release of oxides of nitrogen (NO_x) to the atmosphere. Nitrogen oxides contribute to smog and the total pollution problem. What may be overlooked, however, is that with all other things being equal, the NO_x production is proportional to the fuel consumption. This is of concern since some systems actually produce more contaminants in the form of NO_x than they eliminate in the form of hydrocarbons.

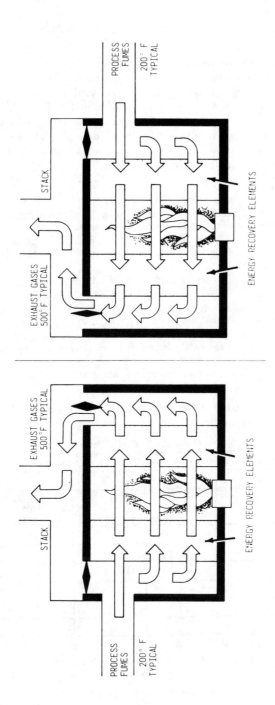

Figure 6-7. Thermal regenerative system.

Available Energy

There is a common and costly misunderstanding that the quoted gross energy of a fuel is also the energy available for work. The determination and quotation of the gross fuel Btu content is based on the use of a calorimeter—yet who has known any plant using fuel under such hypothetically ideal conditions. There is the need to understand why only a percentage of this energy is available to do useful work. The need to determine fuel use efficiencies has become necessary due to the higher cost of fuel and the shortage of low-sulfur fuel.

The following factors exist:

- Available Energy = Gross (Total) Energy-Discharge Losses,
- Fuel costs, calculated (incorrectly) on the gross energy basis, suggest costs substantially lower than actual costs,
- The actual use of fuel based on a 1400 F process exhaust is more than 60 percent greater than one would expect using gross heat calculations,
- Natural gas gross energy is approximately 1000 Btu/cu ft,
- No. 2 fuel oil gross energy is approximately 140,000 Btu/gal,
- Energy costs are often quoted on the basis of the therms required with the result that the measured use (*e.g.,* cu ft of gas) is assumed,
- The equipment operator pays for all fuel on a quantity basis of gas or oil whether the energy is used in the process or whether the energy goes up the stack,
- The lower the process exhaust temperature, the greater the utilization of the fuel energy.

Figure 6-8 shows the energy loss and the energy available on a percentage basis. Variations will, of course, occur with a change in the chemistry of the fuel.

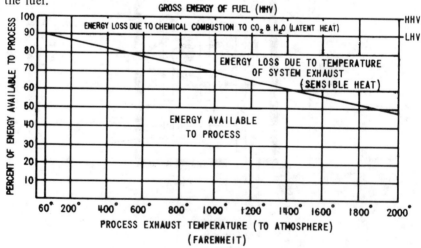

Figure 6-8. Perfect combustion of natural gas, 0 percent excess air.

Table 6-4 provides a detailed breakdown of the fuel costs and the formula used to derive the costs. Comparative fuel costs of $29.82/hr, $15.06/hr, and $5.61/hr clearly show the increasing importance of the fuel available energy as the system exhaust temperature increases.

Table 6-4. Fuel Cost Comparison

Equipment Type	Afterburner	Afterburner with Heat Recovery	Thermal Regenerative System
Thermal recovery efficiency	0%	40%	75%
Process gas volume	16,000 scfm	16,000 scfm	16,000 scfm
Purification temperature	1400 F	1400 F	1400 F
System exhaust temperature	1400 F	920 F	500 F
Energy required for purification (Btu/hr)	21,920,000	13,150,000	5,472,000
Gross fuel energy (Btu/ft^3)	1036	1036	1036
Fuel unit cost ($/1000 scf)	$0.86	$0.86	$0.86
Available energy from fuel	61%	72.5%	81%
Cubic feet of gas required	34,700 scfh	10.808 scfh	6,520 scfh
Fuel cost	$29.82/hr	$15.06/hr	$5.61/hr

$$\text{Fuel Cost} = \frac{\text{(Unit cost)}}{\text{(Gross energy of fuel)}} \times \frac{\text{(Btu required)}}{\text{(\% energy available to process)}}$$

$$\text{Fuel Cost (\$/hr)} = \frac{(\$/1000 \text{ ft}^3)}{(\text{Btu/ft}^3)} \times \frac{(\text{Btu/hr})}{(\%)}$$

Afterburner

$$\$/hr = \frac{(0.86)}{(1000)(1036)} \times \frac{(21.92 \times 10^6)}{(0.61)} = \$29.82/hr$$

Afterburner with thermal recovery

$$\$/hr = \frac{(0.86)}{(1000)(1036)} \times \frac{(13.15 \times 106)}{(0.725)} = \$15.06/hr$$

Thermal regenerative system

$$\$/hr = \frac{(0.86)}{(1000)(1036)} \times \frac{(5.472 \times 10^6)}{(0.81)} = \$5.61/hr$$

The above data show the significance of thermal energy recovery and of lower exhaust temperature.

RELATING FLARES TO AIR QUALITY[129]

Using a computer and common diffusion equations, it is possible to directly relate source parameters for continuous acid gas flares to ambient air quality. The end result is a set of curves for determining the minimum flare height, given the emission and meteorological conditions, such that the ambient air standard for sulfur dioxide is met.

Flares of various kinds are commonly used to combust and disperse waste gases in the petroleum and other industries. However, not much work has been done to predict flare contributions to ambient quality; most research has been confined to large stack sources. Flare emissions, because they are released nearer to the ground than stack gases, might be expected to be important on a local level.

A well known air pollution relationship is the Pasquill-Gifford dispersion equation. This and other equations seem to approximate the behavior of flare plumes reasonably well:

$$\chi = Q \ EXP \ [-H_e^2 \ / \ 2\sigma_z^2] \ / \ \pi\sigma_y\sigma_z u$$

This relationship depicts expected concentrations of pollutants in g/m^3 at ground level beneath the plume centerline. The theory offered here is:

$$H_f = g(Q, \chi, T, ...)$$

It is logical to say that a minimum flare height may be calculated from the dispersion equations, such that the allowable ambient concentration, χ_a, is not exceeded anywhere downwind.

$$H_f \geqslant g(Q, \chi_a, T, ...)$$

Coupled with the dispersion equation is a plume rise equation, many of which can be used. Briggs' equations appear to be well suited for use in flare calculations. For atmospheric stability categories A-D (unstable to neutral)

$$\Delta H = 1.6F^{1\,3} \ (10H_f)^{2\,3} \ / \ u. \ for \ x > 10H_f$$

Briggs' equation is for large stacks, but can be used for flares since the heat release from flares is high, and because field studies seem to warrant its use. For categories E-F (stable)

$$\Delta H = 2.9(F/us)^{1\,3}, \ for \ x > 2.4us^{-1\,2}$$

Evaluation of the dispersion parameters, σ_y and σ_z, may be accomplished readily from Slade's curves (USAEC TID-24190) for given meteorological conditions. However, these curves are not directly applicable to high speed computer analysis.

The table and equations below were derived from least squares analysis of Slade's curves and may be incorporated into a computer program for rapid calculation of the dispersion parameters, σ_y and σ_z:

$$\ln\sigma_z = K_1 + K_2 \ \ln x + K_3 \ (\ln x)^2$$

$$\sigma_y = K_4 \ x \ K_5$$

When all this information is incorporated into a computer program, sufficient output may be obtained in less than 30 seconds. Determined is the required flare height to produce allowable downwind concentrations of waste gases under reasonable conditions. A computer flow diagram is given in Figure 6-9.

Taking the example of a hydrogen sulfide flare, Figure 6-10 provides a quick means of determining the hydrogen sulfide flow rate as a function of H_2S content and total flow rate. The equations are as follows:

$$Q = 60 \text{ V M } (34 \text{ lb/mole}) / 385 \text{ scf/mole}$$
$$Q = 5.3 \text{ V M}$$

From the computer output, Figure 6-11 is constructed to allow rapid determination of required flare height as a function of Q and the heat release rate from combustion of the flare gas fuel mixture. (The curves were drawn for atmospheric stability category C and windspeed 8 mph. Similar curves may be obtained via the computer output for any meteorological conditions.)

A low H_2S gas mixture would require addition of hydrocarbons to promote good combustion, while a high sulfur gas may require no fuel at all. There is, of course, a lower practical heating value below which combustion cannot be supported. This limit will vary with composition of flare gas and fuel, particularly as a function of CO_2 concentration. Careful system design is needed to ensure that the flare will burn reliably. In any event, the net heat released by all the combustibles present contributes to the buoyancy attained by the plume, promoting mixing and dispersion of contaminants.

It should be noted that the value of χ as given in the Pasquill-Gifford equation assumes an averaging time of about 10 min. This fact is rooted in the 10-min averaging time associated with measurement of the dispersion parameters. Therefore, it is necessary to correlate the value of χ given in the equation with the χ_a dictated by the appropriate ambient standard. For instance, some states have a 30-min standard while others have 1-hr, 24-hr or other limits. A useful relationship for extrapolating averaging times is:

$$\chi_a = \chi \left(\frac{t}{10}\right)^{-b},$$

where b is a fraction dependent upon meteorological conditions. Good dispersion may yield a value of about 0.2, and poor mixing about 0.35. Caution should be used, however, when it is significantly longer than a few hours.

In addition to meteorology, topography may also play an important role in the calculations. Care must be taken if the local terrain is very hilly or mountainous. On the other hand, vast areas in the Southwest are flat, and terrain effects are negligible.

An example illustrating the usefulness of these graphs is: Given 500 scfm process gases consisting of 25 percent H_2S, 15 percent light hydrocarbons (assume 1150 Btu/scf), and 60 percent inerts to be flared, what height of flare would be necessary to meet a 24-hr standard of 0.1 ppm SO_2?

Assume C stability and wind 8 mph: From Figure 6-10, Q is about 650 lb/hr H_2S. The heat value of the gas is 0.25(637) + 0.15(1150) = 332 Btu/sec. Entering Figure 6-11 at Q = 650 lb/hr and reading across from a point near the 2500 Btu/sec line, the required flare height is about 110 ft.

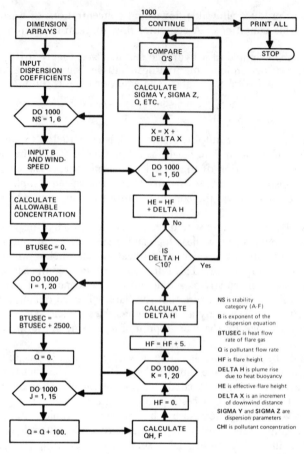

Figure 6-9. Computer flow diagram for determining minimum flare height to meet standard.

χ = Concentration, g/m^3

χ_a = Allowable concentration (ambient standard), g/m^3

f,g = Function of

F = Buoyancy flux factor;
$F = 3.7 \times 10^{-5} Q_h$, m^4/sec^3

H_e = Effective flare height
$(H_e = H_f + \Delta H)$, m

H_f = Actual flare height, m

K = Constants for dispersion parameter equations, dimensionless

Q = Mass flow rate of pollutant, g/sec (Convert from lb/hr in text)

Q_h = Heat flow rate from flare, cal/sec

s = Atmospheric stability parameter;
$s = \dfrac{g\,\delta\Theta}{T\,\delta z}$, sec^{-2}

σ_y = Horizontal dispersion parameter, m

σ_z = Vertical dispersion parameter, m

T = Temperature, K

t = time, min

u = Windspeed, m/sec

V = Volume flow rate of flare gas, scfm

x = Downwind distance from source, m

y = Horizontal distance from plume centerline, m/sec

z = Altitude, m

Θ = Potential temperature, K

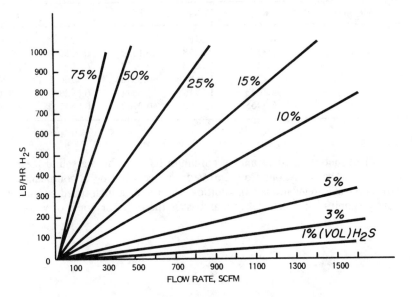

Figure 6-10. Sulfur mass rate as a function of total gas flow, scfm flow rate and volume percent H_2S.

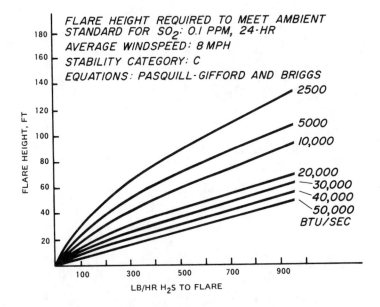

Figure 6-11. Required flare height as function of sulfur mass rate and heat release rate.

Coefficients and Exponents of σ_y and σ_z Equations					
Stability	K_1	K_2	K_3	K_4	K_5
A	8.9796	-3.3185	0.4291	0.3370	0.9160
B	2.8171	-0.8778	0.1717	0.2360	0.9230
C	-3.0171	1.2058	-0.0234	0.1690	0.9230
D	-3.4263	1.2583	-0.0375	0.1200	0.9190
E	-4.8643	1.5846	-0.0621	0.0929	0.9070
F	-6.3765	1.8280	-0.0769	0.0642	0.9010

This method of determining minimum flare height is a sound means of ensuring that emissions from petroleum process will be in compliance with applicable regulations. In addition, the use of a high-speed computer greatly simplifies the calculations, and saves the engineer hours of work.

CHAPTER 7

ODOR CONTROL

ODOR SOURCE INVENTORIES[178]

An inventory of odor sources may be used to predict the scope of odor control procedures needed for abatement, to relate odor sources to effects in the community, or to establish regulatory or enforcement policies. However, there is no standard method for compiling odor source inventories. Any of several methods may be used.

Gaseous Emission Rates and Threshold Data

The basic premise of this method is that a sample of odorous air can be described in terms of the volume to which it must be diluted for its intensity to be reduced to the sensory threshold level. Select any volume of odorous air, V, containing a mass of odorant, m. Then the odorant concentration = $C = m/V$. When the sample is diluted with odor-free air to the sensory threshold concentration, C_t, the volume will increase to V_t, and $C_t = m/V_t$, from which $C/C_t = V_t/V$.

ASTM Standard Method D-1391-57 defines an *odor unit* as 1 cu ft of air at the odor threshold, and *odor concentration* as the number of cubic feet that 1 cu ft of sample will occupy when diluted to the odor threshold. Obviously, odor concentration is equivalent to the dimensionless fraction V_t/V, with volumes expressed in cubic feet. The ASTM Standard defines *odor emission rate* as the number of odor units discharged from a stack or vent per minute. This rate provides one basis for quantifying odor sources; the total emission from a plant or industrial area will be the sum of the odor emission rates. For a nonconfined source such as a lagoon or ditch, some estimate of emission rate can be made on the basis of the odor concentration of the air in equilibrium with the source, and an assumed velocity of transfer of odor to the atmosphere by wind action. Then the total odor emission rate is the emission rate per square foot of surface times the total area in square feet.

Some serious limitations to this approach are:

1. The method does not concern itself with odor quality or objectionability. It assumes implicitly that all odors are to be counted equally in establishing the inventory.

2. The method assumes that it is valid to draw conclusions from threshold data about odor levels to which people respond in problems of community malodors. It is more likely that complaints are related to supra-threshold odors. It has never been shown that the dilution ratio, V_t/V, is a valid measure of the intensity of an odor; in fact, there is evidence to the contrary.

3. The threshold level itself has been shown to be a variable that depends on the response criterion of the observer; it can be manipulated by factors that affect motivation and expectation.

4. The problem of handling odorants in the very low concentrations in which they exist at threshold levels throws doubts on many dilution data. The extreme ranges of odor threshold concentrations reported in the literature support this uncertainty.

A study to determine the extent to which odor travel could be predicted from dilution data was carried out by Wohlers. He compared odor travel from four different plants with stack gas dilutions calculated according to Sutton's equation. Large discrepancies were found. The most extreme sample was odor from a kraft paper mill, for which the average dilution needed to reach threshold was 32:1 and the maximum 64:1. The minimum calculated dilution of stack effluent was 990:1, predicting that the odors would not be detected in the field. Field surveys over a 6-month period showed that the kraft odor could be detected at distances up to 8 miles, where the calculated dilution of the stack effluent was 840,000:1. There were occasional reliable reports of the odor at a distance of 40 miles.

Qualitative Description or
Generating Processes

If it is assumed that different odor sources cannot be measured on an equal basis, then consideration must be given to the role of odor qualities. Odor quality classifications have a long and confused history, and attempts to apply some of the early classification systems to air pollution problems have been unsuccessful.

In a particular plant or industrial area in which the composition of each odor source is known, it is possible to set up a series of standards that are unequivocally recognized in the community. An example is shown in Table 7-1, in which seven odor qualities were defined in a particular area in terms of reference standards taken from or simulating the actual sources. It was shown that panel members could recognize and distinguish among the various standards after very rapid training.

A classification system designed to describe community odors in a general way is more difficult to set up, but perhaps can be attempted. Table 7-2 groups community odor nuisances into ten general classifications.

Table 7-1. Odor Sources, Types and Reference Standards

Odor Source	Odor Type	Odor Reference Standard
Foundry emission	Core oven odor; the foundry cores consist of a mixture of linseed oil, Western bentonite, and corn flour; the material is heated at 600 F for 2 hr	Core mixture made up in laboratory to formula used by the foundry, then kept in a warm sand bath for odor presentation to panel
Bread bakery	Baking oven exhaust odors	Exhaust gases from oven sampled in evacuated stainless steel bomb
Rendering plant	Sour inedible fat from rendering operations	Sour fat from skim tank of rendering plant mixed with a little mineral oil
	Dried nonfat products from rendering operations—protein and bone	Dried inedible protein product from rendering plant
Varnish oven emissions	Linseed oil cooking; oils are heated to about 585 F	Raw aged linseed oil
	Cooking of a mixture of linseed oil and fish oil	Mixture of linseed and menhaden oils, 4:1
Distillation of saponified oils	Fatty acids from vegetable oil	Linseed oil fatty acids

Table 7-2. Community Odor Nuisance Classification

Odor Type	Typical Occurrence
1. Organic nitrogen compounds	Reduction of animal matter, including fish. Typical representative compounds would be trimethylamine and skatole. The animal and fishy odors are different from each other and could comprise two classes instead of one.
2. Phenolic odors	Curing of phenolic resins, creosoting operations and the like. Typical components are phenol, cresols, xylenols and carvacrol.
3. Organic sulfur	Petroleum refinery emissions, emissions from pesticides mfg., gas odorant leakage, and the like. Typical components are mercaptans, sulfides, and disulfides.
4. Organic acidic odors	Emissions of acids like butyric, valeric, and phenylacetic acids.
5. Burnt odors	Burning operations; tarry and asphaltic odors.
6. Fragrant or floral	Manufacturing or compounding of flavors and perfumes.
7. Solvent odors	Dry cleaning exhaust, solvent drying emissions.
8. Camphoraceous	Naphthalene, or paradichlorobenzene odors.
9. Oily	Varnish cooking, foundry core oven emissions. Typical odor of linseed oil.
10. Gassy unsaturated	Pungent, gassy odors like those of acrylic materials, diesel exhaust, etc.

It should be emphasized that this classification does not intend to describe odors in all contexts. It simply represents a large group of odor types that may be experienced from industrial sources and is represented as a suggested approach to the classification of community odor nuisances.

If the classification systems of the type suggested above were to be useful in the analysis of community odors, the descriptive words cited would be insufficient. For each of the types listed, it would be necessary to blend a standard mixture of odors that would most generally represent the indicated type. For example, the organic sulfur odor type would be represented by a standard comprising a blend of mercaptans and organic sulfides and disulfides which would most generally elicit the description of organic sulfur odor. If both odor classification and measurement of odor intensity were needed, then each of the odor type standards could be prepared in a series of dilutions which would represent "weak," "moderate," and "strong" intensities.

How could quality descriptions of this type, either specific or general, be translated into inventories of odor sources? The following procedure is one possible approach:

1. For each odor quality that represents an emission to be inventoried, set up a reference standard whose intensity represents the minimum *acceptable* level of this particular odor. This level would have to be determined by a local survey, not by a trained odor panel. The standard would then necessarily be at a supra-threshold level of intensity, although for highly objectionable odors the level might approach the recognition threshold.

2. An odor inventory might be made in a manner similar to the dilution-to-threshold method, with the modification that each odor quality or type is listed separately and the dilution is carried out to the acceptable rather than to the threshold level.

3. Complaints are usually the driving force behind installation and operation of odor control systems. Some system for logging complaints should be considered that can serve to inventory odor sources, or to relate odor emission rates to their actual effects in the community. Complaint frequencies in themselves are informative only if they can be confidently related to sources. The qualitative descriptions given by uninformed complainants exemplified by adjectives such as "chemical," "obnoxious," and "suffocating," are of little help. What is necessary is to construct a set of reference standards for odor qualities as described above, and to devise a report form on which the respondents can indicate location, odor quality, time, duration, and some estimate of odor intensity. At the same time wind velocity and direction must be recorded continuously. An accumulation of data of this type over a period of months, including a wide variety of weather conditions, would provide an inventory of odor effects which, if correlated with wind data and known sources, would present a valuable overall picture.

4. Social and economic effects of odor on people are difficult to assay, but this factor is crucial toward the objective of evaluating sources. This approach evaluates each odor source by asking the question, "What would be the dollar value to the community to eliminate this source?"

There seems to be little doubt that people consider community malodors to be a significant detriment to the enjoyment of life. In a series of questions in which children living in a slum area were asked to write about their neighborhood, a typical response was, "My block is dirty and it smells terrible." Those effects of malodors to which it might be possible to assign dollar costs include:

a. Decrease in values of existing property. This factor includes not only depression of resale value, but also, for example, increased difficulty in renting out a spare bedroom. Values of both residential and industrial property are included.

b. Lack of use of open land because of the likelihood that offensive odors would be a problem. Some areas of concentrated industrial activity associated with odorous emissions are surrounded by undeveloped "buffer zones." Even if such areas are intended to be recreational rather than residential, any depressed utilization may be interpreted as a dollar cost associated with odors. In many cases even the use of industrial zones are restricted by the presence of a local odor source. Thus, a plant for assembly of electronic parts may avoid locating near a rendering plant.

c. Impairment of use and enjoyment of property. When a back yard cookout is cancelled because of a local malodor, or when the homeowner feels he must close his windows and install air purifiers, or when he operates his air-conditioning system when outside temperatures do not require air cooling, these behaviors may be translated into dollar costs. In fact, the courts often recognize such actions as evidence that odorous emissions are damaging and that compensation should be made by the offender.

d. Direct personal effects. People frequently complain that the effects of malodors include loss of appetite, nausea and loss of sleep. In some cases these sensations result in demonstrable changes in behavior, and it is these changes to which dollar costs may be most reasonably applied. Absenteeism or loss of productivity in work are obvious examples. The latter effect often arises with office workers in a building affected by malodors from a nearby source. In relating odor sources and effects, it must be recognized that the "damage" is not linear with odorant concentration. Decrease of the emission by 90 percent on a material basis will yield considerably less than 90 percent reduction of the effects. As results of social and economic studies of odors become available, it will become increasingly easier to establish relationships between odor sources and effects.

CHEMICAL CONTROL OF ODORS[3]

Odors have many different qualities—permeating, pungent, pleasant, putrid, etc. Even the sound of the word "odor" evokes the memory of a sensation—and all too often that response is of a polluting malodorous emission, instead of the fragrance of a perfume or the aroma of home-made bread.

Odor pollution is becoming a major factor in the total air pollution problem. Because odor is so easily noticed, complaints about obnoxious odors are frequently received by local air pollution control authorities. Much work is being done in the area of odor control to define the major problems, find adequate control methods, develop odor measurement methods, and establish standards for odor emissions.

Studies have been carried out in many cities to establish the chief sources of odor problems. The major offenders have usually been chemical manufacturers (particularly in the plastics industry), pulp and paper manufacturers, rendering plants and other food processing industries, and waste treatment facilities. The chemical compounds which cause the odor problems vary from source to source. Some of the more important sources and causes of odors are shown in Table 7-3.

Table 7-3. Common Industrial Odor Sources

Odor Source	Cause of Odors
1. Pulp and paper manufacturing	hydrogen sulfide organic sulfur compounds
2. Plastics industries	phenols formaldehyde acrolein organic sulfur compound organic solvents
3. Fats and oils-based industries	aldehydes amines
4. Waste treatment facilities	organic nitrogen compounds hydrogen sulfide organic sulfur compounds
5. Rendering and related industries	aldehydes ketones organic acids organic sulfur compounds amines
6. Food canning and processing	aldehydes amines other nitrogenous compounds
7. Pharmaceutical industry	organic sulfur compounds amines
8. Textiles	starch formaldehyde

Odors caused by most of the compounds listed in Table 7-3 are susceptible to control by potassium permanganate ($KMnO_4$). These odors are related mainly to the functional group of the compound—the divalent sulfur of a mercaptan or the carbonyl group of the aldehydes. These functional groups usually can be oxidized rapidly to form different kinds of compounds which are generally odorless or more pleasant.

Given sufficiently vigorous reaction conditions, $KMnO_4$ is capable of oxidizing almost any organic compound to carbon dioxide, water, sulfates, ammonia and nitrates. However, the various types of available pollution abatement equipment suggest that the oxidation reaction be examined in terms of relatively mild conditions. For example, most odor scrubbing solutions are mildly alkaline, contain 1 to 4 percent $KMnO_4$, and are circulated at ambient temperatures.

Under these conditions, potassium permanganate does not react rapidly with all organic compounds. Some odorous compounds which are susceptible, however, to oxidative degradation under these conditions are: aldehydes, reduced sulfur compounds, unsaturated ketones and hydrocarbons, phenols, amines, hydrogen sulfide and sulfur dioxide. Among the compounds which resist oxidation under these conditions are: saturated organic acids and hydrocarbons, ketones and chlorinated hydrocarbons.

The advantages in using potassium permanganate for odor control are easily seen. In general, it is easy to handle, quite stable under normal storage conditions, noncarcinogenic, and noncorrosive toward most construction materials. For odor problems due to a low concentration of a highly odorous contaminant or for odor problems in heavily moisture-laden air streams, scrubbing with this chemical offers economic advantages over incineration methods.

Because potassium permanganate destroys or degradatively alters the compounds responsible for many odors, the odors do not reappear as so often happens with counteractant or masking techniques. Many states have already banned the use of masking agents for odor control.

Application Techniques

The nature of the source of odors determines how potassium permanganate may be applied to abate the problem. Odors emanating from an industrial plant stack are best handled by using a wet scrubber with a solution of $KMnO_4$ as the scrubbing liquid. Odors from cattle feedlots, sludge beds or decaying garbage are easily treated by spraying a solution over the area or by dusting with a dry mixture. Odors in hospitals, auditoriums and other public places can be eliminated by using filters which contain adsorbent materials impregnated with potassium permanganate.

Scrubber Application

Scrubbers are designed to remove soluble gases or liquid mists from the air stream either by absorption or interception. Scrubbing liquids containing potassium permanganate may be used in any scrubber, but most applications have been in cross-flow and countercurrent-flow packed scrubbers and spray chambers.

An important distinction with regard to the use of $KMnO_4$ solutions is whether the scrubbing liquid is recirculated. Potassium permanganate doesn't react until the contaminant is absorbed by the scrubbing liquid. Then the reaction usually continues at a rate determined by the nature of the contaminant and the solution pH, until the contaminant is totally oxidized. If total oxidation takes place, the consumption of this chemical compound generally corresponds to the stoichiometric amount required for degradation to carbon dioxide and water. In some instances, total oxidation is desirable to comply with local wastewater regulations.

Packed scrubbers, both countercurrent-flow and cross-flow, generally use recirculating scrubbing solutions. The use of $KMnO_4$ solutions in these scrubbers is economically favorable for exhaust streams containing small amounts of highly odorous compounds.

On the other hand, "once-through" scrubbers, such as spray chambers can operate at virtually any pH, thus allowing selection of the pH most favorable for absorption and/or reaction. The solutions in these scrubbers contain only enough chemical oxidizer to destroy the odor-causing functional group of the molecule. The odor is removed but the molecule is not completely degraded, and chemical consumption is held to a minimum.

The type of scrubber system used to control odors, of course, depends on individual conditions and on the type of effluent to be treated. Use of a dust collector before an odor scrubber should be investigated. Exhausts from coffee roasting plants can be made odor- and smoke-free by such a combination system. Sewage digestion odors, which are year-round problems in the warmer areas of the country, have been easily controlled with the use of a baffled spray chamber in which the $KMnO_4$ solution is introduced as a mist.

Case Histories

A rendering plant recently installed a system handling 32,000 cfm consisting of a dry cyclone at the outlet of the feather meal dryer, a wet venturi scrubber, and a countercurrent-flow packed tower. Water flows through the venturi at 100 gpm and is pumped to the city sewer system after use. The dilute $KMnO_4$ solution passes through the packed tower and drains to the sewer. All the permanganate is reduced during a single pass through the scrubber. Consumption of the chemical oxidizer is about 60 lb/day.

Potential odor problems were the reason for inclusion of an odor control system in a recently constructed meat cooking plant in Pennsylvania. A countercurrent-flow scrubbing tower 10 ft in diameter and 17 ft high was constructed of fiberglass reinforced polyester. The packed area is four feet deep. Air flow rate is about 25,000 cfm. The scrubbing solution is 1 to 2 percent $KMnO_4$ buffered with borax to approximately pH 8.5 and recycled through the tower at a flow rate of 250 gpm. Tests show a 90 percent reduction in total organic compounds.

An asphalt plant installed a system utilizing potassium permanganate solutions for odor control. Off-gases from the asphalt saturators first pass through a cyclone for removal of bulk particulate matter and then travel through a spray chamber duct about 30 feet long containing 12 spray nozzles, which spray a potassium permanganate solution across the gas flow. Finally the gases pass twice through a $KMnO_4$ solution which is in a chambered solution tank. The spray chamber duct was constructed with a 5-degree slope so that the chemical solution runs back into the tank and is recirculated. A revolving cloth filter continuously removes the MnO_2 sludge from the system. The capacity of this system is 15,000 cfm. About 30 pounds of the chemical oxidizers are used daily.

Several sewage plants in California are using once-through spray scrubbers which fog a $KMnO_4$ solution into baffled chambers through which the gases pass. This treatment has proven very economical and effective in handling the malodorous gases, mostly hydrogen sulfide and mercaptans, from the digestion process. Chemical consumption averages 0.1 to 0.3 lb/hr/ 10,000 cfm.

Each company's odor problems are unique. To effectively and economically use potassium permanganate for odor control, a technical representative of the chemical producer should be consulted.

ODOR MODIFICATION[106]

Increasing awareness and vigilance of public and private sectors tends to produce the general reaction that, whenever there is an odor, something must be done. When gases are toxic and present in sufficient concentration, they present a serious health hazard. In addition, a larger group of effects, resulting from non-toxic odor concentrations, are harmful to the comfort and well-being of the community.

One of the more important drawbacks in pollution engineering is the lack of an exact terminology concerning odor. Degrees of intensity of smell, names, and other classifying details vary from one investigator to another. A comparison of values obtained from different laboratories is often a difficult and complex matter. Human response to sound or light can be exactly defined, because they can be exactly measured. However, exact measurement of olfaction has, at present, not succeeded.

All of the chemical and physical characteristics of an odor can be measured but this does not mean that they allow organoleptic interpretation. One's mental background supplies the higher critical centers of the brain with the decisive information. A person with a pleasant memory of a particular smell will classify this odor as pleasing; another will take a different view.

How do we establish at what level a malodor first becomes noticeable, and is then termed a nuisance? Or, indeed, how do we measure or express the concentration of a malodor at all? Essentially, when we speak of a quantity of air, gas, or its constituents, we measure these relationships in parts per million. The level at which a malodor is recognized or detected in ambient air by an odor panel is termed the odor recognition threshold. Table 7-4 gives threshold values from a recent study. The figures mean that 100 percent of the participants can detect or recognize these odors at the levels indicated.

Table 7-4. Odor Threshold Recognition Values

Chemical	Odor Characteristics	Odor Threshold Recognition Level, ppm
Hydrogen sulfide	Boiled eggs	0.00047
Ethyl Mercaptan	Earthy, Sulfidy	0.001
Pyridine	Burnt, Pungent	0.021

The American Conference of Governmental Industrial Hygienists has also published maximum daily allowable exposure levels for workers. These are given in Table 7-5.

Table 7-5. Hazardous Exposure Levels

Chemical	Threshold Limit Values, ppm
Hydrogen Sulfide	10
Ethyl Mercaptan	0.5
Pyridine	5

Table 7-4 shows the level at which these odors can be detected and adjudged a nuisance, while Table 7-5 shows the danger level of exposure. This comparison represents a very important distinction. While concentrations of these gases at the hazardous level must be eliminated or destroyed at the source, nuisance content can be controlled by other means.

Some of the most common methods of eliminating malodors are: modification, masking, scrubbing, venting or exhausting, combustion, biochemical reduction, oxidation, electrostatic, adsorption and/or absorption.

Masking is based on the premise that people perceive a mixture of smells as a single odor. It is nothing more than superimposing a pleasant odor on top of an unpleasant one. This effort raises the total odor level and creates an overpowering sensation, hopefully more pleasant but usually objectionable.

For example, a sewage-plant operator might try to mask the malodor of a sludge lagoon by spraying a vanilla fragrance in the immediate area. Unfortunately, several things might happen:

1. He might find the sewage odor so intense that the cost of vanilla required would make the process uneconomical.

2. The vanilla fragrance could become pronounced enough to prove objectionable to nearby citizens.
3. The sewage and vanilla odors might combine to produce a new odor which could prove more obnoxious than the original.

A more subtle and scientific method of odor abatement exists based on the principle of modification (compensation). This involves lowering the malodor intensity and mixing it with selected chemical gases or vapors. It has been shown in experiments that when two substances of given concentration are mixed in a given ratio, the resulting odor may be far less intense than that of the separate components.

Eventually it may not be perceptible at all. This modification phenomenon is essentially a decrease of olfactory intensity. The results of these experiments show that modification occurs in the majority of cases where two substances are mixed in uneven proportions.

A pleasant smell and an unpleasant smell will sometimes neutralize (modify) each other so that neither can be recognized. The only possible explanation is that the compensation (modification) is internal, taking place in the central nervous system in the olfactory nerves at the nerve junctions or on the cortex.

There are many instances where a relatively small change of chemical structure will swing the verdict of the observer from agreeable to highly objectionable. Such materials may at one degree of concentration cause repulsion, whereas at another, usually highly dilute, will contribute a pleasant attractive odor. This dualistic effect becomes more interesting and remarkable when applied to giving foods flavor or odor. Concentrated indole, having a high repulsive fecal odor, is used in dilution to give chocolate its "kick." Hydrogen sulfide, at concentrations producing harmful physiological effects, loses its smell of rotten eggs, producing a pleasant odorous sensation.

The implementation of the modification concept has led to positive solutions for problems of nuisance malodors. Pragmatically speaking, there can be no objectionable odor unless there is someone to smell it and complain. This has brought about bathing malodorous air with gaseous odor-modifying chemicals before it reaches the community. The procedure involves setting up a screen of modifying gas between the source of a malodor and the source of complaints. Odoriferous air passes through the screen and mixes with the proper modifier gas. Many of these chemical screen installations are operating successfully in varied applications throughout the country. Waste disposal plants, industrial facilities, feedlots and lagoons have eliminated nuisance odor complaints.

An alternative method of odor modification was applied in the following instance. A major processor of soya oil products was faced with complaints concerning emission of malodorous vapors from a cooling tower. Samples of the liquid waste material were sent to the odor laboratories for analysis. After converting the liquid into a gaseous state and using established odor panel testing techniques, the malodor was described as a fatty, glue-like, rendering type.

The recognition threshold (RT) of the malodor was 0.084 ppm in air. In the odor chamber the RT of 0.084 ppm would be 0.63 mg of fat airborne in 256 cf. The fan output was 570,000 cfm, therefore:

$$\frac{570,000}{256} \text{ cfm} = 2227$$

2227 x 0.63 mg = 1403 mg fat blown per minute
1403 x 60 min = 84,180 mg/hr

$$\frac{84,180}{1,000} \text{ mg} = 84.18 \text{ gm/hr of fat}$$

$$\frac{84.18 \text{ gm}}{28.4 \text{ gm/oz}} = 3 \text{ oz of fat/hr}$$

The selected odor modifier had a recognition threshold of 0.006 ppm or about 1:14.

Tests indicated that 1 sec of the odor modifier sprayed into the chambers at a concentration of 0.01 percent was as follows:

1 sec	= 1.8 gm of 1800 mg
1800 x 0.01 percent	= 0.180 mg active
0.180	= 0.024 ppm

(Factor for Odor Chamber) 7.5
This will modify with:
4 sec spray of fat at 0.05 percent

1 sec	= 1.8 gm
4 sec x 1.18	= 7.2 gm or 7200 mg
7200 x 0.05 percent	= 3.6 mg of active
3.6	= 48 ppm

Conclusion: 0.024 ppm of odor modifier to modify 0.48 ppm of fat or 1.20.

This chemical was incorporated directly into the liquid waste effluent and allowed to vaporize in the same air stream as the malodorous gas effluent.

Admittedly odor pollution is the least directly harmful to health of the environmental nuisances, but it can seriously affect mental attitudes in the home and at work. Malodor conditions inside plants must be corrected to comply with the following rules set forth by the Williams–Steiger Occupational Safety and Health Act of 1970. (OSHA)

Subpart "G": Occupational Health and Environmental Health 1910.93

A. Exposures by inhalation, ingestion, skin adsorption, or contact to any material or substance (1) at a concentration above those specified in the 'Threshold Limit Values of Airborne Contaminants by the Conference of Governmental Industrial Hygienists' shall be avoided.
1910.94 Ventilation (A) (Vi)

(Vi) Clean air. Air of such purity that it will not cause harm or discomfort to an individual if it is inhaled for extended periods of time.

It has been well established that odor modification materials have their proper place in the abatement of malodor nuisances. There are, however, important considerations to think about before selecting this route.

1. No attempt should ever be made to modify or otherwise disguise any gas or vapor that may possibly be toxic or harmful.
2. Odor modification should not be used as a substitute for good housekeeping.
3. Odor control techniques are sophisticated and only trained and experienced personnel should be involved.
4. Where it is feasible and economical the malodor should be eliminated or destroyed at the source.
5. Odor-modifying chemicals must conform to existing specifications and regulations of EPA, OSHA, and other regulatory agencies, whether on a federal, state, or local level
6. Odor-modifying materials should effect a lowering of olfactory intensity.

ODOR CONTROL OF PAPER RECYCLING[13]

Paper recycling by one New Jersey company helped reduce solid waste but created air pollution problems. Although odors could not be quantified or measured definitively, when neighbors complain, a problem exists. The air pollution problem stemmed from small amounts of oil (150 ppm) evaporating into a large volume of air (120,000 scfm) picking up water vapor at the rate of 800,000 lb/day. These pollutants were exhausted from five huge dryers which remove moisture from fiberboard made of paper waste.

One source of oil was solvent-based inks in the newsprint. The oil produced a slight haze and accompanying odor which nearby residents claimed "smelled like wet newspapers." Air pollution, in the form of non-compliance with emission standards, did not exist. This fact was borne out in March 1968 after the New Jersey Department of Health evaluated the company's dryer stack emissions and pronounced them well below the legal limits, and safe for operation.

The following year, however, due to neighbors' complaints, the State served the company with a violation (based on odors) of Chapter 6 of the New Jersey Air Pollution Control Code. During preliminary hearings, the company entered into a "consent agreement" with the State whereby it would "take measures in compliance with Chapter 6."

Evaluating the Problem

In order to be sure that the company was, in fact, the source of the complaint, a consulting firm performed an odor survey around the plant and at neighboring apartments. Wind direction and general weather conditions were recorded. Although other sources of odor were detected, it was concluded that the plant did cause an odor in the area.

To define and measure the intensity of the problem, stack sampling tests were performed, measuring air flow rates, oil concentrations, humidity and temperatures.

The obvious way to solve the problem was by incineration. However, operating costs would have been unrealistic, and the idea was rejected. High water vapor content in the exhaust air would have required large amounts of fuel to incinerate the gas, even with heat reclaiming. Since there were no existing odor pollution standards to be met, all that was needed was for the company to stop annoying its neighbors. As a matter of practicality, the consulting engineers felt that the company did not need 100 percent efficiency to solve the problem. As a result of this thinking, it was projected that a simple, relatively low-cost mist eliminator-entrainment separator system would provide the proper cost/effectiveness combination.

When any business, regardless of its size, is faced with large capital expenditures for which it will reap nothing in return, management looks closely at all possible alternatives before making a commitment. For this reason, and at this time (while the prototype looked good), the company decided to investigate other methods and systems which might offer better performance and cost figures. During this period, it was suggested that the company recycle drying air, thus reducing stack air volume by about 30 percent.

Seven other methods of solving the problem were evaluated. The comparison chart (Table 7-6) shows the systems analyzed, performance and costs. As the chart bears out, the original proposal proved to be the best method. Based on the satisfactory results, five large systems, one for each dryer, were constructed (Figure 7-1).

After the units were installed and operating, another series of stack samplings was performed and test results submitted to the state. The results satisfied the state that the company had made every possible effort to eliminate its odor pollution problem, and a five-year operating permit was issued to the company.

OZONE[196]

Ozone is present in the atmosphere in varying concentrations. It is responsible for the refreshing scent during and following electrical storms. Measurements have been made of concentrations up to 0.3 ppm/v weight at sea level and up to 6 ppm/v at 80,000 ft. Ozone concentrating varying 0 to 3 ppm/v has been measured in the lower atmosphere. There is 0.5 ppm/v to 1 ppm/v in the Los Angeles Basin on a smoggy day.

A maximum allowable limit of 0.1 ppm/v has been set by a federal agency. This is substantially lower than the normal values above described. An area of misunderstanding is the relation of ozone to air pollution. Because of the adoption of the term "ozone count" in some localities in referring to an index of smog intensity, the term has served to create the impression that ozone is synonymous with air pollution. This is comparable

Table 7-6. Comparison Chart of Alternate Control Methods

Method	Performance	Installation Costs	Operating Costs
Low pressure-drop scrubber ($<$10 in. H_2O)	Poor efficiency	Not considered	Not considered
High pressure-drop venturi scrubber (10-50 in. H_2O)	Satisfactory at high pressure drops	High	Operating (power) costs very high; minimum two-year waiting period for necessary power to be installed by publicutility
Wet electrostatic precipitator	Erratic (oil coating of electrodes was suspected)	Not considered	Not considered
High-velocity filtering system	Poor efficiency	Not considered	Not considered
Low-velocity filtering system	Satisfactory (if exhaust gases cooled below water dew point)	Extremely high	Very high with exhaust cooling
Packaged entrainment separator system	Poor efficiency	Not considered	Not considered
Combustion in existing boilers (not tested)	Combustion calculations showed high moisture content of stack gases required special design (new boilers and air volume three times greater than required for boiler capacity)	Very high due to new special boiler requirement	Moderate to high due to raise in temperature of air and moisture from exhaust to boiler stack (300 F in, 450 to 550 F out)

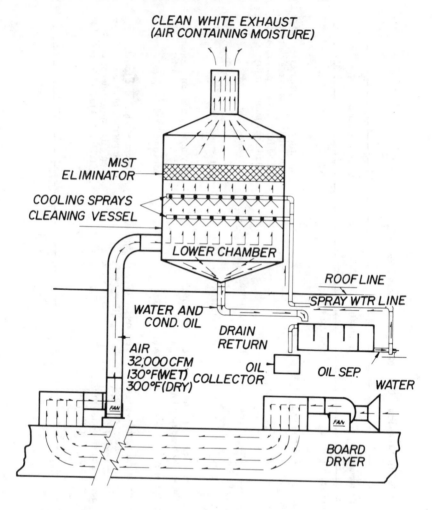

Figure 7-1. Schematic of odor control system for fiberboard drying oven.

to blaming the thermometer for a heat wave. Ozone is a measure, not a cause of smog. The ozone in the city environment is a combination of ozone generated by photochemical reactions in the smog plus that brought down from the upper atmosphere by inversion currents.

There has been much discussion about the effectiveness of ozone for killing odors. Part of the disagreement may be accounted for by the omission of the unit measuring ppm of ozone and lack of controlled standard measurements. Whether ozone is measured in ppm (concentration) by volume in air or ppm by weight can make a significant difference. (1 ppm/wt = 0.61 ppm/v, while 1 ppm/v = 1.65 ppm/wt.) There is a factor of nearly two difference between ppm/v and ppm/wt in air.

Recent lab tests and actual installation data have shown that if the proper amount of ozone is mixed well for the necessary retention time, with controlled temperature and humidity, most common odors can be destroyed. No ozone odor will be sensible.

Ozonated air must be in motion and mix well with the pollutant for effective action. If not, some areas in the same room may smell of the pollutant that the ozone was supposed to neutralize but never reached. In this case a person working in the room would not smell the pollutant anywhere in the room since his nose would have been desensitized by the high mixture of ozone in the cloud. If a second person entered the room, however, he would detect both the odor of the ozone cloud and that of the pollutant as he passed through different areas of the room. To neutralize odors effectively, therefore, it is necessary to have both the correct amount of ozone and a proper mixing factor.

Odor control by neutralization with ozone is a comparatively new application. Among the first and one of the largest sewage odor control installations using ozone is installed at Ward's Island in New York City. Ozone has also been used to combat odors effectively in commercial kitchens, cafeterias, food and fish processing plants, rubber compounding plants, and chemical plants. There is solid evidence that various food odors such as fish, onions, and acrolein from burnt fat can be oxidized by ozone. Tobacco and body odors can also be effectively controlled.

CHAPTER 8

AIR POLLUTION CONTROL
IN SELECTED APPLICATIONS

AUTOMATED AIR POLLUTION CONTROL
IN THE STEEL INDUSTRY[167]

Inland Steel Company has in operation one of the largest installations of electrostatic precipitators in the steel industry—an 11-story-high, 300-ft-long facility—to clean exhaust gases from its Indiana Harbor Works' No. 3 Open Hearth Shop.

Control instrumentation used at this installation is unusual; operators can visualize and regulate the entire precipitator functioning from a room containing a graphic control panel.

Use of a graphic-type control panel added to the costs of instrumentation, but is proving its value. It quickly communicates to operators any changes in conditions of the air pollution control system.

To understand the need for instrumentation at this plant, it is necessary to know how the precipitators are arranged. The graphic panel serves as a visual guide to the system operation. Each of the exhaust stacks from seven open hearth furnaces is equipped with a cap which can be closed by remote control. When the caps are down, exhaust gases flow into the precipitators.

A mixing chamber takes the gases from one or more stacks and distributes the mixture through an inlet manifold to one or more of the ten precipitator chambers where dust particles are collected.

Gases exiting from the precipitators enter an outlet manifold. Automatically controlled dampers regulate the exit flow from each precipitator. Five fans, each with a capacity of 400,000 cfm, draw gases from the manifold under control of separate dampers. Gases enter one common exhaust stack, where the final percent of solids can be checked. Approximately 1.6 million cfm of gas can be cleaned by the system, resulting in some 40 tons per day of dust removed.

293

Basic Control

Essentially, the prime variable affecting efficient performance of a precipitator is rate of gas flow (assuming the electrical fields are maintained and the plates clean). Contaminated gas must flow at a rate slow enough to permit dust particles to be removed. This flow rate can be controlled by the differential pressure across the precipitator. Precipitator chambers are designed basically to provide a reduction in gas velocity so that electrostatic separation can be effective.

On the other hand, too much restriction of flow in the system causes an increased load on the fan motors. If the system were to operate with dampers nearly closed, the motors would draw extra current because of the added load. Therefore, the control system maintains a balance between gas flow rate and fan load.

Another variable in the abatement system process is the number of furnaces in operation. One or more of the seven can be operated at a given time. Thus, load on the pollution control system varies widely. Precipitator operation is arranged so that gas flow is near the optimum rate in spite of the varying input. The system is designed so that nine precipitators are always in use, with one held in reserve. As load increases, the pressure drop or fan load increases.

Reasons for using the instrumentation are relatively secondary, but are nevertheless important—convenience of remote centralized operation, monitoring of variables, and providing the operator with a visible display of actual operating conditions.

The control system is divided into seven parts:

1. Flue pressure control system
2. Off-take and relief damper control system
3. Gas flow balancing system for precipitators
4. Monitors and controls for fans
5. Process temperature and pressure measurement
    a) Off-take temperature and pressure
    b) Precipitator and stack exit temperature
6. Air dilution gas cooling system
7. Other instrumentation, such as the annunciator panel

Most of the instrumentation used in this system, Figure 8-1, is conventional in design. Table 8-1 provides descriptions of instruments and their function in the system.

Flue Pressure Control

The purpose of this system is to maintain a constant negative pressure at the mixing chamber. It uses one differential pressure measurement at the chamber as a basis for control. Regulation of draft comes indirectly from control of the current flow to each of the fan motors. If the draft is not sufficient, the system acts to open the fan dampers.

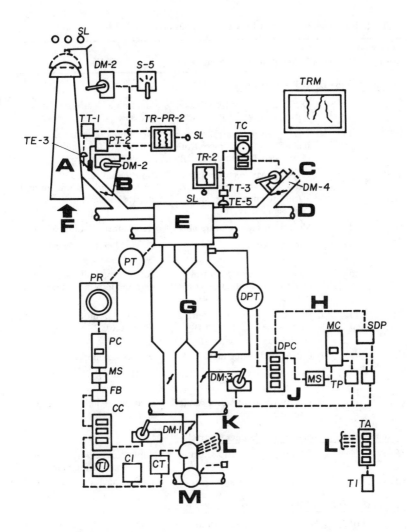

Figure 8-1. Control system instrumentation required for the operation of two precipitators.

A. H. H. stack
B. Off-take damper
C. Dilution air damper
D. Gathering flue
E. Mixing chamber
F. Zinc bearing gases

G. Precipitators
H. Deviation from set
J. Change set point
K. Outlet manifold
L. Motor and fan bearing
M. Stack

Table 8-1. Description and Function of Pollution Control Instrumentation

Diagram Symbol	No.	Instrument Description	Function
System A–Flue Draft Pressure Control			
DPT-1	(1)	Differential pressure transmitter. Range: 0 to -20" water.	Measure low differential pressure
DPC	(1)	Draft pressure control station including 3-function control (Error signal comes from DPR).	Determine set point of current controllers for fan motors, through MS-A.
DPR	(1)	Circular chart recorder (Range: 0 to -20" water) with low alarm to annunciator, and error signal detector.	Continuous record of flue draft pressure. Sound alarm if draft is too low.
MS-A	(1)	Motorized slidewire assembly.	Determine set points of 5 CC.
FB	(5)	Fan bias control	Provide "bias" in set point of each CC.
CC	(5)	Fan motor current control station with 2-function control unit for fan louver drives.	Position fan louvers. Indicate load and louver position.
CT	(5)	Current transducer. Input: 5 amps a.c.; Output: 0 to 10 volts, d.c.	Measure fan motor load. Send signal to CC
CI	(5)	Current indicating meter	Display fan motor current.
Ti	(5)	Timer with electrical contacts	Short out CT when motor is started
DM-A	(5)	Fan louver drive motor with reversing contactor	Position fan louvers to effect draft control.
System B–Off-Take and Relief Damper Control			
SS	(7)	3-Position pistol grip selector switch.	Manual operation of off-take valves and relief dampers.
DM-B	(14)	Drive motor torque.	7 pairs operate by SS.
System C–Gas Flow Balancing System for Precipitators			
PdT	(10)	Differential pressure transmitter.	Measure static pressure across each precipitator.
PdC	(10)	Differential pressure control station with set point and variable indicators	Control static pressure across precipitator by positioning damper on exit line.
MC	(1)	Master controller, including set point meter.	Adjust set point of 10 PdC's.
DM-C	(10)	Drive motor for each precipitator damper.	Position precipitator dampers
SPD	(1)	Deviation from set point meter.	Signals MC about dampers.
Tp	(2)	Pulse timers.	Manual adj. increase of MC set point.

Diagram Symbol	No.	Instrument Description	Function
System D—Smoke Density Measurement			
SmD	(1)	Smoke density detector	Detect and record smoke density.
SmR	(1)	Smoke density recorder	
System E—Fan and Fan Motor Bearing Temperature Monitor			
TI	(1)	Indicating temperature meter.	Indicate any temperature.
TA	(5)	Monitor for four (4) temperature points	Monitor inboard and outboard bearing temps. of fans and motors.
TC-E	(20)	Thermocouples	Detect temperature.
System F—Fan Oil Temperature and Pressure Measurement			
TC-F	(10)	Thermocouples	Measure and record fan oil temperature and pressure.
	(1)	2-pen recorder.	
System H—Air Dilution Gas Cooling System			
T/C-H	(2)	Thermocouples	Measure temperature in gathering flue for dilution.
CC-H	(2)	Control station, including: —Manual set point. —Deviation from set point meter. —Drive motor position indicator. —Auto/Manual station. —3-action control unit.	Position louvers on inlet air line via DM-H.
DM-H	(2)	Drive motor with limit switches	Position louvers in inlet air line.
TR-2	(2)	One-pen recorder. Range: 0 to 1200 F I.C.	Record dilution air temperature.
Other Instrumentation			
ANN	(1)	Annunciator panel with 16 points for panel alarm for failure of excess values of: 1) Screw conveyors 2) Precipitator power 3) Fan oil pressure 4) Fan oil temperature 5) 2400 Volt substation 6) Gathering flue temperature 7) Low flue pressure 8) O.H. stack high temperature 9) I.D. fan no. 1-5 trip 10) Fan bearing temperature 11) Fan bearing vibration 12) Lube oil low level	
SL	—	Various signal lights for indicating rapper operation: precipitator field sequence; precipitator outlet damper positions; fan lower full open or full closed positions; etc.	

The transmitter PT sends a measuring signal to the circular chart recorder PR which determines the set point of one master controller PC for all five fan motors. Individual current control stations (CC) form five control loops— each with its own current measurement (CT and CI) and a motor operator (DM-1) for its fan inlet damper.

The system includes controls for day-to-day operation: (1) means for limiting the maximum motor current (MS); (2) an alarm contact for cases when the flue draft drops too low (PR); (3) indication of individual fan motor currents and automatic-manual control so that individual current set points can be changed manually (PC); (4) indication on each fan damper controller (CC) of current set point, measured current and fan damper position; and (5) an automatic-manual station for manual positioning of fan dampers from the graphic panel.

A time (Ti) for each fan control system is used at startup, putting damper control on manual for a predetermined, timed period. When the fan is stopped, the timer resets and again switches the fan control to manual. Fan motors require a high starting current; the delay time allows the motors to attain full speed before controllers are placed on automatic. Once the fans have reached full speed, the dampers may be opened to draw the exhaust gases, without exceeding the maximum safe motor current.

Off-Take and Relief Damper Control

This system, on each of the seven stacks, determines whether gases pass out the open hearth stack or flow normally to the precipitators. Transfer from one condition to the other is made by the operator, using a pistol-grip selector switch on the console in the control room. This switch sequentially causes drive motors for stack caps and off-take dampers to run to "open" or "closed" positions (off-take opens when cap closes). Indicating lights show damper and cap positions system on the schematic diagram printed on the graphic panel.

Gas Flow Balancing System

This system measures the individual differential pressure across each of the ten precipitators (DPT and DPC), all set by one master control loop It maintains approximately equal pressure drops across all operating precipitators.

Each control loop operates to position its damper motor (DM-C), as dictated by the controller set point, to maintain the differential pressure across its precipitator.

A "floating set point" raises or lowers all precipitator set points. As long as none of the dampers is in its wide open position, the set point of all damper controllers will be slowly raised at a rate determined by a pulse timer. When one or more dampers reaches a wide open position, a limit switch in the damper motor stops the set point increase, and the set point remains at the value reached.

If any precipitator with wide open dampers has a differential pressure below its set point, the pressure set point signal to all precipitators will be decreased at a rate determined by a second pulse timer. A deviation-from-set-point meter with each controller measures the amount by which the differential pressure is below the set point, and permits the pulse timer to function only if the deviation exceeds a fixed amount.

Fan Monitors and Controls

Associated with the fans are two instruments systems: (1) fan and fan motor bearing temperature monitor, and (2) fan oil temperature and pressure recording. The first system indicates either the inboard or outboard bearing temperatures of the fans and their motors.

Process Temperature and Pressure Measurement

For recording off-take temperature and pressure (draft), a two-pen recorder is used for each of the seven stacks. Each instrument contains a high-limit contact to light a signal indicating to the operator which stack has too high a temperature in the incoming gases. Point 8 on the Annunciator Panel, common to all seven alarm contacts, sounds an audible alarm too.

Air Dilution Gas Cooling System

Two safety systems are provided to cool gases with outside air if required. Operating from the same thermocouples used by TR-2, each system has a control station which actuates a damper motor operator (DM-4).

This graphic panel control instrumentation concept can be adapted to systems in other industries. The pollution engineer can specify control panels when purchasing a new system, or even make his own from instrumentation furnished at the time of installation. Measurement and control instruments help operators get top performance from the abatement system and prevent unscheduled shutdowns, as might be caused by an overheated fan motor or a clogged precipitator chamber.

AIR POLLUTION CONTROL IN FOUNDRIES[20]

Air pollution control in metal casting foundries presents unique problems. Figure 8-2 shows a simplified flowsheet of the process operations found in most malleable iron casting foundries. The operation involves two major steps—the manufacture of white iron castings and the conversion of the castings, through heat treatment, into malleable iron products. The charge can be melted either by an air or electric furnace, or by a duplexing operation in which the charge is initially melted in a cupola and the molten metal transferred to a refining furnace. Large foundries favor duplex operations because of economy and ease of operations.

Air pollutants are emitted from many foundry operations—melting furnaces, refining furnaces, sand handling and shakeout systems, core ovens, and heat treatment ovens. However, the major problem is controlling the quantity of contaminants exhausted from cupola melting furnaces.

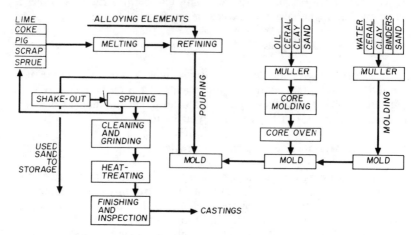

Figure 8-2. Flow diagram for malleable iron casting production.

Effluents from cupola furnaces can consist of gases, dust, smoke and oil vapor. Exhausts are discharged at temperatures ranging from 500 to 1200 F with dust loadings ranging from 0.5-10 gr/scf, in the 50-micron and smaller size. Factors which influence the amount of pollution exhausted from a cupola include method of charging, iron-to-coke ratio, quality of scrap melted, and the amount of infiltrated air.

When designing a collection system for a cupola, the process variables should be evaluated. The takeoff from the cupola can be below the charge door, thereby using the charge as a seal. Or, if the takeoff is above the charge door, the top of the cupola must be sealed. Additionally, the infiltrated air volume will be affected by the dimensions of the charge door and frequency of charging. Use of an enclosure can reduce the infiltrated air but presents some operating problems.

Since the gases are exhausted at temperatures of approximately 1200 F, they must be cooled prior to treatment. This can be accomplished by surface cooling, dilution or flash cooling. Each method has limitations. Direct surface cooling of high gas volumes can require expensive equipment to provide the necessary heat transfer area. Cooling by dilution results in an increase of treatment costs. Flash cooling increases the exhaust volume as a result of the moisture in the gas.

Selection of cupola treatment control equipment is dependent on the degree of cleaning required as dictated by local pollution codes. Since there is a lack of uniformity in the codes from one community to the next, it is important to determine local requirements.

Treatment equipment can be classified according to efficiency. Wet-top stack wasters, simple cyclones and medium-pressure-drop wet collectors, such as orifice scrubbers and wet centrifugal cyclones, can be used where design criteria are not severe. Where rigid control is necessary, high-energy scrubbers, electrostatic precipitators and bag filters should be used.

Baghouse filter systems are versatile for removing particulates, since a system can handle a large variation of micron-size dusts and a wide fluctuation in flow rate. However, a baghouse system must be carefully designed since the fabric bags are subject to abrasive wear and are affected by moisture.

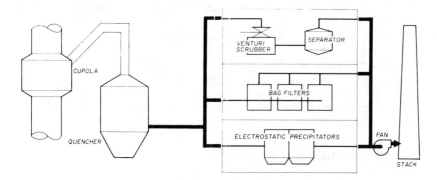

Figure 8-3. Primary methods for controlling air pollutants from cupolas.

High-energy scrubbers are capable of removing vapors as well as dusts from the cupola exhaust. Scrubbers can operate efficiently with alkaline solutions to remove significant amounts of sulfur dioxide. The high-energy scrubbing system has the advantages of handling hot gases, sticky and dry dusts, requiring a small amount of plant space and affording a safe system for fire protection. However, scrubbing systems require some type of clarification equipment to clean the liquid discharge. In addition, they require high energy for fine dust removal and large quantities of water in some operations. Corrosion problems in ducts and emission of a visible steam plume are added problems which must be evaluated.

Electrostatic precipitators can be used for high-efficiency dust collection. However, the characteristics of the dust are important in evaluating the efficiency of this system. Primary collectors such as cyclones are often used to reduce the dust loading of large particles to a precipitator. These systems have the advantage of low pressure drop and the capability to handle moisture and oils. They do, however, have the disadvantage of being sensitive to flow variations, temperature and humidity, and require a means for disposing of collected dry dust.

Afterburners are generally installed in a cupola control system to eliminate the explosion hazard from carbon monoxide concentrations in the exhaust. In addition, combustible pollutants such as oils are burned. Afterburners can be installed as separate units or in the upper portion of the cupola.

Although control of pollutants from cupolas is the primary air pollution problem in a foundry, many contaminants are released from sand handling, shakeout, grinding, chipping and heat treatment operations. Emissions from refining furnaces are usually not a significant problem because of the quality of the melt. Where pollution control is a problem, bag filters, scrubbers or electrostatic precipitators are used. Contaminants exhausted from core ovens consist of organic vapors and smoke discharged at relatively low emission rates. When this emission presents a nuisance, afterburners are used for pollution control. Where control equipment is required to abate emissions from sand handling operations, baghouse filters and wet scrubbers are commonly used.

Pollution from heat treatment processes is usually not a major problem. Control is usually taken care of by proper selection, maintenance and operation of the production equipment. Because each foundry has many operations, pollution abatement facilities should be carefully designed to efficiently and economically control as many processes as possible.

FLYASH EMISSIONS ESTIMATING[159]

Air pollution control regulations are expressed in many different ways by regulating agencies. The nomograph (Figure 8-4), based on EPA emission factors (Public Health Service Publication No. 999-AP-24), provides a quick method for making a rough estimate of the pounds of particulate discharged to a receiving stack or dust collector, and the emission in pounds of flyash per million Btu.

Example

How many pounds of flyash will be discharged to a dust collector with an underfed stoker burning 5 tons of coal per hr having a heating value of 13,000 Btu/lb and 10 percent ash content? What is the emission in lb flyash per million Btu?

Solution

(1) Connect reference point for underfed stokers on **A** scale with 5 tons/hr on **C** scale and mark where line crosses Pivot Line No. 1.
(2) From this marked point line up with 10 percent ash on **E** scale and read 250 lb/hr particulate discharged to the dust collector where the line crosses **B** scale.
(3) Connect 5 tons/hr on **C** scale with 250 lb on **B** scale and extend to Pivot Line No. 2.
(4) From this point connect with 13,000 Btu/lb on **F** scale and read 1.92 lb flyash per million Btu where line crosses **D** scale.

This value is often used to select the type of ash collector needed. For example, if regulations limit the emission to 0.50 lb of flyash per

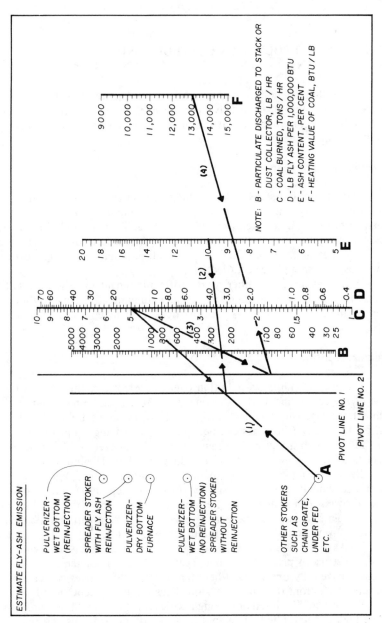

Figure 8.4.

million Btu, the collector efficiency required would be at least:

$$\left(\frac{1.92 - 0.50}{1.92} \right) \ 100 = 73.96\%.$$

MINIMIZING AIR POLLUTION CONTROL COSTS IN OLDER PLANTS[55]

Controlling air pollution in an existing plant requires both study and ingenuity. An older plant usually does not contain equipment designed for easy control of contaminants, and plant layout may make installation of hoods and ducts difficult.

Surveying the Problem

One of the first steps is determining the sources of contaminant emission, quantity, size, physical state, and carrying medium of the pollutants. Emission sources such as process stacks and vents should be sampled isokinetically, and volume flow measured. If flow is intermittent, duration should be measured. When flows are cyclical, peak and minimum flow rates as well as time should be recorded.

A general survey of other sources of contaminant loss should also be made. Raw material unloading and product shipping are frequently areas where contaminants can be released, particularly if materials are handled in bulk.

Minimizing Effluent

One way to minimize pollution control costs is to eliminate the release of the contaminant. Leaky conveyor systems can be modified to eliminate spillage and dusting. Airtight enclosures can be built around conveyors, or cover housings can be equipped with flanges and soft gaskets. Most idler shaft bearings can be redesigned and relocated inside the conveyor housing to prevent dust leakage along the shaft (Figure 8-5).

Before spending money to modify or improve an aged conveying system, consideration should be given to its condition and annual maintenance costs. If a system is old, in poor condition, and has several dusty transfer points, it may be cheaper to replace the system with an enclosed type to convey around corners without transfer points.

Tanks and bins should have air-tight covers and joints sealed with cemented strips of plastic or rubber sheeting. Each bin or tank should normally have only one vent. Where temperature differences can create a natural draft from the vent, other openings should be kept closed.

Where possible, interconnecting a series of vents should be considered. If one tank is being filled while another is being emptied, this will reduce the need for venting air to the atmosphere. It may also be possible to make one small dust collector serve a number of bins.

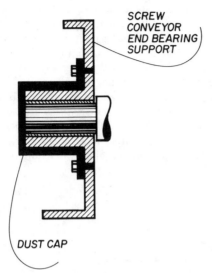

Figure 8-5. Air pollution from screw conveyors can be
minimized by modifying the idler end bearing to
fit it with a dust cap.

When contaminants are a by-product of a production operation, generation often can be minimized by changes in operating conditions. Lower
combustion temperatures will reduce nitrogen oxide formation. Fluorine
compounds released when some ores are heated frequently can be decreased
if water vapor is eliminated from the atmosphere. Thus, pre-drying ore at
low temperatures and use of low hydrogen-content fuels may be beneficial.
While fluoride compounds can be released from the cooling of calcined ores,
the amount released may be decreased if temperature quenching is increased.

When release of contaminants cannot be entirely prevented economically,
use of existing plant equipment should be considered for their partial containment. If contaminant gases are handled by an exhaust fan, it may be
possible to obtain some collection in the fan by adding a water spray to the
inlet. Such a system can collect an appreciable amount of dust and some
fumes and mists. Collection efficiency will increase as the fan tip speed is
increased.

Data indicate that 50 percent of 1-3 micron size particles can be collected in a fan with a tip speed of 16,000 ft/min. A water flow rate of 1.3
gpm/ft of fan wheel circumference is adequate.

Water should be introduced through a solid cone spray nozzle sized
to pass the flow at a pressure of 10-20 psig. The nozzle should be centered
in the fan inlet duct so that an extension of the cone angle of the spray
will just intersect with the diameter of the fan inlet as it enters the fan
casing.

A spray should not be installed in just any fan. The fan casing should be equipped with a bottom drain. Some water will be thrown out the discharge outlet of the fan, so a collection point and drain are needed. Spray systems work best with a fan having a bottom horizontal discharge arrangement. Performance of upblast fans may be hurt by the column of water droplets attempting to return to the fan.

Hooding should be designed to assure complete capture of contaminants, but its openings kept to a minimum with as little air evacuated as needed for complete capture. Figure 8-6 shows a phosphor bronze melting pot which was initially equipped with a large open draft hood. Changing the hood when addition of a fume scrubber was required reduced the scrubber size.

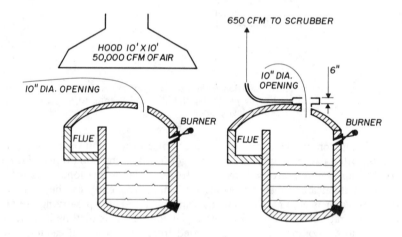

Figure 8-6. The retort at the left shows the common method of constructing a hood to capture fumes. At the right, the modified hooding arrangement permits installation of a scrubber only 1/77th size required formerly for fume abatement.

Reducing Inerts and Recycling

In many direct drying processes where the product is dried by direct contact with hot combustion gases on a one-pass basis, effluent can be reduced by recycling some of the flue gas to temper the incoming combustion products.

Where atmospheric air is used in direct contact with a hot product for cooling, it can become contaminated and may not be suitable for combustion air. In this case the air may be cooled and recycled back to the cooling operation instead of using fresh makeup air.

Volume Reduction by Cooling

Many dust collectors such as bag filters and precipitators are sized more on a volume basis than a mass flow basis. If the effluent is a hot dusty gas, cooling can reduce the volume appreciably. Cooling by addition of cold air is the poorest method cost-wise. Radiation panels, finned surfaces, waste heat boilers, forced-convection heat interchange, and direct spray cooling with water are all possible solutions.

New Equipment

Careful consideration should be given to alternate types of collection equipment. For instance, is it cheaper to reprocess material when collected dry or wet? If the collected waste material cannot be reprocessed or made into a saleable product, what further treatment is required so that it does not become a pollution problem? Disposal should take into consideration the following possibilities:

Dry Collection

1. Can wastes be piled or will surface winds entrain the material?
2. Can rains wash material into surface streams or carry it into ground water as pollution?
3. Is it necessary to surface-treat dumped piles to prevent windage loss?
4. Is burial necessary?

Wet Collection

1. Can material be sewered?
2. Is neutralization required for sewering or for impounding?
3. If impounded, does the pollutant have appreciable vapor pressure so that it will still be lost to the atmosphere from the surface of the pond?
4. Does seepage from the pond into ground water constitute a pollution hazard? Must pond walls and bottom be made water-impermeable?

The processing conditions venting an effluent also must be considered. Is processing done in batches, or continuous? Is the effluent flow steady and uniform, pulsing, or cyclical? Collection equipment should be designed for the required efficiency over the entire range of operating conditions expected.

Apply engineering principles and ingenuity to design of new collection equipment. For instance, in spray chambers, collection efficiency for gas adsorption can be increased at given liquid flow rates by using smaller spray nozzles which atomize the liquid finer. This increases the surface area of the liquid particles in the chamber for absorption. If dust collection is involved, the desired effect is impingement of the spray droplets on the dust particles. Here the maximum number of spray droplets in the 200- to 800-micron size range is the result. Also, use of wetting agents should be considered for better capture of dust particles.

ALUMINUM BONDING OVEN EMISSION CONTROL[153]

For many years a Wisconsin manufacturing company has been producing heat exchangers. In order to metallurgically bond the aluminum fins and tubes together, the mechanically bound units (known as "cores") are wetted by a specially formulated bonding liquid, and then passed through a gas-fired oven. The oven temperature necessary for bonding the fin and tube surfaces is in excess of 1000 F.

As the cones approach bonding temperatures, extremely dense white fumes begin to evolve. Initially, it was theorized that the fume was a particulate aluminum compound, and that a great deal of some type of acidic compound was associated with the particulates. Later extensive testing and detailed studies revealed that the particulate was primarily a submicron aluminum oxide, hydroxide, and/or chloride combination. The associated acid gas was predominantly hydrogen chloride.

In order to cool the oven emission and to scrub the fumes, the gas was passed through a water spray chamber before being vented to the atmosphere. This system did not appear to satisfactorily control the emissions, and the plume continued to throw a bluish-white haze across the countryside. Further analysis and testing continued.

It was then theorized that the fume from the heated cores was sublimed aluminum chloride. It was further assumed that the $AlCl_3$ contacted the moisture in the water sprays, the air, and/or the combustion products and reacted in the following manner:

$$AlCl_3 + 3H_2O \rightarrow Al(OH)_3 + 3HCl$$

This, of course, would explain the high acid content of the gases.

Based on the above facts and assumptions, a fiberglass scrubber was purchased for installation on a prototype operation. It was a high-energy venturi unit designed for 12,000 scfm at 27 in. H_2O vacuum. The fumes were drawn through the unit by two fiberglass fans in series, each capable of 12,000 scfm at 15 in. vacuum. Both fans eventually failed due to the high centrifugal forces involved. The combined power required for the two fans was 100 hp.

A centrifugal mist eliminator was used between the fans and the venturi scrubber. The mist eliminator did not prove adequate, and the fiberglass stack "rained" with entrained moisture. Scrubbing water was fed to the converging section of the venturi at about 100 gpm, and recirculated. A small settling tank removed a great deal of the scrubbed particulates, and the acidity was removed with intermittent lime additions.

Many modifications were made to this prototype scrubber to improve efficiency and to provide more trouble-free operation. The venturi section was originally quite long, and cracked several times. It was replaced with a shorter throat which proved effective. Two mild steel fans replaced the fiberglass units. These corroded frequently due to poor pH control. However, pH control was made more precise and totally continuous, using ammonia as the neutralizing agent. An additional mist eliminator was added, consisting of 2 ft of 1-in. ceramic packing. With the experience gained, a full-scale scrubber system was designed and installed.

Scrubber Design

Figure 8-7 illustrates the flow of both the gas stream and the scrubber liquor. The system handles 21,000 scfm at 28 in. H_2O pressure drop across the scrubber and mist eliminator.

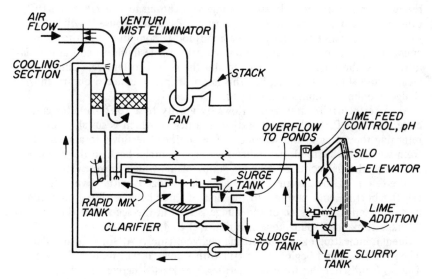

Figure 8-7. Schematic of emission control system for aluminum oven.

Fumes leaving the oven (800 F) are pre-conditioned and cooled by direct contact water sprays (fresh tap water) in an Inconel section of the ductwork. The velocity of the gas is about 3800 fpm upstream of the sprays, and is not slowed by enlarging the ductwork at the cooling section. The temperature downstream of the sprays is about 200 F, and is monitored by a thermocouple. If the temperature should rise above 250 F, a damper is automatically activated which bypasses the hot gases to a "hot" fan and stack without scrubbing. This provides protection for the fiberglass scrubber.

The Inconel cooling section is attached to the scrubber inlet, where an additional 150 gpm of recirculated water is added. The equilibrium water temperature is about 130 F as is the wet bulb temperature of the gas exiting the scrubber. The scrubber manufacturer, in a rather unique design, has provided a structurally stable, easily maintained unit, with an integrated mist eliminator section. The mist eliminator works efficiently with 1½ ft of 1-in. plastic packing material.

The fan is mild steel and handles the total air flow. It is driven by a 200-hp motor, which draws about 150 hp during normal operation. The stack, 40 ft above grade, is mild steel. Fiberglass would have been a better choice, it is now realized. Corrosion is caused by moisture condensing on stack walls, and the amount of entrainment is minute.

Water leaving the scrubber flows by gravity to a rapid mix tank where lime slurry is automatically mixed with the discharged water using automatic pH control. The pH is held at about 10.5 to ensure that the entrained moisture entering the mild steel fan and stack has a pH of at least 6.0 to retard corrosion of the mild steel. Lime was chosen to neutralize the HCl, primarily because it costs less than caustic soda, and because it will cause fluorides to precipitate. There are some fluoride compounds in the oven emission. Ammonia was ruled out since it is not desirable to add ammonia-nitrogen to the wastewater discharge and the odor is offensive. Lime does create more sludge ($CaCO_3$), but this is not of prime concern due to the sludge handling techniques employed.

After neutralization, water flows by gravity to a 16-ft-diameter concrete clarifier. Particulates are settled to a degree which is suitable for returning the water to the scrubber. The clarifier overflows to a surge tank from which the clarified, neutralized water is pumped back to the scrubber.

The clarifier contains a sludge-raking blade, which moves settled sludge to the center of the clarifier. Once each day sludge is withdrawn from the clarifier bottom, and pumped to the first of three large settling ponds connected in series. No trace of the sludge is seen in the second or third pond (average suspended solids concentration is 3 to 4 mg/l in the third pond effluent), indicating excellent settling characteristics. At some future date, when the first pond begins to overfill with sludge, it will require dredging. The sludge discharging from the clarifier is at least 4 percent solids.

Hydrated lime consumption is approximately 70 lb/hr. To avoid a dusty situation, lime is loaded into a vented elevator with filter, which carries the powder to a storage silo. The silo is emptied by a mechanical feeder. The feeder and a tap water line are both actuated by the pH sensing controls. Lime and water are mixed and overflow to the rapid mix tank. A vibrator is required on the storage silo to avoid bridging. Weekly "rodding" avoids plugging of the 1½-in. lime slurry overflow.

Cost and Performance

Table 8-2 lists costs of equipment and installation. The clarifier, lime feeding system, stack, rapid mix, and surge tanks are all designed to handle two scrubbers in the event another oven and scrubber is required.

Both the gaseous influent to the scrubber (upstream of the cooling section) and the scrubber discharge were monitored at full production.

Table 8-2. Capital Expenditures

Scrubber and entrainment separator	$ 9,300
*Circular clarifier, excavation and concrete	7,000
*Clarifier sludge rake	5,500
*Lime feeding system, plus pH controls	6,200
Fan 20,000 scfm at 32 in. H_2O	2,600
Fan Motor	2,000
**Installation, plumbing, electrical, stack, spare parts, mixers, instrumentation, by-pass fan and stack, ductwork, foundations, etc.	37,400
TOTAL	$70,000

*Items are designed to handle two scrubbers.
**Will not require duplication if a second scrubber is installed.

Samples were withdrawn isokinetically. An S-type pitot tube was used to measure gas velocity while simultaneously withdrawing samples.

The sampling train consisted of a stainless steel heated sample probe, a heated cyclone and discharge flask for particulates > 5 microns, a heated glass fiber filter for 100 percent capture of particles > 0.3 microns, and four ice-bath-cooled impingers. Eight sampling points were used in the scrubber inlet, and 16 points at the stack discharge. Table 8-3 summarizes the results of the testing. Both inlet and discharge particulates had essentially the same analysis. Gaseous contaminants were 99+ percent HCl.

Table 8-3. Stack Sampling Data

	Inlet	Discharge	Percent Efficiency
Flow Rate	33,800 acfm	25,900 acfm	−
Temperature	800 F	135 F	−
Moisture	1.3% H_2O (by weight)	8.7% H_2O (by weight)	−
Particulate	0.63 lb*/ 1000 lb gas	.040 lb/ 1000 lb gas	93.5
Gaseous Contamination	0.82 lb HCl/ 1000 lb gas	.0025 lb HCl/ 1000 lb gas	99.5+

*This number is erroneously low. Much of the particulate condensed inside the probe, cyclone and filter inlet. Only that which was captured on the filter was weighed. (The cyclone discharge flask was empty.) A conservative estimate might be 1.2 lb/1000 lb gas or more.

ALUMINUM AND COPPER RECYCLE PROCESS EMISSION CONTROL[105]

The high cost and limited supply of primary copper and aluminum make recycling attractive. However, recycling processes can create air and water pollution problems which, in some cases, are more severe than those associated with the manufacture of primary materials.

Aluminum Reclamation

Aluminum for secondary melting comes from three main sources: aluminum pigs (which can be primary or secondary reclaimed metal), foundry returns and various types of aluminum scrap. Since the scrap contains impurities or extraneous alloy metals, it is usually necessary to flux the molten metal to remove unwanted materials. Fluxing of secondary aluminum is responsible for most of the air pollutants emitted from this process.

Thin-sectioned aluminum scrap is reduced in size by a chipper. If clean, the metal is melted directly in a reverberatory or crucible furnace. It is common practice to melt heavier scrap first and add lighter scrap below the melt surface to prevent further oxidation. Dirty scrap contaminated with oil, grease, paint, etc. is burned clean in a chip dryer prior to secondary melting.

Melting Furnace

Secondary aluminum melting is a batch process of charging, melting, fluxing and pouring operations. A "heat" can vary from 4 to 72 hours. Smaller aluminum melting operations may be carried out in crucible furnaces having capacities up to 1000 lb. However, most medium to large secondary aluminum melting operations take place in reverberatory furnaces. These furnaces are usually gas- or oil-fired and range in capacity from 2½ to 50 tons or greater. A charging well is frequently used on aluminum reverberatory furnaces, which permits chips and other aluminum scrap to be charged and immersed below the liquid level. The charging well can also be used as a point to flux the molten metal with a gaseous halogen flux. It is desirable to use a submerged type hood in the charging well to efficiently exhaust the acid gases generated during fluxing (Figure 8-8).

Fluxing

There are various types of fluxes. Cover fluxes are used to cover the surface of the molten metal to prevent oxidation. These fluxes are usually salts, such as sodium or calcium chloride. A flux containing aluminum fluorides and chlorides is also used. These fluxes cause oxides and dirt to rise to the top of the molten metal where they can be skimmed off as dross. Aluminum chloride and fluorides can be emitted as a sublimed vapor from these fluxes, and are potential air contaminants.

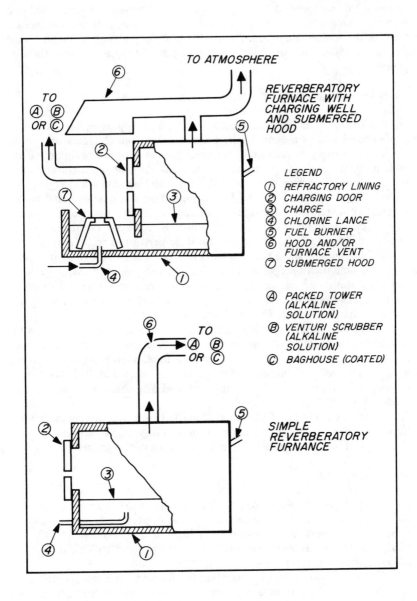

Figure 8-8. Secondary aluminum smelting in reverberatory furnace.

Degassing and demagging fluxes are used to purge the metal of dissolved gases and to reduce the magnesium concentration of the alloy. The most common fluxing agent of this type is gaseous chlorine. Chlorine is fed under pressure through tubes or lances to the bottom of the melt and bubbles up through the molten aluminum metal. While fluxing a typical 50-ton furnace, five or six 150-lb chlorine cylinders can be used during a two-hour fluxing period.

Air Pollutants Generated

If the scrap metal is adequately burned in a chip dryer, smoke and odors should not be present in the emissions from the reverberatory furnace. However, a variety of halogen compounds are emitted during fluxing of the molten metal. Sodium, potassium, magnesium and aluminum chlorides and fluorides can be emitted from both the cover and degassing/demagging fluxes. These emissions vary both in composition and quantity, and are a function of instantaneous conditions during the fluxing operation. Since halogen emissions vary widely, they cannot be accurately estimated from known emission factors. Stack tests must be performed to determine specific amounts of these contaminants generated from any given process.

Aluminum and magnesium chlorides formed from the use of gaseous and cover fluxes are unusual contaminants that change from solid to gas to solid. At high melt temperatures, aluminum chloride sublimes to a vapor, which can later condense and react with water vapor to form hydrochloric acid (HCl) and aluminum oxide particulate. One firm tested its process to estimate chlorine and hydrochloric acid emissions from fluxing operations and found that, for every pound of chlorine used, approximately 15 percent is emitted as gaseous chlorine and 30 percent is emitted as HCl. The remaining chlorine reacts to form aluminum and magnesium salts which are collected in the dross.

The vast majority of particulates emitted from the fluxing cycle are in the submicron particle size range. Consequently, highly efficient air cleaning systems are required to remove them from the gas stream.

Air Cleaning Systems

Chip dryers should be equipped with suitable afterburners or other control devices to control smoke and odors generated from burning dirty scrap. Afterburners should maintain a temperature of 1400 to 1600 F, with exhaust gas residence time of ½ to 1 sec for complete oxidation of all combustibles present.

There are two ways of exhausting the halogen materials generated from fluxing in the reverberatory furnace. If the furnace does not have a charging well, fluxing is usually carried out within the furnace. Here, the halogen gases are mixed with products of combustion and are vented to a control device. This results in a large volume of hot gases passing through the air cleaning system. If a furnace has a charging well and uses an efficient hood (such as a submerged type) to collect halogen contaminants, the combustion gases can be excluded from the air cleaning system. This allows use of a smaller, more efficient air cleaning system.

Caustic packed tower scrubbers can be very effective in controlling the halogen gases generated during fluxing. However, packed towers may plug if particulate loadings are high. Some firms have successfully used venturi-type caustic scrubbers to remove particulates and halogen gases simultaneously.

A new approach for controlling pollution from secondary aluminum melting furnaces is the use of the coated baghouse. These baghouses have been used in the primary aluminum industry for some time. A standard baghouse (fabric dust collector) is coated with a material that adsorbs and neutralizes the acid gases while simultaneously filtering out fine particulates. A Canadian firm has been fairly successful in using a coated baghouse to control its reverberatory furnace fluxing emissions. Their fiberglass filter bags are coated with a powdered alkaline adsorbent at the rate of about 800 lb/week. The coating is applied through the hopper discharge chute when the baghouse is under negative pressure. The baghouse pressure drop is about four in. of water at the start, and increases to a maximum of 14 in. within a week's operation—an increase of about two in. of pressure a day. When the maximum pressure drop is reached the bags are shaken and the contaminated coating is removed. A new alkaline coating is then applied to start another cleaning cycle. Tests appear to indicate that this system can be highly effective in removing both gaseous and particulate emissions.

Gaseous Control Requirements

Since secondary aluminum melting operations vary and no two fluxing operations are alike, halogen gaseous emissions must be verified by stack tests. Emission rate potential is the quantity of gases that could be discharged to the atmosphere in the absence of air-cleaning equipment. Once the maximum emission rate potential is determined, allowable emissions are computed for the respective source rating (Figure 8-9). For **B** rated sources (most secondary aluminum plants) air-cleaning efficiencies in excess of 90 percent are required to control gaseous halogen contaminants.

Particulate Control Requirements

Some metallurgical operations, such as those in the steel industry, can be regulated on a process weight basis. However, since particulate emissions from secondary aluminum processes are a function of the fluxing cycle, which is variable, the State of New York does not feel that the process weight approach is applicable. Instead, they control particulate emissions for B and C rated sources on an allowable concentration basis. Existing sources must meet a value of 0.3 lb of particulate per 1000 lb of exhaust gas concentration. New sources are required to meet 0.05 gr/scf of undiluted exhaust gas. Smoke is limited to 20 percent opacity.

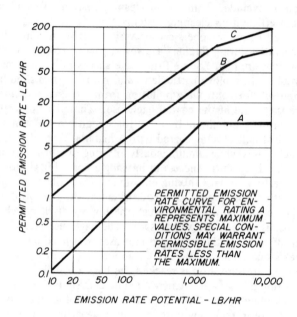

Figure 8-9. Emission rate permitted by New York State for processes, exhaust and/or ventilation systems of gases and liquid particulates (environmental rating A, B, C) and solid particulates (environmental rating A).

Copper Reclamation

The reclamation of copper begins with collection of scrap by a dealer within a given territory. Collection and transportation costs become an important factor for scrap dealers, and for this reason small operations generally confine their activity to within a 100-mile radius. A large portion of reclaimed copper comes from scrap wire. The wire can range in size from large utility power cables down to common household and telephone wire. Often the wire is an integral part of a device such as a transformer, alternator or other electrical equipment. Before the wire can be reclaimed it is necessary to sort out just the copper wire and remove it from the item to which it is attached. Almost universally this operation is accomplished by hand cutting and sorting. Copper in forms other than wire is generally separated by cutting, sawing or melting of soldered joints.

Once the wire is free of the device, it can be stripped by machine if the cable is large, or the insulation can be burned off under controlled conditions. Burning is generally accomplished by use of a multiple-chamber furnace for batch operations (Figure 8-10) or by combustion in a continuous rotary kiln (Figure 8-11).

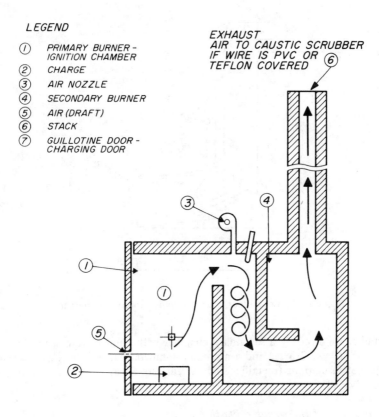

LEGEND

①   PRIMARY BURNER – IGNITION CHAMBER
②   CHARGE
③   AIR NOZZLE
④   SECONDARY BURNER
⑤   AIR (DRAFT)
⑥   STACK
⑦   GUILLOTINE DOOR – CHARGING DOOR

EXHAUST AIR TO CAUSTIC SCRUBBER IF WIRE IS PVC OR TEFLON COVERED

Fiqnre 8-10. Batch wire reclamation multiple chamber incinerator.

The incineration process creates the serious air pollution threat. The magnitude of the problem and the required air pollution controls are directly related to the type of wire being burned. From an air pollution standpoint, wire insulation should be separated into three categories: (1) halogenated plastics, (2) non-halogenated plastics, cotton, silk, paper, and rubbers, and (3) metallics. Combustion of the first group can result in a variety of emissions including HCl and HF. The second group of insulators can be burned to the final end products of CO_2 and H_2O under ideal conditions of sufficient temperatures and residence time. Common values for these parameters range from 1400 to 1600 F and 0.3 to 0.6 sec in the primary combustion

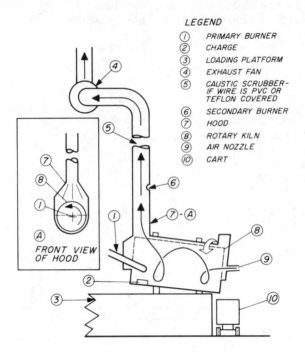

LEGEND

①	PRIMARY BURNER
②	CHARGE
③	LOADING PLATFORM
④	EXHAUST FAN
⑤	CAUSTIC SCRUBBER- IF WIRE IS PVC OR TEFLON COVERED
⑥	SECONDARY BURNER
⑦	HOOD
⑧	ROTARY KILN
⑨	AIR NOZZLE
⑩	CART

FRONT VIEW OF HOOD

Figure 8-11. Continuous wire reclamation rotary kiln burner.

chamber followed by a secondary chamber with good mixing, a residence time of at least 2/3 sec and a minimum temperature of 1400 F. The third group of insulations (metallics such as lear) are always mechanically stripped and salvaged.

Combustion of Halogenated Plastics

Halogenated plastic insulations commonly include polyvinyl chloride (PVC) and some Teflon. Even under optimum combustion conditions, combustion of these plastic insulations results in release of HCl and HF respectively. The quantity of HCl emitted has been reported to be about 58 percent of the weight of the PVC burned. Incomplete combustion can result in release of a wide range of air contaminants in addition to HCl. These include aromatic hydrocarbons and small quantities of aliphatic and olefinic compounds which are known to be photochemically reactive and generate an odor problem. Particulate emissions occur in small quantities, but can be more significant if inorganic fillers are used in PVC.

Air Cleaning Systems

Control of emissions from burning rubber or non-halogenated plastic covered wire is accomplished by careful design of the furnace and/or the use of thermal afterburners.

Emissions resulting from combustion of PVC are not as easily controlled. The potential emissions consist of HCl, small quantities of particulate, and hydrocarbon gases which can cause an odor problem. Several approaches have been tried to control these emissions. In one installation PVC wire was burned and the exhaust gases were passed through wet scrubbers, an afterburner to maintain temperatures above dew point, and finally a baghouse. In this case organic materials from the pyrolysis of PVC coatings were not adequately burned, causing odorous gaseous emissions and sticky particulates capable of plugging the filter media.

The use of a wet caustic scrubber in conjunction with a mist eliminator is another approach; however, this scheme cannot effectively control the hydrocarbon gases which result in odors.

The most effective air cleaning system uses an afterburner or secondary combustion chamber for the combustion of all carbonaceous particulates and residual hydrocarbon gases. At this stage, the particulate and odor problems are solved. The HCl generated can then be effectively removed using a caustic wet scrubber.

Experimental Mechanical Separation

In an effort to eliminate objectionable pollutants associated with wire burning, experimental mechanical separation systems are undergoing tests on a pilot basis. One such system is an inertial separation unit, Figure 8-12, utilizing the difference in densities between copper and jacket-insulating materials. Copper wire complete with jacket was pulverized in a hammer mill to near powder consistency, air conveyed to a series of cyclones and experimentally sized to segregate the copper and the jacket material into two separate bins.

Some of the operating difficulties experienced were:

1. Plugging of the hammer mill used to pulverize the wire. Hand feeding pre-cut lengths of wire was difficult, if not hazardous. Occasionally a length of wire would wrap around the rotor and jam the unit.
2. Blockage of ductwork leading to cyclones, due to buildup of material.
3. Carryover of copper into the transitional cyclone stage of the train, deemed too difficult to remove and landfilled with the jacket material.

While capacity or cost figures were not available, the system did have definite potential as a cleaner method of recycling copper wire. A more thorough engineering approach to the basic design could have resolved most of these operating difficulties.

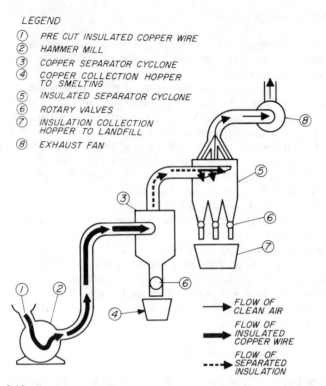

LEGEND

① PRE CUT INSULATED COPPER WIRE
② HAMMER MILL
③ COPPER SEPARATOR CYCLONE
④ COPPER COLLECTION HOPPER TO SMELTING
⑤ INSULATED SEPARATOR CYCLONE
⑥ ROTARY VALVES
⑦ INSULATION COLLECTION HOPPER TO LANDFILL
⑧ EXHAUST FAN

FLOW OF CLEAN AIR

FLOW OF INSULATED COPPER WIRE

FLOW OF SEPARATED INSULATION

Figure 8-12. Copper wire reclamation experimental inertial separation system.

Air Pollution Control Regulations

New York State employs a rule limiting smoke emissions from wire burning to 20 percent opacity. Particulate emissions are generally rated B or C, depending on location of the plant. When halogenated plastics are burned, HCl emissions are usually controlled under a B rating. The HCl emission rate potential can be estimated to be approximately 50 to 60 percent of the PVC burned; however, stack tests are still the only sure method to determine actual emissions. The degree of air cleaning required under a B rating is greater than 90 percent and ranges to 99 percent for emission rate potentials of 10,000 lb/hr or greater.

Odors can be controlled by assigning a C rating and requiring a 90 percent reduction of nonspecific odorous gases. They can also be regulated as a general prohibition of air pollution as defined in Article 19 of the New York Environmental Conservation Law. The key terms in this law are "the presence in the outdoor atmosphere of air contaminants—which unreasonably interfere with the comfortable enjoyment of life."

Each process must be carefully studied to determine its air pollution potential. The use of effective air cleaning systems can enable aluminum and copper recycling processes to be sound investments ecologically as well as economically.

SO$_2$ CONTROL FOR SMALL BOILERS[134]

Caustic wet scrubbing can be geared to the small industrial coal user whose boilers have steam capacities from about 50,000 to 500,000 lb/hr (Figure 8-13). Stack gases containing 0.1 to 0.3 percent SO$_2$, a small amount of SO$_3$, and flyash are passed through a quench chamber for cooling and on through a packed cross-flow scrubber utilizing a caustic scrubbing solution.

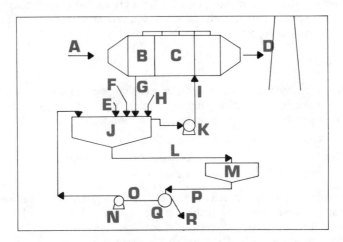

Figure 8-13. Caustic scrubbing of sulfur dioxide and regeneration system.
A. Stack Gas In B. Quench Chamber, C. Cross-flow Scrubber,
D. Scrubbed Gas Out, E. CA(OH)$_2$ Slurry, F. NaOH Make Up,
G. Spent Scrubbing Liquor, H. H$_2$O Make Up, I. Regenerated Solution,
J. Reaction-Settling Tank, K. Recycle Pump, L. Slurry, M. Thickener,
N. Filtrate Pump, O. Filtrate, P. Thickened Slurry, Q. Vacuum Filter,
R. Filter Cake

Precipitation of CaSO$_3$ and CaSO$_4$ from solution occurs. Spent scrubbing liquor is treated with a lime slurry. Caustic is regenerated and returned to the scrubber. The slurry of precipitated solids and flyash from the gas stream is thickened, filtered and removed for disposal. The filtrate is returned to the main body of the scrubbing liquor in order to conserve dissolved salts.

If water is used in the quench, SO$_3$ will be removed and form sulfuric acid. Quenching requires adding water in excess of what is actually needed for cooling. When the water balance on the entire system is critical, caustic scrubbing liquor may be used in the quench for SO$_3$ removal.

Sulfur dioxide is absorbed in the scrubbing solution where the following reactions occur:

$$2NaOH + SO_2 \rightarrow Na_2SO_3 + H_2O$$
$$Na_2SO_3 + SO_2 + H_2O \rightarrow 2NaHSO_3$$
$$Na_2SO_3 + \tfrac{1}{2} O_2 \rightarrow Na_2SO_4$$

Oxidation of some SO_3 occurs in the scrubber each time the scrubbing liquor passes through, resulting in a buildup of SO_4 ion.

Because of the presence of a considerable amount of CO_2 in the stack gas, other chemical reactions in the scrubbing liquid are possible. The degree of CO_3 and HCO_3 formation depends primarily upon the pH of scrubbing liquid. However, the net effect of the reactions is to absorb SO_2.

NO_x in the stack gas can react with NaOH in the scrubbing liquid to form soluble nitrates or nitrites, or with Na_2CO_3 to form nitrates. Cation-anion analyses of extended runs, however, indicate that this does not happen to any appreciable degree.

Regeneration

In the regeneration portion of the scrubbing system a lime slurry is mixed with the spent scrubbing liquor forming $CaSO_3 \times 2H_2O-CaSO_4 \times 2H_2O$ salts which tend to precipitate from solution.

$$NaNSO_3 + Ca(OH)_2 + H_2O \rightarrow CaSO_3 \times 2H_2O \downarrow + NaOH$$
$$Na_2SO_3 + Ca(OH)_2 + 2H_2O \rightarrow CaSO_3 \times 2H_2O \downarrow + 2NaOH$$
$$Na_2SO_4 + Ca(OH)_2 + 2H_2O \rightarrow CaSO_4 \times 2H_2O \downarrow + 2NaOH$$

Results of pilot plant tests indicate that for a given lime addition the formation of caustic depends mainly on the concentration of the NA^+ ion, the temperature of the solution and the reaction time.

Concentration of the Ca^{++} ion in the regenerated solution is of critical importance. Because the solubilities of $CaSO_3$ and $CaSO_4$ have inverse temperature coefficients, precipitation and scaling can occur in the scrubber if the regenerated liquor is delivered at an appreciably lower temperature than that of the scrubber effluent. With a scrubbing system, excellent SO_2 absorption can be obtained with scrubbing liquor temperatures of 125 to 130 F; therefore, there is no need for cooling.

Although the chemistry in the absorption and regeneration stages is complex and the exact mechanisms not well understood, steady-state operation of the system demands little more than adding lime slurry, makeup water, chemical makeup (replacement for sodium salts lost in the liquid portion of the filter cake), and removing filtered solids, all at a constant rate.

Changes in the percent of sulfur in the coal and excess air in a particular boiler result in changes in the concentrations of SO_2, SO_3, flyash, and in the stack gas rate. Therefore, the scrubber used must have flexibility to operate under variable inlet conditions.

Figure 8-14 shows height of a mass transfer unit (HTU) obtained from absorption tests with a scrubber. It is significant that, for the range of gas and liquid rates tested, the HTU and the absorption efficiency remain relatively constant. This simplifies the design of the system. Also, as the gas rate is increased above 1900 lb/hr/sq ft, absorption efficiency actually increases slightly. For a given depth of packing, absorption efficiency as HTU decreases, even though contact time in the packed bed is reduced. This tends to reduce the fixed capital investment for the scrubber. Operation with higher gas rates also increases particulate removal efficiency.

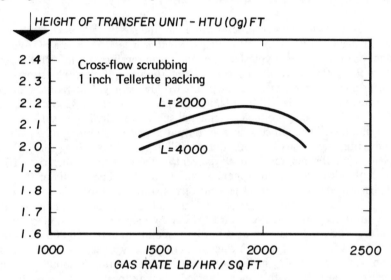

Figure 8-14. Absorption efficiencies of cross-flow scrubber at varying gas and liquid rates. Inlet SO_2 conc. = 0.1%; Scrubbing liquid—NaOH; Liquid temp. = 125-130 F; Outlet liquid pH–7.5; L = liquid rate (lb/hr sq ft).

Advantages

There are a number of advantages associated with this scrubber system:

1. It is a looped system. Chemical losses and water makeup requirements are minimized by recycling liquid in the system. Caustic is continually regenerated (and not replaced) through the addition of lime.
2. SO_2 removal efficiency is high. Caustic solution has a large capacity for absorbing SO_2 without building up an appreciable equilibrium back pressure.
3. Precipitation of solids from solution in the scrubber is eliminated. Regenerated liquor delivered to the scrubber can be treated if the particular operating conditions tend to produce precipitation during scrubbing.
4. The waste product from the system is disposable. $CaSO_3 \cdot CaSO_4$ filter cake can be used as landfill, eliminating potential water pollution problems and dependence upon a market for disposal of chemical products.

The caustic regeneration system may provide a practical and economical means for controlling SO_2 emissions from small-scale industrial boilers. Before the system can be installed in a steam generating facility, however, the fixed capital investment and operating costs may be justified by the difference in cost between low and high sulfur coals, or conversion to other low sulfur fuels. Future availability of the fuel and existing and future air pollution control legislation are also factors to be considered.

SULFUR PLANT TAIL GAS ABATEMENT[4]

A newly developed process can easily reduce the amount of sulfur in sulfur plant tail gases to 250 parts per million or less. This is a 60-fold reduction from current typical values, and means an increase in sulfur recovery from 95 to 99.9 percent or higher. Using the new process, the estimated sulfur dioxide emission of sulfur plants worldwide might be reduced from the present estimated 3000 to 6000 tons per day to about 12 tons per day.

Today, plants producing elemental sulfur from hydrogen sulfide by the modified Claus process have an output above 30,000 long tons daily, and within two years the total output will exceed 40,000 long tons. Sulfur plants typically recover 90 to 95 percent of the entering sulfur; 5 to 10 percent is lost to the atmosphere as sulfur dioxide. Large units of the latest design will recover about 97 percent of the sulfur, and lose about 3 percent.

The composition of the Claus plant tail gas typically contains about one-third SO_2 and H_2S, one-third COS and CS_2, and one third elemental sulfur. The total quantity of sulfur corresponds to about 15,000 parts per million of SO_2 in a typical incinerated tail gas (dry basis). The concentration of equivalent SO_2 in the tail gas before incineration is about 50 percent higher. Consistent with trends observed in many locations, a reasonable goal for concentration of SO_2 in the incinerated gas appears to be 250 parts per million or less. This 60-fold reduction corresponds to increasing sulfur recovery from 95 to 99.9 percent or higher.

The sulfur recovery process is best described by reference to the modified Claus process. Claus discovered in about 1880 that hydrogen sulfide produces good yields of sulfur when mixed with air and passed over iron or bauxite catalyst at elevated temperature. The reaction is highly exothermic but completion is opposed by high temperature, so conversion in a fixed bed of uncooled catalyst is relatively low.

I. G. Farbenindustrie, about 1937, approached the problem of heat removal by first burning one-third of the hydrogen sulfide with air in a pressurized boiler, in which about four-fifths of the overall heat of reaction is removed by generating steam, and some of the sulfur is produced.

After condensed sulfur is separated from the cooled gases, the chemical reaction is carried further toward completion by heating the gases to 400 to 500 F and passing them over a Claus catalyst (bauxite or alumina, usually). This process of free-flame reaction followed by one or more catalytic conversion steps is known as the modified Claus process. The operating pressure is usually 1.0 to 1.5 atmosphere absolute.

The reaction is exothermic and equilibrium favors formation of elemental sulfur at lower temperatures. It is customary to carry out the reaction in a series of two or three catalyst beds, with cooling, condensation and removal of produced sulfur after each bed. The lower practical temperature limit in the condensing steps is set at 260 F by solidification of the sulfur product and plugging of the apparatus. A further barrier is set, when water condensed, by the formation of a solution of sulfurous acid and polythionic acids $H_2S_xO_3$, known as Wackenroder's solution. This solution is very corrosive to the common construction metals, with the possible exception of titanium.

Ultimate conversion in the modified-Claus sulfur plant is set by the reverse of the chemical reaction in which water (mostly that formed from hydrogen sulfide) reacts with sulfur to produce gaseous hydrogen sulfide and sulfur dioxide. In general, plants with two catalytic stages can recover 92 to 95 percent of the potential sulfur; three stages, 95 to 96 percent; four, 96 to 97 percent. In addition to losses in the form of unconverted hydrogen sulfide and sulfur dioxide, some elemental sulfur is lost as vapor, and some as entrained sulfur mist or droplets.

Further losses—0.25 to 2.5 percent of the hydrogen sulfide fed—are experienced in the form of carbonyl sulfide (COS) and carbon disulfide (CS_2). These gases are formed in the flame zone where hydrogen sulfide is burned with air (generally at a temperature in the range of 1500 to 2500 F) when carbon dioxide or other carbon compounds are present. The carbon-sulfur compounds react slowly under the conditions of the sulfur plant, thus often comprise 30 to 40 percent of the sulfur values lost in the tail gas leaving the sulfur plant.

Tail gas leaving the sulfur plant then contains 1 to 3 percent of a mixture of hydrogen sulfide, sulfur dioxide, carbonyl sulfide, carbon disulfide, and elemental sulfur vapor and liquid. This mixture is usually incinerated at about 1000 to 1200 F in an oxidizing atmosphere to convert all the sulfur compounds to sulfur dioxide, their least obnoxious form. The resulting flue gas flows into the atmosphere from a stack.

The new process (Figure 8-15) starts by converting all the sulfur in the tail gas into hydrogen sulfide, then cools the gas to condense out water, removing it from the reaction zone. Conversion to hydrogen sulfide is by both hydrolysis and hydrogenation reactions. Tail gas contains about 30 percent water vapor and consequently the hydrolysis reactions are virtually complete at temperatures of 600 to 700 F when a suitable catalyst is used.

Hydrogenation of sulfur and its compounds proceeds completely according to the following equations:

$$SO_2 + 2H_2 \rightarrow S + 2H_2O$$
$$S + H_2 \rightarrow H_2S$$

Enough hydrogen is usually present in the tail gas to effect the reactions, since both hydrogen and carbon monoxide are formed in the flame

Beavon® Sulfur Removal Process

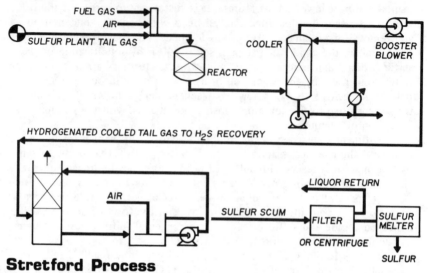

Stretford Process

Figure 8-15. Flow diagram of processes to eliminate air pollutants
from tail gases of sulfur plants.

zone and persist through the Claus plant. The same catalyst is effective for
the hydrolysis of carbon monoxide (the water-gas shift reaction). This
reaction proceeds completely and produces more hydrogen. In addition,
the supply of hydrogen may be augmented by the partial oxidation of
natural gas with air in a simple line burner.

Following the catalytic reactions which convert essentially all the
sulfur compounds in the tail gas to hydrogen sulfide, the gas mixture is
cooled to as low a temperature as is economic, and condensed water is
removed.

Next, hydrogen sulfide is extracted from the hydrogenated, cooled
tail gas using the Stretford process. This process is capable of treating the
gas to about one part per million hydrogen sulfide content.

The Stretford process uses a sodium carbonate solution which reacts
with hydrogen sulfide to form sodium hydrosulfide. The hydrosulfide is
oxidized to sulfur by sodium vanadate also in solution. Subsequently the
vanadium is oxidized back to the pentavalent state by blowing with air,
with sodium anthraquinone disulfonate (also contained) working as an oxidation catalyst. Finely-divided sulfur appears as a froth which is skimmed
off, washed, dried by centrifuging or filtration, and added to the product
from the parent sulfur plant.

Effluent gas from the Stretford plant contains less than one part per
million hydrogen sulfide and small traces of carbonyl sulfide, and does not
require incineration. Recovery of sulfur is virtually complete, and the sulfur
plant is no longer an air pollution problem.

The law of mass action requires that, for one reagent in a chemical reaction to be consumed essentially completely, there must be an excess of another reagent. For example, the Claus reaction must have an excess of SO_2 to consume all the H_2S, or conversely: either condition is undesirable because it means a loss of sulfur to the atmosphere. In the new process, an excess of hydrogen is supplied, driving reactions to completion. The hydrogen is not an air pollutant, hence the use of a small excess is not harmful. Similarly, the Stretford step applies a very large excess oxidation potential at the top of the absorber column, allowing the conversion of H_2S to be driven virtually to completion.

DEGREASER SOLVENT RECOVERY
BY CARBON ADSORPTION[112]

The carbon adsorption method of solvent vapor abatement and recovery offers an advantage which is difficult to ignore. It literally pays for itself.

Vapor adsorption on activated charcoal is a well-known process, utilized for many years in gas masks and industrial gas-handling systems. The new element added to this technology is the completely automated adsorption/desorption/separation cycle, which produces usable solvent with a minimum of secondary pollution. In many cases, the condensed water can be sewered without further treatment after separation. There are none of the disposal problems involving solid wastes or large amounts of heavily contaminated liquid common to scrubbers and other types of pollution control equipment.

Regeneration of an activated carbon by treatment with steam, followed by condensation of the resulting steam distillate and separation of the components from the water, yields high-quality, reusable liquid solvent. The reduction in solvent costs not only pays for the cost of operating the solvent recovery system, it also makes recovery of the original purchase and installation investment possible—often in less than a year. Losses of expensive solvents by evaporation can be reduced by 85 to 95 percent.

Basically, the adsorber unit consists of one or more tanks containing charcoal beds; a blower; a condenser for steam and solvent; a two-layer liquid separator or distillation unit; and automatic controls.

In operation, vapor-laden air is drawn into one of the tanks and through the carbon bed for a pre-set time period. The emerging air, virtually solvent-free, can be vented to atmosphere—or possibly recirculated in the building. The time duration of the entire processing cycle is selected to assure complete removal of solvent vapor from the air. By a comfortable margin, the total amount of solvent vapor passed through the system is always less than the maximum amount of vapor which could be adsorbed by the bed. This is achieved by automatically shutting off the adsorption cycle before the bed saturation point is reached, so that no unadsorbed vapor can pass through the system and contaminate the exhausted air.

At the end of the first tank's adsorption interval, the vapor-laden air brought in by the blower is entirely diverted to the second tank for adsorption. The first tank is then advanced to the desorption portion of the cycle. The desorption cycle consists of passing steam through the bed in the direction opposite to that of the flow of solvent-laden air during adsorption. The steam distillate produced passes through a water-cooled condenser and into a separator. If the solvent is insoluble in water, a separation unit is used, consisting of a tank with siphon outlets for removing the upper and lower layers simultaneously. For water-soluble liquids, a continuous distillation unit is placed in-line after the condenser.

A number of factors are involved in determining the required carbon adsorption capacity of a solvent recovery system. No simple, universal formula can be given to cover every solvent or recovery problem. The type of solvent and its temperature, air flow rates, variations in concentration, and total amount of solvent evaporated per unit of time must all be considered. In practice, recommendations and specifications for a solvent recovery system are preferably left to the adsorber manufacturer, after the pollution engineer has supplied pertinent operating information.

In general, calculations for determining adsorber unit size are patterned after the following method. Solvent consumption records for a given calendar period of work days are used to estimate the hourly solvent loss by evaporation during manufacturing operations. Using the known efficiency or activated carbon for adsorbing the particular solvent, the amount of carbon required to adsorb the solvent vapor can be calculated. The required rate of solvent-laden air flow can then be determined, since it is known empirically that under average bed conditions 100 lb of carbon can efficiently treat 200 cu ft/min of solvent-laden air/hr.

Example

A vapor degreasing line is known to lose approximately 75 lb of trichlorethylene/hr.

Calculation: From Table 8-4 find that the carbon adsorption efficiency for trichlorethylene is 15 percent. This means that the amount of solvent adsorbed by 100 lb of carbon would be 15 lb, so for adsorbing 75 lb in an hour the requirement would be:

$$75 \times \frac{100}{15} = 500 \text{ lb of carbon.}$$

This is the minimum amount of carbon needed in one bed, through which vapors are to be passed for a period of one hour. As noted above, 100 lb of carbon efficiently handles a volume flow rate of 200 cu ft/min. Therefore:

$$500 \text{ lb carbon} \times \frac{200 \text{ cu ft/min}}{100 \text{ lb carbon}} = 1000 \text{ cu ft/min,}$$

Table 8-4. Selected Data of Commonly Used Solvents

Solvent	Ceiling value (ppm)	8-hour time weighted average (ppm)	Acceptable ceiling concentration (ppm)	Acceptable maximum peak concentration		Lower explosive limit (percent by volume) in air	Carbon adsorption efficiency (percent)
				Conc. (ppm)	Time (minutes)		
Acetone		1000				2.15	8
Benzene		10	25			1.4	6
n-Butyl acetate		150				1.7	8
n-Butyl alcohol		100				1.7	8
Carbon tetrachloride	50	10	25	200	5 in 4 hr	n	10
Chloroform		10				n	10
Cyclohexane		300				1.31	6
Ethyl acetate		400				2.2	8
Ethyl alcohol		1000				3.3	8
Heptane		500				1	6
Hexane		500				1.3	6
Isobutyl alcohol		100				1.68	8
Isopropyl acetate		250				2.18	8
Isopropyl alcohol		400				2.5	8
Methyl acetate		200				4.1	7
Methyl alcohol		200				6.0	7
Methylene chloride		500	1000	2000	5 in 2 hr	n	10
Methyl ethyl ketone		200				1.81	8
Methyl isobutyl ketone		100				1.4	7
Perchlorethylene		100	200	300	5 in 3 hr	n	20
Toluene		200	300	500	10	1.27	7
Trichlorethylene		100	200	300	5 in 2 hr	n	15
Trichloro trifluoroethane		1000				n	8
V M & P Naphtha		500				0.81	7
Xylene		100				1.0	10

the minimum rate of air flow required for this quantity of trichlorethylene. In practice, these figures would be optimized to meet the particular recovery requirements of the individual installation.

The initial hardware costs, operating costs and return on investment for a solvent adsorber system will, of course, vary depending on the size of the system, the type of solvent vapor processed, and the volume of solvent used. A typical unit designed to handle 3000 cu ft of solvent-laden air/ minute would cost between $12,000 and $13,000. The installation cost can be estimated to run about 20 percent over and above the price of the unit.

Operating costs have been found to average out at 25¢/hr for electricity, 50¢/hr for water and 90¢/hr for steam, depending on local utility rates. Maintenance on the unit for the first four to five years should come to less than $250 annually.

The rapid return on investment for a solvent adsorber can be seen in the example for a plant using 24,000 gal/yr of a typical chlorinated solvent costing $1.50/gal, or $36,000. At a solvent recovery rate of 75 percent—an extremely conservative estimate and readily exceeded in most cases—the first-year savings would amount of $27,000. At this rate, the cost of the unit plus installation ($15,600—a $13,000 unit plus $2,600 for the installation) could be recovered completely in less than a year. Therefore, the claim that a solvent recovery system based on activated carbon literally pays for itself can be documented.

AIRLIFTING—AN INSTALLATION ALTERNATIVE[188]

An increasing number of contractors and pollution engineers are using helicopters instead of cranes, especially to install dust collectors, scrubbers and other control equipment on rooftops. Helicopters can often get the job done faster and at a much lower cost. Costs on some jobs have been reduced by as much as 50 percent.

Savings can be higher than 50 percent on jobs where the use of a crane would require shoring up a roof and construction of a planking roadway to permit the rolling of heavy units across the roof. Helicopters on the other hand, can set completely assembled units—units weighing as much as 8500 lb— any place on a roof with ease and precision.

Movement of units weighing more than 8500 lb can be accomplished by helicopter with equal ease and economy, if the loads are broken down into several large sections.

Why are installation costs so much lower with helicopters? The reason becomes apparent when one considers that a helicopter can normally do more work in one hour than a crane can do in two days. A helicopter can set units at the rate of one every 2½ or 3 minutes. Labor costs, therefore, are substantially lower.

A contractor who said he saved $1500 using a helicopter on a recent job provided this comparison data: With a helicopter he needed six men for one

Figure 8-16. Helicopter transports air handling unit from ground to top
of manufacturing plant.

Figure 8-17. Ground crew signals helicopter pilot for exact placement of
air handling unit on supporting frame.

hour to set eleven 4000-lb units on the roof of a one-story building. Total man-hours = 6. With a crane, the contractor said he would have required eight men for nearly two days to do the same job. Total man-hours = 128.

A pollution engineer reported he used a helicopter to remove a large dust collector from a one-story plant and set a new unit in its place. The helicopter took only 30 minutes to complete both operations. He estimated it would have taken a crane several weeks to do the job. Rental and labor costs with the crane, including weekend overtime, would have far exceeded the costs of using a helicopter.

However, cost may not be the main consideration for using of helicopters. They may be selected because they are able to get the job done so much faster—without any disruption in plant production or construction activity.

There are jobs, of course, in which the helicopter cannot compete with a crane. It usually is more economical to use a crane if the crane can complete the job in less than a day. However, even on these small jobs, it will probably cost less to use a helicopter, if use of a crane requires special construction to permit its use.

In our investigation, we found Carson Helicopters, Inc., of Perkasie, Pa., to be the largest heavy-lift helicopter operator in the country. They pioneered the use of helicopters in construction. Carson's fleet of 11 helicopters is in constant use everywhere in the country, stringing pipe for gas lines; setting poles for power lines; building ski lifts and bridges; dismantling huge tower cranes; erecting structural steel and airlifting logs from forests on both the east and west coasts. Their main activity, however, is the installation of rooftop units.

Men and Equipment

A helicopter is physically able to work almost anywhere as long as there is a landing area large enough for the particular helicopter being used. For example, on some downtown jobs in large cities, it has been possible to utilize a nearby parking lot.

The types of heavy-lift helicopters used by Carson for rooftop work are the Sikorsky S-55, capable of lifting up to 2000 lb; the Sikorsky S-58, capable of lifting up to 4500 lb and the Sikorsky S-61, which can lift up to 8500 lb.

For the smaller helicopters, those capable of lifting up to 4500 lb, only a pilot is required. The larger helicopters require a co-pilot. A helicopter's ground crew normally will consist of two signalmen; one at the pick-up point on the ground and the other at the delivery point on the roof. The contractor will normally supply two men on the ground to hook each unit to the spreader bar under the helicopter, plus four to six men on the roof to guide the unit into place.

Prior to any helicopter operation, all debris is cleared from the ground and the roof of the building because of the wind velocity created by the helicopter. The helicopter pilot briefs the ground crew before starting an operation. Hard hats and goggles are worn by all personnel involved in the operation, and rubber gloves are worn by all workers who touch the units while they are being lifted.

The helicopter's ground signalman watches the contractor's crew and directs the pilot to lift off after a unit has been attached to the spreader bar. The signalman on the roof directs the pilot to go forward, backward, up or down while the contractor's roof crew moves the unit exactly into place on its curbing.

Case History

The capability of helicopters to install pollution control equipment with speed, economy and precision is pointed up in an airlift by Carson at the Carborundum Company's Monofrax Plant in Falconer, N.Y. The assignment was to set nine air-handling units, weighing between 4000 and 7000 lb each, on the roof of the plant.

These air-handling units replace the high volume of air being removed by a new pollution control system, consisting of nine fabric collectors with more than 7416 bags and a total cloth area of 110,484 sq ft.

The Pollution Control Division of Carborundum, which designed the new control system, had made a detailed study of alternative methods for setting the units before deciding upon the use of helicopters. Cost estimates indicated the use of helicopters would result in a substantial saving, and these estimates were borne out by the actual airlift. Use of a helicopter to set the nine air units saved the company about $12,000 and cut costs by 75 percent.

They said this was the first time they used a helicopter and were "surprised" the helicopter could move a 7000 lb unit to the roof in about three minutes. The total airlift, including preparation and refueling time, took less than 45 minutes. It was estimated that a crane would have taken 2½ weeks to do the job. In addition, using a crane would have necessitated shoring up the roof with steel beams and anchoring the crane so that it wouldn't tip over from the weight of its extra-long boom.

In most large cities, it is necessary to get permission for a helicopter airlift from the department of public safety or the police department. In smaller communities, it is usually only necessary to advise the local police of the airlift.

CHAPTER 9

WORK AREA AIR PROBLEMS

AIR SAMPLING AND ANALYSES
OF CONTAMINANTS IN WORK AREAS[23]

There are generally two categories of pollution control for a manufacturing enterprise: (1) those affecting the environment external to the plant; and (2) those affecting the working conditions within the plant—the controls usually considered necessary for occupational safety and health. It is in this latter area, the updating of safety concepts in line with OSHA requirements, that pollution control becomes part of the program—not only in implementation, but in identification and assessment of problems where they may exist.

Plant safety programs have traditionally emphasized two factors: the guarding of hazardous machinery and providing adequate personal safety protective devices for the worker. Evaluation of occupational exposures has undergone marked evolution in the recent past. By making work injuries—including those resulting from such conditions as fumes, dusts, and noise—immediately and inescapably expensive to employers, workmen's compensation laws and OSHA have done more to promote safety than all other influences combined.

In any program of hazard elimination, air sampling and analysis for contaminants in work areas is part of the broadening structure of plant safety. A sampling and analysis program is initiated to:

1. Evaluate the effectiveness of engineering control,
2. Evaluate process changes,
3. Determine the need for personal protective devices,
4. Evaluate exposure of workers to contaminants arising from and in the course of their work, and
5. Establish compliance with regulations or accepted standards.

335

Air Contaminant Sampling in Work Areas

Work area atmospheric sampling, as in pollution control flue gas sampling, is dependent on an estimate of the amount of contaminant to be found, the sensitivity of the analytical method, and the safety of hygienic standard. There are generally three types of airborne contaminants in work areas. The type of contaminant occurring will, of course, depend on the nature of the business or existing processes. Work area contaminants may, however, be divided into physical-characteristic groups as outlined below.

Odors

In some cases, an air contaminant is immediately detected and identified by odor. Then the sense of smell becomes an analytical and sampling tool, particularly if the amount of material in the exposure atmosphere is so small as to be detectable only by odor.

Particulate Matter

This classification may be divided into solids and liquids, and solids are usually further divided into three classes:

1. Dusts formed from solid inorganic or organic materials reduced in size by mechanical means, such as grinding, crushing or drilling. These particles range in size from visible to submicroscopic. The size below 10 μm is of principal concern to industrial hygienists since these particles are respirable and remain suspended in the atmosphere for long periods.

2. Fumes formed from solid materials by evaporation and condensation. When heated, such metals as lead and cadmium produce vapors that condense in the atmosphere to form metallic oxides. The particles range in size from 1 to 0.0001 μm. Solid organic materials are also capable of forming fumes in a similar fashion, and products of incomplete combustion that form smokes also fall into this class.

3. Liquid particles, which are sometimes classified as fogs or mists. These are produced by atomization or condensation from the gaseous state.

Collectors for particulate matter sampling fall into the following categories: settling chambers, centrifugal devices, impingers, impactors, scrubbers, filters, electrostatic precipitators and thermal precipitators. Collection of particulate matter may be instantaneous or integrated, and the sample itself is subject to identification by accepted chemical and physical methods.

Vapors and Gases

Gases occupy the entire atmospheric area, whereas vapors are condensation products. Examples of gaseous contaminants are hydrogen sulfide, sulfur oxides (SO_x), nitrous oxides (NO_x) and hydrogen chloride. Vapors, in addition to being evaporation products, are usually materials, such as water, that are liquid at normal temperatures.

Vapors and gases are usually collected instantaneously in evacuated containers, such as glass or metal flasks or plastic bags. For integrated gas sampling a collection rate of 0.1 cfm is most commonly used, and the samples may be collected in a solvent with wash bottles, impingers and absorbers; by adsorbents, such as active carbon; or by condensation.

Methods of Analysis

Analytical procedures available in the chemical laboratory today obviously facilitate qualitative and quantitative identification. Because only small quantities of material are often collected, trace techniques are widely used. Here again, the interdisciplinary relationship between pollution control and hazardous substance identification and quantification in process areas asserts itself.

In many cases devices that provide immediate answers for workroom safety appraisal are also required. Direct reading instruments for many parameters are fortunately available and adequate. Such instrumentation includes the halide meter, capable of measuring concentrations of halide to 1 ppm; combustible gas detectors; and thermal conductivity instruments for detection of gases such as carbon dioxide. In addition, small hand-carried direct reading instruments are available for measuring hydrogen and mercury vapor.

Indicator detector tubes have achieved a preeminent place in direct reading and industrial air analysis. They are generally small, light, hand-operated and sensitive over a wide variety of parameters, and give an immediate read-out.

In addition, use of an indicator tube is probably the simplest and most economical air analysis method available for a wide range of air contaminants. Table 9-1 lists the air contaminants that can be analyzed readily by this method and the sensitivity ranges of the available instruments with the prevalently used reagent-indicator systems.

Some Parameters for Evaluating Safety

Table 9-2 lists many common gases and vapors and some significant properties for evaluating plant safety practices. The contaminant gases or vapors are identified by the name in normal use.

Threshold limit values (TLV) are important as these are the concentrations below which daily exposure is thought to be without adverse effect. These values, however, should be used only as a guide in the control of health hazards and should not be regarded as the ultimate dividing points between safe and dangerous concentrations. Also, threshold limit values do not remain constant, but are constantly researched and changed. The values are only valid for a pure substance; in most cases data obtained in measuring working atmospheres will be for components and admixtures. With the present state of knowledge a general method of

Table 9-1. Indicator Detector Tubes Available and Sensitivity Ranges

Contaminant Parameter to be Measured	Reaction Principle (Reagent System)	Measuring Range
Acetone	Dinitrophenylhydrazine	100-12,000 ppm
Acrylonitrile	Via hydrogen cyanide	5-30 ppm
Alcohol	Chromic acid	100-3000 ppm
Ammonia	Bromophenol blue	5-700 ppm
Ammonia	Mercury (1)-nitrate	25-700 ppm
Aniline	Furfurol + acid	1-20 ppm
Arsine	Gold salt	0.05-60 ppm
Benzene	Formaldehyde + sulfuric acid	15-420 ppm
Carbon dioxide	Hydrazine + redox indicator	0.1-60 Vol %
Carbon disulfide	Copper thiocarbomate	13-3200 ppm
Carbon monoxide	Iodine pentoxide + selenium dioxide + fuming sulfuric acid	8-7000 ppm
Carbon tetrachloride	Via phosgene	10-100 ppm
Chlorine	0-Tolidine	0.2-500 ppm
Cyanogen chloride	Pyridine + barbituric acid	2-40 ppm
Ethyl acetate	Chromic acid	200-3000 ppm
Formaldehyde	Xylene + sulfuric acid	2-40 ppm
Hydrazine	Bromophenol blue	0.25-3 ppm
Hydrocarbons	Iodine pentoxide + fuming sulfuric	0.1-1 Vol %
	Selenium dioxide + fuming sulfuric	2-25 mg/l
Hydrochloric acid	Bromophenol blue	1-20 ppm
Hydrogen	Via water vapor	0.5-3 Vol %
Hydrogen cyanide	Mercury (2) chloride + methyl red	2-150 ppm
Hydrogen fluoride	Zirconium-alizarin lac	0.5-15 ppm
Hydrogen sulfide	Lead salt	5-2000 ppm
	Copper (2) salt	0.02-7 Vol %
	Iodine	0.02-7 Vol %
Mercaptan	Copper salt	2-100 ppm
Mercury vapor	Mercury chloride + gold chloride	0.1-2 mg/m^3
Methyl bromide	Via bromine	5-50 ppm
Monostyrene	Resinification (sulfuric acid)	50-400 ppm
Nickel carbonyl	Iodine + dioxime	0.1-1 ppm
Nitrogen dioxide	Diphenylbenzidine	0.5-10 ppm
Nitrous fumes	Diphenylbenzidine	0.5-100 ppm
	Dianisidine	100-5000 ppm
Olefins	Permanganate	1-55 mg/l
Oxygen	Via carbon monoxide	5-21 Vol %
Ozone	Indigo bleaching	0.5-300 ppm
Perchlorethylene	Via chlorine	10-400 ppm
Phenol	Indophenol	5 ppm
Phosgene	(Dimethylanaline-Dimethylaminobenzaldehyde)	0.05-75 ppm
Phosphine	Gold salt	0.1-3000 ppm
Sulfur dioxide	Iodine	20-2000 ppm
Suptox	Gold chloride + N-chloramide	0.5 mg/m^3
Toluene	Sulfuric acid + iodine pentoxide	5-400 ppm
	Fuming sulfuric + selenium dioxide	25-2000 ppm
Trichlorethylene	Via chlorine	10-400 ppm
Vinyl chloride	Potassium permanganate	100-3000 ppm
Water vapor	Selenium dioxide + sulfuric acid	0.1-40 mg/l

Table 9-2. Common Gases and Vapors—Some Significant Properties for Plant Safety

Vapor or Gas	Threshold Limit Value (ppm)	Threshold of Smell (ppm)	Lower Ignition Limit (Vol %)	Upper Ignition Limit (Vol %)	Flash Point (°C)	Boiling Point (°C)
Acetone	1000	-	2.5	13	<-20	56.2
Acetylene	-	-	1.5	32	-	-83.6
Acrylonitrile	20	-	2.8	28	-5	78.5
Ammonia	5	5	15.0	28	-	-33.5
Aniline	5	0.5	1.2	11	76	184.4
Arsine	0.05	-	-	-	-	-62.5
Benzene	25	<100	1.2	8.0	-11	80.1
Benzylbromide	-	-	-	-	-	198
Bromine	0.1	< 0.01	-	-	-	58.8
Bromoform	-	-	-	-	-	149.6
(1,3) Butadiene	1000	-	1.1	12.5	-	-4
n-Butane	1000	-	1.5	8.5	-	-1
n-Butanol	100	-	1.4	11.3	35	117.8
2-Butanol	150	-	-	-	24	99.5
Carbon dioxide	5000	Odorless	-	-	-	-78.5
Carbon disulfide	20	1-2	1.0	60	<-20	46.2
Carbon monoxide	50	Odorless	12.5	74	-	-191.5
Carbontetrachloride	10	70	-	-	-	76.7
Chloral hydrate	-	-	-	-	-	98
Chlorine	1	0.02	-	-	-	-34.1
(Chlorobromo-methane)	200	-	-	-	-	68
Chloroform	50	200	-	-	-	61.2
Ethyl acetate	400	-	2.1	11.5	-4	77
Ethyl alcohol	1000	350	3.5	15	12	78.4
Ethylene	-	-	2.7	34	-	-103.7
Ethylene oxide	50	700	2.6	100	-	10.7
Formaldehyde	5	-	7.0	73	-	-19
Hydrazine	1	3	4.7	100	-	113.5
Hydrocyanic acid	10	2	5.4	46.6	<-20	26
Hydrogen	-	Odorless	4.0	75.6	-	-253
Hydrogen chloride	5	-	-	-	-	-85.0
Hydrogen fluoride	3	-	-	-	-	19.5
Hydrogen sulfide	10	0.1	4.3	45.5	-	-60.4
Kerosene	-	-	0.6	12.0	-	82.4
Mercury vapor	0.1 mg/m³	Odorless	-	-	-	357
Methane	-	-	5	15	-	-161
Methyl alcohol	200	2000	5.5	44	11	64.6
Methyl bromide	20	Odorless	8.6	20	-	3.5
Methylethyl ketone	200	< 25	1.8	11.5	-1	79.6
Methyl mercaptan	10	-	4.1	21.0	-	6
Monostyrene	100	25	1.1	8	32	145
Nickel-carbonyl	0.001	-	2.0	-	<-20	43.2
Nitrogen dioxide	5	1.3	-	-	-	-
Nitrogen monoxide	-	Odorless	-	-	-	151.8
Ozone	0.1	0.015	-	-	-	-
Perchlor-ethylene	100	50	-	-	-	121.2
Phenol	5	0.5	-	-	79	182
Phosgene	0.1	0.5	-	-	-	7.6
Phosphine	0.3	2.7	-	-	-	-88
Propane	1000	-	2.1	9.5	-	-42
Sulfur dioxide	5	3	-	-	-	-10
Toluene	200	50	1.2	7.0	6	110.6
Trichlorethylene	100	50	1.2	7.0	6	110.6
Vinyl chloride	500	-	3.8	29.3	-	-14

calculating threshold limit values of mixtures cannot be recommended. Allowances must, therefore, be made when interpreting analytical results. Values shown are from the American Conference of Industrial Hygienists data.

Threshold smell values are given only as guides, since data from varying sources differ widely. Evaluation of odors is highly subjective and may be considered more qualitative than quantitative.

The significance of ignition limits is that inflammable gases or vapors, mixed with air, are explosive only within given concentration ranges. The data shown in Table 9-2 give the concentration range in volume percentage at 20 C and 760 mm Hg of the gas or vapor, mixed with air, in which ignition by an external source is possible. The minimum and maximum concentrations at which ignition takes place are termed the lower and upper explosive limits. Therefore vapor air mixtures with lower and higher vapor concentrations are not explosive.

The flash point of an inflammable liquid is the lowest temperature (at a pressure of 760 mm Hg) at which, in a given flash-point test apparatus, vapor is produced to such an extent that the vapor-air mixture can be ignited by an external source. The flash point is a criterion of the inflammability of a liquid by an external ignition source.

The boiling point is an indicator of how readily volatile a material may be.

The type of work area or hazardous atmosphere dictates the requirements for safe operation. The classification and designation of the hazard or the provisions for operating under hazardous conditions can be identified by many of the methods in current use for pollution control analysis. It is, therefore, in this latter area that the broadening structure of plant safety becomes closely related.

IN-PLANT AIR CLEANING[175]

Recycling of exhaust air in industrial facilities offers a number of advantages. The management of air use and treatment can result in cleaner plant air and significant savings in operating costs. Air is a fluid that can be directed and controlled. It can be filtered and reused rather than thrown away.

General exhausting of entire plants to eliminate airborne contaminants such as welding smoke, oil mist or fiber dust cannot be regarded as feasible where the cubic volume is great and the distances from contaminant sources to exhaust fan openings are more than a few feet. The velocity required to capture and exhaust particulates varies from 100 fpm, for particles released with practically no velocity in quiet air, to 2000 fpm for particles released at high initial velocity into rapid air motion.

When an exhaust fan is used to remove air from a large space the air moves toward the exhaust opening at a gradually increasing velocity. A series of flow contours can be shown on a diagram (Figure 9-1). The velocities are

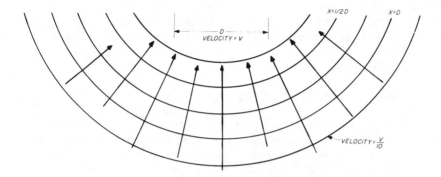

Figure 9-1. Air flow contours near exhaust opening.
At X=D velocity is 1/10 velocity at opening.

calculated by the following formula:

$$V = \frac{Q}{10X^2 + A}$$

where:

V = Centerline velocity, fpm, at distance X from the opening
X = Distance from opening, ft.
Q = Air flow in cubic feet per minute (cfm)
D = Diameter of exhaust opening, ft.
A = Area of opening, or $3.14/4 \times D^2$

A general approximation, illustrated by Figure 9-1, is that at a distance from the opening equal to the opening diameter, $X = D$, the velocity of air movement is only one tenth the velocity at the exhaust opening.

An example will illustrate the futility of attempting to control welding smoke in a production area by means of a ceiling exhaust fan. Assume: (1) Exhaust fan, 15 ft over work area and directly overhead, (2) fan capacity 20,000 cfm, and (3) exhaust opening 5 ft in diameter. Calculate the velocity of air movement at one diameter, or 5 ft from the opening.

$$V = \frac{20,000 \text{ cfm}}{10(5)^2 + \frac{3.14}{4}(5)^2}$$

V = approx. 75 fpm

The minimum capture velocity necessary to control welding smoke is 100 fpm. Yet at a point only 5 ft from the fan opening the velocity has decreased from approximately 1000 fpm to 75 fpm and the welding smoke is still 10 ft away. In this case, however, the thermal convection of the welding smoke, straight up from the source, would assist in carrying the smoke into the area of fan influence. If the fan is located some distance away from the point directly above the source, the smoke and particulate matter will rise on thermal currents and hang in a dense cloud near the ceiling. A duct opening close to the source becomes necessary to achieve adequate capture velocity.

Forced Blowing vs. Suction Exhausting

The removal of airborne particles which are dispersed in a large volume of air is better accomplished by using the forcing characteristics of a pressure jet of air blown from a fan. An air stream blown from an opening of diameter D will travel a distance X equal to 30D before the velocity is reduced to one-tenth of the exit velocity. This suggests that blowing is more effective than suction exhaust for moving airborne particles to an area where they can be removed by filters or exhausted to the outside.

In fact, a horizontal air pattern can be created by locating blowing fans in such positions, and of such size, as to maintain an air velocity greater than the capture velocity. If the exit velocity of each fan air jet is 4000 fpm, at 30 diameters away, the velocity is 400 fpm, which is above the usual capture velocity. Before the velocity of air movement drops below the capture velocity the particles can be filtered out or exhausted to the outside.

The principle of aspiration may also be used in air handling systems. A liquid may be sucked uphill in defiance of the law of gravity by a moving air column. In a similar way, a moving air column can be used to aspirate contaminated air into the exhaust stream.

Recirculation and Filtering

Air recirculation and the filtration of nuisance contaminants can reduce heating and air conditioning costs and allow a smaller make-up air system. A mechanical filtration system, using a paper or cloth filter medium, is generally satisfactory but requires more air-moving capability to overcome the pressure drop across the filter. As the particles to be collected approach 1 micron in size, the filter medium must be very fine and the pressure drop across it goes up considerably. Further, as the medium loads up with particles it becomes more efficient, but the pressure drop increases. This gives rise to a truism in air movement and control—the effectiveness of an air particulate removal system inside a facility is dependent on the filter efficiency and air exchange through the system.

The air exchange factor is too often overlooked or disregarded in design of air filtration and recirculating systems. Consider the following example:

1. Filtration system capable of 90 percent efficiency removal of particulates
2. Air exchange rate through system = 20,000 cfm

3. Volume of air inside facility = 100,000 ft^3 (100 ft X 100 ft X 10 ft)
4. Production rate of contaminants = 0.1 mg/ft^3/hr
5. Efficiency of air pattern = 50 percent
6. 20,000 cfm = 1 air exchange each 5 min

In evaluating industrial plants it has been found in well designed circulation systems that each time the fans have handled a volume of air equal to the cubic capacity of the building space, only one-half of the air has been filtered. The air pattern efficiency is roughly 50 percent.

Effectiveness = filtration efficiency X air pattern efficiency

In the example:

Effectiveness = 0.90 X 0.50 = 0.45

The job of the air cleaning equipment is to remove the contaminants. When the manufacturing operation is started in the morning, the level of contaminants in the room will start to build up at the rate of 0.1 mg/ft^3/hr. The building will eventually reach a level where the rate of contaminant fallout or removal equals the rate of generation. This is the equilibrium condition.

In the example, contaminants are produced at the rate of 0.1 X 5/60 = 0.0083 mg/ft^3 every 5 min, as the 100,000 cu ft of air is moved once. But since the effectiveness of removal is 0.45, the net gain in contaminants each 5 min is:

$$\frac{0.0083}{0.45} = 0.0184 \text{ mg/ft}^3$$

When removal rate equals this figure, equilibrium will have been established. This equals a contaminate level in the building of 0.65 mg/cm.

Another choice of filtering equipment is the self-contained electrostatic precipitator with its own blower system arranged in an air pattern so as to exchange the air in successive sweeps. Although initial cost of such a system may be higher than that of the mechanical filter, operating costs are generally lower and the collecting section is permanent and continuously reusable after cleaning. The small increase in static pressure during use allows smaller motors to power the blower systems. For collection of liquid particulates, a vertical orientation of the collection plates allows a gravity feed down the plates into a sump at the bottom where the liquid can be collected, drained away or recovered. Efficiency of electrostatic filters is usually greater than that of mechanical media for welding smoke and oil mist (Figure 9-2).

A precipitator connected to exhaust ducting and discharging back into the facility is convenient for local sources of contamination such as welding operations (Figure 9-3). The filtration system may be mounted in a stationary position or on a movable cart as shown. With OSHA emphasis on removing dust and oil mists from in-plant environments, installation of electrostatic precipitators provides a highly effective means of meeting compliance regulations.

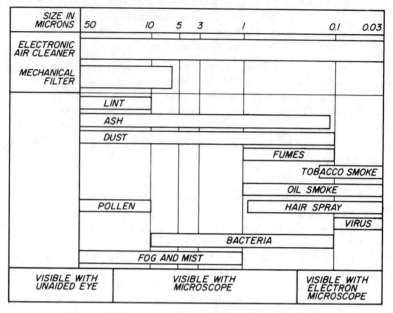

Figure 9-2. Sizes of particulates.

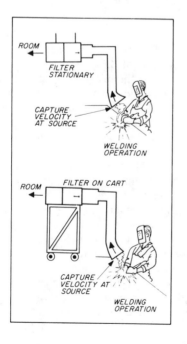

Figure 9-3. Filters for welding operations.

RESPIRATORY PROTECTIVE EQUIPMENT[33]

For many years, only a few industrial occupations were considered a hazard to the respiratory system. In such cases, gas masks were the basic protection device and were used only in emergencies. Today, it is recognized that many occupations involve vapors, dusts and mists which have a cumulative detrimental effect on workers' health. Respirators should be worn regularly under such conditions to prevent long-range lung damage.

Respiratory protective devices are used by many industries. For example, packagers use them as protection against airborne particles in the atmosphere. Sanders and grinders need and use such devices. Painters use them to protect against non-toxic overspray particles. Sandblasters also require their use. Other applications include certain jobs in the chemical industry, steam-cleaning tank cars and cleaning tube bundles where temperatures reach 170 F.

There are two classifications of respiratory protective devices:

1. Air purifiers, which can be subdivided into three categories:
   a. mechanical filter respirators, which filter out contaminants, especially particulates;
   b. chemical cartridge respirators, which remove pollutants through chemical absorption; and
   c. combination respirators, which remove contaminants both chemically and mechanically.
2. Air suppliers, which provide the user with clean air from an outside (or oxygen from a tank).

Air Purifiers

Mechanical Filter Respirators

These respirators cover the mouth and nose, but do not provide eye protection. They must provide a high filtering efficiency and low resistance to breathing. The U.S. Bureau of Mines has established standards for respirators. These types are primarily employed by workers (grinders, sanders, packagers, sandblasters, painters) involved with irritating contaminants in the atmosphere, or with contaminants presenting long-range damage. The contaminants are usually particulate.

Mechanical filter respirators can be classed into the following categories, as established by the Bureau of Mines:

- Respirators which protect against pneumoconiosis-producing dusts. Dusts include those of aluminum, cellulose, cement, charcoal, coal, coke, flour, iron ore, wood, limestone, and others.
- Toxic dust respirators that guard against poisonous dusts (those which are not significantly more toxic than lead). These dusts include arsenic, cadmium, chromium, lead, manganese, selenium, vanadium, and their various compounds.
- Mist respirators which absorb out pneumoconiosis-producing, chromic acid, and other irritating mists.

● Fume respirators which are effective against particulate matter formed by the condensation of vapors; for example, those from heated metals.

Low breathing resistance is of prime importance in a respirator. Breathing resistance must be kept at a minimum even after prolonged use of the respirator under dusty conditions. Excessive inhaling resistance wastes the worker's energy and may cause fatigue or even lung injury if continued. Breathing out must also be considered in the design. This too can prove to be detrimental or tiring to the worker if there is high resistance.

High-porosity filters generally give low resistance to breathing with small filter area. However, they will not stop fine dusts. They merely improve the worker's comfort by arresting the larger particles which are irritating, not dangerous.

Chemical Cartridge Respirators

This type of respirator is made up of a half-mask facepiece which is connected to one or several small, lightweight containers of chemicals. It is very similar to a gas mask. In fact, the chemicals are the same as those found in gas masks, but such a unit is utilized under nonemergency conditions—in harmful atmospheres that present a danger only after prolonged exposure. Chemical cartridge respirators are not used as often as combination respirators.

Combination Respirators

These employ both mechanical filtration and chemical absorption. They utilize dust, mist, or fume filters with a chemical cartridge against multiple exposure. Such a system generally has independently replaceable filters because the dust filter may become exhausted before the chemical cartridge. Painters and welders find use for this device.

Figure 9-4 shows a schematic representation of the three types of personal air purifiers. Note that such devices include an adjustable headband. These units must provide maximum protection but also at sufficient comfort to the user. Thus, they must be lightweight, sanitary, and easily dismantled for cleaning, repairing, and replacing of filter pads or cartridges.

The mechanical type filter is usually one-piece molded rubber with a facepiece designed to fit most facial contours. All three types must assure a close fit around the bridge of the nose without uncomfortable pressure. They must be of rugged construction, able to retain their original shape when subjected to accidents such as being stepped on, or exposed to heat.

Working conditions may require eye protection. In such cases goggles, spectacles, or some type of facelet may be worn. These are generally not considered standard equipment to air purifiers.

Figure 9-4. (A) Mechanical dust respirator provides protection against the inhalation of toxic and non-toxic dusts. (B) Single-cartridge chemical respirator provides protection against toxic fumes. (C) Combination respirator is used primarily against organic vapors, sprays and mists from solvents and thinners, paints, and varnishes. It includes a chemical cartridge to absorb harmful vapors and filter material to remove solids.

Air Suppliers

Self-Contained Oxygen Respirators

These are used under the most severe conditions where the atmosphere is extremely corrosive to the skin and mucous membranes in addition to being deadly. These conditions require a complete suit of special corrosive-resistant clothing that must be considered part of the respiratory unit.

When less severe working conditions exist a complete suit may not be required. An air-supplied hood or mask may be sufficient. In hot dusty work areas the hood design respirator is preferable.

Compressed Oxygen Rebreathing Unit

This unit consists of a mouthpiece and nose clip and may or may not include a full facepiece. This apparatus has a high-pressure oxygen cylinder with reducing and regulating valves. It has a lung-regulating admission valve which supplies oxygen from the cylinder upon inhaling. Most units have a carbon dioxide scrubber and a reservoir breathing bag connected by tubes to the mouthpiece or facepiece. The scrubber purifies the exhaled oxygen and allows the air to be recycled through the unit. With this unit, the seal around the facepiece must be maintained absolutely tight. A rebreathing unit is used under very toxic conditions; however, a full suit generally does not accompany it.

Air Nonbreathing Type

The respirator includes a high-pressure cylinder of oxygen, a cylinder valve, a demand regulator, and a facepiece and tube assembly with an exhalation valve. To operate, the user adjusts the facepiece, turns on the cylinder valve, and breathes in to draw the oxygen through the demand regulator to the facepiece. He must exhale to the atmosphere through the exhalation valve in the facepiece. This type unit can only be used for 30 min at a time, and is considerably less efficient than the rebreathing apparatus. It does not provide a constant positive pressure at the facepiece, which means that perfect facepiece fit is imperative to prevent inward leakage. Beards, sideburns, glasses and sometimes even facial movements may break the seal and allow leakage.

Some units do provide positive pressure. This eliminates the need for a tight fit. However, certain positive pressure units can create other problems such as high noise level, uncomfortable air blasts, and fatigue to some employees, which should be investigated before purchasing.

Oxygen-Generating Unit

This unit includes a chemical canister, a reservoir (breathing bag), a facepiece and tube assembly, along with a relief valve and check valves which regulate the oxygen flow. The chemical in the canister or tank generates oxygen when it comes in contact with moisture and carbon dioxide from the exhaled breath.

This respirator has a tendency to produce more oxygen than is required, so automatic venting is necessary. Most models require a new canister each time the unit is worn. Canisters should be disposed of according to the specific directions on each can.

Many of the air supplier respirators manufactured include personal air conditioning units when a full suit is employed. The wearer can regulate temperature as much as 60 F.

Because of the danger of unit failure, wearers should not be allowed to work in an area alone. Other personnel similarly equipped should be there in case emergency assistance is required.

Air suppliers are generally complicated, so it is necessary to provide a thorough training program. Workers must understand fully the operation and proper handling of such equipment. Some employees may dislike using air suppliers because of some inconvenience or awkwardness. Proper instruction will overcome this.

Care of respiratory equipment also requires proper training and instruction. Blowers and valves to hose masks should be lubricated and periodically inspected. Periodic replacement of the inhalation and exhalation valves is necessary along with regular cleaning of the rubber parts.

Canisters and tanks should be stored as suggested by their manufacturer. In general, they should be stored in a cool, dry area. When a unit with a canister is stored, the seal on the bottom opening of the canister should be unbroken. If it has been broken, it must be resealed before storing.

After cleaning and disinfecting, the apparatus should be thoroughly inspected for defects. An additional check every two months is advisable even if the units have not been used. If the lens or other parts of the hood or facemask need replacing, the unit should be returned to the manufacturer. Employees should not attempt self-repairs which could damage the equipment and even endanger life.

The purpose of lung respirators is to turn hot, dirty, and sometimes dangerous working conditions into safe environments for employees. The choice of the proper unit can only be made after a thorough analysis of the working atmosphere. Table 9-3 provides a guide to selection of this type equipment.

Table 9-3. Guide to Respirator Selection

Type of Respirator	Working Area Atmosphere (Normal Oxygen Concentrations 21 percent)			
	Dust, Fumes Mists (generally irritating but not toxic)	Gas or Vapor Concentration by Volume		
		No Greater than 0.1 percent	No Greater than 2 percent	Any Concentration of Contaminant
Mechanical filter	X			
Chemical cartridge filter		X		
Combination respirator	X	X		
Hose mask with electrical blower	X	X		
Gas mask		X	X	
Self-contained breathing unit	Can be used but not practical	X	X	X

Respirators should not be considered substitutes to pollution control. They can only be used for short periods of time and offer only a temporary solution. However, they are a good alternative under some situations. In general, they increase the worker's safety and productivity by offering maximum protection.

HANDLING AND DISPOSAL OF CHEMICALS[138]

Chemicals—an integral part of manufacturing—pose a problem of handling and ultimate disposal. When any degree of chemical infusion is required for an organized operation, control procedures should be established to oversee all phases of the chemical cycle: purchasing, receiving, use and storage, and disposal. Representatives from all the involved areas should assist the pollution engineer as a chemical control team.

The team should have regular meetings at which problems are discussed, procedures reviewed and innovations suggested. Good chemical control can be maintained only if all personnel involved are fully aware of their responsibilities. To facilitate chemical awareness, regularly held training sessions are vital. Subjects to be covered include: work practices, safety rules, personal protective equipment, packaging and moving requirements, and use of an approved color labeling system to identify the chemical hazard in a container.

Purchasing

All orders for chemicals should require approval by the pollution engineer (Figure 9-5). This provides monitoring of potential build-up of hazardous concentrations or combinations of incompatible items.

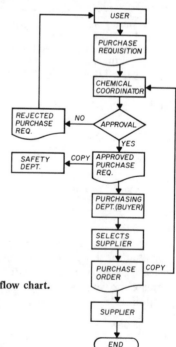

Figure 9-5. Purchasing procedure flow chart.

Investigation of the order may reveal that a nonhazardous chemical can be substituted for a hazardous one. For example, the solvent benzene is inflammable and highly toxic. Methyl chloroform is nonflammable and the toxicity level is considered lower, thus making it a satisfactory substitute.

A chemical data sheet giving the flash point, TLV ratings, and general health hazards is a valuable asset and should be requested from the vendor when a chemical is being ordered for the first time. The pollution engineer should also notify the medical office. Awareness by the proper individuals and departments is vital.

Receiving

Chemical escorts should be designated and made responsible for moving chemicals within a plant (Figure 9-6). When chemicals are received the transporting vehicle should be detained until a chemical escort verifies the order, inspects the shipment for transit damage, and checks for proper labeling and packaging.

To facilitate safe in-house movement, a protective type conveyor (Figure 9-7) should be required. By having chemicals semi-isolated, the transportation safety factor improves considerably.

As deliveries are made to production areas, chemicals should be signed for and the receipt forwarded to the pollution engineer for inventory updating.

Use and Storage

The pollution engineer should take samples of air where hazardous chemicals are used and analyze it for the amount and nature of contaminants. Due to their composition, many chemicals require a controlled environment (CE), with inflammable chemicals requiring fireproof facilities. Thus, storage can become a problem; CE chemicals may require air-conditioned storage and use areas. When such facilities are limited, use of CE chemicals should be avoided whenever possible.

Another storage problem is the incompatibility of chemicals. Accidental contact of some chemicals with others can be disastrous. Some commonly used chemicals causing violent reactions or producing highly toxic gases if allowed to come in contact are listed in Table 9-4.

Disposal

When chemicals have outlived their usefulness they must be either disposed of or reactivated (Figure 9-8). Many times the shelf-life of a chemical can be extended by processing. The pollution engineer should be contacted when chemicals are no longer required for a particular operation or are ineffective.

Disposal processes vary according to chemical types. Inorganic and water-soluble chemicals can usually be treated by a plant's waste disposal

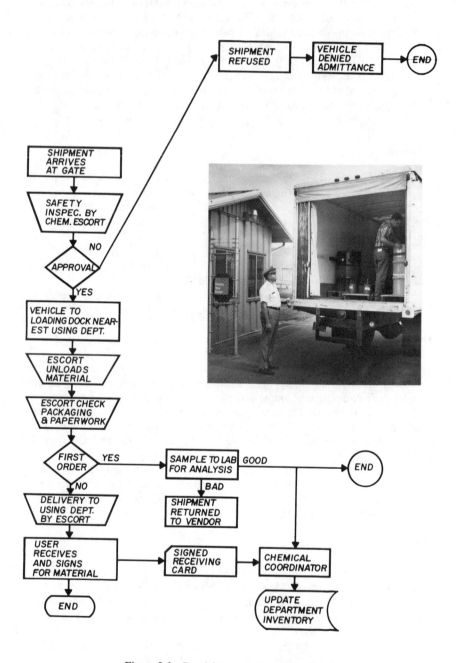

Figure 9-6. Receiving procedure flow chart.

Figure 9-7. Special containers should be built and used to transport chemicals within a plant.

Table 9-4. Substitution of Hazardous Chemicals

Chemicals	Application	Hazard	Substitution
Alcohol	Solder Flux Removal	Flammable	Freon, Alcohol Mixture
Benzene	Cleaning Agent	Flammable, Toxic	Methyl Chloroform
Methyl Ethyl Ketone	Cleaning Agent	Toxic	Isopropanol
Sodium Methoxide in Methanol	Stripping	Highly Toxic	Sodium Ethylate in Ethanol
Toluene	Spraying	Flammable, Toxic	Methylene Chloride
Toluene & Ethyl Acetate	Spraying	Flammable, Toxic	Freon TMC

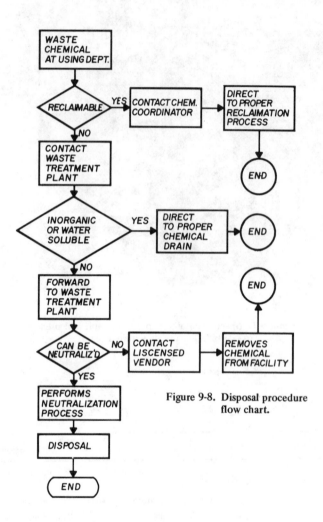

Figure 9-8. Disposal procedure
flow chart.

and treatment area. However, special drains for chemicals must be used. Before chemicals are put in the drains the disposal facility must be contacted to divert the chemical to the proper holding and treatment tanks for solids separation and pH adjustment.

CHAPTER 10

STACK DESIGN AND AIR POLLUTION DISPERSION[22]

INTRODUCTION

The design of a stack for safe effluent release is dependent on many factors. Meteorological factors such as wind and turbulence are the most important. Local topography including both man-made obstructions and natural features also determine stack design. The physical and chemical properties of the effluent must be carefully considered. Adjacent stacks must be accounted for or a multiple-source problem may be created. Certain limiting factors exist which prohibit extremely tall stacks, while local law often dictates minimum stack heights.

An accurate analytical expression for all these factors does not exist at this time. By assuming some of these factors negligible, equations can be used to predict dispersion. Experimental methods are often used to supplement calculations, and computer programs exist which model proposed stacks and the conditions they will face.

METEOROLOGICAL FACTORS

An understanding of basic meteorology is a fundamental requirement for a successful solution to an air pollution problem. Atmospheric dispersion of pollutants is largely controlled by weather factors, the most important being wind.

Wind

Large-scale pressure variations over the earth produce air flows called winds. Wind distribution in an area is determined by the intensity of these pressure variations. Characteristic patterns can be developed for the movement of these pressure systems and diurnal heating effects in a given area. These patterns are plotted on polar diagrams called *wind roses.* Variations of wind speed and direction with season are readily apparent in this useful

355

graph. The prevailing wind is defined as the predominant wind direction in a wind rose. If the wind direction is constant over long time periods, it is called a persistent wind.

Horizontal wind speed and direction vary with height. At high altitudes wind flow is primarily caused by the horizontal pressure gradient (gradient winds). Friction effects nearer the ground slow wind speed and change wind direction. Angular change of wind with height depends on a variety of factors, the most important being thermal stability of the atmosphere and under-underlying terrain changes. Angular change is usually lower during the day than at night and lower over smooth surfaces than rough. The mixing height is the height above the ground surface at which relatively vigorous vertical mixing occurs. It is important that this height be reached by the plume if good dispersion is to be achieved. Mixing height also varies with diurnal periods, ground surface and precipitation.

There are several common terms used in connection with winds. *Drainage winds* occur when a marked slope exists to the land. Winds developing close to the surface are cooled by radiation at night and flow down with gravity. *Valley winds* are caused by the night-time flow of drainage winds toward the valley center from which flow tends toward the lower end of the valley. During daytime the flows tend to travel up the center of the valley. *Land* and *sea breezes* are caused by temperature differences between adjacent bodies of water and land. During the day the air over the land is warmer than that over the sea and breezes circulate toward the land. At night the reverse happens due to the warmer surface of the sea.

Turbulence

Turbulence can be defined as local fluctuations in the wind flow. It is an important parameter since it has a large effect on dispersion. Turbulence can be separated into two classes; *mechanical* and *convective.* Mechanical turbulence is due to the influence of buildings or any objects on the surface which obstruct natural air flow. Turbulence increases with the irregularity or roughness of terrain. Convective turbulence is caused by the difference in temperature between the surface and overlying airstream. These differences can be caused by a large variety of factors including radiation heating of buildings, water masses or land features. The vertical temperature gradient is an indicator of convective turbulence. Temperature gradients can be divided into three general categories: neutral, stable and unstable. Mechanical and thermal influences generally occur simultaneously in varying ratios. One affects the other and it is often difficult to distinguish the type of turbulence. In general a neutral temperature gradient indicates primarily mechanical turbulence, while an unstable gradient adds convective turbulence and a stable gradient suppresses mechanical turbulence.

Stability

Stability is the ability of the atmosphere to resist vertical motion or to suppress or enhance turbulence. It is related to both wind shear and the temperature gradient. Measurement of stability is accomplished through the temperature gradient. As a standard, a small volume of air is considered to rise up within the atmosphere. Assuming no heat exchange between this air and the environment the rate at which the air cools is defined as the dry adiabatic lapse rate (-1 C/100 m). Potential temperature is the temperature this small volume of air would have if its pressure were raised adiabatically to 1000 microbars. A potential temperature increase with height implies stability. The actual vertical temperature distribution is called the environmental lapse rate.

Environmental lapse rates can be divided into five types which are shown in Figure 10-1. Superadiabatic conditions exist on days of strong solar heating. These conditions favor instability and turbulence. When the

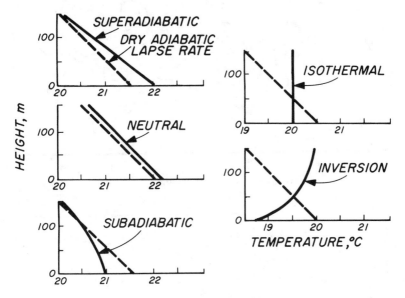

Figure 10-1. Environmental lapse rates.

environmental lapse rate is nearly identical to the dry adiabatic lapse rate, a neutral condition exists. A more gradual temperature decrease with height is known as subadiabatic. If air temperature is constant with height, isothermal conditions exist. When air temperature increases with height inside a stable atmospheric layer, dispersion is restricted. This condition, known as an inversion, can exist at surface level or, in the case of elevated inversions, as high as 600 meters. Inversions can be caused by a number of factors but always result in suppression of vertical motion which will trap a smoke plume.

Plume Types

The geometrical configuration of plume dispersion with the atmosphere has been related to the vertical temperature gradient. Plumes can be classified into six types. Figures 10-2 through 10-7 identify each plume type.

Figure 10-2 shows looping plumes which occur with an unstable or superadiabatic lapse rate on a clear day. Solar heating causes large thermal eddies in the unstable air which periodically bring the plume to the ground. A coning plume (Figure 10-3) occurs during slightly overcast days. The vertical temperature gradient can vary between dry adiabatic and isothermal.

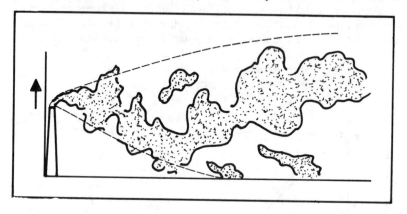

Figure 10-2. Unstable conditions (looping).

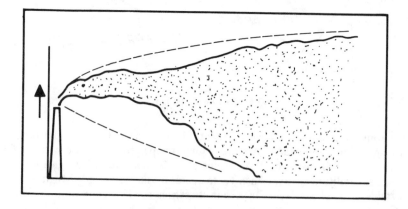

Figure 10-3. Near neutral (coning).

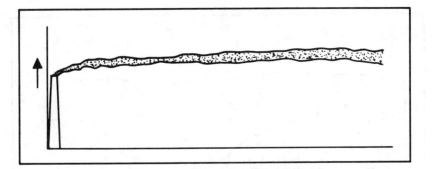

Figure 10-4. Surface inversion (fanning).

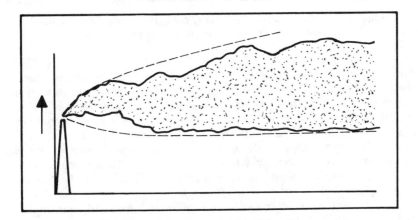

Figure 10-5. Surface inversion below stack (lofting).

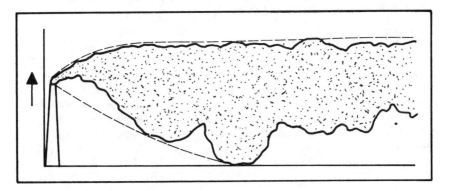

Figure 10-6. Inversion aloft above stack (fumigation).

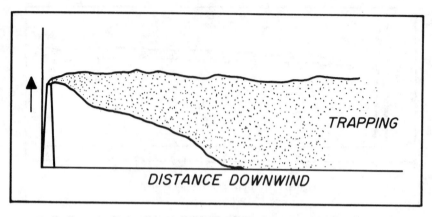

Figure 10-7. Similar conditions exist as in fumigation, but plume does not touch ground midway along its path (trapping).

An inversion prevents vertical diffusion resulting in a thin horizontal plume classified as fanning (Figure 10-4). A superadiabatic layer above an inversion results in lofting (Figure 10-5). The plume will continue to rise and diffuse laterally without reaching the ground level. The opposite of this condition (Figure 10-6), superadiabatic lapse rate below an inversion, results in fumigation. The plume disperses downward toward the ground since it is trapped by the inversion. Trapping (Figure 10-7) exists under the same conditions as fumigation but the plume does not reach ground level along the plume length.

The pollution engineer should be familiar with the conditions causing each plume type. If these conditions can be expected in the stack location, the design must include provisions to deal with them. This is especially important in industries and public utilities which can not be shut down for the duration of adverse conditions.

EFFECT OF TOPOGRAPHY

Surface irregularities range from isolated projections or depressions to dense successions of these features. Collectively these features are called ground roughness and have a great effect on local turbulence as previously mentioned. These irregularities can be man-made buildings or natural terrain exhibiting mechanical, thermal or combined influences over air stability.

Buildings

When a moving air stream approaches an obstacle such as a building, air which would have gone through the object is displaced vertically and laterally. This air displaces adjacent streams of air. The mean velocity increases upward from the ground, rapidly at first but slower at higher elevations. This is called streamline distortion (Figure 10-8). This flow disturbance produces a highly turbulent wake. Within the upper portion of the wake, directly behind the building, a cavity is formed. Within this cavity is a circular flow which will trap effluents. Cavity size is affected by building shape and wind orientation but is primarily a function of the frontal area of the building presented to the wind.

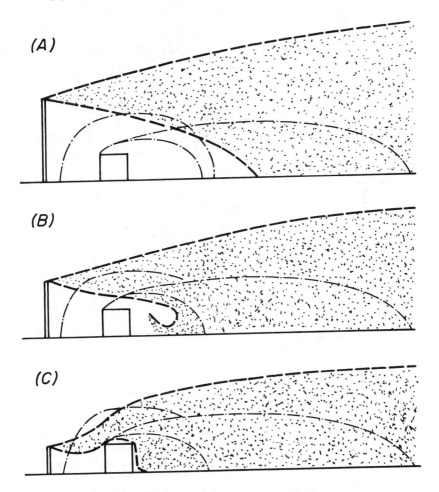

Figure 10-8. Building effect on an upwind stack.

Building effect on plume dispersion will vary with stack location and height. In Figure 10-8 the stack is upwind of the building. A plume from a tall stack will clear the displacement zone but may not clear the wake (depending on atmospheric turbulence). A short stack will cause the plume to be caught in the displacement zone, cavity and wake. A medium stack (plume centerline in vicinity of cavity boundary) will cause the plume to enter the displacement zone, wake, and possibly the cavity.

In Figure 10-9 the stack is supported on the building, thus increasing the physical stack height. Effluent from a tall stack will clear the wake, while a medium stack will clear the cavity. A stack close to the roof will cause the plume to be distributed throughout the cavity and wake.

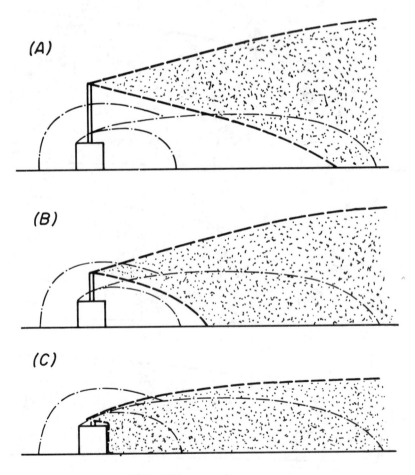

Figure 10-9. Effect on building mounted stack.

Downwind stacks are shown in Figure 10-10. Downwind stacks will not be affected by the cavity, but lower stacks will disperse plumes into the wake.

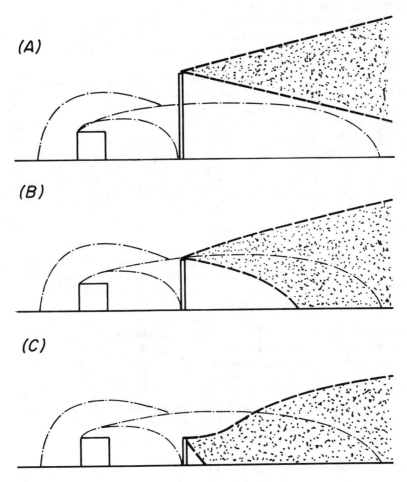

Figure 10-10. Building effect on downwind stack.

Natural Features

Land-water interactions have already been mentioned but are important enough to repeat. Temperature differences exist between the water body and surrounding land. Diurnal variations in temperature cause changes in wind direction and inversions.

The direction of wind flow in a valley will determine plume dispersion. A wind parallel to the valley will result in a normal dispersion confined only by the valley walls. A wind flow from high to low ground will cause vertical distribution of the plume throughout the valley. A reverse wind flow will cause distribution of the plume on high ground.

Under stable conditions stack plumes tend to follow the terrain and resist being forced over hills and mountains. In unstable conditions hills or mountains will create turbulence that can be likened to that of a building causing dispersion within a cavity.

Combined thermal and topographic effects are caused by cities. Heat released from fuel consumption and that absorbed by streets and buildings from the sun creates a warmer environment within the city. Combined with mechanical turbulence caused by the buildings, these heat effects result in rapid vertical mixing of plumes. This phenomenon is known as a heat island.

There are no accurate analytical equations for predicting topographic effects. Estimates can be made using the general rules just mentioned, but differences in topographical elements prove this method somewhat crude. Experimental data can be obtained from scale-model wind tunnel tests.

EFFECTS OF INITIAL PARAMETERS

The two initial source parameters, stack effluent exit velocity and stack effluent temperature, are extremely important to good dispersion. They can be manipulated after stack construction to correct dispersion much more easily than other alternatives.

The exit velocity of a stack effluent must be high enough to prevent downwash, the tendency of a gas to creep down the stack. This speed will depend on local atmospheric conditions and can be determined in wind tunnel tests. Additional elevation of the plume is also supplied by the momentum of the exit gas. It is important to provide additional elevation before atmospheric turbulence becomes controlling. The stack gas is usually much warmer than ambient air, thereby allowing its reduced density to cause a buoyancy effect. The gas will rise until it is cooled by the atmosphere. Pollution control devices could cool the gas and restrict this rise.

EFFECTS OF MULTIPLE SOURCES

The most obvious effect of multiple-source emissions is the increase in pollution. Each source alone may not present a problem, but their combined effects could be above tolerance levels. Legal problems arise as to setting responsibility for corrective measures. Before a stack is designed, consideration should be given to projected combined effects.

When two or more stacks are closely grouped they tend to influence each other. Their plumes combine and will rise higher, but not as high as that of one stack replacing them all. There are no accurate analytical equations to evaluate plume rise or concentration of effluent from combined sources, but an approximation is often used for a very rough estimate.

$$Xmax(N) = Xmax(1)N^{.8}$$

where $\quad$ Xmax(1) = maximum concentration for a single stack
$\quad\quad\quad\quad$ N $\quad\quad$ = number of stacks

Multiple stacks are often in line and when wind direction is also in line, the stack plumes will rise to their highest level. This can be explained by noting that the lifting factor previously mentioned will be at a maximum since all stack plumes will merge. This effect is heightened when the line stacks are placed on a long building which will be in line with the wind.

EFFLUENT CONSIDERATIONS

An understanding of the effluent produced by an industrial process is necessary before a stack can be designed. Consideration should be given to the type of operation. This can roughly be divided into two categories, *batch* and *continuous*. Batch processes may be shut down under extremely bad weather conditions with little equipment damage, whereas a continuous industrial process or power generation unit would be damaged by a shutdown or create a hazardous situation. The sensitivity of the local area to the effluent is another important factor. The local annoyance level may be such that the additional expense necessary for better dispersion may be warranted.

Physical properties of the effluent gas should also be considered. Stack effluent may be composed of gases or a combination of gases and particulates which will require special treatment. This treatment would have a pronounced effect on design as will be mentioned later. The gas density will have an effect on buoyancy rise. The density will be dependent on the temperature of the exit gas. Care must be taken that the temperature of the gas does not fall below the dew point of the gas. This could result in serious corrosive problems within the stack or a precipitation problem in the local area.

Chemical properties must also be considered. Many effluents will react in the atmosphere to cause secondary pollution problems. An excellent example of this problem is the conversion of sulfur dioxide to sulfur trioxide to sulfuric acid in the atmosphere. Particles drop out of the air onto buildings and cars. Deposition or settling of particles could be another problem area.

PREDICTION OF PLUME CHARACTERISTICS

Analytical

The majority of analytical methods for predicting plume characteristics use a two-step approach. The first step calculates the height of the plume above the surface. Step two estimates the diffusion of gas and deposition of particulates. These equations are very limited in that they treat only a single horizontal straight plume. Topography and atmospheric stability are not taken into account. The diffusion equations assume a point source of the plume which cannot be applied for small downwind distances. While these equations are of questionable accuracy they do provide a rough initial estimate of the stack plume.

Experimental

For a specific air pollution problem the available analytical methods may often be inadequate. Two experimental methods which have been widely used are full-scale field experiments and scale-model wind tunnel tests. In a field experiment the actual conditions of stack operation, topography and atmosphere are used. This may often be impractical when buildings are in the design stage or when infrequent atmospheric conditions are to be included in the design. Because of the cost involved limited field experiments are often used in conjunction with other data. Wind tunnel tests are often advisable when dealing with a limited time factor since conditions can be changed as desired.

In field experiments a desirable tracer is released under conditions of stack operation. The tracer should be easily identifiable and dispersed, and should closely simulate atmospheric motions. Fluorescent particles and water-soluble fluorescent dyes are commonly used. At times a gas is necessary to simulate the atmospheric effect on small particles. Smoke has been used primarily as a visual tracer and nuclear techniques have also been developed.

The primary measurements taken of the atmosphere are those most important in dispersion, namely, wind speed, wind direction and wind turbulence. Surface wind direction is measured by vanes, while wind direction aloft is measured by different types of balloons. Wind speed is also measured at ground level and aloft. An anemometer is used at ground level while a balloon is tracked to obtain the speed aloft. Wind turbulence is estimated indirectly from the environmental lapse rate or directly through vanes and anemometers.

Secondary measurements include visibility, humidity, precipitation and solar radiation. The equipment necessary to determine the concentration of effluent will depend on the tracer selected. The ability to carry out wind tunnel tests on a scale model will depend on the availability of data. Detailed information is needed on plant configuration, topography, meteorological characteristics and sensitivity of polluted areas. In the development of a new plant it is often advisable to obtain data on the tolerances of those items in the area sensitive to the pollutant. Plant blueprints and terrain maps should provide all necessary information for construction of the scale model.

Wind speed and direction are easily modeled in the wind tunnel. Air density, thermal and convective turbulence can also be scaled down. Usually the three main stack parameters—stack height, velocity and temperature of the exit gas—are varied for a fixed set of conditions. In this manner an optimum design can be determined. Particulates can also be treated by this same method.

The actual data to be represented in the wind tunnel must be scaled down to the model. This, as well as scale-up of the optimization study, is done with the use of equations derived through dimensional analysis.

Wind tunnel experiments are valuable in design of modifications to existing plants. The addition of nozzles or stacks can be studied as well as changes in operating conditions. As more experience is developed on the effects of building shape and similar factors, it may be possible to modify theoretical analysis to account for these factors. Wind tunnel tests have been effective in investigating atmospheric dispersal of gaseous wastes which did not lend themselves to analytical solutions.

GENERAL GUIDELINES

Over the years several rules-of-thumb in stack design have been developed through experience. These are meant to be taken only as guidelines—not as solutions to air pollution problems. Most of these rules concern stack height. The most well-known calls for a stack to be 2.5 times higher than the tallest building within twenty stack lengths. At this height the aerodynamic effects of the adjacent buildings are no longer a factor in dispersion. In areas of flat country with few large buildings this factor may be scaled down to 1.5. Local laws often govern the minimum height of a stack, which is suggested to be 120 ft by British authorities.

There are a few other guidelines which should be kept in mind. A gas ejection speed lower than the wind speed results in downwashing. It has been found that an ejection speed of 60 fps is high enough to overcome any trend toward downwashing. Through experience, it has also become apparent that any structure added near the top of an otherwise plain circular stack will be detrimental, and a square stack will be inferior to a circular stack.

Practical Stack Design

Ground-level concentration of pollutants can be reduced through modifications of design features in both new and installed stacks. Building height is an important factor because of its effect on turbulence. Building shape may be just as important if not more important than height. The degree of influence of building shape will be determined by wind direction. For example, a long narrow building with a large surface normal to the wind will have a more adverse effect on dispersion than a square building of the same height. It may be possible to favorably modify an existing building, but this would usually require costly additions. Wind tunnel tests can be used to test the effect of any suggested modifications.

The most obvious method of decreasing ground-level concentration is by raising the physical stack height. In general, ground-level concentration is found to vary with the inverse square of the plume elevation. This assumption is based on uniform turbulence; thus additional factors such as buildings will significantly increase this equation.

An increase in gas exit velocity will increase the momentum of the gas, thereby raising the effective stack height. This can be done by increasing the forced draft fan or blower equipment, if practical, or decreasing the exit area of the stack. In any new design a gradual decrease in stack diameter can be incorporated into the stack. In existing stacks, a nozzle can be placed on the top of the stack to reduce the outlet diameter. Care must be taken in nozzle design so that the reduction in stack diameter is not too sudden. The effect of the nozzle is also controlled by the effluent. Again, a wind tunnel can be used to test designs.

A problem is created by processes with varying stack emission rates. In an effluent with large momentum and buoyancy effects, reduction in emission rates will cause greater ground-level concentration. A stack designed for full load will not be satisfactory for a partial load. Special devices which adjust exit area are used so that full load emission rate can be reached with a partial load.

The effluent gas may be physically modified to create lower concentrations at ground level. Heat can be added to a gas to increase its buoyancy effect. Dilution at effluent temperature may be desirable to increase the effluent exit speed. If air is added at ambient temperature, it will have little effect on ground-level concentration because it will lower the density and temperature of the effluent. Since it is costly, heat addition may be reserved for handling especially poor meteorological conditions.

Stack Height Limitations

There are many limitations to the physical height of a stack, the most common being economic restrictions. Construction and material costs are high for tall stacks. Another limiting factor relates to the amount of air traffic in the area. Runway approaches have inclined building height, limit lines increasing from the end of the runway. Wind pressure is also an important factor in stack design. Wind pressure will increase with air density and the square of the wind speed. At higher altitudes wind speed will increase, and stacks must therefore be structurally designed to withstand higher lateral wind speed. Wind forces also form break-away eddies around stacks. These eddies form and break away with a regular periodicity. If this frequency corresponds to the natural vibration frequency of the stack, a resonant vibration is created. Without strong structural damping the vibration may reach critical levels. This phenomenon will only occur, however, in steel stacks.

MATERIAL AND EQUIPMENT CONSIDERATIONS

A stack is composed of two parts, a shaft and a liner. The shaft is the structural element designed to resist wind and temperature and to shield and support the liner. The liner is designed to resist chemical and physical characteristics of the effluent gas. The liner may be independent of the shaft. Costs of the shaft material will vary with height. For example, at lower heights steel will be less costly, while for higher ranges concrete and brick will be cheaper.

The liner material will be dependent on the effluent. For gases containing sulfurous compounds the temperature must be above the dew point, or sulfuric acid will form. Above the dew point steel liners can be used. If removal equipment is necessary, the gas can be reheated before going through the stack. A lining of acid-resistant material can be added when acid formation cannot be prevented. Various types of steel can be used including carbon steel, copper-bearing steel and stainless steel. Polyvinyl steel—polyvinyl chloride bonded to steel—can be used when the gas temperature is less than 450 F. Acid-resistant brick and mortar linings are often used. A radial brick liner is completely independent of the shaft, but exit velocity of the gas is limited to 90 fps to avoid damage to the brick.

An exit gas velocity of 60 fps was mentioned as necessary to prevent downwashing and to ensure good dispersion. A higher velocity is needed to overcome low inversions, by enabling the gas to push through the layer. Natural draft cannot create stack exit velocities greater than 10 fps. Natural stack draft is a function of atmospheric and stack conditions.

DETERMINING WIND CHILL FACTOR[158]

You have often heard your weatherman use the term "wind chill index," and give the actual thermometer reading temperature and the equivalent temperature for the prevailing wind speed. Both temperature and wind speed must be measured to determine the wind chill. People feel colder on windy days, because the greater the wind's velocity the more quickly it carries away natural body heat from skin surfaces.

The wind chill factor has an influence on the demand for fuels such as natural gas, coal and oil. Information concerning temperature and wind velocity is an important factor in predicting this demand. If you know or can estimate the wind speed and atmospheric temperature, Figure 10-11 will provide an easy way for finding the equivalent temperature.

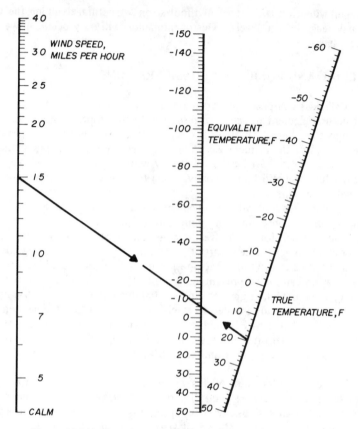

Figure 10-11. Nomograph for determining wind chill index.

Example

If the wind speed is 15 miles per hour and true atmospheric temperature is 20 F, what is the equivalent temperature?

Solution

Connect 15 mph on *Wind Speed* scale with 20 F on *True Temperature* scale, and read -5 F (5 below zero) where line crosses *Equivalent Temperature* scale.

NEXUS—A NEW WAY TO MEASURE[162]

There are many unfortunate examples of residential areas located downwind from heavy polluters. These would not have occurred had a regional land use plan been used, utilizing regional air pollution data inputs.

Regional air quality modeling requires accurate numerical simulation of the effects of advection and eddy dispersion processes on pollutants emitted into the air from both point and distributed sources.

Until now, air pollution dispersion models have been based on an analytic solution of the diffusion equation under very simplified and restrictive assumptions. The most widely used models are based on constant wind and diffusion in a region bounded by the ground and a possible inversion. These models use a Gaussian crosswind distribution together with a downwind plume. Other models use short-time releases or puffs—but still have been restricted to a two-dimensional wind field.

A new tool known as NEXUS (Numerical Examination of Urban Smog) is a mathematical model of air pollution and uses a computer for simulation of transport and diffusion of pollutants in the atmosphere. The techniques employed in NEXUS use the computer to track many parcels of polluted air through a region divided into grid cells. Generality in specifying emissions, winds and dispersion, varying in both time and space is available.

Computer codes predict the flow of pollutants in the atmosphere in three spatial dimensions as a function of time. Thus, NEXUS can be used to:

1. Predict worst-case and nominal air quality,
2. Evaluate the effect on air quality of changes in emissions,
3. Develop optimum land use, industrial site locations, roadway routing and emission controls to minimize pollution levels, and
4. Obtain a regional perspective on air pollution throughout a basin, using graphic data presentation.

Validation of NEXUS as a regional model for three-dimensional air pollution was accomplished by comparing NEXUS simulated carbon monoxide concentrations with monitored air data for the Los Angeles basin. In the original validation, day-long and hour-averaged CO concentrations for 12 widely spaced monitoring stations were calculated by NEXUS and compared to actual observations.

NEXUS also provides qualitative presentation of the results of a simulation in graphic form (Figure 10-12). Such "dot" pictures of polluted air have been used to generate a movie of the pollutant flow. Results of the calculated concentrations can also be presented using computer contouring of isopleths.

NEXUS has been used to study photochemical smog in Los Angeles under a program funded by the U.S. Environmental Protection Agency.

With NEXUS, it is possible to quantitatively determine the dispersion of pollutants from existing sources and—more importantly—from proposed new facilities. With the addition of chemical reaction data to the computer program, interaction of other pollutants with those already present in a region can be taken into account. And, with the addition of differential equations for water vapor and liquid, the effects of cooling towers can be assessed.

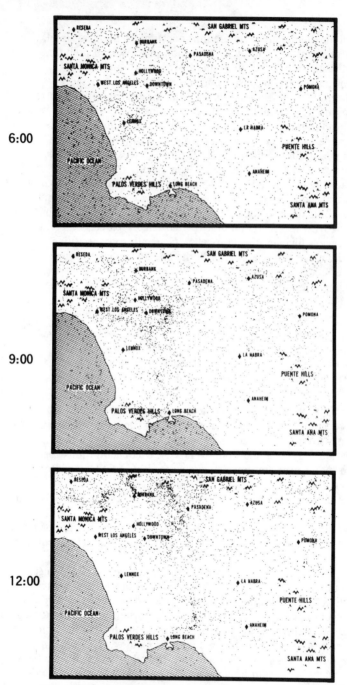

Figure 10-12. Dot representations of carbon monoxide concentrations in the Los Angeles Basin.

The techniques developed for the NEXUS computer code can be incorporated into a plume assessment system (PLASY) computer model and used in micrometeorological applications. In such use, a three-dimensional grid represents the air surrounding the stack. The ambient atmosphere, with a given thermal stratification and wind field, is disturbed by the plume and responds dynamically to it. Turbulence is calculated throughout the atmosphere as a function of local wind shear and thermal stability.

In formulating air quality control plans, a model must be used to ensure that the proposed emission controls will achieve the air quality standards set. NEXUS can provide assurance that the standards will be met, when the model predicts that the controls are effective. PLASY offers a technique for evaluating the local effects of large emission sources which may be most important in complying with air quality standards.

NEXUS can be applied to forecasting air pollution levels throughout a region. Local meteorological forecasts can be coupled with NEXUS calculations to achieve 12-, 24-, or 36-hour forecasts on which emergency procedures can be based.

The impact of a single emission source in an urban multiple-source region is hidden within the many overlapping effects. NEXUS can determine how these emissions interact and where they go. NEXUS calculations with and without the specific source will clearly show the source's impact on regional air quality. They will indicate to what degree regulation of the source is necessary to meet air quality standards. Such indications are inexpensive enough that a company could make the determinations itself and provide the data to governmental regulatory agencies.

CHAPTER 11

WASTEWATER TREATMENT METHODS

INDUSTRIAL WATER TREATMENT[179]

Water supply quality, waste treatment and retreatment of water supply for reuse are especially important to industry, since it ranks, behind agriculture, as the biggest single user of water in this country today. Four industries alone—steel, chemicals, petroleum, and pulp and paper—are estimated to use somewhere between 90 and 95 percent of all water withdrawn for industrial use. The great majority of this water is taken from surface sources, and returned to natural water courses on the surface of the earth.

Requirements for industrial water supplies are greatly different from those of domestic use. Domestic water need only be free from materials which may be harmful or poisonous to human life, occasionally free from scaling or corrosive potentials, and esthetically of a nature which will suit standards with regard to taste, odor and color. Industrial supplies, on the other hand, must meet requirements which vary widely from industry to industry. In the steel industry, water used for once-through cooling often comes in direct contact with the product. The water is used as it comes from the supply source without any pretreatment. The petroleum and chemical industries, on the other hand, utilize most of their cooling water in heat exchangers of one type or another. Treatment of the influent and effluent water is usually required to assure freedom from corrosive attack or fouling.

In more specialized applications, boiler feedwater must have hardness removed to prevent deposit buildup in boilers. Water used for the manufacture of carbonated beverages must be low in alkalinity, to permit effective carbonation and sparkle. Water used in the manufacture of electronic phosphor equipment, such as television video tubes, must be completely free from both mineral and organic impurities, to prevent "poisoning" of the special luminous materials used.

Contrary to the beliefs of many, contaminants can be removed from a water supply of virtually any initial quality, if the proper treatment program is used. The drawback is the price which must be paid. The pollution engineer should thoroughly evaluate the quality of water required by each process before undertaking a program to clean influent water supplies.

Impurities

Characteristics of an industrial water supply depend upon the impurities present in the water (Table 11-1). By removal or modification of the impurities, it is possible to change the characteristics of the raw water to meet the needs of the individual industry and to prevent problems from occurring, or greatly reduce their incidence.

Table 11-1. General Characteristics of Water

Class	Solids Content, ppm	Specific Resistance, ohm/cm
Brackish	over 500	over 10,000
Normal	10-500	200 to 10,000
Pure	less than 2	5×10^5 to 2×10^6
Ultra-pure	less than 50 ppb	18×10^6

Impurities present in raw water supplies can properly be divided into two categories—particulate (suspended) and dissolved. In turbulent streams and rivers, particulate matter can include material such as large sand grains and even small pebbles, while in quiet waters the only suspended matter may be of colloidal size. Larger particles can be removed by simply settling or filtration, but the smaller particles, in the colloidal and near colloidal size ranges, require chemical coagulation under carefully controlled conditions for efficient and economical removal. When necessary, chlorination may be combined with either the coagulation or filtration step, to kill microorganisms. In some cases, soluble materials can be converted to an insoluble form and then removed by precipitation or filtration. An example is the oxidation of iron and manganese with their subsequent precipitation as hydrated oxides.

Certain types of organic matter, particularly those known as the "color colloids," can also be removed by chemical coagulation, sedimentation and filtration. Larger organic particles, such as algae, can be separated out by a filtration technique such as microstraining.

By far the biggest problems caused by dissolved solids in water are those related to hardness—dissolved salts of calcium and magnesium, insoluble salts of these two metals are the principal cause of scale formation in boilers, piping, heat exchangers and many pieces of process equipment; consequently control or removal of hardness is an important part of most industrial water supply treatment programs.

In addition to hardness, the significant metallic ions (cations) contaminating industrial water are iron and manganese. During most uses of industrial water, these materials, originally present in a soluble form, become oxidized and convert to an insoluble which precipitates to cause deposits. They can interfere seriously with water-using processes, since the insoluble forms are highly colored and cause staining or discoloration in such operations as textile dyeing, leather tanning and paper manufacturing.

Alkalinity, which is the total of bicarbonate, carbonate, and hydroxide content, is an important consideration in many industrial applications. Excessive alkalinity can be a cause of foaming and carryover in boilers, and scale formation in cooling systems. In addition, bicarbonate alkalinity breaks down under the influence of heat to release carbon dioxide, a major source of corrosion in steam systems.

Low chloride and sulfate contents are important in water because of their effect on the corrosive potential of the water. Both contribute to total dissolved solids, which may be undesirable.

Silica causes problems because of its scale-forming potential, and its removal is especially important in preparation of high-pressure boiler feedwater. Silica will volatilize with steam and cause deposits on power generation turbines.

Other negative ions (anions) such as nitrate and fluoride generally are not significant in industrial water, unless they are present in sufficient quantity to cause an undesirable increase in the total dissolved solids.

With the exception of some types of cooling operations and various specialized processes, organic matter generally is undesirable in industrial water. Organics can come from various pollution sources, such as oil, drainage, sewage and many industrial wastes.

Dissolved gases, particularly hydrogen sulfide, carbon dioxide and ammonia are significant in industrial water supplies and can cause corrosion.

Treatment

Removal of dissolved solids from industrial water normally is accomplished by chemical reaction which converts them to an insoluble form. The most common technique is the lime softening process, which can reduce both hardness and alkalinity to moderately low levels. Lime, and sometimes soda ash or calcium chloride, is added to the raw water. The process continues through flocculation, sedimentation and usually filtration. Some reduction in the total dissolved solids content is also normally achieved as part of the process. When the process is carried out hot, silica reduction can also be accomplished through the addition of magnesium oxides. Using magnesium oxide or dolomitic lime (CaO-MgO) will improve silica removal, while the use of calcium chloride will remove excess sodium alkalinity.

Soluble iron and manganese can be removed by oxidation and subsequent precipitation by caustic soda. The precipitate is then removed by settling or filtration or both.

Ion exchange systems are also used to remove dissolved solids. The sodium zeolite method exchanges the hardness components (calcium and magnesium) for sodium, thereby producing a soft water with very low scale-forming potential.

By regenerating the ion exchange resins with an acid, all of the metallic ions in the water are exchanged for hydrogen ions, producing a water free of metallic ions, but very acidic. Split-stream softening combines the two processes, whereby part of the water is softened by sodium exchange and part by hydrogen exchange. The two softener effluents are then blended. By proper proportioning, the acid strength of the hydrogen exchange effluent is exactly balanced by the natural alkalinity in the sodium exchange effluent. The result is a neutral pH water.

Anions (chlorides, sulfates, etc.) as well as alkalinity can be removed by base exchange resins. Through the use of hydrogen exchange softener regenerated by caustic soda, water almost completely free from contaminants can be produced. In waters containing a high dissolved solids content (ranging from the brackish through sea water) ion exchange becomes uneconomical, and techniques such as distillation, freeze-separation and reverse osmosis are employed.

Economic Aspects

Costs of treatment vary widely, depending on the quality of the raw water supply and the ultimate purity required (see Table 11-2).

Table 11-2. Representative Water Treatment Costs*

Process	Capital Cost	Chemical Costs/ 1000 gallons
Lime Process (cold)	$25-30,000	2-3¢
Sodium Softening	$ 5- 7,000	4-5¢
Deionization		
A. Two step	$45-50,000	18-22¢
B. Two step plus degasification plus polishing	$55-60,000	20-25¢
Aeration and filtration for iron removal	$10-12,000	No chemicals used

*Basis: 150,000 gpd plant capacity
　　　　Surface water of 150 ppm total hardness
　　　　　　　　　120 ppm total alkalinity.

SURFACE WATER QUALITY[91]

Many extensive studies are now being conducted throughout the world on the quality of water in rivers, lakes and streams. Table 11-3 presents analyses of grab samples collected from a number of major bodies of water. It is important to note that grab sample results are of value only when related to a number of collection factors. The analysis of these samples is for the purpose of general information only.

Table 11-3. Grab Sample Analysis*

Determination—ppm	Lake Ontario, Toronto, Canada, Alexandra Yacht Club 5/15/73	Lake Michigan, Mackinaw City, Mich. W. Side of Bridge 5/14/73	Lake Erie, Fort Erie, Canada 5/18/73	Lake Huron, Parry Sound, Canada 5/14/73	Lake Superior, Gros CAP, Canada 5/14/73	Hudson River, New York, Riverside Park Boat Docks 5/21/73	Rappahannock River, Fredericksburg, Va. 5/25/73	Bay of Fundy, Eastport, Maine 5/19/73	Cape Cod Bay, Plymouth Rock, Plymouth Mass. 5/18/73	Norris Dam-Lake, Tennessee Valley Authority, Knoxville, Tenn. 5/13/73	Chesapeake Bay, Norfolk, Va. Ocean View 5/25/73	Galapagos Islands, Spring at South James Bay 7/12/73	Seine River, Paris, France 6/27/73
Total Dissolved Solids, at 105 C	258	210	218	28	97	2676	42	32,320	34,012	138	21,232	258	360
Phenolphthalein Alkalinity, as $CaCO_3$	0	0	0	0	0	0	0	0	0	0	0	0	0
Total Alkalinity, as $CaCO_3$	105	132	102	5	54	57	30	99	108	96	90	192	204
Carbonate Alkalinity, as $CaCO_3$	0	0	0	0	0	0	0	0	0	0	0	0	0
Bicarbonate Alkalinity, as $CaCO_3$	105	132	102	5	54	57	30	99	108	96	90	192	204
Carbonates, as CO_3	0	0	0	0	0	0	0	0	0	0	0	0	0
Bicarbonates, as HCO_3	128	161	124	6	66	70	37	121	132	117	110	234	249
Hydroxides, as OH	0	0	0	0	0	0	0	0	0	0	0	0	0
Carbon Dioxide, as CO_2	9	23	22	13	35	59	25	51	36	64	30	2	28
Chloride, as Cl	36	3	54	0	9	1200	0	17,700	17,250	39	10,900	9	24
Sulfate, as SO_4	34	26	30	10	4	240	6	2,105	2,240	19	1,425	6	47
Fluoride, as F	0.33	0.31	0.33	0.17	0.17	0.47	0.17	1.5	1.5	0.41	1.3	1.0	0.35
Phosphate, as PO_4	0.8	0.5	0.0	0.5	0.4	0.1	1.0	0.0	0.0	0.5	0.0	7.0	2.2
pH (Laboratory)	7.4	7.1	7.0	5.9	6.5	6.3	6.4	6.6	6.8	6.5	6.8	8.3	7.2
pHs	7.5	7.4	7.5	9.8	8.3	7.8	8.9	6.6	6.6	7.8	6.8	7.5	6.9
Stability Index	7.6	7.7	8.0	13.7	10.1	9.3	11.4	6.6	6.4	9.1	6.8	6.7	6.6
Saturation Index	-0.1	-0.3	-0.5	-3.9	-1.8	-1.5	-2.5	0.0	0.2	-1.3	0.0	0.8	0.3
Total Hardness, as $CaCO_3$	150	144	135	15	42	444	21	5,040	5,340	120	3,221	132	264
Calcium Hardness, as $CaCO_3$	120	102	105	15	33	120	15	900	960	63	630	60	228
Magnesium Hardness, as $CaCO_3$	30	42	30	0	9	324	6	4,140	4,380	57	2,591	72	36
Calcium, as Ca	48	41	42	6	13	48	6	360	384	25	252	24	91
Magnesium, as Mg	7.3	10	7.3	0	2.2	79	1.5	1,006	1,064	14	630	17	8.7
Sodium, as Na	33	11	25	5.7	16	1,140	9.9	16,320	18,000	50	12,720	15	29
Iron, as Fe	0.0	0.2	0.0	0.4	0.0	0.5	0.7	0.0	0.2	0.1	0.9	0.25	0.3
Manganese, as Mn	0.0	0.0	0.0	0.0	0.0	0.0	0.0	0.0	0.0	0.0	0.0	0.0	0.0
Copper, as Cu	0.2	0.0	0.0	0.0	0.0	0.0	0.2	0.2	0.0	0.0	0.0	0.0	0.4
Silica, as SiO_2	2	2	1	3	2	4	10	2	2	9	2	13	11
Color, Standard Platinum Cobalt Scale	0	0	0	15	0	20	0	0	0	0	0	35	15
Odor Threshold	0	0	0	0	0	0	0	0	0	0	0	0	0
Turbidity, Jackson Units	20	5	5	2	0	10	5	2	0	5	15	2	15
Total Organic Carbon	7.9	5.3	5.9	7.3	4.5	7.1	4.7	4.9	3.7	5.8	5.7	1.5	8.3

*To convert ppm to grains per gallon, divide ppm by 17.1 - p.p.m. = mg/l.

Researchers who are making studies of these water bodies can compare their results with those shoreline grab samples, collected without regard to wind direction, tides, etc. Total Organic Carbon content has been included, because it is a pollution parameter and not a standard water analysis.

NEUTRALIZING INDUSTRIAL WASTES[133]

Wastewater is generally considered adequately neutralized if: (1) its attack on metals, concrete or other materials is minimal; (2) it has little effect on the fish or aquatic life in the receiving stream; (3) it has little or no effect on biological matter (*i.e.,* activated sludge system); and (4) pH is in the range of 5.5 to 10.0 (normally accepted standards for discharges to surface streams and municipal systems). Stringent standards or additional treatment requirements might make pH control more critical, however.

Neutralization can be accomplished on either a batch or continuous basis, depending on the volume and rate of flow. Because tanks are used for wastewater neutralization, the size of the treatment facility is entirely dependent on volumes to be treated and rapidity with which the neutralization step occurs. If batch treatment is used, at least two neutralization tanks will be necessary.

Usually, tanks should be sized to provide 2 to 3 ft freeboard. However, if foaming exists, more height will be necessary. The neutralization tanks may be round, square or rectangular. Circular tanks should be equipped with baffles to prevent short-circuiting of waste around the tank, and square and rectangular tanks should be sized to ensure good mixing and good agitation.

Short-circuiting of untreated wastewater through the system can be minimized by installing baffles at strategic positions. Baffles should be arranged at different elevations at the inlet and outlet of the tank.

Tanks should either be constructed of corrosion-resistant materials or lined to prevent corrosion. The inlet and outlet mixing zones should be separated by a baffled section. Each tank should be equipped with a mixer capable of providing thorough agitation of the total contents of the tank; the number and size of mixers needed will depend on the capacity and configuration of the tanks.

Addition of acid or alkali must be controlled by pH measurement. If a batch operation is employed, the operator can take samples, check the pH (titrate the samples to determine the acidity or alkalinity), and add the chemical dosage required for proper pH adjustment. After a suitable reaction period, the pH would again be checked to determine if it is within the acceptable range for discharge. This method requires very little equipment investment. A recording-controlling type pH instrument with field-adjustable control capabilities may be used to sense the pH of the tank contents and feed the neutralizing chemical until the pH reaches a preset value.

On the other hand, if the operation is continuous, wastewater composition and flow rate may be critical factors in determining what chemical control methods should be employed. When flow rate and acidity or alkalinity content are relatively uniform, the feed rate and acidity of alkalinity content are controlled automatically in response to continuous, automatic pH measurement. When flow rate and/or composition are variable, however, both flow and pH should be measured—the flow rate controlling the primary addition of neutralizing material, and the pH measurement controlling the supplementary feed of neutralizing materials. A small pump (1 to 2 gpm) is sufficient to continuously pump a sample of the effluent from the neutralizing tank through the flow chamber equipped with pH sensing probes (Figure 11-1).

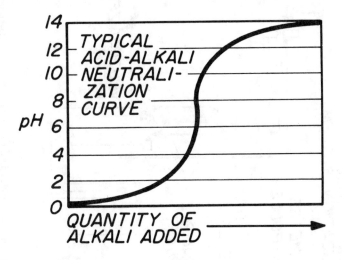

Figure 11-1. As an acid or alkali is neutralized, its pH approaches 7.0, and the effect of the neutralizing chemical increases. If buffers are present in the wastewater, the effect of adding a neutralizing chemical is greatly diminished.

Type of acidity/alkalinity measuring equipment required for continuous treatment is determined by both the magnitude and the nature of contaminants requiring neutralization. In general, if the contaminants are either strong alkalis or strong acids with little or no buffering material present, simple pH measurement will suffice as a basis for automatically controlling the rate of chemical feed. However, if the wastewater contains contaminants that have buffering properties (*e.g.*, iron compounds in acidic wastewater from steel pickling operations), the neutralizing chemical feed requirements should be determined by automatic titration of wastewater samples (Figure 11-2).

Concentrated sulfuric acid solution is generally used to neutralize alkaline wastes, but waste acid can also be considered if it is available as a

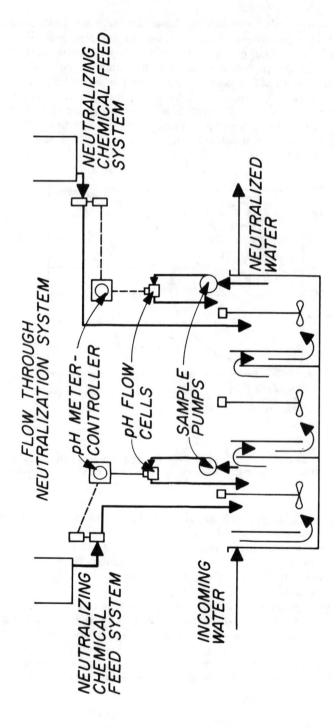

Figure 11-2. Addition of neutralizing chemicals is easily controlled in a continuous system by determining pH at first chemical addition and again just prior to wastewater discharge. Coupling pH meters to chemical controllers automates the system.

plant by-product. Since sulfuric acid is usually purchased in bulk quantities, a large bulk storage tank, constructed of carbon steel or reinforced fiberglass, should be provided. This storage tank should be sized on the basis of acid usage and availability of acid deliveries. Then, a day tank can be filled, preferably by gravity flow, from the bulk storage tank. Acid is pumped from the day tank to the neutralization tank as conditions warrant. The pump used should be an acid-proof, positive displacement type. Acid feed lines and any of the attendant equipment should also, of course, be fabricated of acid-resistant materials.

The selection of an alkaline material for acidic neutralization is based on:

1. Economics,
2. Reactivity of the alkali,
3. Character of sludge formed, and
4. By-product recovery possibilities.

Lime is generally used, but other alkalies may be suitable or preferable. An advantage of lime is that it is widely available in bags or bulk and can easily be shipped by rail or truck. If purchased or stored in bags, they must be kept dry and sealed. Any contact with water will cause the lime to generate heat, posing a potential fire hazard. A dry storage tank will be needed if bulk lime is used.

A disadvantage of lime is that it is sometimes difficult to feed because:

1. Pulverized lime can bridge, arch or hang up in equipment;
2. It settles in mixing pots and clogs openings;
3. It picks up moisture, clogs and cakes;
4. It has low solubility.

The piping through which a lime slurry is pumped should be installed so that it can easily be cleaned with water pressure or rods. Even though problems can be encountered with the use of lime, its use is favored because of low cost and ready availability.

Lime can be fed in a number of ways, each dependent on the amount to be handled during a period of time. It can be placed in a dry hopper and added to water to make a slurry which is then pumped to the neutralization tank. Lime can also be slaked (addition of water to quicklime to form a slurry), diluted with more water following slaking, and then fed. Any lime slurry must be continuously agitated.

Where economics dictate (low amount of alkali needed) or where ease of handling may outweigh other considerations, caustic (either flake or liquid) may be used.

Sludge formation often occurs during neutralization, and settling of this sludge may take place in the reaction tank. If so, it must be removed from the tank and dewatered prior to disposal. This is seldom a large problem in batch operations. In continuous operations, however, neutralization is often followed by sludge settling in a clarifier or sedimentation basin.

As with all other process systems, a preventive maintenance program should be established for the chemical feeders and instrumentation. Failure to do so will result in inadequate neutralization and possible stream pollution.

CHEMICAL WATER TREATMENT[168]

Thousands of tons of chemicals are used each year in treating water. However, all of the chemicals added to water in its treatment contribute their own form of water pollution and must subsequently be handled appropriately.

Boiler Water

Most industrial processing plants require heat obtained from steam or hot water boilers. Whether the boilers are large or small, the water generally requires fairly extensive treatment; or the boilers, steam distribution and condensate return systems will be adversely affected.

In the preparation of water for boiler use, the treatment procedure to be followed is governed by the quality of the water, the type and size of the boiler, the amount of water that the boiler will require, and the temperature of the water in the boiler. The latter is controlled by the pressure. Higher pressure boilers require water of very high purity. Such water usually is obtained by demineralization, which may first have been softened by a conventional process such as the lime-soda process. Very little chemical treatment is added internally to the water in the very high pressure boiler installations.

Common corrosion inhibitors used in closed hot water heating systems are sodium nitrite with appropriate buffering chemicals such as borax and a copper inhibitor like mercaptobenzothiazole. Sodium nitrite functions best as an inhibitor to protect iron and steel in the system. When employing sodium nitrite, it is necessary to maintain a somewhat alkaline condition so that the nitrite will remain stable. For steam boilers where the water would not circulate out into a system containing metals other than iron or steel, it is possible to go to higher alkalinity levels and use sodium hydroxide or caustic soda as part of the boiler water treatment.

In hot water heating systems, another chemical that has been used rather widely in the past is sodium chromate. It functions as a very efficient corrosion inhibitor. However, the problem with this inhibitor is that it has a yellow color and can stain at points of leakage. In addition, since chromate is a toxic chemical, it creates a problem when it must be drained from the system into rivers, lakes and sewer systems.

Another method of inhibiting corrosion in closed heating systems is to remove oxygen in the water by addition of oxygen scavenger chemicals such as sodium sulfite or hydrazine. When used at the proper concentration hydrazine does not contribute to the dissolved solids loading of the water. If leaks occur at pump seals, there will not be any abrasive action which could result from crystallization of solids dissolved in the water.

Turning now to steam boilers, the water treatment possibilities are much more extensive. If an appreciable amount of make-up water is required, and the water has a significant amount of hardness or mineral matter, it is common to remove part or nearly all of the hardness before the water enters the boiler. In many installations, after the water has been softened, it is then degasified by treatment with a controlled amount of sulfuric acid and then heated to remove carbon dioxide and the bulk of the dissolved oxygen in the water supply. Then it is common practice to add an oxygen scavenger such as hydrazine or catalyzed sodium sulfite to the deaerator effluent to remove the last traces of oxygen from the feed water. The catalyst that is frequently used in preparing catalyzed sodium sulfite is cobalt sulfate. This catalyst is required only in trace quantities and serves to speed up the reaction between dissolved oxygen and the sodium sulfite. The reaction between oxygen and sodium sulfite goes to completion at a lower temperature, also in the presence of the catalyst. The end product of the reaction is a salt, sodium sulfate, which is inert in the environment of a boiler and finally is removed in the boiler blowdown.

Within the boiler itself, other chemicals are added such as caustic soda or sodium hydroxide; sodium phosphate to react with any calcium hardness that might enter the boiler because of incomplete removal by the external softening process; and in some boilers, sodium nitrate is added to function as an embrittlement inhibitor. Sodium nitrate is also quite inert in a boiler and is ultimately removed in the boiler blowdown.

In some boilers receiving water of very low hardness, it has been found desirable to treat the boilers internally with chelating agents such as tetra-sodium ethylenediamine tetraacetate, or trisodium nitrilotriacetate. The former compound is sometimes simply referred to as EDTA and the latter as NTA. The advantage of using these organic chemicals is that instead of an insoluble calcium or magnesium precipitate, the reaction products are totally soluble and there is no problem of sludge formation within the boiler. These chemicals also are disposed of ultimately in the boiler blowdown. If a boiler operates at too high a temperature, then the use of these organic compounds, as well as any other organic compounds, are ruled out because of the thermal instability of organic compounds.

Some moderately high pressure boilers do not soften or treat raw water prior to entry. Chemicals in these cases are added directly into the boiler. Removal of the undesirable hardness minerals occurs within the boiler. This practice is referred to as internal boiler water treatment. In such a procedure, the following chemicals are most commonly used: sodium phosphate for removal of calcium, sodium silicate for removal of magnesium if the natural silica content of the water is insufficient to accomplish this, sodium hydroxide to achieve the proper alkalinity in the boiler, and sodium sulfite for removal of oxygen.

Sometimes soda ash or sodium carbonate is used to achieve boiler water alkalinity and also to participate in the removal of calcium and magnesium from the boiler water. At times sodium nitrate is used for embrittlement control. When internal softening is practiced, it is common to use an organic sludge dispersant, so that the precipitates formed within the boiler will remain suspended for removal by the blowdown procedure. The chemicals utilized in this sludge dispersing function are organic in nature. They will form colloidal dispersions and become attached to the inorganic precipitates formed by the chemical reactions in the boiler. The organics encountered are either naturally occurring materials or synthetic dispersants. Among the natural organics, the commonly used ones are tanning which is obtained from quebrancho, starch, sodium alginate and lignin. Lignin is usually added to the boiler in the form of the chemical known as sodium lignosulfonate.

Amines

Other chemicals added to boiler water or introduced directly into streamlines are known as amines. They function to control or inhibit corrosion in steam distribution and condensate return systems. Chemically, amines are divided into two main groups: one group being known as neutralizing, and the other being known as filming amines. In the neutralizing amine group, the most commonly encountered chemicals are cyclohexylamine, morpholine, and diethylaminoethanol.

Filming amines are either octadecylamine, or derivatives of this parent organic compound having varying solubilities in water. Amines serve to neutralize acidity, which would otherwise occur in the steam condensate such as from the presence of dissolved carbon dioxide. In the case of the filming amines, the surfaces of the piping or other equipment will actually become coated with a protective film and thus render the surfaces waterproofed. Corrosion is therefore minimized.

Water Cooling Systems

Corrosion inhibitors are an important part of water treatment for recirculating cooling tower systems. One of the most widely used chemicals for this purpose has been sodium chromate. For many years it has given excellent corrosion control. However, because of its cost and in an effort to reduce the amount required, other chemicals have been used with the sodium chromate to enable the user to get satisfactory corrosion control with a much smaller quantity of the sodium chromate. Chemicals which have functioned in this manner are sodium polyphosphate and zinc sulfate. Some inhibitors are composed of zinc dichromate, which eliminates the extra sodium and sulfate ions in the cooling tower water.

Another problem in cooling water, whether it be in a cooling tower or a cooling pond, is the control of growing organisms such as algae and bacteria. To accomplish the control of these organisms, chemicals known as algicides or biocides are used. In large systems it has been found desirable to use chlorine to control growing organisms. In smaller systems, chemicals such as calcium hypochlorite, sodium hypochlorite, sodium pentachlorophenate, sodium trichlorophenate, quaternary ammonium compounds, iodine compounds, derivitives of tributyl tin oxide, and a host of other chemicals are used. Because algicides and slimicides from cooling towers enter various receiving streams and lakes, the toxicity of the chemicals must be carefully evaluated so as not to damage aquatic plants or animals. The use of mercury compounds, for example, is now eliminated from this application. For many years such mercury compounds as phenylmercuric acetate were used extensively to control the slime-forming bacteria. Mercury compounds have been used extensively too for control of slime in paper and pulp production.

Other chemicals that find application in smaller cooling tower systems for control of the alkalinity are sulfamic acid and sodium bisulfate. Either of these chemicals will reduce the natural alkalinity and control scale formation in the heat exchangers.

A chemical that has been used for a long period of time to control the growth of algae in ponds is copper sulfate, also known as blue vitriol. In sufficient concentration, this chemical, of course, is toxic to fish and must be used in a carefully controlled manner. However, since it is not always easy to carefully control very small quantities of chemicals in water, the use of such toxic materials should be avoided.

WATER AND WASTEWATER DISINFECTION[132]

Disinfection is a unit operation widely practiced in water and wastewater treatment. Its purpose is to destroy harmful or disease-causing organisms. Although such methods of disinfection as boiling water have been known since 500 B.C., disinfection as a means of disease prevention was only fully developed near the beginning of the twentieth century.

Only after 1912, the year of the development of chlorine gas feed equipment, did disinfection practice grow. Chlorination of public water supplies has grown so rapidly since that time that it is now routine practice in potable water and sewage treatment. Because of the widespread use of chlorine as a disinfectant, chlorination and disinfection are frequently used as synonymous terms. However, many other methods of disinfection and types of disinfectants are used.

Many organisms which cause disease are transmitted by water. The major disease producers among these are the causative agents for typhoid fever, the various types of paratyphoid fever, dysentery, cholera and infectious hepatitis. Organisms living in the intestines of man are discharged and usually contaminate water supplies by some sanitary defect.

Frequently, the term "disinfect" is confused with "sterilization." From the viewpoint of definitions, disinfection in water and wastewater treatment concerns itself only with the destruction of harmful or objectionable organisms. It does not necessarily destroy all forms of plant and animal life. The method by which all organisms (bacteria, viruses, etc.), are killed is called sterilization. Organisms kill is dependent upon the type of organism involved. Each organism offers its own relative resistance to sterilization by moist and dry heat with bacterial spores being considerably more resistant than the other organisms.

Physical Characteristics Affecting Disinfection

Various methods are used to disinfect by modifying the microorganism's environment. Physical characteristics such as temperature can be changed, prohibiting active growth, or agents can be added to the environment that interfere with or modify the microorganism's life-sustaining functions. Some of the physical characteristics affecting disinfection are:

1. The nature, type and concentration of disinfectant,
2. Temperature,
3. Contact time and concentration,
4. Nature of the organism,
5. pH or hydrogen ion concentration,
6. Nature of the disinfectant, and
7. Surface tension of the microorganism.

Organism Type

The nature, type and concentration of organisms to be destroyed are important. Microorganisms display varying grades of resistance to disinfection. The destruction of any given microbial population follows a normal distribution curve. Some microorganisms are easily destroyed because they offer little resistance. Others offer a high degree of resistance and are difficult to destroy. Examples of the extreme cases are young vegetative cells which are easily killed and bacteria in spore form which are quite difficult to destroy. Spores are bacteria in a resting state in which all or part of the bacterium is surrounded by several membrane coatings. Young microorganism cultures are more easily destroyed, whereas older cultures offer greater resistance.

Temperature

LeChatelier's principle states that, "If a stress is applied to a system at equilibrium, then the system readjusts, if possible, to reduce the stress." If heat is added to a system at equilibrium, this constitutes a stress on the system. Since chemical equilibrium is a system of chemical reactions, the system will adjust by forming reactants or products, whichever relieves the stress. The rate of most water-related chemical reactions increases as temperature increases. Since disinfection involves chemical reactions, its efficiency generally increases with temperature (Table 11-4). The disinfection time for killing *Staphylococcus aureus*, which is the organism that lives in food and produces a toxin causing food poisoning, decreases as temperature increases for constant phenol concentration. Phenol is a hydrocarbon and benzene derivative which was historically popular, but is now regarded as a pollutant because it is toxic.

Table 11-4. Effect of Temperature on the Destruction of *Staphylococcus Aureus* at Different Phenol Concentrations

Dilution	Percent	Disinfection Time (Minutes)	
		10 C	20 C
1:55	1.82	17.5	5
1:60	1.66	40	7.5
1:65	1.54	70	12.5
1:70	1.43	100	20
1:75	1.33	150	30

Time and Concentration

The rate of kill depends upon the disinfectant concentration and the time period over which this concentration is effective. Chemical concentration is also a critical variable. An insufficient concentration will have little, if any, effect on the microorganisms. Possibilities exist that insufficient chemical concentration will even stimulate microbial growth. Table 11-5 illustrates that increasing phenol concentration results in less time needed to kill the *Eberthella typhosa* organism.

Table 11-5. Disinfection Capacity of Phenol at Different Concentrations Using the Microorganism, *Eberthella Typhosa*

Dilution	Percent	Killing Time (Minutes)
1:70	1.43	5
1:75	1.33	7½
1:80	1.25	12½
1:85	1.17	22½
1:90	1.11	30

Environment

The nature of the organism environment is crucial for disinfection efficiency. Most microorganisms are destroyed because the disinfectant reacts with its associated organic matter. If any organic matter not associated with the microorganism is present in the environment, the disinfectant combines with it and its effective concentration is reduced. This can be exemplified by breakpoint chlorination in water treatment. In this procedure, chlorine is added to oxidize the organic matter present. When all the organics have been oxidized, a residual concentration of chlorine will be obtained in the water. Further addition of chlorine will increase this residual linearly. However, before this linear relationship can be obtained, a very distinct decrease in chlorine residual will be noticed. This decrease is the breakpoint.

pH Factor

Another influential factor is pH. It is perhaps one of the best controls for microbial growth since few organisms can survive below an acidic pH of 4.0 or above an alkaline pH of 9.5. This is a general rule, however, and there are some exceptions. *Thiobacillus*, the bacteria which oxidize sulfite, $SO_3 =$ to sulfate, $SO_4 =$, survive at a pH of 1.0. *Bacterium radiobactor*, a type of bacteria which produces diseases in plants, survives at a pH of near 12.0. For *Escherichia coli*, optimum growth rate occurs near a neutral pH of 7.

Disinfectant Types

Some chemical disinfectants are much more toxic to organisms than others. In water treatment, for instance, different types of chlorination are practiced. Since free chlorine, *i.e.,* chlorine added as gaseous chlorine or liquid hypochlorite, reacts rapidly with oxidizable organic substances, it is used to produce a quick kill. Combined residual chlorination involves reacting chlorine with ammonia to produce monochloramine (NH_2Cl), dichloramine ($NHCl_2$), and nitrogen trichloride (NCl_3). Although about 1/25th as reactive as free chlorine, the chloramines produce a longer lasting residual—a longer period of chlorine availability for disinfection purposes.

Surface Tension

Passing of substances into and out of cells is increased by lowering the surface tension within the individual cell. Chemicals which lower the surface tension increase the penetrating power of the disinfectants. Although a particular chemical may have no disinfecting power, it may lower the cellular surface tension sufficiently to kill the organism or enhance disinfection by other chemicals.

Various Disinfectants

Various disinfectants used for water and wastewater treatment are chlorine and hypochlorites, iodine and bromine, chlorine dioxide, ozone, ultraviolet light, heat, heavy metals, quaternary ammonium compounds, and acids and alkalies.

Chlorine

The most widely used method of disinfection is chlorination because chlorine is relatively inexpensive, efficient, dependable and easy to handle.

The penetration ability of chlorine causes it to attack microorganisms readily. Chlorine is generally used in gas form and as a liquid hypochlorite solution. In water, however, these forms exist as HOCl, hypochlorous acid, and as OCl, hypochlorite ion, the distribution of which is dependent upon pH. Chlorine is used as a disinfectant for water of poor quality or when a short contact period is desired. Increasing temperature and decreasing pH result in a more effective kill. If long contact periods are available, chloramines can be used since a contact period of 100 times the length of that necessary for chlorine is needed using these compounds. However, the chloramines are used to control algae and bacteria in water distribution systems because they are effective for longer periods of time.

Other Halogens

Iodine and bromine, halogen elements in the fluorine group of the periodic table, can also be used as disinfectants. Although bromine has long been used, it is somewhat limited in application because of handling difficulties. Bromine as a liquid burns the skin, and gaseous bromine is corrosive and toxic. Hypobromites are unstable and react to form inactive bromate compounds. Bromine is not as reactive as chlorine and consequently its powers of disinfection are not as good.

While iodine's disinfection properties are as familiar as chlorine, it is usually used in emergencies and primarily as a skin disinfectant. Iodine can be used effectively in swimming pools and for disinfection of small or individual water supplies. It is especially effective for the control of tuberculosis microorganism and bacterial spores. The concentration necessary to do the job does not vary greatly among species. Some of the advantages of using iodine are that pH has little influence, time and temperature are not as important as with chlorine, and it is effective over a broader range of organisms. Important disadvantages of using iodine are that higher concentrations are necessary to produce identical kills with chlorine, iodine is more expensive than chlorine, and iodine produces color problems.

Chlorine dioxide is a more reactive chemical than chlorine. It is used primarily in water treatment plants to oxidize phenols and chlorophenols as well as substances that produce undesirable tastes when treated with chlorine. Because this gas is unstable, on-site generation is necessary. The necessary equipment must be provided at or near the point of application, which makes this method somewhat impractical.

Ozone

A relatively new disinfection method utilizes ozone (O_3), which consists of three atoms of elemental oxygen. In water, ozone breaks down rapidly in the presence of oxidizable matter. It also is used to oxidize phenols in water and wastewater treatment, and can be used to destroy other organics from specialized waste streams. Below a critical concentration

(0.4 - 0.5 mg/l for the microorganism *Escherichia coli*), ozone produces little or no disinfection; however, complete disinfection occurs above the critical point. The major difference between chlorine and ozone disinfection is that ozone disinfects only above a specific concentration while the chlorine kill is concentration dependent.

Ozone can produce water free of tastes and odors which chlorine cannot remove and has a greater tendency to destroy cysts than does chlorine. Reportedly, it can reduce color by 50 percent. Advantages of using ozone are primarily that it leaves no taste, odor, or color problems and disinfects about 300 to 3000 times as fast as chlorine. Additional details on the use of ozone are discussed later in this chapter.

Ultraviolet Rays

Disinfection is also accomplished with ultraviolet light which involves high-frequency wavelengths beyond the violet end of the visible spectrum. A thin film of water (maximum of 120 millimeters in thickness) flows past one or more quartz mercury-vapor arc lamps which emit ultraviolet light. The wavelengths range from 200 to 295 millimicrons, which are shorter than the violet rays. Violet rays are the shortest in the visible light spectrum. Turbidity and color in water reduces the effectiveness in water disinfection. Advantages of ultraviolet disinfection are that no chemicals are added to the water, it is not necessary to satisfy chemical demands before disinfection is accomplished, no tastes or odors are produced, and short contact periods are possible. The design and operation of UV systems is discussed later in the chapter.

Boiling

Boiling of water is another method of disinfection. This method is generally not used in water treatment because considerable expense is involved. However, it can be used in emergency situations if bacterial spores are not present. If water is known to contain intestinal pathogens, boiling for 15 to 20 minutes will kill these organisms.

Heavy Metals

Copper sulfate added to reservoir waters provides a means to algae control. Mercury also exhibits toxicity to algae, but is seldom used. Silver inhibits the growth of *Escherichia coli*. Advantages for the use of silver as a disinfectant are that low concentrations are effective, it is powerful and long-lasting, it inhibits growth of algae and fungi and is easily handled. Disadvantages are that extensive water pretreatment is required, certain microorganisms are resistant, low pH and temperatures affect its reaction, certain anions reduce its action, and long contact periods are necessary. It is an expensive treatment—about 200 times more than that of chlorine gas.

Quaternary ammonium compounds, which consist of four groups of hydrocarbons attached to a central nitrogen atom, are considered as disinfectants since they lower the surface tension of water. This disrupts the cell membrane of the microorganism and permits loss of nitrogen and phosphorus compounds. The possible methods by which this occurs is that the quaternary ammonium compound combines with lipids and proteins of the microorganism's cell membrane. This method suffers from the major disadvantage that it is considerably more expensive than chlorine.

Acids and alkalines can also be used for disinfection. Waters containing high concentration of acids and bases are especially effective if combined with long contacting periods. Apparently, the hydrogen ions attack and destroy the microorganisms' membranes, especially the cell walls of certain bacteria.

Since chlorine is widely used as a disinfectant in water treatment, the harmful effects of excessive concentrations should be considered. Chlorine gas is highly toxic and proper safeguards must be taken when handling. Its solubility in water is about 7300 mg/l at 68 F at 1 atmosphere pressure. In air about 30 mg/l will produce coughing and 1000 mg/l is fatal.

In water, available chlorine (a measure of the oxidizing capacity of chlorine compounds) usually produces no fatalities at a concentration of 0.5 percent. Since no specific clinical toxicity data are available, chlorine solutions with an availability of 4 to 6 percent are probably toxic only in dosages of several ounces if swallowed. However, a one ounce solution at a concentration of 15 percent may be dangerous.

OZONE[173]

Ozone is an allotropic or alternate form of oxygen. Specifically, an ordinary oxygen molecule (O_2) which acquires an extra oxygen atom becomes a molecule of ozone (O_3). It occurs in nature, principally in the upper atmosphere where it is formed as a result of lightning discharges and in atmosphere which is heavily polluted because of the action of sunlight on nitrogen dioxides and organic vapors.

Ozone is a colorless gas which has a peculiar pungent odor in concentrations as low as 0.02 ppm by volume. The property which makes it of particular interest to the pollution engineer is its very high oxidizing potential. A comparison shows that the oxidizing potential of ozone is nearly twice that of chlorine.

Generation

The only practical method of generating ozone which has been developed so far is by an electric discharge. This is created by two electrodes separated by an air gap. A dielectric is placed between the electrodes to act as a stabilizing resistance. When air is passed through the resulting electric discharge, a percentage of the oxygen is converted to ozone. When ordinary air is used, the concentration of ozone produced is approximately 1 percent. If pure oxygen is used, however, the concentration can be as high as 3 to 4 percent.

To protect the electrodes, the air should be at a dewpoint of -40 F or below, which is usually achieved by a combination of cooling and drying.

Actual ozone formation takes place during the migration of electrons across the air space from one electrode to another, which occurs at the moment the critical voltage is reached. The electrons remain at the electrode which they have just reached until the polarity is reversed. When critical voltage is attained, they migrate back across the air space causing the formation of ozone in the air between the electrodes. For this reason, the higher the frequency, the more efficient the ozone production will be. The biggest portion of the energy of the wave is given up as heat. Consequently, commercial ozone generators must be designed with a means of dissipating the heat produced, normally by air or water cooling.

Generators

There are two principal types of commercial generators: the plate type and the tube type. The original plate type generator was designed by Otto in 1905 and is the basis for the design of many ozonators is use today. The electrodes are two plates with a dielectric of glass between them. Some modifications employ hollow plates through which cooling water or oil is circulated. Improvements in the Otto generator have introduced such things as stainless steel electrodes, various alternate means of cooling, and different dielectric materials. However, the basic design has remained unchanged.

Tube type generators consist of a series of stainless steel hollow tubes. Inside these tubes are glass tubes that have been coated inside with a conductor. The concentric tubes are separated by an air space, where the electric discharge takes place and ozone is produced. The outer shells are surrounded by water for cooling purposes. The advantage of the tube type generator is that a greater surface area is available and consequently a higher efficiency is attained. On the other hand, heat removal is more difficult. Only one electrode can be cooled, whereas in the plate type both electrodes can be cooled.

Ozone generators today are compact and efficient compared to the early cumbersome models. Small "table model" style generators are available for commercial odor control and water treatment. These are compact (7½ in. x 20 in. x 12½ in.), air cooled, and operate from a conventional 115-v power source. They will produce approximately 3 to 6 gm/hr of ozone depending on the feed. Larger units, Figure 11-3, are available for industrial applications. These also operate off conventional power sources and require no special equipment. They will produce up to 120 gm/hr for a wide variety of applications. Where higher amounts of ozone are required, pure oxygen may be used as feed rather than air. This will produce concentrations up to double that obtained with air.

Figure 11-3. Industrial size ozone generator.

For extremely large applications, such as water purification or tertiary sewage treatment, special installations are required. These consist of numerous banks of electrode packages, special electrical apparatus, cooling provisions, and cooling and drying equipment. Installation costs can be high—usually more than conventional methods such as chlorine. However, the use of chlorine requires more extensive storage and safety measures than ozone, which narrows the difference of capital costs between the two.

Water Problem Treatment

Ozone water purification was pioneered in France and is still widely used there today. The installation at Nice, France was put into operation in 1906 and treated 5,000,000 gpd. Today it has been expanded and treats over 20,000,000 gpd. Ozonation is also in use in Switzerland and Scotland. Ozone disinfects without leaving any of the unpleasant tastes and odors characteristic of chlorine. In fact, ozone can remove undesirable tastes and odors, as well as color resulting from organic decomposition products.

The Torricelli method, Figure 11-4, introduces ozone countercurrent to the water flow. It also utilizes excess ozone by reintroducing it into the influent tower to preozonate the influent, thus obtaining maximum use of the ozone produced. Dosages one-fourth the usual amounts at retention times one-half normal for conventional contact columns have achieved complete destruction of bacteria in water artificially contaminated with a concentration of 7,000,000 *B. coli* per 100 ml. Ozone is highly satisfactory in the water bottling industry as a means to disinfect both the water and the containers.

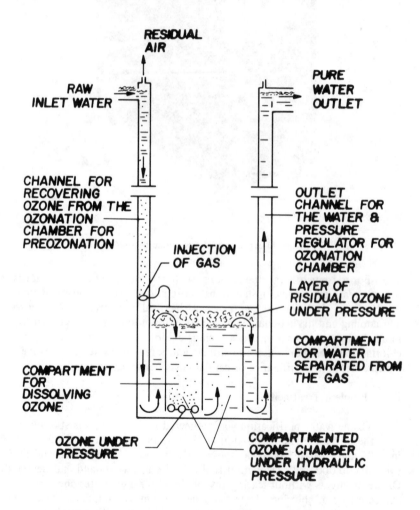

Figure 11-4. Schematic section of Torricelli ozonation system.

Cooling tower water problems may also be solved with ozone. Problems encountered in cooling towers are corrosion, scale build-up, and fouling with organic contaminants such as algae, bacteria, and slime. The current method of treatment is to add chromates or phosphates to control the corrosion and chlorine to reduce algae growth. Recent pollution control regulations are, however, forcing some companies to find alternatives to chromates and phosphates for corrosion inhibition.

A new method of treatment for cooling tower water which avoids the use of contaminating chemicals consists of two operations: controlled blowdown and the application of ozone. The ozone is introduced at the intake to the system to immediately destroy all bacteria, algae, and microorganisms.

The water's tendency to scale or to be corrosive can be determined by the Langlier Index. If the water is corrosive, the pH may be adjusted by adding caustic so that the water will be slightly scaling. It is desired to allow a slight scale build-up to occur so that the system will be protected from corrosion. Proper blowdown procedures can then be used to maintain the scale build-up at suitable levels.

Treatment of cyanide wastes is a continuing problem for the electroplating, steel, and electronic component manufacturing industries. Ozone is a solution which may prove to be more economical and offer a number of advantages over present methods. Approximately eight pounds of chlorine must be applied to accomplish the destruction of one pound of cyanide. The result is an excessive amount of chlorine ions and a free chlorine residual which may be unacceptable in public waterways. Oxidation with chlorine requires the use of large basins or retention tanks as well as expensive handling and storage of large amounts of chlorine.

On the other hand, a maximum of one pound of ozone is required for the destruction of one pound of cyanide, and usually much less if a catalyst such as copper is used. There is no residual, since ozone simply reverts to oxygen. Handling, storage, and freight costs are eliminated due to on-site generation and consumption as produced. Oxidation may be carried out in continuous-flow packed towers, eliminating the need for large retention tanks. Initial investment costs are higher for the installation of the ozone generating equipment, but a lower cost per pound of cyanide treated ($0.25 for ozone vs $1.15 for chlorine) quickly repays the capital costs and results in lower operating costs over the long run.

A potential application for ozone which has generated considerable interest is the tertiary treatment of sewage. Various pilot plants have been built with mixed preliminary results. Costs are difficult to determine because of numerous variables involved. The cost per 1000 gallons will vary with the size of the plant, ozone demand of the material to be treated, efficiency of the contacting system, and power costs. Present indications are that cost for ozone treatment is in the area of $0.077/1000 gallons. This is favorable when compared with activated carbon treatment, but is more expensive than chlorine.

The possibilities of using ozone in tertiary sewage treatment seem to be in conjunction with chlorine rather than in place of it, such as in applications where chlorine is ineffective, where chlorine residuals are unacceptable, and where ozone demand is not too high. Costs for drinking water sterilization, for example, where ozone demand is much lower, are approximately $0.01/1000 gallons which compares more favorably with chlorine.

Toxicity

One objection that has been raised to using ozone is its toxicity and potential danger. While ozone is definitely toxic, there are factors which reduce the immediate danger to individuals working with it. Toxicity of the gas is dependent on two factors: concentration and length of exposure. A maximum concentration over a period of time has been established by the American Council of Governmental Industrial Hygienists as 0.1 ppm by volume of air for continuous exposure under normal working conditions. Exposures at higher concentrations can be tolerated for shorter periods (Figure 11-5).

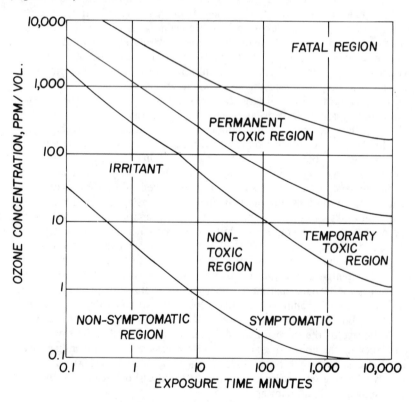

Figure 11-5. Toxicity of ozone.

Thus, while 0.1 ppm can be tolerated indefinitely, 1.0 ppm can be tolerated for eight minutes, and up to 4 ppm can be tolerated for one minute without producing the symptoms of coughing, eyewatering, and irritation of the nasal passages. The concentration at which the odor of ozone becomes noticeable is 0.01 to 0.01 ppm allowing detection long before a critical concentration is reached. Furthermore, once the first threshold is reached, the results are not immediately toxic but merely symptomatic. Since ozone is generated at the jobsite and consumed as produced, large stored quantities of the gas are not on hand to endanger personnel.

Ozone treatment for a wide range of applications appears to be on the threshold of widespread acceptance. The advantages over established methods, particularly chlorine, are considerable.

ION EXCHANGE[126]

Ion exchange depends upon the ability of certain materials to remove and exchange ions from solutions. These ion exchange materials are generally insoluble organic polymers, although some natural inorganic ion exchangers are still used. There are two types—cation exchangers, capable of exchanging positively-charged cations; and anion exchangers, capable of exchanging negatively-charged anions.

Water Softening

The primary function of a sodium cation exchanger is to remove hardness by exchanging the sodium ions on the ion exchange resin for the calcium and magnesium hardness ions in the water. When the water softening capacity of the ion exchange bed is exhausted, the softener unit is taken out of service and regenerated. There are three steps in this regeneration. The first step is *backwashing*, which is done by passing a strong flow of water upward through the softener bed. This action loosens and regrades the resin bed; holds it in a suspended condition; and removes, by washing up and out, any dirt which may have collected in the resin bed during the softener run.

The second step in regeneration is *brining*. This is a reverse of the softening process whereby a solution of common salt is passed through the bed. This salt solution is evenly distributed across the bed and passes down through it. During this passage, the salt reacts with the ion exchange resin, displacing the calcium and magnesium as soluble chlorides and restoring the ion exchanger to its original sodium condition.

The third step in regeneration is *rinsing*. This step consists of washing the calcium and magnesium chlorides plus the excess sodium chloride to the drain with a downward flow of raw water. After rinsing, the softener is returned to service.

Demineralization

What happens if the water softener cation resin is regenerated with an acid? Now, in addition to calcium and magnesium being exchanged, sodium can also be exchanged—all for the hydrogen ion.

This free hydrogen ion in the effluent combines with the anions originally in the raw water. The bicarbonate ion is then converted to carbonic acid, which breaks down to carbon dioxide and water; the chloride is converted to hydrochloric acid and the sulfate ion to sulfuric acid.

Anion exchangers operating on the hydroxyl cycle can remove the strongly and weakly ionized acids present in the effluent of a cation hydrogen exchanger. At the end of each operating run of an anion exchanger, the unit is backwashed, regenerated with caustic soda (NaOH) in the case of strong base resins, rinsed and returned to service.

When this anion exchanger follows a cation exchanger the process is commonly referred to as two-step demineralization (Figure 11-6). But there

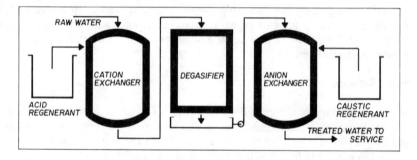

Figure 11-6. In a two-step system a strong base anion exchanger can reduce silica content; with a degasifier preceding the anion exchanger, carbon dioxide can be completely removed.

are other demineralizing systems which find application where varying degrees of purity are required. Some of these systems are:

1. Three-step systems: Cation exchange, intermediate or weakly basic anion exchange, strongly basic anion exchange.
2. Four-step systems: Cation exchange, intermediate or weakly basic anion exchange, cation exchange, strongly basic anion exchange. In some cases, strongly basic anion exchange is used in place of the intermediate or weakly basic anion exchange.
3. Mixed-bed systems: Cation exchange resin and strongly basic anion exchange resin are mixed together in one unit. This creates an infinite series of two-step demineralizers (Figure 11-7).

Mixed-bed demineralizing plants are usually installed where lowest possible electrolyte content in the treated water is desired. With new anion resins in a mixed-bed demineralizer, it is readily possible to obtain treated effluent containing less than 0.25 micromhos conductivity. On the other

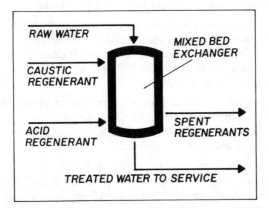

Figure 11-7. Cation and anion resins can be mixed together in one unit creating an infinite series of two-step demineralizers. During regeneration, resins are separated into layers with acid and caustic introduced simultaneously.

hand, most stepwise demineralizers are designed to produce a treated effluent having a conductivity of from 5 to 15 micromhos.

Operating Problems

Cation exchange resins are usually quite stable and require only minor replacement over years of normal use. Anion exchange resins, however, deteriorate more rapidly and require greater replacement. As a result, the problem of anion exchange resin deterioration has received a great deal of attention (Figure 11-8).

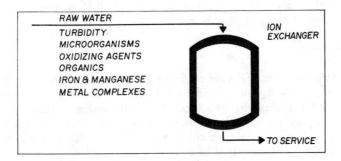

Figure 11-8. Operating difficulties of many ion exchange systems can be traced to contaminants present in the raw water.

The most common contaminants which can foul resins are:

• *Turbidity*, and traces of iron and manganese in the influent water to the anion exchange unit or mixed bed demineralizer. These precipitate in the pores of the resin and interfere with diffusion of anions. A hydrogen cation exchange unit can minimize the amount of these contaminants in the influent water to the anion unit.

• *Sodium hydroxide* and other contaminants already present in the regenerates. A caustic containing excessive amounts of chlorates and iron will attack the anion resin, while chlorides and silica may have some effect on the quality of water being produced.

• *Metal* complexes such as copper, nickel and cobalt cyanides. These contaminants are absorbed in the anion resin and interfere with anion exchange. However, if present in small quantities, these metals can be removed by the cation exchange unit in stepwise demineralizing plants.

• *Oxidizing agents* which attack cation resins can indirectly foul strong base anion resins. Sulfonic acid by-products of cation resin breakdown can irreversibly attach themselves to the anion resin. Chlorine is the most common oxidizing agent responsible for cation resin attack.

• *Industrial organic contaminants* leak high molecular weight carboxylic acids, reactive organic acids, amines, oil and grease.

• *Dissolved organic compounds* of natural origins. They arise from the bacterial decomposition of organic impurities leached from soils, decaying trees and leaves, and can be particularly troublesome.

• *Microorganisms* filtered out by the cation and anion resins. They can grow under proper environmental conditions such as optimum temperature and the presence of nutrients. The contaminating organisms include spore-forming type, iron bacteria, sulfate-reducing bacteria, and even coliform bacteria. Microorganisms are a nuisance, from a health viewpoint and because they cause objectional tastes and odors, slime growths, and general fouling and plugging of resin beds.

Pretreatment

Influent water pretreatment is usually concerned with the removal of turbidity, sediment, color, organic matter and microorganisms (Figure 11-9). Associated with these are taste, odor, iron and manganese hardness.

The principal operations involved in pretreatment are coagulation and filtration. Also important are chlorination, iron and manganese removal and purification by activated carbon.

Coagulation and Coagulants

Coagulants are chemicals which act in one or more of the following ways:

RAW WATER

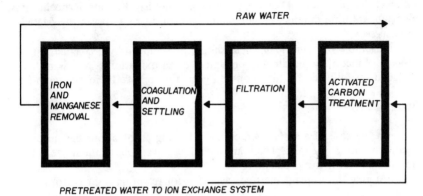

PRETREATED WATER TO ION EXCHANGE SYSTEM

Figure 11-9. Pretreatment for ion exchange systems may consist of any combination of these systems.

1. *Neutralization* of the repulsive forces around the suspended solids. This enables the cohesive force to hold the particles together once they collide.
2. *Precipitation* of sticky flocs such as metal hydroxides. This floc entangles the suspended solids and color.
3. *Bridging* of suspended solids by natural or synthetic long-chain, high molecular weight polymers. The polymer chain bridges between particles. This "bridge effect" tends to take up particles missed by neutralization and precipitation, and toughens flocs.

Commonly used coagulants are aluminum sulfate, ferric sulfate and ferrous sulfate. They are acidic and react with the natural or added alkalinity of the water to produce sulfates of calcium, magnesium or sodium and a gelatinous precipitate. This precipitate is a complex formula of varying composition.

Filtration

Filtration is used in pretreatment to an ion exchange system to remove or reduce suspended solids which may be present initially in the raw water undergoing treatment, or may be the result of a coagulation or precipitation process.

With proper sedimentation following chemical treatment in a clarifier, heavier coagulated particles are removed prior to filtration. Only smaller and lighter particles of floc reach the filter.

Iron and Manganese Removal

In any water supply, iron and manganese can exist in an insoluble or soluble state. Soluble iron and manganese are readily removed by oxidation followed by filtration. Oxidation can be accomplished by air, hypochlorites, chlorine, chlorine dioxide, or potassium permanganate with varying degrees of success.

Insoluble iron and manganese should be removed by a coagulation process. Since they are already precipitated, no oxidation is necessary. Coagulation followed by filtration will generally remove them satisfactorily.

System Management

In the case of a system which has a fouling problem but has no pretreatment, all is not lost. There are techniques available to minimize (but not always solve) the problem.

Among the in-plant methods available for restoring resin performance are:

1. Organic fouling
   a. Hot brine
   b. Hot brine and caustic
   c. Hot brine and hypochloric acid
   d. Hot brine and hypochlorite
2. Iron and manganese contamination
   a. Acid wash
   b. Commercial resin cleaners (phosphates plus hydrosulfite)
3. Oil and grease contamination
   a. Sodium hydroxide wash
   b. Anionic detergents (cation exchanger)
   c. Non-ionic detergents (general use).
4. Sterilization of microorganisms
   a. Hypochlorite (start-up: systems prior to resin fills)
   b. Formaldehyde (systems with resin fills)

ULTRAVIOLET WATER PURIFICATION[121]

A number of factors combine to make the ultraviolet method unique as a means of water purification. Ultraviolet radiation is capable of destroying all types of bacteria. In addition, ultraviolet radiation disinfects rapidly without the use of heat or chemical additives which may undesirably alter the composition of water.

One of the several categories of energy is electromagnetic or radiant energy. Radiant energy travels, in the form of waves, in straight line paths and in all directions from its source. The wavelengths range from very long radio waves to very short X-rays.

The most familiar part of the spectrum is a narrow band of wavelengths visible to the human eye. Another band with wavelengths shorter than those of visible light, and not visible to the eye, is the ultraviolet part of the spectrum.

Ultraviolet radiation can cause changes in living matter. The sun's rays cause sunburn. Rays from a welder's torch burn the unprotected eyes of an observer.

Table 11-6. Ion Exchange System Management

A. Start-up
1. Cleanliness maintained during installation
2. System flushed with quality water prior to use
3. System, prior to resin installation, sterilized with a chlorine base product

B. Maintenance
1. Good housekeeping practices maintained
2. Lubrication points checked for possible introduction of microorganisms
3. Recommended procedures followed during scheduled shutdowns
4. Formaldehyde treatments used for sterilization

C. Analysis
1. Raw water supply completely analyzed at least once a year
2. Resin beds sampled and analyzed at least once a year

D. Records (daily logged items)
1. Raw water hardness
2. Iron and manganese removal system effluent—Fe and Mn
3. Clarifier effluent alkalinity A, B & C
4. Clarifier effluent pH
5. Clarifier effluent chlorine level
6. Clarifier percent sludge concentration
7. Filter effluent turbidity
8. Activated carbon effluent chlorine level
9. Two-step demineralizer—gallons/regeneration
10. Mixed-bed demineralizer—gallons/regeneration
11. Ion exchange softener—gallons/regeneration
12. Pressures losses of filter, carbon and ion exchange units
13. Resistivity-conductivity of demineralizer effluents
14. Backwashes of filter, carbon and ion exchange units
15. Regenerations of ion exchange units
16. Steamings (activated carbon unit only)
17. Sterilizations of ion exchange units
18. Changes of fills (sand, carbon, resins)

The ultraviolet spectrum includes wavelengths from 2000 to 3900 Angstrom units (Å). One unit is one ten billionth of a meter. The 2000 to 3900 Å range may be divided into three segments.

Long-wave ultraviolet—The wavelength range is 3250 to 3900 Å. These rays occur naturally in sunlight. They have little germicidal value.

Middle-wave ultraviolet—The wavelength range is 2950 to 3250 Å, also found in sunlight. Middle-wave UV is best known for its sun-tanning effect; it provides some germicidal action, with sufficient exposure.

Short-wave ultraviolet—The wavelength range is 2000 to 2950 Å. This segment possesses by far the greatest germicidal effectiveness of all ultraviolet wavelengths. It is employed extensively to destroy bacteria, virus, mold, spores, etc., both air- and waterborne.

Short-wave ultraviolet does not occur naturally at the earth's surface, because the atmosphere screens out sunlight radiation below 2950 Å. In order to take practical advantage of the germ-killing potential of short-wave ultraviolet, it is necessary to produce this form of radiant energy through the conversion of electrical energy. The conversion of electrical energy to short-wave radiant ultraviolet is accomplished in a mercury vapor lamp.

Germicidal Lamps

The low-pressure variety of a mercury vapor lamp, which can be referred to as a germicidal lamp, provides the most efficient source of short-wave ultraviolet energy. Medium and higher pressure types are less efficient in short-wave output, consume more power, and may cause purifier overheating.

Germicidal lamps are made of special quartz glass that will allow 70 to 90 percent of the short ultraviolet rays to pass. Ordinary glass is not transparent to wavelengths below 3200 Å (Figure 11-10). The low-pressure mercury

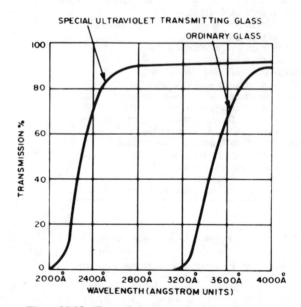

Figure 11-10. Transmission curve for high silica glass.

vapor lamp emits radiation that is predominantly at 2537 Å. This is in the region of maximum germicidal effectiveness (Figure 11-11).

The germicidal lamp works on the following principle: An electric arc is struck through an inert gas carrier (usually proprietary), in a sealed special glass tube. Heat from the arc causes vaporization of the small amount of mercury contained in the sealed tube. The mercury, when vaporized, becomes ionized and in the electric arc gives off UV radiation.

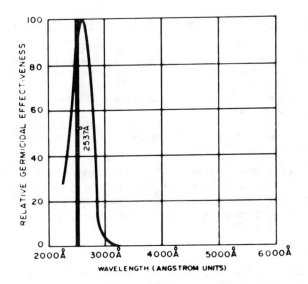

Figure 11-11. Germicidal effectiveness as related to wavelength.

Required Germicidal Energy

Bacteria withstand considerably more ultraviolet irradiation in water than in dry air. *E. coli*, for example, require more UV exposure for their destruction in water than in dry air. In either case, the germicidal radiation must strike a microorganism to destroy it. This implies that the water be clear enough to allow transmission of an adequate quantity of UV energy.

The degree of microbial destruction is a function of both the time and the intensity of the radiation to which a given microorganism is exposed. A short exposure time at high intensity is as effective as a long exposure time at low intensity, provided the product of the time and intensity remains the same. This dosage is normally expressed in microwatt-seconds/sq cm.

The dosages required for common bacteria range upward to more than 20,000 microwatt-seconds/sq cm. To allow for less than 100 percent transmission, a purification system should be designed to deliver 2537 Å energy in excess of 30,000 microwatt-seconds/sq cm.

Any turbidity in the water reduces the range of transmission of UV radiation. Water that is naturally turbid, or that has become turbid from corrosion products formed during storage in steel tanks and lines, should be filtered before UV purification.

Purifier Design

Several design features bear directly on the dosage delivered:

1. Output of the lamp.
2. Length of the lamp—when the lamp is mounted parallel to the direction of water flow, the exposure time is proportional to the length of lamp.
3. Design water flow rate—exposure time is inversely related to the linear flow rate.
4. Diameter of the purification chamber—since the water itself absorbs UV energy, the delivered dosage diminishes logrithmically with the distance from the lamp.

A typical UV water purifier is shown in Figure 11-12 in operation. Water enters the inlet and flows through the annular space between the quartz sleeve (which contains the germicidal lamp) and the outside chamber wall. The irradiated water leaves through the outlet nozzle.

Features to Look For

1. Expandable system—parts should be as uniform and as interchangeable as possible to permit easy expansion later.

2. Sight port—enables visual monitoring of lamp operation; also permits later adaptation to electronic monitor device using same port.

3. Single lamp per chamber—provides greater safety through more accurate monitoring than does a multi-lamp/single-chamber system.

4. Quartz protection sleeve—cold water moving past an unshielded lamp will reduce the lamp temperature and the radiation yield. A protective quartz sleeve will allow the higher lamp temperature required for optimum output of 2537 Å radiation.

5. Mechanical wiper—for cleaning the sleeve surface without shutdown or disassembly of the unit.

6. Optional accessories—flow controls, UV monitors, electronic water shut-off valves, and alarms, should be available to provide fail-safe operation without operator attendance.

Operating Data

For tubular lamps of 12 to 48 in. in length, the following data are typical:

Lamp Length	Power Consumption	2537 Å Output	Rated Effective Life
12 in.	8 W	1.3 W	7500 hours
18 in.	18 W	5.8 W	7500 hours
36 in.	39 W	14.6 W	7500 hours
48 in.	110 W	51.5 W	7500 hours

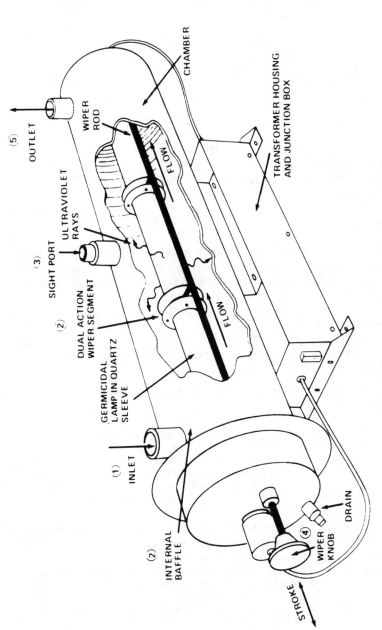

Figure 11-12. Assembly of one type of ultraviolet water purifier.

A single lamp purifier can be designed to handle any flow rate up to approximately 2400 gph. By multiplying purifier units, in series and in parallel, the higher flow rates are obtained. Several flow rate units and approximate costs are:

Flow Rate, gph	Approximate Cost
75	$ 300
250	350
600	400
2,400	800
10,000	3,200
20,000	6,400

The approximate cost per gallon of water treated may range from two mils for smaller units down to one-fifth mil for the larger installations. Lamp replacement and average maintenance is estimated at one cent/1000 gal of water treated.

Applications

The unique advantage of the UV method of sterilization of water is that nothing is added to the water. When chemical methods of treatment are used there may be handling problems, taste and odor problems, and undesirable chemical reactions with substances present in the water.

This difference is most significant when producing water for drinking or swimming, processing foods and bottled beverages, manufacturing cosmetics or pharmaceuticals, use in hospitals and research institutions, and tertiary treatment of municipal or industrial wastewater. The versatility of UV purification includes:

1. UV purification produces germ-free potable water for home, institution and municipal use.

- for application to water wells; bacterial contamination of wells is unpredictable and may occur from seepage of surface water or sewage.
- for installation on outlet side of water cisterns; most cisterns foster the proliferation of bacteria in untreated water.
- for swimming pools; to control bacteria, algae and slime formation. It avoids the undesirable effects of heavily chlorinated swimming pool water by allowing substantial reduction or elimination of the use of chlorine.

2. It provides bacteria-free food process water without the use of germicides, oxidants, algaecides or chemical precipitants; particularly applicable where chlorine adversely affects flavor.

- for the brewery, winery, soft drink, and water bottling industries, where biological purity of the water must be absolutely maintained in order to insure product quality.
- for safeguarding against spoilage of dairy products, *e.g.*, cottage cheese and butter; certain psycrophilic bacteria are resistant to chlorine treatment.

- for sterile washwater; to guard against waterborne bacteria spoilage where vegetables, fruits, meats, fish, and other products must be washed in water before packaging.

3. UV purification is particularly useful in applications where chlorine-free, de-ionized and/or carbon filtered water are extensively employed. Unattended carbon filters and ion exchange tanks act as incubators for bacteria accumulation.

- for electronics; in conjunction with de-ionized and high-purity water systems.
- for pharmaceuticals and cosmetics; strict water treatment standards are necessary for strict maintenance of product quality control.
- for biological laboratories; sterile water is required for testing and research work.
- for hospitals; provides ultra-pure water on demand for maternity labor and delivery areas, pathology labs, etc.

4. In industrial pollution control, it affords an excellent end-treatment for positive protection in wastewater control systems.

- for selective use as a tertiary treatment for bacteria destruction after removal of chemicals and other objectionable ingredients.

Chlorine Versus Ultraviolet Purification

As a tertiary treatment for water, chlorination offers the advantage of continued disinfection after initial treatment, since some chlorine remains in the water with residual germ-fighting action. The ultraviolet method, however, has none of the following disadvantages of chlorine:

1. Chlorine treatment requires operation attention.
2. In small installations, when chlorine gas is liberated from a chlorine cylinder or moistened crystals or pellets, the fumes are extremely dangerous and may even be lethal.
3. Chlorine itself is a highly corrosive and toxic chemical.
4. Chlorine is an additive material which may impart an undesirable taste to the water and a decrease in pH.
5. Chlorine is chemically active and can react with foreign ingredients (*e.g.*, in industrial wastewaters) to form toxic compounds, a matter of increasing concern to the federal government and to many states and municipalities.

- it may combine with ammonia to form "chloramine" which is acutely toxic to fish even at low concentration.
- it may combine with phenol to form "clorophenols," another dangerously toxic compound.

DEEP WELL DISPOSAL OF LIQUID WASTE[95]

Many industries which generate large quantities of concentrated liquid waste are using injection well disposal systems (Figure 11-13). Deep well disposal is a method of pumping or injecting liquid waste into sub-surface formations.

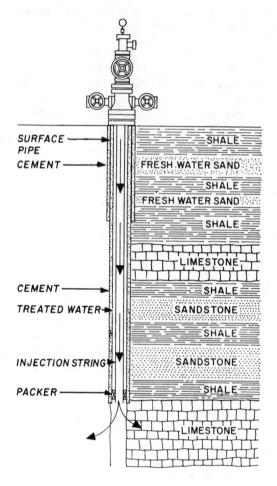

Figure 11-13. Cross section of typical deep well disposal system.
A well should be double-cased to a depth below any water bearing strata.

These deep well systems often have lower investment and operating costs than surface methods. Concentrated toxic liquid waste and odorous fluids can be completely disposed of, where comparable surface treatment could only be accomplished at considerably higher expense and difficulty. Very little surface area is required except for the well pumping equipment.

Properly selected geological disposal zones can accept large quantities of fluid waste. A uniform 100-ft-thick sand strata will accept 500,000 gallons per day, and will permit injected wastes to reach out only 1000 feet from the well in four years of operation.

In the United States, more than 100 liquid waste disposal wells have been drilled to depths ranging to 12,000 ft, handling liquid waste capacities up to 1100 gpm with pH levels ranging from 1.0 to 12.5. Thousands of salt water injection wells are being used in oil fields to improve oil production. Injection wells have also been used to recharge fresh water aquifers.

Type of Injection Zone

Injection wells have been completed in many kinds of geological formations such as rock, shale, shale and sand, and sand.

An effective disposal well has a porosity structure in which pores and voids are interconnected to permit an injected fluid to pass through the media. Porosity of a geologic formation is that portion of the total volume which is occupied by pores or voids. Subsurface formations can vary widely in porosity. Unconsolidated sands usually have 30 percent or more porosity. Shales and clays can hold over 40 percent liquid by volume, but are so fine-grained that they are almost impervious to fluid flow.

Porosities can also be classified according to the physical arrangement of the material surrounding the pores. In clean sand, pores exist between the individual grains. This type of porosity is called intergranular, and is inherent in sand formations. Vugular and secondary porosities may also exist, and are caused by the action of water after the materials are deposited. For example, slightly acidic percolating waters can enlarge and create pore spaces with interconnecting channels. Percolating water rich in minerals may also form deposits which can seal off some of the pores in a formation reducing its porosity.

In sand formations some particles are colloidal size. A characteristic of colloidal particles is that they carry a negative electrical charge, which prevents the particles from clinging together. When injected wastes are highly acidic, ions which are positively charged can neutralize the charges of the colloidal particles causing them to coagulate and form larger grains filling natural voids. This reaction will tend to reduce the voids ratio and decrease porosity of the media. Acidic solutions can also cause some materials such as clay to swell, reducing the pore volume.

Porosity can be used to determine the capacity of an injection zone from which the distance of invasion can be developed. The distance of invasion is the outer boundary of an injected fluid waste in the injection zone or horizon.

Permeability is a measure of the ease with which a formation permits a fluid to flow through it. To be permeable a formation must have some interconnected pores, channels or capillaries (effective porosity). In general, greater permeability corresponds to greater porosity; however, this is not an absolute rule. Some sands have large effective porosities but the grains are so small that paths available for fluid movement are restricted. A geologic sample measured for permeability in a laboratory can vary considerably from the true average for a portion or all of the zone. There are often wide variations both laterally and vertically in a formation, with permeability being several times greater in one plane than another. Permeability measured parallel to a bed of stratified rock is generally larger than in the vertical plane.

Drilling Permits

States may often have several regulatory agencies from which approval must be obtained before drilling a well, such as a department of health, office of the state geologist, board of water development, game and wildlife commissions, or even a railroad commission. Before drilling, a permit normally must be obtained from the state geologist, with reports then forwarded to one or more of these agencies.

Compatibility

It is essential to determine that natural brines in the injection zone will be compatible with the fluid to be injected. If there is incompatibility surface treatment may be required to remove or alter specific contaminants in the injected field. Incompatibility can cause solids to precipitate from the mixture, causing pores to plug.

Geology

Geology of the area into which wastes are to be injected must be thoroughly understood. It should be known if formations are horizontal or slope upward, are bounded by low-permeability rock, and whether there are oil or gas domes in the proposed area. Also, the depth formation that should be selected as the injection zone must be determined. Wells are often drilled one or two zones deeper; if necessary, the lower zone can be plugged off to permit continued well operation at a higher zone.

Bottom-Hole Completion

If a well is completed in a limestone or rock formation, the end of the injection tubing can be left open without a screening device. When the bottom-hole completion is in a shale-and-sand or sand formation, one of several screening methods can be used—gravel packing around a sand screen at the end of the injection tube, or plasticizing sand and epoxy around the sand screen. However, injection of epoxy should not extend more than 4 ft into the sand zone. The permeability of sand treated with epoxy will be reduced about 15 percent from its initial state. Diameter of the bottom-hole arrangement should be sized so that fluid velocity will not permit excessive pressure drop or erosion in the injection tubing or screen.

Injection Pressure

Knowledge of fluid mechanics is essential for determining injection pressures of waste material. When more than one well is to be used and completed in the same formation, it is important to know the effect of one well on the injection pressure and zone acceptance of injected fluid from other wells. Hydraulic gradients will likely increase with time, and this rate of increase must be known when selecting surface equipment (Figure 11-14).

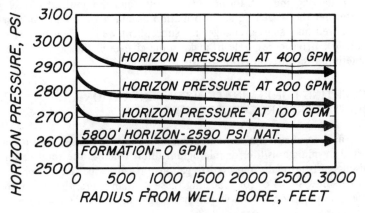

Figure 11-14. The chart shows the effect of an injection pressure rate on a formation at varying distances from the well bore.

Surface Equipment

Surface equipment required for deep well disposal consists of pumps, valving and piping necessary for the injection of the liquid waste. Either centrifugal or positive displacement pumps can be used. If calculations show the well head pressure will be low (less than 400 psig), centrifugal pumps can be used. Positive displacement pumps have an advantage of requiring only the horsepower needed to pump against the well-head pressure. Centrifugal pumps must be designed to produce the maximum possible head pressures even though they may operate at lower rates.

Corrosion-Resistant Materials

Usually liquid waste consists of a variety of chemicals with pH levels ranging from 1 to 12.5. With this range of pH, stainless steel or equivalent metals are suitable for use in pumps. However, if the liquid waste has a high chloride content, stainless steel should be avoided especially where cavitation or eddies may exist.

Costs

Costs for drilling a well will, of course, depend upon its depth, bore size, casing arrangement, geology and type of bottom-hole completion. Costs for drilling through shale and sand can range from $15 to $25/ft for a cased hole. In areas where rock must be drilled through, costs will range from $20 to $40/ft for a cased hole.

To obtain the most economical construction the designer should have a good working knowledge of drilling equipment, loads, drilling operations, mud engineering and well logging instruments, and the ability to develop and interpret the logged data. Information obtained from construction of the first well may suggest changes for subsequent wells. Even if only one is drilled, it may be desirable to make some change in the bottom-hole completion after interpretation of the logged data.

CONVERTING CONTAMINATION CONCENTRATION IN WASTE DISCHARGE TO TOTAL POUNDS[44]

Analyses of contaminants in industrial water discharges are normally reported in milligrams/liter or micrograms/liter, whereas the total volume of discharge is normally reported in gallons or gallons per day. For design of treatment systems or for reporting these contaminants, the values are normally required on a total weight basis, commonly in total pounds of a contaminant.

Figure 11-15 gives the total weight of a contaminant in pounds, when the volume of discharge is given in gallons and the concentration of the contaminant is given in milligrams/liter. The nomograph covers the ranges of values, 0.1 to 100 mg/l of contaminant concentrations; 100 to 1,000,000 gal of discharge, and 0.001 to 100 lb of the weight of contaminant.

Example 1

Plant A discharges 150,000 pgd of water effluents from its operations. Analysis reveals the concentration of oil and grease is 32.0 mg/l. What is the total weight in pounds of oil and grease discharged per day?

Solution: Locate 32 mg/l on the C scale. Locate 150,000 gal on the V scale. Join these two points by a line which intersects the M scale at 40. The weight of oil and grease discharged is 40 lb/day. If the value of concentration falls below or above the range given, the nomograph can still be applied by using a multiplication factor.

Example 2

240,000 gpd of water effluent discharge from plant B contains 50 μg/l of lead. What is the total weight of lead discharged per day?

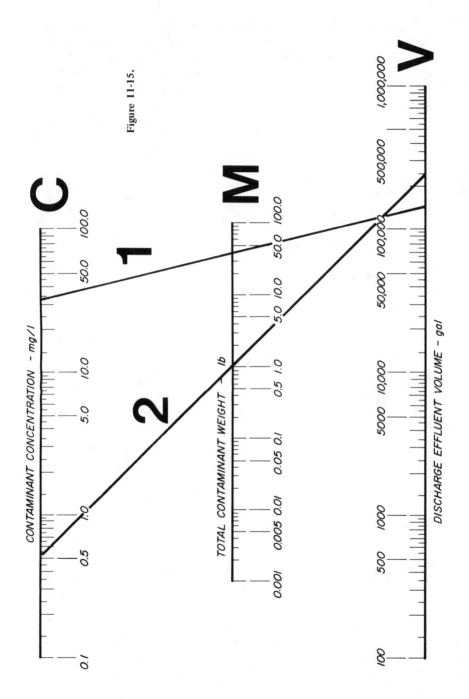

Figure 11-15.

Solution: Locate 240,000 gal on the **V** scale. Locate 0.5 mg/l (500 μg/l) on the **C** scale. Join the two points by a line which intersects the **M** scale at 1.0 lb. Since the actual concentration is 1/10 of 500 μg/l, divide the weight found by 10. Thus, the desired weight of the lead discharged per day is 0.1 lb.

CHAPTER 12

FILTERS AND FILTRATION

WASTEWATER TREATMENT FOR
REMOVAL OF SUSPENDED SOLIDS[137]

Removal of suspended contaminants from water and waste streams is a necessary step to the achievement of economic and environmental objectives.

Water turbidity (suspended solids) in nature generally results from colloidal clay dispersion, and most of the natural color comes from decayed wood, leaves (tannins, lignins, etc.), and organic soil matter. In addition to these contaminants, there are viruses, algae, bacteria, starches, metal oxides, oils and other pollutants that can significantly increase the levels of suspended solids.

Many of these same materials—pigments, tannins, lignins, metal oxides, oils, starches and bacteria cells—are also found in industrial wastewaters, and the concentration is nearly always higher than in natural waters.

Because of the similarity of solids in some natural waters and in waste streams, the same basic factors apply to their removal. They are: (a) particle charge and degree of hydration, (b) particle size, (c) concentration, and (d) contaminant ratios.

Particle Charge

Most solids present in water are negatively charged. This charge is electrostatic in nature, and the degree of charge depends upon the types of solids and the electrolytic environment. When a colloidal particle is dispersed in water, it can ionize, adsorb, and attract low molecular weight ions to its surface. Most of these adsorbed counter ions are held tightly to the colloidal surface (Stern layer). The remainder will be attracted to the particle and extend into the solution (diffuse layer or Gauy-Chapman layer) until electroneutrality is established (Figure 12-1).

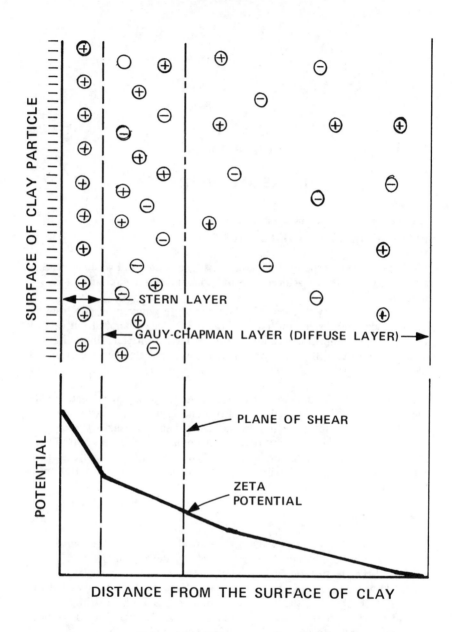

Figure 12-1. Particles in colloidal suspension are sourrounded by ions. Those ions tightly held to the colloid are in the Stern layer; those farther away are in the Gauy-Chapman layer. Outside the plane of shear (zeta potential), the electrostatic charge of the colloid has little effect on the ions in solution.

By placing a colloidal suspension in a direct-current electrical field, one can observe its migration velocity with a microscope. If the mobility is in the direction of the positive pole, the particle has a net negative charge. As this negative particle moves toward the positive pole, certain counter ions will be held tightly to the colloid. These ions are described as existing within the plane of shear. Counter ions outside the plane of shear can be exchanged for other counter ions as the particle moves through the solution. The negative potential that exists at this plane is called the zeta potential.

This net negative charge exhibited by colloidal particles is the strongest force inhibiting their removal. It is the interparticle repulsion that prevents colloids from colliding and forming larger masses. By partial or complete neutralization of this surface charge, colloids can collide through Brownian motion and mixing, and can be attracted to each other by hydrogen bonding and Van der Waal's forces of attraction, enabling them to form larger masses.

When considering charge and its effect on collidal removal, it is of equal importance to consider the degree of hydration of the colloid. Particles that strongly adsorb water (hydrophilic) are much harder to remove than solids that do not hydrate (hydrophobic). Colloidal clay and metal oxide dispersion are examples of hydrophobic colloids. Colloidal material, starches, and certain industrial organic pollutants are examples of hydrophilic colloids.

Natural color in surface water can be either hydrophobic or hydrophilic depending upon its molecular weight and the degree of ionization. So, also, can oil and water emulsions, depending upon particle size and the quantity and ionic character of the emulsifier.

Particle Size

Suspended solids in water can vary in size from 0.001 to 100 microns in diameter (Figure 12-2). As particle size for a given suspension gets smaller, removal becomes more difficult because charge has more effect on the smaller particles. When the solids are large, the ratio between the mass and surface charge is large, indicating that there is insufficient surface charge to cause interparticle repulsion.

Through collision, bonding and physical forces, these particles can coalesce to a size more suitable for removal. As the particles get smaller, the mass-to-surface charge ratios decrease to the point where surface charges are sufficient to cause interparticle repulsion, preventing coalescence.

Concentration

As the solids in water increase, the chemical coagulant dosage required to remove them increases proportionately. However, this only applies to small order of magnitude increases. If solids increase substantially, the amount and/or the type of chemical needed to produce the most effective results may change significantly.

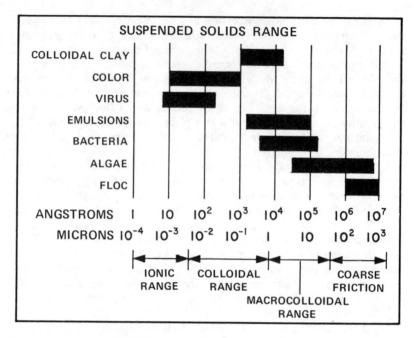

Figure 12-2. Sizes of particles falling within the suspended solids category vary significantly. Chart illustrates sizes of several common materials.

This change occurs because, at lower concentrations, the distance separating the negative colloids is sufficiently large that interparticle repulsion inhibits removal. A cationic material will neutralize these repulsion forces. As solids concentration increases, the distance between colloids is reduced so that they are in proximity to each other. In this case, the interparticle repulsion forces have been dampened by concentration, and a chemical that can bridge this distance may be all that is required for effective removal.

Contaminant Ratios

Theoretically, the removal of turbidity, color, oil, etc. can be examined separately to determine the most economical removal method. However, from a practical standpoint, waste streams are composed of many different (and varying) contaminants. To develop an effective treatment method, one must investigate their combined removal. The development of an effective treatment system for an integrated industrial waste can, therefore, prove difficult. Two examples are removal of solids collected from scrubbing basic oxygen furnace flue dust in the steel industry and removal of wastewater solids generated by a paper mill.

By understanding the nature of suspended solids present in water and the four major factors that affect their removal, the mechanisms that take place when removal occurs can also be better understood. Colloidal solids removal can be achieved by any of the following steps:

1. Destabilization
2. Microfloc formation
3. Agglomeration
4. Physical entrapment

To coalesce colloids, the interparticle repulsion forces must be neutralized. Certain cationic chemicals can be used to decrease the net negative charge. This is destabilization (Figure 12-3).

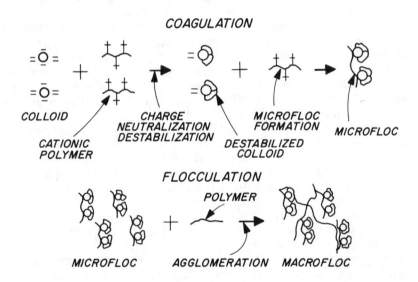

Figure 12-3. Coagulation and flocculation combine to convert colloidal particles into a macrofloc that can be removed from solution. Various steps of colloid removal are shown.

In a hydrophilic colloidal system where colloids are strongly hydrated, it may be necessary to add a chemical that not only neutralizes surface charge, but also forms an insoluble complex with the colloid for destabilization.

Once the colloids have been destabilized, collisions can occur. Through chemical bridging, hydrogen bonding, and Van der Waal's forces of attraction, these particles can then form small microflocs. Continued mixing enables the microflocs to combine to form macroflocs.

Not all colloids undergo destabilization, microfloc formation, and agglomeration. A portion of the colloids are removed by being physically entrapped in macroflocs already formed.

These first three steps are known by the familiar terms of coagulation and flocculation. Coagulation involves destabilization accompanied by microfloc formation. Flocculation does not include destabilization, but does include microfloc formation resulting from interparticle bridging, and macrofloc formation caused by agglomeration. There are a variety of chemicals used to coagulate and flocculate suspended matter. Most of these chemicals can be placed in two groups: inorganic coagulants and organic polyelectrolytes.

Most widely used inorganics are the di- and trivalent metal salts of calcium, aluminum and iron. The principal difference between calcium, aluminum and iron salts is their hydrolysis products. When calcium salts are added to water, the calcium ion is formed. When aluminum or iron salts are added, they form trivalent metal complexes with water.

These complexes contain a number of repeating metal ion units and can more properly be referred to as polyaluminum hydrates or polyferric hydrates. Polymetal hydrates of significant length have been reported. They are cationic and can destabilize a colloidal suspension. They also have a sufficient chain length to bridge the distance between particles. In addition, their strong hydrogen bonding ability enables them to form large macroflocs that can trap other stable colloids.

Other inorganic coagulants that have wide application are bentonite clays and activated silica. These products are used in conjunction with other inorganics such as calcium, aluminum and iron salts. Bentonite clay acts as a nucleus for floc formation and can also function as a weighting agent. Activated silica functions mostly as a flocculant improving macrofloc formation initiated with other inorganic coagulants.

Organic polyelectrolytes can be either natural or synthetic. Natural products, such as starches or gums, have been used for years as flocculants or aids to improve treatment with inorganic coagulants. They act primarily as bridging agents in a similar manner as activated silica, and are still used extensively in ore beneficiation, paper processing and municipal water treatment.

During the past 15 years, the synthetic organic polyelectrolytes have gained acceptance and now play an integral part in solids/liquid separation. They have achieved recognition because they are high-performance products. In many applications, these polyelectrolytes can be used at much lower dosages than the conventional inorganic coagulants and/or natural products, producing equal or superior results at a reduced cost. There are applications where they do not work by themselves, but can significantly improve both the performance and treatment cost when combined with inorganic coagulants.

The high performance of organic polyelectrolytes is a result of their high molecular weight and the variety of ionizable groups that can be placed along the polymer chain. A good working definition for the synthetic organic water-soluble polyelectrolyte is large molecules composed of a number of repeating units held together by covalent chemical bonds where some or all of these units are dissolved or ionizable in water solution.

Synthetic organic polyelectrolytes are classified as nonionic, anionic and cationic. Polyacrylamide is an example of a nonionic polymer. The hydrolyzed polyacrylamides and copolymers of acrylamide and acrylic acid are examples of anionic polymers. One of the principal advantages of anionic polymers is their relatively large available chain length. Nonionic polyacrylamide exists in solution as a random coil.

When the product is partially hydrolyzed, the negative sites repel each other and the random coil unwinds. The anionic materials have comparable molecular weights, but because of the uncoiling phenomenon their chain lengths are larger than the nonionic material (see Figure 12-4). This large available chain length makes a more effective bridging agent when charge neutralization is not an important factor in suspended solids removal.

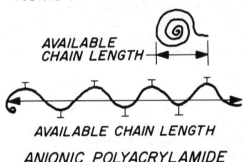

NONIONIC POLYACRYLAMIDE

AVAILABLE CHAIN LENGTH

AVAILABLE CHAIN LENGTH

ANIONIC POLYACRYLAMIDE

Figure 12-4. Chain length of a polymer has a great deal to do with its effectiveness in removing suspended solids. Drawing illustrates that, although two polyacrylamides are similar in formula and molecular weight, they may have very different properties as flocculating agents.

As charge neutralization becomes more important, the nonionic and cationic polyelectrolytes become more useful. A variety of cationic polymers is presently being produced to satisfy the varying solids/liquid separation requirements where charge is an important factor.

Even when the nature of suspended solids, their removal mechanism, and treated water requirements are fully understood, batch laboratory tests are still necessary to arrive at the most effective and economic treatment. The conventional jar testing apparatus, containing four to six mixing paddles tied to a common drive is used extensively to determine application data. Four to six different treatment systems can be compared to each other under the same conditions using this method. Degree of mixing, mixing time and settling time can be varied to simulate plant conditions. In addition, graduated cylinder tests are also widely used to compare various chemical treatments. They do have one limitation in that it is hard to duplicate mixing from one graduate test to the next.

More sophisticated methods employing the stream and current detector and the zeta meter are used and have proved practical in some cases. These tests only attempt to measure surface charge and the product's ability to alter the charge. They do not indicate the degree of floc formation, floc settling ability, or the supernatant turbidity. The conventional jar test must still be used to attain this type of information.

Knowledge of the factors affecting suspended solids removal can help bring about better and more economical water treatment as illustrated by the following examples:

Example I–A large industrial plant clarifies surface water for both drinking and process use. Original treatment with alum and lime produced good quality water at a cost of approximately $31 per million gallons per day (mgd).

Lab and inplant tests were conducted using a cationic polyelectrolyte in an attempt to reduce the cost of treatment and produce better quality water. Tests demonstrated both objectives could be met. As a result, the use of polymers has been incorporated in the treatment procedure. Comparison of the old and new treatment in techniques indicates the savings obtained:

	Original Treatment	Present Treatment
Avg alum used, lb/million gallons	980.0	561.0
Avg lime used, lb/million gallons	263.0	172.0
Avg cationic polymer feed, lb/million gallons	0.0	10.4
Total chemical cost per million gallons	$ 31.00	$ 22.00

Besides the lower cost, the use of the cationic polymer also resulted in turbidity and better and more consistent removal of color. Two additional benefits were less generation of sludge and reduced labor cost, both resulting from a substantial decrease in alum and lime usage.

Example II–A steel mill located in the Ohio Valley operates a gas scrubbing system on its basic oxygen furnace. An integral part of this system is the proper operation of a thickener rated at 1500 gpm. Suspended solids in the thickener influent can vary from 1000 to 20,000 milligrams per liter (mg/l), pH fluctuates between 8.1 to 12.0, and the temperature ranges from 100 to 140 F. Desired results are less than 20 mg/l suspended solids. Extensive lab and field testing with an anionic polyacrylamide indicated it would produce the desired results. The mill has been using this product for over one year.

In this case, the slightly anionic polymer is functioning as a flocculant. The suspended solids are sufficiently large so that they can settle naturally. Agglomeration of the coarse solids is all that is needed to remove them.

DEWATERING SCREENS[6]

Conventional methods for removal of solids from wastewater streams are bar screens, settling basins, vibrating screens, microstrainers, and mechanical clarifiers. However, there are other screening devices available for dewatering.

Screens reduce suspended solids in industrial effluent before discharge into a municipal sewer. In certain industries screens recover sufficient usable materials to pay for their installation.

At the municipal sewage plant influent, screens offer economical removal of cigarette filters, bone, seeds, hair, fibrous material, metal foil, and grease particles. Such materials would impede clarification, clog pumps, and foul digesters. Storm runoff waters carry floating and settleable coarse solids which can be screened out ahead of surge basins.

Static Screens

Static screens have no moving parts. A three-piece multiple-angle screen is shown in Figure 12-5. All static screens use wedge-shaped bars placed horizontally across the wastewater flow. Liquid flows through the spaces between

Figure 12-5. Static screen.

the bars, leaving particulate solids on the screen face. Space openings generally available vary from 0.010 to 0.100 in. The screen spacing provides variations in flow capacity and a choice of the size of the particulate solids to be removed. In the three-screen model, each screen may be varied independently of the others, enabling the unit to be adapted to specific conditions.

The operation of static screens is simple. Wastewater is delivered to the screen head box. It flows over a weir and cascades down the face of the screen. Water flows through the screen while solids travel down the face and fall from the lower edge. Screen unit flow capacities are based upon the capacity of the upper third of the screen surface. The lower two-thirds of the screen protects against flow surges and sudden increases in suspended solids levels.

There is no simple rule-of-thumb for sizing static screens on the basis of hydraulic capacity. Experience has shown that every static screen application has unique conditions which must be considered. Certain general questions need to be answered.

First, what is the nature of the solids to be removed? What are the minimum and maximum sizes of the solids and the approximate percentage of each size? The answers will determine the proper screen openings.

Second, what is the suspended solids load carried by the stream? As suspended solids levels increase, flow capacities for a given screen are decreased because the fixed open area of the screen face is partially blocked by particles. Peak surges in the suspended solids levels must be taken into account.

Third, what are the minimum and maximum flow rates anticipated? The screen must be sized on the basis of the peak flow. Otherwise, a portion of the flow may flood into the solids recovery area.

Typical static screen performance data are shown in Table 12-1. This information should be used only as a rough guide for estimated purposes. The most effective—and, in the long run, least expensive—way to size a static screen for a specific dewatering application is to install a full-size unit on a trial basis. This enables the purchaser to develop accurate data for his particular application and thereby determine the screen best suited to his needs.

Pros and Cons of Static Screens

Suspended solids reductions do not vary appreciably between the different screen configurations. Screens can be designed with bar spacings down to 0.01 in.

Mechanical Reliability: Because static screens contain no moving parts, they are essentially maintenance free.

Blinding and Clogging: As long as there is no grease, fat, or oily solids present in the wastewater, static screens need very little cleaning to maintain efficient dewatering. When such solids are present, as in the food industry or in screening raw sewage, the frequency of cleaning the screen may be every few hours. Manual cleaning is done with brushes. Some static screens used

Table 12-1. Typical Static and Rotating Screen Performance Data

Process	Screen Spacing In.	Typical Static Screen Performance Data		Typical Rotating Screen Performance Data	
		Approximate Flow Capacity Per Ft of Width, gpm	Approx. Percentage of Suspended Solids Reduction	Approximate Flow Capacity Per Ft of Length, gpm	Approx. Percentage of Suspended Solids Reduction
Municipal Pretreatment	0.060	75-125	10-30	400	10-30
Storm Water	0.100	100-140	2-5	550	2-5
Tanning Waste	0.020	20-50	20-30	150	20-30
Canning Waste	0.060	75-100	10-20	400	10-20
Meat Packing Waste	0.040	50-70	10-40	250	10-40
Textile Mill Effluent	0.030	50-90	5-20		
Kraft Mill Effluent	0.020	70-90	5-20		

in food prodcessing and textile plants have been equipped with continuous spray cleaning systems. In a given situation the value of a spray cleaning system can be determined only by observing the cleaning efficiency and the reliability of the mechanical pumping system.

Cylindrical Screens

Rotary dewatering screens (Figure 12-6) eliminate some of the clogging problems associated with static and vibrating screens. Wastewater or process water passes through the top of the slowly rotating screen. Solids which cannot pass through the openings ride over the top of the screen and are removed by a wiper mechanism on the opposite side. The wiper blade channels the dewatered solids away from the screen into a suitable collection system.

Influent water, after passing through the top arc of the screen, falls through its interior, and out the bottom arc. The falling filtrate backwashes away the particles trapped in the screen openings. Even grease, fats, and stringy solids, lodged between the wedge-shaped screen wires and not removed by the wiper, are effectively removed by the backwash. This self-cleaning action has proven to be effective and reliable. Cleaning cycles of six to eight weeks on raw sewage applications are common.

As with static screens, there is no accurate rule-of-thumb for sizing rotary screens. The questions covered earlier are equally valid for rotary screens.

Pros and Cons of Rotary Screens

Performance: Suspended solids reduction of rotary screens is similar to that of static screens.

Figure 12-6. Rotary screen.

Mechanical Reliability: Since the rotary screen rotates slowly (10 rpm maximum), vibration and structural fatigue are not a problem. Wear life of the solids wiper mechanism ranges from 12 to 18 months, and the replacement cost of the wiper is nominal.

Blinding and Clogging: Use of conventional vibrating and static screens is not practical in high grease environments because of continuous blinding and clogging. The rotary screen, however, has proven practical in many such applications. The cleaning cycle in environments such as meat packing, poultry, and tanning applications, varies from once a week to once every two weeks. Rotary screens are easily cleaned by spraying the interior of the cylinder.

Capacity: Because the rotary screen constantly provides a clean screen surface for the wastewater to flow through, its available screen surface is more effective than that of static screens. Floor requirements for the rotary screen are therefore only about one-third to one-half the requirements of static or vibrating screens.

Screen Economics

Typical costs for vibrating, static and rotary screens for a meat packing plant are given in Table 12-2. On the positive economic side, screens offer a means of reducing sewage surcharges by removing suspended solids before they become an expensive problem within the wastewater treatment system.

Table 12-2. Typical Screening Costs for a Meat Packing Plant

Influent: Hair and Manure Water		Flow: 3 mgd 2082 gpm	
Costs	Vibrating Screen	Static Screen	Rotary Screen
Equipment	(3) 4-ft units $ 15,000	(3) 6-ft units $ 16,000	(1) 5-ft unit $ 11,000
Installation	900	900	300
TOTAL	$ 15,900	$ 16,900	$ 11,300
Annual Expenses			
Maintenance	$ 3,000	$ 100	$ 100
Power	360	None	120
Cleaning	3,000	3,000	200
Replacement	1,500	None	75 (wipers)
TOTAL	$ 7,860	$ 3,100	$ 495

Recovery of usable solids is an economic gain. For example, in some meat packing and poultry processing plants, the recovered solids are renderable to useful products. Paper and textile mills recover usable fibers formerly discharged as waste.

GRAVITY FILTERS [125]

With proper sedimentation following chemical treatment, heavier coagulated particles are removed prior to filtration leaving only smaller, lighter floc particles to be filtered. When a freshly backwashed filter is first placed in operation, many of the finely coagulated particles penetrate into the filter bed through voids in the bed surface. As particles lodge between grains of the filter medium, flow is restricted. Coagulated particles then build up on the surface of the filter bed. This coagulated mat acts as a fine filter for smaller particles.

Penetration of the filter medium by coagulated particles normally does not extend deeper than 2 to 4 in. at normal flow rates of 2 to 4 gpm/sq ft. Most filtration occurs at the surface or in the first one or two inches of the bed. The filter medium must have sufficient coarseness so that some penetration of the top few inches of the bed takes place. With no penetration of the bed by coagulated material, head losses would increase rapidly and filter runs would be short.

Filter Media

Filter medium size is important, because the medium should:

1. Prevent passage of suspended matter
2. Hold suspended matter as loosely as possible for easy backwash removal
3. Be able to hold a volume of suspended matter without clogging

To determine if a medium meets these criteria, two parameters, effective size and uniformity coefficient, are used. Measurements are made by doing a screen analysis of the filter media using a series of U.S. Standard Sieves. The results (cumulative percent retained on each screen) are plotted against sieve opening (millimeters).

Effective size of a filter medium is that part which accumulates 90 percent of the sample. However, this does not indicate particle size variation in the bulk of the filter media.

To ensure that variation is not too great, a second measurement must be made. This is the size which accumulates 40 percent of the sample. This size divided by the effective size provides the uniformity coefficient.

Example: Screen analysis showed 10 percent of the sand finer than 0.40 mm and 60 percent finer than 0.64 mm.

Effective size (E.S.) = 0.40 mm
Uniformity coefficient (U.C.) = 0.64/0.40 = 1.6

Media Types

There are three common types of filtering media used for water treatment. They are listed below along with their typical sizes:

1. Fine sand. E.S. = 0.4 mm-0.60 mm. U.C. = 1.6 maximum
2. Fine anthracite (No. 1). E.S. = 0.65mm-0.8 mm. U.C. = 1.85
3. Calcium carbonate. E.S. - 0.5 (0.4-0.7 mm). U.C. = 2

The filter medium can either be supported on graded gravel or heavy anthracite. It can also be used alone in false bottom type underdrains that do not require subfills.

Silica sand is the most commonly used filtering medium; however, it is not used in situations where silica leached from the sand could be detrimental to plant operation.

Anthracite is generally used in cases where silica is leached from sand due to the action of high alkalinities and heat. Anthracite, similar in size to fine sand, is equally effective for turbidity removal. In addition, extended filter runs, less backwash water requirements, and higher filtration rates are possible with anthracite. Its sharp, annular particles produce larger voids than sand providing more dirt-holding capacity for certain types of suspended solids.

Calcium carbonate reduces the corrosive action of low pH waters and can be used to protect plumbing and heating systems. The $CaCO_3$ neutralizes part of the carbon dioxide content of water when the pH of the raw water is under 6.8 while serving as a filter medium:

$$CaCO_3 \quad + \quad CO_2 \quad + \quad H_2O \quad \rightarrow \quad Ca(HCO_3)_2$$

Calcium Carbonate　Carbon Dioxide　Water　(Calcium bicarbonate in solution)

Equilibrium under the recommended conditions is reached at about pH 7.0 or slightly higher with some CO_2 remaining in the water. Calcium carbonate is consumed in this reaction and hardness is added to the water. The filter medium is replaced whenever the bed depth decreases by 10 percent.

Sidestream Filtration

In a cooling water system, a filter can reduce suspended matter in the water by continuously bypass-filtering 1 to 5 percent of the total cooling stream with no chemical treatment. This process is called *sidestream filtration.*

Sands, silt or turbidity normally found in unclarified waters used for cooling purposes can create many problems. Suspended matter can foul heat exchangers, impair transfer rates and accelerate corrosion. More important, it can lead to increased cleaning and maintenance costs of heating and cooling equipment.

Although any pressure or gravity filter will reduce this suspended matter, years of operation have shown the automatic valveless type filter to be well suited to this application because of its ease of operation and minimum maintenance. Ordinarily these filters use a sand or dual media combination or anthracite and sand.

Filter Operation

Water from a constant level source enters through the inlet pipe into the upper part of the filter bed compartment and filters downward through the filter bed. Passing through the strainers into the collector chamber, the filtered water then rises from the chamber up through the outlet to service. During the filter run, the accumulated matter on the filter bed slowly builds up a back pressure which causes a gradual rise of the water in the backwash pipe. When a predetermined level is reached (usually 4 to 5 ft above the level of the filtered water outlet) a self-actuated primer system rapidly exhausts air from the backwash pipe and starts the siphoning action that backwashes the filter.

Since the backwash pipe carries approximately the flow of 20 gpm/sq ft of filter bed area and only 2 to 3 gpm/sq ft flows through the inlet pipe, the remainder is drawn from the backwash storage compartment. This water passes through the ducts leading from backwash storage to collector chamber, and from there up through the filter bed, backwashing it.

Backwashing starts at a higher rate than when it ends because of the diminishing level in the backwash storage compartment. This steadily diminishing backwash rate aids greatly in hydraulically regrading and evenly settling the filter bed.

At the end of the backwashing operation, incoming water from the inlet pipe filters through the filter bed and, rising through the ducts from the collector chamber, fills the backwash storage compartment with filtered rinse water for the next backwashing operation. When this compartment is filled to the level of the filtered water outlet, the water then ceases flowing into this compartment and flows through the filtered water outlet. This returns the filter unit to normal operation with none of the rinse water ever going to service.

Table 12-3 shows operating experiences of five installations using sidestream filtration. Experience at these plants indicates removal of 80 to 90 percent of high turbidity waters are not usual. Low influent turbidities of 10 to 30 ppm still show at least a 50 to 75 percent reduction. In general, average effluent quality of 5 ppm turbidity is not uncommon in sidestream filtration.

Another parameter which will have bearing on the proper sizing of the sidestream filter is the number of turnovers per 24 hr, or the number of total cooling system volumes bypass-filtered per 24 hr. Consider a cooling system which contains a relatively small volume of water circulating at a high rate. A lower percent bypass for this system would give the same type performance as would be obtained in a system containing a large volume of water circulating at a low rate at a higher percent bypass.

Overall, experience indicates that two or three turnovers per 24 hr is sufficient. Therefore, in sizing a filter for this type of service using 2¼ to 2½ bypass, two or three turnovers are desirable. If not, a higher percent bypass may be necessary.

FLOCCULATION[76]

Increasing use of chemicals in earlier stages of industrial and municipal waste treatment is seen for such reasons as: (1) phosphorus removal in primary clarification; (2) greater removals of BOD (Biochemical Oxygen Demand) and suspended solids in primary clarification to better utilize existing biological systems for nitrogen transformation; and (3) proper preconditioning of municipal and industrial wastes for filtration, and carbon absorption or reverse osmosis, including optimum pH (usually close to 7.0). Thus, proper chemical treatment, flocculation, and clarification will heavily contribute to the success or ease of operation of following more sophisticated high-rate processes.

Table 12-3. Comparison of the Operating Experiences of Sidestream Filtration in Five Different Installations

Plant	A	B	C	D	E
Filtration rate gpm/sq ft^2	2.5	3.1	2.5-2.8	3.0	3
Percent cooling water bypass through filters	2.5	1.13	1.7-1.9	4.2	3.6
Water temp. $^\circ$F	110	100	65-75	60-75	80-100
Turbidity-cooling water, ppm	No info	Not tested	5-50 (Avg 10)	No info	10-30
Turbidity-filter effluent, ppm	No info	"Very clear"	Avg 7	No info	1-3
Source of turbidity	No info	Makeup, wind, dust algae, slime	Floc from makeup clarifier, wind, dust, slime growths, glacial silt	Wind and dust	Present in makeup, dust, corrosion
Length filter cycle	Approx 36 hr	7 days	Approx 24 hr	8 hr (summers) 72 hr (winters)	24 hr
Time in service	9 mo	50 mo	38 mo	20 mo	66 mo
Replacement of media	None	None	After 26 mo (bed dirty)	None	Yes-twice 24 mo each
Reduction in cleaning or maintenance costs (annual)	75%	Approx $2600	Slight improvement, difficult to estimate	Yes (no estimate of amt)	Improved operation & reduced maintenance. Cleaning costs reduced as much as 50%
Is turbidity reduction adequate?	Yes	Yes	No	Yes	Yes
Is amount bypassed adequate?	Yes	Yes	Not always	Yes	Yes
Number of turnovers/24 hr	7.5	2.4	0.8	2.3	3.3
Time to recover— i.e., return cooling water turbidity to normal levels after sudden increase		24 hr		24 hr	20 hr

The first step in evaluating the treatment sequence is laboratory bench-scale flocculation plus separation tests where a series of chemicals are screened. Bench-scale studies will determine what chemicals, sequence of addition, blending time, flocculation time and separation rates and detention time are required. The bench-scale test will indicate what effluent quality level can be obtained. In addition, the volume of separated solids can be measured to obtain an estimate of anticipated volumes of sludge. Sludge removal characteristics and sludge dewatering characteristics can also be determined.

In addition to considerations for chemicals, the method used to develop the actual chemical floc is as important as the clarification step. Traditional methods of flocculation such as horizontal and vertical redwood paddle wheels (Figure 12-7) in round or rectangular basis have been supplemented in recent years with a number of alternate and superior methods of flocculation.

Figure 12-7. Conventional horizontal redwood paddle flocculators.

These methods are: (1) vertical turbine flocculators (adaption from center well reactor clarifier units), shown in Figure 12-8; (2) horizontal turbine flocculation (see Figure 12-9); (3) horizontal paddle oscillating flocculators; and (4) air flocculation.

The turbine type flocculator (Figure 12-10) is substantially smaller in diameter than a paddle wheel flocculator. It relies on a pumpage of flow radially outward and a recirculation of flow up to five times within the detention time of the flocculation zone. These units are particularly suited to

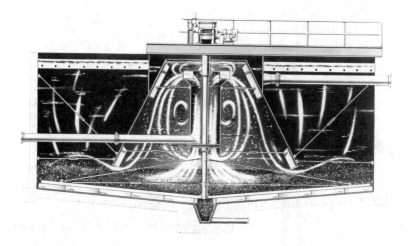

Figure 12-8. Vertical turbine flocculators in high rate reactor clarifier.

Figure 12-9. Horizontal turbine flocculators.

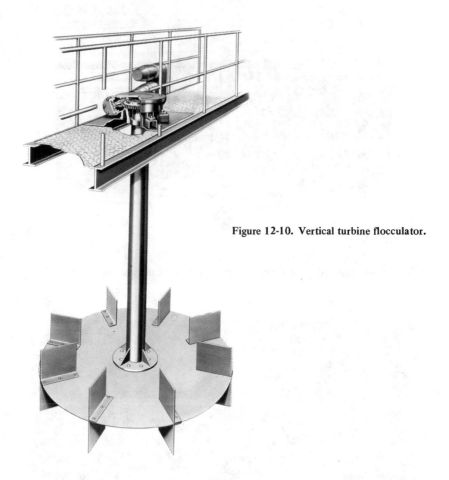

Figure 12-10. Vertical turbine flocculator.

flocculate finer particulate and near-colloidal solids at higher energy levels. Reasonable success in maintaining solids in suspension is achieved. Ferric or calcium hydroxide flocculation would also be appropriate with a turbine type flocculator.

Attempts to more accurately define required flocculation have resulted in the common use of **G** and **Gt** values. These values can be translated into specific criteria for detailed vertical or horizontal turbine flocculator configurations.

Briefly, the velocity gradient, **G**, fps/ft is defined as $\sqrt{\phi/\mu}$, where:

ϕ = dissipation function or work of shear per unit volume per unit time, ft-lb/cu ft/sec; and also expressed as P/V, power input, ft-lb/sec over V, flocculation volume under influence of the turbine flocculator.

μ = dynamic viscosity, lb-sec/ft

Thus $P = G^2 V_\mu$

G values have been defined for flocculation applications identified as color removal, turbidity removal and softening. Table 12-4 shows the range of G values to be used for each application. Note that a range is given and would be controlled by varying the speed of the turbine flocculator. Other variables available during design, such as basin volume and turbine configuration would be fixed during operation. Power input, P, would be based on the maximum G value in the range.

Table 12-4. Velocity Gradient Ranges

Process	Present Max	3 to 1 Speed Range Min	3 to 1 Speed Range Max	4 to 1 Speed Range Min	4 to 1 Speed Range Max
Color removal	15	5	26	5	40
Turbidity removal	30	10	52	10	80
Softening	45	15	78	15	120

Power input may be further defined as follows:

$$P = KN^3 D^5 W/G$$

where
K = a constant for the specific turbine configuration
N = turbine speed (rev/sec)
D = turbine diameter (ft)
W = depth of turbine blades (ft)
g = gravitational constant (ft/sec^2)

In larger applications, the G value is varied through change in turbine configuration to provide a tapered energy gradient. Thus as the floc development progresses, the energy gradient is lowered to maintain uniform suspension, but minimize floc break-up.

Of equal importance is the length of time that the gradient is applied to the developing floc. In other words, exposure to a high gradient for longer periods will produce floc shear conditions. The expression Gt develops an optimum energy input and time combination that will produce satisfactory flocculation. Comparable Gt values for earlier velocity gradient ranges are given in Table 12-5.

Table 12-5. Suggested Gt Ranges

Process	3 to 1 Speed Variation Maximum Range	4 to 1 Speed Variation Maximum Range
Color Removal	46,800 to 62,400	72,000 to 96,000
Turbidity Removal	62,400 to 124,800	96,000 to 192,000
Softening	93,600 to 163,800	144,000 to 252,000

Lighter fragile, voluminous formations, such as alum or biological floc, are better developed with the horizontal, oscillating paddle flocculation (Figure 12-11). Lower speed turbine flocculation has been used, but the slower speed vertically oscillating paddle network more uniformly distributes the energy and avoids floc breakup.

Figure 12-11. Horizontal oscillating paddle flocculator.

Air flocculation offers low energy level flocculation, often adequate in primary sewage treatment applications for addition of chemicals in aerated grit chambers and feed wells of primary clarifiers. In activated sludge plants the blower air is available and represents an economical method for flocculation when coarse solids may cause operational problems for other types of flocculators. No attempts have been made to correlate G or Gt values for air flocculation. Generally, air rates in the order of up to 2 cfm/ft of chamber length develop roll or turnover rates adequate for floc development of maintaining floc in suspension. Danger exists in air entrainment and/or attachment with the floc that would lower the settling rate.

In most cases the flocculator basin should hydraulically be an integral part of the clarifier to avoid floc breakup in transferring from one process to the next. With reasonable baffling a minimum of flocculation energy is transferred to the clarification and sludge compaction zones. This arrangement is readily apparent in circular clarifiers and can be improved at modest expense by designing steeper basin floor slopes. This change will permit a greater depth under the feed well. Substantially lower inlet velocities will exist in the initial stages of the clarification zone.

SLUDGE FILTRATION[68]

Today's rising acreage costs have made sludge disposal by lagooning uneconomical. Sludge concentration before disposal or incineration represents the logical and least expensive approach to investigate.

Concentrating or composting dilute sludges into an easily handled state usually requires a filter system. In cases where small amounts of dilute sludges (1000 to 2000 gpd) must be concentrated, simple gravity-type disposable media filters, or more expensive but compact pressure filters are commonly used. As flows increase much beyond this point, the cost of either type of filter becomes prohibitive and a vacuum filtration system is commonly employed.

In most instances however, a vacuum filter alone cannot be considered a complete solution for a complex industrial liquid-solids separation problem. Though it may be the most critical and singularly the most expensive piece of equipment, its use depends on various pretreatment steps.

Thickening

Basically, a filter is a piece of equipment which uses energy to separate solid and liquid fractions. If a large portion of the liquor can be removed before filtration, then less energy will be expended by the filter. As shown in Figure 12-12, the subsequent filtration rate will be higher because of pre-thickening, which should reduce the size and cost of filtration equipment.

In every application, it is necessary from both process and economic standpoints to concentrate the dilute sludge as much as possible before filtration. Use of an industrial clarifier or thickener is usually mandatory in concentrating slurries as dilute as a few hundred ppm to 5 to 15 percent dry wt suspended solids or higher.

A typical example of effective thickening with a resulting reduction of filter area and a subsequent increase in the filtration rate is the disposal of water-softening sludge. This waste is composed primarily of calcium carbonate with some magnesium hydroxide, and results from the softening of ground or surface waters with lime in municipal or industrial water treatment plants.

An essentially pure calcium carbonate sludge such as might be found in Florida water could not be filtered at a feed concentration less than 6 or 7 wt percent. At a concentration of approximately 10 percent, a filtration rate of 10 to 12 lb dry solids per hour sq ft results. However, after

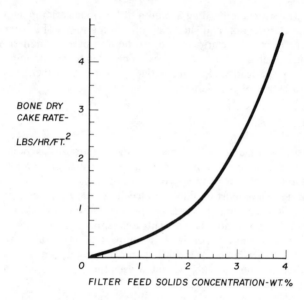

Figure 12-12. When the filtration rate is plotted as a function of the feed solids concentration, the filtering rate increases rapidly as the feed concentration is approached.

thickening the feed to 20 to 22 percent dry solids for filtration, the ultimate design rate of 50 to 60 lb dry solids hour sq ft is achieved.

Design criteria for thickener sizing vary widely for different waste problems. For example, the thickener unit area requirement for a Florida well water sludge is approximately 10 sq ft ton day for a thickener feed of 2 percent yielding an underflow concentration of 20 to 22 wt percent dry suspended solids. The thickening rate depends not only upon the nature of the material, but also on feed concentration, specific gravity of the dry solids carrying liquor, underflow concentration and the overflow clarity required for disposal.

A solids-contact clarifier or separate reaction tank and clarifier may be necessary where effluents such as plating wastes must be chemically precipitated and concentrated to a filterable consistency. This treatment step, along with oil removal, aeration, flotation, and bacteriological removal, may be required to obtain a filterable material.

Flocculation

In almost every case of industrial waste separation flocculants are used to improve the filtration rate and the cake discharge characteristics. Even though the cake may form well, as in a feed containing an excess of 50 percent + 200 U.S. mesh coal particles, the fine fraction usually contains an appreciable amount of slimes which tend to seal off the cake after the first instantaneous deposition. Flocculation to coagulate the fine particles reduces this tendency and the "blinding" effect caused by slimes.

In some instances flocculation is quite complex. For example, in a coal refuse plant flotation tails are treated with a combination of sulfuric acid, starch and a polyelectrolyte. Determination of the most economical system is based on a considerable amount of bench-scale testing, pilot plant studies, and knowledge of the particular field. A single flocculant such as ferric chloride, lime, calcium chloride, or a polyelectrolyte is often sufficient. In a few cases, such as the filtration of water-softening sludge, no flocculation is required since the sludge does not contain a significant amount of fines if it is from a ground water source.

Flocculant requirements vary greatly from one industry to another, but correspond somewhat within a particular industry. In lieu of actual testing of each flocculant, experience with flocculants and similar problems in the particular industry is the only alternative. Occasionally, the situation occurs where the feed to a filter cannot be thickened to a filterable condition. There are many cases where filtrate clarity of less than 100 ppm is mandatory, but cannot be achieved even with a conventional belt-type filter. In these cases, a precoat-type filter has found limited application.

Drum-Type Filters

Since the development of continuous filtration, the conventional vacuum drum filter has been used in a great number of applications. This is primarily due to its flexibility and capability in handling a wide range of slurries. The basic geometry of the drum filter (Figure 12-13) provides a wide range in the percentage of filter cycle time that can be individually devoted to cake formation, optional cake washing, and cake dewatering while minimizing inactive cycle time.

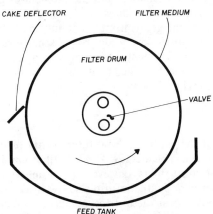

Figure 12-13. Conventional drum filter.

Simultaneously, the conventional drum filter permits discharge of thinner cake in comparison to other types such as disc, pan or horizontal filters. The drum filter principle permits a greater degree of "self regulation" than other units. This feature, when combined with the range in filter cycle time, allows latitude in productivity per unit area and greater ability to handle off-quality feeds as encountered with industrial waste slimes.

A few industrial waste products, such as coarse flotation circuit coal mill effluents containing 50 percent + 200 U.S. mesh (+74 micron) solids, can be handled by the conventional drum-type filter. However, where the feed contains greater than 5 percent - 200 U.S. mesh (- 74 micron) particles, slimes tend to blind or seal off the filter medium, yielding low filtration rates and in some instances a cake too thin for satisfactory discharge.

While the conventional drum filter has been more widely applied than other types of continuous filters, there are major limitations that either prohibit wider use of continuous filtration or reduce its efficiency and increase its operating cost. The major disadvantage is that progressive blinding of the filter medium occurs particularly where extreme fines are in suspension, as in most industrial wastes. Where slow blinding occurs, as in most existing applications, the rate may be severely reduced in several days. This requires a design filtration rate of 70 to 80 percent of the clean medium rate to achieve an economic balance between cloth life, operating time and capital cost. When blinding has progressed to a critical point, the medium must be either replaced or rejuvenated by an acid or alkaline scrubbing or rinse. Blinding generally results from one of three phenomena:

1. Plugging of the interstices within the medium suspended solids in the feed.
2. Chemical precipitation within the filter medium, or
3. Sealing off of the surface of the medium, by the shape characteristics of the feed solids.

Along with medium blinding goes the problem of poor cake discharge. Generally, a ¼-in. minimum thickness of filter cake is desirable, although a 1/8-in. cake can be discharged if it breaks cleanly from the filter medium. When thin, slimy or moist cakes occur, cake discharge can be severely hampered. Also, if multifilament weaves of cotton or other natural fibers must be used, short fibers will become imbedded in the filter cake and tend to hinder cake removal.

Belt-Type Filters

The design of the belt-type drum filter greatly reduces blinding and cake thickness problems (Figure 12-14). This filter consists of a sectionalized drum like a conventional filter; however, the periphery of each drum section contains a soft rubber or synthetic strip slightly raised from the drum surface. The filter medium lays over these strips, sealing the vacuum side from the atmosphere. Since an average filter will operate at 20 to 25 in. Hg vacuum, there is 10 to 12 lb/sq in. of pressure to provide the sealing force.

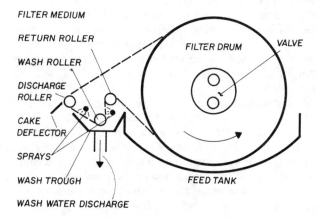

Figure 12-14. Belt-type filter.

The filter medium is an endless belt which travels off the drum to a discharge roll. When the cake is discharged, the belt travels to a wash roller located in a trough which collects the medium wash water. A washing fluid is applied by high-pressure sprays to both sides of the cloth. Occasionally a periodic rinse is required to prevent blinding.

Where the filter medium passes over a small-diameter roller from the drum, the radius curvature of the medium changes abruptly. This action breaks the cake free from the cloth. A scraper blade is not usually required to loosen the cake from the medium, but only to deflect the cake to a conveyor. Cakes as thin as 1/16 in. and less can be continually discharged satisfactorily.

In the case of organic solid, effluent standards for each product are usually determined by the state. Filtrate clarity requirements better than 100 ppm can be easily achieved.

Other Filter Types

A precoat filter is a conventional drum filter with side shims arranged to contain diatomaceous earth or perlite filter aids. After this initial filter aid cake is deposited, the feed slurry is fed to the unit. A small amount of cake residue or slime is deposited in and on the bed, but is discharged every revolution by a programed knife advance which scrapes off a few thousandths of an inch of precoat.

Filtration rates are lower because of the finer filtering bed. Resultant clarities are considerably better·than can be achieved on any other continuous mechanical separator. Whenever fine colloidal particles will not settle or cannot be flocculated, the continuous precoat filter becomes necessary.

Precoat filters have been used for clarifying slop oil, removing TNT fines from wastewater at ordinance plants, and on various laundry wastes. Soap curds, detergents, oil and grease along with a certain amount of silt can be filtered on a diatomaceous earth or perlite filter bed yielding a clear filtrate.

Because of low and variable solids concentration of the filter feed, filtration rates for precoat units are indicated in gph/sq ft. Depending upon the grade of filter aid and the effluent stream, flow rates range on the average from 2 to 50 gph/sq ft. The operating cost of these units is, in general, higher than for conventional drum- or belt-type filters. The cost of the filter aid is usually about 4 to 5 ¢/lb, and is consumed at an average rate of 10 to 15 lb of precoat/gallon of filtrate.

The horizontal top feed belt filter (Figure 12-15) permits extensive washing—even countercurrent staging—to remove objectionable solubles from the cake. This type of unit commonly incorporates a belt-washing feature to eliminate blinding. It has proven somewhat more flexible than any other filter developed, but costs more than a conventional drum- or belt-type filter. Its use has been widespread in industrial waste applications where an extreme range of particle size must be removed, or where cake washing is required.

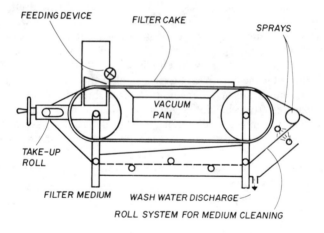

Figure 12-15. Horizontal top feed filter.

REVERSE-OSMOSIS ULTRAFILTRATION[154]

Filtration is one of the most widely used methods for separating particulate matter from fluids. However, it only removes suspended solids, leaving dissolved solids to be removed by other methods. The recent development of semipermeable membrane systems now makes it possible to capture dissolved solids, and allow only water and smaller solute molecules to pass through the membrane.

Action of these membranes can be considered similar to conventional filtration, except that the membrane "pore" sizes are about 10,000 times smaller than the finest mesh filter screens. Water flow per unit area of membrane is substantially less than for normal filters. Water flux rates of 10 to 30 gpd/sq ft are considered within the state-of-the-art for today's commercially available ultrafiltration membrane systems.

Processes using membrane filters to remove dissolved solids are broadly referred to as "ultrafiltration" methods. Hydraulic pressure is applied to the fluid being filtered to overcome the resistance to flow and force the fluid through the membrane. Osmotic pressure is a measure of the natural tendency of water to pass from a dilute into a concentrated solution, across a permeable membrane. It is proportional to the molal concentration difference. Sea water (3.5 percent salt), for example, exerts an osmotic pressure of 350 psi with respect to pure water. The osmotic pressure must be overcome if water is to be made to flow from the concentrated side to the pure water side.

For retaining proteins and other large molecules, osmotic pressures are usually negligible, while operating pressures can be in the neighborhood of 20 to 40 psig. However, when membrane systems are operated to retain small molecules, such as salts or sugars, operating pressures in the range of 500 psig are necessary to drive water across the membrane since osmotic pressure can be high. These high-pressure operations are often termed "reverse osmosis" systems (Figure 12-16).

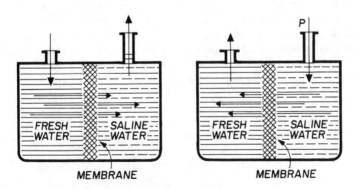

Figure 12-16. Fresh water will pass through a membrane filter into a saline solution by normal osmotic pressure. When pressure is applied to the contaminated solution, "reverse osmosis" will cause water to separate from the solution leaving behind the solid pollutants.

When considering membrane ultrafiltration methods for any process situation, two important membrane characteristics need to be established:

1. *Selectivity or rejection*—membrane ability to discriminate between smaller molecules which permeate through, and larger ones which are retained by the membrane.

2. *Flux rate*—quantity of water passing through the membrane in a unit of time (gallons per 24-hr day per sq ft at a specified pressure).

Development work done at the University of Florida and in California has led to asymmetric polymeric membranes which have good selectivity, yet exhibit flux rates well over a hundred times greater than those of any artificial membranes made previously. This made it possible to conceive and design membrane processes that could compete economically with existing conventional methods such as evaporation for separating dissolved solids from water.

Desalination

Equipment has been designed capable of producing large quantities of pure water from brackish or hard water. Reverse osmosis plants with production costs of $1 to $2 per 1000 gallons of pure water permeate are presently operating.

In desalination use, the main objective of a system is to produce pure water. If there is an adequate supply of feed water, there is little need to operate with high concentration ratios on the feed, or recover dissolved solids. However, for waste treatment applications, it is usually desirable to operate with the highest concentration ratio possible to minimize the quantity of waste material which needs to be processed further.

The more concentrated the "reject" stream, the more problems arise on the concentrate side of the membrane. Concentration polarization (limitations on water passage through a membrane imposed by a more concentrated fluid layer immediately adjacent to the membrane surface) becomes important with high concentration ratios used in solute recovery applications. Experience has shown that membrane plants designed solely for desalination can prove unsuitable when used for water pollution control treatment.

Pollution Control

However, when properly designed, the membrane process can provide an effective new approach to abating water pollution, which compares favorably in certain instances with the more conventional treatment methods. A membrane produces water of at least secondary and usually tertiary treatment quality. Permeate water from a membrane plant is free of bacteria and suspended solids and substantially reduced in dissolved solids. It can usually be reused and is often of better quality than the make-up water received from wells or from municipal supplies. The "pore" size of ultrafiltration membranes can be designed so that nearly 100 percent of the proteins and 95 percent of the sugars and salts of a supply stream can be retained. Thus the permeate from a membrane plant is free of bacteria, BOD may be 99 percent removed, and dissolved solids can be substantially reduced. Even sewage effluent can be converted into high-quality water using an ultrafiltration installation.

However, water removal in a well-designed membrane process is gentle, and little or no degradation of solids occurs. Often valuable by-products can be recovered, particularly in the case of food processing wastes. Recovery of food values by using a membrane process may pay for the operation, and reduce pollution at the same time (Table 12-6). Figure 12-17 provides a simple guide to possible by-product credit.

Table 12-6. Potential By-Product Recovery Credit Values of Liquid Waste Streams

Waste		Percent	Lb/1000 gallons	¢/Lb*	By-product value/1000 gallons	
Whey	Protein	1		5	4.00	7.20
	Sugars	4	320	1	3.20	
Soy	Protein	0.7	60	5	3.00	3.60
	Sugars	0.8	65	1	0.65	
Yeast		1.0	80	1	0.80	
Apple	Sugars	0.2-0.3	16-25	1	0.16-0.25	
Grape	Sugars	0.2-0.3	16-25	1	0.16-0.25	
Potato	Starches	0.2-0.4	16-32	0.5	0.08-0.12	
Citrus	Sugars	0.1-0.3	8-25	1	0.08-0.25	
Fish	Protein	3.5	280	2-3	5.60-8.40	

*Applies to the "wet" solids content, *e.g.*, the solids will be contained at 5 to 20 percent concentration, unrefined.

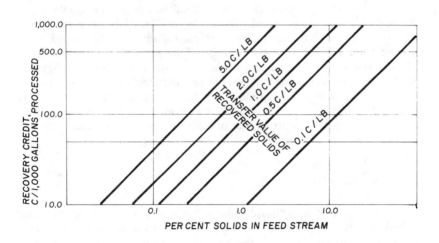

Figure 12-17. Possible by-product credit to be realized from reverse osmosis/ultrafiltration process.

Economics

Any generalized economic comparison between a membrane process and conventional treatment is difficult to develop, and only broadly indicative of the areas where membrane processes should be considered in more detail. Properly designed membrane processes are relatively unaffected by the nature and amount of a contaminant up to at least 5 percent solids, and often to considerably higher levels of dissolved solids. Unlike conventional processes depending upon biological degradation of organics in the waste, the cost of membrane plants per unit of water processed will not vary significantly with the level of BOD or other solids dissolved in the water.

Figure 12-18 provides a comparison of data for an activated sludge process providing 95 percent BOD removal with those of a membrane process. The data indicate that membrane processes become attractive at the higher BOD levels. Membrane processes should first be considered for use on streams with 9000 ppm BOD or higher. However, by-product or water reuse credits may make a membrane process more attractive even for relatively dilute waste streams.

When Are Membrane Processes Used?

Many industrial effluent streams are below the 8000 to 9000 ppm BOD level, but frequently there are individual streams within a plant which carry far more than a proportional share of pollutants. Because membrane plants are compact and are more economical on concentrated waste streams, consideration should be given to installing membrane units on selected streams within the plant. Space requirements are small. Membrane plants can handle 50 to 200 gpd/cu ft of space occupied. This is 100 to 1000 times less space than that required for most biological systems, particularly those removing better than 95 percent BOD. Operation is usually fully automatic. Membrane units can be installed indoors, in an out-of-the-way corner. No odors or off-gases are generated by membrane systems, since no chemical or biological action should take place during operation. Since no living organisms need to be maintained, there is no lengthy start-up and shut-down procedure; neither is the process particularly sensitive to heat or cold. However, membranes cannot be allowed to freeze, and there is some reduction in flux at lower operating temperatures.

Cautionary Guidelines

Because membrane ultrafiltration is a relatively new process, and because pollution control applications are varied, opportunities far outnumber limitations on the process. However, managers should realize that this process is not a cure-all to water pollution problems.

While capital investment and utility requirements for plants using membrane separation processes are substantially less than for evaporators, membrane

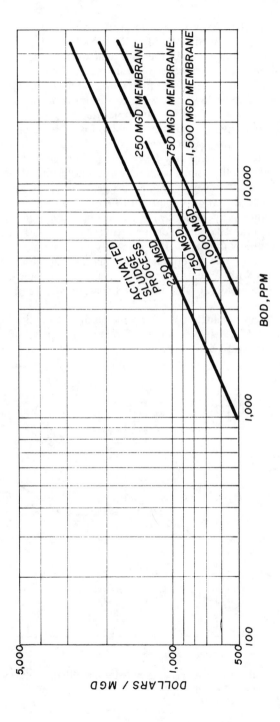

Figure 12-18. Cost comparison between activated sludge processes and membrane filtration system.

replacement costs for large plants may run 10 to 25 percent of the capital investment even for a 1-yr membrane life. Reasonable assurance that adequate membrane life can be obtained with the particular waste stream is essential prior to a commitment to construct a full-scale plant. Design of membrane plants requires careful study of several interacting variables, including operating pressure (should be as low as possible for optimum membrane life) recirculation rate, temperature, and permeation rate per pass.

Membranes are currently available which will operate at pressures up to 800-1000 psi. When applications involve strong brines or concentrated sugars higher operating pressures may be required. Though ultrafiltration in such applications is still feasible, careful design and testing of the membranes and associated equipment is required. Chemical resistance of the membrane material to the waste stream must be explored. For cellulose-ester membranes the optimum operating conditions are between 3 and 5 pH.

In membrane processes used for the recovery of nutrients, the possibility of biological growth exists. This is usually detrimental to both the process and the concentrated by-product, and may also interfere with membrane performance. With systems properly designed for periodic cleaning and the use of a bactericide, trouble from biological growth can be minimized.

As a guide to approximate costs for a membrane plant filtering dissolved solids, Figure 12-19 shows capital and operating expenditures based on a flux rate of gpd/sq ft, and a membrane life of one to two years. The effect of both membrane life and flux rate on capital and operating costs is shown in Table 12-7.

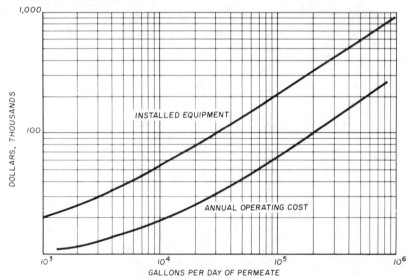

Figure 12-19. Estimated capital and operating costs for ultrafiltration plant with flux rate of 10 gal/day/sq ft and membrane life of 1 to 2 yr.

Table 12-7. Effect of Flux Rate and Membrane Life on
Membrane Process Operating Costs*

1. Membrane replacement (at $1.50/sq ft replacement cost)

Flux rate	$/1000 gallons		
	5 gfd	10 gfd	20 gfd
6 month life	1.64	0.82	0.40
1 year life	0.82	0.41	0.20
2 year life	0.41	0.20	0.10

2. Capital charges (at 25 percent of equipment cost per year)

Flux rate	$/1000 gallons		
	5 gfd	10 gfd	20 gfd
50,000 gpd plant	1.39	1.07	0.86
500,000 gpd plant	0.64	0.48	0.38

3. Pumping power (at 1¢/kwh, and including recirculation pumping power)

Operating pressure	$/1000 gallons		
	50 psi	250 psi	1500 psi
	0.24	0.26	0.38

*Figures represent typical values for industrial membrane processes, and illustrate order-of-magnitude effects of membrane life and flux rate on processing cost (gfd = gal permeate per sq ft of active membrane surface per 24-hr day).

TUBE SETTLERS FOR SEDIMENTATION[180]

Development of the tube settler concept has made a major break-through in shallow depth sedimentation in treating water and wastewater. Shallow tubes meet the basic requirements for ideal settling—shallow depth, a large wetted perimeter, laminar flow conditions and reasonable overflow rates. A 1-in.-diameter tube, 4 ft long, at a flow rate of 10 gpm per sq ft would provide a Reynolds number of 24, an equivalent surface overflow rate of 235 gpd per sq ft, while having a detention time of only 3 min.

There are two basic tube settling systems: essentially horizontal and steeply inclined (Figure 12-20). Operation of the essentially horizontal tube settlers is coordinated with that of the filter following the tube settler (Figure 12-21). Each time the filter backwashes, the tube settler is completely drained. The falling water surface scours the sludge deposit from the tubes and carries it to waste. Water drained from the tubes is replaced with the last portion of the filter backwash water. The tubes are inclined slightly in the direction of the flow (5 to 7 degrees) to promote the drainage of sludge during the backwash cycle. No mechanical sludge removal equipment is required.

It has been observed that sediment in tubes inclined at angles in excess of 45 degrees would not accumulate but would slide down the tube. The

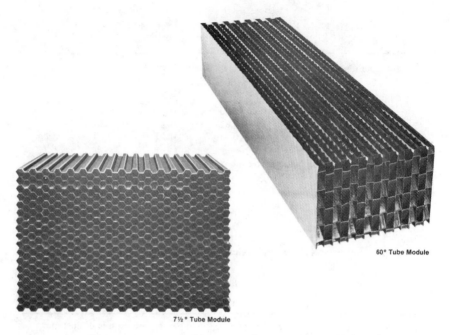

60° Tube Module

7½° Tube Module

Figure 12-20. Settling tube modules
of Nepture MicroFloc Inc.

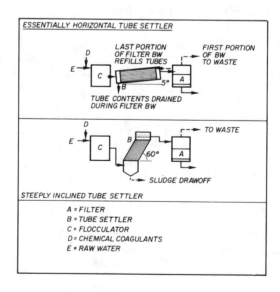

Figure 12-21. Basic tube settler designs.

continuous sludge removal achieved in these steeply inclined tubes eliminate the need for drainage or backwashing of the tubes for sludge removal. An angle of 60 degrees provides continuous sludge removal, though there is some lowering of efficiency.

Tube settlers can be rectangular, circular or any other shape. Circular tubes may not be quite as efficient as rectangular types, since particles entering at the top of the tube have a greater distance to settle than those entering at the sides. Circular tubes also have a void space between them, in a settler array. The chevron design maximizes the settling characteristics (Figure 12-22). The tube settler module is an array of nested tubes 24 in. long with a cross sectional chevron shape. It has the largest surface perimeter of any common shape of the same area. Its V-grooves also promote better sludge flow.

Large plastic module tube settlers are available that can be installed in either circular or rectangular clarifiers. These steeply inclined tubular modules have been used in raw water clarification and in primary and secondary sewage clarification. The tube settlers are capable of being operated at rates of 3-4 gpm/sq ft compared to about 1-1.5 gpm/sq ft in conventional settling basins.

Advantages

The use of tube settlers has been demonstrated successfully on full-scale plant operations. They offer the following advantages:

1. Shallow depths provided by the tubes permit better clarification at detention times less than those used in conventional settling basins.
2. Existing clarifier capacity can be augmented two- or four-fold by installing modules of steeply inclined tubes in the existing clarifier structure.
3. Economics in space and money are significant,
4. Tubes can be installed as an integral part of an aeration basin eliminating separate sludge separation and return systems, and
5. No mechanical sludge removal equipment is required.

Limitations

There has been slow progress in utilizing this technology. More study is indicated in:

1. Inclined tube settlers of different shapes,
2. An inclined tube settler system having a simple alternate sludge removal arrangement, without a sacrifice in system efficiency,
3. The floc build-up and consequent cleaning of the settlers especially in wastewater applications,
4. The application of tube settlers for algae-laden waters, and
5. Tube settlers tested for more industrial process effluents.

Figure 12-22. Chevron tube system of Permutit Co.

The technology has been refined only in the past three to four years and more plant-scale experience would be required before definitive design criteria can be evolved for tube settlers in primary and secondary wastewater clarification.

DIATOMITE FILTER AIDS[113]

Diatomite is a chalky, sedimentary rock composed of the skeletal remains of single-celled aquatic water plants called diatoms. These plants have been part of the earth's ecology since prehistoric times. As living plants, diatoms are basic to the oceanic cycle. They are food for minute animal life, which serves as food for the higher forms of marine life. Literally, the diatom is the "grass of the sea."

The complete diatom consists of the living cell encased in two half-cell walls or valves united by a connecting band. These cell walls are composed of opaline silica, extracted from the water by a mechanism which still remains obscure. Thus the microscopic plants live within a transparent glass-like case.

Typically, the cell walls have lacework patterns of chambers and partitions, plates and apertures in a selection of shapes as varied as snowflakes. The total thickness of the cell wall is only a few ten thousandths of an inch: the internal structure is highly porous on a microscopic scale.

Extreme care is used in the production of diatomite powders to preserve the natural properties of diatomaceous silica. Most diatomite powders have an exceptionally high silica content—as much as 94 percent. Because it is essentially silica, the powder is inert to most chemical reactions and is resistant to extremely high temperatures with a softening point of about 2600 F.

The remarkable physical structure of the individual diatom skeleton is the characteristic upon which almost all of the applications of diatomite functional fillers are based. Whole diatom skeletons and their fragments are characterized by very irregular shapes, spiny structure and pitted surfaces. These rigid, strong particles range from about 5 to 50 microns in diameter.

Loose diatom skeletons in a mass do not pack together. Contact with each other is limited largely to the outer points of each particle involved, like a mass of loose thistledown. This gives great bulking value or low apparent density to diatomite powders—about 10 lb per cu ft.

It has been determined that 93 percent of the apparent volume of diatomite powders consists of a myriad of tiny interconnected pores or voids. This explains its very high absorptive capacity—about 2½ times its own weight. However, in spite of their exceptionally high absorptive capacity, diatomite mineral fillers do not absorb any appreciable amount of moisture from the air.

Diatomite has a high surface area. Only 210 grams of diatomite (less than one-half pound) has a surface area equal to the area of a football field (45,000 sq ft).

How Diatomite Filters Work

Diatomite filtration is a two-step operation (Figure 12-23). First, a thin protective layer of filter aid (the precoat) is built up on the filter septum by recirculating a diatomite slurry (Figure 12-24). After precoating, small amounts of filter aid (body feed) are regularly added to the liquid to be filtered. As filtering progresses, the filter aid mixed with the turbid liquid is deposited on the precoat. Thus, a new filtering surface is continuously formed, the minute diatomite particles providing countless microscopic channels which entrap suspended impurities but allow clear liquid to pass through.

Continuous addition of filter aid is done by feeding filter aid slurry or by dry feeding. Slurry feeding is usually done with plunger or diaphragm pumps. If filtration is a batch process the filter aid can be added directly to the batch. Dry addition can also be done by feeding into a small vessel into which unfiltered liquid is continually being fed. The liquid, with filter aid added, is then educted with the help of a booster pump into the filter feedline. Changes in rate of body feed addition are made by changing the feed from the dry feeder.

The amount of precoat should be 10 to 15 lb of filter aid per 100 sq ft of filter area, the greater amount being used when distribution of flow in the filter is poor, or in starting up new filters. If it is perfectly distributed, 10 lb of filter aid per 100 sq ft of filter area will give a precoat approximately 1/16 in. thick. If distribution is imperfect with 10 lb of filter aid per 100 sq ft, the precoat at the top of the leaves or at the far end of the filter from the precoat slurry entrance may be inadequate. This condition can frequently be remedied by use of baffles.

Precoat slurry concentration will depend primarily on the ratio of filter area to filter and piping volume. The slurries will, however, range from about 0.3 to 0.6 percent. If they are much below 0.3 percent, precoating may be difficult since the formulation of the bridge depends partly on the "crowding" effect of the particles trying to get through the septum openings.

Selection of the proper filter aid is a compromise between high clarity and low flow rate, and low clarity with high flow rate. The best filter aid is that grade which will give the fastest flow rate (or greatest throughput per dollar's worth of filter aid) and yet provide adequate clarity. The correct clarity must be determined and specified by the filter aid user.

Diatomite Filter Types

Filter aids can be classified in three main groups depending on the method of manufacture.

Natural Grade

A natural grade filter aid is selectively mined, crushed, dried and air classified to provide a uniform product. It is not calcined nor does it receive

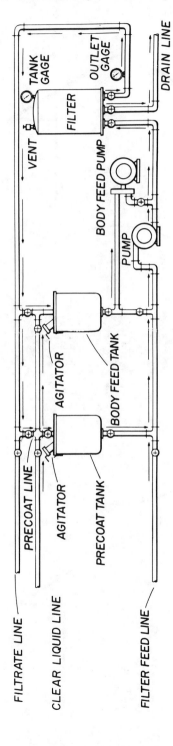

Figure 12-23. Typical diatomite filtration system.

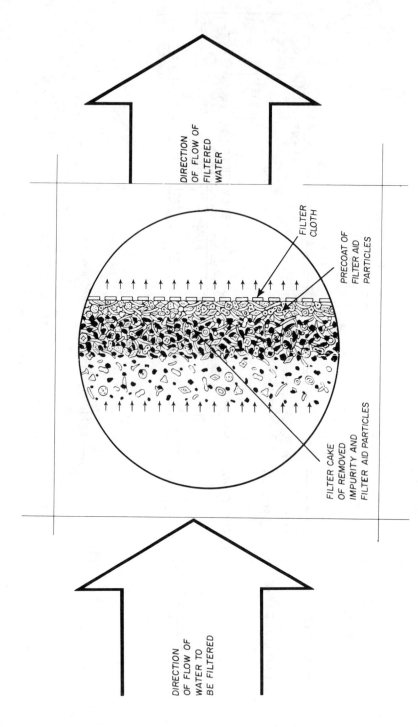

Figure 12-24. The building process of diatomite filtration.

any chemical treatment during its processing. It is the finest grade, giving the highest clarity and the lowest flow rate. Typical filtration applications for this grade are:

1. Process and service water—electronics.
2. Process water—brewing and related industry.
3. High quality—purity for all applications.

Calcined Grades

The calcined grade is produced by the calcination and air classification of the natural grade. After calcining at approximately 1800 F, it is air classified to different particle size gradations. Typical filtrations involving calcined materials include:

1. Quality water production
2. Polishing process water

Flux Calcined Grades

These are produced by adding a fluxing agent to the material prior to calcination. This treatment produces a larger particle and, consequently, a more permeable material. Again, air classification produces products of varying particle size gradations. Typical applications for the flux-calcined grades are:

1. Potable water— turbidity removal
   iron and manganese removal
   lime softening removal
2. Wastewater — tertiary sludge treatment
   deep well disposal
   board mill water reuse.

TRICKLING FILTERS[148]

The trickling filter is a sophisticated duplication of nature's purification processes. It achieves the same cleansing effect produced by natural organisms adhering to the rocky bottoms of waterways. However, its operation is faster, more efficient and compact than a flowing stream.

The process is not so much one of filtration, but rather a method by which contact between liquid waste and the active growth of microorganisms is promoted. The organisms are cultivated on a surface over which contaminated water trickles. The soluble organic pollutants are consumed by the organisms, and converted to CO_2 and water. Thus the process may be described more aptly as attached-growth bio-oxidation. Because of its inherent stability, the trickling filter is not easily upset by shock loads or sudden variations in the quantity of incoming waste.

In early days, the filter medium consisted of rock, tile, coal or slag. Eventually, the effectiveness of rock as a filter medium was investigated. Problems such as plugging, resistance to air flow, low loadings and excessive construction costs have led to the development of preformed materials to replace the rock. In general, these materials are corrugated plastic sheets, welded or bonded together in predetermined configuration. Polystyrene and polyvinyl chloride (PVC) are used in alternately flat and vacuum-formed corrugated sheets.

Each of these products maintains the advantages of the rock medium while overcoming many of its deficiencies. Plastic media possess a greater surface area to volume ratio (approximately 95 percent void space), permitting better airflow. In heavy industrial waste situations, the plastic media can be used to treat high-temperature wastes which exhibit higher oxygen demands than wastes at ambient temperatures. Because of the more uniform shape, the plastic media also allow better distribution of liquid than did the rock filters. Preformed plastic media have eliminated ponding problems which resulted when rock filters were subjected to overloading.

Plastic media units are also used for small treatment plants. In these systems the filter is fed from the bottom, thereby significantly dampening shock loads. This treatment method is actually a hybrid of the trickling filter and the completely mixed activated sludge system, maintaining the advantages inherent in both.

The value of the small plant plastic media trickling filter system is shown in Table 12-8. The Kingsport data show how plastic media reacted

Table 12-8. Effect of Plastic Media to Varying BOD and Temperature

KINGSPORT, TENN.

Day	Influent	Effluent	Percent Efficiency
Monday	561	6	98.9
Tuesday	1235	13	98.9
Wednesday	1064	25	97.9
Thursday	1473	13	99.1
Friday	998	19	98.1
Average	1066	15	98.6

ESCANABA, MICHIGAN

Date	Influent	Effluent	Percent BOD Removal
November	Average	13	93.5
July	BOD	8	96.0
July	200	16	92.4
July	mg/l	15	92.6
July		17	91.7
February		9	95.8
April		13	93.5
May		14	93.2

to varying loads; the Escanaba installation reveals the media's performance
under extreme changes in temperature. As can be seen from the chart, the
Kingsport design for 90 percent BOD removal achieved an efficiency peak
of 99.1 percent, and an average of 98.6 percent. The Escanaba plant treats
domestic sewage. Situated in northern Michigan, this operation must function under extreme variations in temperature; during the winter months,
the receiving stream is usually covered with ice. The system functions well,
illustrating a marked improvement over rock filters, whose efficiency drops
significantly during the winter months.

CHAPTER 13

AERATION[197]

Aeration and other gas transfer operations serve a multitude of purposes in water and wastewater treatment. Aeration occupies a significant place in water quality management and is an important factor in the purification of polluted water. Gas transfer is a physical phenomenon in which gas molecules are exchanged between a liquid and a gas at a gas-liquid interface. It may also be accompanied by biological, biochemical, biophysical and chemical action. These results are often the primary purpose of the gas transfer operation and methods of achieving the desired results may vary. The principal objectives of aeration, however, are usually to add or remove gases or volatile substances from water or to carry out both objectives simultaneously. In the biological process aerators function to transfer the required oxygen and include sufficient mixing to maintain uniformly dispersed oxygen throughout the basin, keeping biological solids in suspension in aerobic basins and the activated sludge process. For high-rate organic loadings, the power required may be determined by oxygen transfer requirements rather than mixing.

Aeration aside from agitation purposes may have the following functions:

1. Addition of oxygen in natural or wastewater treatment and disposal for promotion of biochemical and chemical processes,
2. Addition of oxygen to ground waters to oxidize dissolved iron and manganese,
3. Removal of carbon dioxide, and reduction of corrosion and interference with lime-soda softening,
4. Removal of hydrogen sulfide, odors, tastes; decrease metal corrosion and concrete and cement deterioration,
5. Lessen interference with chlorination.
6. Methane removal, and
7. Removal of volatile oils and odor- and taste-producing substances produced by algae and other microorganisms.

465

AERATORS

The pollution engineer can choose from a wide range of equipment for effective aeration of wastewater. The common classes of equipment most often considered are: diffused aeration systems; submerged turbine aerators; high- and low-speed surface aerators. In addition gravity aerators and spray aerators are available.

Diffused air was the earliest aeration system available, and it is still in use in many plants today. The use of surface and submerged turbine aerators has markedly increased in recent years. The surface aerator is the most recent of mechanical aeration devices and has become increasingly popular in newer system designs. Submerged turbine aerators have found application in many plants where relatively short detention times have been required, and where land is at a premium. These advantages are usually traded for higher horsepower requirements. Submerged turbine aeration systems can also be employed to increase oxygen input where existing diffused air systems have reached the limit of their capacity. Combined systems which use a surface aerator plus a submerged turbine aerator in the same unit are also available. Table 13-1 compares these basic types of aeration for wastewater treatment systems. The areas of application for mechanical aerators are in aerated lagoons, activated sludge processes, aerobic digestion, chemical oxidation, mixing and equalization, lake and stream aeration.

Oxygen Transfer and Aeration

The oxygen transfer process is generally considered to occur in three phases. Oxygen molecules are initially brought to a liquid surface, resulting in a saturation or equilibrium condition at the interface—the rate being very rapid. The liquid interface is of finite thickness, estimated at three molecules thick, composed of water molecules facing the gas phase. In the second phase the oxygen molecules pass through this film by diffusion. In the third phase the oxygen is mixed in the water body by diffusion and convection. As turbulence is increased, the surface film is disrupted and renewal of film or new surface is responsible for increasing oxygen transfer to the liquid body.

Common gas transfer relations therefore support the goals of aerator design and use. Oxygen transfer into water is systemized by: (1) Generating the largest practicable area of interface between a given liquid volume and air; (2) Preventing build-up of thick interfacial films or by breaking them down to keep the transfer coefficient high; (3) Having as long as possible exposure time and maintaining the highest possible driving force or concentration difference for adsorption and desorption. The above, therefore, become the manageable variables. It follows then that the total mass transfer is a function of the eddy exposure time and gas-liquid molecule exchange during that time interval.

Table 13-1. Mechanical Aeration Systems Comparison

System	Oxygen Transfer	Solids Suspension	Advantages	Limitations
Diffused Aeration	Transfer depends on bubble size; fine bubble diffusers higher than large bubble diffusers.	Basin design requires special attention with diffuser location critical for good suspension. Not good for use in deep basins.	Quiet operation. Flexibility through variable gas rates, enabling tailoring of oxygen transfer to system loads.	Fine bubble diffusers are subject to plugging problems, thus often requiring air filtration. Diffused air systems require properly-designed long, narrow basins which may increase construction costs.
Submerged Turbine Aerator	Intermediate efficiency below surface but higher than diffused air.	Can handle very high solids concentrations in deep tanks of 20 ft or more water level. Full hp is applied near basin bottom.	High degree of flexibility in oxygen transfer is available through speed changes and a variable gas rate. Solid suspension in deep basins is good. Not limited by area available. Can be used in high-rate systems or for basins designed for maximum use of premium land.	Higher installed hp due to lower oxygen transfer efficiencies. Submerged piping cannot be installed on floating platforms for shallow lagoon operation.
Surface Aerators	Generally highest of systems described.	High flows produce good suspension of biological solids. Lower impellers or combined units using submerged aeration design techniques are required for deep basins.	High oxygen transfer efficiency. Elimination of submerged piping. Can be float-mounted for lagoon systems. Requires little or no standby equipment for multiunit installations. Submergence adjustment allows changes in oxygen transfer to suit varying loads in the system.	Requires sufficient area for proper aeration. High-rate systems applied in deep basins may have insufficient surface area to utilize full power.
Combined Surface & Submerged Turbine Aerators.	Generally design dependent on split between surface aerator and submerged turbine.	Generally good and adjustable for use in deep basins through proper design.	Aerator speed and gas rate changes are flexible. Advantage of turbine features for good solids suspension in deep basins. Partial oxygen transfer can be maintained with blowers shut down.	Submerged piping required. Combined aeration generally requires a higher installed hp.

Under aeration conditions at high turbulence levels the bubble surface film is constantly disrupted and the renewal of this film is responsible for the oxygen transfer to the liquid body. It follows that the total mass transfer is a function of the eddy exposure time and the molecular exchange during the time interval considered.

Most mass transfer applications in waste treatment are liquid-film controlled. $K_L a$ is an overall transfer coefficient and includes the effects of changes in the liquid film coefficient K_L and varies in the interfacial area **A**. In most aeration applications it is not possible to measure the interfacial area, the overall coefficient $K_L a$ is employed to characterize aeration performance.

The following equation describes the oxygen transfer process:

$$\frac{dc}{dt} = K_L a (C_S - C_L)$$

where
$K_L a$ = overall oxygen transfer coefficient
C_s = oxygen saturation concentration
C_L = initial oxygen concentration
K_L = oxygen diffusion coefficient at the air-water interface
a = A/V, where A is the exposed air-water interface and V is the volume of aerated liquid

Once the equation is integrated:

$$K_L a = 2.3 \log [(C_S - C_1)/(C_S - C_2)]$$

where C_1 and C_2 are the oxygen concentrations at times t_1 and t_2. This equation will plot as a straight line on semilog paper, the slope being $K_L a$. This value can be determined in the field or laboratory by deoxygenating the test liquid, aerating and measuring the oxygen deficit at various time intervals.

In an actual aeration system the absorption rate will be given by the steady-state version of the above equation:

$$N = K_L a (C_S - C) \frac{V}{10^3}$$

where:
N = absorption rate (kg/hr)
$K_L a$ = overall mass transfer coefficient (hr^{-1})
V = total volume (m^3)
C_S = saturation oxygen concentration (mg/l)
C = bulk oxygen concentration (mg/l)

The above equations apply in clean water; however, some constituents of sewage change the oxygen transfer coefficient $K_L a$. Soap, for example, changes the surface tension of water and decreases the rate of oxygen transfer. It is also possible for some substances to enhance oxygen transfer, but this characteristic can only be determined by testing. Thus, in two identical aeration systems the oxygen transfer coefficient differs between clean water and sewage, the ratio of $K_L a$ into sewage to $K_L a$ into clean water is known as the α (alpha) factor. Alpha can vary considerably, but it is commonly between 0.85 and 1.15.

Similarly, the oxygen saturation of a waste may differ from that of pure water. In the case of domestic sewage, the saturation value does not usually differ much from clean water and is rarely less than 90 to 95% of that of clean water. The ratio of saturation value of the waste to that of clean water is usually known as the β (beta) factor. The transfer of oxygen into a waste can be modified as follows:

$$N = \alpha K_L a(\beta C_S - C) \frac{V}{10^3}$$

where:
$K_L a$ = clean water transfer coefficient
C_S = clean water saturation value

Aerator Testing

A number of procedures are available for testing aerators to determine process capabilities of equipment under consideration. As a key step in biological processing of wastewater, good quantitative information is desirable rather than a rule-of-thumb approach. Tests are performed by most aerator manufacturers.

Since there is no standard aeration test established by a recognized authority, there are many procedural variations in methods. The most widely accepted performance evaluation is the non-steady-state reaeration test, which is simple and straightforward. In this procedure a volume of tap water is brought to a state of zero dissolved oxygen by the addition of properly catalyzed sodium sulfite. During ensuing aeration, excess sulfite is oxidized and the liquid bulk increases in dissolved oxygen content over a period of time, until the saturation point is reached. On small-scale tests, bottled inert gas such as nitrogen or helium may be used to remove the dissolved oxygen. This latter procedure is, however, not practical for large-scale testing.

Dissolved oxygen concentrations may be determined at frequent intervals using a galvanic cell oxygen probe. Using this type test procedure it is possible to plot oxygen deficit against time and determine a $K_L a$ value. Knowing the $K_L a$ value, the quantity of oxygen transferred and the coefficient can be computed. Plotting oxygen transfer efficiency against DO (dissolved oxygen), the percent transfer efficiency can be determined for any given DO concentration in a system. Necessary data can be obtained from such tests to show efficiency of particular equipment with respect to oxygen transfer. Tests comparing different aeration devices should also take into account variations in physical as well as chemical conditions. Physical conditions include tank or basin sizes, shapes and surface area.

Mechanics of Operation

As described, aeration equipment commonly employed in the wastewater field consists of air diffusion units, turbine aeration systems in which air is released below the rotating blades of an impeller, and surface aeration units in which oxygen transfer is accomplished by high surface turbulence and liquid sprays. Two principal differences exist in the aforementioned systems. Diffused air and submerged turbine devices accomplish oxygen transfer by bringing quantities of air into contact with the liquid. In other words, the air is the transported or principal phase. In the case of surface aerators, wastewater is the transported or principal phase brought in contact with the air. Various submerged turbine devices operate utilizing both air and water transport in varying degrees of importance to achieve design goals in oxygen transfer.

The mixing requirement is important, since it may influence equipment selection. In the case of diffusers, mixing is accomplished by air rising through the wastewater, the degree of mixing being determined by the gas rate and bubble size. In the case of turbine aerators, mixing is achieved as a direct consequence of water movement from basin bottom to the surface. Proper mixing of basin contents is important for distributing oxygenated liquid and to bring oxygen-deficient waste to the aerator. Good mixing is needed to keep biological life in suspension and intimate contact with dissolved oxygen and biodegradable organic material.

Diffused Air

Air diffusers or injection aerators bubble compressed air into water through orifices, nozzles in air piping, diffuser plates or tubes, or spargers. Diffused aeration equipment can be classed in two general types depending on bubble size generated. Large bubble devices have the advantage of low maintenance over fine bubble devices. A lower adsorption and oxygen transfer efficiency results from large gas bubbles.

Fine bubble devices are generally fabricated of porous media such as carborundum, nylon or tightly wrapped saran. The principal problem encountered is plugging, which may require high maintenance to keep units operative. This problem is overcome by filtering or cleaning the input air. The main advantage over large bubble devices is that greater absorption is obtained due to the increased interfacial area of the relatively small bubbles. Variables affecting performance of diffused aeration units are air flow rate, liquid depth and tank width. Types of air diffusers include simple open pipes for coarse air bubbles to more efficient oxygen transfer baffled devices and porous ceramic tubes and domes for fine air bubbles. Diffusers are located at basin bottoms and spacing depends on type and aeration level required. Greater oxygen transfer can be achieved by locating the air diffuser at a greater depth below the water surface. Optimum balance between oxygen transfer and mixing is usually achieved at an 8 to 16 ft diffuser depth.

Efficiencies indicated in manufacturers' literature show most diffused air systems in the lower range of mechanical aeration devices tested. Fine bubble devices are usually higher than large bubble diffusers, and are competitive with the low end of submerged turbine aeration. Ascending bubbles in wastewater acquire smaller terminal velocities than would drops falling freely in air through the same distance. This increases exposure time. Spiral and cross-current flow lengthens the travel path in the liquid.

Air diffusion is employed in water as well as wastewater treatment. The best known application of wastewater treatment is in the activated sludge process. Floating compressors are also used in raising oxygen content of receiving waters overloaded with waste material that might become septic and destroy water stratification.

Submerged Turbine

Mechanical aerators of this type are widely employed in the wastewater treatment. The design objectives of a good submerged turbine aeration system are to provide, by mechanical and fluid action, sufficient shear to create a fine bubble distribution and maximize the air retention in the system. Here, the mechanical function is important to keep the activated floc in mobile, useful suspension as well as injection aeration for oxygen gas transfer.

While there may be a number of systems that can achieve these goals, the usual arrangement consists of a radial flow impeller located above an orifice sparge ring or an open air pipe. Air rising from the pipe is dispersed by the impeller and distributed throughout the liquid. Balance between air and impeller flow is important. In cases where air rates are too high, the gas may overcome the pumping action of the impeller, and this action results in oxygen transfer efficiency loss. Lower air rates yield good dispersion in a wide range of gas rate, while independently maintaining adequate mixing. Submerged turbines are fixed unit devices.

Submerged turbine aeration devices fall in an intermediate range for gas transfer efficiencies. Oxygen transfer can be varied independently of the mixing and is a decided advantage for these devices, particularly where wide loadings are experienced. The oxygen transfer efficiency of a single-impeller submerged turbine is in the range of 1.5 to 2.0 lb oxygen/hp-hr; a dual-impeller turbine ranges from 2.5 to 3.0 lb oxygen/hp-hr oxygen transfer efficiency.

Turbine aerators can also be installed to augment existing diffused air systems. This offers plants an opportunity to increase oxygen transfer capability at a minimum additional capital investment. Additionally, icing problems associated with surface aerator operation in cold weather do not exist with turbine aerators.

Surface Aeration

During the past several years surface aerators have found increasing application in activated sludge plants and aerated lagoons for wastewater treatment. The surface aerator is a device which brings to the surface the waste in water for contact with air.

A number of designs of surface aerators are in use. The bladed or paddle-surface aerator pumps liquid from beneath the blades and sprays the liquid across the water surface. The brush aerator utilizes a rotating steel brush which sprays liquid from rotating blades with mixing achieved by an induced velocity below the rotating element. A draft tube is employed in some designs. Surface aerators are usually float mounted. These designs can be broken into two classes, namely as to whether the pumping device operates at the liquid surface or substantially below the surface.

The oxygen transfer occurs directly to the waste while being sprayed through the air. Additionally, oxygen transfer takes place with entrained air at the impeller and in the area around the aerator resulting from splashing liquid impinging into the liquid body. Surface aerators generally provide higher efficiencies than other devices; oxygenation efficiencies of low-speed surface aerators range from 3.0 to 3.5 lb oxygen/hp-hr. These units can be controlled for varying oxygen demand requirements by making use of submergence adjustment, cycle timers, and speed control with variable speed motors. High- or motor-speed surface aerators are essentially axial flow pumps. Pumping action occurs using a marine-type propeller on the end of the motor shaft. As the propeller rotates, water drawn up through the draft tube is discharged at a high rate against deflector plates, producing horizontal liquid sprays. In most cases, adequate mixing occurs. Additional improvement such as lower impellers in the case of deep basins may be implemented.

Combined units where a surface aerator is placed on the same shaft with a submerged turbine device have been used. These units are also adaptable to existing diffused air systems for increased oxygen transfer. Efficiencies will vary depending on the designed combination.

DESIGNING A MECHANICAL AERATION SYSTEM[56]

Traditional waste treatment systems, relying on surface aeration, are becoming increasingly outmoded. Purification standards are demanding longer retention times. In urban areas with a high concentration of industry producing organically dirty water, there is often not enough land available for surface aeration ponds. Some form of mechanically induced aeration, therefore, is going to be mandatory in most wastewater treatment processes.

Mechanical surface aeration is an efficient and inexpensive method. As with all aeration systems, its purpose is to increase the water surface area available for oxygen transfer. It has a second equally important purpose—that of mixing the basin contents thoroughly. Mixing keeps the sludge in close contact with the waste material and keeps the oxygen thoroughly dispersed throughout the basin.

In the basic system, clarified wastewater is aerated. Sludge flocculates and settles in a secondary clarifier. The effluent is clear and low in organics.

In the step aeration system, settled wastewater is distributed to different portions of the basin to spread oxygen demand. Aerated water passes to a secondary clarifier to allow flocculation and settling. Sludge is returned to the head of the basin. This system can be combined with one of the other three. In the contact stabilization process, raw wastes are mixed with aerated sludge and then treated by aeration. This is perhaps the most efficient method employing mechanical aeration in terms of space requirements. The extended aeration system handles raw wastes without primary settling at a cost in space requirements. In this system, physical mixing is very important because it evens out variations in load and dilutes concentrations of impurities. Detention time in the aeration tank is usually about 24 hr.

System Design

In all systems, there must be a balance between basin design and aerator performance. New systems wherein basin geometry and aerator selection are matched are most efficient. Older systems can be converted, however, by sizing and placing aerators properly. While it may not match a new system in efficiency, a converted system can be economically feasible.

Aerator selection is perhaps the most important step in arriving at a new system design. Sizes commercially available are shown in Table 13-2, along with important performance data. When selecting aerators, consider the following steps in the design process:

1. Determine the desired retention time in the system.

Table 13-2. Aerator Size and Performance Data

Size (hp)	Transfer Rate (lb/hp/hr)	Zone of Complete Mix (ft)	Zone of Complete Oxygen Dispersion (ft)	Pumping Rate Through Unit (qpm)	Shaft Dia. (in.)	Mooring Cable Dia. (in.)
5	3.8	45	150	3390	1.250	1/8
7.5	3.6	50	160	3780	1.250	1/8
10	3.4	51	142	5060	1.750	1/8
15	3.5	62	200	6140	1.750	1/8
20	3.2	72	230	8320	2.125	1/8
25	3.4	80	255	9830	2.125	1/8
30	3.5	88	280	12570	2.125	1/8
40	3.8	102	325	14000	2.500	1/8
50	3.5	105	330	18560	2.500	3/16
60	3.5	115	350	20560	2.500	3/16
75	3.0	130	380	22550	2.500	3/16
100	3.1	150	440	41000	3.375	1/4
125	3.3	165	490	47500	3.375	1/4
150	3.2	185	530	57000	3.375	1/4

2. Determine the flow rate (mgd) for the system.

3. Calculate the volume in cu ft required in the aeration basin.

4. Referring to Figure 13-1 determine the horsepower required for mixing in the system under consideration. If oxygen mixing alone is required (aerated lagoon system only), use the left-hand band; if solids must be suspended, use the right-hand band.

5. Referring to Figure 13-2 determine the horsepower required for the necessary oxygen transfer.

6. From Table 13-2 select the number and size of aerators required. Use the higher hp figure obtained from steps 4 and 5 above in selecting the aerator pattern. Consider also the mixing dispersion zones for the aerators selected.

If it is important to avoid solid suspension, as in the aerated lagoon, and if oxygen transfer requires the higher horsepower (> 20 hp/10^6 gal), the basin size should be increased. Work backward through Figure 13-1 to determine the size of basin needed. The increased size will add to the retention time within the basin and will probably result in a slightly greater BOD_5 removal.

It is wise to investigate several patterns using various sizes of aerator to arrive at a suitable basin depth.

If the system is subject to cyclic loading, smaller dual-speed **aerators** should be considered. This will allow mixing to be maintained while governing the power consumption of the system. Control of speed can be automatic via dissolved oxygen probe signal inputs, timeclock, or manual. Power consumption can be made to range from full load to approximately 60 percent of full load.

7. Settle on a placement pattern for the aerators. The optimum pattern for a single unit is circular; however, construction economics usually rule out circular tanks in larger sizes. The next best compromise is a square. A rectangular basin is best handled by dividing the overall basin into squares, or rectangles that approach a square configuration.

8. Determine the approximate size of the basin by balancing aerator patterns against depth. Figure 13-3 provides proper operating depths for each aerator size. The basin depth should be such that neither draft tube or anti-erosion assembly is needed.

Choose a pattern that washes slightly the sides of the basin approximated. The above calculations provide "ballpark" dimensions to arrive at a ratio **R** between length and width. Because the basin is normally constructed with sloping sides, the final surface dimensions arrived at below will be larger than those approximated.

9. Select an appropriate side slope **S** (2/1, 3/1, etc.).

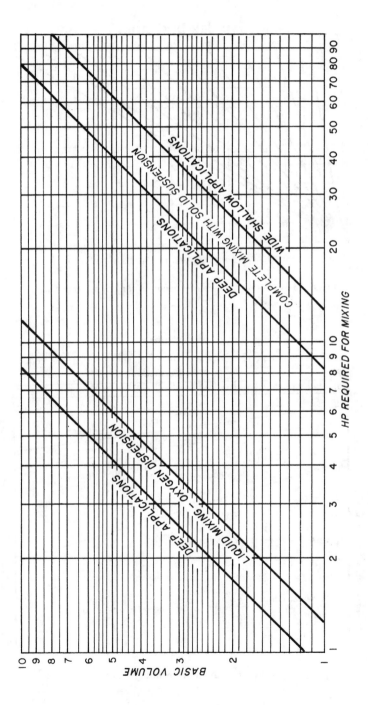

Figure 13-1.

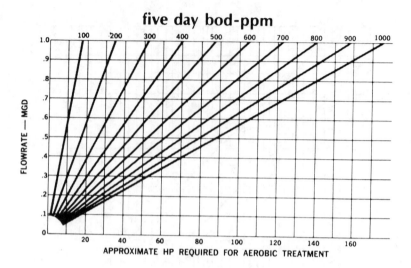

Figure 13-2. Aerator horsepower required for oxygen transfer.

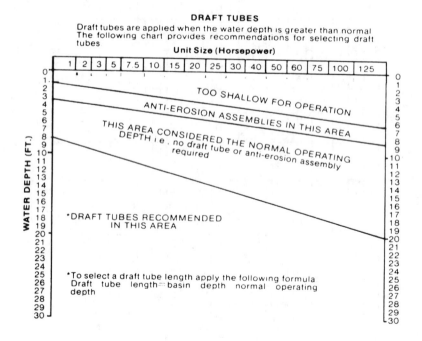

Figure 13-3. Aerator operating depth.

10. Determine the exact length of the basin by solving the following formula:

$$L = \frac{[4R(A) - 6R(SD)^2 + 2(SD)^2]\ ½ + SD\ (1 - R)}{2R}$$

Where: L = basin length

R = $\dfrac{\text{approx. width of basin}}{\text{approx. length of basin}}$

A = $\dfrac{\text{volume (cu ft)}}{\text{average depth (ft)}}$ = average surface area

S = side slope

D = basin depth

11. Determine the basin width by solving the following formula:

$$W = RL$$

where: W = basin width.

12. Obtain bottom dimensions by subtracting **2SD** from both **L** and **W.**

At this point in the design, several details remain to be settled. Piping and liquid transfer provisions should follow industry standards. Aerator mooring must be considered. Aerators can remain stationary or be allowed to rise and fall with the level of the aeration basin. Cables for floating aerators must take into account variations in level. Recommended cable sizes are provided in Table 13-2. (These sizes are for direct-drive aerators only.)

An electric power cable must also be selected. Its insulation must be suitable for underwater use and impervious to the impurities encountered in the waste being treated. It should be a simple continuous piece with no splices. Its size is dependent on the power requirements of the motor, the length of the run, and the available voltage.

If the aerators are to be operated under severe winter weather conditions, they should be provided with removable heat blankets. Heat blankets lace onto the units and protect the exposed motors from damage from icing. They are thermostatically controlled and require very little power.

Converting Existing Systems

When converting existing systems, basin size is seldom a serious problem. Basins tend to be larger than required, rather than too small. Basin configuration may be a problem, however, being too deep or too shallow, or long and narrow or irregularly shaped. Shape problems can be alleviated in part by introducing the waste effluent at several different points to help even out loading. These points should be close to aerators within the basin.

Aerators are selected in the same fashion as for new systems. Two conditions, however, may modify the selection process. First, the configuration of the basin may require several smaller units, rather than one larger one, to assure uniform mixing. Second, in earthen basin applications, aerators must be placed so that high turbulence does not impinge on the earthen walls.

Depth problems are easily solved by adding the proper accessory. If the basin is too shallow, the aerator must be equipped with an anti-erosion assembly to prevent bottom scour. This device consists of a large disc mounted below the intake cone. If the basin is too deep, a draft tube solves the problem. A draft tube is a simple extension which provides for a deeper intake.

AERATED LAGOONS[101]

Lagooning and ponding have long been popular methods in treating industrial and municipal wastes. However, a significant amount of acreage is required for satisfactory treatment. Also, the degree of treatment is unpredictable because the process is very dependent upon weather and climate.

With the advent of surface mechanical aeration equipment, aerated lagoons have now been found to be an economical alternative in biological waste treatment of industrial wastes (Figure 13-4). Other common types of biological waste treatment are the activated sludge process and trickling filters.

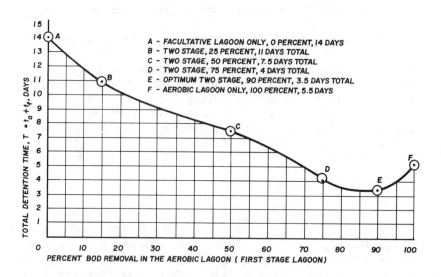

A – FACULTATIVE LAGOON ONLY, O PERCENT, 14 DAYS
B – TWO STAGE, 25 PERCENT, 11 DAYS TOTAL
C – TWO STAGE, 50 PERCENT, 7.5 DAYS TOTAL
D – TWO STAGE, 75 PERCENT, 4 DAYS TOTAL
E – OPTIMUM TWO STAGE, 90 PERCENT, 3.5 DAYS TOTAL
F – AEROBIC LAGOON ONLY, 100 PERCENT, 5.5 DAYS

TOTAL DETENTION TIME, $T = t_a + t_f$, DAYS

PERCENT BOD REMOVAL IN THE AEROBIC LAGOON (FIRST STAGE LAGOON)

Figure 13-4. Comparison of facultative lagoon, aerobic lagoon and optimum two-stage lagooning detention times.

An aerated lagoon has an aerobic environment, where oxygen is supplied by artificial means. The oxygen requirements of the system are usually satisfied with the use of mechanical surface aeration. An aerated lagoon can be further defined as an *aerobic* lagoon or *facultative* lagoon.

In an aerobic lagoon, the mixing level created by the aeration equipment keeps all organic solids in suspension. In a facultative lagoon, the mixing level is low enough to allow organic solids to settle, but high enough to distribute the dissolved oxygen throughout the lagoon.

Economic Comparison

Non-aerated lagoons and ponds are limited to a 3- to 4-ft maximum water depth. Aeration is obtained only by wave action and algae. Aerated lagoons can be constructed to a depth of about 18 ft; this greater depth allows a significant saving in land. Since aerated lagoons mix the volatile suspended solids (a measure of the active biological mass of organisms) and because oxygen is not limiting, significantly shorter detention times are required; thus, an additional saving in land. Also, the mixing scheme in aerated lagoons can allow for a "completely mixed system." Industrial shock loads and toxic loads can be quickly dampened without causing failure of the process.

Since smaller volumes and less surface area are required and because oxygen is added artificially, the effluent quality obtained in an aerated system is more uniform with respect to time. The entire process is far less dependent on climate and weather.

In comparing the two types of aerated lagoons, each has distinct advantages as well as disadvantages. Since the degree of mixing in the aerobic lagoon is high enough to keep solids in suspension, the detention time required for removal of soluble Biochemical Oxygen Demand is significantly less than if a facultative lagoon were used. This reduction in detention time only occurs because a higher equilibrium level of volatile suspended solids (biological mass) is maintained.

Disadvantages of the aerobic lagoon are twofold. First, the horsepower required for mixing in the aerobic lagoon is usually higher than that required for satisfying the oxygen requirements, so much more horsepower is required in the aerobic lagoon than in an equivalent facultative lagoon. Second, a relatively high level of volatile solids are kept in suspension, and they will remain in suspension in the effluent. An aerobic lagoon removes soluble BOD from the influent, but produces an effluent BOD which is associated with the volatile suspended solids in the effluent. A clarifier is required to remove these solids and reduce the suspended BOD. The result is a distinct economic disadvantage.

The facultative lagoon has the advantages of a lower horsepower requirement and a low effluent volatile suspended solids (VSS) level. However, the required detention times are much longer. In practice, the aerobic lagoon is usually employed only as an intermediate biological treatment process where the ultimate treatment scheme is to be extended aeration— that is, solids recirculation from a clarifier.

Optimization of Aerated Lagooning

It has been shown mathematically that a two-stage aerated lagooning system, with an aerobic lagoon as the first stage and a facultative lagoon as the second stage, will require less total detention time than if a single aerobic lagoon or a single facultative lagoon were used. The design indicates that the total detention time using both lagoons is a function of the intermediate soluble BOD level (effluent BOD level from the first stage lagoon). Specifically, the amount of soluble BOD removal in the first stage lagoon (a unique intermediate soluble BOD level) will yield the minimum total detention time for this series lagooning system. With this type of lagooning system, not only is detention time optimized, but also the advantages of each type of aerated lagoon are utilized and the disadvantages minimized.

Oxygen requirements for aerated lagoons can be described by the following equation:

$$O_2 = a'(8.34QS_r)$$

where: O_2 = oxygen required, lb O_2/day
 Q = flow, mgd
 a' = lb O_2 reqd/lb BOD removed, unitless and having a value of 0.9 to 1.1 for aerobic lagoons and 1.1 to 1.4 for facultative lagoons.
 S_r = BOD removed, mg/l

Power requirements for mixing for aerated lagoons can be described as follows:

$$(hp)_m = VP_v$$

where: $(hp)_m$ = mechanical surface aeration horsepower required, hp
 V = lagoon volume, gal
 P_v = required mixing level, hp/1000 gal of volume and having a value of 0.008 to 0.01 for facultative lagoons and a value greater than 0.08 to 0.1 for aerobic lagoons.

Mechanical aeration horsepower required to satisfy the oxygen requirements can be described as follows:

$$N = N_0 \frac{C_{sw} - C_1}{C_{20}} \theta = T - 20_a$$

and

$$(hp)_o = \frac{O_2}{24N}$$

where: N_0 = oxygen transfer rate at standard conditions: 20 C, 0 mg/l DO. 1 atm, and having a value of 2.5 to 3.5 lb O_2/hp-hr depending upon size and type of surface aerator

N   = oxygen transfer rate at field conditions

$(hp)_o$ = mechanical aeration horsepower required, hp

C_{sw} = oxygen saturation concentration of the waste at field conditions, mg/l

C_L = dissolved oxygen concentration of the waste to be maintained, mg/l, usually having a design value of 1.0 to 2.0 mg/l

C_{20} = oxygen saturation concentrate of water at 20 C (9.17 mg/l

θ = temperature correction coefficient, unitless and having a value of 1.024

T   = design temperature, C

a   = oxygen transfer correction factor, unitless, usually having a value of about 0.6 to 0.9.

Using mechanical surface aeration in aerated lagoons provides an economical alternative for industrial biological waste treatment.

COMPRESSED AIR SUBSURFACE AERATION[104]

Aeration of sewage and industrial wastes using compressed air is a common, well documented process. Extensive research has been done to improve upon the basic process of introducing air at some depth within a tank, and allowing the natural rise of the air bubbles and entrained water to provide oxygen transfer. Oxygen transfer depends upon surface area (air bubble size), turbulence (mixing of air and water), and contact time (length of air-water path).

In a typical aeration tank, the length of the air-water path is equal to the tank depth above the air inlet. Turbulence is produced by the drag of rising air bubbles, so that for fixed tank geometry, little can be done to improve upon these factors directly. The method of introducing the air, however, can easily be varied. This controls the bubble size, and indirectly affects turbulence and path length, due to a dependence on the velocity of rise of the bubbles.

Several ways of introducing air have been developed, each with its own particular advantages and limitations. Characteristics of the most common methods are summarized in Table 13-3. The various methods for introducing

Table 13-3. Summary of Compressed-Air Aeration Methods

Type of Diffuser	Examples	Advantages	Disadvantages
Porous (Fine bubbles)	Ceramic plates and tubes Saran pipe Cloth bag	High oxygen transfer efficiency, reduced heat losses to ambient	High maintenance, clogging
Nonporous (Coarse bubbles)	Spargers Nozzles Orifices Shear box Valves	Nonclogging, low maintenance, reduced heat losses to ambient	Low oxygen transfer efficiency

air can be divided between two basic types of diffusers: porous and non-porous. Porous diffusers are high-maintenance devices due to clogging of the holes but they provide efficient oxygen transfer. Nonporous diffusers are low-maintenance devices, but provide less efficient oxygen transfer. No totally satisfactory device has been developed which produces small air bubbles and is maintenance-free.

A new approach to the problem improves upon the compressed air method of aeration by starting with a maintenance-free, nonporous diffuser, increasing the path length, and air-water mixing, and reducing air bubble size after the introduction of air in order to increase oxygen transfer. The device is an in-line, no-moving-part, continuous mixing and processing unit. It is constructed of a number of short elements of right- or left-hand helices. These elements are alternated and oriented so that each leading edge is at a 90-degree angle to the trailing edge of the one ahead. The element assembly is then enclosed within a tubular housing.

When materials are passed through the mixer, two unique mixing actions (flow division and radial mixing) operate simultaneously in the unit. This results in nearly "plug flow" characteristics. When two immiscible fluids are passed through the mixer, shear forces generated by the mixing action disperse one fluid within the other in the form of fine drops or bubbles. Intimate contact, as well as a high degree of interface, is the result (Figure 13-5).

For compressed air aeration the unit is placed in a vertical position above an air inlet. Small bubbles of air are generated, and intense mixing of the bubbles with entrained water takes place. This combination of large surface area and intense turbulence transfers more oxygen. Also, the path of the air-water mixture is lengthened due to the winding channel imposed by the mixer elements.

Basically, oxygen transfer takes place in four steps. The first step occurs during bubble formation at the two orifices used to supply the aerator. The bubbles are formed by high-velocity air jets (100-200 ft/sec) where there is a high degree of turbulence and a high dissolved oxygen deficit. The contact time before the air enters the mixer is extremely short. The orifices—and therefore the bubbles—are large, since this reduces the risk of clogging and provides a high circulation of water. Because of the large bubble size and short contact time, the contribution of this step to the total oxygen transfer is small.

The second step is the actual contacting of the air-water mixture in the device. As described previously, the action of the mixer elements breaks the air into fine bubbles, and completely mixes the oxygen-rich air with the low-oxygen water to give an outlet liquid of high dissolved oxygen content. The length of diameter ratio (pitch) of the elements, and the diameter of the unit are designed to give an optimum combination of dispersion, mixing, path length, and liquid circulation. The length (number of elements) is designed to provide a liquid which is nearly saturated.

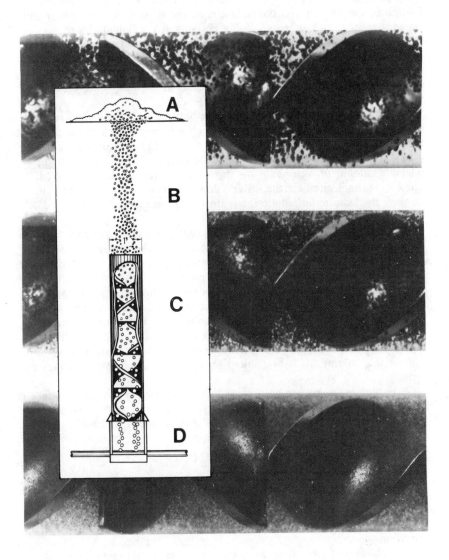

Figure 13-5. Dispersion in static aeration device. Different drop sizes are produced by varying the velocity.

A. Surface turbulence; B. contacting in air-water phase;
C. contacting in the aerator; D. bubble formation.

The third step of oxygen transfer is the rising of the exit air-water column to the surface. As the stream rises, the air slips past the water and expands away from the center. Eddy currents at the boundaries of the rising column slow the rise of the water and actually cause interchange of oxygen-rich liquid with the oxygen-weak surrounding liquid. Additional oxygen transfer results as the rising air bubbles contact the oxygen-weak entrained liquid. Final oxygen transfer occurs in the turbulence created as the rising air-water column breaks the surface.

Aerators are anchored to the lagoon bottom by either a rigid support or a tether line. The elevation of each anchor is predetermined using a transit, and the length of each tether line is determined so that the aerators will be at the same elevation. For large variations in depth, where a considerable amount of water is below the aerators, draft tubes are installed. These extensions circulate the water below the aerators. The aerators lie on their sides during installation, but float to an upright position as the basin is filled.

Air is supplied by blowers at a sufficient pressure to overcome friction losses in the piping and the static head of water in the lagoon. Three types of blowers are commonly used. The positive-displacement rotary-lobe blower operates at constant volume and variable pressure, which permits variation in lagoon depth. Also, the volume can be varied by changing the speed of rotation. The second type is the helical screw positive displacement blower. It has the same operation characteristics as the lobe type, but is more efficient at higher pressures. It is, however, a more costly machine. The third type is the centrifugal blower, which operates at nearly constant pressure and variable volume. It is best suited for applications where air requirement varies.

Extensive testing has been carried out to determine the oxygen transfer and basin mixing characteristics of subsurface aerator systems. Oxygen transfer is proportional to air supply and basin depth. The transfer capabilities of a single unit are given by the equation:

$$N = (.0252)(D-1)(A)(.67)$$

where
 N = Oxygen transfer to pure water (lb/hr-unit)
 D = Depth of the basin (ft)
 A = Air supply per unit (scfm)

In general, great depths and low air supply per unit give most efficient operation.

A single unit will set up a circulation pattern in its immediate area. The general direction of flow is from the bottom of the basin into the mixer, out of the mixer to the surface, across the surface and down to the bottom (Figure 13-6). Strong eddy currents are set up by the rising air-water column, where up to 6 times flow rate through the mixer is entrained and carried to the surface.

Compressed air aeration is best suited for tank applications where the geometry of the tank (the side walls and bottom) can enhance basin mixing.

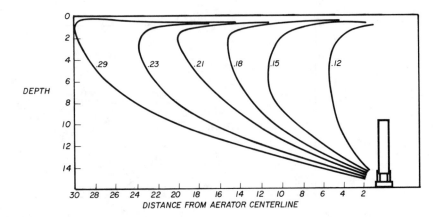

Figure 13-6. Subsurface circulation currents for a single static aeration device operating at 25 scfm in 16 ft of water.

For large lagoon applications, the natural circulation caused by the rising air bubbles is not sufficient to produce complete mixing without excessive power consumption. Circulation patterns must be induced by the placement of baffles, or the use of long, narrow tanks.

The area of influence of a single unit extends for a radius of over 50 ft, as evidenced by measurements of surface currents. Basin mixing is accomplished by a combination of surface and eddy currents. Since both of these effects depend on the depth, the degree of mix varies somewhat with lagoon depth. Figure 13-7 shows the maximum deviation of the dissolved oxygen level from the average as a function of depth.

In all cases, the degree of mixing is more than sufficient for waste treatment applications. Measurements of current velocity in the area of the aerator show that the smallest velocities are at the bottom of the basin. In cases where the units are far apart (aerated lagoons), some settling of solids will take place. In cases where the units are closer together and the basin geometry plays an important part in the circulation pattern (activated sludge), insignificant settling takes place.

Studies have been performed to determine the mixing patterns in aeration basins. Most of the studies have been done with surface entrainment aerators, with the result that the type of mixing (*i.e.,* completely mixed *vs.* plug flow) depends on the length-to-width ratio of the basin and the power input (Figure 13-8). It is believed that small basins with short retention time are completely mixed using sub-surface aerators. For larger basins and longer residence time (aerated lagoon), the basin acts like a plug flow system, and the aeration must be tapered to allow for this.

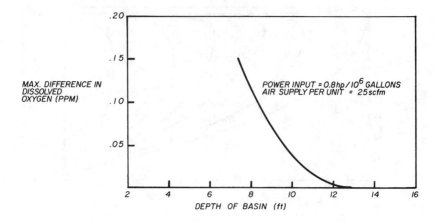

Figure 13-7. Maximum difference in dissolved oxygen level as a function of basin depth.

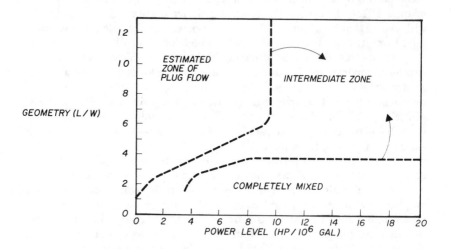

Figure 13-8. Mixing characteristics of aeration basin as a function of geometry and power level.

DISPERSED AIR FLOTATION[170]

Although the term "flotation" is not new to the water treatment process, the definition commonly and traditionally inferred is that of *dissolved air* flotation. For many years dissolved air flotation systems have been used for the removal of oil and suspended solids from waste streams. Although successful, the dissolved air process is now being challenged by a simpler and less expensive flotation process known as *dispersed air* flotation.

The difference between dissolved air flotation and dispersed air flotation is in the way air is generated and introduced into the liquid (Figure 13-9).

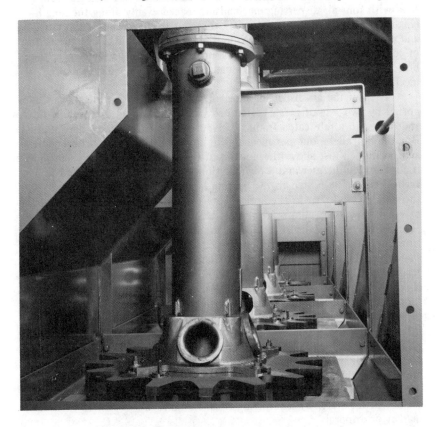

Figure 13-9. Interior of dispersed air flotation machine showing air dispersing mechanisms.

Dispersed air flotation creates a more rapid response. Therefore, a shorter detention time is required to do the job. For a given application, a smaller sized dispersed air flotation machine can be used than is required for dissolved air flotation. In many applications, dispersed air flotation requires a detention time of only four or five minutes vs. about twenty minutes for total flotation response in dissolved air flotation systems. The shorter detention time means that a smaller machine will do the job. Since space is at a premium in most industrial plants, it is an important economic factor to be considered in designing a water pollution control facility.

A four-cell dispersed air flotation machine with a volume of 400 cu ft is capable of processing a million gallons of waste materials per day. Yet, the machine occupies only 100 sq ft of floor space.

Operation

Dispersed air flotation machines basically consist of a single rectangular tank with four air dispersing mechanisms spaced evenly along the length (Figure 13-10). Contaminated water enters at the feed end of the tank, and effluent is discharged at the opposite end. Flotation air, in the form of countless finely divided bubbles, is introduced into the water near the bottom of the rectangular tank by means of the four air dispersing mechanisms.

Each of the air dispersing mechanisms consists of a vertical shaft supported by two sets of grease-sealed ball bearings enclosed in a spindle housing and bolted to a fabricated frame. Drive is by V-belt from standard motors. Rotating impellers, one per mechanism, are attached to the vertical shaft and positioned just above the bottom of the cell. Surrounding each impeller is a stationary diffuser and a recirculation hood, which are attached to a vertical standpipe bolted to the bearing housing. Just above the liquid level the standpipe is open to the atmosphere for admission of air. No compressor, external blowers or pumps are required.

In operation, void is created as the rotating impeller displaces liquid from its center toward the periphery which causes a flow of air down the standpipe. Liquid within the cell circulates to the impeller through four recirculation ports in the stationary hood. Air flowing down the standpipe is engulfed by and mixed with the inflowing liquid by the rotating impeller. Liquid leaving this zone is saturated with fine air bubbles. The degree of aeration is determined by rate of fluid circulation through the impeller and by impeller speed.

Air bubbles provide a surface area to which non-wettable impurities are attracted. Particles of the impurity attach themselves to bubbles and are lifted to the surface. At the surface the bubbles coalesce as a frothy mixture of air, water and contaminant. This froth is then swept off the surface of the water by a revolving froth skimmer and is sent to reprocess, market or disposal.

By adding reagents to chemically condition the wastewater, it is often possible to further improve the effectiveness of the flotation process. Reagents are added to the influent before it enters the flotation tank in order to create non-wettable surfaces on the contaminants and encourage froth formation.

No blower, compressor or pump is required for most dispersed air flotation applications, as the mechanisms generate their own air. However,

Figure 13-10. Dispersed air flotation machine.

where large flow rates are encountered which require high capacity units, low-pressure air from a rotary blower is normally used instead of atmospheric pressure. Gas can be substituted for air in cases where oxygen may not be desirable, or where the contaminant would be more responsive to a specific gas bubble. Gas-tight covered flotation tanks can be employed so that the gas can be recovered and recycled.

Applications

Dispersed air flotation has been used extensively for over 50 years for the separation and selective recovery of minerals. However, only recently have industrial water clean-up applications become widespread. The flotation process has been applied to such tasks as removal of oil from water; paper de-inking; recovery of latex, carbon black, soot, naphthalene, plastics, activated carbon, wood fibers, meat processing wastes, and paints from effluent streams. A side benefit is the thorough aeration of the waste stream as it passes through the flotation machine.

A number of unusual applications have also been explored: processing of waste products from atomic reactors, recovery of sulfur dyes from wastewaters of dye works, reduction of biochemical oxygen demands in domestic sewage plant effluent, cleanup of laundry wastewater for re-use, recovery of silver from photographic film wastes, treatment of "white water" in the paper industry, cleanup of oil-saturated beach sands and recovery of waste porcelain enamel.

DYNAMIC AERATION ECONOMICS[57]

Oxygen transfer in a body of water takes place in three steps. During the first step, atmospheric oxygen quickly saturates the surface film of the water. In the second step, the oxygen penetrates the surface film by molecular diffusion. During the third step, oxygen is distributed throughout the body of water by diffusion and convection. The second step, molecular diffusion, is the slowest and, therefore, governs the rate at which oxygen is made available for the biological cleaning processes taking place in the water.

In passive wastewater treatment processes, large, shallow holding ponds retain the water while it is naturally aerated. With land at a premium and holding times shortened to increase system capacity, oxidation ponds are no longer practical.

Dynamic aeration, the use of agitating aerators, has become one of the most common features in new short loop wastewater treatment systems. Aerators churn water into a spray, multiplying the effective surface area many times. This action speeds the oxygen transfer rate considerably. By increasing subsurface water turbulence, aerators also increase the uniform distribution many-fold. Finally, the aerators provide enough turbulence to keep organic solids suspended in the oxygen-rich water, thereby increasing the efficiency of the biological cleansing processes.

Determining Aerator Requirements

Average and worst-case temperature and barometric pressure conditions are balanced against water condition (BOD) to arrive at a basic system design. All three factors influence the oxygen transfer rate, the first two by determining the oxygen saturation level in the water, and the third by establishing biological oxygen removal rate. In an ideal system, oxygen transfer into the water would always exactly balance the biological removal rate. Thus, the oxygen level in the water could theoretically be kept at any given level, including zero dissolved oxygen, thus maximizing the rate of transfer. Unfortunately, the pollution engineer has no control over any of these factors, so he must resort to a worst-case-plus-safety-factor design for his system.

For a static system, the penalty involved is the size of the holding pond. The extra size needed to accommodate worst-case conditions requires the use of additional land. In a dynamic system, the penalty is extra size in the aeration equipment used, but there is a second penalty as well: the cost of the power consumed by the equipment.

Under any given set of conditions, only a certain amount of oxygen can be dissolved in a given quantity of water. No amount of mechanical aeration will add oxygen to water already saturated. In arriving at an equipment choice, therefore, the pollution engineer should determine exactly what his worst-case needs are and select units no larger than necessary to meet those needs.

Mechanical aerators will furnish 1.5 to 2.0 lb of oxygen/hp/hr under field conditions. The formula used to determine the amount of power needed to provide sufficient oxygen is:

$$hp = \frac{(Q)K(BOD)}{N_C\,(24)}$$

Where: Q = flow, in mgd
K = constant for wt of water (8.34 lb/gal)
BOD = oxygen demand, in ppm (this is 24 hr/day)
N_C = oxygen transfer rate for aerator

Because the aeration tank will probably not have a million-gallon capacity, the formula necessary to arrive at a power-to-volume ratio to determine the total aeration horsepower required is:

$$R_{P/V} = \frac{hp \times 10^6}{V}$$

where: $R_{P/V}$ = power-to-volume ratio
V = volume of aeration tank, in gal

For example, assume a system has a flow of a million gallons/day, BOD of 212 ppm, and an aerator oxygen transfer rate 1.8 lb of oxygen/hp/hr. Further, assume the aeration tank volume to be 250,000 gal. Total aerator hp required would be:

$$\frac{(1.0)\ (8.34)\ (212)}{(24)\ (1.8)} = 41 \text{ hp}$$

Two 20-hp units will provide the aeration necessary for this system. This is an $R_{P/V}$ of 160 hp/million gal and is more than adequate for mixing in a 250,000-gal tank.

Most system designs stop at this point. Unfortunately, the true operating cost savings potential has hardly been touched. Consider the curve shown in Figure 13-11. This is a typical flow pattern for a municipal sewer system waste treatment plant. The amount of oxygen required for biological cleansing parallels this curve. The worst-case situation for which the pollution engineer would design is represented by the heavy flow peaks during the middle of the day.

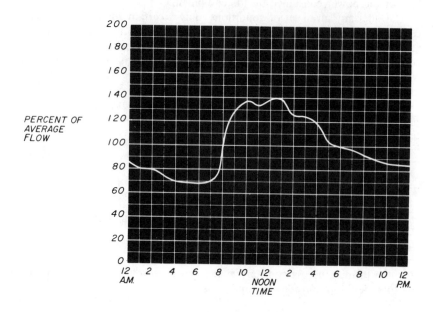

Figure 13-11. Typical variation in flow patterns in a municipal waste treatment plant. Variations in waste strength and oxygen uptake rates parallel this curve.

Ideally the system is designed so that there is a dissolved oxygen residual of about 1 mg/l (enough to support an activated sludge system) in the water at the peak load period. Pragmatically, a slight safety factor is also designed into the system. A residual dissolved oxygen curve (based on Figure 13-11) is shown in Figure 13-12.

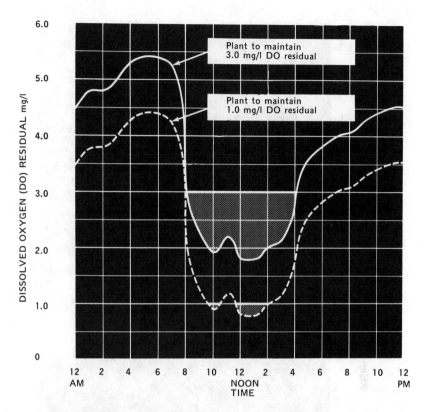

Figure 13-12. Curve of residual dissolved oxygen in an activated sludge system during a 24-hr period. Shaded area represents the time during which full aerator power is required.

Excessive residual dissolved oxygen, however, represents inefficiency in the system, and this inefficiency costs money (Figure 13-13). Figure 13-14 shows the cost of this inefficiency. Traditional aerators produce the excessive residual dissolved oxygen because they run at a constant speed matched to peak-load periods. Their power demands at off-peak periods provide opportunities for reach operating cost savings.

Dual-Speed Aerators

At first glance, a dual-speed aerator might be assumed to be more expensive than single-speed machines. Direct-drive, dual-speed units are considerably less expensive to purchase than radial flow units. They are equipped with dual-wound motors which produce rated horsepower at top speed and 40 percent less power at their low speed.

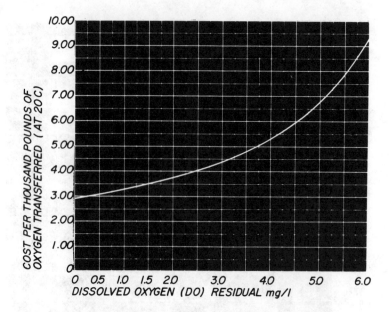

Figure 13-13. Cost of oxygen vs dissolved oxygen residual.

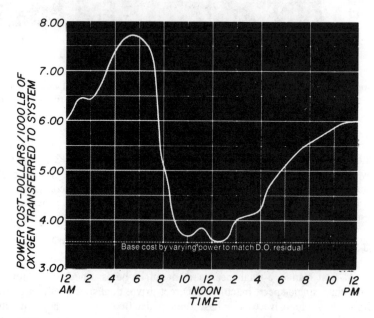

Figure 13-14. Cost of oxygen per pound vs time (at constant power with single speed aerators). Shaded area represents dollars per day lost due to lack of control.

In application, dual-speed units provide a much greater degree of flexibility than their two-speed design would suggest. For instance, an application calling for 50 hp could utilize two 25-hp units, resulting in the following power requirements:

> Both units low speed–30 hp
> One unit high speed, one unit low–40 hp
> Both units high speed–50 hp

As more units are added to a system, the number of possible combinations increases. The units could be switched, one by one, to lower speeds as flow increased. The switching could be manually controlled, or controlled by a timer, to match very closely the varying oxygen transfer demands of the system.

Intermediate horsepower demands are only possible when two or more aerators are operating in the same basin. It is still possible, however, to match aeration to flow, using only one unit. The unit can be controlled by a dissolved oxygen probe. The aerator would operate at top speed when DO residuals are less than 1 mg/l, and drop to the lower speed when the residuals exceeded 2 mg/l. Such a system would automatically match aeration to demand within very narrow limits. Time would be the variable factor in this case, rather than intermediate horsepower demands, but the overall power cost would be similarly reduced from that of constant worst-case aeration.

A DO probe can also be used in systems where the flow pattern is unpredictable. The probe would automate the system, making it unnecessary to monitor conditions closely to assure peak performance. Also, the probe controls can be adjusted to govern at some other point than the 1.0 mg/l and 2.0 mg/l DO residual points mentioned. This capability enables a wastewater system to be tuned to an extremely fine state of efficiency.

The only point which the pollution engineer needs to consider when he is designing a dual-speed system is the power-to-volume ratio achieved at the lowest horsepower condition. As long as this does not drop below 50 to 60 hp/million gal flow, adequate mixing is assured. In most applications, even the lowest hp conditions satisfy mixing requirements, so this point is seldom a problem.

CHAPTER 14

SELECTED APPLICATIONS
OF WATER POLLUTION PROBLEMS

METAL WASTE RECLAIMING[165]

Reverse osmosis serves two functions in solving the pollution problem of the plater and chemical manufacturer. First, a toxic but valuable material is reclaimed for in-plant re-use or for refining outside. Second, the purified water from reverse osmosis is usually of better quality than tap water, and can be re-used in the plant.

Figure 14-1 shows a typical nickel plating line with a reverse osmosis system reclaiming nickel salts and purifying water for re-use. A system for gold cyanide plating process, a copper plating line or a rinse from a chemical reaction would be set up in the same manner.

In the nickel plating line example, the flow of parts is from the plate tank into the rinse tank and then into the next set of plating tanks. Water from the rinse tank is continuously processed through the reverse osmosis machine at 10 gpm. Because of dragout from the plate tank, the rinse is maintained at a concentration of 500 mg/l nickel salts.

Prior to entering the R.O. unit, the rinse is filtered to 50 microns of fine emulsions which contain large organics, a 100-mesh strainer is used to protect the membrane module flow channels from large particles.

The rinse enters the R.O. unit and is pressurized to 450 psig by a stainless steel multistage centrifugal pump. After pressurization, the rinse solution flows over the membranes and purified water is forced through the membranes. The R.O. system is set at 95 percent recovery of the feed as permeate; This means that for every gpm feed, the R.O. unit produces 9.5 gpm permeate and 0.5 gpm concentrate.

Permeate concentration from this system is 75 mg/l. All of the brighteners, which are organics of over 200 molecular weight, are retained in the concentrate. The permeate is recycled to the rinse at the rate of 9.5 gpm. Pure water makeup at the rate of 0.5 gpm is necessary to replace the 5 percent of feed which is retained as concentrate.

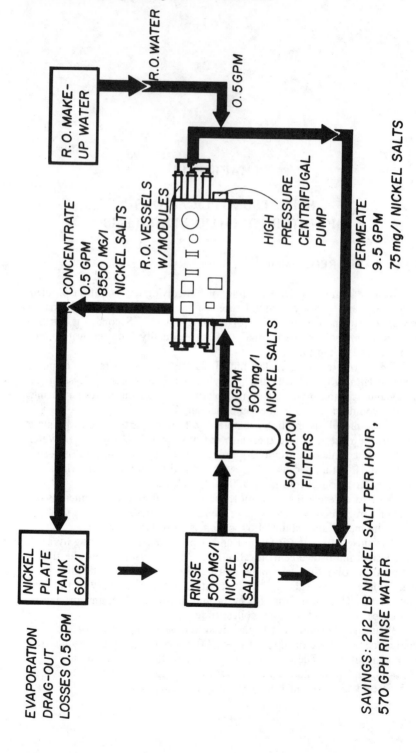

Figure 14-1. Metal reclamation system.

Concentrate is returned to the plate tank to make up for evaporation and drag-out losses. The concentration of the concentrate stream is nearly 20 times the feed concentration, or 8550 mg/l.

On systems where the evaporation and dragout losses are below 0.5 gpm, two- and three-stage systems are installed to increase the concentration of the concentrate while allowing the permeate to remain high in quality.

Economics

Consider, for example, a nickel plating system used for plating decorative trim and ornamental pieces. The nickel plating is followed by chrome plating. Both the nickel and chrome systems use tap water at the rate of 1900 gal/hr for rinsing the plated articles.

Assume the plating line runs 15 hours per day and the total rinse drag-out amounts to 220 lb of plating salts per day. The value of the lost plating solution is:

175 lb NiSO$_4$ @ $0.55/lb =	$ 96.20	
45 lb NiCL$_2$ @ $0.85/lb =	38.20	
1.2 gal #36 Brightener @ $8/gal =	9.60	
0.2 gal #61 Brightener @ $5/gal =	1.00	
Total/day	$ 145.00	

If the plating system uses a two-stage cascading rinse (Figure 14-2), the reverse osmosis unit is designed to process the first rinse after the plate tank. Assume a 2000 mg/l concentration exists in the first rinse and a 100 mg/l concentration in the second rinse.

The water required to dilute 220 lb of nickel salts to a 2000 mg/l concentration is:

$$2000 \text{ mg/l} = 0.2\% = 0.002 \text{ lb of salt/lb of } H_2O,$$

$$\frac{220 \text{ lb of salt}}{0.02 \text{ lb of salt/lb of } H_2O} = 1.1 \times 10^5 \text{ lb of } H_2O$$

which gives

$$\frac{1.1 \times 10^5 \text{ lb } H_2O}{8.3 \text{ lb/gal}} = 13,250 \text{ gallons of } H_2O$$

The hourly flow is:

$$\frac{13,250 \text{ gal}}{15 \text{ hr}} = 884 \text{ gph}$$

Since the R.O. membrane rejects 99 percent of nickel plating salts, permeate concentration of the first stage is 60 mg/l which is below the 100 mg/l limit established for the second rinse. Permeate from stage 1 is thus returned to the second rinse.

Concentrate from the first stage is run into stage 2 for further treatment. The second stage operates at 90 percent recovery. Second stage permeate is

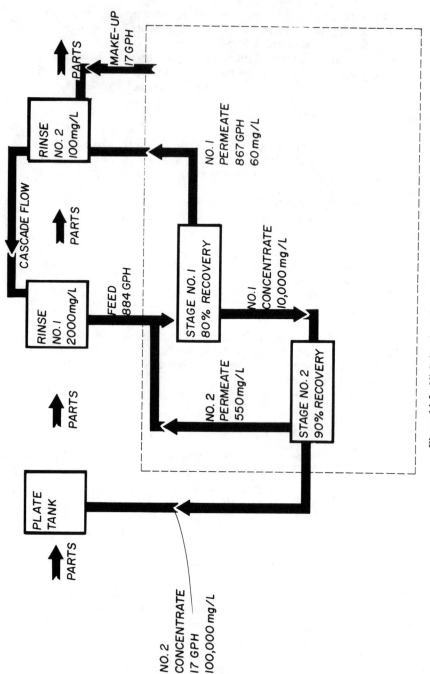

Figure 14-2. Nickel reclamation system.

returned to the feed for stage 1. Concentrate leaving stage 2 at about 100,000 mg/l (10 percent) nickel salt is returned to the plate tank at the rate of 17 gal/hr to make up for water losses due to evaporation and drag-out. Total recovery for this reclamation system is 98 percent of the feed as purified water.

Economics of the nickel recovery system are:

Savings

Plating salt savings: $145.00

Water savings:

At 25¢/1000 gal for city water and 25¢/1000 gal for city sewer,
884 gph x 15 hr = 13,250 gpd less blow-by water,
20 gph x 15 hr = 300 gpd less cooling water,
20 gph x 15 hr = 450 gpd
     Total   12,500 gpd x $ .50/1000 gal = $6.25/day

Costs

Capital: Total system including a water purifier, start-up, internal wiring, etc. $36,000.

Electrical Costs:
Total horsepower = 13.5 hp
13.5 hp x 0.75 kw/hp x 15 hr = 152 kwh/day
    Electricity rate: $0.02/kwh
    Electrical costs: $3.04/day
Note: electrical costs are one-half of the savings in water costs.

Module Replacement Costs:
Assuming a two-year life on the membrane modules,
Total modules: 36 @ $200/module or $7200 every 2 years which is:
$$\frac{\$7200}{24 \text{ months}} = \$300/\text{month}$$
with a 20-day work month:
$$\frac{\$300/\text{month}}{20 \text{ day/month}} = \$15.00/\text{day}$$

Pump Maintenance Costs:
Seals and bearings changed every two years
$$\frac{\$240}{24} = \$10/\text{month} = \$0.50/\text{day}$$

Maintenance:
30 minutes per day average x $5.00/hr = $2.50/day

Savings/Day		Costs/Day	
Plating salts	$ 145.00	Electrical	$ 3.04
Water and sewer	6.25	Membrane modules	15.00
	$ 151.25	Pumps	.50
		Maintenance	2.50
			$ 21.04

Gross savings/day	$	151.25
Less gross costs		21.04
Net savings/day	$	130.21

Using 20 working days per month, the monthly savings are $2604.20. The system can pay for itself in:

$$\frac{\$\ 36,000}{\$2604.20/month} = 13.6 \text{ months}$$

The least expensive method in terms of capital cost for disposing of plating salt is a precipitation system. The capital cost for automated precipitation for this plating system is approximately $18,000, plus daily costs for chemicals, sludge removal and disposal, and large plant space requirements. Therefore, the real economics for the nickel system are:

Capital cost of R.O. system:	$ 36,000
less capital cost of alternative:	18,000
Net capital cost:	$ 18,000
Savings of R.O. system over sewer dumping:	$ 2,604/mo
Extra costs over sewer dumping for alternative system	800/mo
Net savings over alternative for R.O.	$ 3,404/mo

True payback is then:

$$\frac{\$18,000}{\$3,404/mo} = 5.25 \text{ months}$$

Table 14-1. Membrane Rejections

SALTS:

	Cations		
	Symbol	Percent Rejection	Maximum Concentration Percent
Sodium	Na^+	94-96	3-4
Calcium	Ca^{+2}	96-98	*
Magnesium	Mg^{+2}	96-98	*
Potassium	K^{+1}	94-96	3-4
Iron	Fe^{+2}	98-99	*
Manganese	Mn^{+2}	98-99	*
Aluminum	Al^{+3}	99+	5-10
Ammonium	NH_4^{+1}	88-95	3-4
Copper	Cu^{+2}	96-99	8-10
Nickel	Ni^{+2}	97-99	10-12
Strontium	Sr^{+2}	96-99	—
Hardness	Ca and Mg	96-98	*
Cadmium	Cd^{+2}	95-98	8-10
Silver	Ag^{+1}	94-96	*

Anions

Chloride	Cl^{-1}	94-95	3-4
Bicarbonate	HCO_3^{-1}	95-96	5-8
Sulfate	SO_4^{-2}	99+	8-12
Nitrate	NO_3^{-1}	93-96	3-4
Fluoride	F^{-1}	94-96	3-4
Silicate	SiO_2^{-2}	95-97	—
Phosphate	PO_4^{-3}	99+	10-14
Bromide	Br^{-1}	94-96	3-4
Borate	$B_4O_7^{-2}$	35-70**	—
Chromate	CrO_4^{-2}	90-98	8-12
Cyanide	CN^{-1}	90-95**	4-12
Sulfite	SO_3^{-2}	98-99	8-12
Thiosulfate	$S_2O_3^{-2}$	99+	10-14
Ferrocyanide	$Fe(CN)_6^{-3}$	99+	8-14

* Must watch for precipitation, other ion controls maximum concentration.
**Extremely dependent on pH; tends to be an exception to the rule.

ORGANICS:

	Molecular Weight		
Sucrose sugar	342	100	25
Lactose sugar	360	100	25
Protein	10,000 Up	100	10-20
Glucose	198	99.9	25
Phenol	94	***	—
Acetic acid	60	***	—
Lactic acid	90	***	—
Dyes	400 to 900	100	—
Biochemical Oxygen Demand Demand	(BOD)	90-99	—
Chemical Oxygen Demand (COD)		80-95	—
Urea	60	40-60	Reacts similar to a salt
Bacteria & virus	5,000-100,000	100	—
Pyrogen	1000-5000	100	—

***Permeate is enriched in material due to preferential passage through the membrane.

GASES, DISSOLVED:

Carbon dioxide	CO_2	30-50%
Oxygen	O_2	Enriched in permeate
Chlorine	Cl_2	30-70%

CYANIDE PLATING WASTE TREATMENT[66]

Cyanide compounds are widely used in the metal finishing industry in the preparation of plating baths. Cyanide is commonly used in the preparation of zinc, copper, cadmium, gold, silver, bronze, and brass plating solutions, and often as a cleaning solution immediately prior to a plating process. A small quantity of this highly toxic solution will be dragged out of the plating bath as each part is removed and rinsed prior to the next plating operation. The rinse water will also contain cyanide. The problem is to safely dispose of cyanide without creating a water pollution hazard.

Collection

Sewers should be constructed so that only cyanide wastes are discharged to the treatment system. Treatments for other plating room chemicals are not usually compatible with the methods used for the destruction of cyanide wastes. Where large quantities of cyanide are used, it is usually necessary to install separate sewers for dilute and concentrated cyanide wastes. A dual collection system will permit concentrated wastes to be collected and pumped at a controlled rate to the cyanide treatment system.

If a separate collection system is not employed, the capacity of the chemical feed system will probably be exceeded during periods of slug charges.

Treatment

Cyanide wastes can be destroyed by oxidation processes to form carbon dioxide and nitrogen. Among the most widely recognized oxidation methods employed are (1) biological; (2) electrolytic; (3) incineration; (4) radiation; and (5) chemical.

Biological Treatment

Cyanides can be destroyed in combination with domestic sewage through such biological means as the activated sludge or trickling filter processes. Sewage provides the required nutrients for the process, and a biological culture will be produced capable of approximately 99 percent cyanide removal. However, this process cannot tolerate the discharge of slug dosages of concentrated cyanides. Also, the biological population must receive necessary nutrients for growth as provided by domestic sewage. If sewage is not available, the required nutrients must be artificially added to sustain the biological population.

Electrolytic Destruction

This method of cyanide destruction requires long periods of time—2 to 7 days—to reduce cyanide to a 1-ppm level. The anode used in this process can be copper, stainless steel or carbon steel. However, a current density from 30 to 80 amps/ft^2 must be maintained. The cathode employed can be made of carbon steel. Optimum operating temperature for this destruction process is approximately 200 F (93 C).

Incineration

Waste cyanide can be blended with an oil mixture and incinerated at approximately 2500 F.

Radiation

Cyanide solutions can be decomposed by radiation processes. Cyanide can be decomposed to within 90-95 percent completion without excessive radiation dosages. However, to obtain further decomposition markedly higher radiation dosages are required. A solution containing 2 grams per liter of a zinc cyanide-sodium cyanide complex requires 83.6 x 10^6 rads for 95 percent decomposition.

Chemical

The alkaline chlorination oxidation process is the most widely used process for the destruction of cyanide. The oxidation of cyanide by chlorination occurs by two separate and major chemical reactions. In the first stage cyanide is oxidized to cyanate; in the second stage cyanate is oxidized to carbon dioxide and nitrogen. The first stage of the oxidation can be illustrated by the following equations:

$$2Cl_2 + 2NaCN \rightarrow 2CNCl + 2NaCl$$

$$2CNCl + 4NaOH \rightarrow 2NaCNO + 2NaCl + 2H_2O$$

The first reaction is instantaneous and occurs at all pH levels, while the second is one of hydrolysis which converts cyanogen chloride to cyanate and is dependent upon pH. The hydrolysis reaction will not go to completion at pH levels of less than 7.5, but the hydrolysis rate will increase as the pH increases. At a pH of 9.0 the reaction will be completed in approximately 3 minutes. It is important that the hydrolysis of cyanogen chloride to cyanate be completed as rapidly as possible since cyanogen chloride is both volatile and toxic.

The second major stage of chemical destruction is the oxidation of the cyanate obtained in the first stage to carbon dioxide and nitrogen.

$$3Cl_2 + 4H_2O + 2NaCNO \rightarrow 3Cl_2 + (NH_4)_2CO_3 + Na_2CO_3$$

$$3Cl_2 + 6NaOH + (NH_4)_2CO_3 + Na_2CO_3 \rightarrow 2NaHCO_3 + N_2 + 6NaCl + 6H_2O$$

These reactions require an excess of chlorine and caustic. The rate of the reaction will increase as the pH of the solution decreases.

Because a pH level greater than 9.0 is required for the first stage oxidation (of cyanide to cyanate), and a pH of approximately 8.5 is required for the second stage oxidation of the cyanate, it is convenient to perform the oxidation in two separate processes. The prime difficulty in using a single chlorination step process is that a compromise pH condition must be selected and there will be a tendency to have both stages of the oxidation occurring simultaneously.

When solution pH is too low, there will be a tendency for carbon dioxide and nitrogen gases to be formed in the presence of cyanogen chloride, due to the incomplete hydrolysis to cyanate. Nitrogen gas will be swept out of solution, carrying cyanogen chloride with it, and creating a hazardous condition in the immediate work area.

If solution pH is too high, the oxidation of the cyanate will require excessively long detention periods and may never go to completion.

The alkaline chlorination process lends itself conveniently to automatic control. A system can be installed to maintain the required pH level conditions by controlling the addition of caustic. An ORP (Oxidation Reduction Potential) system can be used to control the addition of chlorine. With a properly designed and operated plant it is possible to maintain about a 10-ppm excess chlorine residual in the final effluent.

Treatment System

The configuration and size of a treatment system will, of course, be dictated by local conditions. Figure 14-3 shows a schematic diagram for a typical batch-type treatment process; Figure 14-4 shows a flow-through-type treatment process.

Even though Figure 14-3 shows only one treatment tank in the batch-type system, a well designed treatment plant will normally contain three tanks. One tank will be filling, the second tank treating influent contaminants, and the third tank discharging. A treatment plant designed in this manner allows maximum flexibility for nearly all situations which could arise.

The treatment plant should also provide for a means for the effluent to be discharged into a solids removal system. Final effluent of the treatment plant will normally contain toxic heavy metals employed in the plating processes which normally must be removed to comply with state and local regulations.

PULPING PROCESS TREATMENT[194]

Sulfite pulp and paper mills have dotted the banks of the Columbia River in the state of Washington for years. When the state modified its water quality standards and required a 70 percent reduction in biochemical oxygen demand of the effluent discharged by the mills, the mills had several choices: (1) alter production methods to meet the water quality standards, (2) cut production by 70 percent, or (3) close down.

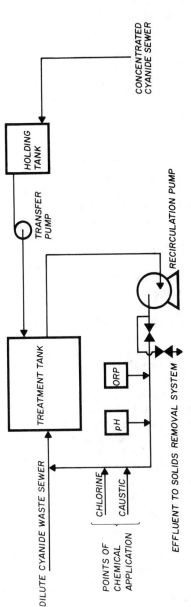

Figure 14-3. Typical schematic diagram for a batch-type treatment process for destroying cyanide wastes.

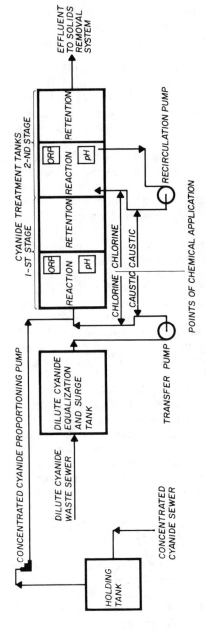

Figure 14-4. Cyanide wastes can also be treated by the flow-through type process as shown in the schematic diagram.

Two mills actually closed. However, one (420 ton/day) mill at Camas met the problem by converting from the traditional calcium base sulfite pulping process to the new Magnefite (Babcock & Wilcox) process.

Basic chemicals in the new pulping process are magnesium hydroxide and sulfur. Sulfur is burned and enters the system as sulfur dioxide. It reacts with magnesium hydroxide to form magnesium bisulfite. This is the cooking chemical used in place of calcium bisulfite.

The raw material, hemlock chips, is unchanged. Digesters cook the wood chips in the magnesium bisulfite liquor at 325 F and 90 psi for 6 hr and 20 min. The pulp is then blown into wooden pits at 50 psi to separate the spent cooking acid ("weak red" liquor) from the pulp. The spent acid then moves through the recovery system (Figure 14-5).

The weak red liquor is concentrated in a multiple-effect evaporator from about 9 percent solids to about 55-60 percent solids ("heavy red" liquor). At this concentration, it will support combustion and is sprayed into the furnace and burned. The furnace produces sufficient steam to supply the cooking and evaporation processes, thus reducing demand for fossil fuels.

Combustion of the heavy red liquor produces a flue gas containing sulfur dioxide and magnesium oxide. The latter is a fine white powder that is separated from the flue gas by a multicone dust collector and washed to remove impurities. It is then dropped into a slaking system where steam and agitation convert the MgO to $Mg(OH)_2$.

Flue gas containing SO_2 is routed from the multicone separator through an induced draft fan and then through an absorption train consisting of four venturi-type scrubbers. Water is added at the first venturi to cool the gas from 425 to 160 F. The gas flows from stage to stage, finally passing out a stack.

Absorption of SO_2 from the flue gas is accomplished by introduction of magnesium hydroxide slurry $[Mg(OH)_2]$ into the suction side of the circulating pumps located under the collection sumps of each stage. Weak cooking acid is formed by the absorption and is routed countercurrent to the gas flow. Weak acid is withdrawn from the No. 2 venturi sump and routed over a fortification tower where it is strengthened further by countercurrent contact with SO_2 gas from the mill's sulfur-burning system. All but trace amounts of SO_2 in the flue gas are absorbed in the venturi system (Figure 14-6).

Compliance and Costs

While meeting water quality standards, the new system has improved pulp quality slightly but also increased production costs. The addition of 13 new jobs is partly responsible for the cost increase.

Ecologically speaking, the new Magnefite system offers at least five significant benefits:

   1. The BOD of the discharge into the river is reduced 70 percent,

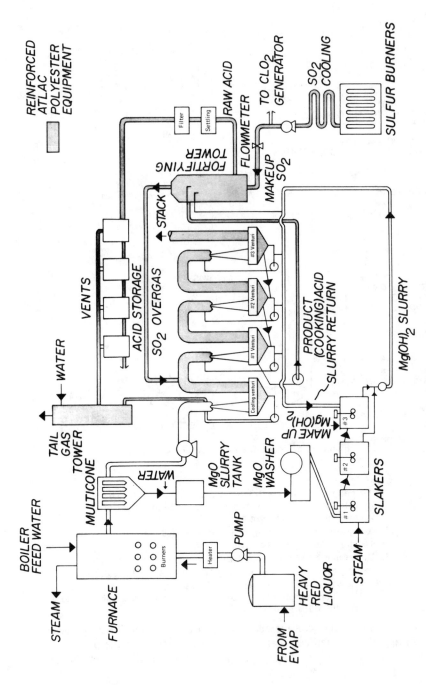

Figure 14-5. Magnefite recovery system (simplified).

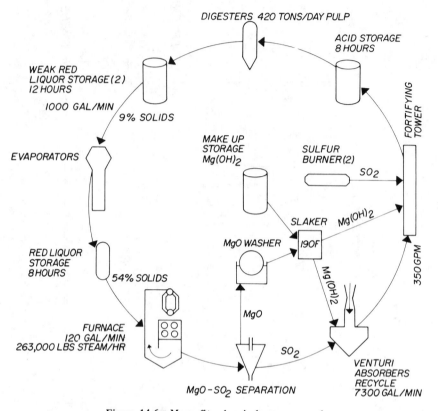

Figure 14-6. Magnefite chemical recovery cycle.

2. Air pollution codes (both federal and state) are met. The plume from the stack is harmless water vapor. All but a trace of the SO_2 is removed in the absorption train,

3. A relatively minor amount of sulfur is needed for makeup in the system because most is recycled,

4. Magnesium hydroxide is similarly recycled,

5. The furnace is capable of burning fuel oil or natural gas along with the heavy red liquor. However, it does so only when steam demand and/or liquor supply requires supplemental fuel. Thus, the demand for gas and oil is reduced.

The mill has had considerable success using fiber-reinforced polyester (FRP) materials. When considering the corrosive nature of the pulping process, the company's engineers specified FRP for major components in the recovery system. The only problem encountered in the FRP equipment to date has been damage to the resin-rich surface in certain areas near the bottoms of the venturis. The slurry moves rapidly in these areas and solids content is quite high. The problem apparently is one of erosion rather than corrosion. These areas have been repaired by adding a stainless plate for impact resistance and covering it with glass and rein.

STABILITY AND REMOVAL OF COMMERCIAL DYES FROM PROCESS WASTEWATER[130]

The total U.S. commercial color production currently amounts to almost 0.5 billion lb/yr. Practically every industry uses dyes and pigments to color its products. Many dyes and pigments are inert and non-toxic at the concentrations discharged into receiving waters. However, some are not so innocuous. In either case, the color they impart may be very undesirable to the water user. This is one of the reasons research has been performed on the stability of dyes to light and water under conditions similar to those encountered when they are discharged to natural streams and reservoirs.

This study is limited to some of the more common dyes used by the textile industry. The total dye consumption of the textile industry is more than 100,000,000 lb/yr. An estimated maximum of 90 percent of these dyes end up on fabrics—the remaining 10 percent goes to the waste stream. Approximately 10,000,000 lb of dye/yr are discharged to waste streams by the textile industry.

Basic Dyes

Basic dyes are cationic organic molecules capable of dyeing wool, polyester, acrylic and other fibers. These fabrics are prepared to contain anionic sites which attract and interact with cationic molecules. The dyes are applied from an aqueous solution containing enough acetic acid to adjust the pH from 4 to 6. The weight of the dyebath will generally be 20 times the weight of fabric being dyed. The solution is carefully heated to 190 F and held for 1 to 2 hr. The bath is then cooled, discharged and fresh water added to rinse surface dye from the fibers.

Basic dyes are the brightest class of soluble dyes used by the textile industry. Their tinctorial value is very high; less than 1 ppm of the dye produces an obvious water coloration. Fortunately these dyes are adsorbed by many minerals and organic matter, so natural processes can generally remove them from a stream without the help of sunlight if given sufficient time.

The basic dye structures studied are the triphenylmethane, phenazine and thiazine types. Fading rate curves for basic dyes are shown in Figures 14-7 to 14-11. All of the basic dyes showed appreciable degradation during their 200-hr exposure to visible and ultraviolet light.

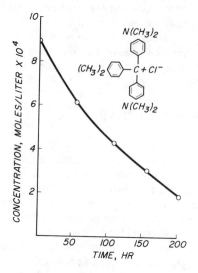

Figure 14-7. Rate of photodegradation of Basic Violet 3 in water at 50 C.

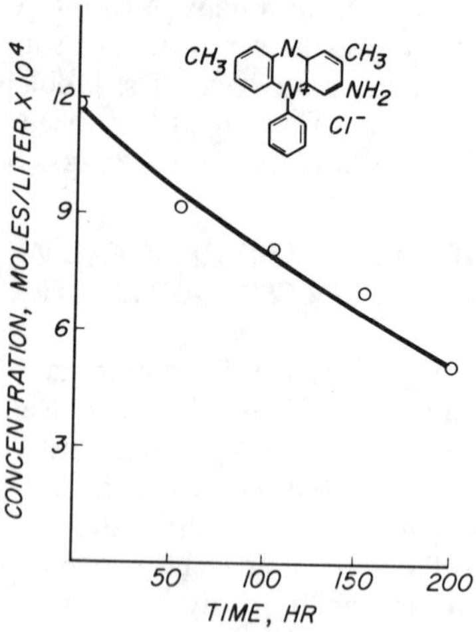

Figure 14-8. Rate of photodegradation of Basic Blue 9 in water at 50 C.

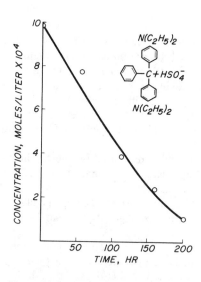

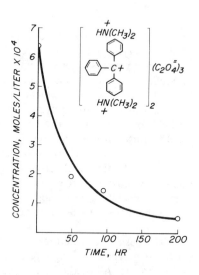

Figure 14-9. Rate of photodegradation of Basic Green 1 in water at 50 C.

Figure 14-10. Rate of photodegradation of Basic Green 4 in water at 50 C.

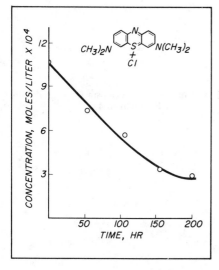

Figure 14-11. Rate of photodegradation
of Basic Red 2 in water at 50 C.

Figure 14-12 compares the effect of sunlight and artificial light on
Basic Green 4. The fading rate in artificial light is at least 10 times as fast.

The degradation products that could be identified by mass spectrometry
and gas chromatography from Basic Green 4 were: Leuco base, p-Methyl-
aminobenzophenone, p-Dimethylaminobenzophenone and p-Dimethylamino-
phenol. The results of this and other studies show that these dyes may
decompose by two principal paths. A proposed mechanism for the degradation of the triphenylmethane dyes is shown in Figure 14-13.

The carbinol form of the dye is converted to the excited state (possibly triplet) by absorption of ultraviolet light. The excited molecule is
converted into products by: (1) fragmentation into radicals and rapid reaction of the radical moieties with oxygen and water to give the products

Figure 14-12. Rate of photodegradation
of Basic Green 4 in water exposed
to sunlight and carbon arc light.

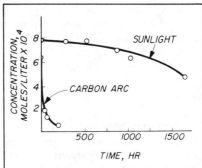

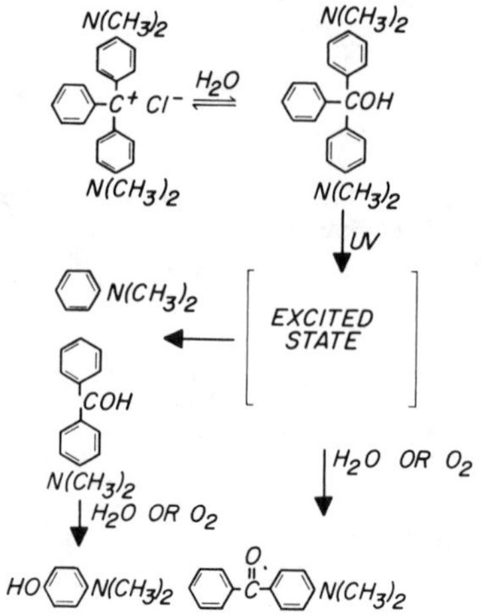

Figure 14-13. Mechanism for degradation of Basic Green 4.

isolated, or (2) concerted reaction of the excited carbinol form with oxygen and water to give degradation products directly. Oxygen seems to be necessary for the formation of the products obtained.

By adding phloroacetophenone to an alcohol solution of Cl Basic Green 4, it has been found that the time required for bleaching the solution was increased from two days to several weeks. This has been attributed to the fact that the adsorption region of the ketone and the dye were similar, so both compounds would be absorbing the same degradative radiation.

Most covalent bonds have dissociation energies which are well within the energy range of natural and artificial daylight. An enclosed carbon arc light produces a spectral wavelength range of 279 nm to 12,000 nm[9] which corresponds to energies of 95.3 to 40.9 kcal/mole, respectively. Since most dyes have adsorption maxima within this region, it is understandable that a number of them are vulnerable to photochemical degradation.

Basic Green 4 has two adsorption bands in the visible region. The band at 422 nm represents an energy of 67.8 kcal/mole, while the adsorption at 619 nm corresponds to an energy value of 46.2 kcal/mole. A third band exists in the near ultraviolet at 318 nm and is equivalent to 89.9 kcal/mole. The leuco carbinol form is a fugitive species and does not absorb visible light. Therefore, it requires ultraviolet light for degradation.

Acid Dyes

Acid dyes have an ancient origin, dating back many centuries to the first use of natural dyes on wool fibers. These dyes are used for nylon, wool and acrylic fibers. The nylon or wool fiber is dyed from a bath having a weight of 20 to 30 times that of the fabric. The fiber is placed into the warm dyebath, and the dye is added. The temperature is raised slowly to

200 F and, if necessary, acetic or sulfuric acid is added to exhaust the dye onto the fiber. If the dye tends to go on to the fiber too rapidly it may be necessary to add a retarding agent such as sodium sulfate. Several retarding or leveling agents are used with acid dyes.

These dyes are comparatively small molecules with one or more sulfonic acid groups attached to the organic substrate. They have good water solubility and may or may not be removed from a waste stream by a conventional biological waste treatment plant. Because of their high tinctorial color value, it is important to be able to estimate their light stability in a natural water supply.

Acid dyes as a class have poor light fastness on textile fibers compared to vat or disperse dyes. This suggests that their photodegradation in water would be quite significant. The curves for their fading rates are shown in Figures 14-14 through 14-19. Of the dyes studied, three showed drastic photo induced degradation.

One generalization indicated is that the acid azo dyes are more fugitive to light than the acid anthraquinone dyes. The only exception is the Acid Black 52, which is really a premetalized azo acid dye and not comparable because of the known stability metal coordination adds to any dyestuff.

The basic reason for the degradation of acid dyes seems to be their susceptibility to electrophilic attack. This has been demonstrated with basic dyes and shown to be applicable to the reactions of oxidizing reagents with other dyes. The more electron-attracting groups that are present on the dye molecule, generally the more stable it is to light.

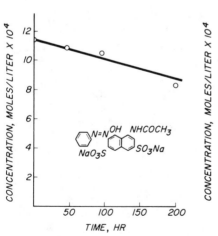

Figure 14-14. Rate of photodegradation of Acid Red 1 in water at 50 C.

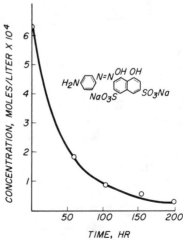

Figure 14-15. Rate of photodegradation of Acid Violet 3 in water at 50 C.

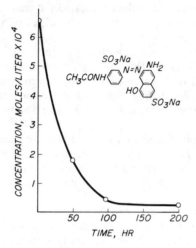

Figure 14-16. Rate of photodegradation of Acid Red 37 in water at 50 C.

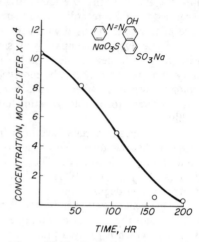

Figure 14-17. Rate of photodegradation of Acid Orange 10 in water at 50 C.

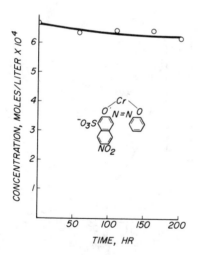

Figure 14-18. Rate of photodegradation of Acid Black 52 in water at 50 C.

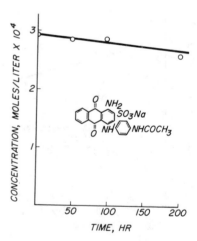

Figure 14-19. Rate of photodegradation of Acid Blue 40 in water at 50 C.

Acid dyes should be one of the best classes to interpret a fading mechanism because they are water soluble and give homogenous solutions. This is not true for the disperse, vat and sulfur dye classes, which can undergo physical changes affecting the absorption of light without undergoing any chemical degradation.

Direct Dyes

Direct dyes make up one of the major classes of dyes used on cellulose fibers. The dyes are applied from a water bath about 30 times the weight of the fabric being dyed. After the fabric and dyes are placed in the bath, it is heated to near boiling and a salt is added to exhaust the dye on to the fiber. The amount of salt required varies from dye to dye, but is approximately 10 percent of the weight of fabric being dyed.

These dyes are more resistant to light degradation than the basic or acid dyes, as can be seen by the fading rate curves shown in Figures 14-20 to 14-24.

Some direct dyes have sufficient affinity for the sludge in a biological waste treatment plant to be absorbed and removed from the wastewater. Others would require chemical or physical treatment for removal. In any case the data show that most direct dyes would be stable and resist photochemical degradation in a treatment plant or receiving water.

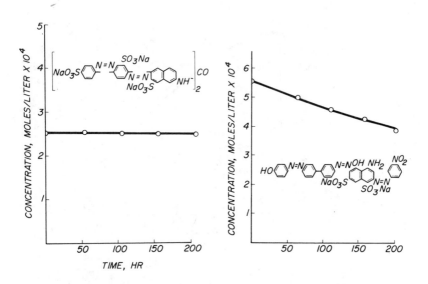

Figure 14-20. Rate of photodegradation of Direct Red 80 in water at 50 C.

Figure 14-21. Rate of photodegradation of Direct Green 6 in water at 50 C.

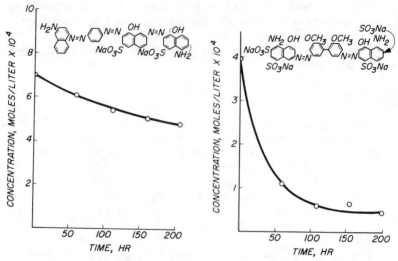

Figure 14-22. Rate of photodegradation
of Direct Black 80 in water at 50 C.

Figure 14-23. Rate of photodegradation
of Direct Blue 76 in water at 50 C.

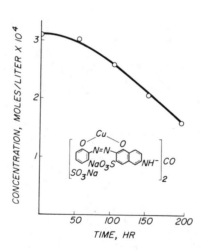

Figure 14-24. Rate of photodegradation
of Direct Red 83 in water at 50 C.

The comparison between degradation by natural and artificial light is shown in Figure 14-25. The data show that the rate of degradation is at least 10 times as slow in natural daylight as in artificial light. This indicates that the direct dyes are very stable in natural waters.

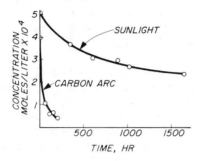

Figure 14-25. Rate of photodegradation of Direct Blue 76 in water exposed to sunlight and carbon arc light.

The dyes discussed in this study were selected from those most used by the textile industry, so that an accurate assessment could be made of the water pollution potential of common commercial colors. Most of the dyes are quite resistant to light degradation, showing an average of 40 percent color loss after 200 hr exposure to artificial light in water. A comparison of artificial light and natural sunlight effects on Basic Green 4 and Direct Blue 76 showed that these dyes degraded at least 10 times slower in natural sunlight. This means that a minimum of 80 days in a natural environment would be required to produce appreciable degradation of the dyes studied.

For the degradation of triphenylmethane-type basic dyes, hydrolysis plays an important part in the degradation mechanism.

Removal Systems

Since most dyes are stable and will remain in the environment for a fairly long period of time, it is best to remove them from waste streams before they are discharged. The type of waste treatment required for their removal will depend on the dye class and chemical composition. If only small quantities of color are present, secondary biological treatment with sludge separation may be effective. The dye can be adsorbed on the sludge in the same way it is adsorbed by plastics and fibers.

If the dye concentration is significant, biological treatment may not remove much of the dye. In this case carbon adsorption, chemical coagulation or hyperfiltration may be needed. Successful removal of color has also been accomplished with carbon and alum coagulation.

However, all dyes cannot be removed by carbon adsorption or coagulation. Some vat, disperse and pigment dyes are not adsorbed by carbon. Some very small molecular weight dyes such as acid and reactive dyes are not removed by coagulation. Each case must be studied and the correct process design formulated. Generally, laboratory and pilot plant data will be needed to ensure the successful operation of the treatment plant.

Hyperfiltration has also been studied for removal of color from textile waste streams. The results look good, but the economics of the process remain to be determined.

No commercial treatment process works in all cases, except reverse osmosis, and operational costs of this process have only been established for a few waste streams.

Several processes are in use or in experimental stages which show distinct promise for effluent decolorization. Although experience is limited in some, laboratory and pilot plant studies have proved their feasibility.

Activated Carbon

Textile dye wastes can be easily decolorized by a single-pass flow through fixed granular activated carbon beds at an average flux of 12 gpm/sq ft provided that the color bodies are receptive to adsorption on the carbon.

Economically, the process is well suited for handling complete treatment of small volume textile wastes (up to 75,000 gpd), and for pretreatment (complete color removal and 50 percent organic removal) of large volume textile wastes prior to discharge to conventional biological waste treatment systems. Activated carbon has an adsorption capacity in excess of 1.6 lb COD/lb of carbon, when the carbon is reactivated only by biological means. The estimated operating cost for decolorizing 1 million gpd is 8.3 cents/1000 gal, not including amortization.

Radiation-Oxidation

Solutions of **commercial** textile dyestuffs can be decolorized by a combined treatment using a chemical oxidant and gamma radiation. The combined treatment is more effective than the effect of the two components applied separately. In a laboratory demonstration of this method, aqueous solutions of several commercial dyestuffs and also a textile dye waste of unknown composition were subjected to irradiation near a Cesium-137 gamma ray source in both the presence and absence of chlorine. Without chlorine, complete decolorization, if it occurred, required a gamma dose of several megarads, necessitating a rather long exposure to the radiation source. In the presence of an excess of chlorine, color was destroyed during an irradiation time of a few minutes.

At a concentration of 0.25 g/l, the transmittance at the wavelength of maximum absorbance of dye solutions is greatly increased by treatment with a radiation dose of 60 kR plus 75 ppm chlorine. Nonoptimized cost estimates indicate $0.31/1000 gal for design treatment, with normal operating costs potentially lower.

Lime Precipitation

Calcium hydroxide in a slurry of constant concentration is mixed with the total process effluent in direct proportion to flow. The mixture is retained in a flocculator for 35 min and then clarified in a center feed clarifier. The colored substances (mainly lignin by-products) are precipitated as calcium salts and are removed from the system in the underflow of the clarifier, together with fiber and other settleable solids.

Operating results show that the color removal system can operate successfully under widely varying conditions to give a relatively constant effluent color in the range of 125 ppm APHA color units at treatment levels of 1000 (± 50) ppm of calcium hydroxide with untreated effluent colors in the range of 1200 (± 200) ppm. Treatment at this level produces a lime cost of $53.73/million gal, with lime at $15.35/ton (90 percent CaO).

An additional advantage of this color removal process is that it preconditions the effluent for biochemical treatment allowing rapid degradation. Foaming is eliminated and phosphorus concentration is reduced, preventing possible eutrophication.

COMBINED TREATMENT FROM POWER PLANTS[189]

All of today's large electric generating plants require enormous quantities of water for condenser cooling. Additionally, large amounts of water are converted to sewage each day by cities. A combined thermal-sewage treatment system has been proposed, however, to treat both these effluents simultaneously.

As shown in Figure 14-26, the combined system incorporates an activated sludge sewage treatment plant with a power plant spray-type cooling channel. The circulating liquid, called mixed liquor, flows through the system in a closed loop path. The mixed liquor is pumped through the power plant condenser after which raw sewage combined with settled sludge is added. Then the mixture is aerated and cooled, and a portion of it drawn off to settling tanks just before re-entry to the condenser.

Raw, screened sewage from a nearby community is introduced into the treatment channel downstream of the power plant condenser location. The raw sewage influent can have a suspended solids content of 300 ppm and a flow rate of 200 cfs for a city of 1 million people.

For a 1000-mW power plant, as much as 2000 cfs mixed liquor will be pumped through the condenser where its temperature is raised approximately 20 F. For maximum treatment of sewage, a suspended solids buildup of 3000 ppm is desired in the mixed liquor. Research is, however, needed to determine the optimum amount of solids concentration that a specific power plant condenser can handle.

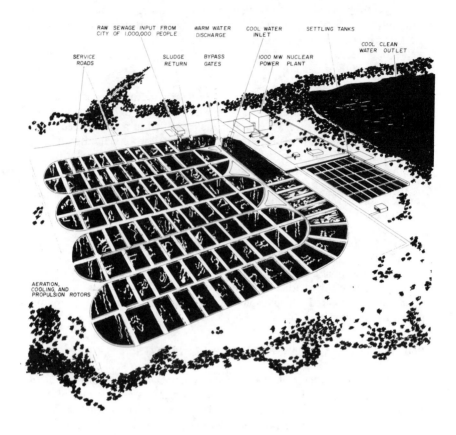

RAW SEWAGE INPUT FROM CITY OF 1,000,000 PEOPLE
WARM WATER DISCHARGE
COOL WATER INLET
SETTLING TANKS
COOL CLEAN WATER OUTLET
SERVICE ROADS
SLUDGE RETURN
BYPASS GATES
1000 MW NUCLEAR POWER PLANT
AERATION, COOLING, AND PROPULSION ROTORS

Figure 14-26. Combined sewage treatment and cooling system.

As mixed liquor flows through the winding channel, it receives aerobic biological treatment and, at the same time, is cooled for return to the condenser. Rotary paddle wheel devices are used to provide aeration as well as motion for the channel. They are alternately spaced across the channel with thermal rotors, like flat spinning disks, which perform the cooling function. These rotors can be automatically controlled for the amount of aeration and cooling needed at any time. Research is also required, however, to determine the optimum oxygenation, cooling, and channel size for the system.

In this type of system, even though the power plant were shut down for maintenance, the sewage treatment in the channel would continue. This is accomplished by special by-pass gates.

Just before the mixed liquor enters the condenser, a portion is drawn off into settling tanks for separation of its suspended solids from the liquid.

Settled sludge, teeming with bacteria at the bottom of the settling tanks, is returned to the mixed liquor channel for continued aerobic treatment and for seeding of the city's raw sewage with bacteria. By continually returning the activated sludge from the settling tanks, a suspended solids concentration of approximately 3000 ppm can be maintained in the mixed liquor channel.

The clean and cool effluent leaving the settling tanks is returned to a river or other natural body of water after chlorination, thereby avoiding any detrimental effect to that body of water.

SOIL EROSION POLLUTION[183]

Each year more than 3 billion tons of soil from rural and urban lands are washed into the watercourses of the United States. Almost a third of this amount is carried from farmlands of mid-America into the Mississippi River and finally deposited in the Gulf of Mexico. This is only a part of the total soil loss from the country's land, unfortunately. The greater part is laid down as a cover over other lower lands or in stream channels and reservoirs.

Soil erosion from rural areas usually remains fairly stable once agricultural land-use practices have been established. This may not be the case in urban areas because of construction. Relatively short-term soil erosion losses from construction sites during land clearing, shaping, and stabilization of new surfaces may contribute significant sediment loads to watercourses. Runoff from bare disturbed construction sites often contains 1000 or more times its normal suspended sediment load, and large construction-site soil losses of from 50 to as much as 150 tons per acre are not uncommon.

Standards

Soil erosion standards in the United States generally have been established on the premise that any rate of erosion which does not harm crops or the crop land is permissible. However, as more and more emphasis is placed on stream quality control, more restrictive standards will evolve. When this occurs, farming, construction, and industrial plant maintenance practices will probably be placed in serious jeopardy.

In many states very restrictive standards already exist, but as yet are not being rigorously enforced. Established water quality standards for permissible suspended solids concentrations in effluent to streams apply to all suspended solid matter regardless of source.

If present standards were rigidly enforced, all farming and construction probably would have to be terminated immediately. This fact is illustrated in Figure 14-27, which relates soil loss rates to suspended solids loads for various amounts of yearly runoff. From the graph it is apparent that, even at a minimal soil erosion rate of 1.0 ton per acre per year, the suspended solid load in runoff to rivers and streams like the Mississippi, Ohio and their tributaries would be about 10 times the maximum allowable concentration.

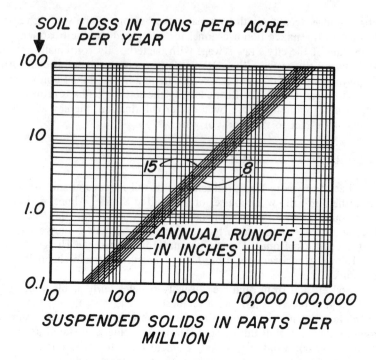

Figure 14-27. Relationship of soil loss rates to suspended solid loads
for varying amounts of annual runoff.

Corrective Measures

Not much attention has been devoted to effectively preventing soil
erosion from denuded surfaces at construction sites and around industrial
plants. However, several methods offer economical solutions. Temporary
embankments or trenches can be designed to capture and retain the silt-
laden runoff. Ground coverings can be used to reduce runoff velocities
below critical levels. These include fibrous matting in either spray-on or
sheet form or fast-growing vegetal covers such as crown vetch. Impermeable
plastic-type sheets or spray-on chemical covers are also available to isolate
bare surfaces from rainfall.

Agricultural soil conservation practices prescribed in the past generally
have not been widely accepted, primarily because the costs could not be
justified on the basis of actual savings to the land owner. Minimum or
zero tillage, a relatively new method now being developed and perfected,
appears to offer far more promise of universal acceptance and eventual
marked reduction of erosion than other methods.

DISSOLVED AIR FLOTATION FOR
TREATING INDUSTRIAL WASTES[102]

Dissolved air flotation is a waste treatment process whereby oil and other suspended matter is removed from a waste stream. Minute air bubbles are attached to the oil and other suspended matter, reducing the effective specific gravity of the suspended particles to less than that of water. The suspended matter rises to the water surface where it can be mechanically removed.

This treatment process has been in use for about 15 years and has been most successful in removing oil from waste streams. Free oil can be removed directly through the use of dissolved air flotation. However, if the oil is emulsified, chemical conditioning is required to break the emulsion and form a floc to absorb the oil. The oil/chemical floc particle can then be removed from the waste stream by dissolved air flotation.

When chemical pretreatment is required, a flocculation compartment and a flash mix compartment (if required) can be designed as integral parts of the dissolved air flotation basin. Figure 14-28 schematically shows how and where chemicals may be added to raw process flow.

A schematic flow diagram of the treatment process is shown in Figure 14-28, and a pictorial cutaway of the process shown in Figure 14-29. From these figures it can be noted that a portion of the clarified effluent is pressurized by a recycle pump. This recycled flow is pumped to a pressure tank into which air is injected. In the pressure tank at approximately 40 psig, the recycle flow is almost completely saturated with air based on the operating pressure. The pressurized recycle flow, containing the dissolved air, leaves the air saturation tank and flows through a pressure reduction valve.

A 40-psig pressure drop occurs at the pressure reduction valve and causes the pressurized flow stream to relinquish its dissolved air in the form of tiny air bubbles. This air-charged recycle flow is then blended with the raw process flow to effect attachment of the air bubbles to the oil and other suspended solids to be removed. The combined flow stream (raw flow plus recycle flow containing the air bubbles) is mixed and uniformly distributed over the cross-section of the basin.

As the incoming flow travels to the effluent end of the basin, separation of the oil and solids from the associated liquid occurs. Solids accumulate at the water surface and form an oily sludge blanket. Clarified liquid flows over the effluent weir and into a wet well. From the effluent wet well, a portion of the effluent is recirculated. The remainder of the effluent flow is removed from the basin for subsequent treatment or discharge.

The floated scum blanket of separated solids and oil is removed from the basin by skimmer flights traveling between two endless strands of chain. Since the influent stream may also contain small amounts of heavy solids, such as grit, which are not amenable to flotation, provision must also be made for solids removal from the bottom of the unit.

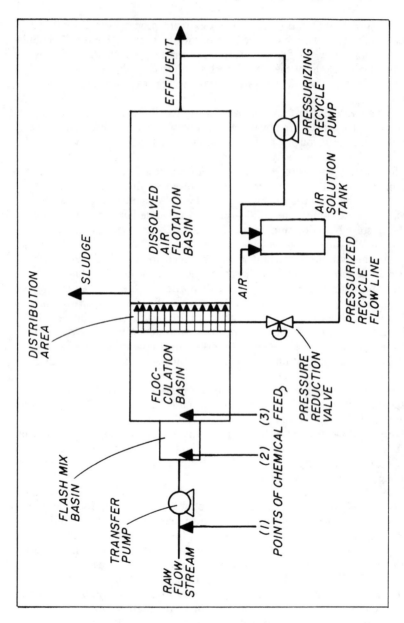

Figure 14-28. Flow diagram of dissolved air flotation system, including points of chemical injection.

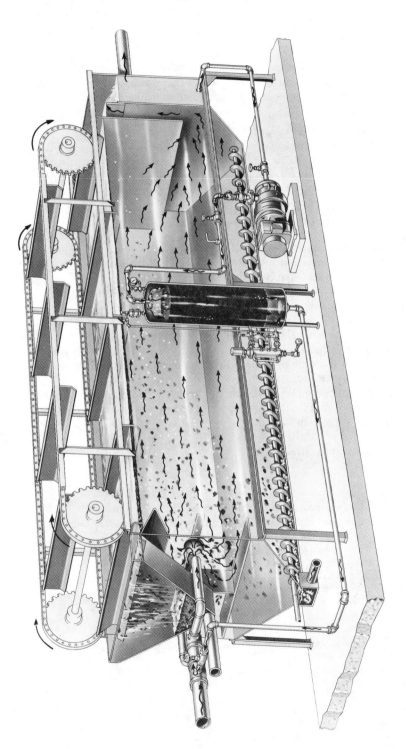

Figure 14-29. Cutaway photo showing major components of dissolved air flotation system.

Treatability Testing

Studies must be conducted on a waste sample to determine if the waste is amenable to dissolved air flotation. Once that has been established, the following design parameters can be established with the use of laboratory bench-scale tests:

1. Rate of rise of the particles or mass
2. Optimum recycle rate
3. Attainable effluent quality
4. Type of chemical conditioning required
   a) type and number of chemicals
   b) optimum dosage of chemicals
   c) point of application of chemicals
   d) flocculation and flash mix detention times

The "Manual on Disposal of Refinery Wastes (Volume on Liquid Wastes)" of the American Petroleum Institute describes these bench-scale treatability test procedures for the dissolved air flotation process.

Design Procedure

From the bench-scale treatability tests, the theoretical mass or particle rise rate V_t can be determined. The optimum rise rate is a function of the recycle rate and the type of chemical pretreatment (if required) as determined from the bench-scale studies.

The design procedure of the actual dissolved air flotation unit is basically the same design procedure as developed for the oil-water gravity separators by the American Petroleum Institute.

1) For the effective length:

$$L = F \frac{V_h}{V_t} d$$

where
- L = effective length of the dissolved air flotation basin, ft
- F = turbulence and short-circuiting factor (Figure 14-30), unitless
- d = effective depth of basin, ft
- V_t = theoretical rise rate of particle (obtained from bench scale tests), ft/min
- V_h = horizontal velocity, ft/min

and

$$V_h = Q_T/A_c$$

where
- Q_T = total combined flow (process raw flow plus recycle flow) cfm
- A_c = effective cross-sectional area of the basin, sq ft

2) Other design criteria and recommendations include

 a. Depth to width ratio between 0.3 and 0.5,
 b. Maximum V_h/V_t ratio of 15,
 c. Maximum V_h pf 3 fpm,
 d. Optimum length to width ratio of 4:1,
 e. Maximum width of 20 ft.

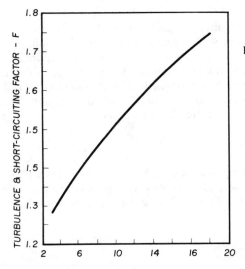

Figure 14-30. Recommended values of F, turbulence and short-circuiting factor.

Applications

  Dissolved air flotation is used to treat the following industrial waste streams. Most of the wastes are pretreated.

Refinery Wastes

  Primary treatment of "general" refinery wastes involves the use of oil-water gravity separators (API separators) to remove free oils. This process is followed by dissolved air flotation to remove the remaining free oil plus emulsified oil if necessary. Typical pretreatment chemicals used are alum, lime and bentonite.

Machine Shop Wastes

  Wastes from machine shops are very difficult to treat, because they are:

1. Highly variable in flow quantity and duration,
2. Very high in emulsified oil, and
3. Extremely variable in the "types" of emulsified oil present. Different concentrations and/or types of chemicals are required for proper treatment.

The treatment procedure for machine shop waste is as follows:

1. Oil-water gravity separation to remove free oils, and/or use of
2. Equalization basin(s) to equalize the flow volumes and types of emulsified oils,
3. Chemical treatment with the use of lime, alum and polyelectrolytes followed by flocculation and dissolved air flotation.

Dual basins are usually required for gravity separation. They also serve to isolate a given volume of flow so that bench-scale treatability studies can be conducted to determine the type and amount of chemicals required for pretreatment. These equalization basins dampen the peak flows, so that a smaller dissolved air flotation basin can be used. Mixing should be provided so that a uniform waste can be fed to the dissolved air flotation unit. Dual equalization can also serve for batch treating a continuous flow.

Meat Packing Wastes

Treatment of meat packing and slaughterhouse wastes serves a twofold purpose: first, to make the waste suitable for discharge, and second, to recover grease by-products (Figure 14-31). Wastes are first sent through a

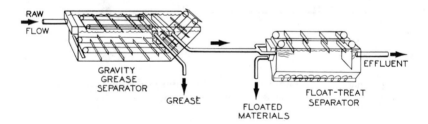

Figure 14-31. Meat packing grease recovery system.

gravity grease separator for removal of heavy quantities of grease and solids, and then through the dissolved air flotation process for more complete removal of grease and fine solids. The U.S. Department of Agriculture restricts the use of chemicals in treating wastes from meat packing and slaughterhouses if the removed grease will be used as animal feed; therefore, chemical pretreatment is limited to certain USDA-approved polyelectrolytes.

Edible Oils Processing Waste

Waste containing light soluble oils can be removed on a continuous or batch feed operation. The process usually starts with a mixing and equalization basin, followed by the addition of two or more chemicals with flash mixing, flocculation and dissolved air flotation. An equalization basin dampens volumetric flow variations. Mixing insures a uniform waste feed concentration for optimum chemical usage and treatment performance.

Railroad and Aircraft Maintenance Wastes

These wastes are first treated with oil-water gravity separators for the removal of free oils, and then sent to an equalization basin for the dampening of peak and maximum flows (Figure 14-32). From the equalization basin, the flow is chemically conditioned and flocculated before being fed to the dissolved air flotation basin. Railroad maintenance waste is usually first passed through a grit chamber before being sent to the gravity separators, since it usually contains a significant amount of gritty material from railroad car washing facilities.

Automotive Production Wastes

Waste from automotive production has a normal treatment process of grit removal followed by an equalization basin for batch feeding the dissolved air flotation unit. Chemical conditioning and flocculation is usually required. Free oil removal is provided for in the equalization basin.

Tank Truck Washout Wastes

These wastes are the result of the internal tank truck washings from clean-out of bulk materials. This waste is treated on a batch process. Chemical treatment is usually required with flocculation before the waste is sent to the dissolved air flotation process. This process removes not only oil and grease, but other types of suspended material washed out of the trucks.

SALT PILING POLLUTION OF WATER SUPPLIES[198]

Chemical pollutants from surface sources are causing serious groundwater quality deterioration in many shallow aquifers throughout the United States. Instances of such pollution are increasing, and widespread contamination of many major aquifers may occur unless stringent measures are employed.

The Illinois State Water Survey was recently requested to determine the cause of cyclic occurrences of high chloride and hardness in groundwater pumped from a shallow unconsolidated aquifer in Peoria, Ill. (Figure 14-33).

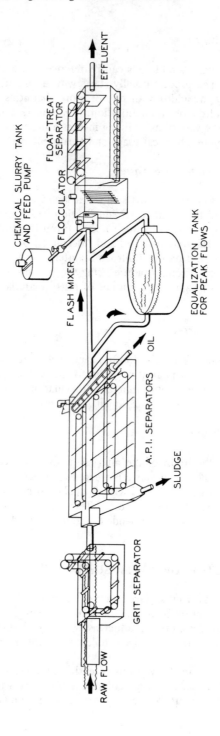

Figure 14-32. Railroad maintenance waste treatment system.

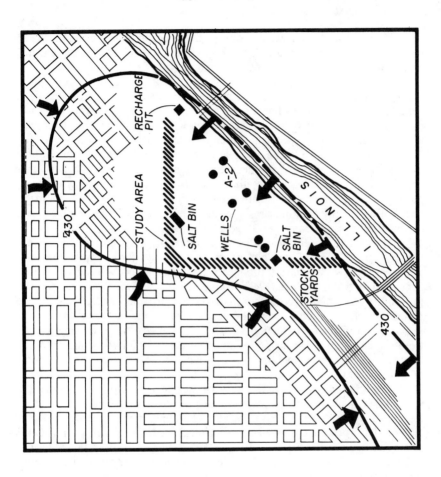

NOTE:

DIRECTION OF
GROUNDWATER FLOW

GROUNDWATER
SURFACE ELEVATION

A-2   AFFECTED WELL
NUMBER(A-2)LOCATION
AND (428.8)
WATER-LEVEL ELEVATION

Figure 14-33.

Methods used to locate the sources of pollution and to evaluate the magnitude of their effects are directly applicable to the solution of similar problems no matter where they may occur.

A serious chloride pollution problem has plagued industrial groundwater developments in this area for many years. Because this was first detected in production wells located in the vicinity of the stock yards, it had been assumed that the pollution was caused by leakage from deep bedrock artesian wells formerly used as a source of brine. After several unsuccessful attempts were made to locate and seal these now abandoned and buried installations, the industries concerned generally accepted the fact that it would be difficult if not impossible to eliminate this source of pollution. Consequently, they attempted to isolate the pollution by adjusting their pumpage distribution and by locating supply wells for chloride-sensitive processes as far away as possible from the suspected source area.

However, in recent years these corrective measures have failed to contain the salt. Formerly low-chloride producing wells, far removed from the stock yards, now periodically produce water that has an even higher chloride content than water from the wells nearest the old suspected source. The probability that these occurrences might be due to a new and different pollution source caused concern, because there is no available property remaining to relocate production wells. Also, the cost of purchasing water from the city supply for high-rate industrial use is rapidly becoming prohibitive. These costs often amount to more than $500 a week for periods of several months.

The aquifer in this area is a very permeable sand and gravel formation about 60 ft thick which is contained in a partially buried bedrock trough associated with the Illinois River valley. The upper part of the water-bearing material lies at or near land surface over much of the southern part of Peoria.

The aquifer is recharged with water derived mostly from precipitation, but a portion also comes from the river through recharge pits and by induced infiltration through the bed of the stream. A significant amount of additional recharge is received from adjoining shallow bedrock aquifers, and from city-street storm runoff through leaks in the old brick sewer system. Groundwater within the aquifer generally is a blend of water from these various recharge sources.

Plots of chemical patterns and graphs of hardness variations showed that groundwater chemical quality within this area has changed considerably in recent years. Much of this change occurred following expansion of one industrial well field (Wells A-8 and A-9) and the further development of municipal well fields.

Changes in groundwater chloride content and water table fluctuations near the center of the study area during the period selected for study are shown in Figure 14-34. Data from this figure for periods when the chloride concentrations were maximum were used to construct a chloride-distance profile (Figure 14-35).

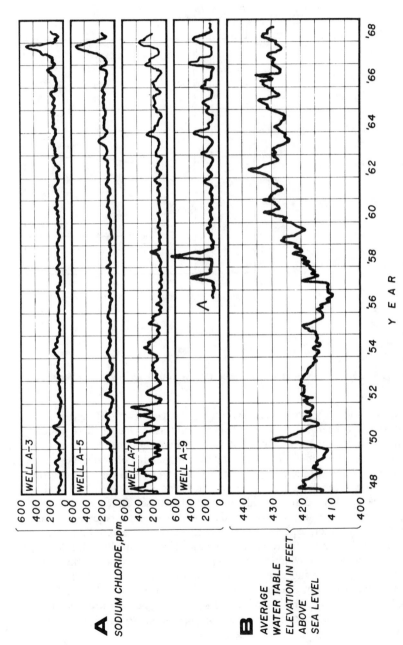

Figure 14-34. Changes in groundwater chloride content pumped from four affected industrial wells are shown in A. The average fluctuations of the water table are shown in B.

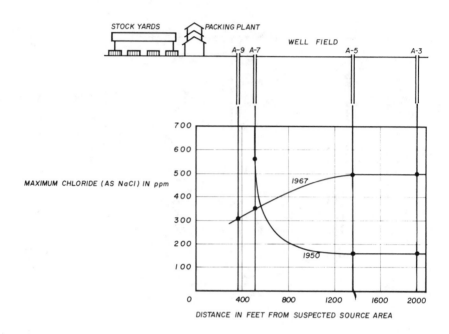

Figure 14-35. Profiles of chloride concentrations in underground water supplies.

From these profiles it was obvious that at least two different sources of chloride had been active in the fairly recent past. One of these seemed to have been centered near the stock yards, whereas the other source appeared to be at some point much further away, in the vicinity of Wells A-3 and A-5.

None of the data seemed to support the hypothesis that the chloride pollution was being caused by underground leakage from abandoned salt wells in the stock yard area. Had this been the cause, evidence of a fairly high, uniform concentration of chloride would have been expected in the affected groundwater, especially during periods of no precipitation when the water table was relatively low. Instead, a series of cyclic chloride occurrences was evident, showing the salt content to be highest when the water table is highest. This suggested that the chloride pollutant was being transported downward from the surface by precipitation recharge to the aquifer.

Investigation of other possible salt sources in the area revealed that the now-abandoned packing plant had for years used large quantities of sodium chloride in meat and hide processing operations. These processes also required large amounts of water. Liquid waste from these operations had been discharged to 48- and 60-in.-diameter brick sewers. It is probable that liquid waste pollutants from this source entered the fresh water aquifer through leaks in the old brick sewer which served the plant.

The salt supply for the packing plant apparently had been stored in the basement of the building which contained the meat and hide processing facilities. During an inspection of the plant, it was observed that several tons of salt had been left in storage after the plant ceased operation. This material was removed to ensure that no further aquifer pollution can occur from this source.

A comparison of hardness and chloride data showed a time sequence of mineralization in affected wells during the winter of 1967-1968. Peak concentrations occurred first in November at Wells A-3 and A-5, whereas comparable peaks did not occur until the end of December in Wells A-7 and A-9 which are located about 1200 ft away. With this time-distance relationship, and the assumption that pollution from the unknown source may have been transported to the aquifer by precipitation recharge, a check was made of climatological data. Rainfall records for the fall and winter months indicated that precipitation during December and November was near average. During October of that year, however, 5.56 in. of rainfall occurred, which is about 2.3 times the mean precipitation value for this month. These data indicated that, if salt pollution was carried to the aquifer from a surface source by precipitation, the most likely time was during October, a month prior to the appearance of salt at Wells A-3 and A-5. The time-distance relationship established for the movement of pollution between Wells A-5 and A-7, and the precipitation-time relationship, indicated that this source of pollution should be within 1000 ft of Wells A-3 and A-5.

A further clue to the new pollution source appeared in the chloride-hardness relationship. Between about 1956 and 1968, chloride increases generally were accompanied by a corresponding increase in hardness. However, this correlation did not appear to hold for dates prior to 1956.

From available data three generalized clues to the source of recent groundwater mineral pollution were evident:

1. It had to be located about 1000 ft west of Wells A-3 and A-5;
2. It could not have been in existence before about 1956;
3. The source material would have to contain both chloride and hardness-forming chemicals.

Within the cone of influence of the affected well field there was only one known source of pollution that satisfied all of these criteria. This was the city street-salt storage facility.

Salt has been stored at this location for about 15 years. Initially, salt was merely piled on the ground; then a concrete slab base and a tarpaulin cover were used to protect the salt piles from the weather. Later, a street department garage was constructed including partially enclosed concrete salt-storage bins covered with a permanent roof. The concrete-slab floor of these bins extended out beyond the roof line by about 5 ft. Between 2000 and 3000 tons of salt were generally stored in the bins for an average of 120 days, from November through February. This material consists of a 4 to 1 mixture of sodium and calcium chloride. In recent years there has been a sizable carryover of salt each year. It is estimated that the recent yearly carryover was 350 tons. Inspection of this storage site showed the ground surface practically covered with salt for a distance of several feet beyond the concrete slab floor of the storage bins. Also, small trickles of a salt-saturated water could be seen flowing from some of the storage bins into the brick-lined storm sewers which underlie the property.

It is suspected that most of the salt pollution which enters the aquifer from the street-salt storage area is transported by rain precipitation recharge downward through the uppermost dewatered parts of the sand and gravel to the water table. If this is the case, pollution from this source will probably continue for some time after this storage facility has been abandoned unless the salt-saturated earth around and underlying the storage area also is removed. The salt pollutant will not be entirely eliminated unless brick conduits within the sewer system are replaced or the practice of street salting is discontinued.

Although it may take several years to completely eliminate this form of pollution, short-term corrective measures may be successful in temporarily containing pollutants so that effects on industrial processes will be minimal.

A rigid pumping control program should be initiated to keep high-chloride groundwater from the street-salt storage area away from wells providing water to salt-sensitive processes. Such a program may require trying several combinations of pumping from production wells closest to proven sources of pollution before an effective cone of influence trap around these wells can be established. Large quantities of groundwater will probably have to be pumped to waste during all of the months that such a program is in operation.

Additional production wells may have to be constructed within the affected pumping cones before effective pumping-center traps of the highly mineralized water can be established. If new wells are drilled, some of these should be located near the recharge pits and the river to obtain a maximum quantity of quality water.

PLASTIC PIPE FOR SEWERS[111]

Deterioration of water and wastewater lines laid many years ago, the need for expanded systems, and meeting new pollution control standards are just a few of the reasons municipalities and industries are turning to high-density polyethylene pipe. This pipe should not be confused with the thin-wall plastics commonly used for pipe liners. It is an engineered pipe with wall thicknesses that make it self-supporting. It is extruded from high-density polyethylene resins in standard outside diameters as small as 1 in. and as large as 48 in. Outside diameters up to 80 in. are fabricated to order.

One of the most significant benefits of polyethylene pipe is its inherent resistance to most chemicals. For most applications, it can be considered inert. This makes it ideal for carrying wastewater as well as for water supply or drainage. Inside diameters are not reduced over long periods of use, so capacities are maintained.

Since enactment of new FWPCA standards on July 1, 1973, sewer collection systems must be free of excessive infiltration as well as leakage. High-density polyethylene pipe meets this requirement because conventional joints—the weakest link in any piping system—are completely eliminated. Sections of pipe are joined by butt fusing. Pipe ends are heated to the melting point and then joined together under pressure. The resultant joint is as strong as the body of the pipe itself and entirely leakproof.

Typical polyethylene pipe weighs only 1/3 to 1/9 as much as other pipe materials. This simplifies handling and substantially reduces the number of cranes and other heavy equipment needed for installation. Weight saving doesn't, however, sacrifice ruggedness. Polyethylene pipe won't break or crack during shipment or handling. It is not necessary to add a breakage factor when estimating total pipe needs.

Like other plastics, high-density polyethylene pipe is flexible. Although it cannot be bent to a right angle, it can be cold bent to a radius of 20 times the outside diameter. Contouring the pipe around obstacles can eliminate the use of fittings.

Inherent flexibility also plays a key role in off-setting the effects of water hammer and transient over-pressure cycles. Instead of propagating the pressure waves, polyethylene pipe tends to expand and dampen the shock waves. There is less need to plan the design to three times rated pressure when selecting polyethylene pipe. The designer can stick to basics— flow, temperature, pressure, chemical resistance and life expectancy.

Polyethylene pipe is basically non-wetting and has a factor 10 to 25 percent greater than concrete, cast iron or other rigid pipes per cross-sectional inch. Such an increase in flow capacity allows the pollution engineer to consider smaller diameter pipe than needed with other materials.

The specific gravity of polyethylene is less than unity, a feature which makes it particularly applicable for water installations. Lengths of prejoined pipe, even lengths of a 1/4-mile or more, can be floated across a waterway and submerged to their resting place on the bottom. Such systems have saved installers thousands of dollars in equipment and services while limiting the disruption of commercial traffic as well.

Limitations

Polyethylene pipe's biggest enemy is temperature. As the pipe body temperature rises about 70 F, strength decreases. Therefore, the pressure which can be tolerated on a long-term basis (50 years) must be reduced for higher temperatures. Slight variations within temperature limits will not affect longevity. Polyethylene pipe will not break if its contents freeze. Normal functions, without degradation, will resume upon thawing.

Polyethylene pipe should be stored and handled in an area relatively free from jagged rocks or ledges. Deep radial scratches are particularly damaging to pipe strength under stress conditions.

Pipe Joining

The recommended method of joining lengths of polyethylene pipe is butt fusion. Mechanical joints defeat the pipe's leakproof properties, and joining with solvents is not possible. The fusion process is simple and can be done by two men. No special skills are required, so joining can be done by personnel with very little training.

The usual practice is to deliver standard 40-ft lengths of pipe to the job site. The most practical approach is to join all of the lengths of pipe into a monolithic structure prior to installation. Continuous lengths of pipe 1000 ft long or longer are not uncommon. But because of the pipe's lightweight properties, they are easy to handle with a minimum of equipment.

Figure 14-36. Even large diameter polyethylene pipes require only a few workmen to butt fuse weld joints into a monolithic leakproof structure.

The joining operation takes only four basic steps using equipment supplied by the pipe manufacturer. First, two pipe ends are aligned and a collar is attached to them. Often, V-rollers are used to simplify alignment and handling. A trimmer is inserted between the two ends which cleans and trues both edges for perfect mating. The trimmer is removed and replaced by a heating ring matching the diameter of the pipe.

Ends of the pipe are then heated above the crystal melt point. The heat ring is removed and the ends of the pipe are quickly forced together under hydraulic pressure and held immobile until completely cooled. The time required for each joint is short. Even large diameter pipes require less than 30 min from start to finish. Because each joint adds 40 ft to the finished length, big jobs go quickly and time required is far less than for conventional installations.

If joint testing is a standard operating procedure, brief exposure under two times normal rated pressure is adequate to verify the integrity of the joints. However, if testing is done while the pipe is subjected to hot sunlight, or if the ambient in the test area is otherwise greater than the expected temperature load, reduce the test pressure proportionally.

Sewer Relining

High-density polyethylene pipe provides an economical, practical method for the relining of deteriorated sewers. Its principal benefit lies in the fact that it can eliminate whole city blocks of sewer excavation that would be required if the deteriorated sewer were dug up and replaced. In most relining jobs involving polyethylene pipe, excavation is required only at entrance, exit and service hookup points. Disruption of traffic and damage to landscaping is minimal. Excavating costs, often many times the price of sewer pipe, are reduced.

A diameter of polyethylene pipe is selected that is slightly smaller in outside diameter than the existing pipe inside diameter. Because of the non-wetting characteristics of polyethylene, predetermined flow factors are often increased.

While excavation takes place at the entrance and exit points, the pipe is assembled into a continuous length at the construction site. When excavations and assembly are completed, a cone-shaped drawing collar is fitted to the head end of the pipe to prevent snagging. A drawing cable is run through the existing sewer and pulling begins. Polyethylene pipe is flexible enough to follow common contours in the original sewer line. In some installations, it is easier to pull manhole-to-manhole, in stages, rather than attempt a single draw the full length of the sewer line.

Continuous lengths of polyethylene pipe exceeding a quarter mile are not unusual although it obviously draws some amazed glances from passers-by as this leviathan begins making its way under the earth. Once the new liner is in place and hookup is completed, refilling and resurfacing at each end of the line is a simple job. Relining sewer pipe instead of replacing it can result in savings of 20 to 40 percent.

New Installations

There are many reasons for using high-density polyethylene pipe in new installations, even though the initial cost of poly pipe exceeds that of conventional pipe material.

1. The waste effluent may affect the life of conventional pipe, whereas polyethylene may be totally immune.
2. Transportation charges for shipping conventional pipe may exceed the cost differential. Polyethylene pipe can weigh as little as 1/3 to 1/9 that of other types of pipe.
3. The gain in flow factor from polyethylene pipe may allow the selection of a smaller diameter. Over the years, encrustation problems (and ID reduction) will be far less with poly pipe than with other materials.
4. Monolithic systems reduce infiltration and exfiltration. This requirement for federal funding is easy to prove.
5. No allowances for breakage or damage are needed.
6. Light weight and longer working length save on manpower and equipment.

MEMBRANE POND LINERS[108]

Increasingly rigid clean water standards have prompted industry, agriculture and government bodies to resort to lined ponds to contain potentially polluting fluids prior to treatment, disposal or re-use (Figure 14-37). Many of the traditional liner materials such as concrete, asphalt, wood, metal and clays have not proven entirely satisfactory in meeting these new regulations because of their vulnerability to fracture, permeation or corrosion, or because they are too costly for large projects. As a result, there is a growing use of flexible non-permeable liners to contain these fluids.

The most important advantages of membrane-type liners are their ability to contain a wide variety of fluids with a minimum of loss from permeation and seepage, their high resistance to chemical and bacterial deterioration, and their relative ease and economy of installation and maintenance.

Major disadvantages of flexible liners are: Their relative vulnerability to ozone and ultraviolet deterioration as compared to some of the hard liners; their limited ability to withstand the stresses of heavy machinery as compared to concrete or asphalt; and their comparative susceptibility to laceration, abrasion and puncture from sharp objects such as metal chips, stones, tree roots, etc. Some membrane materials are prone to crack at extremely low temperatures and to stretch and distort at extremely high temperatures.

Materials used in the fabrication of membrane liners fall into two general classifications—rubber and plastic. There are no clear-cut advantages or disadvantages of either basic material, per se. Formula variations of both materials can produce end products that will be better suited to a specific application than another liner of the same base material. There are many

Figure 14-37. Membrane lined ponds such as this are being increasingly used as an effective low-cost method of containing a wide variety of fluids.

**Table 14-2. Some of the Physical Characteristics
That Should be Considered in Selecting a Liner**

- Material should satisfactorily resist attack from all chemicals, ozone, ultra-violet rays, soil bacteria, and fungus to which it will be exposed.
- It should have ample weather resistance to withstand the stresses of freezing and thawing and seasonal shifts of earth.
- It should have adequate tensile strength to elongate sufficiently and withstand the stresses of installation or use of machinery or equipment.
- It should resist laceration, abrasion and puncture from any matter that might be contained in the fluids it will hold.
- All membrane in a given installation should be of the same material furnished by the same manufacturer to ensure compatibility.
- The liner should be of uniform thickness, free of thin spots, cracks, tears, blisters, and foreign particles.
- It should be of sufficient thickness to insure long term service in a specific application.
- It should be easily repaired.
- It should be of the most economical material that can adequately fill a specific need.

liners made by manufacturers using these basic materials with formula and structural modifications to fill specific needs.

More relevant classifications are as *exposable* and *unexposable* materials. Exposable membranes are formulated to resist ozone and ultraviolet exposure for a longer period of time than unexposable membranes. Costs of the exposable materials are higher, ranging to as much as 80 percent more than the cost of unexposable liners. However, service life of most exposable materials can be expected to range from 20 to 25 years or longer under normal atmospheric exposure, while unexposable liners could only be expected to perform satisfactorily from 10 to 15 years under similar conditions.

In the exposable materials category are *butyl* (synthetic rubber), *EPDM* (ethylene propylene diene monomer), *Hypalon* (DuPont registry and patent), *CPE* (chlorinated polyethylene) and *neoprene* (synthetic rubber).

Unexposable materials include *PVC* (polyvinyl chloride), *polyethylene* and *polypropylene*. While these materials do not resist ozone and ultra-violet attack as well as exposable materials, they can be expected to provide entirely satisfactory service in many applications, particularly if covered by a protective layer of soil or water. Some variations of these materials will last 20 years or longer in some applications if properly protected by a layer of soil or water.

The designation as an exposable or unexposable material does not necessarily indicate that one group or the other has superior physical properties for all applications. An unexposable liner may have chemical resistance qualities that would adapt to a specific application as well as, or better than, an exposable material.

Reinforced Materials

Virtually all flexible liners are available in supported form. This means that reinforcing fabric (or scrim) is laminated between layers of basic materials to increase tensile and tear strength. There are added advantages of better dimensional stability (especially during installation), better puncture resistance, and greater hydrostatic load capacity.

Scrims are available in many materials and strengths. Selection would be determined by the chemical composition of fluids to be contained, slope of installation site, and by the stresses of equipment or other mechanical forces to which the liner will be exposed.

Among the more commonly used scrim fabrics are nylon, Dacron, polypropylene and fiber glass. Scrims are specified in *count* and *denier*. Count refers to the number of woof and warp yards per sq in. (Figure 14-38). Denier is a unit of fineness of a yarn weighing one gram for each 9000 meters. Deniers are usually specified in multiples of 110, 210, 420, 840, and 1000 (1000 being the heavier). In Figure 14-38, assuming the denier is 210 and the fabric is nylon, the correct designation of the scrim would be "11 x 11, 210 denier nylon."

The disadvantages of reinforced liners are low elongation to break, less conformity to ground irregularities, less flexibility, and greater cost than unsupported liner by from 30 to 60 percent.

Following are typical property profiles of some of the more commonly used liner materials. These basic materials may be used by more than one manufacturer to produce a variety of trade-named liners. Each manufacturer may incorporate variations of these materials such as adding plasticizers for greater flexibility, providing scrims (fabric or nylon reinforcing) to increase tensile strength, adding a variety of resins to increase resistance to specific chemicals, or to improve impermeability. Complete information on end products made from these basic materials is available from the manufacturers. It will be helpful in selecting a specific product, for the pollution engineer to know which of the following basic materials best fills his particular requirements.

**Property Profiles of Materials Most
Commonly Used in Liner Manufacturing**

Butyl—A highly reliable synthetic rubber with more than 20 years field service. It has excellent resistance to ozone and ultraviolet rays, is extremely impermeable to water, and retains flexibility throughout its service life with a high tolerance for extremes of temperature. It has good tensile and tear strength, good resistance to puncture, and desirable elongation qualities. Among its disadvantages are a low resistance to hydrocarbons, petroleum solvents, and aromatic and halogenated solvents. It has poor workability and poor sealability requiring a special two-part adhesive and a cap strip, all of which must be applied in dry conditions.

PERCENT COST OF LINER MATERIALS COMPARED TO POLYETHLENE	10-20% GREATER	20-30% GREATER	30-40% GREATER	40-50% GREATER	50-60% GREATER	60-70% GREATER	70-80% GREATER
POLYPROPYLENE	■						
PVC		■					
BUTYL			■				
EPDM				■			
CPE				■			
HYPALON				■			
NEOPRENE							■

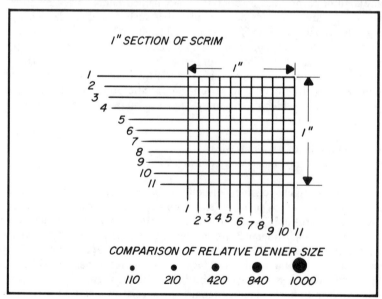

Figure 14-38. Scrim and denier designation
(courtesy of Goodyear Tire and Rubber Co.).

EPDM—A high-strength, flexible compound designed especially for contact with potable water. It is extremely impermeable, has high resistance to weather and ultraviolet exposure, resists abrasion and tear, and has good tolerance for extremes of temperature. It is also resistant to light concentrations of acids, alkalis, silicates, phosphates and brine. It can be expected to last from 15 to 25 years in normal use. It is not recommended for petroleum solvents, aromatic solvents or hydrocarbons. Sealed by a one-step EPDM adhesive.

Hypalon—A widely used synthetic rubber, it provides exceptional weather, ozone and sunlight resistance. It will not crack or fail at extremes of temperature from weather or contents. It is highly resistant to a wide range of chemicals, acids and alkalis and it will not support growth of mold, mildew, fungus or bacteria. Service life can be expected to exceed 10 years without protective covering. Generally supplied unvulcanized, it can be seamed by heat sealing or by solvent welding. Its disadvantages are relatively high materials cost and relatively low tensile strength.

CPE—Chlorinated polyethylene is a very flexible thermoplastic produced by a chemical reaction between chlorine and polyethylene. Its flexibility resembles a soft vinyl. A completely saturated polymer, it has no double bonds, which means it is not susceptible to ozone attack. It has excellent crack resistance at low temperatures and has good tensile and elongation strength.

While CPE has excellent resistance to atmospheric deterioration, it has a rather limited range of tolerance for chemicals, oils and acids. However, it is unique in its compatibility with other plastics and rubbers while retaining most of its desirable characteristics, making it a feasible base material for a broad spectrum of liners designed for specific applications. It has been successfully alloyed with polyethylene, PVC, ABS and several synthetic rubbers. It is widely used to improve stress crack resistance and softness of ethylene polymers and to improve cold crack resistance of flexible vinyls.

Neoprene—A synthetic rubber closely paralleling natural rubber in flexibility and strength. However, it is superior to natural rubber in resisting weathering, ozone and ultraviolet rays. It is extremely resistant to puncturing, abrasion and mechanical damage. Neoprene was developed primarily for retainment of wastewater and other liquids containing traces of hydrocarbons. It also gives satisfactory service with certain combinations of oils and acids for which other materials are not suited for long-range use. It is one of the more costly materials.

PVC—A very popular liner material because of its relatively low initial cost and tolerance to a wide range of chemicals, oils, greases and solvents. While not as resistant to ozone, ultraviolet and weather deterioration as most exposable materials, it can provide satisfactorily long service in many situations if covered with soil or water. Exposed areas may be covered with materials of greater weatherability to improve service expectancy. PVC has high strength-to-weight ratio and good puncture resistance. Exposure to heat causes undesirable deterioration in the presence of some chemicals. It is susceptible to strain from sulfides. Adhesion to metal and wood is poor. It becomes stiff at low temperatures making installation and maintenance more difficult in cold weather.

Polyethylene—A plastic material with good flexibility, tensile strength and resistance to solvents. Excellent low temperature qualities. Initial material cost is low with life expectancy proportionately short. It has poor weatherability and puncture resistance, conditions that can be improved with an adequate protective covering of soil and water.

Polypropylene—While relatively prone to ozone and ultraviolet attack, it has a desirable balance of other physical properties. It is tolerant to many chemicals and to extremes of temperatures, particularly high temperatures. Tensile strength is good and it has low permeability to water. Not recommended for oxidizing solvents.

CHAPTER 15

WASTEWATER

VACUUM TRANSPORT AND COLLECTION[156]

Vacuum sewage transport and collection uses differential air pressure to create flow, as opposed to the gravity-induced flow of conventional wastewater collection systems. In this respect, a vacuum system is essentially a pressure system. However, there are enough differences between vacuum or negative pressure transport and the new technology of positive pressure transport to say they are two different technologies.

To the homeowner, a gravity system is a simple network of underground piping running downhill and eventually arriving at some sewage termination point. Indeed, where the topography is appropriate, some systems are that simple. However, unless the sewage is allowed to outfall into a natural body of water, it first empties into a pumping station, and then into a treatment facility. Generally, gravity systems are not that simple and incorporate lift stations along the route, avoiding deep trenching normally necessary with flat or undulating terrain (Figure 15-1). Therefore, the great majority of gravity sewer systems are also mechanized systems. Pipe sizes usually start at 8 in. dia. and get larger.

The new technology of positive pressure sewers eliminates the need for laying pipe to hydraulic grade lines. However, in this instance, a pump is required at every input point to pump sewage into the network of collection lines (Figure 15-2). The collection lines eventually empty into a larger pumping station which is used to feed the treatment plant. The positive pressure system eliminates lift stations of a gravity system and substitutes small-diameter plastic pipe for large-diameter tile and concrete pipe. However, there is the expense of requiring electrically operated mechanical equipment at every sewage input point to the collection system.

A vacuum system requires a main collection station or pumping station similar to that of the other two systems (Figure 15-3). Unlike the other two methods, however, it also requires vacuum pumps in this station to maintain a vacuum on the collection lines feeding the station. The collection

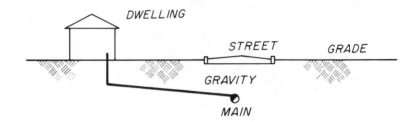

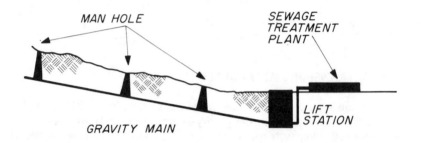

Figure 15-1. Conventional gravity sewer system with lift station.

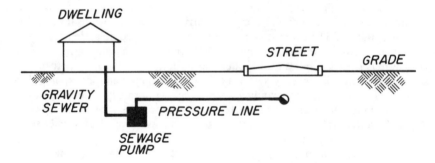

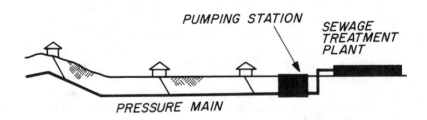

Figure 15-2. Positive pressure sewage transport system.

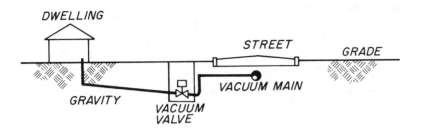

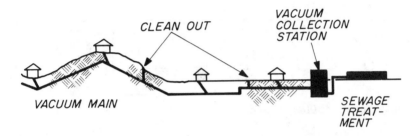

Figure 15-3. Vacuum (negative pressure) sewage transport system.

line is small-diameter plastic piping laid independent of hydraulic grade lines.

The system requires a normally closed valve at each sewage input point to seal the vacuum collection lines so that vacuum can be maintained. This valve opens automatically when a given quantity of sewage has accumulated on the upstream side. It then admits the sewage and closes. The valve is entirely pneumatic in its control and operation. The differential pressure between local atmospheric pressure, and the vacuum pressure on the immediate downstream side of the valve, controls and operates it automatically.

A vacuum sewage collection system closely resembles a water distribution system, except the flow is in reverse (Figure 15-4). The analogy would be complete if the sewage valve was manually operated by the homeowner, much as he manually opens a water faucet in the home. With proper design, proper equipment selection and proper installation, a vacuum system can be made to approach a water distribution system in dependability.

Collection Station

The collection station consists of a collection tank which is the equivalent of a wet well in a gravity pumping station (Figure 15-5). It

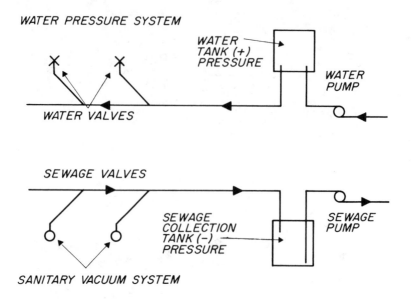

Figure 15-4. Comparison of components in pressure and vacuum systems.

can be either fiberglass or a welded steel tank properly coated internally with material such as an epoxy cold tar coating. Sewage and air are drawn into this tank. It is always maintained under vacuum through one or more vacuum main inlets, each inlet having its own shutoff valve.

Directly connected to the collection tank is a pair of nonclog dry pit sewage discharge pumps. They are installed in duplex fashion, each capable of pumping the design peak sewage flow for the collection system, and alternately pumping every other cycle. The pumps are controlled by liquid level sensors in the collection tank. The lowest, or #1 liquid level sensor, is the pump shutoff switch. The #2 level sensor is the end pump onswitch. The #3 level sensor is the lag pump onswitch, and #4 is an alarm switch.

Should the sewage rise to alarm level within the tank, the situation would indicate that both sewage pumps were inoperative. Consequently, the power supply to the vacuum pumps would be interrupted to prevent any more sewage being collected. A remote alarm would be turned in to those responsible for maintaining the station. The discharge piping incorporates shutoff valves to isolate either pump for repairs. Also, each pump is effectively doublechecked with weighted lever swing check valves as insurance against previously discharged sewage being drawn back.

Vacuum pressure is maintained on the collection tank by a pair of vacuum pumps also installed in duplex fashion. These pumps may be of any type as long as they are compatible with pumping moist air containing

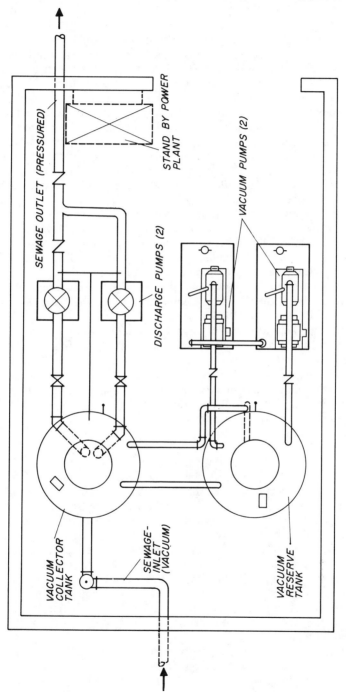

Figure 15-5. Plan view of a vacuum collection station.

some sewer gases. The general choice to date has been liquid ring-type pumps, although other types have also been used. The vacuum pumps are controlled by a vacuum pressure switch set to turn the pumps on and off at fixed vacuum levels. A usual setting range is 16 to 20 in. of mercury vacuum. The vacuum switch is located on the vacuum reserve tank.

The vacuum reserve tank is materially the same as the collection tank and is installed in series between the collection tank and the vacuum pumps. Its purpose is primarily to act as pneumatic reservoir to decrease the frequency of vacuum pump operations. The greater the volume of the reserve tank, the greater will be the time duration between vacuum pump starts for any given rate of air flow into the collection station. The suction lines to the vacuum have check valves to prohibit atmospheric air from bleeding back into the vacuum tanks during vacuum pump operation.

The collection station will include power and control panels to automatically operate the pumps. Any additional auxiliary equipment such as a ventilator, heater, or dehumidifier would also be included. Finally, a standby power generator would be part of the collection station. In the event of a power failure the vacuum sewage collection system would continue to operate. This could be accomplished since only the collection station would require electrical power.

Vacuum collection stations may be located above or below grade or both. Also, tanks may be buried or housed within the walls of the station. Above-grade exterior construction could be varied to suit the locale to the extent of blending in harmoniously in residential areas. In reality, a vacuum collection station is very similar to a gravity pumping station from the standpoint of equipment and operation. The only difference is the inclusion of vacuum pumps for the vacuum system.

Transport Lines

The network of transport piping which includes vacuum mains, branches and laterals is normally chosen to be either PVC or ABS thermoplastic pipe, the reasons being: sealability, excellent flow characteristics, low cost, and ease of installation. Class 200, or Schedule 40 DWV pipe, is usually used. It may be either solvent-welded or gasketed, bell-joined pipe. However, all gasket joint pipe is not satisfactory for vacuum service. Gaskets are normally designed to seal against leakage from internal pipe pressures. Consequently, some types will not seal against external pressures. Common pipe sizes are 3, 4 and 6 in. laid relatively parallel to grade with minimum cover. Depth of cover is dictated by frost depth or load conditions. Within limitations, this includes upgrade transport. The strength of PVC and ABS is largely affected by temperature.

Manholes, common to a gravity system, are replaced by capped cleanouts equivalent to the trunk line pipe diameter brought to grade. A unique feature of the piping is the installation of pockets which are basically built in low spots in the pipe. Coupled with this is the feature of laying the pipe in the ground in roller coaster fashion, whether going upgrade, downgrade or in level transport (Figure 15-6).

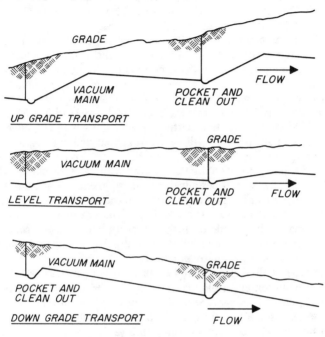

Figure 15-6. Method of installing vacuum sewage mains for uphill, level and downhill transport.

Vacuum Sewage Inlet Valve

The sewage input points to the system are normally the same as in gravity systems. However, in the usual gravity service lateral running from the dwelling to the main in the street, an automatic operating, normally closed valve is installed. This valve provides the interface between the atmospheric gravity flow or sewage on the upstream side and the vacuum transport piping on the downstream side.

Building fixtures, plumbing, and vents are conventional. Sewage flows from the dwelling by gravity and accumulates upstream of the interface valve. A sensor, installed in a tee fitting on the upstream piping, senses the hydrostatic head pressure created by the accumulating sewage. The sensor, which is basically a pneumatic switch, can be adjusted to close anywhere within a range of 3 to 30 in. water column.

When it closes, due to the accumulated sewage, it signals the controller which in turn operates a three-way air switch. Atmospheric air is then exhausted from the topside of the valve air operator into the downstream vacuum line, thereby opening the interface valve. The valve will now stay open for an adjustable time period and then close.

The time **adjustment** is in the controller and can be varied from 3 seconds to a full minute. The timer is usually set to hold the valve open for a total time equal to twice the time required to admit the sewage. In this manner, atmospheric air is allowed to enter the system behind the sewage for an equal amount of time. For a usual sewage input per valve operation, which is 10 to 15 gal, the total time of operation is generally less than 10 seconds. The time setting depends on where the valve is installed since vacuum pressure will vary throughout and govern the rate of flow of sewage into the system.

Usually, one valve would be associated with every dwelling. However, where possible, it is sometimes advantageous to gravity flow a number of dwellings into a small holding tank serviced by a single valve. The decision is chiefly one of economics. The location of the valve could be in the dwelling immediately outside or at the property line. Valve location also depends on the type of system, whether commercial, industrial, public, or residential and who owns and maintains it.

The type of fluid flow is unlike other types of pressurized fluid flow with which pollution engineers may be familiar. First, it consists of two components, air and sewage. Second, the flow is intermittent.

Consider a simplified system of a collection station, a long length of pipe laid out in roller coaster fashion with a vacuum sewage valve at the end of it. As long as that valve is closed, no flow is occurring. The two fluids occupy the line such that the low points and rises are liquid filled, and the high points and down slopes are air filled.

For example, the valve opens and lets in a plug of sewage. Behind that, there is an equally sized plug of atmospheric air which also enters before the valve closes. Remember, the sewage plug is being pushed into the line by atmospheric air pressure. As long as the entering plug of sewage remains integral, it acts as a liquid piston compressing the air in front of it. In turn, this air plug starts the next sewage plug moving, and so on down the line.

Motion starts at the valve and progresses downstream toward the collection station, but progressively ceases in the same direction. Motion will start to cease when the entering plug of sewage breaks down due to shear forces. The reason is that atmospheric air, having entered behind it, has expanded to an equilibrium vacuum pressure within the line. Having broken down, the sewage will flow by gravity to the lowest point within the line and there will reform a plug.

In most cases, a valve will open and close before the effect of its opening is noticed a few hundred feet downstream. The installation of the piping in the ground in a rolling fashion is thus a device to separate the two components into many alternate plugs of liquid and air.

One might ask at this point, "Why let air into the system in the first place? Why not close the valve before air arrives? Why not operate the system exactly as the reverse of a water distribution system?"

The prime reason is that scouring or cleaning velocities cannot be maintained, especially in the gravity and vacuum house laterals. The factors that account for this are (1) the limited differential pressure available, 7 to 10 psi compared to the much higher pressures of water distribution systems, and (2) the limited quantity of sewage admitted per valve operation. Increasing the quantity would mean a larger holding capacity, longer holding time, and increased septicity.

Using the same example as for the air-liquid system, when a valve opens on a full line, the entire column of liquid has to start moving. This represents, in most cases, a considerable inertia which cannot be made to reach a scouring velocity in the time it takes the first 10 or 15 gal. of sewage to pass through the valve. On a system where the valves served exclusively high-flow, large-volume sewage input, a full bore system could operate effectively. Such a system would serve scattered apartment buildings, motels, schools, or high density housing where units could be gravitied to a common holding tank, there to be picked up by vacuum.

Cost Savings and Options

It is obvious from the foregoing discussion that a vacuum system does have limitations, because there is a limited pressure range within which to work. However, even with its limitations, it can do everything a gravity system can do and then some.

The choice of a gravity system or a vacuum system for any given application is one of economics. Obviously, a vacuum system requires a cost item not found in a gravity system, namely, automatic valves installed in a valve box at every input point. Where the terrain is natural for gravity, the vacuum system cannot compete.

However, in the majority of cases, the terrain for a sewer system is not natural. Then, the vacuum system starts to save dollars. Lower piping costs are associated with easily installed small-diameter plastic pipe, with shallow, narrow width trenches, and with elimination of costly manholes and lift stations. These savings can more than offset the additional cost of the automatic valve.

This becomes increasingly true in areas of normally high trenching costs. These have high water tables, rock conditions, unstable sandy soils, underground or aboveground obstacles, or utilities. The obvious examples are around lakes, the seashore, and in existing communities.

One not so obvious example is in the separation of sanitary sewage from process waste in manufacturing facilities. Regulations are now requiring such separation. To install a gravity system in a plant is a formidable and costly task. A myriad of underground utilities usually exist.

All of this is not to say that any given collection system should be all gravity, all vacuum, or all pressure pumping. The marriage of any two or all three is possible if it will result in the best system at the best overall cost (Figure 15-7).

The remaining item in a vacuum system is the collection station. The cost of this item can vary considerably depending on the size and capacity of installed equipment and building construction. However, experience has shown that, even with the additional vacuum pumps and controls, it will cost little or nothing more than a factory built pumping station of the same capacity.

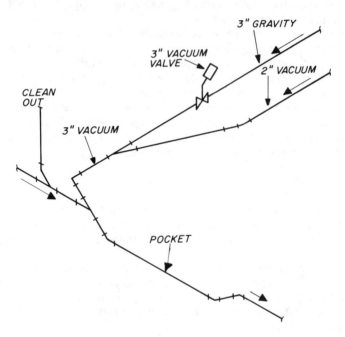

3" GRAVITY

3" VACUUM VALVE

2" VACUUM

CLEAN OUT

3" VACUUM

POCKET

Figure 15-7. Typical piping system combining vacuum and gravity transport.

NEIGHBORHOOD TREATMENT PLANTS[81]

An entirely new concept for treating domestic wastewater is being demonstrated at a new housing development at Freehold, N.J. The concept places a community's waste disposal facilities inside a neighborhood house erected on the recreational set-aside land in the housing project.

The plant will serve as the Freehold development's interim municipal sewage treatment plant until regional facilities become available. At the same time, it will be a demonstration project to prove the feasibility of a small, local, highly effective treatment system now versus the regional system sometime in the future.

In the system, the "house" contains four major unit processes for wastewater reclamation (Figure 15-8). In the first process, the raw

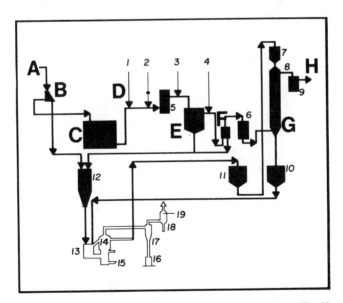

Figure 15-8. Community wastewater treatment plant. (1) coagulant, (2) pH control, (3) polymetric flocculant, (4) magnetic additive, (5) mixer, (6) adsorber feed, (7) carbon feed, (8) chlorine, (9) chlorine contact, (10) spent carbon, (11) regenerated carbon, (12) sludge hold, (13) sand filter, (14) fluidhearth reactor, (15) blower filter, (16) ash bin, (17) dust cyclone, (18) scrubber, (19) spray water.

sewage **A** entering the system via standard underground sewer lines flows through a wedge-wire screen **B**, which captures the large and intermediate size solids. The solids removed in this process, as well as those removed in the successive processes, are fed directly into an incinerator unit. From the screen the waste stream, with gross solids removed, flows into a surge tank **C** which levels the flow through the remainder of the treatment system.

Emerging from the surge tank at an even flow, the stream is treated successively with inorganic coagulants **D** and precipitants, and a polymeric flocculant, to precipitate the phosphate and coagulate the remaining solids which are subsequently settled in a clarifier **E**.

The clarified stream is treated with a few parts per million of powdered magnetic iron oxide which combines with the remaining suspended solids. This combination is subsequently removed in a magnetic filter **F**.

Following magnetic filtration, dissolved organics are removed by adsorption on granular activated carbon **G**. (This system is an upflow configuration with pulsed countercurrent operation.) The stream is then chlorinated for the discharge **H**.

The solids removed in each unit process are fed into a dewatering and incineration system. This same unit is used periodically to thermally regenerate the spent carbon from the adsorption process.

CHAPTER 16

INSTRUMENTATION—ANALYSIS OF WASTEWATERS

SAMPLING AND MONITORING
INSTRUMENTATION FOR WATER POLLUTION CONTROL

Water pollution control programs depend on a variety of analytical techniques. Many water treatment systems are equipped with advanced instrumentation, and procedures employed are often highly specialized. Wastewater sampling, monitoring and control instrumentation systems are similar to those used in other chemical process work even though the nature of a wastewater plant's purpose is somewhat different.

There are generally two types of instrumental analysis for water quality: (1) on site, *in-situ* measurements; (2) procedures which remove water from its environment for off-site analysis. In this latter method, collection must be made to ensure a representative sample. Generally, the grab sample is relatively expensive on a cost per sample basis and is statistically questionable.

Automatic sampling and analyses are being more frequently used to monitor industrial plant discharges as well as water quality in rivers, streams, fully automated and requiring little or no operator attention or service are commonly used to measure such parameters as flow, temperature, pH, dissolved oxygen, turbidity, hardness, chlorides, residual chlorine, redox potential and specific conductance. Equipment for use in measuring water quality parameters has become increasingly available in recent years. Many of the new instruments represent state-of-the-art techniques in wet chemical, electrochemical, electrical design and mechanical construction, and all are superior to those available a few years ago.

Some of the important reasons for sampling and monitoring are:

1. Ensure proper operating conditions in the process plant.
2. Provide a continuous, permanent record of process and wastewater handling.
3. Ensure efficiency and reliability of manufacturing operations and waste processing.

561

4. ʼAs a basic tool for pollution control and determining requirements for control equipment needs.
5. Ensure that environmental standards set by regulatory agencies are met.

Continuous analysis and recording of important parameters over an adequate time period representative of a plant's manufacturing cycle provides valuable information for subsequent selection of treatment methods. Continuous analysis should be augmented by composite samples proportional to flow, and grab samples collected automatically whenever one of the parameters which is continuously monitored exceeds the norm. Sampling conducted in this manner provides:

1. Continuous 24-hour record of flow and important parameters,
2. Discrete samples for off-limit values and a time record which shows when off-limit values occurred,
3. A 24-hour composite for average daily analysis.

Often a decision to be made is whether to buy or rent required monitoring and sampling equipment. An economic decision based on the amount of use equipment will be put to, in addition to reporting requirements (internal or to regulatory agencies), will prevail. Where only one or two parameters need to be monitored to establish concentration, it is generally more economical to measure these directly using either a submersible sensor nest or individual submersible sensors. Where the purpose of an industrial wastewater analysis is exploratory and analyses are performed only occasionally rather than on regular or continuous monitoring and control basis, then adequate samples may be sent to professional testing laboratories for analysis.

Measurement System Design

A measurement system for water quality determination should establish: (1) why the analysis is needed, (2) what are the parameters of the analysis, and (3) the methods of analysis.

After establishing the analysis objectives it is possible to proceed with design of the measurement system. Parameter choice for analysis depends upon the type of information required. Parameters for identification from various types of industries associated with industrial wastewaters are shown in Table 16-1.

Methods of analysis will be determined by: (1) accuracy of method, (2) sensitivity required, (3) interferences present, (4) number of samples, (5) sampling site, (6) availability of instruments and personnel and (7) methods accepted as standard.

Sampling Program

There is no general procedure for establishing a satisfactory sampling program applicable to all situations. Water or wastewater composition is

Table 16-1. Typical Industrial Waste Parameters Sampled

SIC 2818 - Industrial Organic Chemicals

Color	Fluoride	Copper
Chlorinated hydrocarbons	Sulfate	Iron
Oil and grease	Sulfide	Lead
Phenols	Aluminum	Magnesium
Total organic carbon	Cadmium	Manganese
Chloride	Chromium	Mercury
Cyanide	Cobalt	Zinc

SIC 332 - Iron and Steel Foundries

Color	Oil and grease	Cadmium
Turbidity	Fluoride	Chromium
Hardness	Arsenic	Iron
		Mercury

SIC 26 - Paper and Allied Products

Color	Sulfite	Lead
Specific conductance	Phenols	Mercury
Bromide	Total organic carbon	Nickel
Chloride	Chromium	Sodium
Sulfate	Copper	Zinc
		Total coliform

SIC 283 - Drugs

Color	Phenols	Total coliform
Chlorinated hydrocarbons	Turbidity	Chloride
		Sulfide

---

dependent on many variable factors. A few samples may suffice in some cases, while in other cases frequent sampling at many locations may be necessary. The primary consideration is to have the sampling cover a complete cycle. Obviously, data are much easier to collect and interpret during periods of steady flow as opposed to periods of highly variable flow.

Many water analyses are performed on grab samples collected in buckets or sampler bottles. Samplers programmed to collect on a volume of sample at regular intervals are also available, producing a composite sample. If a pump is used, flow rate and tubing size should be chosen to ensure a minimum sample time in the tubing. All sampling procedures must be capable of providing a valid sample. Complete information on the source and conditions under which the samples are collected is paramount.

There are six major types of sample, classified by their method of collection—the individual grab sample, simple composite, sequential composite, continuous sample, hand proportioned composite, and automatically proportioned composite.

The individual or grab sample is retained as a separate entity in its own container.

The simple composite requires that all samples taken over a specified time interval be deposited in a single container. If the composite is to be analyzed for oily materials, it must be divided into two containers. This is also true if different preservation methods other than refrigeration are required for proposed analyses.

The sequential composite is the collection of a series of individual samples per container, each container representing a specific time period, usually one hour. This procedure is particularly useful where the character of the waste may vary significantly from hour to hour, where batch dumping is experienced or where self-cancelling conditions occur, such as alternating high-low pH which would not be apparent in a simple composite.

In a continuous sample a small amount is collected in continuous flow. It is useful in feeding monitors or pilot scale processes. The method is usually not used in cases of high suspended solids unless provision is made to avoid settling or line pluggage.

The hand proportioned sample is readily obtainable where flow charts are available. Individual or sequential composite samples are manually composited in proportion to flow to obtain this representative type sample.

The automatically proportioned composite requires additional equipment. Sampling in proportion to flow is achieved by varying the size of individual aliquots in proportion to flow volumes.

Sample preservation is important in cases where required analyses are lengthy and require specialized equipment off the sampling site. It may therefore be necessary to store samples for some period of time. Sample changes may result because of storage times required. Some sample parameters change so rapidly that only analysis at the sampling site is valid. For example, temperature and pressure changes will cause variation of the concentration of dissolved gases such as O_2, CO_2, H_2S and CH_4. Fixing of some parameters through chemical treatment may be required. Sulfide may be stabilized by formation of mixed zinc sulfide-zinc hydroxide precipitate. Carbonate or sulfide equilibria shifts will cause pH changes. It is therefore recommended that pH measurements be made at the sampling sites. Heavy metal ion analysis may be acidified to a 3.5 pH with glacial acetic acid which minimizes precipitation and adsorption onto the walls of the container. Acetic acid stimulates mold growth which may be overcome by addition of small amounts of formaldehyde as a sample preservative.

In the analysis of water systems, it is frequently necessary to improve sensitivity or remove interferences. The most commonly employed separation and concentration methods include carbon adsorption, ion exchange, freeze concentration, chromatography and liquid-liquid extraction.

Instrumental Analysis

A wide variety of instrumental methods for the chemical analysis of water constituents exists.

Atomic Absorption Spectrophotometry

This is a technique of great popularity in the analysis of wastewater. It is a combination of emission and absorption phenomena and resembles flame photometry. The flame excites the elements in a sample to produce an emission spectrum; only a small percentage of the atoms are excited. Atomic absorption increases the sensitivity of the flame technique by utilizing the unexcited atom in the flame. In atomic absorption the sample solution is atomized into the flame producing atomic vapor of the elements in question. A monochromatic light from a hollow cathode tube containing the desired element and emitting light of the same wavelengths as that of the desired element is passed through the atomic vapor of the sample in the flame. The atoms of the desired element in the vapor are mainly in the unexcited or ground state in the flame and absorb radiation from the light source. The amount of light absorbed is proportional to the amount of element in the sample.

Atomic Fluorescence Spectrophotometry

This is a technique relative to atomic absorption spectrophotometry using atom fluorescence. In some cases, it is possible to increase sensitivity of measurement with atomic fluorescence spectrophotometry.

Infrared Absorption Spectrophotometry

A molecular absorption spectrometric technique, it involves the dispersing of a polychromatic infrared beam of light using a suitable prism or defraction grating. Wavelength bands of infrared light are isolated by a slit and allowed to pass through the sample. Analysis is run by passing a polychromatic light through the sample which absorbs portions of the light. The remaining light then passes through a prism or grating where it is dispersed into its combining wavelengths. The resulting spectrum is scanned by a detector which records intensity at each wavelength. The solution sample will preferentially absorb specific wavelengths of radiation. Various functional groups composed of definite atomic configurations absorb strongly at characteristic wavelengths. Infrared absorption analysis provides a means for characterizing organic compounds by means of functional group analysis.

Ultra-Violet Absorption Spectrophotometry

This is similar to the infrared technique, except that a polychromatic ultraviolet source is used.

Electrochemical Analysis

Electrochemical analysis utilizes electrode systems and electrochemical techniques, which are used routinely in onsite analysis and continuous monitoring of wastewater effluents. Electrochemical methods can be classified as based either on the passage of Faradaic current, which is classical polarography, or on electrode equilibrium-potentiometry.

Chemical Spectrophotometry

This technique utilizes the absorption of portions of the electromagnetic spectrum. Atomic emission or absorption spectra consist of sharp lines of specific spectral wavelengths. The energy of polyatomic molecules consists of electronic energy, involved with the electrons in the atom, rotational energy involved with the rotation of a molecule and vibration of the atoms relative to each other along internuclear axes. The overall result is a broad band spectra at specific wavelengths.

Molecular Absorption Spectrophotometry

This technique is based on absorption of radiation by molecular species. Its principal use in wastewater analysis is for metal ions based on reacting the metal ion with various organic reagents to form colored compounds which may be determined spectrophotometrically. A complex reaction between the metal ion and the organic molecule is usually involved.

Molecular Fluorescence Spectrophotometry

This technique is based on spectral measurements of fluorescence radiation from luminescent compounds upon excitation by incident radiation.

Continuous Monitoring Systems

Monitoring equipment can establish analysis continuously on a 24-hour basis. Continuous monitoring systems are generally located in fixed positions, and afford excellent coverage of fluctuations in water quality due to their continuous operating capability. Monitoring systems are usually established after complete sampling programs have defined the types of waste and their characteristics. Generally speaking, there are two types of continuous monitoring systems.

1. Measurement systems without chemical modification of the sample;
2. Systems which make measurements after a suitable chemical reaction has been affected in the sample.

Systems not requiring reagent additions are available which automatically measure and record such parameters as pH, temperature, conductivity, turbidity, dissolved oxygen, oxidation reduction potential, and chloride. These parameters may also all be measured by portable instruments without use of monitoring stations which are generally fixed. Monitoring instruments are usually modular in design and contain a sampling module with sensors, a signal conditioning module, and a transmitter or data logging module.

Continuous flow of sample material is provided to the sensors by a submersible pump. Caution must be exercised to ensure that no air pockets are present so that measurements such as dissolved oxygen will be reliable.

Physical and Chemical Properties of Water

The major physical parameters in water pollution are: temperature, electrical conductance, density, turbidity, viscosity, volatile, particulate, and dissolved solids. Temperature is the principal property which influences all physical, chemical and biological action in water environments. Changes in temperature are always of primary interest in any water pollution control system. There exists a wide variety of temperature measuring devices. Methods of temperature measurement, aside from liquid in glass thermometers, include:

1. Temperature measurements by bimetallic strip sensors based on the physical property and expansion of two metals. The sensor is composed of two interconnected sheets of different metals which have different thermal expansion coefficients.

2. Radiation pyrometers are temperature sensing devices based on measurement of radiation energy. All bodies emit radiant energies at the rate that increases with temperature. Temperature measurement can be made by measuring the emitted radiant energy without making physical contact with the sample.

3. Electrical thermometers including resistance thermometers, thermistors and thermocouples. Temperature measurement by resistance thermometers is based on dependency of the electro-resistance of most metals to temperature. By accurate measurement of the resistance, the temperature can be determined. The higher the temperature coefficient of the resistant, the better the metal is for thermo-electrical purposes and the higher the sensitivity. Resistance measurement is usually done by Wheatstone bridge circuits or modifications. Thermistors are transducers used for temperature measurement as they are extremely sensitive to temperature change. Thermistors are commonly made of sintering mixtures of specially prepared metallic oxide, modified with platinum alloy wire leads in a circuit whose electrical resistance varies with temperature. Thermistors are used frequently for *in-situ* temperature measurement and find wide application in temperature control. The thermocouple is a temperature transducer consisting of two wires of different metals joined together at both ends. When the two junctions are at different temperatures, the electromotive force generated is the result of the temperature difference.

Electrical conductance is used to give an overall estimate of the ionic strength of the base solution. Conductivity measurements are used for *in-situ* and continuous-type analysis. Factors affecting such measurements include temperature and pressure.

Density determinations are usually done for overall water characterization. Commonly, density is measured with a hydrometer at the temperature of the sample with appropriate corrections to 20 C. Density determinations in the laboratory may be made by measuring the weight of an exact volume of solution at a given temperature. Density measurements are significant for relating concentrations on a weight basis to concentrations on a volume basis. At low concentrations variations will not affect density appreciably.

Viscosity is a direct measurement of the resistance of the liquid to flow or fluidity and is of interest in industrial wastewaters of high solids content, waste slurries and sludges. Results are usually expressed in centipoise units at 20 C. Viscosity of pure water is taken to be equal to 1 at 20 C.

Turbidity signifies a visual response to the absorption and scatter of light by suspended matter in a given water sample. Turbidity is measured in terms of light scattered or absorbed by matter suspended in water. Measurements are usually done in reference to a standard suspension of fine silica. The standard procedure for the measurement of turbidity is based on the use of a Jackson Candle Turbidity. The turbidity of a given water is a function of both the amount of light absorbed and scattered by the sample.

Organic Pollutant Analysis

The chemical oxygen demand (COD) method is principally used as a measure of the organic pollution load of wastewater. It is based on the principle that most organic compounds are oxidized to CO_2 and H_2O by strong oxidizing agents and acid conditions. The measurement represents the amount of oxygen that would be required from the receiving water, if oxygen could oxidize the organic material to the same end products in the absence of microorganisms. The method determines total oxidizable organics; however, it does not discern between those that are biodegradable and those that are not. For this latter reason it is difficult to correlate between COD and BOD values. The test involves reacting a standard dichromate in acid solution with the sample containing organic matter until the oxidation is complete. Excess dichromate is measured by titration with freshly standardized ferrous ammonium sulfate using Ferroin indicator. The Total Carbon Analyzer allows a total soluble carbon analysis to be made directly on aqueous sample.

Separation and identification of organic pollutants may be specifically affected by gas-liquid and thin-layer chromatographic procedures. These methods have been quite useful in the detection of pesticides and phenols, as well as compounds encountered in the control of biological sewage processes such as digester gas, volatile and fatty acids.

Metal Pollutant Analyses

Specific metal analyses are of particular importance because of the toxicity of many of the substances and their compounds. Methods of analysis for metal ions include atomic and molecular absorption spectrophotometry, emission spectroscopy, activation analysis, and electrochemical analysis.

Analysis of Non-Metal Inorganics

Non-metal inorganics include compounds of nitrogen and phosphorus, usually classed as nutrients. Separation and concentration of inorganic anions in wastewater samples are carried out by various techniques. These include evaporation, precipitation, ion exchange, and partial freezing. Use of chromatography has also been made. Sulfate, sulfite, thiosulfate, and sulfide ions can be separated by anion exchange chromatography.

Analysis of inorganic anions is also made by techniques utilizing absorption spectrophotometry. All these techniques depend upon the displacement of ligands in a metal complex or chelate. Fluoride ion analysis can be made by displacement of a chelate dye anion from a zirconium complex. Numerous photometric methods are employed. Indirect UV spectrophotometry and atomic absorption methods have been developed for phosphates and silicates. Such techniques are based on the selective extraction of molybdophosphoric and molybdosilicic acids followed by ultraviolet molecular absorption spectrophotometry and/or atomic absorption spectrophotometry.

Electrochemical methods of analysis have also been employed in such wastewater testing. Direct potentiometric techniques have been used for the analysis of chlorides or sulfides. Ion-selective electrodes are used in electrochemical analysis of anions. Such electrode systems are primarily solid state or precipitate ion exchange membrane electrodes. Cyanides, nitrates and chloride ions may also be determined by potentiometer membrane electrodes.

The most widely used method for analysis for ammonia is the Nesslerization reaction which is based on the development of a colloidal yellow brown color on addition of Nessler's reagent to an ammonia solution. The Standard Method and the ASTM reference test recommend separation of the ammonia from the sample by distillation prior to the Nessler reagent reaction. However, for rapid routine determinations direct Nesslerization is most often carried out. In some industrial wastewaters, it is often required to distinguish between free and fixed ammonia. The former may be estimated by a straightforward distillation and the residual liquor then treated with an alkali and distilled to determine the fixed ammonia.

Determination of nitrites in water is based on forming diazonium by diazotization or sulfonilic acid by nitrite under strongly acidic conditions and coupling with alpha-napthylamine hydrochloride to produce a red-purple color. Spectrometric measurement of the color is performed, or comparison

Table 16-2. Analysis Methods Employed for Various Parameters

Parameter	Method
Alkalinity	Electrometric titration or AutoAnalyzer
Color	Visual comparison
Turbidity	90° scatter photometer
Solids	Gravimetric
Specific conductance	Wheatstone bridge
Biochemical oxygen demand	5 day, 20 C
Chemical oxygen demand	Dichromate reflux
Ammonia	Distillation-Nesslerization or AutoAnalyzer
Kjeldahl nitrogen	Digestion-distillation or AutoAnalyzer
Nitrate	Brucine sulfate or AutoAnalyzer
Nitrite	Diazotization or AutoAnalyzer
Total phosphorus	Persulfate digestion or AutoAnalyzer
Phosphorus-ortho	Single reagent, stannous chloride, or AutoAnalyzer
Bromide	Colorimetric
Chloride	Mercuric nitrate titration or AutoAnalyzer
Cyanide	Silver nitrate titration or pyridine pyrazalone
Fluoride	SPADNS with distillation or probe
Hardness	EDTA titration; AutoAnalyzer; atomic adsorption
Sulfate	Turbidimetric or AutoAnalyzer
Sulfide	Titrimetric or methylene blue colorimetric
Sulfite	Iodide-iodate titration
Surfactants	Methylene blue active substance
Phenols	Colorimetric
Oil and grease	Hexane soxhlet extraction
Total organic carbon	Combustion-infrared
Pesticides	Gas chromatography
Aluminum	Atomic absorption
Arsenic	Colorimetric
Cadmium	Atomic absorption
Chromium	Atomic absorption
Copper	Atomic absorption
Iron	Atomic absorption
Lead	Atomic absorption
Manganese	Atomic absorption
Nickel	Polarographic
Zinc	Atomic absorption

color standards may be used. This method is known as the Griess-Ilosvay method.

Colorimetry, UV spectrometry and polarography have been used for nitrate determinations. The phenoldisulfonic acid and the brucine method are two colorimetric procedures frequently used.

Gases

Dissolved gases in wastewaters can usually be separated by vacuum degasification or by various stripping techniques. Stripping is essentially a gas-liquid extraction in which an inert gas carrier is bubbled through a sample to carry off the dissolved gas for further separation concentration or detection. Gas exchange separation can be carried out either on batch or on a continuous basis. The gas stripped from a wastewater sample may be separated into its various components by gas chromatography.

COMPARISON OF WATER POLLUTION INSTRUMENTATION[195]

The detection and measuring of contaminants can be divided into two general categories: (1) instrumental analysis, and (2) classical, laboratory "wet" analysis. In either case, moderate to highly trained personnel are required and small organizations may find it more economical to sublet such work to established laboratories. However, before an analysis can be performed, samples must be obtained. Standard procedures have been developed and published for obtaining the sample, so that representative sampling of the source will be meaningful.

Since sampling equipment is relatively inexpensive, much can be done by even small installations. A high percentage of sampling equipment is portable. Table 16-3 lists some water samplers, their relative costs and applications.

Instrumental Analysis

When selecting an analyzer, instrument characteristics such as stability, sensitivity, response time, collection efficiency, response to interfering substances, cost and portability all need to be considered (Table 16-4).

Infrared and *ultraviolet instrumentation* provide a very precise qualitative and quantitative analysis of pollutants. Both instruments are in the medium price range and rapidly indicate what is in the water.

Another instrument for qualitative and quantitative analysis is the *gas chromatograph.* This tool is versatile, low in cost and can be used with great precision for trace concentrations.

Optical microscopy is a basic economical means used to determine particle size and, in many cases, to identify solid pollutants, often without going to expensive secondary devices. In fact, an amazing amount of information can be gathered from a polarizing microscope, especially since chemical microscopy is not limited to inorganic crystalline substances. For example, organic fibers and other organic resinous materials represent a broadened aspect of this low-cost analytical tool. The microscope's only disadvantage is that it requires a highly competent microscopist.

The *atomic absorption analyzer,* though not a new instrument, is not a familiar device in most laboratories. It is a good analytical tool for

Table 16-3. Comparison of Water Sampling Equipment

Type	Application	Portable Water Sampling Equipment			Remarks
		Collection Time (Average)	Intake Room	Cost*	
Constant rate	Where volume and character of the liquid is constant—for BOD sludge, turbidity and dissolved pollutants.	1-8 days	50-200 cc/min with sample size of 50-2000 cc	Medium	Portable and nonportable models. Obtains samples with or without refrigeration. Automatic collection features available. Operates with pressure or vacuum applications.
Flow proportioning	Where volume and character of the liquid is not constant. Sample obtained in direct proportion to the volume of flow. For BOD sludge, turbidity and dissolved pollutants.	1-8 days	50-200 cc/min with a sample size of 50-2000 cc.	Medium	Portable as well as nonportable with or without refrigeration.
Bottle	BOD sludge, turbidity, etc.	2 min- 24 hr	200-500 cc	Low	Features available to prevent air entrainment.

*Low Cost – up to $500
Medium Cost – $500-$2500
High Cost – over $2500

Table 16-4. Comparison of Instruments for Water Analysis

Pollutants	Analytical Tool	Pollution Type	Analytical Time	Sensitivity to Concentration Trace	Medium	Major	Reproduction of Results	Price	Personnel Required
I. ORGANIC									
1. Detergents	Optical microscopy	II	Medium	Poor	Ex	Good	Good	Low	Highly skilled microscopist
2. Dyes									
3. Pesticides, etc.	Gas chromatography	I	Rapid	Ex	Ex	Good	Good	Low	Skilled technician
4. Resinous waste	Emission spectroscopy	II	Medium	Ex	Good	Poor	Good	High	Skilled technician
5. Other dissolved organic compounds	X-ray fluorescence	II	Medium	Poor	Good	Poor	Good	High	Skilled technician
	Ultraviolet absorption	I	Rapid	Good	Good	Poor	Good	Medium	Skilled technician
II. INORGANIC	Infrared absorption	I	Rapid	Poor	Ex	Good	Good	Medium	Skilled technician
1. Mineral acids	Atomic absorption	II	Medium	Ex	Good	Poor	Good	Medium	Skilled technician
2. Alkalies	X-ray diffraction	II	Medium	Poor	Good	Good	Good	High	Skilled technician
3. Metals and metalloids	Mass spectroscopy		Long	Ex	Good	Good	Good	High	Highly skilled technician
4. Dissolved gases	Nuclear magnetic resonance	II	Medium	Poor	Ex	Good	Good	High	Skilled technician
5. Siliceous material	Classical wet analysis	I,II	Long	Poor	Good	Good	Good	Low	Skilled technician
6. Dissolved carbon									

LEGEND

TIME: Rapid—15 min-1 hr
Medium—1 hr-8 hr
Long—8 hr-40 hr

PRICE: Low—$1000-$3000
Medium—$3500-$10,000
High—$10,000-$100,000

SENSITIVITY: Trace—less than 0.1 percent
Medium—0.2-10 percent
Major—10-99 percent

detecting trace elements. Its sensitivity range of 0.01 to 100 ppm makes it an excellent tool for monitoring pollutants. Table 16-5 presents the atomic absorption sensitivity for various metals.

The *emission spectrograph* has been used as an analytical tool for a long period of time. This device is capable of detecting 0.01 percent, or even less, of most of the metallic ions and of certain metalloids such as arsenic, silicon, phosphorus and boron. It is an excellent tool for monitoring pollutants in the inorganic field. Sample size can be in the milligram range. The high cost of this device prohibits it from being a familiar laboratory tool.

Table 16-5. Atomic Absorption Sensitivities

	Concentration in ppm		Concentration in ppm
Sb	0.5	Mo	0.5
Ba	5.0	Ni	0.15
Be	100.0	Pt	0.7
Bi	0.5	K	0.03
Cd	0.03	Rh	0.3
Ca	0.1	Se	5.0
Cs	0.15	Ag	0.05
Cr	0.05	Na	0.05
Co	0.2	St	0.15
Cu	0.1	Te	0.5
Au	0.3	Th	0.03
Fe	0.1	Sn	5.0
Pb	0.3	Zn	0.03
Mg	0.01		
Mn	0.05		
Hg	5.0		

Laboratory "Wet" Analysis

The classical analytical techniques for the measurement of impurities are well documented. A trained technologist can obtain accurate measurements of pollution—whether the contaminants be organic or inorganic—using gravimetrical, electro-analytical, oxidation reduction, acidimetry and alkalimetry methods to detect and monitor contamination.

Quantitative analysis by strictly a classical "cookbook" technique has many advantages. Sample size is of utmost importance. A small vial of "gunk," 10 milliliters or less, represents quite a technical challenge to accurately obtain good repeatable information. Trace concentrations of contaminations (substances of less than 0.1 percent of the total makeup) are not easily discerned or measured. Another serious drawback to the conventional techniques is the length of time required to obtain accurate information. The cost, however, is quite low.

Colorimetric devices do, however, provide an inexpensive and rapid means of analysis. By using proper test reagents, various pollutants such as alkalinity, carbon dioxide, chlorides, chlorine, chromates, copper, fluorides, calcium, hydrogen sulfide, iron, manganese, nitrates, nitrites, dissolved oxygen, pH, orthophosphate, metaphosphates, silica, sulfates, and others, can be analyzed.

pH CONTROL[186]

Poor neutralization of waste streams can be harmful in terms of economic waste, poor public image, polluted environments, and even legal censure. However, it is possible to avoid these harmful ramifications by applying advanced control systems to waste neutralization. Feedforward control shows particular promise in solving many of the problems usually encountered with severe pH applications.

With feedforward control the neutralization system does not rely on large expensive holding tanks to average out changes in either flow or influent pH. A neutralization tank is constructed only large enough to permit the chemical reaction to occur. Acidity, or hydrogen-ion concentration of a solution, can be measured continuously with a sensitive electrode. This electrode develops a voltage proportional to the concentration of hydrogen ions in solution.

Figure 16-1 shows a plot of the reagent required for neutralization as a function of influent pH. Near a pH of 7, very small amounts of reagent will cause large changes in effluent pH, whereas larger quantities of reagent

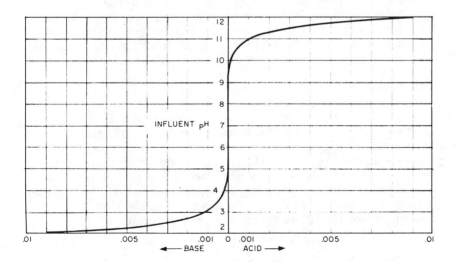

Figure 16-1. Mols reagent required per liter influent waste stream.

are needed to make small changes in the pH further away from neutrality. The logarithmic shape of the curve illustrates that the required reagent will change approximately tenfold for each unit change of influent pH.

This log shape puts severe accuracy requirements on a control system. If an influent pH of 2 is to be neutralized to a pH of 7 ± 1, required accuracy is 10 parts in 10^5, or 0.01 percent. A reagent deficiency of 0.01 percent gives a pH = 6.0 and an excess of 0.01 percent gives pH = 8.0.

Control problems are further magnified when it is necessary to neutralize an entire spectrum of influent materials, often varying in pH from 2 to 12. Problems develop when the system rangeability exceeds the capacity of a single valve, feeder or pump. If the influent pH can vary from 2 to 5 and the flow changes by a factor of 5, the pH change, because of its logarithmic nature, gives a rangeability of 1000 to 1. This, multiplied by the flow change, gives a system rangeability of 5000 to 1.

Feedforward Control

In a conventional feedback control system, a measurable error must exist before a restoring force is applied. Therefore, perfect control is not obtainable. The feedback controller does not know what its output should be, so it changes its output until measurement and set point agree. It has only sufficient information to solve the control problem by trial and error. This often results in oscillatory response. Difficult processes, subject to frequent disturbances, may never settle.

With feedforward control, on the other hand, the control problem is solved directly. Principal factors affecting the process are measured and, with the set point, are used to compute the correct controller output for the current conditions. Whenever a disturbance occurs, corrective action starts immediately before the controlled variable is affected—effluent pH. Feedforward control is theoretically capable of perfect control, its performance limited only by the accuracy of measurements and computations.

Mathematical Model

To design a feedforward control system, a mathematical model of the process is first derived. This is an equation which shows how the controlled variable, effluent pH, responds to changes in influent flow, composition and reagent addition. The required reagent flow **B** with known hydroxyl-ion concentration can be calculated by measuring the influent flow rate **F** and the influent pH expressed as hydrogen-ion concentration pH_1. The amount required is proportional to the influent flow multiplied by the difference between the influent pH and the set point, in this case, neutrality.

$$B \text{ [OH-]} = k \times F \times C \times ([H^+] \, pH_1 - [H^+] \, pH_7)$$

The term **k** is a dimensional constant and **C** represents the buffering capacity of the stream.

This equation is then converted to a logarithmic form since the pH measurement is logarithmic:

$$\log B = \log k + \log F + \log \Delta [H^+] - \log [OH^-] + \log C$$

To apply this equation, the required reagent flow must be varied logarithmically with respect to influent pH by an equal-percentage valve. The reagent flow rate is converted to an equivalent valve opening for control purposes. The equation for an equal-percentage valve is:

$$\log B = (M - 1) \log R + \log B_{Max}$$

where
R = rangeability
M = fractional valve position
B = flow rate
B_{Max} = max flow rate

If neutralization is to be complete, the equations must be set equal to each other so that the valve opening (controller output) produces the required reagent flow.

Solving for the valve position or controller output necessary for neutralization, and converting it mathematically to a form usable by the feedforward computers:

let

$$F = F' \times F_{Max}$$

$$f(F) = 1 + \frac{1}{\log R} (\log F')$$

and

$$r_B = \log \frac{F_{Max}}{B_{Max}} [OH^-]$$

then,

$$M = 1(F) + \frac{1}{\log R} (\log k + \log C + r_B + \log \Delta [H^+]$$

where r_B is the feedback trim adjustment which will change variations in reagent strength, etc.

Control Systems

Figures 16-2 through 16-4 show a series of pH control systems. The feedback control system shown in Figure 16-2 is used when flow rate is unstable and the pH variation is within the rangeability of the reagent addition system. Figures 16-3 and 16-4 show feedforward systems. The system in Figure 16-3 controls only one side of neutrality. If both sides are encountered, the system, except for flow, must be duplicated. Increasing reagent rangeability is shown in Figure 16-3, with two sequenced valves, and Figure 16-4 with six sequenced valves.

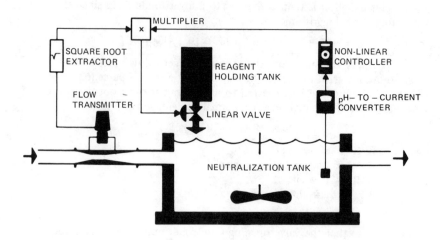

Figure 16-2. Medium range feedback system for unstable flow rate.
Rangeability 10 to 20:1.

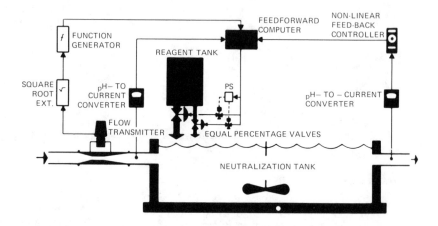

Figure 16-3. Feedforward system controlling on one side of neutrality only.
Rangeability 1500:1.

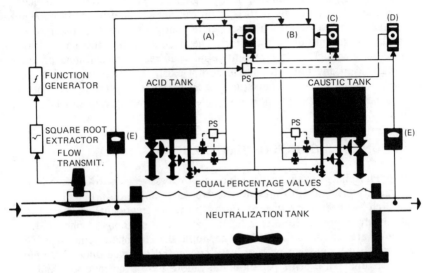

(A) ACID FEEDFORWARD COMPUTER
(B) CAUSTIC FEEDFORWARD COMPUTER
(C) NON-LINEAR FEEDBACK CONTROLLER (P&R)

(D) FEEDBACK CONTROLLER (P&D)
(E) pH– TO – CURRENT CONVERTER

Figure 16-4. Feedforward system controlling both sides of neutrality. Rangeability 250,000:1.

While feedforward control can compensate for changes in influent flow and pH, the tremendous sensitivity of the pH curve in the region of neutrality makes feedback trim an essential component in the system. A nonlinear controller offers distinct advantages over a conventional linear controller in a feedback trim application. A linear controller avoids limit cycling in the highly sensitive neutrality region with an extremely wide proportional band. This means that recovery from a large upset will be slow.

Considerations

A number of considerations affects the success of any pH control system. Effluent pH measurement must be made at a point where the reaction is complete. The use of lime as a reagent is not advisable because it dissolves slowly, making the system unpredictable. It is usually best to use more soluble reagents which dissolve, disperse and react immediately.

Thorough mixing is also essential in accurate control. The extremely high process gain encountered in the region of neutrality can be balanced by a control system only if the system can detect the error early enough to take corrective action. The sooner the corrective action is taken, the smaller the magnitude of the error. The treated stream must be held in a vessel long enough for the reagent to mix and react, so that areas of unreacted reagent and untreated waste do not exist. Under some conditions, particularly in unbaffled tanks, individual particles may follow parallel circular paths around the tank almost indefinitely, mixing very slowly or not at all.

pH MONITORING OF PHOSPHATE REMOVAL[97]

Phosphates are nutrients which abet excessive growth of algae in receiving waters, and thus present a severe water pollution problem. Much research has been done in recent years on waste treatment processes for phosphate removal. Results indicate that several methods of phosphate removal are feasible and relatively economical. Phosphate removal of greater than 95 percent with effluent concentrations well under 1.0 ppm at costs under five cents per thousand gallons have been reported, and full-scale plants are now operational.

Figure 16-5 shows a combined chemical-biological treatment. In this process lime is added with pH control in the flash mix tank. Calcium

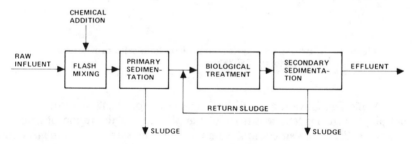

Figure 16-5. Combined chemical-biological wastewater treatment plant.

phosphate precipitates are settled out in the primary sedimentation tank. The degree of removal is dependent on pH level and lime concentration. A relatively small amount of lime (about 100 mg/l) reduces phosphates to approximately 7 mg/l, but an additional 150 mg/l is required for further reduction to about 2 mg/l, so this scheme suggests removal below 7 mg/l by biological means.

Biological uptake is efficient at low levels of phosphate concentration. Adjustment of pH with lime to a level of about 9.5 ensures the proper calcium level. CO_2 production in the secondary treatment process reduces pH to levels readily acceptable for discharge.

pH has customarily been measured in treatment plants of medium and large size. Frequently the measurement has been made in raw influent and occasionally in settled sewage after primary sedimentation. The reason for the measurement has been to forewarn of pH change which could affect both anaerobic and aerobic processes. pH control has not been widely practiced. In many instances, measurement was difficult and maintenance requirements high. Problems resulted from the coating of sensing electrodes with grease from the sewage.

Coating of electrodes results in erratic, erroneous pH readings which, if not corrected by cleaning of the electrodes, soon result in off-scale readings. Most serious problems result at the reference electrode. If one assumes the reference electrode is functional, the pH system will continue to operate until the coating completely isolates the pH-sensitive glass from the stream.

In the sewage environment there is no guarantee that any electrode will totally resist coating. Steps should be taken, however, in selecting amplification systems which will minimize the effect electrode coating can impart.

MEASURING OXYGEN DEMAND[11]

Oxygen demand is different from other kinds of pollution—it takes something away. It removes dissolved oxygen that is needed to support animal and plant life. If oxygen removal is complete, nothing can live in the stream but anaerobic organisms in the bottom slime.

Oxygen-demanding pollutants consume oxygen in two ways. One is by direct chemical reaction. Common constituents of industrial wastes such as ammonia, nitrites and sulfites are readily oxidized in a stream. Pollutants can also indirectly consume oxygen by supporting the growth and multiplication of organisms which use up oxygen. The effluent from sewage treatment plants, laden with organic matter, is typical.

How should oxygen demand be measured? Historically, the first problem undertaken was that of sewage. At the time, it seemed realistic to simulate actual oxygen depletion by incubating a water sample over a long period with typical microbes and measuring oxygen consumption versus time. The shape of this curve varies greatly with the type of wastes involved. The domestic sewage curve reaches its upper limit (ultimate biochemical oxygen demand), in 20-30 days, while the paper waste demand curve continues to rise beyond 200 days.

The importance of specific time points on the oxygen demand curve varies by location. A short-term demand (1-3 days) determines local oxygen sag conditions. The long-term demand is of great importance for discharges anywhere above slow-moving estuaries and rivers or lakes and reservoirs. Nevertheless, tradition has established the use of the five-day biochemical oxygen demand (BOD_5) as an index for the entire demand curve, ignoring the fact that BOD_5 represents only about 60 percent of the ultimate oxygen demand for domestic waste and no more than 5 to 10 percent of ultimate.

Furthermore, the five-day BOD analysis fails to account for nonbiological consumption of oxygen as well as for the differences between conditions in the stream versus the sample bottle. It is not easy to simulate riverine micro-ecology in a test tube. Analysis results are often inconsistent; replicates can vary by 25 percent or more, particularly for industrial effluents.

A newer approach to demand measurement has been to use non-biological techniques. These methods approximate total oxygen-demanding constituents in order to get a more reliable index of the entire demand curve than the five-day biological analysis. There are indirect analytical methods which measure the amount of carbon present (TC) or the amount of organic carbon present (TOC), on the assumption that this is in constant proportion to the oxygen demand.

Another alternative to the BOD_5 analysis is to react an effluent sample at elevated temperatures with a strong oxidizing chemical. The intent here is to measure *all* oxidizable substances in the sample and thus obtain a value somewhat higher than (but correlated with) ultimate BOD. The chemical oxygen demand (COD) analysis is a considerably more repeatable laboratory procedure than BOD. However, it still takes hours to get the answer, and the technique is not as readily adapted to field automation or continuous analysis. It is subject to interference (notably chlorides) and to low values through failure to react or loss of organics by volatilization.

Still another method of analysis uses combustion of all the oxygen-demanding matter. Total oxygen demand (TOD) analysis reacts the sample with oxygen gas at about 900 C, and the oxygen consumption is measured. The results are available in a few minutes and replicate samples give closely similar answers—within 2 to 5 percent. There are interferents, notably nitrate; nitrate interference varies from a relatively small effect to a highly significant one, depending on the design of the specific TOD instrument.

It may be felt that the reaction conditions of TOD are not remotely like those of oxygen depletion in a river. However, TOD may prove to be more realistic than it appears, just as the biochemical test proves to be less so. There is now a large body of evidence to support this view. TOD measured in actual effluents is correlated with COD, but more convenient to determine. Both TOD and COD are considerably larger than BOD_5 (by factors of 2 to 10); despite this, they may be better correlated with real *in-stream* oxygen demand for several reasons:

1. BOD_5 may not be a good measure of biochemical demand (wrong species of microbes, effluent sample may be toxic to microbes, five days may represent a tiny fraction of ultimate demand).
2. Biochemical demand may not measure the direct chemical demand, which can be large.

Figures 16-6 and 16-7 show correlations between TOD and either COD or BOD, for two effluents, one with low oxygen demand and one with high demand. For both effluents the correlation of TOD with COD is approximately 1:1. In contrast, neither BOD correlation is 1:1 and the two are widely different. In Figure 16-6 TOD:BOD is about 3:1; in Figure 16-7

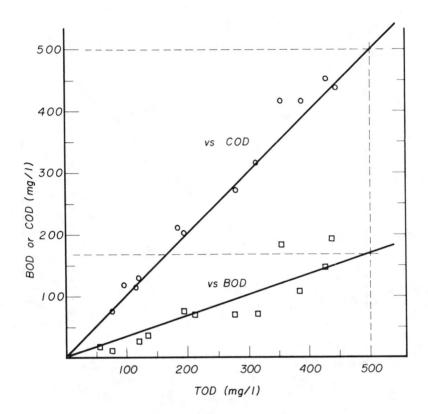

Figure 16-6. Correlations between TOD, COD and BOD
for effluent with low oxygen content.

it is 1.9:1. However, the correlations between the two are reasonably consistent (within the 25 percent or more variability encountered in industrial effluent BOD determinations).

In both correlations there are 14 comparisons over a concentration range of about 8:1. This means that once the correlation has been established for a particular plant, TOD could be used to monitor BOD—test results will be available *on-line* rather than days later. The significance of this on-line readout for preventing serious spills cannot be overestimated. A number of major industrial plants have already demonstrated that it is possible to eliminate almost all spills with on-line monitoring. Many pollution control agencies currently accept TOD correlations with BOD or COD on a case-by-case basis.

There are two types of TOD instruments on the market being used by industries, waste treatment plants and enforcement agencies. In one instrument, a continuous flow of 3.5 ml/min of waste sample enters the reactor

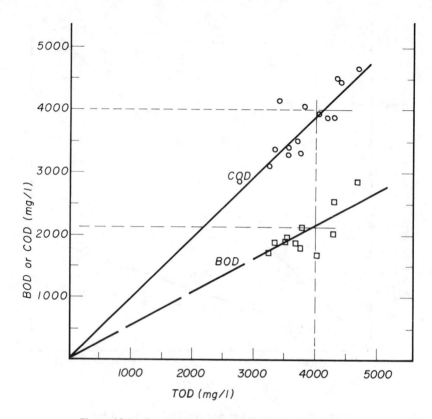

Figure 16-7. Correlation between TOD, COD and BOD for effluent with high oxygen content.

from a metering pump. The reactor also receives a metered flow of air, which reacts with the sample at 875 C in a self-cleaning fluidized bed. The hot gases are cooled in a condenser and pass to an oxygen sensor which continuously measures the amount of oxygen used up by the sample and records it as TOD on a strip chart. The instrument has an automatic provision for waste overload. If the TOD of the sample increases so that all the oxygen in the intake air might be used up, a Hi/Lo alarm switches the air supply to a higher rate. When the TOD returns to the normal range, the air flow is automatically restored to the low level.

In the other type of instrument, discrete filtered samples of about 20 microliters each are drawn from the sample stream at 3- to 5-min intervals by a metering valve; the sample volume and frequency may be changed to suit conditions. The sample meets a carrier stream of nitrogen gas containing a controlled low concentration of oxygen (200 ppm). The flow passes into a fixed bed catalytic reactor at 900 C where the oxidizable components react with and consume some of the oxygen. The gaseous mixture is analyzed by an oxygen-measuring fuel cell which shows a steady reading on a strip chart until the sample combustion gases cause a temporary drop (a "negative peak") which is proportional to the TOD of the sample.

Readout from both types of instrument is obtained within a few minutes after sampling.

Effluent monitoring is often thought of only in terms of keeping within environmental regulations. But a monitor—especially if it has an on-line readout and an automatic alarm—can spot trouble quickly and save expensive product wastage or damage to process equipment. TOD instruments have paid for themselves several times over: one chemical producer claims savings of $6 million while avoiding hundreds of spills. A recent spill occurred at a large brewery. A leaking valve would have let thousands of gallons of beer run to waste but for the vigilance of a TOD analyzer. In addition to the wasted product, the brewery's biological treatment plant would have been put out of action for some days.

However, TOD measurement is only applicable presently to certain industries. In addition to sewage treatment plants the following industries represent potential users of TOD monitors: brewing and distilling, beverages, chemical and plastics, dairy, food processing, paint, petroleum, pulp and paper, pharmaceuticals, rubber, soap and detergents, sugar, tanning and textiles.

PLANNING AND AUTOMATING FOR
POLLUTION CONTROL: A CASE STUDY[118]

It was found that wastewaters from a recently developed facility contained such materials as zinc, acids, alkyl benzyl sulfonate (ABS), oils and hydrocarbons and settleable solids. The treatment process that was designed to remove these contaminants is shown in the simplified schematic, Figure 16-8.

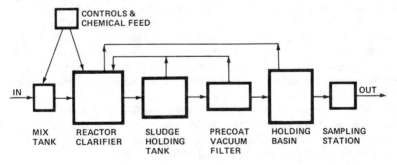

Figure 16-8. Schematic diagram of industrial waste treatment process.

Waste Treatment System

A three-million-gallon, stream-fed reservoir supplies water to the plant for cooling and fire protection. City water is used for in-plant processing. As the used process water is discharged from the plant, neutralizing chemicals are added. This wastewater is held in a reactor clarifier where precipitated materials are settled out. The clarified water overflows into a holding basin before being discharged to the reservoir or stream. The key elements in the system are as follows:

Mix Tank—The total industrial effluent enters the head of the treatment process at a mix tank where neutralization and precipitation occur. Electrodes monitor effluent character to determine chemical feed requirements, and proper flow of chemicals into the tank is automatically activated. The entire plant flow is retained in this basin for about five minutes and mixed vigorously with metered amounts of lime slurry.

Controls and Chemical Feed—The control system includes a series of pH meters present to demand proper chemical feed to bring pH of wastewater back into a predetermined, regulated range. Chemicals normally added are:

    Lime—to neutralize acids and precipitate zinc salts.

    Acids—sulfuric acid to neutralize excess alkalinity.

    Auxiliary chemicals—chemical agents such as cationic surfactants and
        filter aids are added when required to precipitate surfactants or to
        increase floc size.

Lime treatment with one vacuum filter was chosen for economic reasons. This approach affords the opportunity at a later date to continue with one vacuum filter, install an additional filter, or change to caustic treatment if hauling costs prove to be excessive.

Reactor Clarifier—Flocculation takes place in a 65-ft-diameter reactor clarifier after the previously neutralized waste is brought into contact with recirculated slurry in the center well. A turbine impeller recirculates the incoming waste and lime slurry at rates up to ten times the incoming flow.

The pretreated effluent is fed into the center of the clarifier where final pH adjustment is made. The material precipitated out, such as zinc is concentrated at the center of the clarifier. Clear and neutralized water overflows from the outer edge of the clarifier into the holding basin. As a positive check, the pH of overflow water is again monitored and recorded at control panels.

Effluent from the outer section of the clarifier or from the vacuum filter flows into the holding basin, which acts as a settling pond to trap any further fallout. Water underflows from the holding basin through a secondary metering and sampling station into Sinking Creek.

Sampling Station—Water discharged from the holding basin into the creek is sampled and controlled at this station. Chemical tests conducted on a daily routine basis include: pH, flow rate, temperature, zinc content, oil content, surfactant content, dissolved solids, total hardness, alkalinity and suspended solids. The system is automated to constantly monitor effluent quality, and atomic absorption spectrophotometry is used to determine zinc concentration.

As a double check on the effluent discharged into the creek, the station has the responsibility of taking and testing upstream and downstream samples. A monthly summary of these test results is submitted to the state pollution control agency.

Lime vs. Caustic Economics

Once experimental results indicated that a precoat vacuum filter was the practical way of filtering out the precipitated waste, the next decision was whether to use lime or caustic soda as the neutralizing and precipitating agent. The choice of caustic or lime was a critical one because it would determine the size of clarifier and vacuum filters needed.

Five cost alternatives were investigated and evaluated:

1. Lime—2 vacuum filters, large clarifier and thickener.
2. Caustic—1 vacuum filter and small clarifier.
3. Lime—(a) 1 vacuum filter and large clarifier;
   (b) 2 vacuum filters and thickener.
4. Lime—1 vacuum filter and large clarifier
5. Caustic—1 vacuum filter and large clarifier.

If caustic were chosen as the precipitating agent, a smaller clarifier could be used (Option 2), but the yearly chemical cost would be more than 50 percent higher. In addition, no future option of using lime would then be possible. The choice of lime would require a large clarifier and the possibility of two vacuum filters, Options 1, 3(a) or (b). Utilizing the larger clarifier would leave open the option of using either lime (1, 3a and 3b, 4), or using caustic (5).

There was a strong indication that one precoat vacuum filter could handle the total projected waste flow even if lime were used. Therefore, Option 4, lime with a large clarifier and one precoat vacuum filter, was chosen. This alternative enabled the company to start up with the lowest annual operating cost while still maintaining enough flexibility to go to future plans, which would involve Options 3a or 3b and 5.

Instrumentation

An instrumentation schematic (Figure 16-9) shows a flow scheme of control equipment required. A feed-on-demand system was selected with a built-in capability for feedforward operation, if desired in the future.

As the schematic indicates, the signal from the flow-through electrode assembly in the influent is converted into an electropneumatic signal, which is recorded at the indicator/recorder. The primary controller **A** positions the valve openings on the large and medium chemical feed lines. A feedback trim signal from the submersible electrode assembly in the clarifier positions the small valve in the chemical feed line through the secondary controller **C**. It then resets the control set point at **A** through the auxiliary control **B** with alarms and relays completing the system.

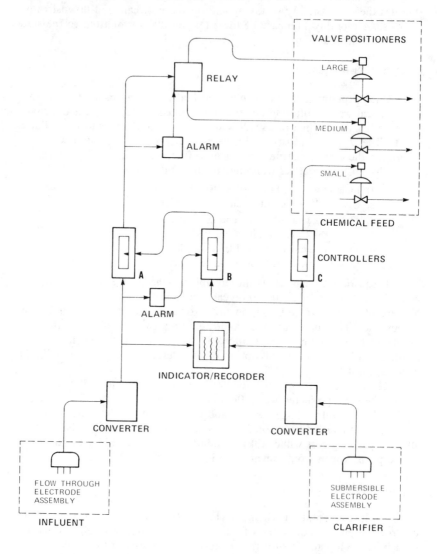

Figure 16-9. Schematic diagram of instrumentation system for water pollution control.

RADIO SYSTEM MONITORS[199]

Northern San Diego County covers 1100 square miles of coastal, hilly and flat land and comprises a dozen communities. Many of the county's water pollution control facilities are operating at near capacity. Constant checking must be done to ensure that they are operating effectively at all times. Because of hilly terrain, a number of lift stations are required to pump wastewater to the treatment facilities. Any prolonged failure at the pumping stations would cause the wastewater to overflow.

The Sanitary Engineering Division, Department of Special District Services, has turned to modern electronics. Using a radio alarm reporting system, the Division is able to keep an eye on all eight pumping stations with one man at the County Operations Center. Some of the stations are located more than 40 miles away from the center; however, if a failure occurs, an operator can be dispatched immediately.

Prior to installation of the radio system, an alarm system was used within each pumping station. However, it was necessary to have someone physically check the station's control panel daily to see if the system was operating. This procedure was time-consuming and costly. Since installation of the radio system, a continual physical check has not been necessary. Now, district pollution engineers know exactly what is wrong at what station when an alarm is received, 24 hours a day.

A typical pumping station includes a power source, a dry well containing the pumping system, and a wet well to receive and serve as a suction pit for the wastewater flow. The station also contains a radio transmitter, antenna and an alarm encoder. The radio transmitter is tuned to the same frequency as the radio dispatcher room at the County Operations Center in San Diego. Each alarm station is set up to provide an immediate alarm as to location of the pumping station, primary power failure, wet well and dry well problems.

Information received from each station is printed on a digital readout printer at the center (Figure 16-10). Each station is assigned a 2-digit address of 12. Any change in the address number other than the second digit indicates a malfunction at the station. The second digit is assigned to the test transmission. When a test is made, the second digit should always print as "0." A test signal from a properly functioning system would print out: 10, 12 12 12. Should an alarm from station 12 be transmitted due to a malfunction, the test indicator (second digit) will print as a 2 but another number in the row will be printed as "0."

The standard code assigned to each digit on the printout is:

1. Power failure alarm on first digit
2. Test alarm on second digit
3. Wet well alarm on third digit
4. Dry well alarm on fourth digit

The last four digits are reserved for future use. If station 12 reports a malfunction at the dry well, the printout will read: 12 10 12 12. A wet

Figure 16-10. A read-out printer at the county's operations center
receives signals from each pumping station.

well alarm will read 12 02 12 12. A printout for a primary power failure
will read: 02 12 12 12. To assure that the alarm gets through, the
message is transmitted five times at one minute intervals.

In every case where the digit "0" replaces any particular address digit—
with the exception of the second digit position—a malfunction has occurred
and an alarm is transmitted. Should two malfunctions occur at the same
time at a station, the printout will reflect it by replacing each address digit
with an "0" digit. If an additional alarm is detected during the cycle of the
five transmissions, the equipment will send five transmissions after the last
alarm is detected. In the event an alarm is being received during the five-
minute cycle and the condition at the station corrects itself, the code will
reflect the correction by printing the proper sequence of digits in a subsequent repeat of the signal. This feature saves the district engineers from
running off on any false alarms.

To check the system, Sanitary Engineering Division personnel artificially
actuate each of the alarms at the station on the first Wednesday of the

month. The alarm system is actuated either from a sensor located and connected to the pumping equipment in the dry well, through a sensor in the wet well, or, in case of a power failure, by a sensing circuit inside the alarm transmitter (Figure 16-11). The sensors, in turn, actuate a transmitter at

Figure 16-11. Radio technician checks sensor for alarm circuit.

the station, sending the signal to receivers on Mount Cuyamaca and on North Peak overlooking San Diego. The signal is then sent over microwave to the operations center where it is connected to the visual read-out system. The alarm transmitter normally operates from a-c power but, in case of a power failure, is operated by a self-contained battery. A sharp increase in population and industrial development is expected soon in the area. This increase will necessitate additional pumping stations. With each addition, the radio alarm reporting system will continue to reduce costs of checking the treatment facilities, which would otherwise be added to the operating budget.

TREATMENT EFFICIENCY TEST FOR PLANT UPSET[96]

In activated sludge waste treatment processes, influent waste is mixed with return sludge. This sludge contains the active mass of bacteria which,

in the presence of oxygen, feeds on organic matter in the waste and biochemically oxidizes this matter to harmless compounds and additional
cellular matter. If the waste is relatively constant in nature, the bacteria
become acclimated to those organics in the waste.

If a strange organic enters with the waste, the active mass may not
readily oxidize this compound. However, if the organic persists, the bacteria
may eventually become acclimated to it and biochemically oxidize the compound at a much faster rate. Therefore, until the sludge becomes acclimated,
it may stabilize the organic at a relatively slower rate than the normal organic content. As a result, the dissolved organic carbon concentration in the
aeration tank effluent will increase.

Waste treatment plants have a finite volume. If influent volume increases, the hold-up time within the aeration tank will decrease. Decreased
retention time generally results in less completely treated waste. If the influent contains the same level of biodegradable organics at high flow rate
as at low flow rate, the organic carbon concentration in the effluent increases.

However, if the waste is diluted by water containing little organic carbon, as might happen with storm water, then the effluent might contain
less soluble organic matter than at lower flow, yet still be incompletely
treated.

A method is suggested for assessing waste treatment efficiency. It depends not only on the level of dissolved organics in the effluent, but also
the rate of change of dissolved organics in the effluent when held in a batch
reactor to add retention time. The method presupposes that the waste
stabilization process taking place in the plant aeration tank will continue in
a batch aeration tank. It also presumes that samples taken from the plant
to the batch reactor will maintain approximately the same concentration of
mixed liquor suspended solids.

Figure 16-12 shows a generalized curve of the progress of waste stabilization taking place in a plant treatment process. If, with the prevailing flow,
effluent leaves the aeration tank at time T, then the reaction will continue
as shown, but with that portion of the reaction after time T taking place
in the batch reactor.

The slope of the curve, as determined from the change in dissolved
organic carbon divided by time, indicates the rate of biodegradability of the
waste by that biota present in the mixed liquors with its prevailing active
bacteria.

Both the biodegradable organics remaining in the plant effluent and
the refractory organics exert an ultimate oxygen demand on the receiving
water. The effects on the receiving water can only be qualitatively estimated,
but the amount of biodegradable organics and the rate of biodegradability
give an indication of the oxygen depletion which will take place in the
receiving stream.

In this procedure a sample is taken from the process and held in a batch
reactor at room temperature. Periodic measurements of the dissolved organic
carbon concentration are made until a constant plateau level is obtained.

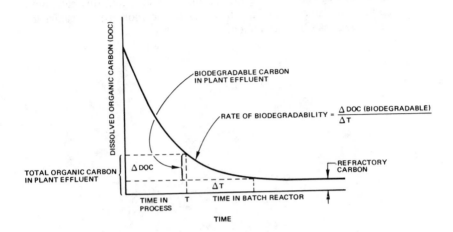

Figure 16-12. Water stabilization in sewage plant treatment process.

The procedure can be modified to speed biodegradation by increasing the temperature of the liquors (the reaction will be about twice as fast at 35 C as at 20 C).

Recommended Procedure

1. Set up a batch reactor, and provide an agitator for it.
2. Collect a sample of about one liter of mixed liquors from the effluent end of the process aeration tank.
3. As quickly as possible take the sample to the laboratory batch reactor. Keeping the sludge in suspension by frequent shaking of the sample container, add 500 ml of the mixed liquors to the batch reactor.
4. Begin agitation. Agitation must be sufficiently vigorous to keep mixed liquors oxygenated from entrained air.
5. Collect a 2-ml sample in a 2-ml syringe. Filter through a 0.45-micron filter.
6. Collect filtrate in a small test tube.
7. Add 5 microliters of concentrated HCl solution by syringe to the sample.
8. Gently sparge sample with nitrogen for 2 min.
9. Take 200 ml of sample into spring-loaded syringe and run analysis on the total channel of a total organic carbon analyzer. The instrument should have been previously standardized for 0-20 mg/l sensitivity full scale. Duplicate analyses should be run.
10. Repeat sample from batch reactor every 15-20 min. Repeat sample preparation and analysis until it has been ascertained that there is no further decrease in dissolved organic content.

11. Plot results as dissolved organic carbon vs time. From this plot, both the biodegradable carbon and the refractory carbon from organics in the effluent can be determined.

CHAPTER 17

POLLUTION FROM OIL SPILLS

DISPOSAL OF OIL WASTES[107]

One problem which concerns many engineers is the proper disposal of unwanted waste oils and wastewaters from lubrication operations. There are many varieties of lubrication wastes which may contain any combination of oils, emulsifiers, acids, metals, dissolved and suspended solids, organic and inorganic salts.

Oil substances can be deleterious in water supplies. Free oil and emulsions can act on the gills of fish to interfere with respiration. They destroy algae and other plankton, thereby removing a source of fish food. Settleable oily substances can coat the bottom of a water body, destroy benthal organisms, and interfere with spawning areas. Organic materials can deoxygenate waters sufficiently to kill fish. Soluble and emulsified materials ingested by fish can taint the flavor of the flesh and water or have a direct toxic action on fish or fish food. If films of free oil are present they can interfere with the natural processes of stream reaeration and photosynthesis.

Oil emulsions are particularly troublesome to control. Even when emulsions are treated and separated from the water, the effluent can contain substantial quantities of soluble oil.

At-Source Reduction

There are several methods which can be used to reduce pollution and overall abatement costs before end-of-pipe treatment. The best opportunity to reduce wastes is during engineering design. It is easier and more economical to incorporate abatement measures than to modify facilities at a later date. For example, all lubrication systems contain bacteria, which can attack emulsions, increasing the size of the oil droplets. This can reduce oil performance, oil life, and cause discoloration and odors.

When used in large concentrations, cresylic acid is quite effective in controlling bacteria. However, with extended use or at moderately low

Table 17-1. Waste Sources and Contaminants by Industry

Industry	Waste Sources	Fat and Grease	Free Oils	Emulsions	Dissolved Hydrocarbons	Dissolved Solids	Suspended Solids	Acids	Toxic Compounds	Scums and Foams
Metal products:										
Forming	Mill scale, oils and grease	XX	XX	X		X	XXX	XXX	XX	X
Finishing	Dissolved metals, spent acids	X	XX	XXX	X	XXX	XX	XXX	XX	XX
Plating	Pickle liquor		XX	XX		XXX	XX	XXX	XXX	XX
Petroleum	Oil sludges, brine, spent caustics	XXX	XXX	XXX	XXX	XX	XX	X	X	X
Chemicals	Paints, adhesives, plasticizers	XXX	XXX	XXX	XXX	X	XX	X	X	X
Foods	Meat products	XXX		XXX	XXX	X	XX			X
Textiles:										
Natural fibers	Scouring, treating	XXX	XXX		XXX	X	XX			XX
Man-made fibers	Sizing, processing		XXX	XX	XXX	X	XX			X
Aircraft	Washracks depainting	XXX	XXX	XXX	XXX	XX	XX	XX	XX	XX
Automobile	Soluble oils, emulsifier paints, metal finishing		XXX	XXX	XXX	X	XX	X	XX	X
Railroad	Washing	XXX	XXX	XXX			XX	XXX		XXX
	Service depot	XXX	XX	XX						
	Railway works		XX	XX				XXX	XXX	
Machine shop	Stamping, grinding	XXX	XXX	XXX	XX		XXX	XXX		XXX

XXX—Major problem
XX—Moderate problem
X—Minor problem

cresylic acid concentrations, certain bacteria can actually thrive on this compound. Several new compounds, such as dichlorophenol and dichlorophen, have in some cases increased emulsion life fivefold. A visible in-line cartridge could be installed for dispensing bactericide directly into the recirculating coolant.

In-plant Measures

Once a facility is in operation, pollution control becomes much more difficult and expensive. First review the operation to determine if design changes can reduce waste and improve operation. Before a waste is dumped into a sewer, recovery and reuse should be considered. While the use of several emulsions may be necessary in a single operation, separate sewer systems for similar materials can facilitate recovery, reuse, or more effective treatment.

After all possible design and operating improvements have been explored, several pretreatment steps should be considered. Most lubrication wastes are disposed of in batches. Therefore, equalization or intermediate storage should be provided.

Recovery and Reuse

The following physical and chemical methods are employed in waste oil recovery:

Gravity differential separation—Separation alone may remove water and settleable solids, with free oil being recovered.

Vacuum filtration—Precoat vacuum filtration incorporates a layer of filter aid on a vacuum rotary drum and is a proven method for treatment of petroleum refining cuff.

Acid treatment—Sulfuric acid treatment can break emulsions and separate the saturated naphthenic and paraffinic molecules.

Temperature change—Heat treatment can be used alone or in conjunction with gravity separation and/or acid treatment to break emulsions and separate oil.

Electrostatic cleaning—Electrostatic separators remove brine and sediments from crude oil. A new technique has been developed for application of this technique to waste oil recovery.

Centrifugation—Recent improvements have made centrifugation an attractive method for materials separation.

Chemical treatment—Chemical methods of emulsion breaking are widely used and include the reaction with salts of polyvalent metals, salting out of the alkali soap, and destruction of the emulsifing agents.

Flocculation and sedimentation—Chemical flocculation can break emulsions and a recent study indicates that the addition of finely divided solids to emulsions will greatly enhance separation. Materials such as clay, flyash, and even wastewater treatment sludges have been used in conjunction with a coagulant to effectively separate emulsions.

Dissolved gas flotation—When used in conjunction with chemical flocculating agents, flotation can effectively remove oil and finely divided solids from emulsions.

Extraction—Certain compounds can be separated by selective extraction. Many times, though, the properties of the extracted oil are altered. For example, pentane may precipitate materials which are actually dissolved in the oil rather than suspended, while benzene may hold in solution materials which are actually insoluble in the oil itself.

Agitation—Emulsions may be broken simply by vigorous agitation.

Ultrasonic vibration—Subjection of the emulsions to high frequency sound vibration has been investigated, but data is not available to permit assessment of this technique.

Ultimate Disposal

Land disposal of oily wastes is the most widespread method used. For several years a petroleum refiner has employed a unique technique for land disposal of petroleum sludges and stable emulsions from tank bottoms, oil-water separators, and ship ballast water. The area used consists of approximately seven acres divided into four sections. Earthen dikes were constructed to retain the sludge, and 20-in.-high terraces were graded with openings on opposite ends to allow oily materials to travel over the surface area. Oily sludges and emulsions are applied to a depth of 6 inches, and a bulldozer with a scraper blade mixes the sludge with 6 inches of soil. This mixing is repeated two to four times per month until the hydrocarbons have been consumed by bacterial action.

The four sections are rotated so that as one is filled, another is opened to receive sludge, and the two remaining sections are "working."

Decomposition requires three to nine months, depending upon ground temperature, soil moisture content, and type of hydrocarbon. The overall rate of decomposition varies from 5 to 60 lb of oil per cu ft of soil per month. This method of disposal is, of course, not suitable where soil conditions would permit contamination of underground water supplies.

Incineration of oily substances has many advantages, and in some cases, is the only proven method of ultimate disposal. Incineration may be the most economical method available, if the fuel value of the waste is sufficient to maintain combustion.

There are several incineration processes available—multiple hearth, fluidized bed incinerators, and wet air oxidation. Rotary kilns can be used for plastic, tacky, and viscous materials including filter aids and sludges.

Incineration processes require relatively high capital expenditures, and unless the quantities of waste oil are sufficient, the economics of incineration can be marginal or prohibitive. Cooperative efforts with other industries can often reduce disposal costs and provide an acceptable means of ultimate disposal.

The importance of a step-by-step approach to solving oil waste problems cannot be overemphasized. Risks of improper design include: corrosion; air, water, and land pollution; safety hazards; increased cost; limited effectiveness; and bad public relations. It is wiser to define problems thoroughly and develop treatment processes through proper investigation.

CONTROLLING OIL SPILLS[163]

"Spill!"—that word indicating the contact of oil with water—is one of the most misunderstood words in the ecological vocabulary.

Oil is in constant transit around the globe in a variety of forms. There have always been spill problems at sea. Ocean-going vessels have dumped petroleum products into the seas for many years. More than 4000 tankers, many with capacities of more than 300,000 tons, transport oil from one country to another.

Yet, pollution on the high seas is not the major oil spill problem. The major cause of oil pollution is the average day-to-day spill that occurs in harbors, rivers, lakes and streams. Two-thirds of all oil spills occur during routine operations as the result of human error. Whether the captain of a barge has fallen asleep, allowing a tank to overflow during transfer operations; or whether a valve has sprung a leak at a refinery, thus seeping oil into the river; or whether a hose has burst; the main cause of the oil pollution problem remains the same: lack of attention to details, lack of follow-through, and a lack of planning at the local level.

Oil pollution is a complex problem. It must take into account the interaction of oil, water and wind. The problem requires an overall systems approach to its solving: planning, equipment and training.

Planning

Whether at a drilling site of one of the 8000 offshore wells which surround this country, or in a tanker passing through narrow straits or during transfer operations, there exists the possibility that a spill will occur.

During transfer operations, faulty valves and burst hoses are always sources of possible danger. On a ship or barge, pumping, running aground or tanks overflowing can be problems. The pollution engineer should seek out the places where he is likely to have a spill.

Equipment

After a survey has been completed of a facility, the next consideration is the equipment needed to control an oil spill. There are a number of devices on the market designed to control or contain spills. When one takes into consideration the water current directions and speed, wind and weather, the real problem is controlling—rather than merely containing—the spill. The slick must be controlled so that the oil can be recovered effectively.

Each of the available "controllers" roughly conforms to what is generally called an oil "boom" (Figure 17-1). A boom is a floating device with

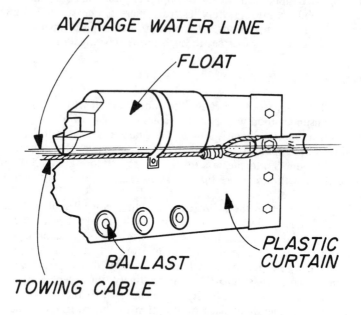

Figure 17-1. Major components of an oil boom.

an underwater curtain. Booms go under a variety of trade names—curtains, shields, guards—but all are basically designed to hold the oil for recovery, Figure 17-2. Many are not adequately designed. As a result, deployment results in failure—the pressure against the curtain has not been estimated correctly, or the product is made of improper material or doesn't float properly.

Commonly used materials for the floating sections are synthetics, either polyethylenes or polyurethanes. Styrofoam wooden logs and even steel barrels are used by some manufacturers. However, when purchasing a boom it is important to determine:

1. Will it float?
2. Will it follow wave contours?
3. Will it withstand the forces of nature?
4. Will it maintain physical integrity?
5. Will it control the oil spill?

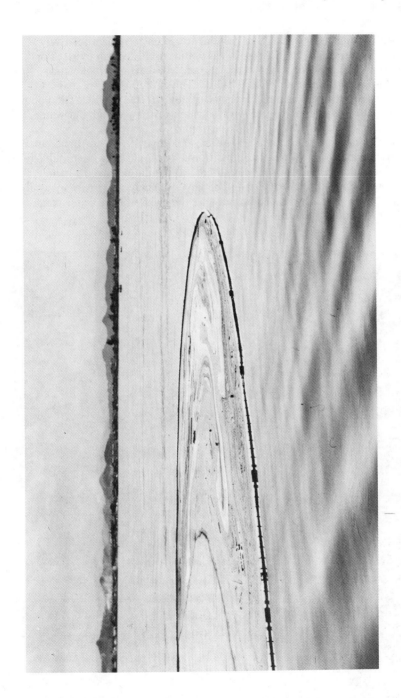

Figure 17-2. An oil boom is a floating barrier for containing oil and floating objects.

Most booms have subsurface curtains weighted to remain relatively perpendicular to the water's surface. Here, too, problems can exist. For example, it must be determined if the curtain will remain perpendicular under the force of given water conditions. The curtain must be at the proper depth and made of a material that is not affected by oil. The boom must be adequately reinforced to withstand the forces of wind and water— (forces often exceeding 100 lb per linear foot).

The boom must be adequately balanced so that it will not break apart or tip over under unusual weather conditions. It should have been publicly test-engineered under actual spill conditions. It should be easy to handle, easy to store safely (Figure 17-3), and deployed quickly and easily by available manpower. Finally, the boom must be the right size for water conditions surrounding a facility.

Figure 17-3. An oil boom should be constructed so that it is easily stored.

Once the proper boom has been selected, the problem of recovering the oil is next. Recovering oil from water is a tricky procedure. It involves a knowledge of what type oil is spilled, whether it can be recovered, and what the per gallon cost will be to recover. Specific gravity, viscosity and water temperature all affect the recovery process.

Some crude oil can be recovered by straw. However, Number 6 fuel oil is difficult to recover, especially when cold. Other problems include who is to deploy the straw; once deployed, how to clear it from the water; once cleared from the water, how to recover the oil from the straw; once separated from the oil, how to dispose of the straw? Burning straw can create serious air pollutants; burying it can contaminate the ground. It is estimated that using straw to recover oil can cost up to $20 per gallon.

Skimmers can recover oil at 40 cents per gallon. They come in a variety of shapes, sizes and capabilities, and are generally of two types: floating weirs and suction heads. Floating weirs collect more water than oil. Suction heads are more efficient and can be used in shallow water along shorelines (Figure 17-4).

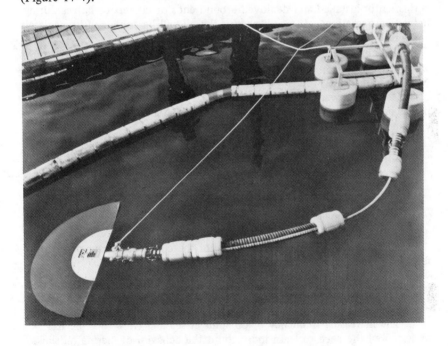

Figure 17-4. Skimmers are used to remove oil trapped by an oil boom.

Getting the oil into the recovery equipment without breaking the surface tension of the oil on the water is the problem.

Training

When purchasing any oil pollution control equipment it is necessary to find out if the manufacturer will train your employees in what to do in the case of emergency. Training involves more than equipment deployment and storage. In the case of oil pollution, a manufacturer should provide your crews with some basic engineering knowhow about what happens when oil meets water.

First, a survey of your facility should be conducted to determine wind directions and water currents. It is not generally known, but oil will be more influenced by wind direction than by water current. It is entirely possible for water to be flowing south, and an oil spill to be influenced northward by the wind. It is therefore necessary to know the prevailing wind conditions, concurrent with the water directions.

Secondly, the crew must be properly versed in deployment of the equipment (Figure 17-5). It must be done rapidly with very little disturbance of the oil slick. Some companies have solved this problem by keeping booms permanently installed and deployed around ends of tankers or barges during unloading operations. If a spill occurs, the oil will be contained immediately and will only have to be recovered.

Training involves time, but it is well worth it and is inexpensive insurance. Recently, a spill estimated at over 600,000 gallons occurred at a Northeastern utility storage facility. The slick, up to 5 in. deep in some places, headed toward Long Island Sound. Fortunately, two utility companies had booms available. The Coast Guard initiated action using this equipment, and the oil never escaped the harbor.

This spill, estimated by some to be the largest in the history of the United States, did not get out of hand for three very important reasons:

1. Local officials, Coast Guard, and neighboring companies had planned for the possibility of a slick.
2. Proper equipment was available from a number of sources.
3. Responsible persons were trained in the proper use of equipment, so that the spill was quickly controlled and recovered.

However, because of the efficiency of the control program procedures, few people ever knew that the spill occurred.

OIL SPILLS MEASURED[77]

The increasing frequency of oil spills at sea and close to shore has underscored the need to learn more about the behavior of marine oil slicks. Many basic questions about oil spills and their environmental impact are still unanswered. There is still much to be learned about the spreading behavior of oil slicks and the estimating of flow rates and quantities of pollutants involved.

Estimates of a Tampa Bay spill varied by a factor of two. For the Santa Barbara oil spill, spillage estimates varied by a factor of ten. One

Figure 17-5. Crews should know how to deploy a boom, and practice for
rapid efficient handling.

source reported a quantity of 10,000 gallons for the Tampa Bay spill, and
also claimed that 100 square miles were covered by oil. Based upon these
figures, the resulting oil film would have been so microscopically thin that
it would have been barely detectable. The Santa Barbara spill was at one
time reported to be covering 800 square miles.

Estimates of oil quantities in a marine oil spill are based on estimates
of the area of sea-surface affected. However, gross uncertainties can result
from inaccurate estimates of the area. An effective system for monitoring
oil spills can be devised by combining operational state-of-the-art remote
sensing techniques with improved "ground truth" procedures.

First, a primary sensor system must be selected. Since one of the basic problems is finding the areal extent or distribution of the pollutant, an imaging sensor system is preferable to an analog recording system, and an areal map is a ready by-product of an imaging recorder.

The region of the energy spectrum for the imager is the next choice to be made. Oil on sea water does not photograph well on panchromatic photographic film emulsions. Contrast between an oil patch and the background water is too low for repeatable, positive recognition. Favorable sun angles can accentuate reflectance from oil films, but reflectance is difficult to distinguish from sun-glitter on oil-free waters. On black-and-white photographic prints made during a number of conventional aerial mapping system overflights of the Santa Barbara oil spill, the tone of the oil slick was almost identical to that of the adjacent waters even near the actual spill site at Platform A.

Color photography is, as a rule, just as ineffective. The color contrasts between oil and the background water is again very low. Except for areas where thick concentrations of oil register as dark gray, or for very thin films of oil producing iridescent reflectance, oil-polluted waters show a grayish green color which an observer would accept as a normal sea surface tone (Figure 17-6). The normal-appearing sea surface is actually completely covered by a film of oil. The only clue for the interpreter is the boat's wake, which shows the lighter tone of the underlying water.

Infrared Imagery

These problems of object-to-background contrast do not affect thermal infrared imagery. Oil-covered areas show up clearly and distinctly. The oil shows a mottled and streaked pattern, both lighter and darker than the background sea surface.

While emissivity characteristics of oil on water are important, the immediate concern is pollutant detection by an imaging system. A thermal infrared mapper shows distinctive responses for oil-covered areas, and allows them to be readily distinguished from the less-polluted background.

After the image is obtained, its contrasts must be enhanced to aid the interpreter. Densitometric color enhancement maximizes tonal differences to distinguish oil slicks from "background" water.

It is at this stage of the procedure that a trained interpreter's judgments introduce subjective control. Color enhancement has to be manually adjusted to a setting and judged to be optimal for separating the "object" (oil) from the "background" (water). No machine will do this automatically.

The film reader system presents density values of a photographic image as analog voltage levels of a video signal which can be displayed in color. A high-speed analog-to-digital converter continuously changes the analog voltage levels to discrete stepped values. Each single value is presented to a number of digital-to-analog converters (three for each color) that produce the red, green, and blue voltage signals for the three electron guns of the color monitor display screen.

Figure 17-6. Without the contrasting tone image of the boat's wake, the oil covered sea could easily be mistaken for a normal sea surface using black and white or color photography.

The three colors can be mixed on the display screen to produce any desired color hues. Colors can be shifted to assign any hue to any selected density level. Quantitative density readings are obtained through a calibrated photographic step-wedge which can be included as a reference. The same colors will appear in all areas of the image having the same density levels.

Color enhancement of an image can automatically increase object-to-background contrast. This enhancement permits identification of oil patches, slicks and films which show even minor density differences from the background water, which emits almost a uniform density.

Figure 17-7 is a cartographic representation of a color-enhanced image. Pattern areas "A" and "B" are increasing step increments of density, representing oil patches and concentrations of varying thickness. The essentially oil-free sea surface is shown as the unpatterned area "C." This representation is an isopleth map of the affected sea surface, showing the distribution of oil pollutants in various degrees of concentration. Ground truth measurement of average thickness within each of the mapped concentration areas would have permitted the assignment of thickness ranges.

A digital planimeter will provide a readout of specific areas of the color density. Area calculations can be automatically expressed as percentages of the total image area. A pushbutton selector enables an operator to quickly measure the area of one or more combination of ten basic colors. The percentage of the image area depicted in each of the selected colors is indicated by a digital voltmeter.

Of the area shown in cover illustration, approximately 62 percent is within the density range interpreted as water. Barges, boats, and drilling platforms represent an additional 0.5 percent of the area. Approximately 37.5 percent of the sea surface is covered by oil spill pollutants.

Allowing for linear distortion in the infrared scan, the total area depicted in this image is calculated to be 2.51 square miles. Approximately 1.6 square miles are in the density range interpreted as water, and 0.9 square miles are within the density range interpreted as oil pollutants. If a value of 0.01 in. is assumed as the average oil thickness in this part of the spill, approximately 155,000 gallons of oil are depicted on this one image.

If the actual amount of crude oil spilled must be determined, other variables must be investigated. The pollutants shown are not pure petroleum but actually an emulsion of crude oil and sea water, which may vary in concentration over the spill area. The volume estimate should be reduced to reflect the accepted percentage value for water in the emulsion. On the other hand, oil can evaporate as much as one-fifth of its volume within an hour under properly turbulent conditions.

The question of amounts of polluting agent spilled, apart from areal extent of the pollution, may well become one of great legal and financial significance in the future when violators are required to pay for cleanup or damages. But, for determining pollutant quantities, the areal extent is the starting point. Thermal infrared imaging, color enhancement and digital planimetry can provide basic information economically, semi-automatically, repeatedly and accurately.

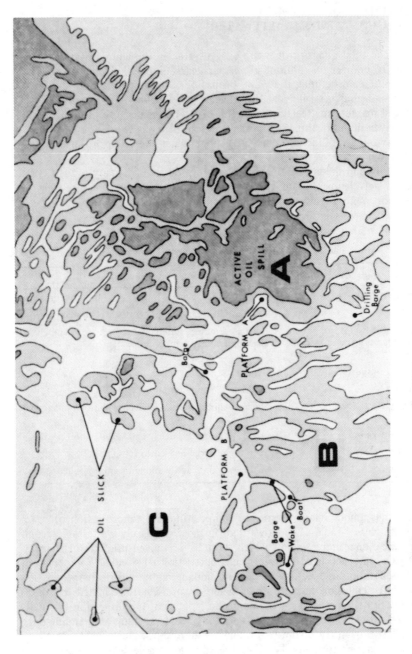

Figure 17-7. Cartographic rendition of a color enhanced image. Oil-free surfaces are represented by C, "lightly" oil-covered surfaces by B and thickly covered surfaces by shades A.

TREATMENT OF OILY AND
METAL-CONTAINING WASTEWATER[109]

The total quantity of waste oil generated in the United States is estimated to be over one billion gallons per year. Since 1965, the market for reclaiming or re-refining the waste oil has been declining because of: (1) disadvantageous tax rulings, (2) packaging and labeling requirements, (3) reduced markets for reclaimed mineral oil, (4) increased use of additives and synthetic lubricants making reclamation difficult, (5) increased transportation and refining costs, and (6) submarginal operation of outdated technology.

Oily and metal-containing wastes are produced by many industries, such as petroleum refining, steel, automobile, shipbuilding, marine transportation, railroad, aircraft manufacturing, miscellaneous metal-working, food processing, textiles, and tanning, Table 17-2. Each of these industrial

Table 17-2. Sources of Oily and Metal-Containing Wastes

Industry	Major Waste Sources
Petroleum refining	Spent caustics, general oily water, tank bottoms
Steel	Rolling mills, pickling and plating, coke plant
Automobile, shipyard, aircraft manufacturing, and machine shops	Cutting, grinding, stamping and heat treating, cleaning/plating, paint spray system
Marine transportation	Bilge water, ballast water
Railroad	Tank car cleaning terminals
Military installations	Machine shops, washing/cleaning, photographic and plating shops
Food processing	Meat, fish, poultry, dairy, and vegetable oil processing
Textiles	Scouring of wool and cotton
Tanning	Soaking, liming, and washing of animal skins

wastes has unique characteristics in composition and concentration of contaminants. Table 17-3 shows that, in most cases, oily wastes are accompanied by undesirable metals which should be removed prior to discharge.

Oil emulsion is an important characteristic that should be understood, since separation or breaking of oil emulsions is required for satisfactory disposal. An emulsion can be formed when oil and water are mixed together under agitated conditions. The emulsion is, in fact, the dispersion of finely divided droplets of one liquid into another liquid medium. Depending on the relative quantity of the two liquids, either oil-in-water or water-in-oil emulsion can be formed. To make the emulsion stable, a third substance called an emulsifying agent is required. Common emulsifying agents include soaps, sulfated oils and alcohols, sulfonated aliphatics and aromatics,

Table 17-3. Characteristics of Selected Wastewaters

Contaminants	Separator Effluent mg/l	Coke Plant Waste Liquor mg/l	Electroplating Wastewater mg/l	Automobile Metal Cleaning & Spray Painting Wastewater mg/l
Oil	60	30	---	900
BOD$_5$	160	4,000	---	---
COD	340	9,000	20-90	3,100
Suspended Solids	80	70	5-60	4,300
Total Solids	---	11,100	---	5,100
Phenol	20	1,600	---	---
Ammonia	60	600	---	---
Cyanide	---	20	4-26	---
Chromium (as Cr)	1.2	---	0.4-23	47
Iron (as Fe)	4.5	10	---	170
Copper	0.2	---	1.0-7	0.45
Cadmium	0.1	---	0.2-25	0.19
Zinc	---	---	1.2-6	14
Silver	---	---	---	0.44
Molybdenum	---	---	---	6.2
Nickel	---	---	0.6-12	0.6

quaternary ammonium compounds, non-ionic organic ethers and esters, and various solids (fine particles). Emulsifying agents enhance the strength of the interfacial film around the droplets of the dispersed liquid by providing layers of electrical charges, thus increasing the stability of emulsions. Stability is created, because the finely divided droplets are prevented from coalescing or merging together into larger droplets.

Oil emulsions, used in machine shops for cutting and grinding and in steel mills for cold rolling, are often oil-in-water emulsions of two to eight percent of long-chain fatty acids with soaps or detergents as the emulsifying agents. Bactericides and antioxidizing agents are often added to increase the life of the emulsions. Nevertheless, the emulsions will become rancid or degraded by biological oxidation and eventually will have to be discarded from the circulation systems.

In refineries, oil emulsions in the wastewater can be formed through agitation in the sewer line, discharge of spent caustic into the sewer, presence of solids, or pumping of API separator influent streams. In hot-rolling steel mills, oil emulsions can be formed from pumping of the scale pit effluent or clarifier effluent for reuse of spray cooling purposes, or from the presence of silt from river water or iron oxides from rolling operations.

Treatment Methods

Selection of a specific treatment process depends upon the characteristics of the waste and pollution control regulations. Alternatives available for control of oil- and metal-containing wastes can be broadly classified into four categories: in-plant measures, pretreatment, effluent treatment, and ultimate disposal of oil and sludge.

In-Plant Measures

1. Proper design of oil emulsion coolant or lubricant recirculation system to reduce the quantity wasted,
2. Use of bactericides such as cresylic acid, dichlorophenol, or dichlorophen to extend the life of coolant or lubricant in the recirculation system,
3. Use of vacuum dehydration in the lubricant recirculation system, to reduce bacterial growth by control of water content,
4. Use of cartridges, bags, or vacuum filters, or of solid bowl and/or disc centrifuges to separate solids and water from oil,
5. Use of solvent extraction for recovery of oil or other organic compounds,
6. Use of a countercurrent rinse water or wash water system to reduce wastewater quantity,
7. Chemical precipitation of selected process waste streams to remove metals at the source,
8. Containment and segregation of spent caustic to avoid formation of oil emulsion, or
9. Segregation of highly emulsified oil, acid, alkaline, or toxic waste, to facilitate pretreatment.

Pretreatment

(For discharge to joint municipal/industrial treatment system, or for further on-site effluent treatment).

1. Gravity separation to remove free flotable oil and settleable solids,
2. Treatment with steam, acid, alum, or iron salts to break oil emulsions,
3. Use of an equalization facility to reduce variations in waste flows and characteristics,
4. Use of chemical coagulation followed by sedimentation or dissolved air flotation to remove suspended solids and emulsified oil,
5. Treatment of hexavalent chromium with acid and a reducing agent to convert it to trivalent chromium for subsequent precipitation,
6. Chemical precipitation of metals with lime or caustic,
7. Alkaline chlorination of cyanide-containing waste,
8. Use of electrolytic methods to remove selected dissolved metals, or
9. Neutralization of excess acidity or alkalinity.

Effluent Treatment

1. Biological treatment such as activated sludge, trickling filter, or aerated lagoon to reduce or remove biochemical oxygen demand (BOD), soluble oil and solids,

2. Use of deep-bed filters (such as sand filters, dual-media filters, or multi-media filters), to remove solids, oil and insoluble BOD,
3. Use of activated carbon to adsorb biodegradable as well as non-biodegradable dissolved organic material, or,
4. Use of reverse osmosis, ultrafiltration, ion exchange, or evaporation to remove metals and dissolved solids.

Ultimate Disposal

1. Re-refining to recover oil for reuse,
2. Incineration in furnaces to recover heat values (need scrubbers to control air pollution),
3. Incineration in multiple-hearth furnaces, fluidized beds, or special liquid waste incinerators to destroy oil and putrescible organic matter (need scrubbers to control air pollution), or
4. Disposal of dewatered sludge or incinerated residues in approved sanitary landfills.

Table 17-4 summarizes some of the available methods for treatment for oily wastewater in the petroleum refining industry. Table 17-5 summarizes the applications and limitations of treatment processes for removal of heavy metals and toxic substances. Treatment efficiency will, of course, be highly variable depending upon the initial concentrations of raw wastes.

Table 17-4. Available Levels of Treatment for Oily Wastewater in the Petroleum Refining Industry — After Initial Treatment by API Separator

Treatment Processes	Removal Efficiencies, Percent					
	BOD$_5$		Oil		Suspended Solids	
	Range	Typical	Range	Typical	Range	Typical
Dissolved air flotation	5-35	25	20-70	40	10-60	35
Chemical coagulation and dissolved air flotation	20-70	45	50-90	75	60-85	70
Stabilization pond	50-90	70	50-95	80	20-70	45
Aerated lagoon	50-85	72	50-80	60	40-65	50
Activated sludge	70-92	84	55-95	75	40-80	60
Chemical coagulation, dissolved air flotation, activated sludge	80-93	87	80-99	90	40-80	62
Filtration, activated carbon adsorption	80-95	87	95-98	96	70-97	84
Equalization, chemical coagulation, dissolved air flotation, activated sludge, filtration & activated carbon adsorption	90-96	94	95-99	97	80-98	88
Ultrafiltration	High	100	100	100	100	100
	Effluent Concentrations, mg/l					
Dissolved air flotation	50-300	120	10-70	40	20-100	50
Chemical coagulation and dissolved air flotation	40-200	85	5-35	20	10-60	25
Stabilization pond	20-150	50	5-50	15	25-100	45
Aerated lagoon	10-120	45	5-60	24	10-80	40
Activated sludge	15-60	26	5-30	15	10-60	35
Chemical coagulation, dissolved air flotation, activated sludge	10-30	20	2-15	6	10-60	30
Filtration, activated carbon adsorption	5-20	15	2-20	10	3-60	12
Equalization, chemical coagulation, dissolved air flotation, activated sludge, filtration & activated carbon adsorption	5-15	10	1-5	2	5-15	10
Ultrafiltration	Varies	Varies	0	0	0	0

Table 17-5. Treatment Processes for Removal of Heavy Metals and Toxic Substances

Contaminant	Treatment Processes	Possible Effluent Concentration after Treatment[a], mg/l
Arsenic	CC (with iron salt) + S + F	0.05-0.5
Barium	CP (with sodium sulfate) + S + F	1-2
Cadmium	CP + S + F	0.1-1
	CP (with lime) + S + RC + F + AC	0.002
Chromium (trivalent)	CP + S	0.06-4
Chromium (hexavalent)	R + CP + S	0.7-1
	IE	0.03
	CP (with lime) + S + RC + F + AC	0.02
Copper	CP + S	0.15-2.5
	CC + S + F	0.5
	CP (with lime) + S + RC + F + AC	0.05
Cyanide	Chl (Alkaline)	0.0
Fluoride	CC (with alum) + S	1.0
	A (with hydroxyapatite)	0.5-1.5
Iron	CC + S + F	<0.3
Lead	CP + S	0.1-1.4
	CP (with lime) + S + RC + F + AC	0.02
Manganese	CP (with lime) + S + F	<0.05
	O (with permanganate) + F	<0.05
Mercury	IE	0.002-0.005
Nickel	CP + S + F	0.1-1.9
Selenium	IE (cation and anion)	NA[b]
	CP (with lime) + S + RC + F + AC	0.002
Silver	CP (with chloride) + S	1-3
	Chl + CP (with FeCl$_3$ and lime) + S	0.1
	IE	Trace
Total Dissolved Solids	RO	<500
	El	<500
	Ev/D	<25
	IE	NA[b]
Zinc	CP (with lime) + S	0.5-2.5
	CP (with lime) + S + F	0.1-0.3
	CP (with lime) + S + RC + F + AC	0.4

Legend

AC	– Activated carbon adsorption		F	– Filtration
A	– Adsorption		IE	– Ion exchange
CC	– Chemical coagulation		O	– Oxidation
CP	– Chemical precipitation		R	– Reduction
Chl	– Chlorination		RC	– Recarbonation
El	– Electrodialysis		RO	– Reverse osmosis and ultrafiltration
Ev/D	– Evaporation/Distillation		S	– Sedimentation

[a]Treatment efficiencies are variable depending upon initial concentrations of raw wastes
[b]NA – Data not available.

CHAPTER 18

COOLING AND COOLING TOWERS

INDUSTRIAL COOLING TOWERS— A USE PROFILE [124]

Increasing concern about water conservation and thermal water pollution is causing industry to search for more efficient ways to cool water. However, government thermal water pollution regulations are changing rapidly. The cooling system designed to meet needs this year may not economically or even feasibly answer next year's requirements if proper care is not taken in selecting equipment.

Removal of heat in a cooling tower is a combination of sensible heat transfer and evaporation between the hot water droplets and the main body of air which is drawn from the atmosphere. For a cooling tower to operate, air flow must be created in the tower. The early water cooling towers used by industry were called *atmospheric towers.* Air flow through the tower was established by local winds and the natural draft effect. These towers were long, narrow and high. Decks were placed at intervals within the tower to break up the water.

The basic advantage of the atmospheric tower was the absence of mechanical equipment costs. However, pumping costs were relatively high because of the tower's length and height. Another problem was the tower's dependence upon natural air flow and local winds to achieve the performance requirements. This restricted its use to geographic areas with high relative humidities and where prevailing winds could be consistently depended upon for air flow through the tower.

An evolutionary step in water cooling towers was the introduction of *mechanical draft towers* in the late 1920's. This kind of tower used fans to induce or force air movement, and overcame the geographic limitations and operational stability problems of the atmospheric tower. Added mechanical equipment costs were a trade-off for reduced pumping costs, increased flexibility, reduced tower height and off-tower temperature.

As a result of design improvements, mechanical draft towers are the most widely used industrial water cooling towers in this country. There are two basic types of mechanical draft towers.

The *counterflow tower* (Figure 18-1) was the earliest type of mechanical draft tower. Both forced-draft and induced-draft fan systems have been used in counterflow tower designs. Current designs use induced draft fans located at the top of the tower. Air is drawn through the louver situated at the tower base and flows upward through the tower fill material. Water is pumped to the top of the tower plenum and is sprayed downward over the fill material in a direction counterflow to the air movement.

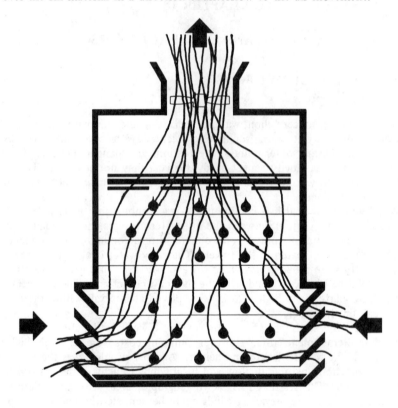

Figure 18-1. Counterflow cooling tower.

The *crossflow tower* (Figure 18-2) is a more recent development in mechanical draft towers. It also uses an induced draft fan system. However, in the crossflow tower, air is drawn horizontally across the fill section. Water is pumped to hot water basins at the top of the tower and flows vertically downward through the fill.

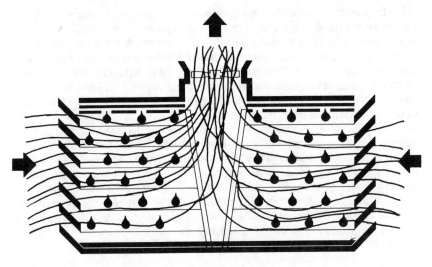

Figure 18-2. Crossflow cooling tower.

In selecting efficient cooling tower packings, the most important characteristics necessary for a thermally efficient system include:

- *Water dispersion rate*—The more droplets produced, the higher the air-water interface surface will be.
- *Retention time*—The longer the air and water droplets are in contact, the better the heat transfer.
- *Minimum resistance to air flow*—Fan horsepower requirements are reduced.

Most commonly used cooling tower packings in the United States are splash-type packings. These depend primarily on water breakup to create the required transfer surface area between water and air. A second method of creating a transfer surface uses a film-type fill, which consists of thin sheets of material spaced closely together and suspended vertically. Water is distributed over the top of the sheets and runs over the sides to create water film.

In principle, the counterflow tower is the most efficient from the standpoint that a lower air flow is required for a given set of thermal conditions. This advantage becomes more pronounced as the cooling requirements become more difficult. However, most engineers purchasing cooling towers are only interested in the initial cost of the equipment and the energy consumption of the tower to produce the air flow.

The crossflow tower, while less efficient thermally, offers less resistance to air flow. It can also operate effectively at higher air velocities than can a counterflow tower. Therefore, lower total horsepower requirements and smaller cell sizes are sometimes possible with a crossflow tower.

In practical application, counterflow and crossflow mechanical draft towers each have distinct advantages and limitations which should be surveyed as a primary step in system evaluations. The *approach* for a cooling tower is defined as the difference between the cold water temperature leaving the tower and the wet bulb temperature of the atmospheric air. This is a key parameter in tower design. *Range* of a cooling tower is defined as the difference between the hot water temperature entering the tower and the cold water temperature leaving the tower.

Mechanical draft towers can be economically designed for systems requiring a low approach (cold water temperatures 5 F above wet bulb temperature are not unusual). They may also be designed to handle a broad range of water flow rates. This is accomplished by simply increasing the number or size of the cells. Mechanical draft towers are a fairly reliable and accurate method of controlling the cold water temperature even though climatic conditions change. Use of multiple cell mechanical draft towers offers a tremendous amount of versatility and ability to respond to changes within the operating system by proper operation of the fans.

Hyperbolic Towers

The natural draft hyperbolic tower has been used in Europe for a number of years, but has found wide acceptance in the United States only recently. Air flow is established by the natural draft principle. Temperature differences between the atmospheric air and exhaust air from the cooling tower inside the hyperbolic shell create a density difference which induces air flow through the system.

Both counterflow and crossflow natural draft hyperbolic towers are currently being offered in the United States. In the counterflow arrangement, the fill is located inside the hyperbolic shell directly above the air inlet (Figure 18-3). In the crossflow design, the cooling tower section forms a ring around the base of the shell and a canopy is used to seal the system. The hypoerbolic shell is constructed of reinforced concrete and can typically measure as high as 500 ft with a base diameter of 300 ft. These towers are designed to handle extremely large heat loads, typical of large power plant installations. Water flow rates in excess of 450,000 gpm are possible with this tower design.

There were no hyperbolic towers in use in the United States until construction of the Big Sandy power plant in Louisa, Kentucky in 1963. However, by the end of 1969, 16 hyperbolic towers were operational in this country. Current projections indicate there will be 32 to 40 more towers constructed in the next 2 or 3 years.

SELECTING AND SIZING COOLING TOWERS[114]

Optimization is the key to selecting and sizing an industrial cooling tower. However, since 100 or more tower configurations might meet the

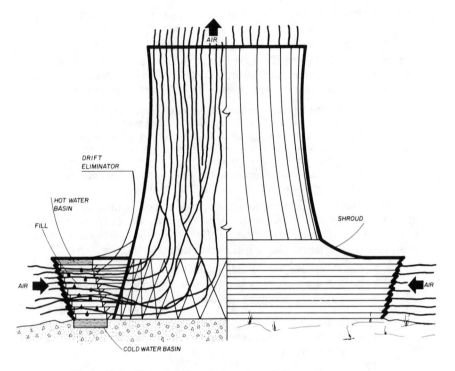

Figure 18-3. Natural draft crossflow hyperbolic cooling tower.

cooling specifications for a particular application, the pollution engineer
must establish selection criteria which will help him arrive at that optimization.

There are two major categories of information which must be identified
and considered to reach financial optimization in tower selection. The first
group of facts includes such items as the specifications relating to water temperatures, GPM of water and environmental realities. This data can be called
the *design conditions* needed to provide sufficient information to select a
tower.

The second group of selection criteria are the *evaluation factors.* These
will include such variables as the costs of fan horsepower, pump horsepower
and the cost of money. Other criteria coming to the forefront in recent
years are aesthetic considerations as they relate to tower architecture and
site location.

The following design conditions are essential factors in proper cooling
tower selection:

   1. **Input Temperature**—The hot water temperature as it enters the tower.
      This will vary according to industry and application.
   2. **Output Temperature**—The temperature required as the cold water
      comes off the tower and is recirculated back through the system.

3. **Wet Bulb Temperature**—This will vary according to geographic location. Exact wet bulb is available from local weather bureaus.

The design figure is usually stated in a percentage frequency, such as 1, 2½ or 5 percent. Normally a six-month summer period is considered, and the percentage is based upon the number of hours during the summer period that the temperature can exceed the wet bulb temperature.

4. **Gallons Per Minute**—The water flow relates entirely to the specific installation and should be based upon forecast peak capacity.
5. **Elevation of Site**—Air density and enthalpy vary with elevation. However, using a computer makes it feasible to develop exact rating solutions at any elevation.
6. **Water Impurities**—Impurities in the water to be cooled can vary the latent heat of evaporation and affect tower size.

When the above information is supplied, a cooling tower can be selected to adequately cool the water needed. However, the configuration might be far from the optimum tower financially. The optimum system can be determined only after considering these additional evaluation factors:

1. **Fan Hosepower Rate**—usually supplied in dolar rate per kw hour.
2. **Pump Horsepower Rate**—usually supplied in dollar rate per kw hour.
3. **Basin Rate**—basin construction rate per square foot includes local factors as difficulty of excavation, labor cost, land cost and availability for tower site.
4. **Cost of Money**—rate of amortization normally practiced by the purchaser.
5. **Initial Equipment Cost**

Example

A hypothetical problem will help demonstrate the importance of this kind of evaluation.

Assume a system requirement of 40,000 gpm at design conditions of 107 F hot water, 88 F cold water and a wet bulb temperature of 75 F. A great number of cooling tower designs are available with varying initial and evaluated costs which will fulfill the requirements.

Power costs and basin construction costs for the two apparently best towers could be compared for an operating period of four years.

Tower "A"   3 cells @ 30$'$ L x 42$'$ W x 38$'$ Fill Height

Drive equipment (3 @ 130 brake hp)	150 hp
First cost	$184,000
Evaluated cost, 4 years	409,000

Tower "B"   3 cells @ 36$'$ L x 42$'$ W x 34$'$ Fill Height

Drive equipment (3 @ 115 brake hp)	125 hp
First cost	$191,000
Evaluated cost, 4 years	399,000

In the example, Tower A has the advantage of first cost savings of $7,000 over Tower B. On the other hand, Tower B has an evaluated cost of $10,000 lower than Tower A when evaluated for four years of operation.

Since both towers have the same design conditions, it is the evaluation factors which can guide engineering management to a choice between towers. Those factors making the important differences in this example are:

	Tower "A"	Tower "B"
Total basin size	$90' x 42'$	$108' x 42'$
Fan electrical costs	390 brake hp	345 brake hp
Pumping costs	39 ft pump head	35 ft pump head

The engineering task of estimating tower configurations and costs has been greatly diminished by the use of modern data processing techniques. To manually examine only 10 of the most promising configurations would require possibly 32 man-hours of effort. A sophisticated computer program will, however, select and analyze the 20 or 30 most economical designs to meet the required specifications within 30 minutes.

Thermal water pollution requirements will probably never relax from the restrictions now in force. Changes can be expected only in the area of even tighter controls. While legislation varies from state to state, maximum allowable temperature rises in the range of 1 and 2 F are common on many waters. Potential legislative changes may outdate many "once through" systems, thus requiring expensive modification. Therefore, consideration should be given toward design of closed systems.

Additional sizing and selecting criteria which should be given careful consideration are related to the building site itself. A checklist of restrictive items to be reviewed by the prospective tower user should include local building codes, cost of land, non-expandable sites, proximity to urban areas and tower orientation with prevailing winds.

Aesthetic considerations can also come into play in selection of exterior building materials used, height of the tower and the relationship of the tower to other structures on the building site. While these concerns have been of relatively small importance in the past, more and more industries are turning their attention to the appearance of their towers and especially plume dispersion.

COOLING TOWER REPAIR AND MAINTENANCE[131]

After the optimum cooling tower has been selected and installed, proper repair and maintenance practices are necessary to assure continued performance. Maintenance is normally less of a problem than are repair considerations. Once a maintenance schedule is set into practice it usually becomes routine. On the other hand, cooling tower repairs can plague plant operations and budgets if they are handled incorrectly. This section will discuss guidelines to proper maintenance scheduling, problem-solving concepts and specifics of cooling tower repair.

Inspections may be weekly, monthly, quarterly, semi-annually or annually based upon practical experience with the components of cooling towers. Weekly inspections should include the cold water basin, drive shaft, fan, speed reducer, motor operating temperature, vibration cut-out switch and sump screen. Monthly inspections should cover drift eliminators, header and laterals. Quarterly inspections should include the distribution system. Semi-annual inspections should provide a detailed review of all plenum components. The annual inspection should thoroughly cover every part of the tower system. This will include all items previously mentioned plus filling, hardware and all safety devices, railings, stairs and access walkways.

Establishing a pre-work analysis plan is vital to obtaining economical, up-to-date repair work. All current and future process and capital expenditure considerations should be evaluated. Repair and modernization priorities should be coordinated and recommended to fit into the scope of the plant's present and future requirements.

Plants with multiple cooling tower installations require a more complex review to determine structural need, performance requirements, fund availability and total plant priority rating. Typical performance-improvement results for seven actual repair case histories are shown in Table 18-1. These tower users gained maximum value of their repair dollars by undertaking planned repair of the items checked.

Annual Inspection

Inspection methods and standards are usually the least understood yet most important single factor connected with cooling tower repair. Following are some of the key considerations for developing inspection procedures.

Regular inspections should be performed by a cooling tower manufacturer on a one- or two-year basis depending upon the size, location and thermal performance of the tower. Inspections can be performed in water-washed sections even while the tower is in full operation. For this kind of inspection, the inspector is outfitted in a diver's wet suit and mask.

However, it is recommended that the cell being inspected have its mechanical equipment and water circulating systems deactivated during the inspection. In addition, the pollution engineer should accompany the repair inspector to provide verification of needed repairs.

If the cell to be inspected must remain in operation, the inspector has two options. Inspect the plenum, exterior and air intake areas with the tower in operation. This is a difficult, dangerous and incomplete method. However, due to operational necessity, this arrangement is used on perhaps 10 percent of all inspections.

The second method is to deactivate the fan, but leave the water distributing system in operation. This method will permit the inspector to carefully review the most important and likely areas for defects to occur— the plenum and fan deck/stack areas. Review of distributing systems and fill can usually be made to determine if a total shutdown is necessary.

Table 18-1. Performance-Improvement Case Histories

Plant	Before GPM	T_1, °F	T_2, °F	Twb, °F	After GPM	T_1, °F	T_2, °F	Twb, °F	Fill	Distribution System	Mechanical Equipment	Fan Stack	Frame Height
A	13,200	120	90	80	20,000	120	90	80	X	X			
B	1,250	125	75	65.2	1,850	125	75	65.2	X	X	X		
C	183,600	128.8	94.2	75.1	191,000	122.5	90	77	X	X			
D	1,800	95	85	79	4,200	95	85	79	X	X	X		
E	3,100	105	85	77	3,880	105	85	77	X	X	X		
F	4,500	110	85	78	6,500	103	85	78	X	X	X		
G	13,200	120	90	80	22,200	120	90	80	X	X	X	X	X

Plant	Results
A	51.5 Percent Increase GPM
B	48 Percent Increase GPM
C	Colder Water
D	133 Percent Increase GPM
E	25 Percent Increase GPM
F	40 Percent Increase GPM
G	68 Percent Increase GPM

Time needed to perform an inspection is relatively short. To inspect the plenum area of a 30 ft x 30 ft cell would require from one to two hours. The plenum of water-worked areas of a 30 ft x 60 ft cell will take about two to four hours to inspect. Inspection time would be from four to eight hours for a complete one-cell tower and would average two to four hours per cell for complete multi-cell towers.

Spare Parts

The approach to spare parts stocking can be determined by reviewing how important the tower is to the plant operation, and what effect a 25 to 50 percent loss of capacity will mean in lost process revenues from the loss of one set of air moving equipment. The loss of one set of mechanical equipment due to failure of one component in a two-cell tower could approach a 50 percent reduction in total capacity. One set lost in a three-cell tower could approach a 33 percent reduction in total capacity.

Additional guidelines for stocking spares should include the manufacturer's recommended list of spare parts. If these are not stocked in the plant, delivery delays of several weeks can often slow down emergency repair. Normal parts recommended for in-plant spares stocking are: fan, gear, coupling and respective assembly components.

If these parts are not carried in the plant inventory, the pollution engineer should obtain availability quotations from the manufacturer to ensure adequate spare parts protection. It is also recommended that spares be upgraded to the latest series available whenever possible.

Repair

Following inspection, a detailed inspection report should be prepared which includes an itemized checklist to indicate the conditions for each component. The pollution engineer should require modifications utilizing the latest designs and components such as fiber glass fans and fan stacks: PVC drift eliminators; fiber glass, PVC or coated steel piping; PVC or polypropylene filling; PCV or CAB sheathing and louvers; fiber glass, stainless steel or coated steel connectors, and pressure treated (ACC) redwood components.

When planning repairs, the time required to accomplish the work is an important consideration. An average job will require six to twelve weeks for jobsite delivery of materials after receipt of the order. A normal job will take one to two weeks work per cell.

The most common arrangement for performing repairs is on a partial tower shutdown basis. Usually one or two cells are shut down at a time. Total shutdown is very infrequent on repair projects involving multi-cell towers.

When purchasing tower repair work, "spotty" or half-complete work should be avoided. It is better to delay repairs until the job can be approached properly. Necessary budget restrictions can be better handled by doing most urgent items first.

FRP FOR COOLING TOWERS[9]

A number of years ago it was discovered that fiberglass reinforced plastic (FRP) offered exceptional resistance to the corrosion, weathering and breakdown induced by long periods of exposure to or saturation with water. FRP does not support fungus growth and is impervious to chemical attack. It retains its strength and structural characteristics over a wide temperature range. These features, combined with its high strength-to-weight ratio, make it desirable for use in cooling tower construction. In addition, FRP's availability in colors appealed to designers and architects and provided image-conscious corporations with a material with which they could brighten or color-harmonize plant facilities.

Although it was originally used for casings and louvers, FRP currently fulfills many functions in cooling towers. Most of the crossflow towers built in the past few years use an FRP fill hanger system, and some towers use FRP fill in place of the traditional redwood. A number of new towers incorporate drift eliminator systems fabricated from FRP components. Independent tests, as well as those conducted by tower manufacturers, show that FRP fill provides performance well within the range of conventional materials. Drift eliminator sections fabricated from FRP have shown more effective results in eliminating water from the tower exhaust than more widely used materials.

Figure 18-4.

A considerable amount of FRP in flat sheet form has been used as a lining material for shallow tower basins. Fastened in as few points as possible to the wood basin structure, the FRP is not subject to the shrinking and swelling of the wood and makes for a more water-tight receptable. FRP has also been widely used in fan rings, velocity recovery stacks, piping, fan blades, structural members and cell partition walls.

For years, the FRP industry has countered the "high initial cost" objection with their quite valid "greatly reduced maintenance costs" and the "lower cost per year of satisfactory service" concepts. Now FRP also competes favorably on initial dollar costs.

Resin Formulations

The largest single use of FRP in cooling tower construction continues to be that of the corrugated material used as casing and louvers. Initially, the material used was of a standard resin formulation with a few installations using fire-retardant resins.

For years the standard 60 percent polyester, 40 percent styrene resin system was the type most generally used for cooling tower FRP panels. These panels are readily available in a variety of stock colors from the larger FRP manufacturers. Other panels employing acrylic-modified, fire-retardant or noncombustible resins are available and have been used in varying amounts. Since these other materials are generally considered as non-stock items requiring from one to four weeks for production and delivery, they have found limited use.

In 1957, an FRP panel using an acrylic-modified resin system was introduced. The resins incorporated a methyl-methacrylate (MMA) monomer of 20 percent as a direct substitute for one-half of the original styrene monomer with a resulting resin formulation of 60 percent polyester, 20 percent styrene and 20 percent MMA. This panel was found to offer greater resistance to outdoor weathering with a longer retention of surface gloss and appearance. In applications such as residential awnings where these qualities are a prime requisite, the acrylic-modified panel has gained wide acceptance and become the norm. Variations in acrylic content from 1 to 100 percent were tested and it was found that a 10 to 20 percent content gave optimum results. Acrylic content of 15 percent is probably the most widely used today. Increasing the percentage of acrylic above 20 percent has a weakening effect on the panel, while increasing production costs due to lengthening the cure time.

The past several years have seen increased use and interest in fire-retardant panels. When first introduced, sheet manufacturers offered panels with flame spread ratings of 60 to 75 which were just under the high limits of the self-extinguishing category.

Through the use of newer hot-acid resins and variations of pigments and additives, the flame spread ratings have been gradually reduced to an area of 30 to 40 last year and now to the 25-and-under range. Since 25-

and-under ratings are classified as noncombustible, future development efforts in this area will be pointed toward cost reduction and product improvement rather than lower flame spread ratings.

It should be noted that there are cost differentials for the acrylic-modified, fire-retardant and E. I. du Pont's Tedlar clad noncombustible panels. The 10 to 20 percent acrylic-modified panel costs approximately 15 percent more than the standard panel, the fire-retardant about 65 percent more and the Tedlar clad noncombustible about 125 percent more. Since the casing is only a small part of the total cost of the tower, the panel cost differentials result in a relatively small increase in the total price of a tower and can usually be justified by the lengthened service life.

Surface Finishes

The lower flame spread rated product and the new noncombustibles are adversely affected by solar radiation which causes color deterioration and loss of surface gloss. It is recommended that panels fabricated with these resins and used in outdoor exposure applications be manufactured with a Tedlar film applied to the outside surface. A panel of this type offers the lowest possible flame spread with the best possible, long-weathering surface.

Most FRP sheet materials furnished for cooling tower use have smooth finishes on both sides. Occasionally, material is furnished with a pebble or textured surface on one or both sides. The textured surface is desirable on FRP used for fill material since it increases the surface area and water retention capacity of the material. A pebble or textured surface on the outside of a casing sheet will add to its normal weathering characteristics, since the resin-rich high spots will serve as additional protection against surface erosion.

Weight/Thickness/Color

An important factor in FRP panel specification is the thickness or weight of the sheet. Most corrugated sheets are sold by weight while most flat sheets are sold by thickness. The standard for the cooling tower industry has been the 4.2-in. corrugated, 8 oz/sq ft panel. This is used almost exclusively for tower casing while a variety of shapes and weights are used in louvers, fill, eliminator sections, partition walls and fan stacks. FRP panels are commercially available in weights ranging from 4 to 12 oz/sq ft and thicknesses ranging from 0.030 to 0.125 in.

The 4.2-in. corrugated panel combines the features of greater load-bearing or spanning capabilities than most standard shapes and is available in a variety of colors. There are several panels which will span greater distances and/or carry greater loads; however, the present column spacings in most cooling towers are such that these stronger shapes are not required.

A final consideration in using FRP materials for tower casing is color. Most panel manufacturers offer three to six colors which can be shipped from stock. Usually these colors will include the gray and brown panels which have been the most widely used. Lately there has been an emphasis on medium and lighter pastels. Virtually any color of the rainbow can be special-ordered where the total requirement will exceed 3000 sq ft. Special-ordered colors usually require a slight surcharge and additional time to produce.

SPRAYS FOR COOLING [139]

For several years, a continuous direct-spray water cooling system has removed heat instantaneously from the Pion Focusing Magnet of the 12.5 billion-volt atomic accelerator at Argonne National Laboratory. Heat generated every 2.4 seconds by the atomic beam has to be removed immediately. This installation demonstrates that adequate preplanning for maximum cooling with recycled process water will save money, compared to a system engineered after startup. This approach can readily be applied to other in-plant industrial cooling and wastewater treatment process problems.

Argonne National Laboratory has four separate ZGS cooling water piping systems supplied by raw water from several on-site deep wells. Raw well water is roughly filtered, screened, and treated with corrosion inhibitive chemicals and biocides. Total dissolved solids are reduced from 1040 ppm average to a range of 390 to 430 ppm. Limited ion exchange and line rough filtering with secondary treatment is provided. Chemicals are added and controlled to adjust pH value at varying water temperatures from 55 to 160 F.

Additional water treatment varies according to final process uses of particular cooling water tapped from a system. All systems usually receive some secondary treatment. Reduction and elimination of finely suspended and dissolved solids which carry over takes place in main water systems.

For example, the main system of chilled water (40 to 45 F) circulates through piping to cool the series of Accelerator Ring Magnets, auxiliary power and electronic triggering equipment. Total solids carried in from well waters are therefore treated by the above chemical, biological, and electrolytic (base-ion exchange) actions to provide low conductivity water with resistance in 10^5 range for serving most ZGS processes.

Cooling water circulating in a process loop is ultra-fine filtered with disposable cellulose fiber filter elements (3 to 10 micron particulate retention). Microstrainers using barrier basket-type element of monel metal are placed downstream to rough filter (about 130 to 150 micron size), and protect against high pressure surges in circulating water.

Slip Stream—Spray Cooling

In direct-spray cooling of the focusing magnet, as little as 100 to 120 gpm recirculates from the deionized chilled water system, within a closed process piping loop. This chilled water recirculates through a 250-gallon storage tank with a 5000 grain/hr mixed-resin bed demineralizer to complete deionization and demineralization. The system produces nearly ultra-pure water of uniform quality with a constant pH value and most of the remaining dissolved and finely suspended particulates removed.

Some carryover of fine particulates and contaminates will take place in circulating water, since the demineralizer bypasses a percentage of the flow. Therefore, 3-10 micron cellulose filters are used to remove any fines that might block nozzles or pressure control and solenoid valves. Water conductivity is lowered appreciably, further purifying the circulating water for continued process reuse, with a resistance of one to two megohms electrolytic resistance.

Loop piping circulates the purified water to a supply manifold-heater. Solenoid controlled pressure flow actuated valves meter the water through tube runouts feeding 17 symmetric tubing rings with nozzles. Each ring tube holds five equi-spaced nozzles, set into the piping harness assembly. This ring harness is anchored to the inside diameter of the primary of eight magnet sections of the round transformer. The transformer primary is built of eight doughnut sections to which the tubing harness assembly is fastened.

Generally, off-the-shelf, commercially available components were used to simply and economically fabricate this spray water cooling system. Runouts were fabricated from heat-resistant Acrylonite-butadiene-styrene thermoplastic (ABS) tubing, which extends the full length of the metal secondary and connects the series of 17 circular nozzle rings. Each ring is strategically located, so that the five nozzles are focused to allow the water droplets to strike the entire surface.

Spray water tests showed that heat could be removed at a rate of 167,000 Btu/hr (about 120 kilojoules per 2.5 second pulse) by this controlled spray action using 100 to 120 gpm water supplied at 40 to 58 psig.

Pressurized ultra-pure spray water from each set of five nozzles cools the cone secondary after each energy pulse from the extracted atomic beam. Atomized water droplets up to about 80 micron size are sprayed at nearly 80 ft/sec stream velocity. This action produces maximum heat and mass transfer, achieving Reynolds number in order of 2.6×10^5 for water-metal-film interface along the throat portion of the aluminum secondary.

Spray Pattern and Timing

Water timing of the paired on-off solenoids is critical. Overall scatter time between solenoids for the 85 nozzles on-off spray action is in the order of 140 msec. A ball check valve in the branch feed pipe kept scatter between nozzles to a minimum of 60 msec. All nozzles operated during the 2.5-sec pulse period, except within the off-time of 100 msec that the horn cleared water for the accepted ZGS beam (the delay time between the last nozzle shutoff and the first nozzle-on).

Shortening the delay time was the main advantage of using paired on-off solenoids in each inner cone branch cooling water circuit and nozzle bank runout. Using a single solenoid to open and close water flow to nozzles would have markedly increased the delay.

Hydraulic Circuit

The entire cone spray water hydraulic circuit was made of these symmetrically spaced runouts of 5/8-in. diameter ABS thermoplastic tubes. Each ABS plastic tube with 17 selectively set stainless steel nozzle groups included: 7 nozzle banks set in the throat section of each runout, proving 1.0 to 1.2 gpm capacity per nozzle with 15 to 25-deg spray angles (35 total cooling the throat), Figure 18-5; 3 nozzle banks of the same capacity

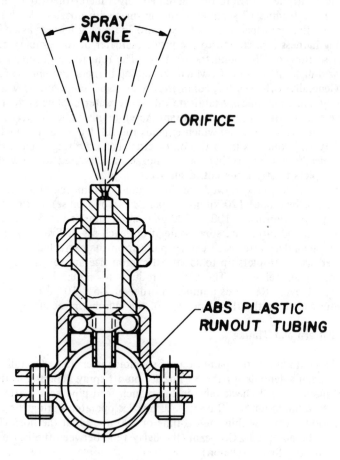

Figure 18-5. Flat spray nozzle assembly on ABS tubing.

set toward the horn left end of each runout (15 total cooling the left section of the cone); and 7 cooling nozzle groups variably spaced toward the horn right side, ranging from 1.0 to 1.6 gpm with 25 to 80-deg spray angles (35 total); or 5 circuits, total of 85 nozzles providing about 90 gpm. Each 5/8-in. plastic tube extension was epoxy sealed and welded to one of 5 shop-fabricated stainless steel tees set 140 cm from the left end. Fifteen to 22 gpm of 42 F water, adjusted from 50 to 80 psig pressure, flowed into each of the 5 equi-spaced runouts from special stainless steel tees. These tees were welded to two 3/4-in. stainless steel pipe extensions to the horn outside shell and water fed through five 1-in. reinforced neoprene hose connections from the main 3-in. cooling water supply header. The spray water drained by gravity from the horn returns to a 250-gallon lined storage tank.

Conservatively calculated, 1 to 2¼ seconds was required between pulses for spraying the cone and draining and returning water through the shield outlets, including time for solenoid jitter and signal response. Less than .013 sec delay time was required to shut off nozzle flow by dropping water pressure acting on the servo-acting ball valves when solenoids closed off flow.

Off-the-shelf standard spray nozzles were used in this system. However, careful analysis of stream cooling action along the secondary's surface perimeter was carried out to specify the proper nozzle.

Study of this cooling-process development, engineering computations, and tests indicate that high heat and mass transfer rates overall can be achieved fractionally in one second or less. Tests proved that continuous action of pulsed droplet stream cooling is reliable and adequate to remove heating instantaneously. In addition, the less than two percent vaporization to steam which occurred at the throat of the secondary mainly took place without noticeable rise of internal pressure within the transformer enclosure.

SPRAY NOZZLES FOR POLLUTION CONTROL[169]

Pollution control involves a wide variety of processes and equipment, and it would not be unusual to encounter spray nozzles in any particular system. The structural simplicity of many nozzles does not adequately reflect their complex functional characteristics. Therefore, a basic understanding is required to design and operate equipment in which they are used.

The link between nozzles and pollution control systems becomes apparent on reviewing the main purposes of liquid spraying:

1. To **atomize** or break up a liquid into small droplets.
2. To **disperse** liquid in a specific pattern, and with sufficient penetration, kinetic energy or impact.
3. To **regulate** the flow of liquid discharged through the nozzle.

Nearly all atomizers generate a broad spectrum of sizes. Typical sprays comprise many fine particles (often submicron), as well as larger droplets that may range up to a few hundred microns in diameter. They can be fully represented by means of curves (Figure 18-6), or equations defining the complete distribution function. In practice, however, it is often convenient to express droplet size by a single value such as the mass median, arithmetic mean or Sauter mean diameter. (The latter is a hypothetical droplet whose ratio of surface area to volume is equal to that of the entire spray.)

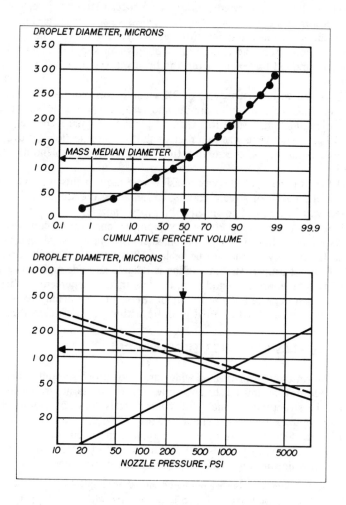

Figure 18-6. Droplet size and flow data for typical spray drying nozzle. Upper graph represents size distribution for 40 gph and 300 psig liquid pressure. Lower curves show effect of nozzle pressure on flow rate, mass median and Sauter mean droplet diameters.

Droplet size often changes after the spray forms at the nozzle. Evaporation and burning are obvious causes in many applications. In spray drying, the size distribution of the dried particles seldom matches that of the liquid spray. Though evaporation tends to reduce size, the heated particles may puff up and become larger. Still another effect is coalescence of liquid droplets that collide in flight.

Choosing optimum nozzles for certain installations can be a challenging, if not exasperating, engineering problem. A major difficulty is understanding exactly what happens to a spray within a combustor, dryer, scrubber, or reactor. However, it is helpful to know how spray characteristics are affected by atomizer design and operating conditions. Nozzles may be classified by the form of energy supplied to them—*hydraulic, pneumatic,* or a combination of both.

Simplex Pressure Atomizers

A common type of hydraulic atomizer is the simplex centrifugal nozzle in which the energy of pressurized liquid is converted into a high-velocity swirling film that collapses into ligaments and droplets outside the nozzle (Figure 18-7). Hollow or solid conical patterns can be produced; spray angles are generally between 30 and 120 degrees. Nozzles with elliptical or slot orifices have been developed for fan patterns.

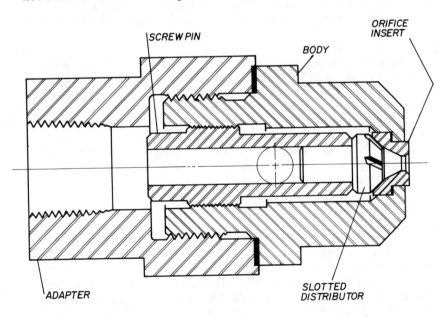

Figure 18-7. Simplex pressure nozzle: distributor slots impart swirl to liquid, resulting in hollow conical spray pattern.

Operating pressures with hydraulic nozzles cover a wide range. Whereas 30 or 40 psig is often adequate for aeration or pesticide spraying, metal descaling or spray drying of viscous materials may require pressures in excess of 2000 psig. With a given simplex nozzle, flow is nearly proportional to the square root of pressure. Droplet size varies approximately as the -0.3 power of pressure. Coarser droplets are associated with narrow spray angles, high capacities, and liquids having high viscosity or surface tension.

Two-Fluid Nozzles

Nozzles that use air, gas, or steam for liquid breakup are classified as pneumatic or two-fluid atomizers. They are attractive when it is expensive or inconvenient to pressurize the liquid, when the flow rate is very low, when viscosity is a problem, or when extremely fine droplets (aerosol range) are required. In such nozzles, droplet size largely depends on the air-liquid ratio and the relative air velocity. Because there are many designs of two-fluid atomizers, it is risky to estimate droplet diameters without reference data for a particular type.

Some designs provide for air-liquid mixing within the nozzle (Figure 18-8); others direct air on the liquid as it issues from the orifice. In the former, air and liquid pressures must be balanced to prevent throttling or backflow, and positive liquid metering is often recommended. Two-fluid atomizers are sometimes specified because their relatively large fluid passages can often handle suspended solids or contaminants that might plug or erode a pressure nozzle.

Like hydraulic nozzles, some pneumatic atomizers operate at very low pressures—others at high pressures. With ample air volume (scfm), good spray quality is possible at pressures approaching 0.1 psi air ΔP. This has led to some interesting developments in burners and turbine combustors where pressures are limited by existing blowers and engine compressors. On the other hand, stationary compressors in the 15- to 100-psi range are needed for processes where moderate or large quantities of liquid must be finely atomized. In certain devices, high air pressures also produce sonic or ultrasonic radiation that may enhance droplet evaporation, burning, or chemical reactions.

Other Types

Some atomizers rely on both hydraulic and pneumatic energy. One example is the air-assist nozzle which uses air at the low end of the flow range, but functions as a pressure atomizer as flow increases. Certain types of combustion systems operate in the reverse manner: at low flow, pressurized fuel is sprayed through a pilot (primary) injector until a turbine compressor generates enough pneumatic energy to handle the larger fuel rates in the main nozzle. Some atomizers use liquid and air pressure at all operating points.

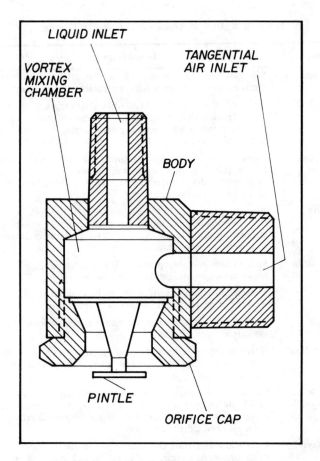

Figure 18-8. Two-fluid atomizer: air and liquid mix in vortex chamber, then pass through annular orifice where pintle determines shape of spray.

These designs are often specified when it is necessary to modulate flow over a broad range while maintaining good spray quality. If air is not used, similar performance is possible with hydraulic nozzles such as the dual-orifice (duple) or bypass (spill).

Table 18-2 shows how the characteristics of sprayed droplets relate to functional requirements in several anti-pollution processes and briefly indicates why atomizers play a vital role in diverse applications.

Though nozzles are relatively small and inexpensive components, one should not underrate their contribution to the efficiency and success of pollution control systems.

Table 18-2. Uses of Atomizers in Pollution Control

Application	Description
Cleaning and Rinsing	Water or cleaning solutions are sprayed, often at high pressure, to clean contaminated industrial or agricultural equipment.
Wet Scrubbing	Spray nozzles in scrubbers and other dust recovery equipment produce water droplets that impact with and remove hazardous particulates emitted in foundries, chemical plants, mining operations, and other industrial processes.
Spray Drying	Liquids or slurries are sprayed into a chamber where contact with hot air quickly evaporates most of the water. It is therefore possible to recover as dry powder waste materials that might otherwise contaminate the environment.
Evaporative Cooling	It is often necessary to protect electrostatic precipitators or baghouses from thermal degradation by hot stack gases. By injecting finely-atomized water into the gas, it is cooled by transfer of latent heat to the evaporating spray.
Spray Aeration	Entrainment and absorption of air in spray ponds is a common technique in many water purification systems.
Effluent Water Cooling	Thermal pollution of waterways can be reduced by dispersing the water as drops to permit cooling by evaporation.
Oil Slick Removal	One method of cleaning up oil spills is to spray dispersants on contaminated areas of water.
Spray Irrigation	Nozzles or sprinklers are often mounted on large mobile booms. Water is uniformly distributed to enhance the productivity of large areas of air land.
Ground Application of Pesticides	Conventional fan-spray nozzles are often used at low pressure to avoid fine droplets that are likely to drift. Some recent systems utilize air-aspirating nozzles and spray adjuvants to produce stable foam for drastic drift reduction.
Aerial Spraying	Insecticides and other chemical agents are disseminated by conventional or special atomizers. Careful control of output and particle size is necessary for effective coverage with minimum drifting. Certain nozzles have been designed for improved droplet-size uniformity, while others are capable of generating foam for low-altitude spraying.
Electrostatic Painting	Wasteful overspray is reduced by using special nozzle tips and guns that impart an electrostatic charge to the paint spray.
Noise Attenuation	The ability of water sprays to absorb sound has been utilized to reduce objectionable noise (e.g. rocket silos).
Water Injection in Engines	Recent experiments indicate that emission of nitrogen oxides can be reduced by injecting water, both in gas turbine combustors as well as in the inlet manifold of automotive engines.
Fuel Injection in Gas Turbines	Precision nozzles control fuel flow and spray quality for optimum combustion efficiency and reduction of visible smoke, unburned hydrocarbons, and nitrogen oxides. Many types of custom-designed fuel injectors have been developed, both for aircraft and stationary turbines.

Table 18-2, continued.

Application	Description
Fuel Oil Burning	Various types of pressure or two-fluid atomizers are capable of efficient and quiet combustion in residential and commercial heaters, industrial boilers, and utility powerplants. Proper nozzle design can help minimize smoke and nitrogen oxides.
Incineration	Liquid waste or toxic materials are sprayed into a chamber and burned. In solid waste disposal systems, fuel may be injected for ignition or for secondary burning of exhaust products to reduce smoke.

CHAPTER 19

CORROSION

CORROSION RESISTANCE OF PIPING AND CONSTRUCTION MATERIALS[43]

Corrosion is deterioration or decay occurring when a material reacts with its surroundings or the fluid being transported or contained. It may be either uniform, where the material corrodes at the same rate over the entire surface, or localized with only small portions affected. There are twelve types of corrosion.

Uniform corrosion occurs over the entire metal surface at the same rate. It can be controlled by proper selection of materials and using protection methods such as coatings.

Galvanic corrosion occurs when two different metals are in contact in a conductive solution. An electrical potential exists between the different metals which serves as a driving force to pass current through the corrodent. The result is more corrosion of one of the metals in the couple. The more active metal becomes the anode, and corrodes at a faster rate than the cathode. When, for example, joints and valves of two different piping materials come in contact, the possibility of galvanic corrosion exists.

Erosion corrosion occurs from movement of a corrosive over a surface, increasing the rate of attack due to mechanical wear and corrosion. Erosion is attributed to the stripping or removal of protective surface film or adherent corrosion products. Erosion appears as smooth-bottomed shallow pits. Attack may also have a directional pattern which is attributed to the corrodent path moving over the pipe surface. Rate of corrosion is increased under high velocity conditions, especially during turbulence or impingement. Fast moving slurries containing hard or abrasive particles are likely to generate such corrosion. There are several methods which can be used to prevent this type of corrosion. First, of course, is the selection of a more resistant material. Erosion can also be reduced in transport applications by increasing pipe diameters which decreases velocity and turbulence. Also, flared tubing can reduce problems at the inlet of tube bundles. Generally, erosion occurs sporadically.

Cavitation corrosion is a special form of erosion corrosion. Cavitation is produced by the rapid formation and collapse of vapor bubbles at a metal surface. High pressures that are produced can deform the underlying metal and remove protective films. Smooth pipes reduce sites of bubble formation and lessen cavitation corrosion.

Fretting corrosion is another form of erosion corrosion. Fretting occurs when metals slide over each other, producing mechanical damage to their surfaces. Vibration generally causes sliding. Corrosion products cause continued exposure of fresh surface that actively corrodes. The use of harder materials reduces friction.

Crevice corrosion is common at gaskets and lap joints, and comes from dirt deposits and corrosive products. This type of corrosion can be attributed to one of three things: (1) changes in acidity in the crevice; (2) lack of oxygen in the crevice; or (3) the buildup of a harmful iron species or the depletion of an inhibitor. Alloys are generally less susceptible to crevice corrosion than pure metals.

Pitting corrosion is the formation of holes on an unattached surface with the shape of the pit responsible for continued growth. This is generally a slow process, taking months or sometimes even years before first traces are apparent.

Exfoliation corrosion begins on a clean surface, but spreads below it. The attack has a laminated appearance, and entire areas can be eaten away. Exfoliation is marked by a blistered or flaky surface with aluminum alloys most commonly attacked. It is combated by heat treatment and alloying.

Selective leaching or parting corrosion is the removal of one element in an alloy. An example is the leaching of zinc in copper-zinc alloys (dezincification). Leaching is detrimental since it adds a porous metal to the effluent, and is combated by utilizing nonsusceptible alloys.

Intergranular corrosion involves an attack upon grain boundaries. When molten metal is cast, it solidifies at randomly distributed nuclei, and grows in a regular atomic array to form grains. The planes of atoms in neighboring grains do not match up because of the random nucleation. This area of mismatch between the grains is known as the grain boundary. The atomic mismatch offers an ideal place for segregation and precipitation. Corrosion takes place because the corrodent attacks the grain-boundary phase.

Under severe conditions, entire grains are dislodged due to complete deterioration of their boundaries. A surface that has undergone intergranular corrosion appears rough and feels sugary. The grain-boundary phenomenon that produces intergranular corrosion is heat sensitive. Such an attack is generally a by-product of a heat treatment, welding, or a stress relieving operation. The cure is another heat treatment, or selection of a modified alloy.

Stress-corrosion cracking is due to the combined action of tensile strength and a corrodent. It is the most serious of all corrosion problems because many alloys will undergo stress-corrosion cracking. Stresses that cause cracking are due to residual cold work, welding and thermal treatment, or may be externally applied by mechanical injury.

The cracks are generally in intergranular or transgranular paths, and such corrosion usually takes a long time. Preventative measures include: stress relieving; removing the critical environmental species; or proper selection of a more resistant material.

Corrosion fatigue is a special form of stress-corrosion cracking. It is caused by repeated cyclic stressing, and occurs in the absence of corrodents. It is common in structures which are subject to continued vibration. The presence of a corrodent increases susceptibility to fatigue.

Most construction materials are expected to undergo some type of corrosion. It, therefore, becomes important to determine what effects chemicals in an environmental system will have on materials. Careful analysis must be made of effluents, and existing piping and construction materials should also be examined and compared. The following factors influencing the extent of corrosion should be considered: (1) concentration of major constituents being handled; (2) pH of effluent; (3) temperature of effluent; (4) degree of aeration (limited aeration may enhance certain types of corrosion; (5) velocity of the fluid stream in the transport system; (6) inhibitors; and (7) startup and downtime procedures.

Construction materials for pipe and tanks are available in numerous materials, metallic and nonmetallic. Physical properties should be thoughtfully examined before any final selection is made. The following tables have been compiled as a general guide to proper selection. Tables 19-1 and 19-2 are divided into metallic and nonmetallic construction materials. A use guide is provided for many materials that may be encountered in pollution control service at ambient temperature.

ANALYZING ATMOSPHERIC CORROSION[119]

Structural metals deteriorate by reaction with moisture, gases and pollutants in the atmosphere; but in order for corrosion to take place, water must be present. The water can exist as a moisture film, condensed droplets or flowing rainfall. Corrosion is accelerated by sulfur dioxide, salt, dust, soot and soils in the atmosphere. Solids will attract and retain moisture and promote the formation of tiny corrosion cells on the surface. At increased temperatures the reaction rates are accelerated, but so are the drying rates which remove the water and stop the reaction. Rain provides water for the reaction but it also washes away solids, thus reducing the effect of these contaminating materials.

Metals are protected from corrosion by providing a barrier to the moisture, such as paint or another metal, or by alloying to produce a more

Table 19-1. Metals

R = Recommended
M = Moderate Service
L = Limited Service
U = Unsatisfactory
Blank = No information

	Carbon Steel	Cast Iron & Ductile Iron	304 Stainless Steel	316 Stainless Steel	347 Stainless Steel	Nickel Resist Iron	Carpenter 20; Durimet 20	Worthite	Durriron-Durichlor	Monel	Inconel	Hastelloy B	Hastelloy C	Hastelloy D	Chlorimet 3	Aluminum and Alloys	Copper and Cu Alloys	Brass	Lead	Nickel
Aluminum Chloride	U	U	U	U	U	L	R	U	R	U	L	R	M	R	R	U	L	L	U	L
Aluminum Hydroxide	U		R	R			R	R	M	M					R	R	R	R		L
Aluminum Sulfate	U	U	L	L	R	L	R	R	R	L	L	R	R	R	R	R	L	L	R	L
Alums, Dilute	R	R	R	L	L	L	R	R	R	R	R	R	R	R	R	R	L	L	R	R
Amines (various)	R	M	R	R	R	R	R		R		R	R	R	R	R	L	L	L		R
Ammonia Gas	M	U	M	M	R	R	R	R	R	M		L	R	M	R	U	U			R
Ammonium Carbonate	U	R	U	L	L		R	R	R	M	M	M	M	M	R	L	L	L	L	M
Ammonium Chloride	R		R	R	R	R	R	R	R	U	M	L	M	U	R	U	U	U	M	U
Ammonium Hydroxide	U		R	R	R	R	R	R	R	R	R	M	M	R	R	L	U	U		
Ammonium Nitrate	R	M	L	L	M	R	R	R	M	M	L	U	R	M	R	U	U	U	L	M
Ammonium Sulfate	U		R	R	R	R	R	R	R	R	R	M	R	R	R	L	L	L	M	M
Benzene	M		R	R	M	L	M	L	R	M	M	R	M	M	R	R	R	R	R	R
Calcium Carbonate	R	R	L	L	R	R	M	R	R	R	M	R	R	R	R	U	R	R	U	R
Calcium Chlorate			L	R			L	L	M	R			L	U	M	L	L	L	L	R
Calcium Chloride	R	R	R	L		L	R	R	R	M	R	M	M	M	R	L	M	M	L	M
Calcium Hydroxide	L		L	R	U	R	L	R	M	R	M	R	R	R	U	U	M	L	U	R
Calcium Hypochlorite	L		R		L	L		L	R	L	M	U	R	R	R	L	U	U	L	R
Calcium Sulfate	L		L	R			R	R	R	U	M	R	M	R	R	R	R	R	U	R
Carbon Dioxide (dry)	R	R	R	R	R	R	R	R	R	U	U		R	R	U	L	L	R	R	R
Carbon Dioxide (wet)	L	L	R	R	R	L	U	R	L	U	R	U	R	U	R	R	L	L	L	U
Chlorine (wet)	U	U	U	L	U	U	M	R	R	U	U	U	R	U	R	U	U	U	M	L
Chromic Acid Solution	L	M	U	L	U		M	R	R	U	L	U	R	R	M	M	U	U	U	U
Copper Chloride	U	U	U	U	U	U	U	R	U	U	U	R	R	R	L	U	U	U	U	U

L	L	U	U	U	R	R	R	L	L	R	R	L	R	R	L	U
R	U	L	L	R	R	R	R	R	R	R	R	U	R	R	R	U
U	U	U	U	U	U	U	L	U	R	U	U	U	U	U	R	U
U	L	U	L	M	M	M	M	R	M	R	R	R	M	L	U	U
U	U	L	U	U	U	L	U	U	R	U	U	L	R	R	U	U
L	L	L	U	U	U	L	R	U	L	U	U	U	R	U	U	L
L	L	U	U	R	R	L	L	U	L	U	U	U	L	U	L	R
L	U	U	U	R	R	M	M	R	M	R	U	R	R	R	D	U
R	R	R	R	R	R	R	R	R	U	R	R	M	R	R	R	M
R	L	U	U	U	L	R	L	L	L	L	L	R	U	L	U	U
R	U	U	R	R	L	L	M	L	R	L	M	R	L	R	R	M
R	R	U	M	R	U	U	L	L	M	U	U	R	R	R	L	R
U	R	U	R	U	D	D	R	U	R	L	R	R	L	L	R	R
U	U	U	U	U	U	U	R	U	R	U	U	L	U	L	U	L
U	U	U	U	U	U	U	R	U	R	U	U	L	R	R	L	L
U	U	U	U	U	U	U	R	D	U	L	R	U	R	R	L	L
U	R	U	R	M	M	M	M	M	M	M	M	R	R	R	R	M
M	U	U	U	L	M	U	U	L	L	R	U	M	R	R	R	U
U	L	U	L	M	M	M	L	R	U	L	U	U	R	R	M	M
L	R	R	R	U	L	L	M	R	R	L	R	L	R	R	M	M
L	L	L	R	R	L	M	M	L	M	R	R	R	R	R	L	U
M	M	M	M	R	M	M	M	M	M	M	M	M	R	M	M	M
U	U	L	L	U	R	R	R	U	U	L	R	L	R	R	L	M
L	R	R	L	R	R	R	U	L	L	U	U	R	R	R	R	L
M	M	U	U	E	U	U	U	R	R	R	R	R	R	E	E	U
R	R	R	R	R	R	R	R	R	R	R	R	R	L	R	L	M
M	M	M	M	R	U	R	U	M	R	R	R	M	R	R	L	M
U	R	L	M	R	R	R	L	L	M	L	L	L	M	M	R	R
M	L	L	M	R	L	U	M	R	R	R	R	M	R	M	M	M
M	R	L	L	R	L	R	L	R	R	R	R	L	R	R	M	M

<table>
Copper Sulfate
Fatty Acids
Ferrous Chloride
Ferrous Sulfate
Hydrochloric Acid (conc)
Hydrochloric Acid (dilute)
Hydrogen Chloride (dry gas)
Hydrofluoric Acid
Hydrocarbons (aliphatic)
Hydrogen Peroxide (conc)
Hydrogen Sulfide (dry)
Hydrogen Sulfide (wet)
Nitrating Acid (>15% H_2SO_4)
Nitrating Acid (<15% H_2SO_4)
Nitrating Acid (<15% HNO_3)
Nitrating Acid (<1% acid)
Nitric Acid (conc)
Nitric Acid (dilute)
Nitrous Acid
Phenol (conc)
Phosphoric Acid (100%)
Phosphoric Acid (hot >45%)
Phosphoric Acid (cold >45%)
Phosphoric Acid (<45%)
Potassium Carbonate
Potassium Chlorate
Potassium Chloride
Potassium Permanganate
Sodium Bicarbonate
Sodium Bisulfate
Sodium Bisulfite
Sodium Carbonate
Sodium Chlorate
Sodium Chloride
Sodium Hydroxide (conc)
Sodium Hydroxide (dilute)
</table>

Table 19-1, continued

R = Recommended
M = Moderate Service
L = Limited Service
U = Unsatisfactory
Blank = No information

	Carbon Steel	Cast Iron & Ductile Iron	304 Stainless Steel	316 Stainless Steel	347 Stainless Steel	Nickel Resist Iron	Carpenter 20; Durimet 20	Worthite	Duriron-Durichlor	Monel	Inconel	Hastelloy B	Hastelloy C	Hastelloy D	Chlorimet 3	Aluminum and Alloys	Copper and Cu Alloys	Brass	Lead	Nickel
Sodium Hydrosulfite	U	U	L	L	L	L	R	L	R	U	U	L	R	L	R	R	L	L		U
Sodium Hypochlorite	R	U	R	L	R	L	L	L	R	U	U	U	L	U	R	U	U	U	U	U
Sodium Nitrate	R	M	L	R	M	L	L	R	R	M	R	U	M	U	R	R	M	M		M
Sodium Phosphate	R		R	R	M	R	R		R	M		R	R	R	R	U	L	L		M
Sodium Silicate	R	R	R	R	R	R	R	R	R	M	M	R	R	R	R	L	L	L	R	R
Sodium Sulfate	M	R	L	L	R	R	R	R	R	M	R	R	R	R	R	R	R	R	L	R
Sodium Sulfide	M	R	L	R	L	R	R	R	U	L	M	R	R	R	R	U	U	U	R	R
Sodium Sulfite	L	R	L	L	L	R	R	R	R	L	L	R	R	R	R	L	L	L	L	L
Stearic Acid	L	R	R	R	R	L	R	R	R	U	R	R	R	U	R	L	L	L	R	R
Sulfur Dioxide (dry)	R		R	R		R	R		L	U	U	U	R	R	R	R	R	R	R	U
Sulfur Dioxide (wet)	U	L	L	R			R		L	U	U	R	R	U	R		U	U	L	U
Sulfur Trioxide	R	M	R	U		R	R		U	U	U		R	R	R	L	L	L	R	U
Sulfuric Acid (fuming 98%)	L	L	U	U	L	L	M	R	R	M	U	R	R	R	M	R	U	U	R	U
Sulfuric Acid (hot, conc)	U	U	L	U	L	L	R	R	R	U	L	U	R	R	R	L	U	U	L	U
Sulfuric Acid (cold, conc)	M	M	L	U	L	L	R	R	R	L	L	R	R	R	R	D	U	U	R	
Sulfuric Acid (75-95%)	M	L	L	U	L	L	L	R	R	L	L	U	R	L	R	U	U	U	R	U
Sulfuric Acid (10-75%)	U	U	L	L	L	L	L	R	R	U	U	R	M	R	R	U	U	U	R	U
Sulfuric Acid (<10%)	U	U	R	R	L	U	R	R	R	R	R	R	R	R	R	L	U	U	L	L
Sulfurous Acid	U	U	R	R	R	R	R	R	R	R	R	R	R	R	R	L	R	R	R	R
Toluene	R	R	R	R	R	R	R	R	R	R	R	R	R	R	R	R	R	R	R	R
Water (fresh)	R	U	R	R	R	U	R	R	R	U	R	R	R	M	R	R	R	R	U	U
Water (distilled)	U	U	R	R	R	L	R	R	R	R	R	M	R	R	R	R	R	R	L	L
Zinc Chloride	U	U	U	L	R	L	R	R	R	R	R	M	R	M	R	U	R	R	U	R
Zinc Sulfate	U	U	M	M	M	R	R	R	R	M	M	M	M	M	R	L	M	M	M	M

Table 19-2. Nonmetals

R	=	Recommended
M	=	Moderate Service
L	=	Limited Service
U	=	Unsatisfactory
Blank	=	No Information

Chemical	Carbon and Graphite	Glass (Pyrex)	Chemical Porcelain	Natural Rubber	Neoprene	Butadiene	Asphalic Bitumastic	Acrylic (Lucite, Plexiglas)	Polyethylene	Polyvinylchloride	Saran	Kel-F	Teflon	Penton	Polystyrene (Styron)	Haveg 41	Heresite	Molded Phenol Formald. (Durez)	Epoxy Resins	Nylon	Durcon 6	Woods—Maple, Oak, Pine
Aluminum Chloride	R	R	R	R	R	R	R		R	R	R	R	R	R	R	R	R	R	R	R	R	R
Aluminum Hydroxide	R	R	R	R	R	R	R	R	R	R	R	R	R	R	R	R	R	R		R	R	R
Aluminum Sulfate	R	R	R	R	R	R	R	R	R	R	R	R	R	R	R	R	R	R	R	R	R	R
Alums, Dilute	R	R	R	L	R	R		R	R	R	U	R	R	R	U	R		R	R		R	
Amines (various)	R	L	R	L	L	R	R		R	L	R	R	R	R	R	R		R	R		R	
Ammonia Gas	R	R	R	R	R	R	L	R	R	R	R	R	R	R	R	R	R	R	R	R	M	
Ammonium Carbonate	R	R	R	R	L	R	L	R	R	R	R	R	R	R	R	R	R	R	R	U	R	
Ammonium Chloride	R	R	R	L	L	R	R	R	R	R	R	R	R	R	R	R	R	R	R	R	R	
Ammonium Hydroxide	R	L	R	R	R	R	R	U	R	R	L	R	R	R	R	R	R	U	R	R	R	
Ammonium Nitrate	R	R	R	R	U	R	R		R	U	L	R	R	R	R	L	R	R	R	R	U	
Ammonium Sulfate	R	R	R	R	L		U	U	L	R	R	R	R	R	R	R	R	R	R	R	R	
Benzene	R	R	R	R	R		R			R	R	R	R	R					R	R	R	
Calcium Carbonate	R	R	R	R	R	R	R	R	R	R	R	R	R	R	R	R	R	R	R	R	R	
Calcium Chlorate	R	R	R	R	R	R	R		R	R	R	R	R	R	R	R	L	R	R	U	R	
Calcium Chloride	R	R	R	R	R	R	L	R	R	R	R	R	R	R	R	R	L	R	R	R	R	
Calcium Hydroxide	R	R	R	R	U	R	R		R	R	R	R	R	R		R		R	R	U	R	
Calcium Hypochlorite	R	R	R	R	R	R	R		R	R	R	R	R	R	R	R	R	R	R	R	R	U
Calcium Sulfate	R	R	R	R	R	R	R	R	R	R	R	R	R	R	R	R	R	R	R	R	R	
Carbon Dioxide (dry)	R	R	R	R	R	R	R	R	R	R	R	R	R	R	R	R	R	L	L	R	R	
Carbon Dioxide (wet)	R	R	R	R	R	L	U		R	L	L	R	R	M	R	L	L	L	L	U	U	
Chlorine (wet)	U	R	R	U	U	U	U		R	L	R	R	R	L	U	U	R		U	U	U	
Chromic Acid Solution	R	R	R	L	R	R	U	U	R	L	R	R	R	M	U	U	L	U	U	U	R	R
Copper Chloride	R	R	R	L	R	R	R	D	R	R	R	R	R	R	U	R	R	R	R	R	R	

Table 19-2, continued

R = Recommended
M = Moderate Service
L = Limited Service
U = Unsatisfactory
Blank = No Information

	Carbon Ceramics			Rubbers			Plastics																Woods
	Carbon and Graphite	Glass (Pyrex)	Chemical Porcelain	Natural Rubber	Neoprene	Butadienne	Asphaltic Bitumastic	Acrylic (Lucite, Plexiglas)	Polyethylene	Polyvinylchloride	Saran	Kel-F	Teflon	Penton	Polystyrene (Styron)	Haveg 41	Heresite	Molded Phenol Formald. (Durez)	Epoxy Resins	Nylon	Durcon 6	Woods-Maple, Oak, Pine
Copper Sulfate	R	R	R	R	R	R	R	U	R	R	R	R	R	R	R	R	R	R	R	U	R	
Fatty Acids	R	R	R	U	U	R	U		U	L	R	R	R	U	R	R	R	R	R		R	
Ferrous Chloride	R	R	R	R	R	R	R			L	L	R	R	R	R	R	R	R	R	U	R	
Ferrous Sulfate	R	R	R	R	R	R	R	R	R	R	L	R	R	R	R	R	R	R	R	U	M	
Hydrochloric Acid (conc)	R	R	R	R	L	R	R	R	R	R	R	R	R	R	U	R	R	L	R	U	R	
Hydrochloric Acid (dilute)	R	R	R	R	L	R	R		R	R	R	R	R	R	R	R	R		R	U	R	
Hydrogen Chloride (dry gas)	U	U	R				U	R		R		R	R					L			R	U
Hydrofluoric Acid	R	U	U	U	U	R	U	R	R	L	U	R	R	M	R	U	U	L	R		R	R
Hydrocarbons (Aliphatic)			R	U	R	R		R				R	R		R	R	R	R			M	
Hydrogen Peroxide (conc)		R	R	L	U	R	L			R		R	R	R	U	R	R	R	R		R	
Hydrogen Sulfide (dry)	U	R	R	R			R			R		R	R		R					U	R	
Hydrogen Sulfide (wet)	U	R	R							R		R	R	R					R	U	R	
Nitrating Acid (>15% H_2SO_4)	U	R	R	U	U		U					R	R			U	R				R	
Nitrating Acid (<15% H_2SO_4)	U	R	R	U	U		U					R	R			L	R				R	
Nitrating Acid (<15% HNO_3)	U	R	R	U	U	U	U					R	R			L	R				R	
Nitrating Acid (<1% acid)	L	L	R	L	L	R	U		U		L	R	R	U	U	L	R	R		U	M	
Nitric Acid (conc)	R	R	R	U	R	R	U	R	L	U	R	R	R	M	R	U	R	L	U		R	
Nitric Acid (dilute)	M	L	R	L	L	R	U			R		R	R	R		L	R				R	
Nitrous Acid	R	R	R	L	L	R	U	R				R	R	M				L	U		R	
Phenol (conc)	R	R	R	R	L		R		R	L	L	R	R		U	L	R	U	U	U	R	
Phosphoric Acid (100%)	M	R	R	L	L		U		L	R	L	R	R		R	R	R	U		L	R	
Phosphoric Acid (hot >45%)	R	R	U	L	L		R		L	R		R	R			R	R				R	
Phosphoric Acid (cold >45%)	R	R	L	L	L		U		R	R	R	R	R	R	U	R	R	R	R	U	R	
Phosphoric Acid (<45%)	R	R	L	R	L		R		R	R	R	R	R	R	U	R	R	R	R		R	

Potassium Carbonate	R						R		R			R	L	R	R	R	R	R	R
Potassium Chlorate	L	R				R	R		R	U	R	R	R	R	R	R	U	R	R
Potassium Chloride	R	R	R	L	R	U	R		L	R	U	R	R	R	R	R	R	U	R
Potassium Permanganate	R	R	L	R	R	R	R	R	R	R	R	R	R	R	R	R	U	R	R
Sodium Bicarbonate	R	R	R	R	R	R	R	R	R	R	R	R	R	R	R	R	R	R	R
Sodium Bisulfate	R	R	R	R	R	L	R	R	R	R	R	R	R	R	R	R	R	R	R
Sodium Bisulfite	L	R	R	R	R	U	R	R	R	R	R	R	R	R	R	R	R	L	R
Sodium Carbonate	R	R	R	R	R	R	R	R	R	R	R	R	R	R	R	R	R	R	R
Sodium Chlorate	R	R	R	U	R	R	R	R	R	R	R	R	R	R	R	R	R	R	R
Sodium Chloride	U	L	R	R	R	R	R	U	R	R	R	R	R	R	R	U	R	U	R
Sodium Hydroxide (conc)	L	L	R	R	R	R	R	R	L	L	R	R	R	R	R	R	L	R	M
Sodium Hydroxide (dilute)	L	L	R	R	R	R	R	L	L	R	R	R	R	R	R	R	U	R	R
Sodium Hydrosulfite	U	R	R	R	U	R	R	R	R	R	R	R	R	R	R	U	R	R	R
Sodium Hypochlorite	R	R	R	L	R	L	R	R	R	U	R	R	R	R	R	L	R	U	U
Sodium Nitrate	L	R	R	R	R	R	R	R	L	R	R	R	R	R	R	R	R	R	R
Sodium Phosphate	L	L	R	R	R	R	R	R	R	R	R	R	R	R	R	R	R	R	R
Sodium Silicate	R	R	R	R	R	R	R	R	R	R	R	R	R	R	R	R	R	R	R
Sodium Sulfate	R	R	R	U	R	R	R	R	L	R	R	R	R	R	R	R	R	R	M
Sodium Sulfite	L	R	R	U	R	R	R	L	L	R	R	R	R	R	R	R	R	R	R
Sodium Sulfide	R	R	R	U	R	R	R	L	L	R	R	R	R	R	R	R	R	R	R
Stearic Acid	R	R	R	R	R	R	R	R	R	R	R	R	R	R	R	R	R	R	R
Sulfur Dioxide (dry)	R	R	R	R	R	R	R	U	L	R	R	R	R	R	R	R	L	R	R
Sulfur Dioxide (wet)	U	R	R	R	R	U	R	L	L	R	U	U	U	L	R	U	U	R	R
Sulfur Trioxide	R	R	R	R	R	R	R	U	U	U	L	L	L	R	R	L	R	U
Sulfuric Acid (fuming 98%)	U	R	U	U	R	R	U	U	U	L	L	M	R	U	L	L	R	R
Sulfuric Acid (hot, conc)	U	U	U	U	U	U	U	U	L	R	R	L	L	R	M	M	U	R	U
Sulfuric Acid (cold, conc)	R	R	R	R	R	R	R	U	R	R	R	R	R	R	R	R	R	R	R
Sulfuric Acid (75-95%)	R	R	L	R	R	L	R	R	R	R	R	R	R	R	R	R	R	L	L
Sulfuric Acid (10-75%)	R	U	U	R	U	R	U	R	R	R	R	U	R	U	R	L	U	R	R
Sulfuric Acid (<10%)	R	R	R	R	R	R	R	L	L	R	R	L	R	R	R	U	R	R	R
Sulfurous Acid	R	R	R	R	R	R	R	R	R	R	R	R	R	R	M	R	R	L	L
Toluene	R	R	R	R	R	R	R	R	R	R	R	R	R	R	R	R	R	R	R
Water (fresh)	R	R	R	R	R	R	R	R	R	R	R	R	R	R	R	R	R	R	R
Water (distilled)	M	M	R	R	R	R	R	R	R	R	R	R	R	R	R	R	M	R	R
Zinc Chloride	R	R	R	R	R	R	R	R	R	R	R	R	R	R	R	R	R	R	R
Zinc Sulfate	R	R	R	R	R	R	R	R	R	R	R	R	R	R	R	R	R	R	R

corrosion-resistant metal. Steel is protected with zinc, cadmium, aluminum, lead, tin or nickel. In decorative applications chromium is deposited over the nickel to provide a stain-resistant surface. Pitting corrosion of steel is reduced by alloying with small amounts of copper, and general corrosion is greatly retarded by alloying with large amounts of nickel and chromium.

Conventional metals use known successful finishes such as galvanized steel, anodized aluminum or chromium/nickel plated zinc die castings. However, the life of all of these is still limited by the environment. Good economics will estimate the life of parts that are: to be exposed to a severe marine or industrial environment, difficult to recoat or service, expensive to replace, or critical to a function.

Corrosion Rates

The American Society for Testing and Materials has reported several extensive programs on atmospheric corrosion of metals. When an outdoor corrosion problem exists, the ASTM data provides an opportunity to take advantage of widespread observations that have been made at many places and with many metals. Report STP 435 is particularly useful. Forty-five locations were calibrated by measuring the weight losses of 4x6-in. specimens of steel and zinc exposed outdoors for two years. The data are recalculated to mils per year (MPY) in Table 19-3.

Steel and Zinc

Corrosion rates of steel and zinc at 45 test sites are charted. The sites are arranged as numbers 1 to 45, with increasing corrosion rates for steel from top to bottom. It is apparent that zinc corrodes at only a fraction of the rate of steel, and that zinc does not respond progressively as steel does. Table 19-3 also lists the environments as rural (R), urban (U), industrial (I), marine (M) or a modification of these.

Comparison

Figure 19-1 compares corrosion at a number of well-known sites. The scale consists of blocks with a range of corrosion of 2 to 1 within each block. For example, the State College, Pa. site falls in the H6 block with limits on the block of 0.8 to 1.6 MPY for steel and 0.031 to 0.062 MPY for zinc. Areas of the figure are also defined arbitrarily as mild, moderate and severe.

The corrosion rates at various types of sites can be easily compared. Normal Wells and Phoenix, although much different in climate, show low corrosion rates because each is dry and rural. Saskatoon, also a rural site, is cold and dry in the winter but a little more corrosive to steel. The next blocks above these are classed as moderately corrosive. Vancouver Island is the most moderate and Monroeville, Pa. the least moderate. Industrial sites approximately follow a 20 to 1 ratio, but the differences in corrosion rates

Table 19-3. Corrosion Rates of Steel and Zinc Panels Exposed Two Years

No.	Location	MPY		Env.	Block
		Steel	Zinc		
1.	Norman Wells, N.W.T., Can.	0.06	0.006	R	D3
2.	Phoenix, Ariz.	0.18	0.011	R	E4
3.	Saskatoon, Sask., Can.	0.23	0.011	R	E4
4.	Vancouver Island, B.C., Can.	0.53	0.019	RM	G5
5.	Detroit, Mich.	0.57	0.053	I	G6
6.	Fort Amidor, Panama C.Z.	0.58	0.025	M	G5
7.	Morenci, Mich.	0.77	0.047	R	G6
8.	Ottawa, Ont., Can.	0.78	0.044	U	G5
9.	Potter County, Pa.	0.81	0.049	R	H6
10.	Waterbury, Conn.	0.89	0.100	I	H7
11.	State College, Pa.	0.90	0.045	R	H6
12.	Montreal, Que., Can.	0.94	0.094	U	H7
13.	Melbourne, Aust.	1.03	0.030	I	H5
14.	Halifax, N.S.	1.06	0.062	U	H6
15.	Durham, N.H.	1.08	0.061	R	H6
16.	Middletown, O.	1.14	0.048	SI	H6
17.	Pittsburgh, Pa.	1.21	0.102	I	H7
18.	Columbus, O.	1.30	0.085	U	H7
19.	South Bend, Pa.	1.32	0.069	SR	H7
20.	Trail, B.C., Can.	1.38	0.062	I	H6
21.	Bethlehem, Pa.	1.48	0.051	I	H6
22.	Cleveland, O.	1.54	0.106	I	H7
23.	Miraflores, Panama C.Z.	1.70	0.045	M	I6
24.	London (Battersea), Eng.	1.87	0.095	I	I7
25.	Monroeville, Pa.	1.93	0.075	SI	I7
26.	Newark, N.J.	2.01	0.145	I	I8
27.	Manila, Philippine Is.	2.13	0.059	U	I6
28.	Limon Bay, Panama C.Z.	2.47	0.104	M	I7
29.	Bayonne, N.J.	3.07	0.188	I	I8
30.	East Chicago, Ind.	3.34	0.071	I	J7
31.	Cape Kennedy, Fla. ½ mile	3.42	0.045	M	J6
32.	Brazos River, Tex.	3.67	0.072	M	J7
33.	Pilsey Island, Eng.	4.06	0.022	IM	J5
34.	London (Stratford), Eng.	4.40	0.270	I	J9
35.	Halifax, N.S.	4.50	0.290	I	J9
36.	Cape Kennedy, Fla. 180 ft.	5.20	0.170	M	J8
37.	Kure Beach, N.C. 800 ft.	5.76	0.079	M	J7
38.	Cape Kennedy, Fla. 180 ft.	6.52	0.160	M	K8
39.	Daytona Beach, Fla.	11.7	0.078	M	K7
40.	Widness, Eng.	14.2	0.400	I	L9
41.	Cape Kennedy, Fla. 180 ft.	17.5	0.160	M	L8
42.	Dungeness, Eng.	19.3	0.140	IM	L8
43.	Point Reyes, Cal.	19.8	0.060	M	L6
44.	Kure Beach, N.C. 80 ft.	21.2	0.250	M	L9
45.	Galetea Point, Panama C.Z.	27.3	0.600	M	M10

R	rural	SI	semi-industrial	M marine	IM industrial-marine
RM	rural-marine	SR	semi-rural	I industrial	U urban

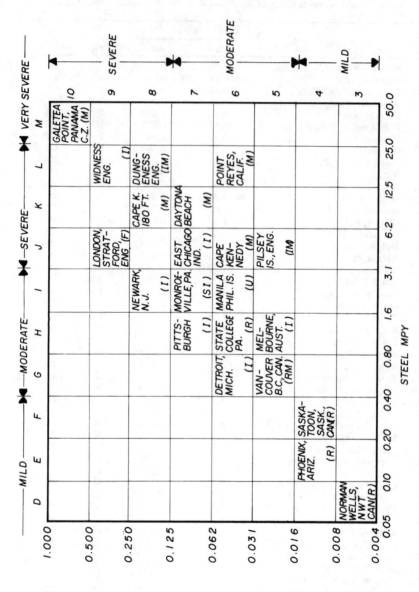

Figure 19-1. Comparison of corrosion rates of zinc and steel in major areas of the world.

are quite substantial. The Pittsburgh atmosphere is more corrosive than Detroit both for steel and zinc.

Block designations are given in Table 19-3 to relate similar sites. For example, block H6 indicates that corrosion at several sites in Pennsylvania and Ohio are similar to Halifax, Durham and Trail, B.C. The block is central to the chart and it contains rural and urban sites and a semi-industrial site. It is therefore logical to classify the industrial sites of Trail and Bethlehem as moderate industrial.

There are enough sites given in the industrial areas, Great Lakes region and various marine and arid regions to indicate what corrosion rate might be expected. Where several sites are reported, such as those in block H6, the data can be applied with confidence to a similar area. Where there are great differences such as sites No. 6, 23, 28 and 45, all in the Panama Canal Zone, then obviously more has to be known than the general locality.

Corrosion of Other Metals

The steel-zinc data are valid only for two years exposure. The corrosion of zinc is approximately linear with time, but the corrosion rate of steel is not. If other steel data are compared to this, it should be calculated from the corrosion for two years.

If a corrosion problem exists with other metals such as aluminum or magnesium, it should not be assumed that the steel-zinc data estimate the severity of the environments for alloys of these metals. However, if outdoor testing is done with a specific metal or alloy, it will be useful to test zinc and steel at the same time, or to test the metal in question at a site that has been calibrated in order to take advantage of a possible comparison with the existing data.

Solving Corrosion Problems

Outdoor corrosion can be stopped by isolating the metal surface from the moisture. Or, it can at least be inhibited to an acceptable degree by restricting the rate at which it reaches the surface. This is done with impermeable corrosion-resistant metals (nickel/chromium), slightly permeable coatings (paints) or sacrificial coatings (zinc or cadmium). Specifications should be developed by the pollution engineer for acceptable coating thicknesses and quality testing assurance to protect abatement systems.

Unfortunately, most system designs cannot wait for securing long-time data. When such is true, then existing data as close to expected service as possible should be examined. In the meantime it is advisable to start an outdoor testing program.

MODERN PROTECTIVE COATING TECHNOLOGY[110]

Paints and coatings are usually of last concern both in specification and actual job site practice. They are usually the first area from which the dwindling overall capital budget is cut. As a result, they are often the most obvious area of failure in a waste or water treatment plant or air pollution control facility.

A small pump, which might control the operation of a facility, can fail and not be noticed until too late. Moreover, when paint begins to chip and peel and rust appears, poor housekeeping is readily evident to everyone. Also, unless frequent recoating is done, equipment, tanks, and other steel must be replaced prematurely.

Paint and coating technology has advanced rapidly in recent years. Unfortunately, most of the specifications which are written today reflect methods five to twenty years old. This outmoded protection results in high installation costs, excessive maintenance expenses, and early failure. Proper selection of paints and coatings, carefully worded specifications, skilled application, and close field inspection greatly enhance plant performance. At the same time, they reduce both initial capital expenditures and ultimate maintenance costs.

Selection of Paints and Coatings

There are many conditions which must be considered when making a specific end use recommendation for a protective coating system. It is important to remember that the conventional system is not necessarily optimal either in price or performance.

The first area of consideration should be the type of service involved. Will it be immersion, splash, spillage, fumes, weathering, or an environment of salt, air, or industrial chemical fallout graded as to severity? The substrate (steel, aluminum galvanizing, concrete, masonry or wood) over which the coating will be applied must be considered. Then, a decision must be made on the type of surface preparation. It must be decided whether the work will be done in the fabricating shop or at the job site. The new environmental laws, as outlined under the OSHA regulations, will have a bearing on this.

The physical and mechanical abuse to which the coating will be subjected must be known. For floor coatings, the severity of foot traffic, steel-wheeled traffic and rubber-wheeled carts affects the coating choice. The temperature to which the coating will be exposed as well as thermal shock and extremes of both heat and cold are important.

Another area of consideration is the wear factor. This is becoming especially important in pipe and tank linings. There will be abrasion resistance requirements for a coating inside a pipe, slurry tanks, precipitator scrubber, or clarifier mechanism. This relates to the size and nature of the particles, chemical and physical makeup, and the flow rates. In general, it relates to what will be the wear phenomenon which would reduce the life of the coating.

Finally, after all of these performance characteristics have been defined, the cost of the protection enters into the equation. This includes both the initial cost in terms of capital dollars and the ultimate yearly maintenance costs. Maintenance costs are becoming more important, especially in view of the recent OSHA regulations. In a few years, sandblasted surface preparation for maintenance, while desirable, may no longer be allowed because of health hazards. This would encourage providing for maximum new steel surface preparation in the fabricating shop.

Knowledge of all of these basic criteria prepares the pollution engineer for his decision on what coating systems to use for his new project.

The selection of protective coating for steel surfaces, especially in new construction, is governed by the degree of surface preparation *allowable* or tolerable. It is desirable in new construction to specify the form of surface preparation, making certain directions are followed.

The life of a coating system is proportionate to the quality of the surface preparation over which it is applied. It is unwise to apply paints or coatings over mill scale-bearing steel. The steel under the mill scale will be in contact with water and oxygen. Technically, with conventional paint systems, mill scale absorbs water which has penetrated the coating. The mill scale is "popped" from the steel when iron oxide forms. The protective coating also breaks loose. Corrosion undercutting then proceeds at a fast rate.

The Steel Structures Painting Council estimates that a coating applied over a blast-cleaned surface will last approximately three and one-half times longer than the same coating applied over power tool-cleaned steel. Therefore, in order to minimize future maintenance costs, it is always recommended that blast cleaning be used wherever possible.

Over 500 fabricating shops have installed automatic blast cleaning equipment and mechanized steel pickling lines which make mill scale removal very economical. Many of these shops also have automatic spraying equipment which greatly reduces the cost of priming steel prior to shipment. The economics of shop blast cleaning or pickling compared to field sandblasting or hand blast cleaning in the shop are approximately as follows (figures do not include overhead and profit):

Method of Cleaning	Cost/Sq Ft Commercial Finish	Cost/Sq Ft White Metal Finish
Modified acid pickling	2-5¢	3-7¢
Automatic blast cleaning	3-8¢	4-10¢
Hand blast cleaning (shop or in field)	15-30¢	25-50¢

Note that hand blast-cleaning figures do not include costs of scaffolding and removal of sand necessary if this were done in the field. From this, it is apparent that automatic blast cleaning or modified acid pickling is the most economical method of preparing steel.

When shop blasting is specified, priming of the steel is essential. The primer must be capable of protecting the steel during fabrication, and through the final erection and topcoating phases of the project. It must withstand physical abuse during shipment and erection of the steel. Today, primers are readily available which permit the primed steel to be welded or bolted with high-tension bolts. They have excellent corrosion resistance and require negligible touchup in the field. These materials can save up to 50 percent in painting costs when properly specified and applied.

Selection of Shop Primers for Steel

Steel corrodes through an electrolytic cell reaction. In order for this to happen, there must be an anode, a cathode, and an electrolyte which become a small electrical cell. Anodes and cathodes may themselves be found on the surface of the steel. Variations in amounts of iron, carbon, or other materials in an alloy, placement of iron oxides as mill scale, or surface contaminants cause this. When water as an electrolyte is present, an electric cell is completed and corrosion begins.

One common way of protecting steel is to remove the water by forming a barrier over it. Protective coatings are a barrier. Another way is to protect it cathodically which is similar to galvanizing. Zinc, the most commonly used metal for this type of protection, will sacrifice itself to protect the steel underneath. Inorganic zinc primers, which are used widely, rely on sacrificial protection.

Protective coatings generally are of two basic types, organic and inorganic. Organic coatings include alkyd (often called red lead, zinc chromate, or iron oxide), epoxy polyamide, vinyl, chlorinated rubber and organic zinc-rich. Inorganic zinc primers, which protect cathodically, fall into the second category. Inorganic zinc primers are classed as alkali silicate inorganic zincs, or ethyl silicate-based inorganic zincs. Of these, the alkali silicates, which are often called water-based, are less desirable. The ethyl silicate-based inorganic zinc primers have a reduced tolerance for surface preparation. Most alkali silicate-based inorganic zincs require a minimum of a near white metal blast cleaning for adequate adhesion over steel. Ethyl silicate-based inorganic zincs can be applied over mill scale, brush-off blasted steel, or commercially blasted steel.

As mentioned earlier, shop applied primer must protect the steel during shipment and construction phases prior to topcoating. A properly chosen primer will require minimum touchup prior to topcoating and have a long protective life. The following table is a summary of comparative test data collected on various primers. All were applied over a commercially blasted and cleaned steel surface (SSPC-SP6-63).

Primer Type (2½-3 mils Dry Film Thickness)	Untopcoated Primer on Steel Time in Months Before Rusting Occurs	
	Gulf Coast	Inland (not desert)
Alkyds (red lead or zinc chromate)	2-3	5-6
Epoxy polyamide	5-6	8-10
Vinyl	3-4	6-8
Organic zinc-rich	8-10	10-15
Inorganic zinc	24-120	48-180

The data show that inorganic zinc primers are, indeed, the best choice from a performance standpoint. In fact, they are used untopcoated in many areas and provide adequate steel protection. Inorganic zinc primers have other advantages. They can be used in contact with surfaces for high-tension bolts. This is specifically allowed in the ASME code for bolting using Type A-325 or Type A-490 high-tension, high-strength bolts. Inorganic zinc primers are available which may be easily welded through. Welds made over the primed steel are as sound as those made over bare steel. There are few, if any, porosities in the weld. Burnback will normally be less than 1/16 in. as compared to 1 to 4 in. for common organic primers.

Untopcoated inorganic zinc primers withstand temperatures ranging from minus 40 to plus 750 F. With appropriate topcoats, the range is minus 40 to 1000 F. Typical organic primers have a useful temperature range of minus 20 to plus 150 to 250 F. These advantages demonstrate that the optimum choice for a steel primer is inorganic zinc normally applied at 3 mils dry film thickness.

A specification for shop priming should read as follows:

Steel in Non-Immersion Areas—Temperature minus 40 to 250 F.

Surface Preparation—Blast clean the steel surface to a commercial finish in accordance with Steel Structures Painting Council Specification SSPC SP6-63 or National Association of Corrosion Engineers Standard NACE #3, or pickle the steel to a commercial finish in accordance with SSPC SP7-63, using modifications indicated by the manufacturer.

Shop Prime and Field Touchup—Apply one coat of an ethyl silicate based, partially hydrolyzed inorganic zinc primer to a dry film thickness of 3.0 mils (nominal). The inorganic zinc primer should be as defined in SSPC-PS 12.00, containing not less than 75 percent zinc in the cured film.

Selection of Field Topcoats

The major physical and chemical properties of inorganic zinc primers have been covered up to this point. It should be stated that inorganic zinc primers may be used in virtually all areas of a water or waste treatment plant or an air pollution control facility. In order to complete the coating systems, a decision must be made on topcoats to be used over the inorganic zinc primer. These general recommendations serve as a guide for topcoat selection:

Non-Immersion Service

There is a variety of topcoats which may be used over inorganic zinc primers. However, for enclosed steel surfaces, topcoats are not required since aesthetics and direct exposure to chemicals are not considerations.

a. General weathering—coastal or inland or areas where architectural colors are desired. Where resistance to corrosive fumes is not required or in a mild fume environment, an acrylic latex is an excellent topcoat over inorganic zinc. This type has excellent weathering and chalking resistance, and retains both color and gloss. It is water-based and available in a wide variety of architectural colors.

b. Acid or caustic fumes—In areas of acid and caustic fumes, a high-build vinyl or chlorinated rubber topcoat is recommended over the inorganic zinc primer. These have been successfully used for a number of years in wastewater treatment facilities. However, they have normally been used over their own generic type primers. This has proven to be costly due to premature failure of the primer and high cost of replacement.

c. Solvent spillage—In solvent spillage areas, it is recommended that the inorganic zinc primer be topcoated with an epoxy polyamide topcoat. These topcoats are resistant to mild acids and alkalies to heavy solvent conditions. However, these catalyzed materials are more inconvenient to use than single component vinyl and chlorinated rubber topcoats commonly used in municipal water and waste treatment facilities. The epoxy polyamide high build materials have greater acceptance in industrial waste treatment facilities.

d. Severely corrosive environments—These exposures exist in advanced wastewater treatment facilities of the tertiary type or in industrial waste treatment facilities. For recommendations, it is wise to consult with the protective coating suppliers. Generally, modified phenolic and modified epoxy phenolic coatings, applied over inorganic zinc primers, perform well when exposed to acids, alkalies and solvents. However, conventional painters have difficulty applying them, and they are more costly than vinyls, chlorinated rubbers, or epoxy polyamide materials.

High Temperature Steel Coatings

Steel, when operating at temperatures above 250 F normally does not require coating since water will not condense at this temperature. Unfortunately, continuous operation above this temperature cannot be guaranteed. Most high-temperature stacks, incinerators, and furnaces cycle up and down in temperature. Weekend shutdowns of incinerators are common, causing a cool-down of the entire plant. During thermal cycling and periodic plant turnarounds, the steel cools down, moisture condenses on the steel, and rust forms. For this reason, the steel must be protected.

Surface preparation consists of blast cleaning the steel to a near white metal finish in accordance with Steel Structures Painting Council Specification SSPC SP 10-63 or National Association of Corrosion Engineers Standard NACE #2. Next, the steel should be shop primed with 3 mils of an ethyl silicate-based, partially hydrolized inorganic zinc primer. A silicone topcoat is applied in the field. Pure silicones are used for temperature ranges between

1000 and 1200 F. Silicones may be modified, but the degree and type will determine the temperature range, performance, and cost. Some silicones are modified for use between 750 and 1000 F, others for use between 500 and 750 F, and still others for use between 250 and 500 F. Since costs increase in proportion to the quantity of silicone resin in the formulation, the system used should be recommended for the temperature range specified.

Immersion Service

In immersion service, a white metal blast-cleaned surface conforming to SSPC SP-5-63 or National Association of Corrosion Engineers Standard NACE #1, or pickling to a white finish in accordance with SSPC SP 8-63 with manufacturer's modifications should be specified. This is followed with a 3-mil partially hydrolyzed, ethyl silicate-based inorganic zinc primer. Topcoats may then be selected for a given service.

a. Potable Water Service—The lining of potable water tanks is commonly done with four or five coats of low solids, vinyl materials. While these materials hold up very well in service, they have poor abrasion resistance. During winter months, ice chunks fall down from the tops of tanks and damage the bottoms. They are very costly because of the labor of applying multiple coats.

A better recommendation is high film build, high abrasion resistant, hard durable coatings for the interior of potable water tanks and process equipment. These are catalyzed epoxy materials, either epoxy polyamide or epoxy amine. Normally, they are applied in two coats at dry film thicknesses between 4 and 6 mils per coat. They are applied over inorganic zinc primers to take advantage of shop surface preparation and shop priming. Epoxy polyamides are preferred to epoxy amines as topcoats over inorganic zinc primers.

b. Wastewater and Process Water—Here, the topcoat recommendation is a coal tar epoxy material. As a rule, coal tar epoxies are applied in two coats, 7 to 9 mils each, over the inorganic zinc primer. One of the more commonly specified coal tar epoxy materials, with excellent performance characteristics, is the Corps of Engineers C-200 formulation (when properly manufactured).

c. Other Environments and Corrosives—It is recommended that a quality coating manufacturer be contacted for recommendations on specific services other than those outlined.

Topcoats for Scrubbers

Flyash for both fossil-fired electric generating stations and incinerator waste service systems are acidic and abrasive conditions which must be considered when selecting systems. Most organic tank lining materials used for scrubber installations are not recommended above 250 F. If higher temperatures will be encountered, materials such as acid-proof brick or mortar linings should be used. Such conditions are generally encountered prior to the quenching of scrubber gases. For gases that have been quenched, it is generally possible to line scrubbers with organic coatings.

If organic zinc primers are to be used in scrubber installations, the primed surface should be brush blasted prior to the application of lining material. This will leave a very thin coat of inorganic zinc over the steel, or none at all. Preferably, the scrubber lining material should be applied over practically bright white metal steel. A thin film (approximately ½ mil of inorganic zinc) would be allowable as a primer in order to hold the blasted surface. The surface should be cleaned prior to application of the protective lining system.

Materials which have found their greatest application in scrubber lining and demister lining work have been bisphenol A fumerate, flakeglass polyester lining materials. These are normally applied either by spray or trowel methods. The trowel-applied material must be rolled to align the glass flakes after they are trowel applied. Spray–applied materials do not require this extra step, and are normally less expensive to install. Application generally consists of two coats, 20 mils per coat. Some manufacturers have insisted upon a third coat to yield a dry film thickness of 60 mils for additional safety in scrubber linings and demister lining application areas. The same cautions and design parameters as taken in tank linings should be observed in this application.

Scrubber Ductwork

The ductwork of a scrubber system and the stacks coming off it present a different problem from the normal scrubber. The duct stacks are generally made of very light gauge steel. This steel is subject to rapid thermal shocks and severe flexing. Because of this, it is necessary to use a chemically resistant, abrasion resistant, elastomeric material rather than a rigid flakeglass polyester coating. The selection of this elastomeric material is very critical, since there are many on the market which have experienced failures. Fortunately, there are some which are specially formulated for this service and have proven successful.

The most successfully used elastomeric material for steel ductwork has been an abrasion-resistant, chemically-resistant, polyurethane lining. This lining is applied in two coats (20 mils each) over an epoxy polyamide tie coat and inorganic zinc primer. This system has now been accepted and is performing satisfactorily on scrubber installations. Earlier installations of rigid materials, such as flakeglass polyester, had failed in both stacks and ductwork lining areas.

Galvanized Steel Surfaces

The procedure has always been to passivate the galvanized steel by pretreatment with a material such as two-package wash primer, phosphoric acid, or a vinegar wipe. Then, it is primed with an inhibitive primer, followed by a tie coat and a finish coat. This four-step operation is costly and, in fact, has only marginal performance. The galvanizing itself protects the steel very well because of the cathodic protection of the zinc. It is painted, essentially, to improve the appearance.

There are available on the market today self-priming vinyl copolymers with high film build characteristics. They are applied at a dry film thickness of 3 mils, and only one coat is necessary. Surface preparation consists of solvent wiping the galvanized surface to remove grease and contaminants. These one-coat systems perform far better than the traditional multiple-coat systems. Over well-aged galvanizing, acrylic latex coatings show excellent performance characteristics. Here, the state of the coating art has progressed from an expensive four-step to an easy one-coat system. The vinyl system may also be used over brass, bronze, and aluminum with the same reduction in cost as experienced with galvanized surfaces.

Concrete Surfaces

Concrete and masonry surfaces are generally divided into three categories:

1. *Walls and ceilings*—These surfaces are painted for decorative and sometimes for protective reasons.
2. *Floors*—Floors are coated to protect the substrate and, where desired, for appearance and non-slip properties.
3. *Concrete tankage and sumps*—Concrete in immersion service is coated for protection from either the corrosives or abrasives in the liquids or slurry which is flowing through.

Before specifying coatings for concrete and masonry surfaces, the nature of the substrate over which these coatings are to be applied should be checked. Certain properties are common to all concrete and masonry surfaces. Practically all concrete, regardless of how well it is reinforced, will crack. Small hairline cracks, often referred to as after-cracking, are common with all concrete structures. In cinder block or concrete block wall constructions, the foundation generally settles. As it settles, cracks show up in the mortar joints between the blocks. Thus, if a coating is to protect concrete, it must have the ability to bridge or tolerate these small hairline cracks. If used to improve the appearance, it does not have to resist the hairline cracks, and this should be taken into account when selecting the type of coating.

Concrete is the most commonly used substrate in a pollution control facility. It is important to note that a crustlike formation, concrete laitence, forms on the outer surface during curing. This is much more apparent on horizontal slabs than on vertical surfaces. Concrete laitence is hard, crusty, and not tightly bonded to the concrete. Most protective coatings adhere well to laitence. Consequently, if a coating system is to perform satisfactorily over concrete, it is necessary to remove the laitence prior to applying the protective coating system.

After etching or brush blasting, a careful check should be made to determine that concrete laitence has been thoroughly removed. This is most easily done by wirebrushing the concrete. If this results in a shiny surface or polish on the concrete, laitence is still present and the process should be repeated. After etching or brush blasting, the surface of the concrete should be blown with clean, dry air to make sure that contaminants have been removed.

When coating concrete for purely aesthetic reasons, extensive surface preparation and care are not normally needed. It is generally adequate to blow the concrete surfaces down with clean dry air to remove surface contamination. It should be noted, however, that this marginal surface preparation will greatly reduce the life of any coating system.

Decorative Concrete Coatings

Decorative coatings for concrete may be of several varieties. Most common are acrylic latex, chlorinated rubber, or vinyl materials generally applied in two or three coats. They merely provide a color, or colored and textured finish, to the concrete or masonry surface. These are the more common wall paints and decorative finishes usually available in a wide array of colors. They echo the appearance of the substrate, but add color to its appearance. Sand or other inert materials may be added to these coating materials to give them the textured appearance sometimes desirable.

Becoming very popular are a wide range of tilelike finishes. These generally consist of a concrete surfacer of either acrylic latex and sand or epoxy polyamide. From experience, the epoxy polyamide surfaces are more desirable. They do a better job, with much less labor, of sealing the concrete surface even though the material cost is slightly higher. The labor cost for epoxy polyamide surfacers is generally one-third to one-half that of acrylic latex materials. This results in a great cost savings to the applicator and owner.

The surfacer materials in tilelike finishes are generally coated with a basic epoxy polyamide resin with or without a high gloss characteristic. This is followed by a second epoxy polyamide finish or a polyester glaze coat to give the appearance of virtrous tile. The better the surfacing characteristics on concrete, the better the tilelike appearance will be. These surfaces are easily cleaned, generally FDA approved systems, and find wise use both in waste and water treatment facilities.

The final decorative finish (for hiding hairline cracks) is elastomeric. The system consists of an epoxy polyamide base coat followed by approximately 20 to 40 mils of a urethane elastomeric material. A final topcoat of a thin weather resistant elastomeric material completes the system. This type system has sufficient elasticity to bridge hairline cracks in concrete and stucco walls. Thus, it will have an aesthetically pleasing appearance and a surface not marred by foundation settling or concrete cracks.

Concrete Floor Coatings

The selection of floor coating for concrete on a properly prepared floor will be based upon traffic and chemical spillage presented in an area. There are many varieties of protective floorings available today. The most common is a two- or three-coat epoxy polyamide, or other catalyzed epoxy system having good chemical and normal usage resistance. It is also rather inexpensive to apply. If a non-skid surface is desired, silica sand or other inert material may be broadcast in an intermediate coat. The result is a skid resistant surface.

Also falling into this category are a variety of chlorinated rubber flooring paints which may be applied successfully in light foot traffic areas. In heavier foot traffic areas, it is best to use a silica-filled epoxy polyamide or other catalyzed epoxy materials. These types have better resistance to chemical spillage and abrasion than their unfilled counterparts. They are quite often applied very heavily by spray or trowel at thicknesses from 1/16 to 1/8 in. There are also urethane flooring systems available having excellent abrasion resistance, and fairly good chemical resistance. The elastomeric properties of urethane give it the ability to bridge hairline cracks in the flooring substrate. These materials, however, must be applied at a minimum thickness of 25 mils in order to be effective. Coating manufacturers should generally be consulted for the optimum recommendation in floor applications.

Concrete Coatings for Immersion Service

Traditionally, in immersion service on concrete, thick materials such as coal tar epoxies or epoxy polyamide mastics have been used. These coatings generally serve as a very adequate barrier coating. However, if the concrete begins to crack, they do not have the ability to elongate. At this time, they will crack with the concrete substrate. This creates openings for the corrosive media or slurry to reach and attack the concrete.

Modern technology has developed elastomeric materials, mainly polyurethanes, which are applied over epoxy polyamide primers on concrete. When these are used, the concrete should be surfaced with an epoxy polyamide surfacer prior to application. The surfacer should provide as uniform a substrate as possible. The elastomeric coatings should be applied at a minimum of 25 mils dry film thickness, preferably 40 mils, to take advantage of bridging characteristics.

Elastomeric polyurethanes are usually chemically resistant and have abrasion resistance to slurries, typical in scrubber operations, and clarifier mechanisms. It is felt that while they are the optimum engineering choice, their rather high material cost will retard their general acceptance. They will be slow in replacing the more traditional coal tar epoxy and epoxy polyamide materials. These perform adequately in most services where crack bridging is not a problem.

There are two areas which are often overlooked in immersion service coating. The first is the requirement to brush blast walls as opposed to acid etching. The second is the requirement to waterproof below-grade concrete tank exteriors. Most wastewater treatment facilities are located close to rivers, and the water table is rather high in these areas. Because of this, groundwater pressure will quite often creep in and permeate concrete. If the groundwater pressure is high enough, it will literally pop the coating from the concrete walls. When this water pressure is anticipated, it is necessary to waterproof the exterior of all concrete vessels and concrete walls which will be below grade.

Traditionally, very rigid materials have been used for waterproofing. These have been coal tar, asphaltic, bitumen, and felt materials or coal tar epoxies. All of these methods are rigid. They will not bridge hairline cracks which develop in concrete. Once the foundation begins to settle, the waterproofing material cracks with the concrete and cannot do an effective job of waterproofing.

Available today are elastomeric bitumen-modified urethanes which are economical and proficient at waterproofing exterior, below-grade foundations. There are two commonly used bitumen urethanes. The first is the single-component, moisture-cured urethane which is applied at a minimum of 60 mils. It is very sensitive to humidity and temperature during application, and this creates problems. The second is the catalyzed urethane bitumen. By contrast, it is less sensitive to moisture conditions and temperature during application. It has better general application characteristics and ability to form an adequate waterproofing barrier at 40 mils dry film thickness. It has more acceptance in the waterproofing membrane field today than its single-component counterparts.

It is interesting to note that the two-component elastomeric membrane types have been used successfully in waterproofing below-grade structures, and in replacing rigidly built-up roofing systems.

CORROSION-RESISTANT LININGS FOR STACKS AND CHIMNEYS [135]

Power plants and industries discharging gases from sulfur-bearing fuels have always had a need for corrosion-resistant linings in chimneys and stacks. With lower operating temperatures, changes in fuels, and growing use of wet scrubbers to meet environmental regulations, corrosion resistance is even more critical.

In the past, independent liners constructed with acid-proof brick or steel have proven very successful. With the decrease in flue gas temperature and an increase in moisture content, condensation is higher inside the chimney and corrosion is accelerated. Unprotected steel liners are no longer considered suitable. Independent brick linings will still meet these new requirements. A new generation of corrosion-resistant mortars has been developed for use in power plants, refineries, steel mills, incinerators, and chemical plants.

Mortars based on soluble silicate comprise some of the original corrosion-resistant cements used. The first silicate mortars were simply mixtures of fillers, such as silica, quartz, gannister, clays, or barytes, and sodium silicate solution. Mortars of this type harden by loss of water and require exposure to air or heat to set. Construction with such a mortar is extremely slow. Although thin joints are used, the fluid mixture squeezes out if more than three or four courses of brick are laid at one time. In most cases, not over 6 ft of brickwork per day can be installed. Very careful drying is also necessary. A 30-day period is usually recommended before putting the structure in service. Air-drying mortars are no longer used for brick linings due to these drawbacks and the development of improved mortars.

Chemical Setting Mortars

In the 1930's, chemical setting sodium silicate mortars were developed. These utilize a setting agent which reacts with the soluble sodium silicate to cause the mixture to harden. The setting agent may be either an acid or a compound which will decompose and liberate acid to accelerate the cure. Typical setting agents are: ethyl acetate, zinc oxide, sodium fluorosilicate, glyceral diacetate, hexamethylene tetramine, formamide, and other amides and amines.

Chemical setting mortars are supplied as two-component systems. They consist of liquid sodium silicate solution and filler powder incorporating selected aggregates or as one-part systems in powder form to be mixed with water. Chemical setting mortars take initial sets in 15 to 45 min, and final sets in 24 to 96 hr or longer, depending on the temperature. Continuous bricklaying is possible because of chemical reaction, and does not require exposure to air or heat.

Large quantities of chemical setting sodium silicate mortars have been successfully used in industry for the past 40 years. Mortars of this type are still employed today for many types of acid service.

Potassium Silicate Mortars

In the early 1960's, new chemical setting mortars were developed using potassium silicates instead of sodium silicates. Several fundamental properties of potassium silicates combine to make them preferable to sodium silicates. Potassium silicate mortars have better workability due to their smoothness and lack of tack. They do not stick to the trowel nor run or flow from the joints of the brickwork. They possess greater resistance to strong acid solutions as well as to sulfation, and have greater refractoriness. Moreover, they do not effloresce or bloom and have less tendency to form hydrated crystals in the hardened mortar.

Potassium silicate mortars are supplied in two parts—the silicate solution and the filler powder. Mortars are available which utilize organic or inorganic setting agents or a combination of the two. The properties of the mortar are determined by the setting agent used. Such properties as absorption, porosity, strength, and water resistance are affected by the choice of setting agent. For example, organic setting agents will burn out at low temperatures, increasing porosity and absorption. The organic setting agents are water-soluble and can be leached out if the mortar is exposed to steam or moisture. Due to crystal structure formation, the mortars take a longer time to gain strength, remaining in a plastic state for 96 hr or more.

Comparative data for both are given in Table 19-4, plus sodium silicate mortar utilizing an organic setting agent. A typical potassium silicate mortar with an inorganic setting agent is designated as Mortar A and, with an organic setting agent, Mortar B. Sodium silicate mortar with an inorganic setting agent is designated as Mortar C.

Table 19-4. Comparison of Typical Chemical Setting Silicate Mortars

Property	Mortar A	Mortar B	Mortar C
Compressive Strength (7 days) ASTM C-396—psi	4000-4800	2900-4200	3000-3800
Bond Strength, ASTM C-321—psi	175	150	125
Shrinkage, ASTM C-531—percent	< 1	> 3	< 1
Working Life @ 70 F, ASTM C-414—min	45	30-40	30
Final Set @ 70 F—hr	24	120	24
Max. Recommended Service Temp.—F	2000	1850	2000
Effective pH range for Use*	0.0-7.0	0.0-7.0	0.0-7.0

*These mortars are not resistant to hydrofluoric acid or acid fluorides.

At first glance, it would appear there is no significant difference between the mortars with the two types of setting agents. However, closer investigation discloses several distinct advantages provided by the potassium silicate mortar with the inorganic setting agent. These include resistance to water and moisture immediately upon setting and without special treatment. The mortar with the organic setting agent requires multiple acid washes after it has set to impart some degree of resistance.

Figure 19-2 shows specimens of both type mortars which were subjected to a 2-hr water boil. Specimens on the left-hand side of each mortar were not acid treated while those on the right were. The results clearly indicate the superiority of Mortar A, since there is no apparent difference in the two specimens.

Mortar A takes a final set in 24 hr at 70 F permitting continuous construction without danger of brickwork slipping. Mortar B requires 120 hr at 70 F to attain a final set. Mortar A also has lineal shrinkage of less than 1 percent in 7 days as compared to over 3 percent for Mortar B.

Figure 19-2. Specimens of two potassium mortars subjected to a 2-hr water boil.

Modified Silicate Mortars

As previously mentioned, the one-part powder form chemical setting silicate mortars have been commercially available for some time. They have not been used extensively because of their higher cost. These one-part silicate mortars are similar to the two-component mortars except they have somewhat lower physical properties. Research and field testing have recently produced new modified silicate mortars with characteristics not previously available in either one- or two-part systems. These new modified silicate mortars utilize different classes of setting agents. The properties of the mortar are dependent upon the setting agent selected.

In Table 19-5, comparative data for two proprietary mortars of this type are given. One of the advantages that new modified silicate mortars provide is greater pH range of service than other types of chemical silicate mortars. They are resistant to most acids except hydrofluoric and acid fluoride salts. Their resistance to most alkalies is from pH 0.0 to 9.0, and in some cases to pH 14.0, depending on the alkali. They are highly resistant to sulfation, blooming and efflorescence. Modified silicate mortars have excellent adhesion to brick, concrete and steel.

Table 19-5. Comparison of Typical Modified Silicate Mortars

Property	Mortar D	Mortar E
Compressive Strength (14 days) ASTM C-396—psi	3500-4000	2500-3600
Flexural Strength ASTM C-580—psi	1360	1200
Shrinkage ASTM C-531—percent	0.75	3.75
K Factor, ASTM C-117 (Btu's/ft^2/hr/F/in.)	8.3	7.25
Coef. Thermal Expansion ASTM E-228 (in./in./F)	7.5×10^{-6}	6.8×10^{-6}
Modulus of Elasticity ASTM C-580—psi	1.4×10^6	1.4×10^6
Working Life @ 70 F—min	35	40
Final Set @ 70 F—hr	24	192
Maximum recommended service temperature—F	1750	750
Effective pH Range for Use	0.0 to 9.0*	0.0 to 8.0

*pH Range may be extended to 14.0 for certain alkalies. These mortars are not resistant to hydrofluoric acid or acid fluorides.

Figure 19-3 shows typical compressive strength values for 1, 3, and 7-day periods for all types of silicate mortars.

Monolithic Linings

Many chimneys and stacks have linings of calcium aluminate cement or refractory concrete applied by Guniting, cast-in-place, or troweling. Calcium aluminate cement linings are unsatisfactory if the pH of the acid condensate is below 3.5. The alumina gel common to these cements dissolves rapidly in such an environment. Refractory concrete is not acid-resistant, and its additional cost is not warranted when operating temperatures are below 500 F.

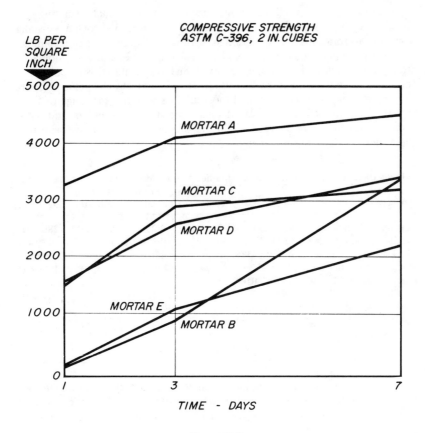

Figure 19-3.

Monolithic corrosion-resistant lining materials are available and have been used successfully to restore and repair brick linings in concrete chimneys and steel stacks. These monolithic linings may be applied to both new and existing steel liners by Guniting.

The original acid-proof concrete lining is an inorganic silicate composition which resists all acids except hydrofluoric, water, oil, most solvents, and temperatures to 1750 F. It is recommended for use over a pH range of 0.0 to 7.0. Supplied in two parts, the powder and liquid are mixed together when used. The liquid is substituted for the water normally used in Guniting.

Modified silicate-base cements designed for use as monolithic linings, to be applied by the cement gun process, are now also being supplied. These are single-component systems which produce a high-strength acid-proof lining when mixed with water. They have extremely good adhesion to concrete, brick and steel, and require only a minimum of surface preparation. These linings are not affected by acids, except hydrofluoric and acid fluoride salts, from the lowest pH to 9.0.

There is also a modified inorganic silicate-base thermal insulating chemical-resistant monolithic lining which is recommended for new installations and restorations. Since it is lighter in weight than most other linings, it reduces the dead load, provides high insulating value, and exceptional strength at low initial cost. This lightweight lining material is in powder form and mixed with water when used. It resists most acids, except hydrofluoric and acid fluoride salts, and most alkalies within a pH range of 0.0 to 9.0. It is not affected by water, steam, moisture, or weather, and withstands temperatures to 1700 F. Table 19-6 provides comparative data for these monolithic lining materials.

Table 19-6. Comparison of Typical Modified Silicate Monolithic Lining Materials

Property	Acid Proof Concrete	Modified Silicate	Light Weight
Compressive Strength (7 days) ASTM C-396, psi	1800-2500	3500-4000	2800-3200
Tensile Strength (7 days) ASTM C-307, psi	420-450	300	400-450
Flexural Strength (7 days) ASTM C-580, psi	450	1360	550
Modulus of Elasticity ASTM C-580, psi	2.32×10^6	1.4×10^6	1.06×10^6
Coef. of Thermal Expansion ASTM C-228, in./in.F	7.1×10^{-6}	7.5×10^{-6}	7.1×10^{-6}
K Factor ASTM C-177, Btu/ft^2/in./F/in.	10.56	8.3	4.9
Water Absorption ASTM C-413, percent	6.3	1.25	1.25
Shrinkage, ASTM C-531, percent	0.4	0.75	0.36
Density, lb per cu ft	168	135	98

The proper selection of lining materials is mandatory, even though the percentage of the total project cost may be relatively low. Downtime, maintenance expense, and service life are all items the pollution engineer cannot ignore.

CHAPTER 20

SOLID WASTES HANDLING METHODS AND MANAGEMENT

INDUSTRIAL SOLID WASTE HANDLING AND DISPOSAL EQUIPMENT[49]

Disposal of industrial solid wastes is a formidable problem. Industrial wastes are generated in excess of 100 million tons annually. Additionally, there are wastes collected by pollution control equipment such as wet scrubbers, cyclones and precipitators, and sludge from wastewater treatment processes.

The general approach to disposal of industrial solid waste includes:

1. Transfer by a disposal contractor off-site, and
2. Internalized handling of the wastes to:
   a. Reduce volume and/or weight of the refuse handled for economic reasons, and
   b. Reduce risks of hazardous and toxic materials.

There are a number of methods widely accepted for reducing weight and volume and negating many hazardous properties of wastes. The principal methods used are shredders, compactors and balers for volume reduction and incineration for weight and volume reduction. A wide variety of equipment is available from manufacturers ranging from single hardware units to complete systems for solid waste handling and disposal. The need for such equipment is increasing because of higher handling and removal costs due to increased waste generation and spiraling labor, land and equipment costs.

Waste Handling Methods

The physical nature of a waste may determine its handling characteristics. The following are common handling methods based on physical classification:

1. Solid, semi-solids, some wet materials, sticky or tarry substances may be handled by front-end loaders or buckets,

671

2. Viscous liquids may be pumped when hot or transported by special pumps,
3. Liquids are handled by normal pumping equipment,
4. Packages may be handled in combustible boxes or cartons, and
5. Some materials are handled in 55-gal steel or fiber-pack drums.

The Incinerator Institute of America has established a classification system for wastes based partly on content and partly on Btu value and moisture.

Type 0—Trash: Mostly paper, cardboard, wood, etc. This type of waste contains 10 percent moisture, 5 percent incombustible solids, and a heating value of 8500 Btu/lb.

Type 1—Rubbish: Paper, wood scrap, foliage, floor sweepings. This type of waste contains 25 percent moisture, 10 percent incombustible solids, and has a heating value of 6500 Btu/lb.

Type 2—Refuse: An even mixture of rubbish and garbage. This waste contains up to 50 percent moisture, 7 percent incombustible solids, and has a heating value of 4300 Btu/lb.

Type 3—Garbage: Animal and vegetable wastes. This waste contains up to 70 percent moisture, up to 5 percent incombustible solids, and has a heating value of 2500 Btu/lb.

Type 4—Pathological waste: Animal remains. This waste contains up to 85 percent moisture, 5 percent incombustible solids, and a heating value of 1000 Btu/lb.

Type 5—Industrial gaseous, liquid or semi-liquid residues.

Type 6—Industrial solid wastes.

Waste types 5 and 6 are undefined because of the wide variety in materials. The only information given is that the characteristics of each waste vary and must be determined individually in order to engineer an effective disposal system. However, more data are needed in a solid wastes classification system than Btu values, moisture content, and ash content in order to evaluate the waste.

Loose Refuse Haulaway

The simplest and easiest way to get rid of solid wastes is to remove them from the plant for disposal by other means such as landfill, composting, central incineration or ocean dumping. Hauling away general refuse as loose material is common practice even though littering, odor and other unsanitary conditions are created. Loose hauling may superficially appear inexpensive, but in reality its actual cost is high because rates charged by refuse collection companies are based on volume of refuse instead of weight. Obviously, uncompacted refuse occupies greater volume, resulting in higher removal costs. Even where local refuse contractors have enclosed trucks with hydraulic rams to pack loose refuse, charges are usually based on loose volume.

Advantages and disadvantages as well as cost comparisons with compacted haulaway systems and on-site disposal methods are covered later in this chapter.

Solid Waste Shredders

Solid waste shredding machines can convert rubbish into a form more easily and economically handled for processing. Hammer mills, in one adaptation or another, are the most commonly used size reduction machine. The machines are variously and/or commonly called shredders, crushers, pulverizers, mills and hoggers. The most frequently used term is shredder. A wide range of capacities is available with the largest waste shredder capacity exceeding 100 tons/hr. Units are available that can nuggetize two complete automobiles per minute.

The most common shredder design consists of a welded steel frame protected on the inside with abrasion-resistant steel alloy liners. Within the frame is a power-driven rotor with rows of pivoted hardened manganese steel hammers. For oversized bulky wastes or hard refuse the hammers are free to pivot back under impact for overload relief. A series of sizing grates is positioned below the rotor. The hammer action reduces the material to a size that will pass through openings in the grates with maximum product size controlled by installing desired sized grates.

Shredder size should be selected to allow for surges in material feed rate. The machine should be sized to accommodate the largest pieces expected. Power consumed is proportional to the feed rate and degree of reduction. Shredder output of final product size can range from 1 in. for composting to 10 in. for landfill. Arbitrarily specifying a small final product means increasing the shredder size, cost, power requirements and maintenance costs. For most solid waste applications, a motor is selected to provide 12 to 20 hp/ton refuse handled per hour.

Shredders can be used to reduce the cost per ton of handling in cases where reduced bulk and improved handling characteristics result. Shredders can help a waste incineration plant operate more effectively, particularly in cases involving large objects which could not normally be fed to the incinerator. Shredded materials improve combustion due to increased surface area for burning.

Major types of shredders for primary or secondary shredding are large heavy-duty equipment. Design and construction features include:

1. Heavy welded steel construction frame or housing,
2. Oversize bearings and rotor parts,
3. Hydraulic system to raise parts for ease of servicing interior parts,
4. Statically and dynamically balanced rotors, and
5. Convenient access to shredding chambers.

Portable refuse shredding plants are available in smaller sizes. They are fed by portable steel apron feeder conveyors discharging onto a portable belt to the transfer truck. These small portable units have a capacity range of 15 to 100 tons/day, while shredders for stationary installations range in capacities from 5 to 100 tons/hr.

Compactors

Refuse can be compacted in many operations of refuse disposal. Compacting units are used on-site at industrial plants, institutions, supermarkets, schools and apartment buildings, where large amounts of refuse are generated. Collection vehicles may be equipped with compactor units. Heavy construction equipment and special vehicles are used to provide compacting at the disposal site.

Compacting of refuse is accomplished by equipment available from a large number of manufacturers. Hydraulic or pneumatic cylinders exert forces as high as 50 tons, reducing the original volume of refuse by 60 to 80 percent. In many cases refuse does not retain its compacted shape and must be restrained by keeping it under pressure either by bagging, baling, or some other means.

Of the several basic systems for plant on-site compaction, some may utilize the pneumatic ram to compact the refuse into paper sacks or plastic containers. The principle of operation is simple—refuse compressor and paper bag holder are mounted below a refuse chute. As refuse enters the bag, it triggers a mechanism that shunts the bag under the pneumatic ram, places a clean bag under the chute and compresses the collected refuse. The partly filled bag stands ready to be shunted under the chute until full. The paper bag compactor has the following advantages:

1. Provides a sanitary method of collection,
2. Reduces volume of refuse by two-thirds,
3. Reduces refuse to packages easy to handle,
4. Reduces the amount of on-site storage required, and
5. Reduces the need for compaction equipment on the collection vehicle.

Some of the disadvantages of such a system are:

1. High pressure storage during compaction,
2. Power costs,
3. Bag costs, and
4. No reduction in weight of refuse.

The other principal system of compaction in general use consists of a horizontal ram with a compression chamber connected to a wheeled detachable container. Volume reduction of refuse in such a unit is approximately 75 percent. When the detachable container is fully loaded, a signal is energized, so that the unit does not accept additional refuse. A typical standard detachable container holds approximately one-half ton of compacted refuse, which can be wheeled to the area for a contract disposal pick-up. It has all the advantages and disadvantages of the bag compactor.

Additionally available are a number of systems in which heavy-duty crushers or disintegrators chew up refuse. Crusher or disintegrator refuse is then delivered to a baling machine, where it is wrapped and sealed for pickup by compactor-type collection vehicles. There are over thirty manufacturers of compactor vehicles. Presently about three-quarters of all refuse trucks in operation are of the compactor type.

Maintenance problems associated with stationary compactors are those typically found in simple mechanical and hydraulic equipment systems. Problems often involve misalignment of moving parts. Oil levels in hydraulic reservoirs must be checked regularly to prevent costly damage to hydraulic power units. Installation of automatic low-oil-level alarms can help ensure maintenance of proper oil level. Some compactors have no ram guides, which may cause the ram to rub back and forth on the charge box floor and sides. In cases where plastic guides are used, small pieces of metal may become imbedded in the surface, and then act as a cutting tool on the plastic. Other guide problems involve materials of construction. For example, some manufacturers have bronze guides which wear rapidly. Some packers even use wood guides, which swell or split. Iron runners riding on replaceable steel wear strips, regularly lubricated, assure trouble-free longevity. Wear affects only replaceable parts.

Balers

Balers, like standard compactors, are made in varying sizes for different jobs. Balers are primarily used to bale paper, corrugated and rags. They are also used to bale scrap metal and nonferrous material, complete automobiles, as well as municipal refuse. The paper baling operation usually consists of throwing waste paper or corrugated onto a feed conveyor. The material then passes a hogger, which tears the scrap into small pieces and throws it into a high-velocity air stream created by the exhauster of the hogger. It then blows into a cyclone separator located directly over the baler feed chute. The separator allows air to escape and scrap to drop into the baler feed chute.

As scrap drops into the feed chute and builds up to the height of an electric eye in the chute, a hydraulic ram moves horizontally compressing the scrap into the bale. When the bale has been formed to the desired length, an operator may insert tie wires to retain the bale shape as it moves forward.

A popular size of horizontal baler for industrial disposal of paper and corrugated is rated at 12 tons/hr. There are also small paper balers on the market today which do not require grinding or hogging. These are principally used by large retail grocery stores.

Cardboard boxes are thrown in a chute and a signal ram travels down until full pressure is reached. Weight of these bales is usually about 300 lb compared to 1200 to 1400 lb for larger industrial-type bales. Hydraulic presses used in the scrap metal industry are entirely different from the paper balers. Triple-compression, heavy-duty hydraulic presses produce high density bales weighing 2 to 4 tons.

Incineration Equipment

Solid and chemical wastes are often disposed of through incineration. Incinerators for industrial-type wastes are often more specialized than those used in municipal or general refuse service. Some chemical industrial wastes may be hazardous, toxic or corrosive and require equipment specifically engineered for them. Incinerators can be classified into categories based upon burner chamber or fuel bed.

Burner Chamber Type

1. Fixed bed incinerators
   a. Open pit burning or incineration (unacceptable under existing air pollution codes).
   b. Closed chamber burning
      (1) Single-chamber
      (2) Multiple-chamber
   c. Tray furnace incinerators
2. Moving bed incinerators
   a. Rotary tube or kiln
   b. Fluid bed
   c. Moving grate

3. Pyrolysis-type incinerators. Not truly incineration—but thermal decomposition without oxidation or in an oxygen-deficient atmosphere. Yields products that may have some intermediate value for re-use.

Solid Stationary Hearth

As shown in Table 20-1, this type of system is used for solids incineration. The primary maintenance problem encountered is that refuse charges on the solid hearth tend to accumulate in a pile. The waste burns only on the surface; complete combustion is difficult to achieve. Industrial solid wastes normally require constant agitation to allow oxygen to reach all areas for complete combustion. Agitation is often done manually in an unsafe and tedious manner.

Ash removal from the solid stationary hearth is usually a hand batch operation. This is often unsafe and disrupts any attempt toward smooth operation of either the combustion operation or pollution control equipment. The combustion chamber must be properly sized to allow flame space for complete combustion, refractory protection, and adequate temperature control over the desired feed range. Proper utilization of the air pollution control system and fuel and air controls can improve the operating turndown ratio.

Solid Mono Hearth (Rotary Hearth or Rotating Rabble Arms)

This system is not only used for incinerating solids, but the rotary hearth can incinerate essentially any liquid waste capable of being fed to a

stationary liquid tar burner. The combustion chamber must be properly
sized to allow flame space, complete combustion, refractory protection, and
adequate temperature control over the desired feed range. Liquid burners
are normally positioned to aid combustion of solid wastes. An adequate
supply of solids on the hearth is required for flame impingement protection,
which is a major problem. Protection of the rabble arms is also a major
consideration, since they may break off if the ash or feed consists of oversized unbreakable material.

No air for combustion or turbulence passes up through the hearth.
All air must be supplied from above through the rabble arm plows. Even
with the use of rabble arms, turbulence and air contact is limited. The
burning rate may range from 8 to 15 lb/sq ft/hr, depending on the solids
being incinerated. Rabble arms are usually air cooled to protect them from
heat damage and to supply some of the combustion air requirement. Rabble
arms are a high maintenance problem and need periodic replacement. The
solid waste must be free of large heavy items such as metal drums or rings,
which will damage the arms.

Rotary hearth are commonly used to incinerate waste sludges and
granular material. These materials will not fall through the hearth, as they
would with grating. If the solids form a melt or liquid phase, some material
may flow through the center discharge before complete incineration.

The **stationary** hearth with rotating rabble arms is sometimes simpler
and cheaper to construct, but the rotating hearth has a certain noteworthy
advantage. A rotating hearth with a ram feed device will allow solids to
move away from the feed area and partly burn before contacting the rabble
arms. Stationary rabble arms can be of simpler, stronger design with lower
maintenance costs, time and frequency.

Grate Hearth

Traveling or reciprocating grates work well with raw municipal refuse.
However, many industrial wastes tend to fall through the open gratings.
Plastics and other industrial wastes which form a melt phase tend to flow
through and around the grating. This can jam the grate drive mechanism
or cause high temperature damage, as the wastes burn directly on the grates.
The high temperatures and abrasive action on the moving grate increases
maintenance costs. Drive mechanisms and grates require periodic replacement.

Rotary Kiln

The rotary kiln provides the design flexibility for incineration of a wide
variety of liquid and solid industrial wastes. Any burnable liquid capable of
being atomized by steam or air through a burner nozzle can be incinerated
concurrently with a wide range of industrial solids. Heavy tars may be fed as
solid waste in packs or metal drums. The kiln may be designed to receive
55-gal drums, or a feed mechanism can be designed to empty the drum. It is

Table 20-1. Comparison of Incinerators by Types

Type	Uses	Advantages	Limitations
Solid Stationary Hearth	Solids incineration	1. Low capital 2. Tight air control 3. Can be designed to include liquid incineration	1. Slow burning rates 2. Limited to batch operations 3. Requires manual ash removal 4. Does not lend itself to good air pollution control 5. Does not provide turbulence, mixing or aeration
Solid Mono Hearth (Rotary Hearth or Rotating Rabble Arms)	1. Used for solids incineration The rotary hearth can incinerate essentially any liquid waste capable of being fed to a stationary liquid tar burner 2. Rotary hearth without rabble arms also used for tire destruction 3. Rotary hearths used to incinerate sludges and granular material	1. Has continuous ash 2. Capable of incinerating waste solids independently or liquids and solids in combination. 3. Good maximum to minimum operating range 4. Incinerating materials will not fall through hearth 5. Can be readily incorporated with a gas scrubbing system	1. Has limited turbulence and air contact 2. Requires rabble arms or plows—thus additional maintenance 3. Susceptible to rabble arm damage 4. Partly combusted materials may flow out ash discharge
Grate Hearth	Solid wastes	1. Provides under fire air to aid combustion 2. Allows ash removal through grating through the incineration system 3. Can be designed to forward solids through the incineration system 4. Does not require extensive fuel preparation, such as shredding	1. Limited turbulence for air contact 2. Solids may fall through grating before complete burn out 3. Plastics or melt phase materials may damage grates

Type	Application	Advantages	Limitations
Rotary Kiln	Used for incinerating a wide variety of liquid and solid wastes	1. Capable of receiving liquids and solids independently or in combination 2. Not hampered by materials passing through a melt phase 3. Feed capability for drums and bulk containers 4. Wide flexibility in feed mechanism design 5. Provides high turbulence and air exposure of solid wastes 6. Long inventory time for slow burning refuse 7. Continuous ash discharge 8. No moving parts within the kiln 9. Can be readily incorporated with a wet gas scrubbing system	1. High capital installation for low feed rates 2. Cannot use suspended brick in kiln 3. Operating care necessary to prevent refractory damage 4. Not normally practical for very low feed rates 5. Airborne particles may be carried out of kiln before complete combustion 6. Spherical or cylindrical items may roll through kiln before complete combustion. 7. Combustion air tends to channel down center of rotary kiln requiring relatively high excess air for complete combustion 8. Drying or ignition grates used prior to the rotary kiln can cause problems with plastic melt plugging grates and grate mechanisms
Fluid Bed	Used for incineration of a moderate range of liquid and solid wastes	1. Rapid heat transfer from gas to solid 2. High combustion rate, high turbulence and air exposure 3. Low excess air requirement 4. Large heat sink to smooth out fluctuations in feed rate or fuel value	1. Requires fluid bed preparation and maintenance 2. Feed selection must avoid bed damage 3. May require special operating procedures to avoid bed damage 4. Incineration temperatures limited to a maximum of about 1500 F
Stationary Liquid Waste Burner	Capable of incinerating a wide range of liquid wastes	1. May use suspended brick 2. No continuous ash removal system required other than air pollution controls	1. Must be able to atomize tars or liquids through a burner nozzle 2. Heat content of liquids must maintain adequate temperatures 3. Must allow for flame space, complete combustion, and refractory protection

also capable of handling pallets, plastics, filter cakes, and other solid chemicals passing through a liquid phase before combustion.

Rotary kilns provide a maximum of turbulence, agitation and surface air contact to achieve a complete burn. Complete combustion of slow-burning refuse is aided by a relatively long inventory time in the combustion chamber. Ash discharge is continuous. Roll-through of spherical or cylindrical items is normally prevented by the other solid refuse being incinerated. Normal kiln operation would not be expected to incinerate such items as metal drums. A metal drum may be melted or deformed, depending primarily on its contents. The ash conveyor system must be designed to remove such items.

Since the drive mechanism is outside the kiln, maintenance is low. There are no internal moving parts such as rabble arms, grates or plows. Care must be exercised in sizing the kiln to accommodate solid wastes and to maximize refractory life. As the kiln size decreases, the unit becomes increasingly sensitive to excessive heat release, and therefore temperature control is more difficult.

Since the rotary motion of the kiln precludes the use of suspended brick, the refractory is more susceptible to thermal shock damage. Continuous operation should be maintained as much as possible. Rebricking of the hottest part of the kiln must be done on an annual basis. It is often advisable to maintain an inventory of kiln refractory in protected storage.

Airborne particles present another operating problem. They may be carried out of the kiln before complete combustion. A high temperature secondary combustion chamber with intimate flame contact is normally required for complete burnout. The fuel for the secondary combustion chamber can be dependable high-quality waste liquid or commercial fuel.

Fluid Bed

Fluid bed incinerators are designed to expose wastes to several feet of hot bed with high turbulence and good air contact for rapid complete combustion. Burnout may be accomplished with as low as 20 percent excess air, which will provide operating economy of low horsepower requirements and less air-to-heat.

Fluid bed incineration appears most advantageous when the bed can be formed as a natural product of the refuse being incinerated. This is especially true if the refuse has a high ash content. Otherwise the bed must be frequently replaced. Low ash, highly volatile compounds such as wet coffee grounds appear to have good incineration application.

Fluid bed particles may be temperature and composition sensitive. Eutectic mixtures may be formed, which will destroy the bed fluidization. Some beds may be very susceptible to caking during shutdown.

Stationary Liquid Waste Burner

A wide range of industrial liquid wastes may be incinerated provided the heating value is sufficient to maintain temperature for complete combustion.

When a low heat value liquid is incinerated, it must be blended with high quality liquid or the use of auxiliary commercial fuel will be required. Tars must be atomized through a burner nozzle by air, steam, or mechanical means. Mechanical atomization is normally avoided because of the high pressure requirement and the wide range of liquid viscosities. Since there are no moving sections, suspended air-tempered brick is utilized. This contributes to longer life and lower maintenance.

The ash is essentially all gas-borne particles which presents the largest problem. Ash must be removed by appropriate air pollution control systems. Because a certain amount of particulate will drop out within the incinerator, occasional shutdown and cleanout is necessary. Depending on the tar burned, cleanout may be required at six-month intervals.

COMMON INDUSTRIAL PRACTICES FOR SOLID WASTE DISPOSAL[51]

In most solid waste disposal methods, the primary objective is to treat the refuse in such a way as to render it safe and sterile so that upon returning it to the environment, it will not pollute the air, water or land. These methods do not actually destroy the refuse, but merely convert it into safe by-product materials which will eventually decompose into various elemental substances to make new organic or inorganic materials—thus completing the ecological cycle.

There are presently only three disposal methods which are practical for most industrial applications: haulaway loose, haulaway compacted, and on-site incineration. Plants with very large amounts of homogeneous wastes may find some other methods feasible, but these apply only to special cases.

Haulaway Loose

One of the easiest and simplest ways to eliminate solid wastes is to remove them from the plant for disposal by other means, such as landfill, central incineration, or composting.

Hauling away or exporting the refuse as loose material has been popular for quite some time. There are, however, several disadvantages. Littering, odor, and other unsanitary conditions may be created. In most cases, loose hauling is a very expensive method, because rates charged by local refuse collection companies are based on volume of the refuse instead of weight. Since uncompacted refuse occupies greater volumes, much higher removal costs are required. Although local refuse contractors use enclosed trucks with hydraulic rams to pack the loose refuse, the charges are based on the volume of the loose refuse prior to compaction in the trucks. Volume is usually measured by noting the volume of refuse containers.

Usual rates charged by local refuse contractors range from $2.50 to $4.00 per cu yd. The higher figure is equivalent to about $1.00 per 55-gal drum or $0.60 per 33-gal drum.

About the only advantage this system offers is the low capital investment required. The only equipment needed is a supply of steel drums and/or plastic liners to collect and store the refuse until the local scavenger truck arrives. If the refuse generated is less than 2 cu yd per day, this method is hard to beat economically.

If, on the other hand, the refuse is greater than 2 cu yd per day, other disposal methods (such as haulaway compacted or on-site incineration) should be examined. The savings offered by these other methods may be enough to offset in a short time the additional capital required.

Figure 20-1. Smaller versions of the traditional "garbage truck" are becoming popular for collecting and hauling industrial wastes. Similar containers, mounted on trailers, are suitable for use inside buildings.

Haulaway Compacted

If the amount of refuse generated on the premises is substantial, or if present handling methods give rise to nuisance conditions, refuse compaction should be investigated.

In this method, compacting equipment is used to reduce the volume of the refuse, thus reducing the haulaway costs proportionally. Refuse is compacted into commercially available metal containers ranging from 2 cu yd to 40 cu yd. In the larger units, the whole box is hauled away on special rollup truck rigs. The smaller containers are emptied into the regular refuse trucks used by scavengers. There are even very small compactors that pack refuse into steel drums, paper sacks, and plastic bags.

Figure 20-2. Market for waste compactors is flourishing, and hundreds of makes and sizes are available. Compacting refuse before hauloff reduces transportation costs.

Compacted refuse in the containers is much easier to handle than loose material. Compacting and containerizing the refuse is, in essence, the same as preparation for shipment of other materials from the premises. Containers should be sturdy and troublefree, and be shipped by the lowest available rates. In shipping compacted refuse, the lowest rates are obtained by reducing the volume as much as possible.

Available compacting equipment usually reduces the volume of loose refuse by 3-5:1. Higher compaction ratios could be developed, but the additional equipment needed is not usually justified.

Capital costs of compactors range from $3000 to about $10,000, depending on the size of the unit and its degree of automation. Small units have a capacity of ½ to 5 cu yd per day and a compaction ratio of about 5 to 1.

Larger units are suitable for operations that generate about 10 to 50 cu yd of refuse per day. The compaction forces developed by these units are high, but the overall compaction ratios (volume reductions) are rarely over 3 to 1 because the refuse tends to bridge inside the containers.

Compacting and hauling the refuse away has become a very popular waste disposal method for industrial installations. The idea of just dropping the refuse in the compactor hopper and pressing the load button has appealed to plant personnel. It offers a number of desirable advantages—easier handling of the refuse; less volume, maintenance, and littering; and lower costs than hauling it away loose.

Even with compaction, refuse transportation costs can be quite high for plants with large volumes of solid waste; that is, over 50 cu yd of loose refuse per day. For instance, if a plant generates about 100 cu yd per day of loose refuse, the compacted amount would be about 30 cu yd per day. Since most contractors charge at least $3.50 per cu yd for compacted refuse, the refuse removal costs would be about $105 per day. Based on 250 working days per year, the annual cost of refuse removal would be $26,250.

There can be another disadvantage to refuse compaction—the "juices" or liquids that tend to drop out the end of the container as the refuse is squeezed. Deodorizers and drip-pans can minimize the odor and unsightliness, but controlling this problem should be a major consideration when planning a compaction system, large or small.

On-Site Incineration

On-site incineration of industrial wastes is generally accepted as a good all-around method for disposing of solid wastes. Compact units are available for burning industrial wastes efficiently and economically without polluting the air.

The capacities of industrial incinerators range from 200 to 2000 lb per hour. Incinerators with less than 200-lb-per-hr capacity, in most cases, are considered domestic units. Units with higher capacities than 2000 lb per hr are usually special units or municipal-type incinerators.

In the industrial capacity range, there are two basic types of incinerators—the multichamber design and the newer controlled-air system. Both of these types are acceptable methods for incinerating solid refuse without emitting pollutants.

Multichamber Design

In the multichamber design, the refuse is burned on top of metal grates located in the main combustion chamber. Smoke and other combustion gases are diverted into a secondary chamber where temperatures of about 1400 F (maintained by auxiliary burners) complete the oxidation of smoke or unburned, volatized hydrocarbon gases. The oxidized gases or products of combustion (CO_2 and H_2O, etc.) are then routed through a scrubber where particles of flyash and dust entrained in the gas stream are removed.

The design of multichamber incinerators is such that large quantities of air are moved through the unit at substantial velocities. As a result, a lot of dust, smoke and flyash is entrained in the air stream as it moves from chamber to chamber. Baffles or impact walls and 90-degree turns reduce the velocities of the moving gases and help remove airborne pollutants.

If proper operating procedures are followed, the particulate emissions of this system can be reduced to less than 0.20 grains/standard cubic foot (scf). Lower emissions such as 0.10 grains are not easily attained unless special high-efficiency scrubbers are installed.

On major drawback of the multichamber design is the inability to control the combustion air at all times. In most units, the air is introduced into the combustion chamber below and above the grates. Air is drawn into the incinerator chambers by the stack effect and/or by an induced draft fan. The fan is usually needed to overcome the large pressure drop through the water scrubber. With or without the fan, the ideal fuel/air ratios for complete combustion cannot be maintained at all times.

Proper operation of a multichamber unit is dependent to a large degree on the competence of the operator. Since the combustion air is more or less manually controlled, the air port openings should be adjusted from time to time during the burning process. Also, the unit must not be overloaded, and it should be cleaned regularly to ensure proper movement of air through the unit. Feeding large chunks of highly flammable material, such as plastics, should be avoided because they tend to clog the unit and generate excessive smoke.

Controlled-Air Systems

In the starved-air or controlled-air incinerator, the design and engineering principles are quite different from those utilized in the multichamber unit. Combustion air is introduced into sealed chambers by blowers, making it possible to control the air required for combustion.

By introducing just the right amount of air in the combustion chamber, thus reducing considerably the amount of excess air, the combustion chamber is always hot and conducive to proper combustion or oxidation of the organic refuse. In addition, with the forced-air system, the air velocities are kept low so dust particles or flyash are not entrained in the moving air stream. Design of the afterburners is such that any smoke particles emanating from the combustion chamber will be oxidized in the secondary chamber where 2000 F temperatures are maintained.

Controlled-air incinerators attain the lowest possible particulate emissions. Tests have indicated that particulate emissions as low as 0.02 to 0.05 grains/scf are possible in units fitted with special high temperature, cyclonic-type afterburners. In fact, some incinerator manufacturers now guarantee emission limits to be less than 0.10 grains.

As a result of previous haphazard methods used to incinerate solid wastes, incineration has been labeled a major polluter. Modern incineration methods, to the contrary, do not contribute significantly to air pollution, and the possibility of their being outlawed is not likely.

ANALYZING THE COST OF
SOLID WASTE DISPOSAL [50]

Determining the most economical waste disposal method for a particular plant requires an analysis comparing all of the investment and operating costs of the alternative systems. To prepare such an analysis, a substantial amount of information is needed.

Some of the factors related to solid waste disposal for industrial installations are indicated in Table 20-2. These variables may be used as a checklist in evaluating specific waste disposal systems.

**Table 20-2. Factors Influencing the Economics of On-site Incineration
On Commercial, Industrial and Institutional Installations**

Factor	Estimated Range	Ideal Requirement For Incineration
1. Amount of Refuse	0 to 2000 lb/hr	500 lb/day and up
2. Type of Refuse	Type 0 to Type 6	Type 3 or less
3. Space Requirements	As needed	As needed
4. Compaction Ratio	3-6:1	3:1 or less
5. Haulaway Costs	$2.4/cu yd	$2.50/cu yd and up
6. Labor Requirements	0.1-0.4¢/lb	0.1¢/lb or less
7. Combustion Reduction	90-98 percent	95 percent and up
8. Gas Consumption	1-3 cu ft/lb	1 cu ft/lb or less
9. Electric Consumption	5-40 kwh/ton	20 kwh/ton or less
10. Maintenance Costs	0.1-0.2¢/lb	0.1¢/lb or less
11. Heat Recovery Amounts	0-60 percent	30 percent and up
12. Capital Investments	$20-$40 per hourly capacity	$30 or less per hourly capacity

A sample cost analysis (Figure 20-3) will illustrate how these variables are interrelated and how each contributes to the overall disposal costs. The estimated costs indicated in the analysis are typical, but it is suggested that prevailing local conditions and costs be used wherever possible.

Amount of Refuse Generated

In order to determine the capacity of the equipment needed to dispose of the wastes, the amount of refuse generated in a particular building must be known. Daily or hourly amount should be computed both in pounds and cubic yards. The reason for measuring the daily weight as well as the daily volume is that incineration is measured in pounds per hour, while haulaway is measured in cubic yards per day.

For example: If refuse is generated at a rate of 8000 lb/day, and the waste is to be incinerated during one 8-hr shift, it is clear that an incinerator with a capacity of 1000 lb/hr is needed. Assuming the waste weighs 160 lb/cu yd, cost of haulaway can be computed using a 50 cu yd/day rate.

SOLID WASTE DISPOSAL COST ANALYSIS

Basic Data

1. Type of wastes generated (circle one) 0 ① 2 3 other _____

2. Amount of wastes generated:
   a. By volume: __6__ cu yd/hr, __50__ cu yd/day, __350__ cu yd/wk, __1,500__ cu yd/mo
   b. By weight: __1,000__ lb/hr, __8,000__ lb/day, __56,000__ lb/wk, __240,000__ lb/mo

3. Plant hours of operation: __8__ hr/day, __7__ days/wk, __30__ days/mo

4. Hauling service requirements:
   a. Investment costs: $ __8,000__ (COMPACTOR-3:1 RATIO)
   b. Average charge for hauling service: $ __3.50__ /cu yd
   c. Electric consumption: (__5__ kwh/ton) x (__120__ ton/mo) x ($__0.02__ /kwh) = $ __12__ /mo
   d. Labor requirements: O
   e. Maintenance costs: (__0.03__ ¢/lb) x (__240,000__ lb/mo) = $ __72__ /mo

5. Incinerator requirements:
   a. Capacity required: __8,000__ lb/day
   b. Unit recommended: _____ Installed cost: $ __35,000__
   c. Capacity of unit: __1,000__ lb/hr   Daily operation: __8__ hr/day
   d. Electric consumption: __80__ kwh/day
   e. Gas consumption: __1,000__ cu ft/hr, __10__ therms/hr, __80__ therms/day
   f. Labor requirements (to charge and clean): __3__ hr/day
   g. Combustion reduction ratio: __5__ % of original volume
   h. Estimated Maintenance costs: __0.08__ ¢/lb of refuse

6. Other data:
   a. Unit cost of electricity: __2__ ¢/kwh
   b. Unit cost of natural gas: __7__ ¢/therm
   c. Average cost of labor: $ __3.50__ / hr

Operating Costs with Refuse Hauling Service

1. Hauling costs:
   a. Loose: (_____ cu yd/day) x (_____ day/mo) x ($_____ /cu yd) = $ _____ /mo
   b. Compacted: (__50/3__ cu yd/day) x (__30__ day/mo) x ($ __3.50__ /cu yd) = $ __1,750__ /mo

2. Other costs: (maint. #72 , elect. #12 , labor _____) = $ __84__ /mo

3. Total Costs: $ __1,834__ /mo

Operating Costs with Onsite Incineration System

1. Gas: (__80__ therms/day) x (__30__ days/mo) x (__7__ ¢/therm) = $ __168__ /mo

2. Electricity: (__80__ kwh/day) x (__30__ days/mo) x (__2__ ¢/kwh) = $ __48__ /mo

3. Labor: (__3__ hr/day) x (__30__ days/mo) x ($ __3.50__ /hr) = $ __315__ /mo

4. Maintenance: (__8,000__ lb/day) x (__30__ days/mo) x (__0.08__ ¢/lb) = $ __192__ /mo

5. Ash removal: (__1,500__ cu yd/mo) x (__5__ % ash) x ($ __2.00__ /cu yd) = $ __150__ /mo

6. Total costs: $ __873__ /mo

Additional Investment for Incineration

Incinerator cost   Hauling cost
($ __35,000__) – ($ __8,000__) = $ __27,000__

Monthly Savings for Incineration

Hauling service   Incinerator Operation
($ __1,834__) – ($ __873__) = $ __961__ /mo

Payback of Incineration System

Payback = (Additional investment for incineration) / (Monthly savings for incineration) = ($ 27,000) / ($ 961) = (__29__ months) = (__2.4__ years)

Figure 20-3. Solid waste disposal cost analysis.

Existing records of wastes generated and disposed are helpful in determining the overall amounts. If such records are not available, then a survey
must be made to determine these daily amounts.

Type of Refuse Generated

Choosing the appropriate type of equipment for disposing of the wastes
efficiently without polluting demands taking into account the type of refuse
generated. For instance, it is much more difficult to dispose of plastic wastes
and wet garbage than it is to dispose of paper wastes or rubbish. This is
especially true if incineration is being considered.

Very often it is not easy to classify rubbish accurately because the
mixture of the refuse varies from day to day. However, exactness is not
really necessary if most of the major constituents are known. Amounts of
paper, wood, cardboard, plastics, moisture content, etc., are most important,
and each should be estimated as a percentage of the total mixture. Existing
records of waste disposal are helpful in determining the amount and type
of refuse generated, but if existing records are not available, then a survey
should be made.

In the sample analysis, the type of refuse seems to fall between Type
0 and Type 1. Unit weight of the refuse is estimated at about 6 lb/cu ft
or about 160 lb/cu yd, and the heat content is assumed to be about 7500
Btu/lb.

Space Requirements

Of major importance in a waste disposal program are the space required
and the space available. Refuse must be collected and temporarily stored
until disposal time. There is great concern for health and safety in all waste
disposal areas.

If refuse is to be hauled away loose, it should be stored in containers
in order not to litter, be a fire hazard, attract vermin, create odors, or contribute to other nuisance and hygienic hazards. Adequate space should be
available to store the containers.

If refuse is to be hauled away compacted, additional space should be
allocated for compaction equipment. Also, provisions should be made for
the compacted refuse to be moved out of the plant. Deodorizers and drip
pans should be used to control odors and refuse "juices."

If refuse is to be incinerated, adequate space should be provided for
the equipment and for temporary refuse storage. Ideal stack and breeching
locations should be sought, as well as gas, electricity and water outlets.
Width of doorways and hallways should also be measured in order to determine if existing space will accommodate passage of incineration equipment
to area of installation. All of these factors tend to contribute considerably
to the installation costs of an incineration system.

In the sample analysis, it is assumed that outdoor space will be made
available for either a compactor or an incinerator.

Compaction Ratio

The main purpose of a refuse compactor is to reduce the volume of loose refuse to the lowest possible amount, thus reducing the haulaway costs by a corresponding ratio.

The ability of a compactor to reduce refuse volume is indicated by the "compaction ratio." Usual range of compaction ratio is between 3:1 and 5:1. Higher ratios are available from some compactor manufacturers at increased expense. In the sample analysis, a 3:1 compaction ratio is used.

Haulaway Costs

If the refuse is hauled away from the plant either loose or compacted, some operating costs will be incurred. Usually, local refuse removal contractors transport the refuse to secondary disposal process centers.

Rates charged by local scavengers range from $2 to $4 per cu yd, depending on the type and amount of refuse and local conditions. Lower costs are usually for dry and loose refuse; higher rates are for wet or compacted refuse. Rates are likely to increase as refuse amounts and costs of secondary waste disposal processes increase.

In the sample analysis, the haulaway costs are $3.50/cu yd for compacted refuse.

Labor Requirements

There is a considerable amount of labor required in collecting, storing, handling and disposing of solid wastes. The amount of labor needed depends largely on the type of refuse, type of treatment, amount of waste generated, degree of automation, and other variables intrinsic to a specific location or installation. Labor required to collect and store refuse is the same whether the wastes are incinerated or hauled away. In disposing of refuse, however, labor requirements vary with the different disposal methods.

If the refuse is to be hauled away, the labor requirements are related to collecting the waste, storing it in containers, and loading it into refuse trucks. The exact amount of labor in this method is not easily established.

If the refuse is to be hauled away compacted, the labor requirements for collection would be the same as for loose wastes. Containerizing and pickup labor is reduced, but labor needed to maintain the compactor equalizes the labor requirements of these two methods.

If the refuse is to be incinerated, labor is required for loading and maintaining the incinerator, in addition to collecting and storing the refuse. The exact amount of labor required in each method is not easily obtained, but a comparative cost analysis can be simplified by eliminating the labor requirements that are common to all methods. Only the additional labor requirements of a method should be noted.

In the above three methods, zero labor would be allowed for both haulaway methods and some labor would be used for incineration. The exact amount of labor required is based not only on the type and size of the incineration unit but also on the loading system and degree of automation. If it cannot be established from job conditions, a cost figure between 0.1¢ and 0.4¢/lb is sufficiently accurate. In the example, a 0.13¢/lb labor cost is used.

Combustion Reduction

Efficiency of an incinerator is measured not only by its hourly combustion capacity but also by its "combustion reduction" limits: that is, how thoroughly the combustibles are burned. This is determined by comparing the amount of ashes removed from the combustion chamber to the amount of combustibles placed in it. Combustion reduction ranges from 90 to 98 percent—an indication of the amount of ashes that must be hauled away. A 95 percent combustion reduction is used in the sample analysis.

Gas Consumption

The amount of gas required for incineration of solid wastes ranges from about 1 to 3 cu ft of gas per lb of refuse. Lower amounts are for dry refuse and controlled air incinerators, and the higher amounts are for wet refuse and multichamber design units.

In the example analysis, 1 cu ft of gas per lb of refuse is used. The unit cost of gas varies from 5¢ to 9¢ per therm (1 therm = 100.000 Btu = 100 cu ft natural gas), depending on local conditions. In the example, 7¢/therm is used.

Electric Consumption

The method and the degree of automation utilized determines the electric consumption. In "haulaway loose" systems, there is no electricity consumed, either directly or indirectly. In "haulaway compacted" systems, electricity is consumed by the electric motor operating the hydraulic ram or compactor. The exact amount cannot be readily obtained, unless watt-hour meters are installed on the system for a short period of time. Electric consumption has been estimated at 5-10 kwh/ton of refuse. The sample analysis uses 5 kwh/ton.

In on-site incineration systems, electricity is consumed to operate loading devices (if any), blowers, fans, pumps and other control devices. Here, too, the amount of electricity is not easily determined, but various published reports indicate electric consumption varies between 5 and 40 kwh/ton. In the sample analysis, 20 kwh/ton of refuse is used.

Unit cost of electricity varies from about ½¢ to 2½¢/kwh, and in the sample analysis 2¢/kwh is used.

Maintenance Costs

As with other mechanical equipment, waste disposal systems require adequate maintenance to perform properly. Amount of maintenance needed depends on the degree of automation, the amount of usage, and care or abuse of the system.

In the "haulaway loose" system, there is no maintenance involved since no equipment is used other than drums or containers.

In the "haulaway compacted" system, some maintenance is required for the compaction unit. Most of these units utilize hydraulic systems which are composed of electric motors, pumps, three-way valves, flexible pipes, switches, relays, etc. Since it is quite difficult to come up with an exact amount for maintenance costs, an estimated value of 0.03¢/lb, or 60¢/ton of refuse can be used. In the example, the total funds allocated for maintaining a compaction system are about $864/year (8000 lb/day x 360 days/year x $0.0003/lb).

In the past, not much emphasis was placed on maintaining waste disposal systems, especially incinerators. Now, however, waste disposal systems require adequate maintenance, not only for efficient operation but also for control of emissions.

In the "on-site incineration" system, adequate funds should be allocated for necessary maintenance, also. Some of the items that should be upgraded, repaired or routinely maintained are: the refractory bricks, grates (if any), stack, burners, controls, scrubber (if any), and loading devices (if any). Here, again, an estimated amount of 0.08¢/lb, or about $1.60/ton, can be used to cover almost all maintenance contingencies. In the sample analysis, a total amount of about $2000/year is allocated (8000 lb/day x 360 days/year x $0.0008/lb).

Heat Recovery

The amount of heat released by refuse depends on the particular type burned. Range is from 8500 Btu per lb for Type 0, to 6500 Btu per lb for Type 1, to 4300 Btu per lb for Type 2, etc. All of this heat energy, however, is not, and cannot be, recovered because the exhaust gases are neither very clean nor very hot. Efficiency of heat recovery in incinerators is not expected to be over 60 percent, and a more realistic amount is between 30 and 50 percent.

For instance, if 50 percent of the heat released by Type 1 refuse is recovered, this would represent about 3000 Btu/lb (6500 Btu/lb x 50 percent). If 10,000 lb of refuse are burned per day, the heat recovered would be about 30,000,000 Btu/day. This amount of heat could produce about 30,000 lb of steam or about 30,000 gal of hot water, equivalent to the energy of about 300 therms of natural gas. With a gas price of 7¢/therm, this represents a savings of about $20 per day or about $5000 to $7000 per year.

However, state-of-the-art for heat recovery from industrial refuse incinerators has not advanced substantially. Heat recovery is not considered in the sample analysis.

Capital Investments

In disposing of solid wastes, certain types of equipment are required to render the refuse safe and sterile. The cost of such equipment depends largely on the process employed and its degree of sophistication.

In the "haulaway loose" system, very minimal equipment is needed, such as metal or plastic cans, bags, or boxes. In the sample analysis, the capital investment for equipment is assumed to be zero.

In the "haulaway compacted" system, compaction units are needed to pack the refuse into metal containers. The costs of these compactors vary considerably but, in general, are related to the size of the unit's charging hopper and the degree of automation. In the smaller units, with hopper volumes of about 10 cu ft, the price is about $4000. In medium size units, with 20 cu ft hoppers, the costs are about $6000, and in larger units, with about 50 cu ft hoppers, the price is about $10,000. As indicated, the costs range from $20 to $40 per cu ft of hopper volume. In the analysis, a 1-cu yd compactor system is assumed, and the installed costs are estimated at about $8000.

In the "on-site incineration" system, the investment costs depend on the type and size of the units as well as the loading mechanisms and control devices. Costs of incineration units vary from $20 to $40/lb of hourly capacity. Lower prices are for manually loaded units with a minimum of pollution control devices. Higher costs reflect the prices of highly automated systems with modern pollution control devices and an automatic loading system. Cost of this unit is estimated at about $35/lb of hourly capacity, or about $35,000 total.

SAFELY HANDLING SOLID WASTES[187]

Solid wastes can produce many undesirable effects by biological, chemical, physical, mechanical, and even psychological means. Simple examinations reveal that pathogens (disease-causing microorganisms) in human waste provide biological threats; industrial wastes create chemical hazards; flammable materials have the potential danger of fires and/or explosion; and broken glass or other sharp-edged materials create mechanical hazards. These hazards, plus unsightliness, costs of waste disposal, and threats to property, contribute to the possibility of psychological disturbances.

To directly relate human disease to the generation, storage, or treatment of solid wastes is not a simple task because complete proof is necessary to make such an association. The harmful or potentially harmful agents must be found in the waste, shown to develop within or in close proximity to the wastes, and proven responsible for the disease.

Disease Transmission

Disease agents must have access to the body if they are to have an opportunity to cause illness; therefore, the agent and the victim must have a direct or indirect environmental association. Direct contact can occur in the handling of waste materials. Indirect contact can occur through transportation of disease agents to the victim by means of a biological vector, such as a fly, mosquito, flea, or rodent.

Any living agent which transports, directly or indirectly, a disease agent (bacteria, viruses, rickettsia, nematodes, protozoa, etc.) is termed a vector. The carrier (vector) may be either a true "host" of the disease or serve only in its transportation. Domestic, commensal, or wild animals which produce infectious solid wastes serve as links in a chain of infection, ultimately ending in man.

Flies have been shown to transport many diseases which can infect man. When an accumulation of fly breeding media is permitted, the potential for human infection via fly-borne pathogens is great. Flies breed in large numbers in human and animal excreta as well as in food wastes and sewage sludge. Many species are highly adaptable and breed in whatever medium is available.

It has been shown that flies can enter a garbage container through openings as small as one-eighth of an inch in diameter and deposit their eggs. Many of these eggs are carted away when the refuse is collected, but during warm weather large numbers of larvae migrate from the can before collection. To minimize fly breeding in garbage and other putrescible waste, two domestic collections per week should be instituted. Currently, many commercial and industrial refuse collection services use dump-type containers. Use of these containers offer many advantages. However, since disposal rates are in many cases based on container size, service companies tend to provide oversized units. These units require less frequent collection; however, they allow prolific fly breeding during the extended period.

A proper disposal facility will carefully handle refuse to prevent any further fly protection. One of the most important reasons for burying garbage and mixed refuse is to control fly breeding. Flies will emerge from as much as five feet of uncompacted cover over refuse. Yet, only six inches of compacted cover is sufficient to prevent their eventual emergence. Because of this, health authorities stress a thorough soil covering and compaction at the end of each day's landfill operation.

Recent investigations have shown fly larvae are destroyed by shredding refuse prior to landfilling. This accounts for a reported reduction in requirements for cover material. Since shredding also disperses potential fly breeding media, less proliferation has been demonstrated. Studies have also shown that flies will travel up to 20 miles between food sources and therefore are capable of functioning as vectors of human disease.

Some pathogenic fungi occur normally in the soil and will multiply through the nutritive effect of some wastes. Disturbance of infested soils in preparation for solid waste disposal provides the possibility of the fungi causing respiratory and other diseases in man. This can be done by the releasing of spores and dust for inhalation or by their contact with minor wounds. Therefore, sanitary landfill roads should be kept from drying up in warm weather, and landfill machine operators should wear masks or have air conditioned cabs.

Chemical Hazards

Solid wastes often contain flammables which can in the course of biological and chemical degradation, give off explosive, poisonous or asphysiative gases. These gases, as well as those normally produced by refuse in decomposition, can travel through the soil and create hazards. A normal waste product of anaerobic bacteria (found in landfilled refuse) is methane gas. This gas is colorless, odorless and lighter than air. If mixed with air in the proper proportion, it is highly explosive. Methane unable to escape vertically from a landfill will migrate laterally through porous subsoils and vent wherever a means of escape is available. This could be the basement of a home or plant. A pilot light or faulty switch activated by an unsuspecting person can cause devastating results. Many explosions directly attributed to refuse decomposition have been documented.

Even incineration of refuse at average temperatures of 1200-2000 F has been found to allow large numbers of microorganisms to pass through the process unharmed. The reason for this is that care has not been taken to make sure all of the refuse reaches the recorded temperatures. Resulting residue and quench water can contain organisms. Bacteria can even pass through air pollution control equipment including scrubbers, then directly out the stack. Dust around handling areas such as tipping floors, charging areas and residue storage areas can be prime sources of infectious agents when not properly controlled.

With the current emphasis on recycling of waste materials, solid waste handling projects should not be overlooked as processes containing inherent health hazards. Skin rashes and irritations can be traced to soiled clothing, while cuts and injuries from sharp cans and exploding aerosol cans are more dangers. Heavy gloves, overclothes and safety goggles should be used by workers handling refuse.

To avoid hazards associated with the handling of solid waste, personal hygiene and sanitary housekeeping are fundamental requirements, but it is most important that awareness of such hazards is made through a continuous safety program. Table 20-3 outlines supplemental equipment and precautionary measures which are recommended for a successful program.

SOLID WASTE HANDLING CONVEYORS[152]

A systematic engineering approach is as necessary for planning a solid waste material handling system as it is for any other process. Waste materials

Table 20-3. Precautionary Equipment and Measures for Solid Waste Handlers

Hazard	Sanitary Landfills	Treatment Facilities Incl: Incinerators, Balers, Shredders, Transfer Stations, Salvaging Depots	Domestic Storage Areas	Conveyor Systems	Collection Vehicles
Mechanical	Safety boots, uniforms, goggles and helmets, first aid kit	Face shields, helmet and goggles, gauntlets, uniforms, guard rails, first aid kit	Goggles, gloves, first aid kit	Face shields, helmet and goggles, gauntlets, uniforms, first aid kit, safety shoes	Uniforms, goggles, safety shoes, first aid kit
Fire	Fire retardant overclothes, stored cover material, water under pressure, venting of methane	CO_2 extinguishers, self-contained breathing equipment, pit water sprays	CO_2 extinguishers, water under pressure	Water sprays, scuba gear	Two-way radio equipment, hand racks, CO_2 extinguishers
Electrical	Not applicable	Rubber mats where required, proper grounding UL approved equipment	Not applicable	Proper equipment, grounding, UL approved equipment	Not applicable
Disease Agents	Personal hygiene, inoculations	Personal hygiene, inoculations, ventilation, masks	Personal hygiene	Personal hygiene, inoculations, ventilation, masks	Inoculations,
Chemical	Segregation, respirators, safety boots	Respirators, scuba gear w/face masks	Careful handling, notify collecting agency	Respirators, scuba gear, face masks	Special handling, segregation
V E C T O R S — Rodents	Continued extermination and baiting, covering	Continued extermination, routine baiting	Metal containers with tight covers	Not applicable	Not applicable
Flies	Fogging w/larvicide, 6" compacted cover	Fogging with larvicides, pest strips	Metal containers with tight covers or plastic bags	Not applicable	Routine washing of vehicles
Mosquitos	Proper drainage, remove occasional ponding	Not applicable	Remove occasional ponding	Not applicable	Not applicable
Noise	Not applicable	Ear covers	Not applicable	Ear covers	Not applicable

produced at various points within a plant must be accumulated and then arranged to follow paths of flow to the final point of disposal.

The type and volume of solid waste, the amount of accumulation required, the plant arrangement and size, and final method of disposal selected (compaction, incineration and/or product recovery) must be given consideration. In order to better understand the overall waste handling problem, a system can be broken down into *collection* and *process*. The collection segment can then be subdivided into *containerized* and *conveyorized* methods.

Why Use Conveyors

When a product is continuously produced by a process it is a likely candidate for handling on a conveyor. If the volume produced is large a conveyor may be mandatory. Similar reasoning can be applied to the solid waste products produced by a process.

The main purpose for using a conveyor for handling solid waste is to minimize the need for manpower and reduce space requirements. A waste product could be collected in cans, tote boxes, etc., and then manually moved to a central collection point. If the volume of solid waste material to be handled is large the numbers of containers required and the manpower involved can be extensive. If a conveyor could be used to collect and move the solid waste to the collection point, it would eliminate the need for manpower and could reduce the amount of space required for the operation.

Conveyors are applicable to the handling of most forms of industrial solid wastes. The problem is to determine when a conveyor should be used and what type of unit is most applicable to the circumstances involved.

When to Use Conveyors

A large percentage of industrial waste is of a uniform character. It is independent of the specific industry involved, and consists of shipping waste, plant trash, and office waste. Some industries have wastes peculiar to the industry involved, such as the food and chemical industries, metalworking or glass plants. Determination of the character and quantity of waste requires a detailed study of the individual application.

Assume that an overall analysis of a particular solid waste handling problem leads to the conclusion that some form of conveyor could be used to collect the waste product and deliver it to a selected process method. The two major steps then required to solve the solid waste conveying problems are: (a) analysis of the conveying job to be done and (2) evaluation and final selection of the equipment which will do the job most efficiently.

Analysis of the Conveying Job

The first step is to analyze the system requirements:

1. **Compile data on the material to be conveyed.** Maximum size to be handled, weight and other material characteristics need to be determined to select the conveying method. In the case of ferrous or nonferrous metal scrap, glass, etc., those characteristics are known and readily available. Other forms of waste are not so easily definable and Table 20-4 has been provided to help identify various types of solid waste with respect to their bulk density and moisture content.

Table 20-4. How to Identify Wastes

Type	Description	Lb Cu Ft	Moisture Content (Percent)
0	Trash, highly combustible paper, cardboard, wood boxes, sweepings; up to 10 percent plastics, rubber, etc.	8 to 10	10
1	Rubbish; combustible paper, cardboard, wood, sweepings, etc.; up to 20 percent food waste, but no plastics, rubber, etc.	8 to 10	25
2	Refuse; approx. even mixture of rubbish and garbage	15 to 20	50
3	Garbage; animal and vegetable food wastes	30 to 35	70
4	Pathological; human and animal remains	45 to 55	85
5&6	Special industrial, chemical wastes (survey required)	–	Varies
7	Magazines and packaged paper	35 to 50	Dry
8	Loose paper	5 to 7	Dry
9	Scrap wood and sawdust	12 to 15	Dry
10	Wood shavings	6 to 8	Dry
11	Wood sawdust	10 to 12	Dry

*Weights are general and based on material usually expected in refuse collection.

2. **Establish objectives for the conveying job**—set down volumes of material to be moved and conveying distance.

3. **Other factors** which determine the selection of conveyors for waste handling include:

Material Transfer—The manner in which material is loaded upon and discharged from a conveyor.

> *Loading*—The method by which the material is transferred or loaded upon the conveying medium. Method and severity of loading, such as impact or shock, are important factors in selection.
> *Discharge*—The method by which the material is transferred or discharged from the conveying medium.

Profile—The path over which the material handling must be done. When handling solid wastes, particular attention should be given to deviations from the horizontal travel.

> *Angle of Inclination*—The steepness of the slope up (or down) at which the material must be conveyed.
>
> *Complexity*—The variety of directions the material is to be conveyed (as compared to a straight line).
>
> *Horizontal Carry*—The distance the material must be conveyed horizontally. Equipment with longer centers requires stronger components and structures.
>
> *Vertical Lift*—The height the material must be moved vertically. The vertical-lift factor is often critical.

Material Properties—The "as handled" properties and characteristics of bulk materials that affect the selection of conveying equipment.

> *Material Size*—When handling solid waste material, special consideration should be given to maximum-sized item to be handled and the percentage of the large items in the total product to be handled. Unbroken cardboard boxes, cracked glass gallon-sized jars, or metal turnings which intertwine into bundles can be a most important factor in selecting a conveyor unit.
>
> *Flowability*—The ease with which a material will flow through a hopper opening or down a chute. In other words, flowability might be termed "broken solids viscosity." The angle of repose is normally an approximate measure of flowability for most bulk materials; however, the flowability of waste materials is not easily defined and may require a test program for final determination.

The following material characteristics should also be considered, since they determine the materials used in constructing the conveying equipment selected.

> *Temperature*—As a general rule, material temperatures under 150 to 200 F can be handled by conventional designs.
>
> *Corrosiveness*—Material in the 1 to 7 pH range may require special consideration.
>
> *Abrasiveness*—Usually defined in terms of MOHS scale. Material with a MOHS rating of 6 or above requires special consideration.

Evaluation and Selection of Conveying Equipment

A conveyor design suitable for handling waste products is generally chosen from the following:

- Belt conveyors
- Hinged metal belt conveyors
- Vibrating conveyors
- Drag chain conveyors
- Screw conveyors
- Pneumatic conveyors

Table 20-5 is a quick reference guide for selecting a conveyor to meet the conditions as determined in the basic analysis.

Table 20-5. Selection of Solid Waste Handling Conveyors

SELECTION OF SOLID WASTE HANDLING CONVEYORS	Material Characteristics													Profile					Type of Action			Typical Material & Conveyor Selections									
	Size					Flowability				Abrasiveness												Trash Type 0	Rubbish Type 1	Refuse Type 2	Wood shavings & sawdust	Glass cullet, bottles, etc.	Glass cullet, plate	Metal chips	Metal turnings bushy	Die cast scrap	Stamping scrap
Conveyor Type	Very fine	Fine	Granular	Lumpy	Irregular, stringy	Very free flowing	Free flowing	Avg. flowing	Sluggish	Non-abrasive	Abrasive	Very abrasive	Very sharp	Horizontal	Inclined-declined	Vertical	Horizontal & inclined	Horizontal & vertical	Carries material	Pushes material	Drags material										
Belt Conveyor	●	●	●	●	●	●	●	●	●	●	●	●		●	●		●		●			●	●	●	●	●	●				●
Hinged Metal Belt Conveyor		●	●	●	●		●	●	●	●	●	●	●	●	●		●		●			●	●	●				●	●	●	●
Vibrating Conveyor		●	●	●	●	●	●	●	●	●	●	●	●	●	●				●						●	●	●		●	●
Drag Chain Conveyor		●	●	●			●	●	●	●	●		●	●	●		●				●						●			
Screw Conveyor	●	●	●			●	●	●	●	●	●			●	●	●				●					●					
Pneumatic Conveyor	●	●				●	●			●				●	●	●	●	●		●					●					

Belt Conveyors

A belt conveyor is an endless rubber or treated fabric belt which carries the solid waste material directly upon it. Terminal pulleys are provided for changing direction, driving the belt and adjusting tensions. Rollers, troughing idlers or wooden or metal beds can be used to support the belt and its load.

One of the main problems with this type of conveyor is the confusing variety of cross-sections available. The inexperienced individual can easily choose an uneconomical and troublesome arrangement. Belt conveyors can be furnished in four basic designs (Figure 20-4). All of the cross-sections can be modified to suit individual applications.

Figure 20-4c represents a very suitable arrangement. First, it is economical. Second, there is no need for adjusting skirtboard rubber seals and no possibility of side spillage. Surprisingly, very little material will work under the belt, since the open spaces between rollers causes a downward pressure which tends to seal the belt to the slider plate portion of the design. Material that does work under the belt usually works its way clear. It falls on the return belt and can be collected by return belt plows.

Conveyors for solid waste having a high temperature (+250 F), or which contains sharp objects, require special designs. Carry-back of material can never be completely eliminated on belt conveyors; therefore units must be designed to operate under this condition. Belt scrapers, self-cleaning return and snub rollers and winged tail pulleys minimize the problems resulting from carry-back.

Standard designs suitable for solid waste handling can be provided by several manufacturers in widths from 12 through 48 in. These units can be operated at inclines up to 25 degrees.

Inclined Cleated Belt Conveyor

In those cases where inclines are greater than 25 deg, special belts with cleats fastened to the carrying surface can be used. Depending upon the type of solid waste, angles up to 60 degrees can be used. Spillage will usually occur at the tail and discharge points of the unit, and daily clean-up is a common requirement.

Hinged Metal Belt Conveyors

This unit consists of a series of overlapping metal pans mounted on chains which operate over terminal sprockets. The pans can be provided with side wings to form a metal trough. This unit can be provided with a horizontal section for loading, with the discharge section inclined up to 45 degrees. The unit can be designed to withstand heavy impact loading, such as is encountered in municipal transfer stations, feeder conveyors for shredders and in landfill projects.

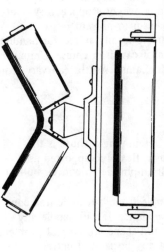

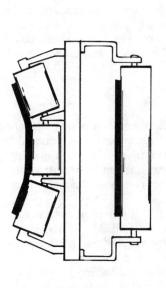

A. Conventional three-roll idlers are provided in widths from 18 through 72 in. They are commonly spaced at 4- to 5-ft intervals on channel or truss frames. These conveyors designs are normally used for high capacity installations. Continuous skirtboards can be provided to enable the conveyor to carry large pieces.

C. The combination trough slider and roller bed construction of this conveyor is ideally suited for handling waste material. Sideboards can be increased in height to handle large pieces.

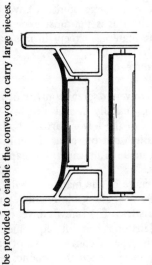

B. Two-roll carriers are generally spaced at 2½- to 5-ft intervals. Rolls are inclined at 35 deg, and can be supported on a formed bed to protect the return bed from spillage. Two-roll carriers provide medium duty service at moderate cost.

D. A one-piece formed, troughed slider bed is the lowest cost design of those shown. It is used for handling relatively light, nonabrasive waste materials.

Figure 20-4.

When handling typical solid waste products, the discharge should be carefully designed to eliminate possible material carry-back. If fine materials or wire particles are contained in the waste, this material can work its way into the unit's moving parts, causing wear and conveyor stoppage. A safety clutch can be incorporated in the drive mechanism to prevent possible damage to the equipment. Hinged metal belts are ideally suited for handling most types of metal scrap, with the exception of cast iron chips. Units can be furnished with 2½, 4, 6, 9 and 12-in. pitch chains. Widths can vary from 18 to 72 in.

Vibrating Conveyors

The form of mechanical conveyor most resembling a chute is the vibrating conveyor. This unit is a simple trough, flexibly supported and vibrated at relatively high frequency and small amplitude to convey bulk material or objects.

The vibrating conveyor is one of the simplest, most trouble-free conveyor types. It is limited to straight line movements. It will handle any material which is not sticky, and particle size and shape are no real problem. Vibrating conveyors can be arranged in almost any horizontal pattern, feeding into each other at any angle to collect waste material and deliver it to a central point. As a practical matter, most systems are kept under 80 ft in length. Multiple units which feed each other are utilized for longer lengths.

Vibrating conveyors are extremely adaptable for the handling of all types of metal chips and cullet resulting from the manufacture of plate glass. Because of the low maintenance, these conveyors also are an excellent choice for location in pits under presses and other types of process machines which have a low discharge point. Conveyor widths vary from 10 to 60 in. Unfortunately, this is the most expensive form of conveyor and cannot be used to elevate materials except in rare cases.

Screw Conveyors

A conveyor screw consists of a steel helix mounted on a shaft suspended in bearings, usually in a U-trough. As the shaft rotates, material is moved by the thrust of the lower part of the helix, and is discharged through openings in the trough bottom or at the end. Screw conveyors can be operated with the path inclined upwards, but capacity decreases rapidly as the inclination increases.

Because of the manner in which the material is moved in a screw conveyor, the size, allowable loading, and speed are controlled by material characteristics. The screw conveyor is among the simplest and most versatile types of material handling equipment—also the cheapest.

This type of conveyor is not easily adaptable to handling solid waste material, except when the product size is less than 4 in. Stringy material can wrap around the shaft and helix. Abrasive and corrosive material, which is in intimate contact with all the working parts of the conveyor, can cause wear. If not properly selected, this type of conveyor can be most unreliable, and have a high operating cost. Standard units can be furnished in widths from 6 through 24 in. Usually, for handling waste product, sizes of 12-24 in. are used.

Drag Chain Conveyors

This type of conveyor has an endless chain which drags materials in a trough or pan. Drag conveyors combine the features of easy accessibility to the moving parts, minimum-sized trenches, total containment, no extra wear due to fines, and maximum flexibility of conveyor layout.

Drag conveyors fall into three general categories: double chain over and under, single chain side pull and single chain en masse.

The over- and-under type of conveyor is preferred in applications requiring a conveyor to run in a straight line with or without an elevation at the discharge end. The side pull drag conveyor can be arranged in a loop and can pick up from more than one point. The unit can be installed in a very shallow trench, with access from removable floor plates.

Drag conveyors operate with very little wear when handling metal chips, when there is also coolant present. Even when coolant is not present, if designed properly, rate of wear can prove to be extremely small. Widths of these units can range from 12 in. through 5 or 6 ft.

Pneumatic Conveyors

There is a variety of pneumatic conveyor systems which have been designed to handle solid waste products. The most common applications are in the handling of paper and fiberboard waste products, and handling wood chips and sawdust waste. Some of these systems are over 50 years old.

In many industries where paper products such as bags, boxes, etc. are produced throughout the plant, a pneumatic conveying system can be provided with pickup chutes at various operating points. From these pickup points, the paper product passes through a positive displacement blower which provides the conveying air for the system, and shreds the conveyed material. The shredded paper stream then passes to a cyclone separator. The paper drops into a baler and/or incinerator and the air stream passes to a dust collector, where fines are recovered.

The units summarized here are those most commonly used for the handling of solid waste materials. Many other units are available, including enclosed en masse conveyors, harpoon conveyors, skip hoists, elevators, etc. Other types of units will undoubtedly be developed in the near future.

SOLID WASTE MANAGEMENT–
LEGAL ASPECTS[8]

Many types of solid waste management problems are coming before the courts. Solid waste management involves health, property rights, zoning, the constitutionality of ordinances, and many other areas of legal importance. The types of situations which are coming into our courts for adjudication and the principles upon which decisions are being based are illustrated by the following recent cases.

P will be used to indicate the **Plaintiff**. This is the party or parties who brings the action and claims that a law was violated, or he was injured. **D** will indicate the **Defendant**. This is the party or parties who is claimed to have caused the injury or to have acted illegally.

Case I

This was an action by **P**, a firm in the business of collecting garbage, against **D**, Township officials. The purpose of **P**'s action was to declare void an ordinance adopted by **D**, regulating and prohibiting garbage disposal areas in the Township.

A firm which we'll call Smith, Inc. had collected garbage and operated a disposal in the town for many years and was using the latest approved methods. A previous court order had required Smith, Inc. to use these methods. **D** adopted the ordinance in question, allowing Smith, Inc. to continue in business but prohibiting any other disposal operation in the Township.

P said that it was disposing of garbage by the same sanitary landfill methods as Smith, Inc. and that it was entitled to continue business in free competition. **P** said the ordinance discriminated and was unconstitutional.

The Trial Court dismissed the action and **P** appealed.

The Appeals Court said:

"The problem is less one of the need to [dispose of] garbage or even the competitive right to do so than it is the power of the local authorities to control it. Considered in such terms there can be only one answer, for the public health is closely involved and the police power imminent.

"The Legislature has put the right to regulate or prohibit the disposal of garbage into the hands of the Townships. . . . The ordinance in suit passed under the aegis of this Act, limits garbage disposal to Smith, Inc., presumably because it has been doing it so long and because it is acting under a Court order which effectively compels him to bury garbage by sanitary landfill or trench burial, the best and most modern method. . . . **P** urges that this is discriminatory and is a denial of the equal protection of the laws.

"We are . . . not dealing with a nuisance . . . neither the Act nor the Ordinance describes it so, and the Court below did not find that **P** were conducting their business as a nuisance. However, it is an activity that could rapidly become one and be a menace to health. Hence we cannot say that the Ordinance does not have a reasonable relationship to public health, morals, safety, and general welfare."

The Appeals Court affirmed the finding in favor of D.

It is important to note in the above case that the Township was given the right to regulate solid waste disposal by the Legislature. Also solid waste disposal is a matter involving public health and safety. An ordinance properly enacted in such an instance is presumed valid. It will usually be upheld in the Courts unless it can be proven that its enactment was arbitrary and un-reasonable or that it had no relationship to protecting public health and safety.

The following case illustrates a pollution situation where an individual was punished although he had no intent to violate the law or knowledge that it was being violated.

Case II

D, a landfill operator, was charged with violating a statute which pro-hibited the pollution of water by allowing deleterious or poisonous substances to enter streams in quantities sufficient to be destructive to wildlife. **D** claimed no knowledge of the pollution and no intent to pollute.

The Trial Court found **D** not guilty because there was no proof of guilty knowledge.

This was appealed and the Appeals Court said:

"The landfill contained large quantities of organic matter and the pollution was the most intense in the area closest to the landfill. . . . the landfill was a continuing operation and, from that continuing oper-ation and by the manner in which it was conducted, the pollution of the adjoining waters resulted.

"However, the lower court . . . concluded that there could not be a conviction because there was no proof of guilty knowledge.

* * *

"A wide variety of social and economic problems has impelled the adoption of numerous 'strict liability' penal statutes by both Congress and State Legislatures. Violations thereof result in penal sanctions regardless of moral culpability.

* * *

"Proof of guilty knowledge [in this case] is not a prerequisite to a finding of guilt . . . the decision of the [Trial Court] is reversed and I find **D** guilty."

D was fined $500.

Case III

D, a collector of discarded material and waste, was convicted of violation of an ordinance, of the city, **P**. Under the provisions of the ordinance, the City reserved to itself the exclusive right to collect garbage and refuse in the City. **D** said the ordinance was unconstitutional.

The ordinance, in part, read:

"Terms used in this Ordinance shall have the meaning herein given to them.

" 'Garbage' shall mean solid wastes from the preparation, cooking and dispensing of food and from the handling, storage and sale of food or food products.

" 'Refuse' shall mean all waste and discarded materials, including rubbish and debris, waste and discarded food, animal and vegetable matter, waste paper, cans, glass, ashes, night soil, offal and boxes.

" 'Waste' shall not include materials manufactured into by-products."

The evidence showed that the waste which **D** carried was inorganic and consisted of cardboard boxes, wood pallets and steel bands.

D had a permit issued by the State Public Service Commission which gave him a right to operate as a carrier of industrial and trade waste and refuse in the City and surrounding County. This was issued under an act to make the highways safe and regulate shippers.

On appeal, **D**'s conviction was upheld. The Appeals Court saying:

"[This State is] among the great majority which hold valid ordinances which give the governmental body itself the exclusive right or privilege of operating garbage or rubbish removal services.

"The mere fact that the particular refuse picked up and disposed of by the **D** may not have been injurious to the public health does not mean that the City could not reasonably decide that the control of the disposition of such material was necessary for the protection of the public health and sanitation. It is a matter of common knowledge that inorganic refuse is frequently mixed with organic refuse. The legislative body of the City could reasonably determine that the possibility of such mixtures renders it advisable that all refuse whether innocuous in itself or not be dealt with in a controlled and uniform manner. This is sufficient justification for the Ordinance regulating the disposition of inorganic as well as organic refuse.

"In addition, the City could reasonably find that the manner of disposing of such refuse was a matter of serious public concern, affecting the public health and well being and that the public interest required that such disposition be conducted under controlled conditions.

"It was necessary for one who desires to handle [garbage and refuse] . . . to obtain a permit in order to operate his vehicle upon the highways, but the permit so obtained does not give him the right to operate his vehicle in violation of other State and local laws."

The Ordinance was constitutional.

The above case again shows that it is difficult to prove that a properly enacted ordinance is invalid. It is, of course, necessary for the governmental unit to follow all steps required by law in enacting the ordinance such as giving publicity to the proposed ordinance, holding hearings or whatever else might be required. If some of the required steps are omitted, this could be grounds for a court's declaring the ordinance invalid.

Case IV

This was an action by property owners, **P**, to prevent the County, **D**, from interfering with a sanitary landfill operation which they conducted on their property. **P** further demanded that **D** issue necessary license and permits for the sanitary landfill.

D contends the sanitary landfill can be prohibited by a state statute covering nuisances; also that the operation is not in accordance with County Zoning laws.

The Trial Court found for **P** and this was upheld on appeal, the Appeals Court saying:

"In a statute regarding open dumps, . . . the legislature made sanitary landfills and incinerators an exception by express statement. Since the statute prohibits open dumps everywhere, it is a legislative recognition of the fact that a sanitary landfill is a different thing and is not objectionable. The use of the word 'sanitary' by the legislature in connection with the word landfill clearly indicates a legislative recognition that a landfill which includes garbage or other material also properly subject to incineration, is not a nuisance.

"By reason of the recent origin of the sanitary landfill method of disposal and by reason of the subsequent legislative recognition of the same as unobjectionable, we cannot conclude that the [State] nuisance statute . . . was or is intended to apply to a sanitary landfill.

"We are brought to the question whether the provision of the . . . County Zoning Ordinance forbidding the use of this particular property as a sanitary landfill for the disposal of waste including garbage are in that respect arbitrary and therefore unconstitutional and void.

* * *

"The issue is whether the ordinance bears any real and substantial relation to the public health, safety or welfare when it denies permission to operate a sanitary landfill upon this property and there has been refusal of a special permit."

The Appeals Court then discussed the area of the sanitary landfill which was on a plot of 187 acres surrounded mostly by farms. Portions of the testimony submitted in the case stated:

"On the basis of current population projections, it is estimated that 150 million cubic yards of refuse will have to be collected [in this area in the near future]. Existing incinerators could burn approximately one-third of this amount leaving a balance of 105 million cubic yards of refuse that must be accommodated at disposal sites. The volume of refuse would cover an area of 50 square miles to a depth of 2 feet."

The Appeals Court went on to say that the true test of the zoning regulation was whether it "bears some real substantial relation to public health, safety or welfare."

The Appeals Court said it did not and that the sanitary landfill was not objectionable from a zoning standpoint.

We see in the above case that the sheer magnitude of the problem of solid waste disposal is influencing courts. It is of course likely that statistical testimony such as this will be used more frequently. It should be noted that it was not shown that this sanitary landfill posed any threat to the environment.

CHAPTER 21

INCINERATION DISPOSAL OF SOLID WASTES

INCINERATOR EMISSIONS:
UNITS, CORRECTION AND CONVERSION[1 2 3]

In recent years, incinerator air pollution testing procedure and expression of results have been greatly standardized. However, to understand the need for such standardization requires a knowledge of conversion formulas and units of contamination concentration. The pollution engineer may find the need of comparing current results with those of older tests based on different sampling techniques.

Using the old nomenclature, the result of a contaminant emission test was referred to as a dustloading. The only emission sampled was particulate matter. The units of the dustloading were either pounds of dust per 1000 lb of flue gas or grains of dust per cu ft. Grains are a weight unit, where 7000 gr equals 1 lb. Present test procedures measure not only particulate emission but also condensible matter which is vaporized at flue gas temperatures. The condensibles are usually organic in nature and account for 40 to 60 percent of the total emission for a typical refuse consisting primarily of cellulose. Consequently, current test results are substantially higher than former results for a comparable installation. While present procedure is much more thorough, the units of emission rate are the same, and in fact the results are still sometimes erroneously referred to as a dustloading.

Pollutant concentration is first determined at experimental conditions of temperature and pressure. The required correction formula for standard conditions of 68 F and 29.92 in. Hg (ASME standards) is:

$$D_1 = D_0 \left[\left(\frac{t + 460}{68 + 460} \right) \left(\frac{29.92}{P} \right) \right]$$

where D_1 = contaminant level, gr/scf, and D_0 = contaminant level, gr/cu ft at sampling conditions of t degrees Fahrenheit and atmospheric pressure P in. Hg.

The formula to convert gr/scf to pounds of contaminant per thousand pounds of flue gas (lb/1000 lb of flue gas) is:

$$D_2 = D_1 \left[\frac{1544}{(M)\,(29.92)} \right]$$

where M is the molecular weight of the flue gas, and D_2 is the pollutant concentration in lb of contaminant per 1000 lb of flue gas.

Combining the two formulas enables the pollution engineer to proceed directly from gr/cu ft (D_0) at sampling conditions to lb contaminant/1000 lb flue gas.

$$D_2 = D_0 \left[\frac{3.12\,(t + 460)}{(M)\,(P)} \right]$$

The inverse of the above formula which is required to convert lb contaminant/1000 lb flue gas (D_2) to gr/cu ft (D_0) is:

$$D_0 = D_2 \left[\frac{(.320)\,(M)\,(P)}{(t + 460)} \right]$$

When M and P are not precisely known, the typical values of M = 29.5 and P = 29.92 in. Hg may be used.

All of these conversion formulas, while presented without derivation, merely represent a density conversion based on the universal gas law with the appropriate numerical factors for conversion of units.

Correction Formulas

There are two principal standards of correction; the 50 percent excess air correction and the 12 percent CO_2 correction. There is a need for a standard unit of correction. When incinerator refuse or any fuel is being oxidized, some minimum air supply is stoichiometrically required for combustion. For incineration processes, however, far more air is supplied than is actually required on a purely chemical reaction basis. The percentage of excess air is defined as:

$$P = 100 \left[\frac{\text{lb air supplied/lb ref.}}{\text{lb air needed/lb ref.}} - 1 \right]$$

where P = the percentage of excess air.

The more excess air which is admitted during the combustion process, the more diluted become the flue gases. Consequently the contaminant concentration is a function of the percentage of excess air. To eliminate this as a variable, the contaminant concentration is specified at 50 percent excess air. The accepted procedure for such a correction is:

$$D_C = D_U \left[\frac{\text{lb f.g./lb ref (actual)}}{\text{lb f.g./lb ref. (50\% e.a.)}} \right]$$

where D_C is the corrected contamination concentration, and D_U is the un-corrected contamination concentration. The units of D_C and D_U are arbitrary but must be the same for both.

Under certain assumptions the ratio lb flue gas/lb refuse (actual) $\div$ lb flue gas/lb refuse (at 50 percent excess air) may be reduced to a function of the Orsat volume percentages. The refuse being burned must be of the general form $C_X H_Y O_Z$. If elements other than carbon, hydrogen and oxygen are present in substantial quantities, the formula is no longer applicable. Furthermore, the formula is derived on a dry basis; that is, all moisture content has been neglected, whether resulting from oxidation of the refuse or from ambient air of combustion. This is not an uncommon assumption, since Orsat readings give dry volume percentages. Finally, the presence of CO has not been considered. Efficient incineration yields only trace amounts of CO; consequently ignoring this gas component does not introduce significant errors.

With these assumptions in mind, a practical formula for the contaminant concentration correction on a 50 percent excess air basis is:

$$D_C = D_U \left[\frac{CO_2 + .25(O_2) + 1.75}{2.88 - 13.9(O_2) - .136(CO_2)} \right]$$

Care must be taken to use only decimal percentages of O_2 and CO_2 in the formula.

The terms $0.25\ O_2$ and $0.136\ CO_2$ are very small and may be neglected without introducing an error over 2.5 percent. The simplified formula is then:

$$D_C = D_U \left[\frac{CO_2 + 1.75}{2.88 - 13.9(O_2)} \right]$$

A second standard of correction which is becoming universally accepted is the 12 percent CO_2 correction. When burned at 50 percent excess air, certain fuels create a flue gas which is 12 percent CO_2 by volume. Common refuse, which consists primarily of cellulose, does not fall into this category; the 12 percent CO_2 correction has been nonetheless retained for incinerator applications.

The formula for this standard of correction is very straightforward, which is one of the reasons it is so widely accepted:

$$D_C = D_U \left[\frac{.120}{CO_2} \right]$$

CO_2 is the decimal volume percentage of carbon dioxide.

T-6 Method

A simplified method of incinerator testing is the T-6 method. This approach is ideal for quickly obtaining approximate results under difficult

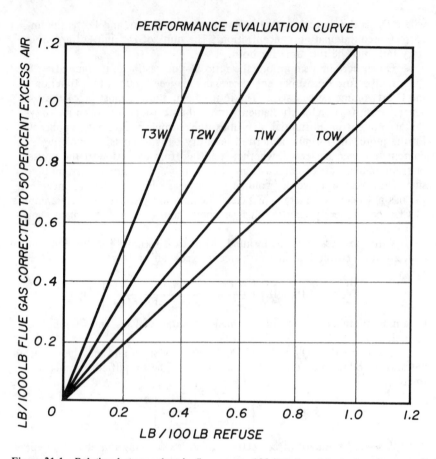

Figure 21-1. Relation between dust in flue gas per 100 lb refuse (abscissa) and corrected dust loading of flue gas (ordinate)

TOW = Type 0 waste   T2W = Type 2 waste
TIW = Type 1 waste   T3W = Type 3 waste

Test Result	Multiply by	Resultant Dustloading	Type Waste
Pounds of dust per 100 pounds of refuse	0.937	Pounds of dust per 1000 pounds of flue gas corrected to 50 percent excess air	0
	1.20		1
	1.71		2
	2.58		3

field conditions. Although the test procedure is not usually accepted by air pollution agencies, T-6 test results still occasionally appear in the literature. In its simplest form, the T-6 method measures only particulate matter and the results are a real dustloading figure. No Orsat readings are taken nor are any volume measurements of sampled gas made.

By measuring the weight of dust collected and monitoring the incinerator burning rate (lb/hr), the quantity lb dust/100 lb of refuse is known. By multiplying this value by an appropriate conversion factor, the dustloading in lb dust/1000 lb flue gas at 50 percent excess air may be determined. The illustration summarizes the conversion procedure.

SOLID WASTE INCINERATORS— DESIGN AND OPERATION[54]

Incinerators can be clean and economical if properly designed and operated. Smoke, odor, fumes and flyash are all potential air pollutants from an incinerator, and to avoid this pollution potential, certain design criteria must be employed.

Sizing Requirements

Determination of the theoretical minimum internal volume of the incinerator is a basic consideration. This volume of the incinerator is a basic consideration. This volume is a function of total heat released per hour from the burning of refuse. Table 21-1 provides average heating values of typical types of refuse. The internal volume requirements, excluding the

Table 21-1. Heating Values of Common Wastes

	Btu/lb
Type "O" Waste (cardboard, wood, paper, oily rags, 10% or less plastic)	8,500
Type "1" Waste (cardboard, wood, paper, foliage)	6,500
Polyethylene Scrap	21,407
Polystyrene Scrap	18,400
Phenol-Formaldehyde Scrap	15,100
Urea Formaldehyde Scrap	7,600
Wood or Sawdust	7,300 to 8,000
Asphalt Tar	17,100

ash pit, can be calculated using Formulas I and II. If auxiliary fuel (gas or oil) will be burned, its heating value should be added to the total heat release determined by Formula I.

Formula I: (lb refuse burned/hr) X (average heating value/lb) = total heat release Btu/hr

Formula II: $\dfrac{\text{(total heat release Btu/hr)}}{\text{(25,000 Btu/cu ft/hr)}}$ = minimum theoretical internal volume

Combustion Air

Combustion air in proper quantities is essential to complete burning and smokeless operation. To determine this quantity of air, first calculate the theoretical amount based on total heat release. Formula III shows this calculation which utilizes the total heat release from Formula I.

Formula III: Theoretical combustion air lb/min =

$$\frac{(7.5 \text{ lb dry air}) (\text{Total heat release Btu/hr})}{(60 \text{ min/hr}) (10,000 \text{ Btu})}$$

The theoretical combustion air represents the amount of air required for a stoichiometrically balanced reaction. Since this reaction is impossible to achieve in an incinerator, an excess of combustion air must be supplied, which is referred to as actual combustion air.

Actual combustion air is calculated by multiplying the theoretical air by an excess air factor of 2.0 to 4.0. The exact factor used depends upon the properties of the waste volatility, heating value, bulk density, etc. Generally, low heating value-low volatile waste has an excess air factor of 2.0. As heating values and volatility increase, so will this factor.

Formula IV: (2.0 to 4.0) (Theoretical combustion air lb/min) =

Actual combustion air lb/min

(Actual combustion air lb/min) (13.35 cu ft dry air/lb @ 70 F) = actual cfm

Combustion air within an incinerator undergoes two reactions. The first takes place with the refuse in the primary chamber to form a flue gas. Flue gas entering the mixing chamber usually contains incomplete products of combustion which require the secondary air reaction to complete the burning process. This gas is now considerably greater in volume than the combustion air for two reasons. The first increase is caused by the products of combustion which represent a 6 percent volume gain. Second increase is approximately a 300 percent volume gain, caused by thermal expansion as the gases heat to 1600 F in the mixing chamber.

Meeting demands of air pollution control emission laws requires that the flue gases flow at certain speeds within the unit. An adequate retention time is also necessary for good performance. Since the time function of flow within an incinerator is traditionally expressed in seconds, the flue gas must be converted from cu ft/min (cfm) to cu ft/sec (cfs) and at the same time corrected for volume changes, Formula V.

Formula V: Flue gas flow @ 1600 F in cfs = $\dfrac{(\text{actual cfm}) (1.06) (4.0)}{60 \text{ sec/min}}$

The flame port and mixing chamber should have a cross-sectional area that generates flue gas velocities of 25 to 35 fps (see Figure 21-2). Area under curtain wall should be large enough to slow the gases to 9 fps

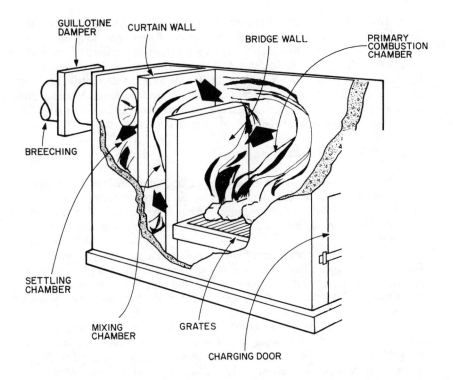

GUILLOTINE
DAMPER

CURTAIN WALL

BRIDGE WALL

PRIMARY
COMBUSTION
CHAMBER

BREECHING

SETTLING
CHAMBER

MIXING
CHAMBER

GRATES

CHARGING DOOR

**Figure 21-2. A cutaway portion of a typical in-line incinerator showing its major
components and the air pattern.**

or less. Finally, the cross sectional area of the settling chamber should allow
the gases to flow at a speed of 10 to 15 fps.

Total distance from the secondary burner to the breeching entrance,
measured along the path the gases will travel, should be of sufficient length
to require a travel time, or retention time, of 0.2 to 1.0 sec at the velocities
mentioned above. The exact retention time required depends upon the
nature of the material being burned.

Refuse that burns in the primary chamber with little or no smoke
emission needs about 0.2 sec retention time for odor elimination. However,
material which produces large amounts of smoke or fumes will probably
require as much as one second or more. Minor variations in these calculations
will exist, of course, and local air pollution codes should be consulted prior
to making extensive calculations.

Negative Pressure Problems

The actual combustion air requirement is the minimum amount which must be constantly brought into the incinerator room as makeup air. Negative pressure is a condition which causes a shortage of combustion air and results in many serious incinerator problems. This condition can be caused by a shortage of makeup air for combustion or by exhaust ventilation without provision for equivalent makeup air, which puts the entire building under a negative pressure. In extreme cases, a building may become so starved for air that a chimney may actually begin to reverse flow.

Pollution Control Devices

When purchasing a new incinerator or upgrading an existing system, accessories are an important consideration, the most important being air pollution control devices. Afterburners (secondary burners), flue gas washers and flue gas scrubbers are becoming necessities in many areas having stringent emission limits.

The function of secondary burners is to consume smoke, odor and fumes. This is accomplished in two ways: first, the flame itself consumes substances by direct contact, and secondly, the burner should raise the temperature of the flue gases to 1600 F or higher, which will destroy smoke with sufficient retention time.

A flue gas washer or scrubber's primary function is to remove particles from the gas stream. A general rule of thumb is that if emission limits are between 0.4 and 0.8 lb of dust per 1000 lb of flue gas (corrected to 50 percent excess air or 12 percent CO_2), then a gas washer should be used. If the limit is less than 0.4 lb, then a scrubber will probably be necessary. A simple distinction between a washer and a scrubber is the relative static pressure drop across the device.

Washers generally operate at a pressure drop of 1 in. water column or less, and have a good efficiency down to a particle size of 5 to 10 microns. A scrubber usually operates with a pressure drop of 4 in. W.C. or more, and has efficiency down to 1 micron in size. Some scrubbers with high pressure drops have good efficiencies on submicron size particles.

Cost

Cost of air pollution control appurtenances is an important consideration for any installation. Both initial and operating costs will vary greatly with different models and locales of operation. The following examples show some typical expenditures:

Example I	Secondary Burner	
Capital Investment	5,000,000 Btu/hr Gas Burner	$ 300
	Installation	200
		$ 500

Operating Cost   5 therms per hr at 8¢ therm = 40¢ per hr

Example II   **Secondary Burner**

Capital Investment	1,000,000 Btu/hr Gas Burner	$ 850
	High Power Flame, Installation	225
		$1075

Operating Cost   10 therms per hr at 8¢/therm = 80¢ per hr

NOTE: Costs for oil burners will be similar

Example III   **Washer**

Capital Investment	Flue Gas Washer for a 1000 lb/hr	
	Type "0" Incinerator	$1800
	Induced Draft Fan	1100
	Installation	250
		$3150

Operating Costs/hr	(Without Recirculation)	
	800 gal water per hr	$ 0.30
	Electricity	0.12
		$ 0.42/hr

Example IV   **Scrubber**

Capital Investment	Flue Gas Scrubber for a 1000 lb/hr	
	Type "0" Incinerator	$2750
	Induced Draft Fan	2000
	Installation	250
		$5000

Operating Costs/hr	(With Recirculation)	
	400 gal water per hr	$ 0.15
	Electricity	0.35
		$ 0.50/hr

Operators

Often, little consideration is given the selection of personnel assigned to operate an incinerator. The practice of giving the job to anyone who is available, or to the lowest paid laborer, is poor policy. A competent, conscientious person should be placed in charge on a permanent basis. Incinerators have much in common with production machinery—quality performance and low maintenance are directly related to the operator. As with other equipment, troubleshooting charts can be a valuable aid to the operator in the rapid correction of problems. A sample chart is provided in Table 21-2.

If all of the foregoing design criteria are met and discrimination used in the choice of operators, incineration need not be equated with pollution.

Table 21-2. Typical Incinerator Trouble Shooting

Problem	Cause	Remedy
Smoke & Odor	1) Retention time too short	1)(a) Reduce total amount of refuse fired per hour (b) Fire smaller amounts of refuse per charge (c) Enlarge unit
	2) Mixing chamber temperature too low	2) Install secondary burner or increase size or number of existing burners
	3) Total heat release per cubic foot is too high	3) See remedy 1 (a&c) above
	4) Refuse volatilized too rapidly	4)(a) Shut off primary burner (b) Increase overfire air
	5) Insufficient combustion air	5)(a) Open damper* fully (b) Increase air supply to the room (c) Increase air intakes in incinerator
	6) Improper location of combustion air entry	6)(a) Reduce underfire air (b) Overfire air must enter above refuse bed (c) Secondary air should be increased
	7) Incinerator, in general, not designed to burn the material being fired	7) Consult manufacturer for redesign information or consider a new unit of proper design
Flyash	1) Dirty unit interior	1) Ash must be removed regularly
	2) Velocity under curtain wall and in settling chamber is too high	2)(a) Reduce draft by partially closing damper* (b) Open adjustment on barometric control
	3) Unnecessarily heavy stoking of fuel bed	3) Heavy stoking shouldn't be necessary to burn capacity in a properly sized unit
	4) Agitation of ash bed by air jets or burner	4)(a) Raise burner (b) Use a washer or scrubber in conjunction with air jets
	5) Damper left open during cleanout	5) Damper* should always be fully closed during cleanout
	6) Miscellaneous unavoidable causes	6) Install a washer or scrubber
Smoke back into room	1) Negative pressure in room	1) Increase air supply to room
	2) Overfiring	2) Reduce amount of refuse burned per hour
	3) Low draft (in absence of room vacuum)	3)(a) Open damper* fully (b) Reduce restrictions in breeching and/or chimney (c) Clean out unit (d) Increase diameter of breeching and/or chimney (e) Eliminate downdraft by raising chimney height (f) Use induced draft fan in breeching

Table 21-2, continued

Problem	Cause	Remedy
Over-heating	1) See smoke causes 3 & 5	1) See smoke remedies 3 & 5
	2) See smoke-back causes 1-3	2) See smoke-back remedies 1-3
Grate burnout	1) Normal wear	1) Periodic replacement
	2) Insufficient underfire air	2) Increase underfire air supply
	3) Ash built up under grates	3) More frequent cleanout
	4) Material burning under grates	4) Install grates with finer openings or a hearth
Brick or refractory	1) Overheating	1) Reduce capacity or install higher temperature duty lining
	2) Mechanical abuse during stoking and cleanout	2) Operator must be instructed to be more careful
	3) Corrosion	3) Refrain from burning materials producing corrosive gases or install corrosion resistant lining

*Damper referred to is type shown in Figure 21-1, not a barometric damper control.

INCINERATION ECONOMICS[147]

Old ideas die hard, as witness a recent statement in a national consumer magazine: "Landfill will be the most popular solid waste disposal method because of incineration costs, not only for new construction, but for upgrading old units to meet clean air standards." Prior to this statement, inexpensive incinerators were being sold, reducing the cost of solid waste disposal hauling to landfill sites by as much as 90 percent. These same incinerators were below the permissible EPA air pollution levels by 50 percent and were well below the standards set by any state.

Yet this statement is still true if applied to conventional incineration based on high-draft design augmented by wet scrubbers, heavy-duty blowers, baffles, electrostatic precipitators, large-quantity natural gas combustion, and other features added on to get rid of smoke. Rather than considering the old conventional design, whose economies are unfavorable, take a close look at those based on the starved or controlled air design.

To compare the hauling of loose or compacted waste to landfill with other methods of disposal, pollution engineers must determine the total annual volume of solid waste disposal in cubic yards. One year's data will provide a sufficient period of time for comparison purposes. The cost for hauling compacted as well as loose waste should also be calculated. Since compaction generally reduces the volume of most Type 0 and 1 waste by a factor of 40 to 60 percent, there is a theoretical cost reduction of 40 to 60 percent. However, this figure is lessened by the cost of either renting or purchasing a compactor.

Allowance must also be made for the continuous increase in hauling costs due to increased labor and equipment costs, longer distances to sanitary landfills, congested traffic problems increasing time and cost of hauling, and the rise in operating and maintenance costs of the trucks. Thus, it is important to compare incineration with the hauling of both loose and compacted waste, putting all methods into proper perspective.

There are other costs associated with solid waste disposal. One is wages and fringe benefits for labor needed to handle the waste. This includes cutting up scrap and putting it into boxes, cleaning the collection area, and segregating materials that can be economically sold for salvage. Rarely will more than a small percentage of industrial-commercial-institutional waste be sold for more than the cost of segregation. Baling is a feasible solution to the solid waste problem when the waste is economically reclaimable and saleable.

Solid waste disposal costs also include the rental charge of depreciation expense of owning either a compactor, shredder, incinerator or other device. Table 21-3 presents a typical industrial analysis for a capital expenditure. It represents a good discipline for accounting purposes, and places all capital considerations on the same footing.

Table 21-3. Capital Appropriation Evaluation Sheet

| Appropriation Request No: | Date: |
| Equipment Description: | Plant: |

Item	Annual Costs Present	Annual Costs Proposed
A. Direct Labor		
B. Outside Contractors		
C. Indirect Labor		
D. Fringe Benefits		
E. Material Costs		
F. Maintenance Costs		
G. Utilities		
H. Supplies		
I. Total Operating Costs		
J. Equipment Cost Including Freight and Installation		
K. Estimated Life		
L. Depreciation (S/L Method = J ÷ K)		
M. Taxes and Insurance (2½% of J)		
N. Total Equipment Operation Costs (L+M)		
O. Total Annual Operation Costs (I+N)		
P. Annual Savings (Present - Proposed) (Line "O")	XXXX	$
Q. Payout Time (J ÷ P) = _____ Years _____ Months _____		

In one recent case, a major facility generated 12,000 cu yd/yr of waste. Hauling costs averaged $9600/yr. Estimated labor costs were $5000 for cutting up all corrugated boxes so they would fit into the hauling containers. This came to an overall total of $14,000 annually. Conventional incineration had already been banned by the city as violating the antismoke ordinance.

The owner then looked at a starved-air incinerator which met city, state, and federal standards, and found that he could have it installed, with a mechanized feeding system, for $21,345. This cut the disposal hauling bill from $14,000 to $480 per year. After further deducting $432 for gas to the afterburner, the net annual saving would be $13,568 before depreciation expenses—a saving that would recur every year.

Tax Benefits

Under present tax laws, there are significant tax benefits that should be factored into economic comparisons so the pollution engineer will get the most for his invested dollar. If a taxpayer elects to use the rapid amortization available for pollution control facilities (a smoke-free incinerator qualifies), he cannot also take the investment credit. He must choose one or the other.

Rapid Amortization

The law states that 60 month amortization, or 5 years, is allowed only for the portion of the cost of a facility attributable to the first 15 years of normal useful life. Many incinerators should fall in this category since the useful life of an incinerator is not expected to exceed 15 years. The law further states that the taxpayer has the right to start amortizing the eligible equipment within the first month that it's placed in use. He may also do so at the beginning of the taxable year following the date in which its installation is completed. Under this method, the asset costs are completely amortizable. In other words, there is no need to maintain a residual value. The asset can be depreciated to zero.

Investment Credit

Should the company use the investment credit, then the useful life established for depreciation purposes will be the same one used for investment credit. Assuming life is in excess of 7 years, the taxpayer is eligible for a 7 percent of total cost credit against his tax for the year in which the facility was put into operation. In addition to the investment credit, he is allowed to normal depreciation charge for the asset. However, under normal depreciation charges, a residual value must be maintained and the asset is never depreciated to zero. Only equipment, 50 percent of whose basis is attributable to value added within the U.S., qualifies for the investment credit.

PLANNING INCINERATION WITHOUT AIR POLLUTION[64]

Between the time that a community runs out of suitable land for the satisfactory operation of a sanitary landfill and the time when it can afford to construct a municipal or regional incinerator, on-site incineration inevitably plays an important role in the community's waste disposal operations.

On-site incineration, for industrial plants, stores, and other commercial buildings is attractive because of (1) the expense of hauling long distances to a municipal incinerator, (2) the high cost of scavenger services for removal of solid waste from the premises, and (3) more rapid disposal vector control purposes. It must be remembered, however, that there are several problems inherent in on-site incineration—air or water pollution control, and the disposal of noncombustibles and ashes.

Many current air pollution regulations specify that multiple-chamber units be installed to meet the particulate emission standard of 0.2 gr/cu ft or less. Some state and local air pollution control agencies are even passing regulations lowering the particulate emission limit to 0.1 gr/cu ft (Figures 21-3 and 21-4).

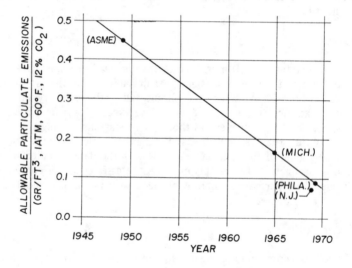

Figure 21-3.

Many new and existing incinerators may not meet these regulations without installation of some type of control device. The only air pollution control device which has proved to be economical and effective on small package-type incinerator units up to 1000 lb/hr capacity is the wet collector. Knock-out chambers and settling chambers have not solved the problem of smoke and flyash emissions.

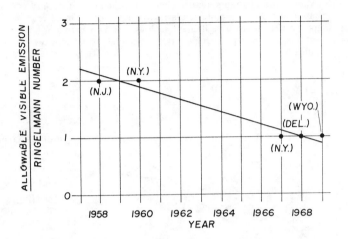

Figure 21-4.

Two types of wet collectors are used. The first is the spray chamber in which the gas velocity is reduced and water is added to speed up the settling and removal of particulate matter. Unfortunately, these units do not effectively remove enough particulate matter to meet the more stringent air pollution control regulations. To meet the lower emission standards it is necessary to use a more efficient scrubber such as a venturi. This high energy device will meet the new requirements, but has a high initial, operating and maintenance cost. For a small installation the scrubber may exceed the cost of the incinerator (Figure 21-5). For a 350 lb/hour or smaller unit, it may not be desirable to consider a scrubber. In the case of package incinerators, however, the convenience of on-site disposal will have to be considered in relation to other methods of disposal, *i.e.*, scavenger disposal or compaction.

Pre-planning is essential for designing an incinerator installation. Included in this plan should be an analysis of:

1. Methods of collecting and charging the refuse;
2. Space around the incinerator for charging, stoking, ash handling and general maintenance;
3. Adequate air supply to the incinerator room at the stoking and charging levels;
4. Effects of air-conditioning equipment, ventilating fans, etc., upon air supply or the draft available from the draft-producing equipment;
5. Adequate draft (negative pressure) to assure safe operation and complete combustion at reasonable temperatures (draft-producing equipment should be capable of handling all theoretical and excess air requirements);
6. Location of the chimney top or stack with respect to ventilation intakes and penthouse or other obstructions;

7. Examination of the immediate environment, and local, county, state and federal air pollution laws, to determine the need for appropriate auxiliary equipment such as flyash collectors or washers, pyrometers, burners, draft gages, and smoke-density indicators.

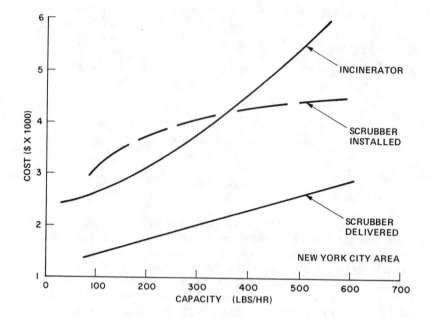

Figure 21-5.

SLUDGE INCINERATION[79]

For years the classical approach to sludge disposal has been dumping. This practice is rapidly becoming impossible because of high land cost, inaccessibility of sites to industry, odor, vermin and unsightliness. Sludge is a liability to waste treatment processes as there is no way to handle it at a profit. Treatment of a disposal sludge from an industrial plant makes up 25 to 50 percent of the total wastewater treatment cost.

Incineration of sludges is usually more expensive than land dumping due to both capital and operating costs. Also, supplemental fuel is required. Incineration of sludge must be relatively complete to reduce the ash volume, and controlled to prevent air pollution.

Six methods of sludge incineration are:

1. Multiple hearth furnace
2. Flash-drying and incineration system
3. Cyclone type

4. Rotary kiln type
5. Fluid bed reactor
6. Wet air oxidation

Some principles are common to almost all of these systems. First, the heat value of a sludge may be estimated from its ultimate analysis. The Du Long formula is:

$$Q = 14,600\ C + 62,000\ (H - O/8)$$

When   Q = Btu/lb of dried sludge
                C = Carbon, percent
                H = Hydrogen, percent
                O = Oxygen, percent

This calculated value must be reduced if inorganic coagulation is used in the thickening process, because the chemicals are inert and lower the heating value. Further, if hydroxides are present, the value is reduced again by their heats of dehydration. The same is true for calcium carbonate which decomposes endothermically to calcium oxide. Figure 21-6 shows the auxiliary fuel costs versus percent volatiles in the sludge to be burned.

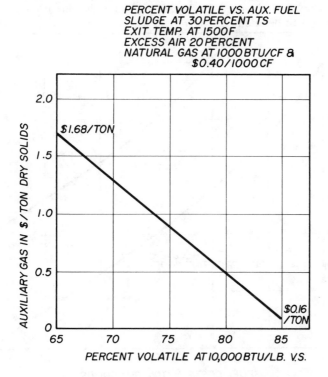

Figure 21-6. Cost of auxiliary fuel vs the percentage of sludge volatiles to be burned.

Two factors affect the amount of auxiliary fuel required for incineration: the heat value of the sludge, and the temperature required for complete, odorless combustion. The temperature, for normal residence times, is generally accepted to be a minimum of approximately 1400 F. It has also been determined that to assure complete combustion in five of these processes, excess air of 50 percent minimum over the stoichiometric value is required. Auxiliary firing or air preheating may be eliminated only under the rare combination of circumstances where: (1) the excess combustion air may be maintained at the minimum; (2) all inorganics have been eliminated; (3) moisture content of the cake is exceedingly low; and (4) volatiles in the sludge are in excess of 70 percent.

The effect of air preheating on fuels cost is shown in Figure 21-7. The effect of excess air is shown in Figure 21-8. Figures 21-6, 21-7 and 21-8 are based on a fluid solids incinerator, but are indicative for other methods as well.

Most sludges will not combust autogenously because of their water content. The effect of total solids on auxiliary fuel requirements is shown in Figure 21-9, which is average vendor data. The combustion temperature versus total solids with and without air preheat is also shown in Figure 21-10. The data emphasize the need for dewatering and preheating if odorless combustion is to be attained.

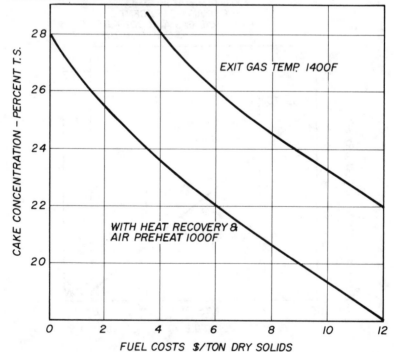

Figure 21-7. Effect of preheating on fuel costs.

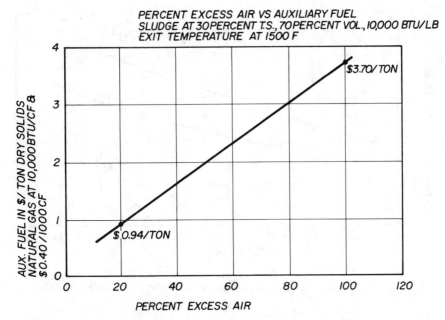

Figure 21-8. Impact of excess air on auxiliary fuel in sludge combustion.

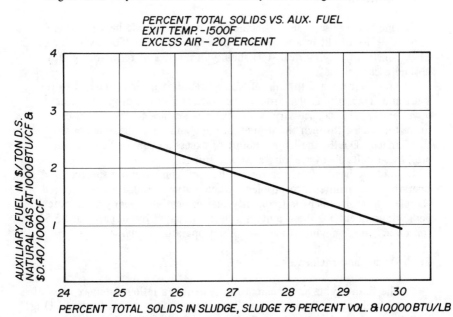

Figure 21-9. Effect of moisture content on cost of sludge combustion.

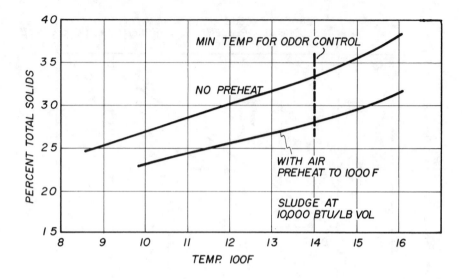

Figure 21-10. Effect of combustion temperature **vs. the percent of total solids.**

Multiple Hearth

One widely used type of incinerator is the multiple hearth. These units have been built in a wide range of sizes; *i.e.,* 500-2500 lb/hr dry solids. They consist of a cylindrical steel shell with a series of refractory grates around a central shaft.

In a multiple hearth unit, air-cooled rabble arms are attached to the shaft and scrape the sludge around the grate and into a drop to the next lower grate. This is necessary to expose new surface to the hot gases and move the sludge through the drying, burning and air preheating zones of the furnace. Due to the large amount of excess air required for cooling, etc., these units always require auxiliary firing.

Multiple hearth furnaces are not heat efficient and are subject to operating and maintenance problems when a sticky sludge is encountered. Rabble arm replacement is frequently due to heat and wear on the outer ends. Investment for these units has been evaluated by McLaren at $5-$10/ ton on a 25-year payment, and $4-$7/ton operating costs.

Flash Drying and Incineration

The flash drying and incineration system is a rather complex mass of pneumatic piping, cyclones, cage mill, conveyors, fans, and a furnace. Originally, units were designed to dry sewage for fertilizer and burn only the excess. Their chief advantage, therefore, is linked to the demand for this type of soil conditioner. They are not competitive with other furnace units for incineration alone.

Cyclonic Reactor

The cyclonic reactor is designed for small capacities (500 lb/hr or less). High-velocity preheated air is introduced tangentially into a cylindrical combustion chamber maintaining intensely heated refractory surfaces. The sludge feed is sprayed radially toward the hot walls where it is caught up so rapidly by the hot cyclonic gases that it is essentially combusted before it reaches the refractory walls. The sludge spray is fed by a progressive cavity-type pump. The ash is carried out with the exit flue gases. Sludge detention time in the reactor is less than 10 seconds. Operating temperature is above 1400 F for odor control. These units occupy a space approximately 4 ft x 6 ft x 6 ft and weigh 2 tons, or less. Cost of a 200 lb/hr unit is $120,000, exclusive of dewatering.

Rotary Kiln

The rotary kiln incineration unit consists of a waste feeder, a sludge feeder, a rotary kiln, a burner package, an afterburner, combustion air inlet, ash bin, effluent scrubber, and draft fan (Figure 21-11). The units are

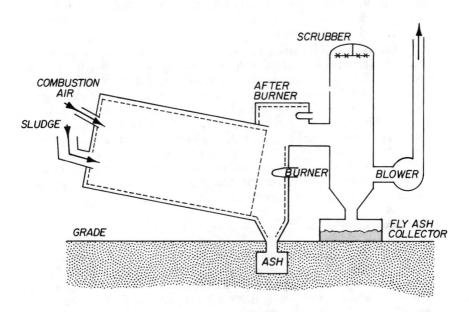

Figure 21-11. Rotary tumbler burner system.

available in 12 size modules with sludge burning capacities of approximately 40 lb/hr to 2400 lb/hr. Dimensions of the incineration package are 15 to 34 ft long by 5 to 14 ft wide by 5 to 15 ft high. The tumble burner has a 75 deg "tumble center" rotating waste at 6 in./sec.

Investment is relatively low, ranging from $1000/lb/hr of feed for small sizes, down to $500/lb/hr of feed. The kiln does not need chains, knockers, or other standard kiln anti-stick equipment. One unit is reported to have operated 1½ years with no maintenance problems. This type unit is designed for continuous operation.

For the majority of sludges in the 500-2400 lb/hr feed range, this unit should be investigated for competitive price. The flyash system appears adequate for air pollution. This leaves the liquid disposal problem of the flyash collector water, which is common to all incinerators.

Fluid Bed

The fluid bed incinerator system uses an air blower, an upright refractory-lined cylinder with a grid plate at the bottom, a burner in the side and a flyash scrubber on the effluent gases. The bed is composed of gradiated silica sand which is heated by the preheat burner and fluidized by the combustion air. Influent air is preheated by the exit gases. Sludge is fed into the fluidized mass which is maintained at approximately 1400-1600 F. Many sludges do not require supplementary heat and volatile ones may not require air preheating. Published operating data show that only 20 percent excess air is required for combustion.

The process advantages are: (1) excellent mixing of air, sludge, and hot sand, which reduces excess air requirements; (2) no moving parts in the reactor; (3) the feed nozzle is the only equipment in contact with high temperature; (4) operation is at 0-2 psig; (5) there are no heat exchange surfaces to scale; (6) ash removal is by the exit gas; (7) odor control is automatic at bed temperature; and (8) this unit may be batch-operated with assurance of combusting the residue after feed shutoff and of a short reheat time, due to the large heat sink afforded by the sand bed.

The disadvantages are flyash separation and disposal. Investment cost is higher than the other true combustion processes. Actual data on operation have placed cost of incineration at $25.32/ton dry solids.

Wet Oxidation

The wet oxidation process (Figure 21-12) is not strictly incineration, but is oxidation in the presence of water at 250 to 700 F. This usually requires steam injection. Operating pressure is between 150 and 300 psig. The advantage of this process is that no dewatering is required. Sludges having 3 to 6 percent solids have been treated by this process. The degree of oxidation achieved depends, in addition to temperature and pressure, upon the solids content of the entering sludge and on the retention time.

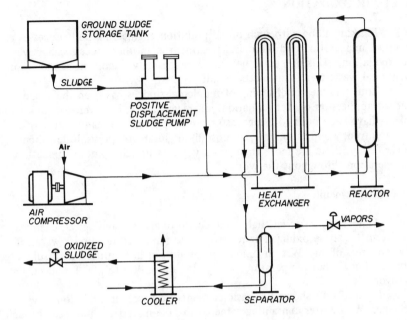

Figure 21-12. Wet oxidation system.

The oxidized solids are equivalent to those from incineration—sterile and separable by settling, centrifugation, or vacuum filtration. The liquid layer generally contains dissolved organics and requires treatment. The cost of sludge disposal by this process is high. It is the one process, however, that has no flyash problem.

Flyash Handling

Assuming the combustion temperatures have been adequate (1400-1600 F) during incineration, odors should have been destroyed. However, air pollution from particulates (flyash) in the exit gases will occur unless there is further treatment. Two means of reducing particulate emission are mechanical collection and wet scrubbers.

Wet scrubbers are generally employed in controlling flyash because they are cheaper than bag filters or precipitators. Scrubbers must be well designed internally to prevent excessive emission of water vapor. Also, the effluent waters which contain the flyash must be concentrated. Generally, due to its high settling rate, a 2-hr retention period will produce 99.9 percent settling of the ash. The overflow water is recirculated and the ash concentrate is withdrawn.

WET AIR OXIDATION[75]

Ranking along with such broad pollution engineering unit processes as thermal incineration and biological treatment is another process called wet air oxidation. This process is becoming increasingly important as more conventional waste treatment methods fail.

Thus far, the conditioning of municipal sludges has been the single largest application of wet oxidation. In recent years, however, wet oxidation plants have been built to treat wastes resulting from the manufacture of acrylonitrile, wood pulp, glue, monosodium glutamate, polysulfide rubber, etc. Today, there are more than seventy plants utilizing wet air oxidation in operation throughout the world.

Operating Principle

Almost any material that is subject to oxidation—biological, chemical, or thermal—can be oxidized at relatively low temperatures, under water, in the presence of air. Wet air oxidation should not be confused with submerged combustion in that temperatures rarely rise above 400 to 600 F and no flame is involved.

Figure 21-13 shows the basic elements of a wet air oxidation flow scheme. Wastewater containing some oxidizable material is pumped through

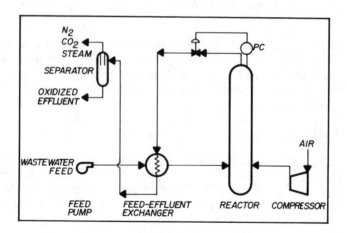

Figure 21-13. Wet air oxidation basic flow scheme.

an exchanger. There the temperature is increased to a point at which the reaction between the oxidizable material and the air will proceed autogenously. The temperature of the waste is further increased to the desired reaction temperature by the heat of combustion, which is released as the oxidation takes place. Heat is economized by exchanging the reactor effluent with the incoming waste stream. Spent air, carbon dioxide and steam are then separated from the oxidized effluent in a pot separator.

System pressure is controlled to maintain the reaction temperature as changes occur in feed characteristics, *i.e.,* organic content, heat value, temperature. The mass of water in which the reaction takes place serves as a heat sink. This prevents a runaway reaction in the event that a slug of concentrated organics is fed into the system.

Biological Sludge Treatment

Treatment and disposal of sludges resulting from biological processes such as trickling filters or activated sludge units can be difficult and costly. Such sludges typically contain 95 to 99 percent water and frequently represent as much as 1 to 5 percent of the total volume of influent to the biological treatment plant. Dewatering of sludges to decrease the volume and increase the solids content to a reasonable level is made difficult by the gelatinous nature of the material. In fact, dewatering processes such as vacuum filtration or centrifugation are almost useless without some prior treatment of the sludge.

One of the more common sludge pretreatments is that of chemical conditioning; that is, the addition of massive doses of lime, ferric chloride and polymers to break down the gelatinous structure. Wet air oxidation and thermal conditioning are competitive processes which eliminate the need for costly chemical conditioning.

In low pressure oxidation (LPO), sludge is reacted with air at about 350 F at 500 psig. Only a 5 to 10 percent reduction of COD takes place. However, physical characteristics of the sludge are completely changed. The vacuum filter area required to dewater a wet oxidized sludge is roughly 10 to 20 percent that required to dewater a chemically conditioned sludge.

Complete disposal of sludges can be accomplished by high pressure oxidation (HPO). In this case, the reaction temperature and quantity of air is increased. COD reductions of 80 to 90 percent and more can be accomplished with a corresponding 95 to 99 percent reduction in the volume of the dewatered sludge. The solid material which remains is essentially an ash, which can be readily land-filled.

Table 21-4 indicates the relative economics of chemical conditioning, low-pressure oxidation and high-pressure oxidation for a hypothetical sludge-handling facility processing 5 tpd (dry basis) of waste activated sludge. A credit has been taken for the nutrients which could be recycled to a nutrient-deficient industrial biological plant.

Direct Oxidation of Industrial Wastes

Wastewater streams that can be conveniently treated by some other means are generally not considered good candidates for wet air oxidation. It is an inherently expensive process in terms of initial capital investment.

Table 21-4. Comparative Economics of Complete Sludge Handling Facility
Chemical Conditioning vs. Wet Air Oxidation

	Chemical Conditioning	Low Oxidation	High Oxidation
Equipment Included	Settler Filter—420 sq ft Incinerator	Flotation, Thickener Filter—54 sq ft (LPO) Unit Incinerator	Flotation, Thickener (HPO) Unit
Installed Cost	$320,000	$503,000	$775,000
Operating Costs[a]	105,000/yr	38,000/yr	37,000/yr
Amortization Cost[b]	33,000/yr	51,000/yr	79,000/yr
Total Annual Cost	$138,000/yr	$ 89,000/yr	$106,000/yr

[a]Includes labor, maintenance, utilities and chemical

[b]8 percent money for 20 years

It can rarely compete with deep well injection. An economic comparison with biological treatment will favor wet oxidation only when the organic content of the wastewater is extremely high (greater than 10,000 mg/l BOD), the waste is toxic or marginally biodegradable, or the real estate required for bio-treatment is prohibitively expensive.

Wet air oxidation will compare favorably with incineration only when the incinerator would require elaborate air pollution devices or the volumetric throughput is so high that fuel costs become a factor. Theoretically, a wet oxidation system can be made thermally self-sustaining in an organic concentration of about 0.5 percent, whereas thermal incineration requires 25 percent or greater organic content in a waste stream. Figure 21-14 is a plot of energy input required versus organic content of the wastewater feed for thermal incineration without heat recovery and for wet air oxidation.

Energy values contained in the wastewater stream can be conveniently and economically recovered through wet air oxidation. This can be done by using the reactor effluent to raise steam or to generate power (Figure 21-15). Both schemes have been successfully proven in full-scale commercial applications.

The inherent nature of wet air oxidation minimizes any air pollution problems. Since the oxidation takes place in the aqueous phase, the usual troublemakers such as sulfur, phosphorus and chlorinated hydrocarbons are simply oxidized to their respective acids. Neutralization or cooling of the effluent stream prior to pressure let-down and discharge to the atmosphere almost completely eliminates any volatile acid components. If necessary, the small amounts of unoxidized light hydrocarbons contained in the off-gas can be removed from the vent gas by means of a low-temperature catalytic vapor oxidation unit. Nitrated compounds fortunately do not form

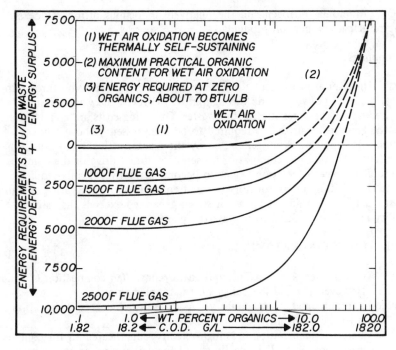

Figure 21-14. Energy requirements vs. organic content, and thermal
oxidation vs. wet air oxidation.

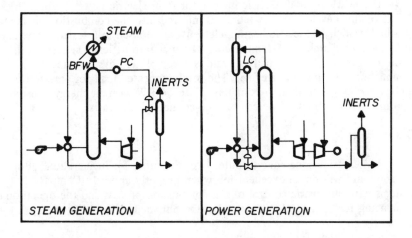

Figure 21-15. Systems by which reactor effluent can generate steam or electric power.

NO_x. Ammonia, nitrogen and the relatively innocuous nitrous oxide are the principal combustion products.

Economics

The capital investment for wet air oxidation is dictated by three principal factors: (1) volumetric throughput, (2) COD of the wastewater and (3) oxidation characteristics of the waste. Operating costs are greatly affected by the amount of air that is required to be compressed. Above 500 compressor horsepower, steam generation by the reactor effluent can usually be justified. Above 1500 compressor horsepower, direct drive of the compressor and/or power generation begins to become attractive. Plants requiring more than 2500 compressor horsepower frequently show a negative operating cost; that is, the value of power or steam generated exceeds the value of utility and labor required to operate the plant.

ELECTRIC INCINERATION[190]

Auxiliary heat is normally required for controlled combustion of solid waste. However, gas and low-sulfur oil for supplemental energy may be scarce in certain areas of the country. As a result, all-electric incineration has come to the forefront.

All-electric incineration uses high-velocity, superheated air for combustion of waste products. It is ideally suited for destruction of compact and dense waste, such as confidential documents, ledgers and blueprints. In most cases, however, electric incinerators still require the use of a wet scrubber to comply with local air pollution codes.

The heart of an all-electric incinerator is an electric power burner. By means of this device, superheated air is injected into the combustion chamber at a high velocity. Electric heating elements are housed in a sealed refractory-lined plenum chamber. A remote control panel automatically sequences the entire operation. Ignition of the waste is accomplished by spontaneous combustion after a preheat cycle. The total connected load for an electric incinerator ranges from approximately 22 to 84 KW with an energy consumption in KWH of about 65 percent of the connected kilowatts.

Gas vs. Electric

Why consider one energy source over the other for incineration? Both are good and will do an excellent job when properly applied. Comparable systems with automatic control of the combustion process and the operational sequencing for control of emissions may be obtained using either energy source.

Reasons then for using electricity are:

1. Electric power is readily available in many parts of the country where natural gas is not.

2. In certain areas gas for industrial or commercial use is on a temporary demand basis, and can be terminated with very little warning from the utility.
3. In an all-electric building, the cost for bringing in a gas line and meter are prohibitive to operate just an incinerator.

Economics

The cost picture for using one energy source over another is complicated. However, the installed cost of an all-electric incinerator is generally in the neighborhood of a 10 to 15 percent premium over an equivalent gas-fired system. Cost of operation of either type of incinerator will vary widely with location and prevailing energy rates. On an average, however, cost of operating an electric system could be half that for a gas-fired system. This is due primarily to the fact that considerably less Btu/hr of auxiliary heat is required for an electric system because of the more efficient utilization of uniformly distributed high-velocity superheated air.

Other factors which can offset the lower installation cost of a gas-fired system are insurance regulations requiring the use of FM or FIA safety controls for gas burners. When plans are being formulated for processing solid waste, it is well worth the time to closely examine energy rates and consider all-electric incineration as an acceptable method of on-site disposal.

FLUIDIZED BED REACTORS[46]

Fluid bed reactors have been used in many process applications such as catalytic cracking of petroleum, drying and roasting. Now, the fluid bed process for pollution control is becoming increasingly important with tightening regulations and must be considered in controlling processes emitting solid waste, air, and water pollution.

When a gas passes upward through a bed of finely divided solid particles, four general class conditions may exist, depending upon characteristics of the particles, geometry of the bed, and velocity of the gas. These situations are: (1) fixed bed, (2) dense-phase fluidization, (3) two-phase fluidization and (4) pneumatic transport.

In a fixed bed at low velocity, the passage of gas up through the bed causes a relatively low pressure drop. The solid particles are essentially undisturbed and the gas merely percolates through the void spaces in the bed. Particles remain in place in the fixed bed as long as the pressure drop is less than the weight per unit area of the bed (Figure 21-16). Further increase of the gas velocity causes the bed to expand slightly and the particles gain some freedom to move.

At a flow rate indicated by G_{mf}, the pressure drop becomes equal to the weight of the bed per unit area and the particles are suspended in the gas stream. At this point the bed is said to be in the *fluidized* state. This gas-solid system has liquid-like characteristics. Intense and violent mixing

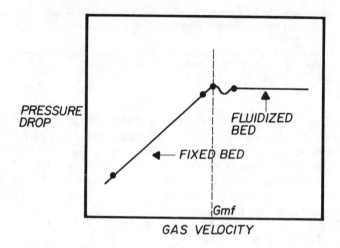

Figure 21-16. Variation of pressure drop with velocity in fixed and fluid beds.

between the solids and gas occurs and results in completely uniform conditions as to temperature, composition and particle size distribution throughout the bed.

Fluidization is dependent on fluid and particle properties. The maximum gas velocity that can be allowed in a fluid bed is the thermal velocity G_t of the particle. These velocities are used in pneumatic transport. The limits of fluidizing velocity is $G_{mf} < G_f < G_t$. Depending upon the operation, the actual gas velocity is chosen between the two extremes.

Recent concern about the evaluation and development of new pollution abatement processes has led researchers to investigate the advantages of fluid bed technology in various pollution control applications.

Water Pollution Control

Carbon Adsorption

The use of carbon as an adsorbent to remove odors or taste from water to make it potable has been well developed in fixed bed technology. This operation requires that the maximum utilization of the carbon surface area is achieved. This criterion can be achieved easily in a fluid bed where intimate mixing of solids and liquids can occur.

Carbon Desorption

Adsorbed carbon is regenerated by passing water vapor through it at a temperature of about 1000 F. This adsorption is achieved in a fluidized bed of sand which is heated by means of combustible gases from the plenum burner (Figure 21-17). Adsorbed carbon is fed to the fluidizer by a screw conveyor at the bottom of the bed. Steam or nitrogen is then passed through the fluid bed to regenerate the carbon. Desorption of carbon takes less than two minutes. Carbon collects at the top of the bed because of the difference in fluidizing velocities of carbon and sand. Regenerated carbon is drawn off the fluid bed from an overflow.

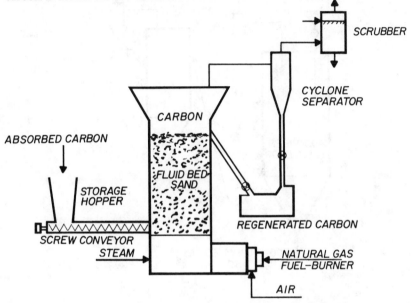

Figure 21-17. Fluid bed carbon regeneration unit.

Air Pollution Control

Absorbers

Sulfur removal during combustion of coal or low-temperature coal char is a problem facing the power plants. The SO_2 produced during such combustion can be fixed as $CaSO_4$ by reaction with dolomite. This is accomplished in a fluid bed at fluidizing velocities of 1.5-3.0 ft/sec and residence times between 1-2 sec. The temperature usually will be 1700 to 1900 F at an operating pressure of 8 psig.

A similar application can be found in desulfurization operations. It is difficult to eliminate hydrogen sulfide and sulfur dioxide at low concentrations other than by adsorption. It has been found that high percentage desulfurization is achieved in a fluid bed of iron oxide (Figure 21-18). Typical operating conditions for desulfurization are: space velocity 1000 gas volumes/spatial volume/hr, particle size 52-85 mesh, bed temperature

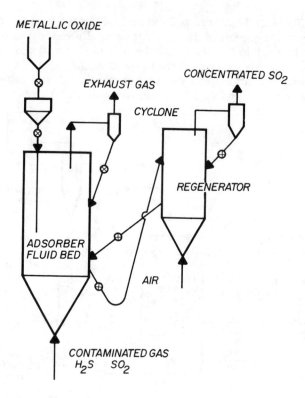

Figure 21-18. Continuous fluid bed desulfurization unit.

30 C. The ferric oxide is converted to ferric sulfide during the process. This can be easily regenerated to produce concentrated sulfur dioxide, which can be used in sulfuric acid production.

Contaminated gas is passed through a fluid bed of the metallic oxide (usual gas velocities of about 10 ft/sec). Clean gas is passed through the cyclone and vented to the atmosphere. Spent oxide is withdrawn from the absorber and pneumatically conveyed to the regenerator, which is also a fluid bed. Metal oxide is regenerated by producing concentrated sulfur dioxide which can be scrubbed with caustic, soda ash, or ammonia solutions to form salts or converted to sulfuric acid.

Fluid Bed Absorbers

These devices are also known as turbulent contact absorbers, a gas-liquid-solid contacting tower. It is characterized by the use of an essentially nonflooding packing (½ in. hollow polyethylene balls), placed between two retaining grids sufficiently far apart to permit turbulent and random motion of the spheres, thus allowing both high gas and liquid rates at modest pressure drop.

This equipment is essentially nonclogging and can be useful when a solid is present in the gas stream or is formed by reaction of contacting fluids. Typical applications can be found in treating sulfur dioxide with sodium hydroxide or magnesium hydroxide. A high overall mass transfer coefficient of 540 lb mole/hr cu ft/min can be obtained. Gas velocities as high as 1200 ft/min could be attained in this type of tower.

Solid and Liquid Waste Disposal

Municipal sludge after dewatering (25 percent solids) can be fed into a fluid bed reactor by means of a screw conveyor or high pressure pump (Figure 21-19). The fluid sand bed is initially heated to about 1500 F and air is passed through the bed at a velocity of 4 ft/sec. About 50 percent excess air is used to ensure complete combustion.

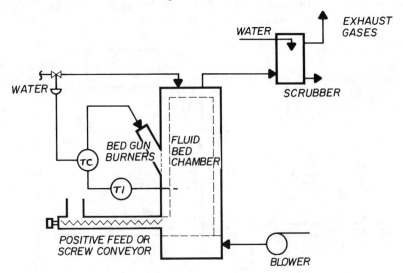

Figure 21-19. Fluid bed incinerator.

Temperature of the fluid bed is controlled by either adding or removing heat. Bed gun burners are provided to add heat to the fluid bed and water is sprayed on the top of the bed to remove the heat. For the incineration to be self-supporting, a minimum of 28 percent dry solids is required in the sludge. Garbage incineration is normally a highly exothermic reaction and heat usually must be removed in the burning operation. Therefore, it is advisable to mix municipal sludge and garbage together for burning.

Since incineration does not give any useful by-product other than heat, pyrolysis is receiving increasing attention, particularly for economic potential. Pyrolysis operations can be carried on in an inert atmosphere or under controlled oxygen conditions (under partial oxidation or combustion).

Pyrolysis of solid waste yields carbon, methane and hydrogen. Pyrolysis of garbage can be carried out either in a fluid bed or in a fixed bed reactor.

Pyrolysis has been applied in the pulp industry. Two-stage pyrolysis incineration units have been developed for the disposal of solid waste. The first stage of incineration is carried out at a temperature with indirect heating at an optimum point where pyrolysis takes place. Pyrolysis gases issuing from the second stage may contain free sulfur and can be used without cooling in a stoichiometric combustion zone to produce heat. Gases containing sulfur products that can create a nuisance are cleaned up by absorption.

Municipal refuse contains such things as aluminum and glass, which cannot be removed by a magnetic separator, but may be eliminated in the pyrolysis unit. This can be done in a fluid bed where advantage is taken of the differences in fluidization velocities of sand, aluminum and carbon (Figure 21-20). Aluminum cans and glass can be removed from the bottom, while carbon is removed from the top of the unit.

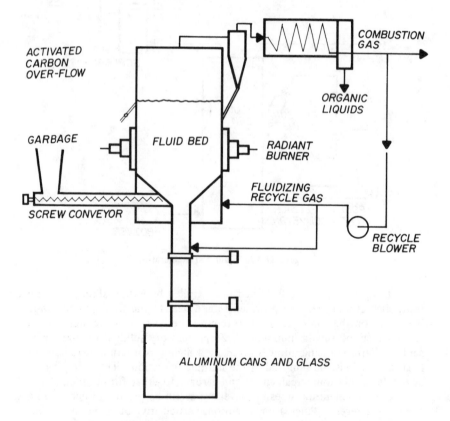

Figure 21-20. Pyrolysis system for handling municipal garbage.

Recent developments in coal gasification have yielded the conversion of municipal solid wastes to pipeline gas. A fluidized bed of sand acts as a thermal flywheel, and by cycling between combustion and pyrolysis zones it supplies the heat of reaction to allow solid waste to pyrolyze. The by-product of the chemical decomposition is solid activated carbon char which is carried out with the pyrolysis gas and can be used to purify and reclaim liquid and gas streams and/or as fuel for the combustion.

REMOVING COATINGS WITH FLUID BED REACTORS[146]

Fluidized bed reactors have been used for many years in the chemical and ore handling industries with good results. Such installations can be economically justified for large volume, bulk, continuous processes. The hard goods manufacturing industries, however, need a relatively smaller, stop-and-go, type of equipment. Recently a new approach in heating fluidized beds has made possible the design of equipment which can even be used for stripping of paint from parts.

When an electric motor is repaired, the old winding in the stator slots are tightly bonded, and difficult to remove. Conventional burn-off ovens incinerate the insulation under controlled conditions, but not without problems of warpage. Several hours are required to heat up and gradually cool the part. Electric repair shops can now immerse a number of stators in a basket into a 850 F fluidized sand bath for about 15 to 30 min to do the same job better (Figure 21-21).

Figure 21-21. Small fluidized bed reactor for industrial cleaning applications.

Another example of the use of the technique is for stripping of encrusted paint from hangers that carry parts through electrostatic paint booths and bake ovens. Hangers, hooks or parts are placed into a basket in an 850 F fluidized sand bath for several minutes until the paint has been burned off. Better electric grounding contact of a part with a clean hook reduces scrap loss in the painting department.

The new design fluidized sand bath uses the same burner that heats the sand to incinerate the fumes coming off the bath, in an insulated enclosure. The design is simple, using off-the-shelf components such as gas burner, fans, automatic temperature control, and approved safety devices. Basically, the unit resembles an industrial oven (Figure 21-22). A unit containing about 5 tons can be heated from 70 to 850 F in about 2 hr. When shut down over a weekend, the sand, being a good insulator, will still be at about 600 F on Monday morning. The operator can push a button, and in about an hour can start processing work. During the day the unit can be started and stopped any time.

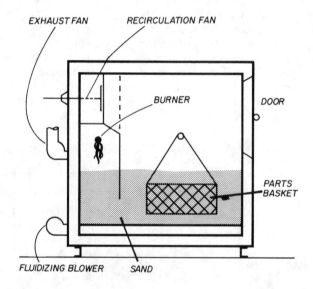

Figure 21-22. Reactors can be constructed using off-the-shelf components.

Canadian air pollution authorities require fluidized sand bath-type paint strippers be used in the following manner: Fluidized sand is first heated to 500 F before parts are immersed for treatment. As soon as the parts and the sand bath are up to 850 F, the heat from the burner is diverted from the sand bath, whereupon the fluidizing blower cools the sand bath, but the fume incineration continues. When the bath has cooled to 500 F the parts are taken out. This process requires about an hour per basket load. Conveyor hangers, hooks, and spray booth floor grating are cleaned in this manner.

Other industrial applications for this small fluid bed reactor are for heat treatment of coiled springs, stress relief after hardening and oil quench, removal of core sand remnants from castings after shake-out, and re-conditioning of foundry resin bonded core chunks.

PYROLYSIS[117]

One of the more promising solid waste disposal processes currently being developed is pyrolysis. Sometimes referred to as destructive distillation, pyrolysis is the process of chemically decomposing an organic substance by heating it in an oxygen deficient atmosphere.

Pyrolysis of most organic materials results in the formation of three classes of products:

1. a solid residue, char, composed of elemental carbon and ash;
2. a condensable liquid, water plus mixed organics;
3. a gas with some heating value.

Except for the water and ash, all the products can be used as fuels with minimal air pollution. In addition, the possibility exists for recovery of saleable by-products such as acetic acid, methanol and mixed solvents. Table 21-5 shows the typical composition of municipal refuse used during one pyrolysis study and the results of the pyrolytic operations on the refuse.

Table 21-5

Chemical Analysis of Raw Refuse	
	(Percent by weight)
Moisture	20.00
Carbon	29.83
Hydrogen	3.99
Oxygen	25.69
Nitrogen	0.37
Sulfur	0.12
Ash and metal	20.00

After Pyrolytic Operations	
	(Percent by weight)
Pre-Pyrolysis	
Inorganics removed	9.00
Moisture removed	20.00
Pyrolysis products	
Charcoal	11.25
Liquids	40.80
Gas	7.95
Ash and metal	11.00

The U.S. Bureau of Mines started work on pyrolysis in 1929 with a research unit to study the carbonization of coal. Rubber from shredded tires was the first waste material processed. Gas with a high Btu content, oil and solid residue with a high heating value were obtained. Other waste products including plastic battery cases and shoe soles have also been tested.

The BuMines unit (Figure 21-23) is a nickel-chromium resistor furnace with a recovery train that acts as a trap. Shredded refuse is sealed in the furnace and heated. Gases evolved pass through an air-cooled trap to condense out tars and heavy oils. Remaining vapors are passed through a series of water-cooled columns to remove additional oils and aqueous liquors. Oil mist is removed by electrostatic precipitators and finally the gas is passed through acid and caustic scrubbers. The gas is disposed of by flaring (99 percent) and gas sampling (1 percent).

Table 21-6 shows a typical yield from 1 ton of municipal refuse as determined by BuMines. Actual values depend on reaction conditions. Estimates for plant operation costs also are shown in Table 21-6. These figures take into account revenue from fuel sales.

1. Pyrolysis is feasible as a method for processing municipal solid wastes.
2. The energy content of the products of pyrolysis, char and hot vapor, is more than sufficient to sustain the process once brought up to temperature.
3. Sulfur content in tars of crude vapor indicated combustion of the vapor would not produce stack emissions exceeding the level set by the San Diego County APCD.
4. Use of pyrolysis by a community with solid waste similar to San Diego (16 percent noncombustible, 74 percent combustibles by volume) would result in significant reduction in sanitary landfill space requirements.
5. The optimum temperature for minimum operation time is 1500 F for material weighing 15-20 lb/cu ft.
6. Twenty-three compounds were isolated including: methanol, ethanol, isobutanol, *n*-pentanol, tertiary pentanol and acetic acid.
7. The carbonaceous residue contains approximately 5 to 6 million Btu/ton of typical San Diego combustibles.
8. The carbonaceous residue, if properly handled, becomes a superior "activated charcoal."
9. Most plastics depolymerized into carbon and gas at 1500 F.
10. Glass softens at 1500 F, but does not become fluid.
11. Iron and steel are not affected. Aluminum cans behave much like glass at 1500 F.

While most of the emphasis in pyrolysis of solid waste has been on resource recovery, Monsanto's Enviro-Chem Systems, Inc. attacks the problem from the standpoint of disposing of solid waste. They have developed and field tested a pyrolysis plant called the Landgard System.

The system (Figure 21-24) is a totally enclosed and self-contained operation. Refuse is delivered to an enclosed (negative pressure) receiving area and dumped into a pit. Enclosed conveyors deliver the refuse to shredders where it is reduced to particles 3 to 4 inches in diameter.

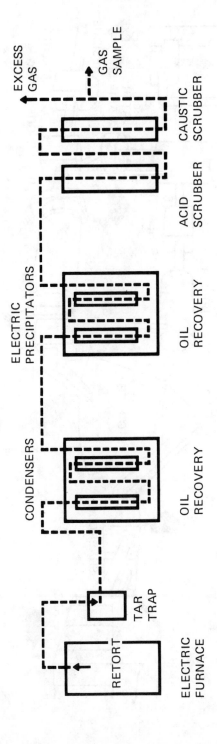

Figure 21-23. Diagram of the U.S. Bureau of Mines pyrolysis system.

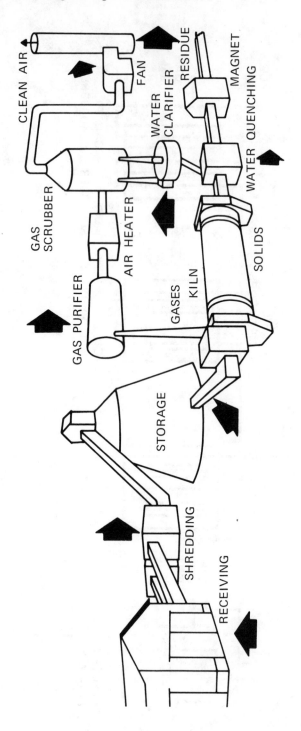

Figure 21-24. Landgard system.

Table 21-6

Yield and Typical Thermal Values	
Char residue	154-230 lb, 8-13,000 Btu/lb
Tar and pitch	0.5-5 gal
Light oil	1.2-2 gal 150,000 Btu/gal
Gas	11-17,000 cu ft 500 Btu/cu ft
Aqueous liquor	80-133 gal
Ammonium sulfate	18-25 lb

Disposal Costs vs. Plant Size	
500 ton/day	$6.00/ton
1500 ton/day	3.00/ton
2500 ton/day	2.00/ton

Another enclosed conveyor moves the shredded refuse to a storage silo. From the silo the refuse is conveyed to an inclined rotary kiln lined with refractory material. Since there are no grates or moving parts in the kiln, the problem of clogging is greatly reduced. The charred residue falls by gravity into a water quenching unit. After being water cooled, the residue is passed through magnetic separators to remove ferrous metals. The wet residue, now only 6 percent of the original trash volume, is hauled away to a landfill. The cooling water is recirculated after passing through sedimentation basins.

Gases formed in the kiln during pyrolysis are passed through a combustion chamber to oxidize the hydrocarbons. An afterburner is used to insure complete combustion. Product gases (CO_2, N, etc.) are passed through an adiabatic spray-scrubber and released to the atmosphere. Steam plumes are eliminated by heating the stack gases. Acid or caustic scrubbing can be added if it is necessary to remove hydrogen chloride or sulfur dioxide and metal separation can be performed before or after pyrolysis, as desired.

Results of government and private investigations in the use of pyrolysis for solid waste disposal indicate that this is definitely a process to be considered by governments desiring to extend the life of their sanitary landfills or to replace open dumps and incinerators.

HIGH ALLOY CAST INCINERATOR COMPONENTS[191]

In an incinerator, rabble arms, plow shoes, access doors, and grates must withstand high heat plus combustion gases, abrasion, and thermal cycling. However, to upgrade the solid waste disposal capability, many incinerators are being operated at elevated temperatures causing parts to be short lived.

Materials such as cast iron cannot withstand the high temperatures required for complete combustion of waste materials. Now, many pollution engineers are specifying heat resistant cast high alloys to solve existing and future problems of high temperature incinerator operation.

A case in point is the increasing use of multiple hearth furnaces to dispose of sewage and waste solids. The furnace consists of a stacked array of circular hearths and rotating arms. The waste is moved alternately toward the center, then toward the periphery in successively lower and hotter passes across the hearths. The heat generated by the burning sludge is approximately 3000 Btu/lb. Only 0.11 gallons of fuel oil are required to reduce a ton of sludge to ash.

The waste material is moved by plow-like shoes affixed to rotating rabble arms. The plows keep turning the waste over, moving it across the several furnace hearths. Since the furnace may operate at temperatures up to 2000 F, the shoes and rabble arms must retain strength and resist corrosive attack (scaling).

High alloys offer useful strength and good resistance to corrosive scaling to temperatures as high as 2200 F. Because parts operate in the region of plastic (rather than elastic) strain, considerations of time are vital. For this reason, conventional short-time high-temperature mechanical strengths may be less important than other characteristics.

Two criteria are often used for industrial furnace design. The first is limiting-creep-stress which is the stress resulting in a creep rate of 0.0001 percent/hr. The second is stress-to-rupture in 100, 1000, 10,000 or 100,000 hr, which is the stress required to cause failure after a specific time interval. Comparative data for cast iron are difficult to obtain because few high-temperature studies include this material. Consequently, limiting-creep-stress values for ductile iron and for two cast heat-resistant alloys are compared in Table 21-7. These high alloys demonstrate markedly superior properties in the 1400 to 2000 F temperature range typically encountered in a multiple hearth furnace.

Table 21-7 provides representative limiting-creep strengths and stress-to-rupture strengths of the heat-resistant high alloys at 1400, 1600 and 1800 F. For comparison with class 40 iron containing chromium and molybdenum, the entry at the bottom has been provided showing the stress-to-rupture for iron at 1200 F. No values are shown for iron above 1200 F because of complications introduced by material growth at higher temperatures.

Two factors contribute to the phenomenon of growth. When internal oxidation occurs, as often happens with repeated cyclic heating, it causes internal expansion. This produces a fine surface cracking or crazing, and results in dimensional changes. Unalloyed gray iron, for example, also displays a permanent increase in volume when heated above 900 F.

To avoid objectionable changes in critical dimensions and associated diminution of strength, alloying elements are added. Chromium, particularly, has a marked effect on the growth characteristics of iron and steel; it prevents decomposition of iron carbide and provides corrosion resistance too.

Comparisons of scaling rates of alloy compositions indicate that maintaining low scaling rates above 1200 F requires large additions of oxidation-resistant elements. These amounts bring the materials into the high alloy field. The pollution engineer's choice, between high-carbon heat-resistant

Table 21-7. Representative Long Time Elevated Temperature Strength[a]

Nominal Composition Percent		LCS[b], psi	SR[c], psi	LCS[b], psi	SR[c], psi	LCS[b], psi	SR[c], psi
Cr	Ni						
28	4 max	1,300	3,300	750	1,700	360	1,000
28	5.5	3,500	10,000	1,900	5,000	900	2,500
28	9.5	4,000	11,000	2,400	5,300	1,400	2,500
21	10.5	6,800	13,500	3,900	7,200		
26	12.5	6,000	14,000	3,900	6,800	2,100	3,200
28	16	6,600	13,000	3,600	7,500	1,900	4,100
26	20	10,200	15,500	6,000	9,200	2,500	4,750
30	20	7,000	15,000	4,300	9,200	2,200	5,200
21	25		18,000	6,300	11,000	2,400	5,600
15	35	8,000	15,000	4,500	8,500	2,000	4,500
19	39	8,500	10,000	5,000	8,000	2,200	4,500
12	60	6,000	13,000	3,000	6,000	1,400	3,600
17	66	6,400	4,000	3,200	6,700	1,600	3,500

[a]Values represent constant-temperature operation. If alloys are exposed to cyclic temperatures, lower values would apply.

[b]Stress-for-creep rate of 0.0001 percent/hr.

[c]Stress-to-rupture in 100 hr.

[d]Alloy with more than 2.5 percent Ni and 0.15 percent N.

[e]Wholly austenitic (ASTM B190, type II).

[f]Class 40 iron containing Cr and Mo. Stress-to-rupture is for 1200 F. No value is shown for LCS because of complications introduced by growth temperature above 1200 F.

alloy irons and low-carbon heat-resistant alloy steels, is based on hot strengths and toughness, rather than scaling resistance. It is well known, though, that heat-resistant cast iron is more prone to cracking than heat-resistant steel.

For operation above 1400 F, scaling rates less than 0.1 inch per year (ipy) are considered satisfactory. Research has established how much chromium and nickel are necessary.for cast high alloys to resist corrosion in air and flue gases containing sulfur. These studies show that increases in either chromium or nickel effectively reduce the rate of attack at low sulfur levels. However, at high sulfur levels (about 100 grains of sulfur per 100 cu ft of gas), the chromium-predominant alloys are superior to the nickel-predominant types.

Service Parameters

For incinerator service, selection of materials may be considered either from the viewpoint of existing equipment or for future installations. Plants where gray iron castings have short life due to scaling or growth may be concerned with improved resistance obtainable from high chromium alloy types (HC and HD). Where failure is due to lack of strength in low alloyed irons, high-strength austenitic grades (HF or HH) will prove beneficial. Where thermal shock or fatigue results from cold loads or rapid cyclic heating, a nickel-predominant grade (such as HT) may provide improved service life.

Incinerator rabble arms and plow shoes are complex shapes which lend themselves well to casting. The rabble arms are hollow tapered cylinders with socket-type fittings for attachment to the central rotating shaft. Along the bottom, there is a flanged track to support the plow shoes. This would be a difficult shape to form by any other process, but can be readily cast. The final one-piece units have large grain structures that are stronger than wrought steel and have uniform physical characteristics. Moreover, through casting, metal can be located exactly where stresses and loadings require.

Incinerator shoes, too, can be readily cast. Their shape must act like the moldboard on a plow and bring the bottom waste material to the top in a continual turning action. The curved contour and the attachment fittings can easily be produced by casting.

TEEPEES[82]

Wigwam refuse burners have been one of the most economical means for disposal of wood wastes, cotton gin waste, peanut hulls, and even some municipal waste. But, they have also been the source of many localized pollution problems and nuisances. Adequate, safe waste destruction is possible, however, for materials with heating values between 4000 and 10,000 Btu.

There are many problems with the old, conventional wigwams. Except for enclosing the waste pile and blowing air under it, no effort has been made to control excess air in the past. Low temperatures hamper efficient combustion and produce smoke (usually white or gray). Higher velocities inside the shell carry embers, unburned waste and ash out the top of the unit.

A few wigwams are different. Until Oregon State University began a research program on wigwams, no one fully understood why they were different. Investigators found that wigwams in good repair, burning at relatively constant high feed rates, could operate with little smoke and flyash emission. Tests showed that wigwams with adequate underfire air systems (relying less on natural draft air) operated at lower excess air values, usually in the range of 300-500 percent.

If the wigwam is properly fired, low-velocity circular movement of the combustion gases help control particulate emissions by a settling effect. With high temperatures, smoke particles and fines can be destroyed prior to emission. However, higher temperatures coupled with excessively high velocities in the unit carry out coarse particles and sparks. This condition can be prevented if a unit is properly sized and air leakage into the shell is reduced.

To accomplish these objectives, the owner of an existing wigwam might examine his unit and answer these questions:

1. Is the unit in good repair to prevent excess air from entering? (No rust holes in the shell, access doors close well, no air leakage around foundation or piers.)
2. Is the unit cleaned out regularly? (Prior to ashes interfering with forced air, or fire damaging walls.)
3. Is there a reasonably steady feed rate of waste deposited in the center? (Discharge end of chutes and cyclone tailpipes should be as low as possible to prevent waste from burning in suspension.)
4. Is the feed rate high enough to maintain good temperatures? (Usually 750 to 900 F near top of shell.)
5. Are synthetics or rubber being burned that would produce black smoke?

Scientists tested 19 units with emissions under 0.20 gr/scf. These teepees had no special controls, other than maintaining the five points mentioned.

Control Systems

Two types of controls or modification can be applied to wigwam units to decrease air pollution. The first is gas cleaning equipment; the second is basic modifications to the combustion process.

Gas cleaning devices are relatively expensive but necessary in some instances. Spray towers are the most common; they remove coarse particles prior to gas emission. Afterburners have also been installed for treating wigwam discharge gases. The initial and operational cost for afterburners becomes prohibitive for wood waste installations, but may prove to be within the economic range for municipal or industrial wastes. Other gas cleaning devices, such as venturi scrubbers, centrifugal scrubbers and electrostatic precipitators have also been operated with some success.

The most successful and economical method of pollutant control to date has been by basic modification to the combustion shell and techniques to control the fuel and air mixture. Two manufacturers use the following:

1. Close natural draft ports, relying on seam leakage to cool the shell.
2. Add controlled high velocity overfire air at the perimeter of the waste pile by means of cast iron pipe directional outlets.
3. Provide controlled underfire air.
4. Add a dampering device or "heat shield" at the top of the shell to control loss of heat in excess air.
5. Control air volume by electronically modulating dampers on fan inlets as a function of temperature.

The overfire air outlets, usually cast iron pipe elbows, are directed over the pile in such a manner as to produce an area of high cyclonic turbulence over the pile. The circular velocity of the gas stream decreases toward the walls of the wigwam, allowing particles to settle and be deposited on the floor of the unit. This is a major improvement over conventional wigwam construction, which allowed air to enter the shell wall tangentially.

Control by this method (Figure 21-25) is relatively inexpensive to install on new or existing units, provided the five basic points of good operation can be maintained. Operating and maintenance costs are only slightly higher than in conventional units.

Modulating Systems

Tests show that the ideal temperature range for particulate and smoke control for wood waste is from 750 to 900 F, as measured 18 to 36 in. from the shell wall just below the outlet.

Upon daily startup and lighting of the fire (usually manual), the temperature gradually moves up to the "control set point," say 800 F. During this time, full forced air volume is being passed to the burning waste pile. As the temperature approaches 800 F, the primary air control will start reducing the air volume by partially closing the fan inlet dampers. Upon reaching the set point, the control will vary the inlet air volume to maintain this temperature, usually within 40 degrees of the set point, so long as in-feed is relatively constant.

Low feed rates make it impossible to reach an efficient temperature, while high feed rates can cause damage to the metal structure. Fluctuating rates combine these two effects. For estimating purposes, Figure 21-26 shows an "allowable feed rate deviation" chart, when using an air modulation or proportioning system and dome damper. This chart may be applied to most wood waste fuels. Since a wigwam is commonly rated on a 1-hr average feed rate, the chart is based on computing to the average 1-hr capacity rate.

With temperature controls and dome damper set, in-feed can vary by some 5 percent either side of the hourly average without affecting emissions

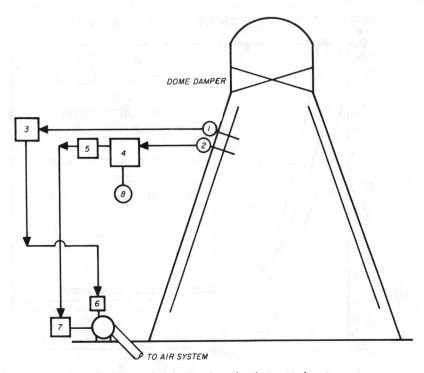

DOME DAMPER

TO AIR SYSTEM

Figure 21-25. Typical heat and combustion control system.

1. Thermocouple for air proportioning circuit
2. Thermocouple for high limit safety circuit
3. Proportioning control with temperature indicator
4. High limit control and recording pyrometer (shuts down fan motor and sounds alarm)
5. 24-hour fan motor timer
6. Proportioning motor on fan inlet damper
7. Fan motor control
8. Alarm horn

Examination of a typical wigwam will usually find the following design features:

1. Reinforced shell of 16 or 14 gage mild steel or aluminized steel sheet
2. Inner shell of light gage corrugated sheet on some units
3. Base diameter approximately equal to height
4. Outlet diameter approximately 1/3 the base diameter
5. Air for turbulence and combustion mechanically produced, and introduced under the waste pile
6. Openings on the periphery of the shell to introduce excess air tangentially for cooling and to promote circular motion

PER CENT DEVIATION ABOVE NOMINAL CAPACITY

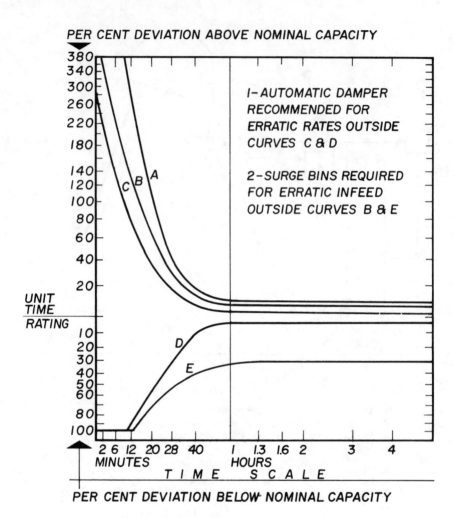

I-AUTOMATIC DAMPER RECOMMENDED FOR ERRATIC RATES OUTSIDE CURVES C & D

2-SURGE BINS REQUIRED FOR ERRATIC INFEED OUTSIDE CURVES B & E

PER CENT DEVIATION BELOW NOMINAL CAPACITY

Figure 21-26. Chart for determining allowable deviation from unit time nominal capacity for burning efficiency.

or combustion efficiency. By adjusting the dome damper to retain or disperse heat through excess air control, the efficiency of the unit can be operated above the normal rating by approximately 10 percent, or below by about 30 percent without affecting pollution control efficiency.

Wider variations in high, low or no feed rates can be allowed for short time periods, based on the accumulated deviation not exceeding the allowable "1-hr" deviation.

The allowable deviation for a shorter time, such as 1 min, is allowed. Without resetting the dome damper, it is possible to overload by nearly 300 percent (400 lb/min) or lose feed completely, if the total hourly rate deviation was not greater than 10 percent or for any time period allowance in between. The curves are for estimating purposes only, and will vary somewhat with Btu value and ease of ignition for specific wastes.

Nearly 100 installations are known to be successfully operating with modulating controls and dome dampers, burning wood residues, peanut hulls and municipal waste. For wigwams with modulating controls only total equipment costs are on the order of $3500 to $6000 per ton in the capacity range of 20 to 60 tons per 8 hr.

A wigwam or teepee incineration system can be operated in compliance with most air pollution regulations, particularly when burning wood wastes. However, the wigwam cannot be installed, the fire lit and forgotten. It must be treated as any other piece of finely tuned equipment to remain in compliance.

CHAPTER 22

VOLUME REDUCTION OF SOLID WASTES

SOLID WASTE SHREDDERS[90]

As received, most solid waste has a low bulk density and is composed of a wide variety of objects in all sizes and shapes. Solid waste shredding machines are not capable of destroying waste matter, but only of converting it into a form more easily and economically handled for processing.

Hammer mills in one adaptation or another are the most commonly used size reduction machine. However, nomenclature is far from standardized. The machines are variously called "shredders," "crushers," "pulverizers," "mills," and "hoggers." Shredder is the most frequently used term.

At one time, both capacity and durability posed important limitations to the application of waste shredders on a large scale. This is no longer true. Today, shredders are in service which can nuggetize two complete automobiles per minute (Figure 22-1). Capacities of the largest waste shredders exceed 100 tons/hr.

Design and Operation

In this country, the most common shredder design consists of a welded steel frame protected on the interior with abrasion-resistant steel alloy liners.

Within the frame, there is a power-driven rotor with rows of pivoted manganese steel hammers (Figure 22-2). For handling OBW (oversize bulky waste) and other "hard refuse," hammers are free to pivot back under impact to give relief from overload or shock. A series of sizing grates are positioned below the rotor. The violent hammer action reduces the material to a size that will pass through openings in the grates. Maximum product size can be controlled by installing grates with large or small openings.

Characteristics of shredded waste differ somewhat from those of other materials which are processed in shredders and hammer mills. With many materials, it is possible to arrive at a screen analysis or size degradation which tells the size of the particles produced, as well as the relative quantity

759

Figure 22-1.

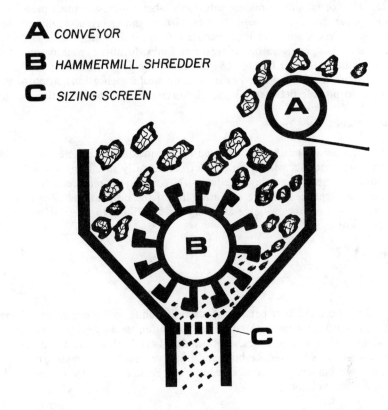

A CONVEYOR

B HAMMERMILL SHREDDER

C SIZING SCREEN

Figure 22-2. Pivoted steel hammers rotate to shred waste to any desired size.

of the various size particles. However, because of the variety of materials contained in solid waste and seasonal variations which can be expected, a quantitative analysis of size degradation is almost impossible.

Capacity

Shredder size should be selected to allow for anticipated surges in the rate of material feed. In addition to processing the hourly tonnage of materials, a refuse shredder must be sized to accommodate the largest pieces anticipated.

Power Requirements

Size reduction machines consume power in proportion to the feed rate and to degree to which material is reduced.

Little or no shock loading will occur when processing small pieces of wood, paper, corrugated board, bottles, plastic, and assorted organic matter. However, where a shredder must handle oversize bulky waste such as rubber tires, mattresses, refrigerators, stoves, tree limbs, furniture, packing crates, and demolition lumber the power source and the drive train must be designed to withstand shock loading. For most solid waste applications, a motor is selected to provide between 12 and 20 horsepower per ton of refuse per hour.

Final Product Size

Shredder output is characterized by the maximum particle size though much of the product will be far smaller. Arbitrarily specifying a small final product will mean increasing shredder size, cost, power requirements, and cost of maintenance over the operating life of the installation. Shredder output can range from as large as 10 in. for landfilling to as small as 1 in. for composting.

Applications

Transfer Stations

To reduce the cost per ton of waste handling, many communities have installed transfer stations. At the transfer station, small payload collection trucks are unloaded and quickly returned to neighborhood route service. Refuse is shredded to reduce bulk and improve handling characteristics. Compaction immediately after shredding can reduce the waste to one-third of its original volume.

Incineration

Shredders can help a waste incineration plant operate more effectively. They can be used to reduce large objects which in the past were not incinerated because of the difficulty in feeding them to the incinerator. Even when physical size of the waste is not the limiting factor, firing theory indicates that more efficient combustion will occur when solid waste material is first shredded.

Under current investigation is an incineration process where shredded waste is used as part of the fuel supply for a coal-fired steam power plant. Though preliminary, feasibility studies indicate that the economics of this scheme are attractive due to the savings in coal requirements. One study indicates that 3 tons of shredded refuse will provide the same heat as 1 ton of coal.

Composting

Modern high–capacity composting operations would not be possible without the use of shredding machines. In these plants, shredders reduce waste to a size which can be quickly decomposed by bacterial action. Composting plants operate best with a relatively fine particle size. The process is speeded if fibrous material is also opened. To produce the required fineness at high material flow rates, composting operations usually employ two shredders in series.

Following bacterial decomposition, a third shredder may be used to thoroughly mix the compost and to break up any agglomerates not destroyed in the digester tank. Output size from the secondary refuse shredder operation will usually be on the order of 1 in.

Sanitary Landfilling

Recent studies conducted by the federal government have shown that advantages are gained when shredding is done in conjunction with sanitary landfilling operations. When deposited in a landfill, waste shredded to a maximum size of 6 to 10 in. will not support combustion, will not support vermin, will not produce odor, and will not provide a breeding ground for insects.

Reduced volume of the shredded waste leads to prolonged life of the landfill site. Materials such as rubber tires and demolition lumber which could not previously be effectively compacted into a landfill present no problem after shredding. The effective life of the landfill site is prolonged from two to three times that normally expected, and the expense for covering the site daily with topsoil is eliminated.

Economics

Costs of a shredder, allied equipment, and maintenance are important considerations when evaluating a shredding system. Unfortunately, good data are difficult to obtain. Often, it is difficult to extrapolate for differing job conditions. Maintenance costs will vary with the relative quantities of hard and soft refuse. Conveyors, motors, controls, site preparation, and concrete foundations must be added to determine a total system cost.

MILLING REFUSE–A CASE STUDY[69]

Extra handling and processing of refuse costs money. Until total recycling is possible, refuse is just that--refuse. Why should any more money be spent processing it than what is ordinarily required for landfilling?

Because—space for landfills is becoming increasing scarce around most communities. Cost of operating solid waste disposal systems is steadily mounting. Efficient methods must be developed for municipalities to dispose of waste rapidly, economically and safely.

Since 1961 when the city of Madison, Wisconsin stopped open burning operations, efforts have been made to continually upgrade operations at landfill sites to a point where they are sanitary landfills in a true sense.

Despite a continual increase in the amount of equipment and numbers of employees involved in actual operations, Madison found a truly sanitary landfill difficult to attain.

In areas which have adverse climatic conditions, such as Wisconsin, it is almost impossible to properly operate a refuse disposal site every day refuse is collected. Equipment used is earthmoving and roadbuilding machinery, generally run by equipment operators with limited training in landfill operations. Earthmoving and roadbuilding operations are not practical when temperatures fall well below zero, or after a steady downpour of rain. A sudden blizzard that dumps 6 inches of snow simply shuts down such operations.

Demonstration Program

A demonstration project to assist Madison in solving a portion of its solid waste disposal problem was recently undertaken in a classic example of cooperation between governments, educators and industry. Involved were the city of Madison; the U.S. Department of Health, Education and Welfare; the Heil Company and the University of Wisconsin. Under the Solid Waste Disposal Act of 1965, a solid waste milling project was partly financed by a three-year matching fund. One dollar of matching funds was provided for every two dollars supplied by federal grant. The total three year budget was $555,000. The federal government provided $370,000, the Heil Co. $116,000 and Madison $69,000.

The city of Madison furnished project management, project site, mill site preparation, utilities, plant operating personnel, refuse to run the mill and landfill area for depositing milled refuse. The Heil Co. furnished the mill, technical services and operating advice, matching funds for participation by the University and matching funds for construction and erection of the pulverizing system. The University of Wisconsin was responsible for scientific and technical research and personnel needed to evaluate the system. They provided reports at various stages in the work.

The pilot program of pulverizing or milling refuse was designed to provide four areas of information:

1. Evaluation of the milling system;
2. Comparison of sanitary characteristics of milled and unmilled refuse;
3. Study of the physical characteristics of milled refuse deposited in a landfill; and
4. Comparison of a landfill with milled refuse to one containing unmilled refuse.

During early analysis of the refuse milling, rubbish and garbage were combined in the collection system. Service included single and duplex residences, apartment buildings, commercial buildings, filling stations, office buildings, and grocery stores.

Packer-type route trucks are used for refuse collection, while separate dump trucks are assigned to collect brush and tree trimmings and bulky

Table 22-1. Costs for Operating 1-4 Mills, 1 Shift

	Number of Mills		
	1	2	4
A. REDUCTION PLANT			
Operation			
Labor	$23,300	$32,300	$43,300
Mill Maintenance	5,890	11,180	21,740
Power for Mill(s) & Compactor	4,980	8,940	16,140
Lighting	2,000	3,400	6,000
Gas (heat)	1,430	2,430	4,580
Water	270	550	1,090
Truck Operation & Maintenance	1,060	1,430	2,860
Compactor Maintenance	120	120	240
Front-End Loader Operation & Maintenance	1,950	1,720	2,580
1. Sub-total–Operation	$41,000	$62,070	$98,530
Depreciation			
Foundations & Building	3,000	5,260	11,280
Scale	840	840	840
Front-End Loader–Case	——	2,140	2,140
Grinders & Conveyors	7,320	14,640	21,960
Huge-Pac Compactor	890	890	1,780
Containers	1,000(2)	1,000(2)	2,000(4)
Truck with Uni-Haul	2,120	2,120	4,240(2)
Hold Bin & Shuttle	1,000	1,000	2,000
Front-End Loader–Caterpillar 950	1,950	——	——
Erection of Each Feed Conveyor and Mill	1,200	2,400	4,800
2. Sub-total–Depreciation	$19,320	$30,290	$51,040
B. LANDFILL			
Labor	$11,300	$11,300	$11,300
Compaction Equipment– Operation & Maintenance	1,950	4,600	4,600
1. Sub-total–Landfill Operation	$13,250	$15,900	$15,900
Depreciation–Compaction Equip.			
Caterpillar 950	$ 2,100	——	——
Michigan CS-70	——	4,000	4,000
2. Sub-total–Depreciation	$ 2,100	$ 4,000	$ 4,000
C. TOTAL OPERATING COST– INCLUDING DEPRECIATION	$ 75,670	$112,260	$169,470
D. TOTAL ANNUAL TONNAGE	27,300	54,600	109,200
E. COST PER TON	2.77	2.05	1.55

waste. The dump trucks deposit their cargos at the landfill where the refuse is crushed and buried in the conventional manner. All other material collected in the trucks is deposited on the operational floor of the shredder-transfer station (Figure 22-3). A front-end loader feeds refuse onto a conveyor which carries it into the charging chute of a vertical shaft sorter-pulverizer (Figure 22-4).

Inside the machine, hammers of the pre-breaking section begin the pulverizing process. Refuse is cut up and readied for further action. The throat of the mill narrows just below the pre-breaking hammers. A clearance of 1½ in. exists between the hammers and the machine walls. Hammer speed chokes off fires and an ejection chute discharges anything not capable of further reduction. Ejected objects can be combined with milled refuse later or sent to a separate station. Magnetic separation is not necessary. Partially milled refuse enters the grinding chamber after it passes the throat of the pulverizer, where it is ground uniformly. Machine maintenance is minimal.

Results

In any landfill the purpose of confining the refuse in the smallest volume is to save landfill space and control fires by reducing voids containing oxygen. The main purposes for providing a daily earth covering are: to adequately control insects and rodent vectors, to eliminate blowing paper, debris and dust, to eliminate odor problems, to minimize accidental fire hazards, to prevent open access of refuse to the general public, and to reduce the amount of leaching to ground water.

Milling refuse overcomes local operational problems in a manner that will not only improve operation during normal times, but actually produce a superior method of landfill under all conditions. It is an important part in the full program of solid waste management, because milling makes possible consistent results which were unobtainable previously. Milled refuse has characteristics which might eliminate the need for earth covering at the end of the work day.

Rat tests were conducted by offering rodents special diets of aged milled refuse and freshly milled refuse. Rats surviving on aged milled refuse registered weight loss by the fourth day. Five days later one animal was cannibalized. Another on the sixth night; two on the seventh and one each night for the next five days. By the end of the 15th day the last rat was weak but alive; he died the next day.

Professor Robert Ham, University of Wisconsin, observed that although fresh milled refuse contained some food, it was not enough to sustain life. It was concluded that if rats don't have food and can't dig burrows, they won't remain around the disposal operation.

Experiments were also conducted regarding the breeding of flies. It was found that when optimum conditions exist, breeding of flies is possible in a milled refuse operation. However, under the environmental conditions

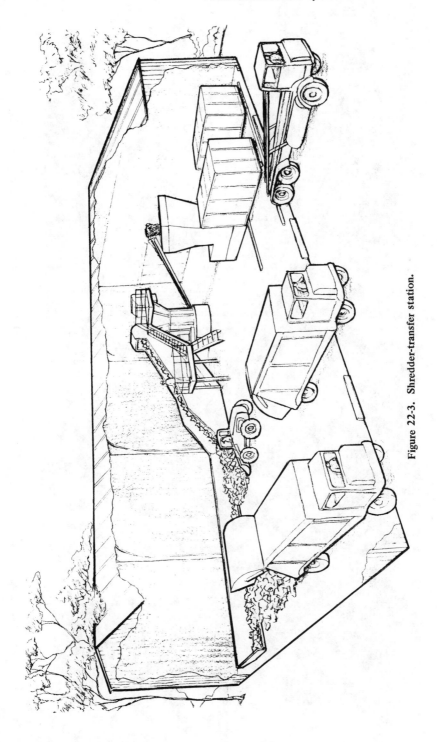

Figure 22-3. Shredder-transfer station.

Figure 22-4. Vertical shaft-type shredder.

at the Madison site, successful breeding was not prominent. Though weather conditions could potentially affect this conclusion, it is not considered likely. It was found that milled refuse did not readily attract flies. Fly populations did not build up as they do in a raw refuse landfill site, where earth cover is provided daily. Large percentages of fly maggots normally found in trash are actually killed by a pulverizing system.

The Madison Fire Department conducted tests on fire potentials. It was found that little fire hazard is present in milled refuse of a year or more in age. Fire is possible in freshly milled refuse because of a large degree of combustibles, but fire penetration is uniform. Because of a lack of oxygen, rapid spreading is retarded, and fires are easily extinguished.

A host of other benefits came from milling refuse. The public has lost interest in probing about the site. Uniform settlement is now attainable. Existing settlement decreases because of higher density. Less landfill space is used. Landfill operators can work closer to definite grades and contours. Frozen cover material and wet weather problems are eliminated. Trucks can operate on milled material without becoming bogged down in mud which results where earth cover is present. A financial saving also results in that earth cover does not have to be acquired.

MOBILE MULCHER[98]

Destroying confidential material has long been a headache to private businesses as well as governments. At San Diego, the problem of classified material disposal caused considerable expense, time, and security problems—as well as air pollution—until the Navy Public Works Center installed a mobile mulcher for the destruction of classified paper. The Public Works Center, funded by the federal government, provides services for Navy activities on a non-profit basis, but is organized in all other respects like a large private business with an annual financial statement totaling $26 million.

At the outset the NPWC evaluated the various types of approved destruction techniques available for disposal of classified material. Most methods, such as chemical destruction, were eliminated immediately because of high cost. Burning seemed the logical choice, since it is the most frequently used method of destruction of classified paper.

However, the city of San Diego passed an ordinance that authorizes burning only in officially approved incinerators. An approved incinerator is expensive. A typical model with a capacity of only 200 lb/hr was estimated at $50,000 to $75,000 per unit plus $10,000 per year in maintenance charges. Incineration, therefore, was discounted because of the high cost and low capacity.

Shredding or pulverizing methods were next considered. Although both shredding and pulverizing are cleared by the Navy for destruction of classified material, only pulverizing is authorized as an adequate method both for classified and top secret or cryptographic destruction. Since several activities generate both types of classified material, pulverizing was selected to avoid the time and expense of physically separating the two types of paper.

For centuries, men have torn confidential material into tiny pieces to destroy it. With today's technology machines perform the task more thoroughly and rapidly, through either a dry or wet process. Of the two methods, the dry process was eliminated for two reasons: the considerable volume produced, and the severe dust problem created during operation of the machine.

Pulverizing is usually accomplished by high-speed rotating metal blades forcing material through a steel mesh screen, ripping it into pieces smaller than a centimeter. The blades create an air current which can whirl small particles through a funnel and out of the machine, spewing a fine dust over a widespread area. The remaining, slightly larger particles, form a more solid mass, similar to cotton in texture and appearance. This mass is nearly triple the volume of the original paper, causing disposal problems, either in providing an unattached separate storage container large enough for the destroyed material, or by increasing the frequency of disposal if the container or dumping facility is attached to the machine.

Wet process machines offered little improvement over the dry system because of high cost and low capacity. A typical wet process machine with a capacity of 400 lb/hr costs $25,000. Greater capacity was essential to the naval activities in the San Diego area which generate over 20,000 lb of classified material per week.

The NPW Center considered designing modifications for a high-capacity dry process machine which would produce a compact wet product. Several dry-process pulverizing machines acceptable for destruction of classified material were available. Although most of these machines were stationary units, some were mobile or could be installed on a truck trailer bed. Mobility is essential. One of a security officer's nightmares is the fear of a traffic accident while moving classified material over streets and highways to a safe disposal area.

A mobile dry mulcher for $24,042 had an attached storage facility for 2000 lb of waste, but it was eliminated because of the high cost and low storage capacity of the disposal container attached to the machine. Several less expensive dry process pulverizers were available, which could easily be converted to mobile units.

A machine originally designed to grind silage, with a capacity output of 1500 to 2000 lb/hr at a low cost of $3,460, was finally selected by the Center. A water pump, filter and belt were installed on the unit to create the wet process. In the modified system developed by NPWC personnel, classified material is fed into a large hopper where it is fed to the pulverizer. From there is proceeds to a cyclone separator, and then through a tunnel where a fine water spray wets it into a mulch form. The waste is then carried by a porous belt through a series of pressure rollers which compact it into a papier-maché consistency. The porous belt allows water in the mulched material to drain off easily as it passes through the pressure rollers (Figure 22-5). Originally, a dewatering system was planned that would compress the mulch into a brick using hydraulic pressure. In use, however, this plan was found unsuitable.

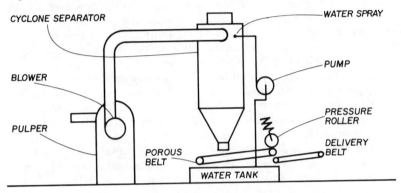

Figure 22-5. Schematic drawing of mobile mulcher.

Although it was thought that the mulcher would not need an external water supply, it became necessary to add a garden hose hookup to replenish the machine's water supply. Other modifications included more efficient filters and new screens.

Since the mulcher has no storage capacity, the naval activity served must supply a container for the mulched material. The machine, which rests on a 35-ft low-bed truck trailer, is driven next to a refuse container during operation. Because of the machine's size, water and container requirements, and noise, a large open area such as a parking lot is required for efficient performance.

During early demonstrations of the mulcher, it was thought that the machine would be unsafe if metal or glass objects were inadvertently inserted at the feed point. Although the material would be destroyed, small projectiles would be ejected at high speed from the feed point. To avoid this danger, the Center redesigned the feed point area and installed a chain guard.

The machine has an average capacity of 50 standard burn bags in 25 min or approximately 1500 lb/hr. It is presently on the road 5 days a week, serving 29 naval activities with an estimated waste total of 10 to 15 tons per week. Thirty-two other activities are awaiting funding and scheduling.

Since the mulcher began operation, it has generated increasing interest and, correspondingly, more business for NPWC. Not only is the mulcher three to four times faster than burning or shredding, it is much cheaper than chemical destruction or disposal by available wet-process pulverizers. Bringing the disposal system to the doorstep of an activity removes the headaches caused by the need to transport classified material to a destruction area.

SOLID WASTE COMPACTION ECONOMICS[73]

The shortage of adequate disposal sites and spiraling labor costs, have caused solid waste handling expense to skyrocket. Recent studies show that collection and disposal costs in some parts of the nation now exceed $20 per ton. Local air pollution regulations in some areas have all but eliminated the effective burning of industrial and consumer wastes. The cost of updating an incinerator with emission control devices such as scrubbers or electrostatic precipitators often is prohibitive.

Frequently, the only solution for those facilities and municipalities faced with a solid waste disposal problem is compaction. Compaction has the virtue of conforming to pollution control laws and is practical as well as economical. It should be pointed out, of course, that there are certain conditions under which compaction should not be used. For example, rubber and similar resilient materials cannot be effectively compacted.

There are many compactor types available and a variety of optional equipment for them. Basic units can range in price from $4000 to $18,000. Beyond the basic unit price, there are other cost factors as well: options, maintenance, and operation.

Equipment Costs and Options

The initial cost of a compactor depends on:

1. **Unit capacity,** which generally can range from 1 to 4 cu yd. Capacity can determine the size of the compaction chamber, in which case a proper selection of compactor unit will take into account the size dimensions of the waste items to be handled.
2. **Compactor hopper style, i.e.,** whether the unit is to be window, side, or ramp loaded. This, in turn, depends on where the compactor is to be installed.

The day-to-day cost of running a compactor is fairly low, with electricity the chief factor. Electrical rates vary greatly, depending on the region of the country. Even within the same power company's territory, the rate for electricity is a function of the amount used. In any event, the 20 amps that practically all compactors draw, on 400 volts, amounts to pennies a day regardless of the rate.

Another factor in the category of operating costs is service, maintenance and repairs. Most compactor models are of sturdy, heavy-duty construction. Properly serviced and maintained, the need for repairs should be very low. Many distributors who sell compactors also service them with service contracts available at approximately $300 a year. This type of contract will usually include monthly visits and a semi-annual change of hydraulic fluid, but will not include the cost of the fluid itself. Different models have different size hydraulic fluid reservoirs ranging from 5 to 30 gal. At about 50¢ per gal, the cost of hydraulic fluid will vary between $2.50 and $15, depending on unit size.

As a rule of thumb, maintenance costs of compaction are about 5 percent per year of the acquisition cost, for the first 10 years of operation. For example, a $4000 compactor should average $200 in maintenance costs, while an $18,000 compactor will average about $900 in maintenance costs.

Compaction and Economy

In most cases, compaction pays for itself in the first year or two of operation. In a study prepared by the U.S. General Accounting Office, the GAO reviewed comparative 10-year life-cycle costs for five waste disposal systems—compaction, shredding, pulping, incineration, and conventional or loose waste handling. The study considered equipment and equipment materials costs; personnel costs, including wages and fringe benefits; floor space and maintenance costs; and operating costs including utilities and outside contracted services.

Table 22-2, derived from the GAO study, shows that compaction has the lowest life-cycle cost of all the systems compared. For that reason, compaction is used as the base percentage to compare the various systems. These comparative figures are based on data from health care facilities, such as hospitals and nursing homes. However, there is no reason to believe that the findings are not applicable to any type of facility generating a wide spectrum of trash, including industrial and commercial facilities.

Table 22-2. Waste Disposal Systems
Life Cycle Costs (10 Years) With Compaction as Base (100%)

Discount Rate	5%		7.5%	10%	
Inflation Rate	5%	7.5%	2.5%	5%	7.5%
Waste Disposal Method					
Compaction	100.0%	100.0%	100.0%	100.0%	100.0%
Shredding	126.2	126.2	126.4	126.4	126.3
Pulping	147.3	144.7	152.9	152.7	149.9
Incineration	159.5	151.6	176.8	176.4	167.7
Unprocessed (Loose Trash)	200.1	202.1	197.5	197.6	199.2

Case History

In December 1972, a manufacturer installed a NSWMA-rated 2.08-cu yd solid waste compactor. An engineering cost analysis was made on the unit over an 11-month period of cost savings plus operating expenses. Table 22-3 provides the actual cost analysis:

Compactor—Depreciation cost per month on the system.

Installation—Total installation cost of compactor, construction of hopper and all electrical work.

Table 22-3. Waste Removal Cost Analysis Using Compactor

	Jan	Feb	Mar	Apr	May	Jun	Jul	Aug	Sep	Oct	Nov
Compactor Cost	$ 22.82	$ 22.82	$ 22.82	$ 22.82	$ 22.82	$ 22.82	$ 22.82	$ 22.82	$ 22.82	$ 22.82	$ 22.82
Installation Cost	3.45	3.45	3.45	3.45	3.45	3.45	3.45	3.45	3.45	3.45	3.45
Safety Door										3.33	3.33
Total Depreciation											
over 15 yr	26.27	26.27	26.27	26.27	26.27	26.27	26.27	26.27	26.27	29.60	29.60
Maintenance Cost	78.00			15.00				48.00			
Operating Costs											
Hours of Run Time	11.7	13.9	13.4	13.0	14.3	15.2	11.9	13.9	13.7	14.0	12.9
Electric Service Cost											
(based upon 22½¢/hr)	2.63	3.13	3.02	2.93	3.22	3.42	2.68	3.13	3.08	3.15	2.90
Cost to Remove Trash											
Number of Dumps	6	6	6	6	7	8	5	6	8	9	8
Cost @ $45/dump	270.00	270.00	270.00	270.00	315.00	360.00	225.00	270.00	360.00	405.00	360.00
Man hours to load container	11.7	13.9	13.4	13.0	14.3	15.2	11.9	13.9	13.7	14.0	12.9
Cost @ $4/hr	46.80	55.60	53.60	52.00	57.20	60.80	47.60	55.60	54.80	56.00	51.60
TOTAL COST	423.70	355.00	352.89	366.20	401.69	450.49	301.55	403.00	444.15	493.75	444.10

Safety door—Door was installed to comply with new OSHA standards.

Total depreciation—Total of all the above depreciated over 15 years.

Maintenance cost—In January there was a $78 charge for labor and material to fill the reservoir tank with hydraulic fluid. In April there was a $15 charge to repair a leaking hydraulic hose. In August there was a $48 charge to replace two sheared bolts on ram.

Operating Costs

Hours of run time—Actual hours that compactor operated.

Electric service cost—Total cost for electric service to operate the 15 hp (440v 3-hp) motor.

Cost to Remove Trash (work performed by outside contractor)

Number of dumps—Actual number of times a 31-cu ft container was removed from the facilities.

Cost to remove container—The outside contractor charged a hauling fee of $45 per load.

Man-hours to load container—Based on the hour run time, since janitor loads unit when it is in operation to do away with any delay time.

Total cost—Total operating cost for the compactor unit, plus labor and maintenance per month.

Engineering cost analysis shows that installation of the compactor has saved $7100 for the company in 11 months. This was due to the reduction in man hours, and also the reduction in dumping fees.

COMPACTOR SIZING AND SELECTION[60]

In many facilities, on-site processing may be used to reduce the volume of solid waste collected. Compaction is one of the alternatives for on-site processing of these wastes. Volume and type determine whether a compactor is less expensive than an incinerator with the control devices required to meet air pollution regulations. [Particulate emission from incinerators in Maryland is restricted to 0.03 gr/scf (a) 12 percent CO_2.] However, the basis upon which fees are charged by waste haulers will normally resolve whether compaction is attractive or not.

Compaction reduces the waste volume by 6 or 10 to 1, but at the same time increases the unit weight. If a scavenger charges by the volume of waste hauled, then a compactor is usually very attractive. When charges are based upon weight, compaction may be an expensive choice.

Most compactors consist of a container unit and a compaction unit, together referred to as a compactor. Compactors are limited as to the types of solid waste they can handle efficiently. Trash and loose refuse can be taken care of effectively. However, wastes with tar, sludges, or materials having high moisture content are difficult and often impossible to handle.

Frequently, drainage is required to remove water from compressed wastes in the vicinity of the compactor. A sanitation aid, often used where organic or putrescible materials are compacted, is a masking agent or counteractant for odor reduction. Most of these deodorizers are added by intermittent sprays or drip mechanisms.

The majority of compactors used in industrial applications are stationary units. Others, the bag, console, and rotary types have special applications.

A wide variety of **bag compactors** exist. Some of them have been proven by use, while others are barely beyond the conceptual stage. Variations among types include: horizontal ram, vertical ram, single bag, continuous and multi-bag, pre-shredding, optical controls, and sonic controls.

Bag-type compactors can be chute fed. Manufacturers claim productive equipment capacities ranging from 7 to 44 cu yd/hr. The production of any of these machines is dependent upon the time and attention given by personnel. Single bags must be removed when full and replaced by empty ones. Continuous multi-bags must be tied off, removed and replaced; filled castered containers must be replaced. All of this emphasizes the necessity for matching equipment to anticipated daily volumes and the availability of operating personnel.

Compaction ratios from 4:1 to 8:1 and package densities from 18 to 60 lb/cu ft are claimed by manufacturers, depending upon the composition and mix of solid wastes. Based upon a density of 6 lb/cu ft, and a compaction ratio of 3 or 4 to 1, a density range of 18 to 24 lb/cu ft may be expected. Containerized packages weighing as much as 200 lb are claimed by one manufacturer, but do require the use of more than one man for removal and transport.

Console compactors employ a vertical compacting ram which may be mechanically, hydraulically, or air operated. The unit is usually hand fed. Chute fed models are in the development and testing stage. These units compress waste into a corrugated box container or a plastic or paper bag.

Models are available to process, side by side, one or two containers housed within the same cabinet. Also available are in-line compactors. Some of these units will accommodate up to eight containers, but they are not within an enclosed cabinet. The containerized packages are about 3½ cu ft in volume, and models by one manufacturer produce packages from 5 to 6 cu ft. The densities of containerized packages range between 12 and 30 lb/cu ft. Manufacturers claim compaction ratios as high as 10 to 1. Manufacturers of this type of compactor include Auto Pak, The Tony Team, and Trans World.

Rotary type compactors are sometimes called carousel types. They consist of a ram mechanism which packs loose wastes into paper or plastic bags held in open positions on a rotating platform. When the bag under the packing ram is filled, the platform indexes one position, moving the full bag and replacing it with an empty one. The bag is held in place within a compartment which confines and prevents it from rupture during packing. These compactors are made in standard models of 8 or 10 bag compartments, but are available to accommodate 20 or 30 bags.

Stationary compactors are frequently referred to as stationary packers. Both manufacturers and users employ the terms interchangeably. Usually, it is a compaction unit having a hydraulically operated ram which moves in a horizontal direction. Wastes are fed into a receiving hopper and the ram, when actuated either manually or by optical or sonic devices, compresses the wastes into a steel container. These containers, generally box-like in appearance, usually have an opening in the lower half of one end of the box. This opening allows the compactor ram to operate inside the container during the compaction cycle. This loading end is hinged as a tailgate to swing fully open to permit dumping of refuse at the point of disposal. The container is also equipped with a pair of cables which are used to retain the compacted load during transport to the disposal area.

When a container is full, the compactor ram is entirely withdrawn. Retaining cables and a tarpaulin are put in place, and the container is winched onto a tilt-frame hoist truck to be hauled away. An empty container is then strapped or locked to the compactor, and the entire unit is ready again for operation. The sides and top of the container are tapered slightly to facilitate refuse slideout when dumping.

Standards

With over 50 producers of stationary compactors and more than 150 models from which to select, it can be difficult for the pollution engineer to compare technical and performance characteristics of equipment. The National Solid Wastes Management Association has, however, been formed to help establish rating criteria for compactors. The NSWMA Stationary Compactor Rating Criteria is shown below with references to Figure 22-6.

National Solid Waste Management Association
Stationary Compactor Rating Criteria

Base Size Rating—the theoretical volume of material moved by the ram within the confines of the stationary compactor in a single stroke. This volume displacement is calculated in cubic yards using the distance from the ram face in the retracted position or the rear wall of the charging chamber, whichever is less, to the front of the breaker bar (Dimension B), inside wall-to-wall dimension of the charging chamber (Dimension E), and the height of the ram head (Dimension J).

Clear Top Opening—the minimum free fall clearance from the top to the bottom of the charging chamber. It defines the dimensions (length and width) of the largest object that can be placed in the chamber. Its length is measured from the ram face in a retracted position to the nearest point on the breaker bar of the compactor or the minimum throat (Dimension A). If the ram retracts behind the rear wall of the charging chamber, the minimum free fall clearance, measured from the rear wall to the nearest point on the breaker bar (Dimension A) will be used to define the length of the charging chamber. The width of the clear top opening is defined as Dimension D, or E, or F, whichever is less.

Chamber Length—the linear displacement of the ram within the compactor. It is measured from the ram face in the retracted position or the rear wall of the charging chamber, whichever is less, to the face of the machine (Dimension B). If the ram retracts behind the rear wall of the charging chamber, chamber length is measured from the rear wall to the face of the machine.

Ram Stroke—the maximum linear displacement of the ram in the compaction phase of a cycle, forward stroke (Dimension C). It includes the distance that the ram penetrates into the container during a compaction stroke. It does not include any retraction distance of the ram behind the rear wall of the loading chamber.

Base Unit Weight—weight of basic compactor with standard power unit less hydraulic fluid. It does not include weight of hoppers, support stands, ramps, container guide tracks, and any other optional equipment.

Cycle Time—the actual measured time, stated in seconds, that is required for a complete ram cycle. A ram cycle is defined by the displacement of the ram from full retraction to full extension to full retraction again.

Force Rating—the pounds per square inch (PSI) on the ram face (full width and height of ram) at stated pum pressures. Both the normal and maximum force ratings will be used. These ratings are determined by dividing the normal and maximum cylinder forces by the total area of the ram face in square inches. Normal forces are stated at the recommended operating pressures available for continuous service. Maximum forces are available for final compaction (pin-off-jogg-packout) and are governed by the relief valve setting of the power control unit.

Ram Face—the specification in terms of the full width and height of the ram in inches (Dimensions E and J).

Machine Volume Displacement Per Hour—the number of cubic yards per hour of material that can be processed by the machine at 100 percent utilization and continuous cycling of the machine. It is calculated as the NSWMA rating times the number of cycles per hour of the machine as determined by measured cycle time.

Discharge Opening—the specification in terms of the overall width and height of the lead-in flange of the compactor (Dimensions G and M). It defines the minimum dimensions for the mating container. Where a partial lead-in flange is used, that is, a top or bottom flange only, the height dimension should be stated as if a full flange was intended (both top and bottom of discharge opening).

Motor Size—pump motor horsepower defined in terms of NEMA rating.

Ground Height—height from ground to bottom of discharge opening (Dimension GH).

Penetration—the difference between ram stroke (Dimension C) and chamber length (Dimension B). It represents the portion of the forward ram stroke available for final compaction of the waste load in the container.

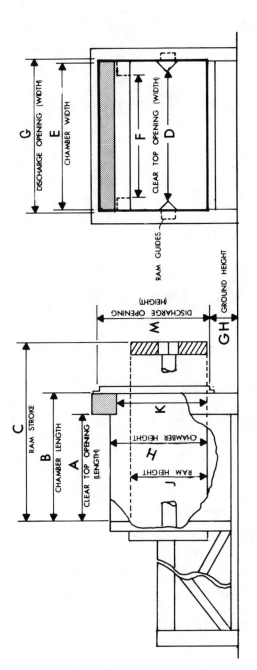

SIDE VIEW

END VIEW

NSWMA Rating Criteria

NSWMA base size rating = B X E X J (cu yd)
Clear top opening = A X (D or E or F, whichever is less)
Chamber length = B (in.)
Ram stroke = C (in.)
Penetration = C - B (in.)
Ram face = E, J (in.)
Discharge opening = G, M (in.)

Ground height = GH (in.)
Cycle time = measured time (sec.)
Force rating = ram pressure at stated pump pressure (psi)
Machine volume displacement per hour = at 100% utilization
(cu yd/hr)
Base unit weight = basic compactor weight

Figure 22-6.

CHAPTER 23

SANITARY LANDFILLING

APPLICATIONS AND LIMITATIONS[150]

Landfilling can be the most economical method of disposing of all types of normal solid waste, including large and bulky items. A properly designed landfill will cause neither air nor water pollution.

A sanitary landfill is not only a refuse disposal operation, but it is also a construction project which results in the reclamation of land. Therefore, careful planning is required in the site selection, design and operation of the landfill.

Distance of the landfill site from the waste generation points should be studied carefully. Sites which require truck hauling distances greater than 10 miles can be economically unsound. However, if closer sites are not available, strategically located transfer stations should be considered.

Land Requirement

An important factor in the selection of a landfill is volume of land required (Figure 23-1). The volume of usable land must be known rather than its area since topographical characteristics vary. In some cases, a few acres may be enough if valleys and ravines can be filled. On the other hand, if the land is fairly level and is to be raised only a few feet, more land would be required.

Cover Material

Refuse in a landfill operation must be covered daily with a layer of soil or other suitable, inert material. Most state laws require at least 6 in. of daily compacted cover. Therefore, the site should be inspected to determine the amount and type of cover material available. A sandy loam is a desirable cover material, with rocky material, pure sand or gumbo clay being the least desirable.

781

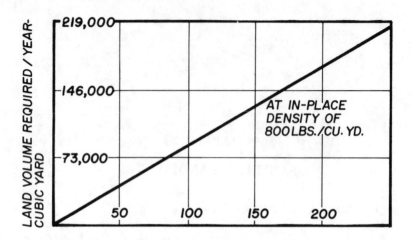

REFUSE GENERATED – TONS PER DAY

Figure 23-1. Amount of land required for a sanitary landfill operation.

The purposes of the daily cover are: confine and keep the waste from scattering, promote proper drainage of surface water, eliminate problems associated with insects, vermin and animals, prevent scavenging and provide a clean, finished appearance.

Access Roads

A landfill operation must be readily accessible to refuse trucks. Roads must be capable of withstanding the loads and constant traffic imposed by the collection vehicles. Preferably, roads should be of a hard surface, all-weather design.

Water Supplies

To ensure that water supplies will not be polluted, the site should be carefully examined for water table depth and the existence of ponds, creeks or underwater springs. A landfill must be operated above the water table, and be designed for proper surface drainage.

Utilities

Most state codes require that utilities be available at the site. Water is required for fire protection along with electricity and telephone service.

Since installation of these utilities can be a major cost factor, they should be considered in an overall assessment.

Land Cost

Initial cost of a parcel of land may immediately rule out its selection as a landfill site. Generally, an engineer should look for inexpensive land that needs reclaiming—land that can be put to a constructive use. He should also investigate the possibility of renting or leasing land, or using public land.

Proximity to Residential Dwellings

Public acceptance and approval of a site for a landfill is many times the most crucial factor in its establishment. Many residents are unfamiliar with the modern methods of sanitary landfilling and will undoubtedly be influenced by the old unsanitary "dump" operations. To alleviate this problem, it is best to locate the site away from residential housing. Consideration should be given to locating in industrial areas or taking over an existing, unsatisfactorily operated dump.

In any case, a program of public relations and education is important in the early planning stages. In some areas, proper zoning will have to be obtained for the site, and public opinion can greatly influence the zoning board's decision.

Sanitary Landfilling Methods

Two basic methods are considered the most effective for sanitary landfilling—the area method, and the trench method. A third, the slope or ramp method, is a variation which can be combined with either of the first two.

Area Method

This method (Figure 23-2A) is best suited for flat or gently sloping land, but can be used in valleys, ravines and areas with land depressions. Basically, it consists of spreading and compacting refuse with a heavy piece of equipment such as a bulldozer, then covering the refuse with an even layer of soil.

Trench Method

This method (Figure 23-2B) is best suited for flat or gently sloping land where the water table is not near the ground surface. Initially a trench is cut in the ground and refuse is dumped into this opening. A bulldozer spreads and compacts the refuse in the trench; an even cover of earth is then placed over the refuse and compacted. Normally, the earth taken out of the ground to form the trench is used as the cover material.

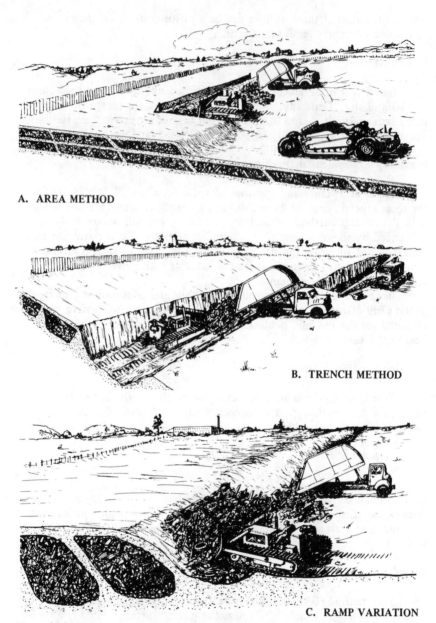

A. AREA METHOD

B. TRENCH METHOD

C. RAMP VARIATION

Figure 23-2. In all methods of sanitary landfilling, refuse is formed into compacted cells
totally surrounded by cover material. The Area Method (A) is used to restore
large areas of land to proper grade, while the Trench Method (B) can be used for
filling ravines and ditches. The Ramp Variation (C) can be employed with either
the area or trench methods. Solid wastes are spread and compacted on a slope.

Ramp or Slope Method

This variation (Figure 23-2C) is generally suited to all areas and is commonly used with the area and trench method. Refuse is dumped on the side of an existing slope, and spread out over the slope and compacted. The cover dirt, usually obtained just ahead of the working face, is then spread over the refuse and compacted.

A landfill operation has some disadvantages. Suitable land is scarce in highly populated areas and there is often public opposition to the operation. Structures on the completed landfill site require special design and construction, due to settling of land. Gases produced from the decomposition of the buried refuse may become a hazard or nuisance.

Costs of sanitary landfilling vary, but generally fall between $.75 and $1.50 per ton for large operations depending on the volume of refuse handled and the particular operation. The cost per ton of refuse disposal in a small operation of 50,000 tons per year or less may vary from $1.25 to $5.00, due mainly to the low efficiency derived from the crew and equipment.

Many factors influence the cost of operating a landfill:

1. Initial land and equipment purchase or rental costs.
2. Site preparation and engineering fees.
3. Rate at which refuse will be deposited at landfill.
4. Useful life of landfill.
5. Payroll costs and fringe benefits.
6. Maintenance costs.
7. Utilities and fuel.
8. Taxes, insurance and interest.
9. Performance bond.

These cost factors lend themselves to computer application and analysis. All of the variables can be programmed to give an accurate cost per ton to be charged at various profit levels. Computer programming can be done in the planning stages before the landfill is started to determine the feasibility of a landfill at any given location. Life expectancy of the landfill at various input volumes can also be obtained. Once the landfill is in operation, more exact data can be supplied to the computer to obtain updated information.

HIGH DENSITY LANDFILLING[99]

Cedar Hills, King County's major solid waste disposal site, is located approximately 12 miles southeast of Seattle. The area is well suited for a disposal site; it is remote and screened from public view with native trees. The site is situated 300 ft above the valley floor and has little run-off or pollution problems. The difference between this landfill project and others is that a Mole buries the refuse.

The Mole is a machine 47 ft long, 9 ft wide, and 27 ft high. Its shape is rectangular with one opening in the top to receive refuse and another to extrude the compacted refuse. It travels like an earthworm in a rigid body.

1. COMPRESSION CHAMBER FILLED

A. Compression plate retracted
B. Refuse fills hopper & compression chamber
C. Loose backfill piled over backfill compactors
D. Backfill compactors cycling
E. Containers emptied into machine as they arrive

2. COMPRESSION STROKE BEGINS

A. Compression chamber closed
B. Excess material sheared off
C. Cover plate seals bottom of hopper

REFERENCE LINE

1

2

3. COMPRESSION STROKE FINISHED

A. Refuse compressed 10 to 1
B. Compression reaction advanced machine in trench
C. Loose backfill compacted as it fills void

4. COMPRESSION CHAMBER FILLING

A. Compression plate retracted
B. Remaining refuse falls into compression chamber
C. Loose backfill piled over backfill compactors

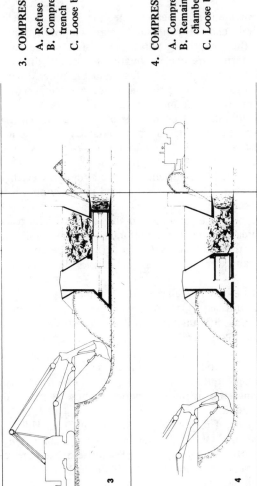

Figure 23-3. Mole Sequence of Operations

That is, everything that goes in its hopper extrudes out its end. The force of the extrusion against the solid immovable back trench propels the machine forward as a ditch in front is relieved.

The Mole operates with the function of two additional machines. One is a 2-yd-capacity, hydraulic backhoe and the other is a crawler-type bulldozer. The backhoe must dig a trench for the Mole to follow, and the bulldozer must keep the extrusion covered and the resistance solid to the rear. The backhoe is positioned to the side to charge the Mole's hopper as each truck-trailer drives up. The Mole rides suspended approximately 2 ft above the surface of the ground. It can be turned for positioning into a new trench and moved forward while it lowers itself down into the pre-dug excavation. The machine slides forward easily on its own compaction machinery behind the backhoe.

Mole is operated from a cab to the rear of the refuse feeding hopper. The cab is entirely enclosed with safety glass windows. Activities of all equipment and the compacting plunger are visible to the operator.

Primary controls are two modulating levels, one driving a large piston rearward and into the compacted refuse, and the other driving its return retraction. Two gages indicate the oil pressure developed by the hydraulic system due to the resistance of the refuse and the resistance to the forward portion of the Mole in the trench. Therefore, it is known that 3000 psi oil pressure is the maximum reaction to be obtained from the machine and its maximum thrust is 700 tons. Pressure on the 7-ft-sq piston and the refuse at this maximum condition is 200 psi.

A bulldozer smooths out and compacts the clean earth backfill as the Mole moves forward. Once refuse is placed in the hopper it is not exposed to the surface again.

The disposal area is on a section and a half of land totaling 920 acres. Access to the area is more than a half-mile from the public thoroughfare. However, use of the site is restricted to 620 acres because of zoning regulations and the many electrical power transmission lines that traverse the property.

Soil at this location is glacial till. The material is well graded, ranging in size from silt and clay fines up to large gravel, cobbles, and occasional boulders. It is very difficult to excavate. Almost like poor concrete in consistency, it softens considerably and turns to muck when excavated and saturated with water.

Due to the impervious nature of the material in its natural state, no subsurface drainage strata exist at Cedar Hills. When water is trapped on the surface during rains, it must evaporate or flow away. It does not penetrate the ground.

Two methods of disposal are employed at Cedar Hills; both are sanitary landfill systems. One is the standard trench system using standard equipment bulldozers, dragline, wheeled compactor, self-propelled and tractor loading cans. The other system is the high density landfill recovery program using the Mole.

Refuse operations serve all of King County's 2298 square miles except the city of Seattle, about one-half million people. Twenty-nine trailers are used to haul refuse from seven transfer points around King County. Containerized 8x8x19 ft refuse boxes are transported on the trailers.

To load and unload containerized refuse loads a hydraulic shovel was modified with a dumping yoke. The shovel lifts the 10-ton containers and swings 180 deg. Time required to empty an 84-cu-yd trailer box from initial approach of the hoist to back-on-road motion is approximately 5 minutes.

Economic Aspects

The Mole was constructed as part of a demonstration grant from the federal government for solid waste experimentation of recognized costs of $446,008.

Cost of operation, including amortization of equipment, is well within the cost range of the average sanitary landfill methods of $.75 to $1.50 per ton of disposal. The Mole's cost for disposal including backhoe, dozer and labor is $1.20 per ton of refuse.

Considering that the Mole places refuse in a sanitary landfill occupying 1/2 to 1/3 the ground-filling requirements of any other known compaction system, then land cost factors are also important. If land costs $3000 per acre, the Mole can easily save 1/2 of that figure, or $1500 per acre land replacement value.

COMPOSTING[78]

Composting is a fermentation process. It differs from the fermentation in the yeast brewery and alcohol industries, however, on two significant points. Composting is conducted in a mass of solids and depends on a great many different types of microorganisms, while the other processes are conducted in a liquid and depend on a single species of microbes with great care having to be taken to keep out all other types.

The basic principles for composting were established after many years of studies and experiments in India by the late British scientist, Sir Albert Howard. It is interesting to note that he predicted the need for applying his ideas to large-scale composting of city wastes, expressing the hope that modern pollution engineering would solve problems of adaption. It seems though that many composting systems have ignored both the highly developed industrial fermentation technique and Sir Albert's more specific teachings.

Practically all organic waste material mixtures are suitable for composting. Very coarse materials like sugar canes should be broken up. Sir Albert accomplished this in his experiments by placing them on the access roads for the ox carts to run over and crush. The moisture in the piles should be between 50 and 60 percent, and the carbon-nitrogen ratio approximately 35:1.

Therefore, attempts to compost common garbage-rubbish with a carbon-nitrogen ratio of 80:1 or higher are bound to fail from the very start. Some material rich in nitrogen must be added to supplement the garbage rubbish. A logical selection of material is sewage, treated or untreated, wet or dry, the untreated variety being preferred.

Sir Albert demonstrated that the composting process starts out with a highly thermophilic phase of fermentation with temperatures in compost piles reaching 160 to 170 F in less than a week. After another three weeks, microbial activities have used up the oxygen in the piles. Turning of the material is required to prevent anaerobic development and resulting petrefaction.

After the turning, the temperature of the material is brought down towards ambient. The piles gradually recover lost heat but will not reach their previous highs—they usually reach 130 to 140 F, which represents moderately thermophilic microbial development.

After another month the oxygen is used up again, requiring a second turning. In this third stage, the temperature will rise considerably less than after previous turnings—usually to about 100 to 110 F—thereby providing the desired mesophilic climate for the final stage, at the end of which the temperature will **fall** to the ambient. This temperature drop signifies the end of a proper composting process and a well-decomposed, stable end-product.

The all-important principle of efficient composting is that the highest temperatures are produced during the initial stage. This constitutes an essential preparatory step for softening up the materials for subsequent activities by other types of microorganisms which thrive at lower temperatures. The high initial temperature is also important to assure that all pathogens in the raw materials are destroyed. Health authorities have set 140 F as minimum temperature for such purpose.

Most compost plants have either ignored Sir Albert's temperature standards or failed to attain them. Maximum temperatures developed near the end of the digester processes rarely reach the 140 F level. Proper aeration or turning of the piles is often neglected. Anaerobic developments are then inevitable, resulting in bad odors and fly and rodent problems.

Sir Albert stressed that piles should be kept loose for easy entrance and circulation of air in the mass. In modern digesters, aeration must be provided with forced air. But here again his teachings have too often been completely ignored, *i.e.*, that composting is a multi-stage aerobic process, with greatly varying air requirements as to both quantity and quality for each stage of the process.

The usual practice has been to force atmospheric air through the entire mass by a turbo blower. This precludes proper control, since air delivery varies with the resistance encountered. Similar climatic conditions are created throughout the mass, which prevents the stagewise microbial developments essential for a proper process and a well-decomposed product. In particular, it prevents the vitally important, highly thermophilic initial stage of development.

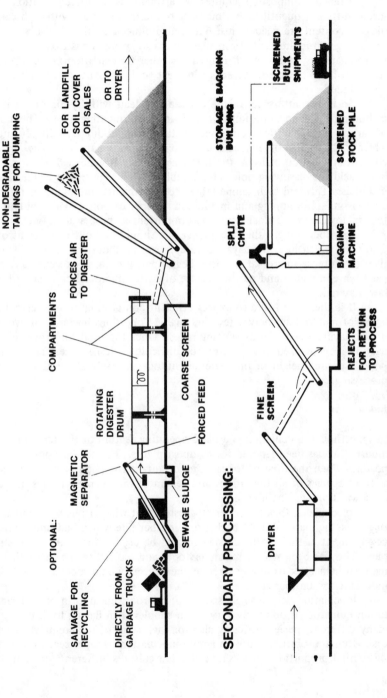

PRIMARY PROCESSING:

OPTIONAL:

SALVAGE FOR RECYCLING

DIRECTLY FROM GARBAGE TRUCKS

MAGNETIC SEPARATOR

ROTATING DIGESTER DRUM

COMPARTMENTS

FORCES AIR TO DIGESTER

NON-DEGRADABLE TAILINGS FOR DUMPING

FOR LANDFILL SOIL COVER OR SALES

OR TO DRYER

SEWAGE SLUDGE

FORCED FEED

COARSE SCREEN

SECONDARY PROCESSING:

DRYER

FINE SCREEN

REJECTS FOR RETURN TO PROCESS

SPLIT CHUTE

BAGGING MACHINE

STORAGE & BAGGING BUILDING

SCREENED BULK SHIPMENTS

SCREENED STOCK PILE

Figure 23-4. Composting system for rubbish and sewage in operation in Big Sandy, Texas and Des Moines, Iowa.

Efficient composting requires better methods for forced aeration. This means different quantities and qualities of warm, moist air with high carbon dioxide content are needed, just like that produced by the microbes to suit the changing requirements from stage to stage. Fresh air is toxic to the microorganisms, and where it is used for other fermentation processes it has to be properly buffered. Direct or indirect sunlight is also toxic and should be excluded.

An important aspect of the climate not dealt with by Sir Albert is the carbon dioxide concentration of the air in the fermenting mass. The amount of carbon dioxide varies from stage to stage, from about 1 to 20 percent, which is 25 to 500 times greater than that of atmospheric air. The highest concentration is normally in the first stage, where it is needed to form carbonic acid for dissolving non-water soluble microbial nutrients in the raw materials. This and high temperature are fundamental prerequisites for rapid composting. Developing and maintaining proper carbon dioxide content in the air, lowest in the final stage, is as important as proper aeration.

Sir Albert taught that a compost pile should be built to a height of 5 to 6 ft, which in a few days would compact to about 4 ft or less. Only then can intense microbial activities begin, partly because excessive air space has been eliminated and partly because of better microbial contact with the raw material.

It is therefore futile to expect efficient composting in digesters where layers of material are only a few feet high. Conditions become still worse if such layers are stirred by moving rakes. Once the need for proper compaction is recognized, it should not be difficult to design digesters which provide desired height of the material without adversely affecting the processing.

Inoculation

Although inoculation is a basic operating feature of the fermentation industry, it has been ignored for compost plants. For old-fashioned composting—when the time factor is unimportant—it is obviously of little benefit as some species of all the groups and genera of the microorganisms needed are always present in rubbish.

Recognizing this, Sir Albert suggested that when a pile was built or turned, some material, held back from another pile which had undergone development in that particular stage, could be used as inoculant for the new pile. This has been widely misunderstood, and has led to the belief that material held back from any stage of the process would serve as an efficient inoculant for another stage.

Inoculation with a single species of microorganism as in other fermentation industries is obviously very much simpler than having to deal with many different species, differing also for each stage of the composting process. But the solution is not to ignore inoculation. Proper seeding with substantial proportions of fresh, acclimated cultures, different for each stage and introduced at their own optimum climates, is a prerequisite for speedy composting and a well-decomposed product. Once the process has been properly established it will produce its own seed requirements, continuously, economically, and indefinitely, much like other fermentation processes.

AIRBORNE SENSING FOR LANDFILL SITE EVALUATION[172]

Local governments are faced with a rapidly growing problem of finding acceptable landfill sites. Growing concern over pollution problems is creating tighter specifications on acceptable sites. A great deal of information about possible landfill sites can be collected by airborne remote sensing techniques using methods such as multispectral aerial photography and aerial thermal mapping.

These methods can be valuable during initial reconnaissance and analysis of surface and near-surface geology, which in turn affects leaching and drainage from landfill sites. Remote sensing methods provide data on features such as:

- Surface and subsurface stone
- Subsurface cavities (potential sinks)
- Near-surface fault structures
- Soil moisture and drainage patterns
- Soil types
- Topography
- Ground cover classifications

Remote sensing can be used for initial rank-ordering of potential sites in area-wide preliminary studies. For individual sites, it can be used to aid in a more meaningful placement of cores. Remote sensing data may also reduce the total number of core samples which might eventually be required. The combination of a large-area reconnaissance capability plus reduction of ground-based engineering costs can make the remote sensing methods attractive on both a technical and a cost-benefit basis.

Thermal Mapping

Measurement and interpretation of surface temperature patterns can contribute to the interpretation of surface and subsurface geology, soil types, drainage patterns, and related matters. These surface temperature patterns are mapped by utilization of airborne infrared line scanning (thermal mapping) equipment operating in either the 1-5.5 or the 8-14 micron region of the thermal infrared spectrum. This portion of the spectrum is not visible to the human eye nor is it directly measurable by photographic techniques.

A thermal mapper scans lines perpendicular to the direction of aircraft flight. The scan rate is 120 lines per second with a maximum field of view of 120 deg. The maximum geometrical resolution is approximately two milliradians, and the limiting temperature sensitivity is approximately ±0.25 C.

Thermal energy emitted from the terrain or water system surface is focused through the optical scanning system onto a liquid nitrogen cooled electronic detector. The detector converts the thermal signal into an electrical signal which is recorded on data tape in flight. Gyroscopic reference signals to compensate for aircraft roll also are recorded on the data tape.

During subsequent playback of the data tape, black and white film imagery can be prepared. The imagery can be considered as a thermal map of the area, with the various grey scales corresponding to variations in surface temperature and emissivity.

The system records variations in temperature rather than absolute true temperatures. By use of a few calibration points, the system can also be used for temperature measurements. However, for landfill site application, relative temperature information generally is sufficient.

Differential rates of heating and cooling during a 24-hr period will cause noticeable changes in the relative temperature shown in the imagery. Normally at least two thermal mapping surveys of an area should be done during a 24-hr cycle in order to make use of both the temperature patterns and the diurnal changes in the temperature patterns during subsequent interpretation and analysis of the imagery.

Figure 23-5 is a block diagram of thermal mapping equipment. A black and white "thermal imagery" is generated by ground-based operations

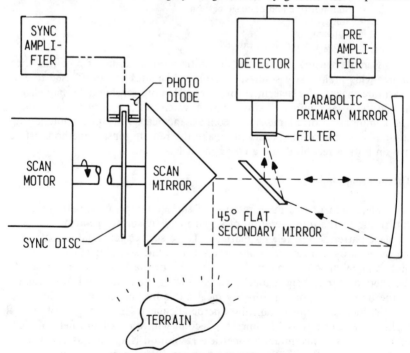

Figure 23-5. Thermal mapper diagram.

from an original data tape. The map produced of the terrain is a visualization of the topographic features. In reality, these are surface temperature patterns which are related to subsurface cavity structures.

Near Infrared

Ektachrome infrared or black and white infrared films are available which extend the photographic range from the visible spectrum out to almost 1 micron in the infrared. For terrain analysis Ektachrome infrared will make this portion of the spectrum visible to the human eye because the emulsion layer is color coded with cyan. Thus, the color seen by the human eye in viewing the film has been shifted from those portions of the spectrum originally seen by the film. (Hence the frequent use of the term "false color infrared.")

Color infrared photography has two important uses for landfill evaluation. First, it is sensitive to soil moisture and thus supplements the information contained in the thermal mapping imagery. Second, it is very sensitive to chlorophyll-containing plants and to slight variations in their condition.

Over reasonable periods of time, moisture and drainage patterns produce vegetative effects which may not be visible to the human eye but which are clearly visible through false color infrared photography. Drainage patterns producing variations in soil nutrients also produce observable differences in the vegetation. For these reasons, near infrared (false color infrared) photography is a valuable tool.

Ektachrome Photography

Ektachrome aerial photography provides imagery with a color balance to which the eye is normally accustomed. While Ektachrome photography is less sensitive to variations in moisture and vegetative character than color infrared photography, it does provide a very useful element in the overall data processing and analysis operations.

It is important to recognize that the airborne remote sensing techniques are of value in providing total area coverage, improving the accuracy and dependability of total site evaluation, and in reducing subsequent costs attributable to improper site selection or missed problem areas. Other applications of these methods include evaluation of sewage lagoon sites, construction sites, highway corridor studies, leakage from reservoirs and earthfill dams, and other problems related to near-surface geology.

CHAPTER 24

SOLID WASTE SALVAGE AND RECOVERY

PROCESSING AND RECLAIMING[151]

Many environmental managers feel that the only lasting solution for the solid waste disposal problem lies in recycling and reuse of wastes (Figure 24-1). Since our natural resources may neither be created nor destroyed, they must be, to the greatest extent attainable, reused to complete the cycle.

It is easy to assume that incineration, deep-well disposal or other such methods are the most straightforward and, therefore, the most economical means of disposal. However, this approach may overlook the economic value of recovering components of the waste for reuse or sale through a reprocessing technique (Figure 24-2). The reprocessing system might cost more than the direct disposal of the waste, but the ultimate value of the recovered product might prove this approach to be the most economical solution.

This does not mean, necessarily, that the value of the recovered material from the waste product will pay for the reprocessing system. But it could change the economics of the reprocessing system in contrast to direct disposal methods, so that the annual totals of the fixed and operating costs are less. This is especially true of wastes generated by large industrial plants where there is a large percentage of reclaimable materials involved.

The first step in the evaluation of any waste disposal problem should be a consideration of possible reuse or sale of the waste or its components. The goal of a processing plant is to economically recycle the highest percentage of this waste at high volumes. Everything that enters a plant must leave in one form or another. By means of a processing plant usable materials can be removed from industrial waste and processed into salable products. The remaining waste residue can be further processed for beneficial usage. Waste processing results in the following advantages:

1. Added revenue is obtained from processed paper products.
2. Less waste at a high density is transported from the plant to the disposal site; therefore, savings in transportation costs are realized.

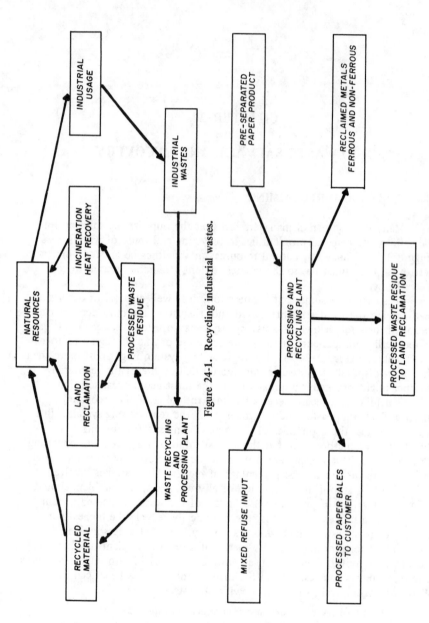

Figure 24-1. Recycling industrial wastes.

Figure 24-2. Material flow to and from a processing plant.

3. Less waste has to be disposed of at the sanitary landfill or incinerator; therefore, dump fees are lower.
4. Processed residue waste is put into a form which makes it suitable for land reclamation.

Containerization

Good materials handling principles dictate that travel time for individual waste collection trucks be kept to a minimum. Therefore, site selection of a processing plant is important. It must be easily accessible to the plants being served. If low-payload collection vehicles travel beyond a certain distance, they reach a point of diminishing returns. Beyond this distance it is more economical to use the transfer station concept. In this method each truck empties its load at the centrally located processing plant instead of making the long haul to the disposal area. The transfer concept allows the short-range trucks time for more profitable work on their individual routes.

Plant Description

A recycling plant uses standard equipment such as conveyors, shredders, compactors, compaction containers, container hauling equipment and a specially designed pulverizer to receive, separate, process and dispose of industrial solid waste materials (Figure 24-3). The process consists of initially receiving the raw industrial waste and conveying it to a salvage-separation area.

The salvageable materials are then separated from the waste material to be further processed by means of a secondary processing system.

Incoming loads from large industrial plants might arrive in compaction containers, since there is a growing trend for plants to install stationary compaction systems. Each compacted load is dumped parallel to the primary receiving conveyor. The unseparated industrial waste is fed onto the recessed primary receiving conveyor by means of a front-end loader tractor. This material is conveyed to the pulverizer conveyor which is at a right angle to the primary receiving conveyor. The speed of the primary conveyor is slightly less than the separation conveyor speed which results in a more even distribution of the material on the separation conveyor and aids the manual separation process. Material is then conveyed to the separation area, where both ferrous and nonferrous metals can be manually picked and deposited into open collection containers. Cardboard and paper products can also be removed in this area and placed onto the paper salvage conveyor for entry into the paper processing system.

Remaining unsalvaged waste residue is conveyed to the main pulverizer unit. Pulverized residue is discharged directly onto a belt takeaway conveyor which conveys the material to a reversible cross conveyor feeding either of two stationary compactors.

A dual compaction system should be installed so that continuous material flow may be maintained during the interval when full containers

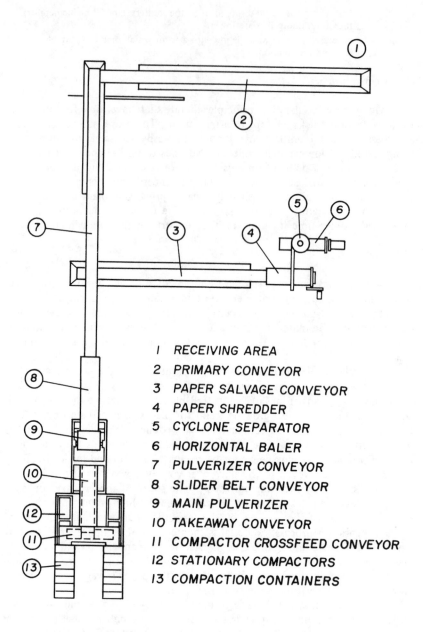

1   RECEIVING AREA
2   PRIMARY CONVEYOR
3   PAPER SALVAGE CONVEYOR
4   PAPER SHREDDER
5   CYCLONE SEPARATOR
6   HORIZONTAL BALER
7   PULVERIZER CONVEYOR
8   SLIDER BELT CONVEYOR
9   MAIN PULVERIZER
10 TAKEAWAY CONVEYOR
11 COMPACTOR CROSSFEED CONVEYOR
12 STATIONARY COMPACTORS
13 COMPACTION CONTAINERS

Figure 24-3. Suggested layout for a processing plant.

are being removed and replaced with empty containers. A single switch reverses the direction of the compactor crossfeed conveyor and controls compactor operation.

Pulverization

A serious problem exists at sanitary landfills in the form of unstable ground conditions due to the poor compaction characteristics of bulky industrial wastes. It is almost impossible to eliminate voids created by the material which will not completely collapse during initial compaction. The pulverization process eliminates these voids and greater in-place densities of each individual cell can be obtained.

It is expected that the pulverization of refuse will permit establishment of a greater number of close-in landfills due to the nonoffensive nature of the fill material. Sites previously discounted for use as potential landfills because of proximity to residential areas can now be reconsidered for possible reclamation areas with the utilization of pulverized residue waste.

Economic Evaluation

Return on Investment

The following calculations compare the invested capital to profits before taxes for a processing plant. Three major cost items are included:

Total Capital Expenditures—Accumulation of all cash outlays made to construct the plant, including facilities, equipment, engineering fees and contingencies.

Total Annual Operating Expense—All capital expenditures after conversion to an annual expense. This is accomplished by multiplying the actual cash expense for each item by a capital recovery factor (crf). The capital recovery factor takes into consideration such values as the economic life, salvage value and cost of money. Labor, direct operating expenses, working capital, general and administrative expenses are also included as a part of the total annual operating expenses.

Total Annual Income—Income from all sources, including income for accepting industrial waste and income from the sale of salvaged material.

For example:

Capital expenditure	$650,000
Salvage value (after six years)	350,000
Annual operating expenses	500,000
Annual income	650,000
Annual profit	150,000

Cost comparisons are based on present worth. This compares the capital expenditure with expected income over the economic life of the installation. Consideration should be given to the time value of money; for instance, the profits obtained after the first year of operation are worth more than those received after the second year, etc. At the end of the economic life of the project, salvage value is considered.

Capital Expenditure–Present worth of annual income over life of project + present worth of salvage value. The return on investment calculations will have to be developed by a method of trial and error using various rates of return. The rate of return which makes both sides of the equation equal is the rate of return for the project.

Present Worth–Using a 10 percent cost of money, six-year life:

$$\$650,000 \overset{?}{=} \$150,000 \ (pwf - 10\% - 6) + \$350,000 \ (pwf^1 - 10\% - 6)$$
$$\$650,000 = \$150,000 \ (4.355) + \$350,000 \ (0.5645)$$
$$\$650,000 \neq \$853,000$$

where (pwf - 10% - 6) = present worth factor for a uniform series of payments with interest at 10 percent for six years

(pwf^1 - 10% - 6) = present worth factor for a single payment with interest at 10 percent six years from the present.

The higher value indicates that a 10 percent return on investment figure is too low; therefore:

Present worth using 20 percent cost of money, six-year life:

$$\$650,000 \overset{?}{=} \$150,000 \ (pwf - 20\% - 6) + \$350,000 \ (pwf^1 - 20\% - 6)$$
$$\$650,000 \neq \$617,000$$

Lower value indicates return on investment is too high; therefore, the correct return on investment is between 10 and 20 percent. Using straight line interpolation we have:

Return on Investment –

$$10\% + \frac{\$203,000}{\$203,000 - (-\$33,000)} \ (20\% - 10\%) = 18.6\%$$

Breakeven Analysis

The breakeven analysis for a plant should compare operating cost with varying income levels. The income levels are a direct function of input tonnages. Breakeven is the point where income equals expenses. In a recycling plant, income can be realized from the dump charges for incoming loads and for the sale of reclaimed materials. The reclaimed materials consist basically of paper products separated from the industrial waste and paper product loads that already have been separated. To determine the varying levels of income, an equation must be derived for the types of income. This equation must relate income to tonnages.

Income from Separated Paper (I_{sep}) = (Price received per ton)(number of tons) or

$$I_{sep} = 12.5y; \text{ where } y = \text{number of tons of separated paper}$$

Income from Salvaged Paper (I_{sal}) = (price received per ton)(number of tons). For salvaged waste the price received per ton consists of two elements:

1. Price received for disposing of waste material
2. Price received for the sale of salvaged paper and metals

The number of tons of salvaged product is expressed as a percentage of the total tons received. In this case, 150 tons of salvageable refuse is received, of which 20 percent is salvaged. Therefore:

$$I_{sal} = 7x + (0.2)(27.50)x, \text{ where } x = \text{number of tons of salvaged refuse}$$

$$I_{sal} = 12.5 \ x$$

In order to combine the two equations, y and x must be expressed in common terms.

$$z = x + y, \text{ where } x = 150 \text{ and } y = 50, \text{ or}$$

$$y = \frac{50z}{200} = 0.25z$$

$$x = \frac{150z}{200} = 0.75z.$$

The breakeven equation becomes:

$$\frac{\text{Operating cost} = \text{income}}{\$500,000 = I_{sal} + I_{sep}} = 12.5x + 12.5y$$

$$= 12.5 \ (0.75z) + 12.5 \ (0.25z)$$

$$= 12.5z$$

$$z = \frac{500,000}{12.5} = 40,000 \text{ tons per year}$$

Operating cost:

$$OC_{sep} = \frac{50}{200} (500,000) = \$125,000$$

$$OC_{sal} = \frac{150}{200}(500,000) = \$375,000$$

Breakeven:

$$OC_{sep} = I_{sep}$$

$125,000 = 12.5y \qquad y = 10,000 \text{ tons per year}

and

$$OC_{sal} = I_{sal}$$

$375,000 = 12.5x \qquad x = 30,000 \text{ tons per year}

Figure 24-4 shows graphical representation of these results.

RECLAIMING PAPER WASTES[149]

Though many items can be salvaged from industrial wastes, one of the most common and profitable is paper products, such as corrugated materials, paper, paperboard and packaging materials. Most industrial solid waste normally contains a high percentage of these items. Basically, two major types of paper waste reclamation systems have been developed and can be economically employed:

Baling systems form paper materials into dense bales retained by baling wires or straps. Shredding the paper materials prior to baling will result in denser bales.

Containerized systems compact and contain wastes in large, steel containers which may either be hauled directly to the buyer or transported to a centrally located processing plant or transfer station.

Baling

A complete baling system may be arranged as shown in Figure 24-5. The major components are shredder, baler, and conveying systems.

Salvaged material to be processed is placed on shredder feed conveyor and delivered into the shredder hopper. After shredding, the small pieces are discharged out the bottom of the shredder into an air handling system. The pieces are air-conveyed into a cyclone collector, where inertial separation of the pieces from the main air stream takes place. Air is discharged into the atmosphere and the shredded paper and cardboard settle to the bottom of the cyclone separator. Shredded particles drop by gravity into the baler charging chamber. The baler then compresses the shredded material into bales which are tied with wire or steel strap.

A shredder added to the processing system will usually pay for itself in a short period of time due to the resultant denser bales or loads. Shredded material allows the production of high-density uniform bales, which saves handling costs, freight and storage space (Figure 24-6). When selecting a particular shredding machine the following items should be considered:

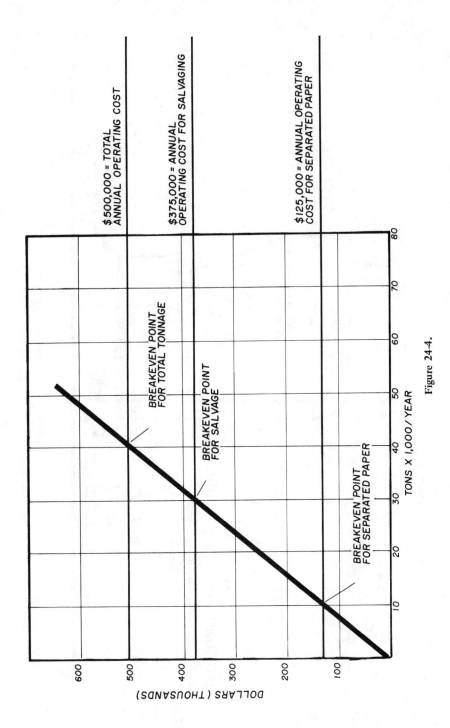

Figure 24-4.

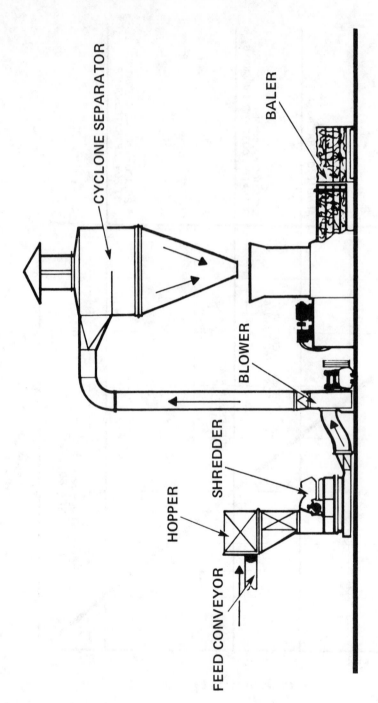

Figure 24-5. Shredding and baling system.

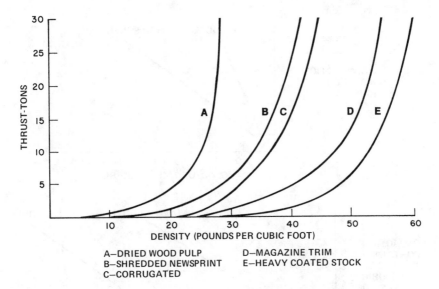

DENSITY (POUNDS PER CUBIC FOOT)

A—DRIED WOOD PULP   D—MAGAZINE TRIM
B—SHREDDED NEWSPRINT   E—HEAVY COATED STOCK
C—CORRUGATED

Figure 24-6. Pressure-density graph of waste paper products processed in a baling system.

1. Volume and type of material to be shredded.
2. Systems required to work in conjunction with the shredder, such as feed and takeaway conveyors.
3. Initial and installation costs.
4. Maintenance costs.
5. Reduction in transportation costs to be gained with increased payloads of shredding operation.

Compaction

Major components of this system are: a feed conveyor, stationary compactor and compaction container. When paper wastes are shredded prior to compaction denser loads, higher payloads and fewer trips are possible (Figure 24-7).

A stationary compactor is a large, heavily reinforced baler. One end contains a cylinder-actuated ram blade, which is the only moving part of the machine. The other end has a large opening to which a compaction container is attached. Paper is fed through a top opening and the compactor blade pushes it into a removable container.

The compaction container is a closed, heavily reinforced storage container. When a compaction container is packed to capacity, the driver detaches it from the compactor and hoists it aboard a truck with a tilting frame. Once aboard the truck, the container and contents are transported to the buyer.

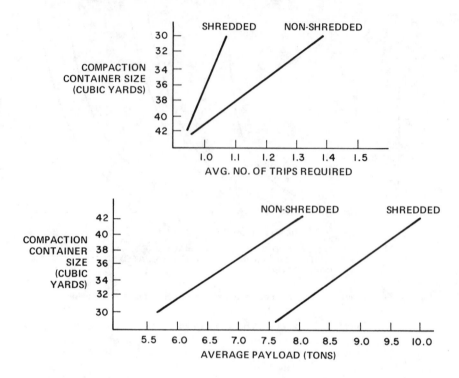

Figure 24-7. Trip and payload analysis of shredded versus non-shredded materials processed in a compactor-container system.

MAGNET CONTROL[14]

Magnets have long served industry to speed materials handling, promote the recovery and beneficiation of ores and remove tramp iron or fine iron contamination in processing operations. However, they can also serve a major role in cleaning industrial and municipal waste.

Wastewater Control

In the water pollution field, magnetic forces are also used to eliminate— or greatly reduce—metallic contaminants in wastewater treated in settling tanks, clarifiers and lagoons. Usually, water is pumped into a tank and the contaminants are simply allowed to settle out.

Magnetic forces can now increase settling rates to make the process more economical and permit smaller settling basins. The system is called *magnetic flocculation* (Figure 24-9). When the contamination in wastewater has a high iron content, the particles can be magnetized. Once magnetized,

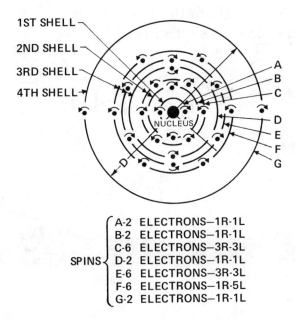

1ST SHELL
2ND SHELL
3RD SHELL
4TH SHELL
NUCLEUS

A
B
C
D
E
F
G

SPINS
A-2 ELECTRONS—1R-1L
B-2 ELECTRONS—1R-1L
C-6 ELECTRONS—3R-3L
D-2 ELECTRONS—1R-1L
E-6 ELECTRONS—3R-3L
F-6 ELECTRONS—1R-5L
G-2 ELECTRONS—1R-1L

D/r must be 3 or more to create ferro-magnetism. This condition occurs in iron, cobalt, nickel and rare-earth groups. Physical concept of the inner structure of a ferromagnetic atom showing the electron arrangement necessary for the creation of magnetism.

The uncompensated, or off-balance, planetary spin of the electrons in the third incomplete quantum shell, together with specific dimensional characteristics creates a magnetic moment, or force.

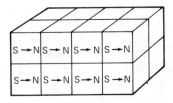

Application of an external magnetic field causes magnetism in the domains to be aligned so that their magnetic moments are added to each other and to that of the applied field.

With soft magnetic materials such as iron, small external fields will cause great alignment, but because of the small restraining force only a little of the magnetism will be retained when the external field is removed.

With hard magnetic materials such as Alnico a greater external field must be applied to cause orientation of the domains, but most of the orientation will be retained when the field is removed, thus creating a larger permanent magnet, which will have one North and one South pole.

Figure 24-8.

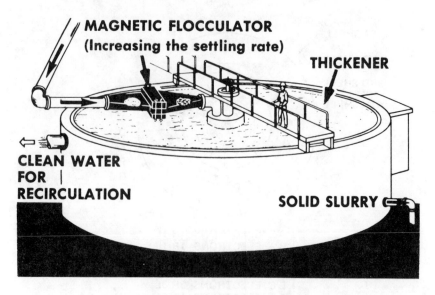

Figure 24-9. Magnetic flocculator.

they agglomerate into flocs which settle at a much faster rate. Chemicals are normally used to promote agglomeration but, when the two processes are combined, a settling rate five times faster than normal has been achieved. Reduction of ferrous contamination from 600 parts per million to a low of 40 ppm is usual.

To magnetize particles, a magnetic flocculator is installed at the inlet pipe to the clarifier or thickener. The flocculator is so shaped that it forces wastewater to spread out over a wide area. Powerful permanent magnets above and below the water flow subject the ferrous particles to a split-second exposure of their magnetic fields. The magnetized particles then attract each other into flocs and begin settling almost at once.

Not only do the magnetic flocs settle at a much faster rate, but they tend to entrap a substantial number of nonmagnetic particles in their descent, thus speeding up the whole settling process. Chemical-magnetic flocculation systems have been used to treat blast furnace dust, cupola dust, flyash, nonmagnetic clay and other solids.

At the Lackawanna, N.Y. plant of Bethlehem Steel Corp., wash water from wet scrubbers on the basic oxygen furnaces is treated. The magnetic flocculator is installed in a 30-inch diameter slurry feed line leading to the primary thickener, which handles 12,000 gpm.

In the metalworking industries, several types of magnetic equipment are used to clean ferrous chips, grindings, turnings, borings and other machining waste from liquid coolants. Armco Steel Corp. in Middletown, Ohio, uses four magnetic wet drum separators in its new strip mill to remove ferrous particles from a cooling mixture of oil and water that is recirculated around the rolls on a five-strand rolling mill. The coolant enters the separator near the bottom of a revolving stainless steel drum capable of handling feed rates up to 600 gpm (Figure 24-10).

Figure 24-10. Magnetic drum separator.

Caterpillar Tractor Co's parts plant in Davenport, Iowa, uses a magnetic chip and parts conveyor to remove chips and fines from cutting oils in facing and tapping operations. The conveyor is shaped like a dogleg, and has a series of powerful permanent magnets running beneath a nonmagnetic stainless steel surface. It has a liquid-tight housing with no external moving parts. The lower section is installed right in the sump so that the cutting liquid and all fines must pass over it. The magnets collect chips and fines from the oil and carry them up the surface of the conveyor to a waiting hopper. The cutting liquid, cleansed of ferrous material, can be recirculated for a considerable time before discharge.

Solid Waste Disposal

Magnetic separating equipment is coming into widespread use for recycling cans and other metal containers. At special collection centers, the containers are dumped onto conveyors fitted with magnetic head pulleys capable of handling 500 12-ounce cans per minute. Aluminum cans passing over the pulley fall freely from the conveyor into weigh hoppers while ferrous containers are trapped and held by the pulley until they are carried outside the magnetic field and dropped. Once separated, the containers are processed for reuse.

Magnetic technologies have developed improved ways to extract valuable metal from junked cars. Junked autos—glass, upholstery, plastic and all—are compressed in a giant squeeze box, then chopped on a shearing machine and thoroughly macerated in a heavy-duty grinder. The shredded pieces fall onto a revolving drum separator. Nonmagnetic materials drop to a conveyor belt below for further processing. Similarly, magnetic separating equipment is used to remove bead wire from old tires in rubber reclamation processes, to eliminate tramp iron in waste paper salvaging, to produce ferrous-free cullet for glass recycling operations and to separate ferrous trash and fine iron contamination from aluminum foil scrap in the production of powdered aluminum.

Incinerator residue is another waste product which can be processed through magnetic separation equipment. Of an estimated five million tons of incinerator ash produced annually in the U.S., some two-thirds is valuable metal. Using magnetic drums and suspended electromagnetic separators, ferrous materials can be reclaimed for scrap. Once separated, the ash increases in density for landfilling.

Each installation, and all the variables that apply to it, should be considered separately. Among the factors involved are the volume of the waste flow, density, composition, temperature, velocity, the amount and magnetic susceptibility of ferrous material present, and the end result desired (Figure 24-11).

SCRAP STEEL MAGNETIC REMOVAL[80]

Until a practical method was developed to ensure a consistent physical and chemical content of scrap steel for reclamation, the removal of piles of scrap auto bodies, farm implements, and large appliances from the landscape was a slow process. In fact, the piles usually grew because it was unprofitable to bring the material to processing areas for the stripping and preparation required before remelting. Now, scrap car bodies and other metallic wastes can be systematically processed by shredders. Separation is effected by feeding the shredder discharge to a magnetic drum designed and constructed specifically for the system. The drum separates steel from all the nonmagnetic components. Magnetic and nonmagnetic charges then fall from the drum onto separate take-away conveyors.

The key to whether the end product, shredded steel, is the cleanest possible lies in correct use of the magnetic drum. This is the stage that determines the system's profitability because the cleanest steel scrap will command the highest price.

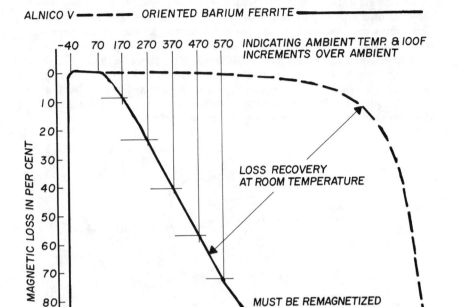

Figure 24-11. Magnetic material temperature chart—
temperature change vs. magnetic loss.

Drum Design

The magnetic "scrap drum" or "shredded steel upgrading drum" is a reverse application of the more familiar product purification drum. A product purification drum is used to remove small quantities of iron or steel contamination from food products, chemicals, wood chips, and other free-flowing materials. The iron or steel contaminate is a very small quantity of the flow—perhaps 1 percent or even less. Handling and upgrading the discharge from a shredder, however, presents the opposite problem. The ferrous (iron or steel) content of the product flow to the drum can be 90 percent or higher. Obviously this product analysis dictates a particular drum arrangement, a specific magnetic circuit, and heavy-duty construction, all tailored to the characteristics of the material to be handled.

Viewing a magnetic drum from its end with rotation clockwise, the shredder discharge should be fed to the drum from an 8 to 9 o'clock position. Feed must be uniform without excess surges, so that nonmagnetic items will not be entrapped in the steel pieces as they jump to the drum surface. A vibrating feeder pan should be placed to receive the discharge from the shredder and carry it to the drum. Other feed methods can be used (from a belt conveyor head pulley, from an inclined chute), but the main consideration in drum feeding is uniformity. The ultimate degree of steel purity depends on a smooth level feed to the drum.

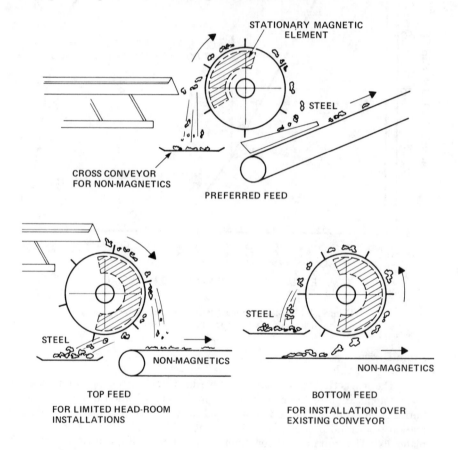

Figure 24-12. Methods of feeding a magnetic separator.

Drum Construction

Maintenance of a magnetic drum must be minor, because any downtime of a drum will shut down the entire processing system. For this reason all drum components should be heavy-duty and selected for continual use. The drum shell, which is the only surface of the drum experiencing any impact or wear, should be abrasion-resistant manganese (nonmagnetic) steel. Wiper angles should be used on the outside of the shell to move the steel along so it does not bunch up in any area on the shell. Drum end flanges should be heavy ribbed castings or weldments.

Shaft and thrust bearings should be lubricated from outside the drum. They can be contained in a separate bearing cap assembly which can be removed with the drum remaining in place. With the bearing cap removed, all bearings and seals can be exposed for examination, service or replacement.

The magnetic element itself must be uniform in intensity across the drum face with its maximum strength at the pickup or feed area. Magnetic strength gradually decreases through the arc of the element to the release point. Only the drum shell should rotate. The magnetic element is stationary behind it and is anchored to the nonrotating shaft by heavy hanger plates.

Loosening the shaft clamp blocks and revolving the external shaft through a small arc will move the magnetic element the same amount. The pickup or "grabbing" area of the magnetic element can be put in the best radical position to suit the specific feed method and the exact nature of the product handled.

Magnetic scrap drums have been installed in diameters from 3 to 8 ft and with face widths up to 8 ft. Most installations have one separation stage (one drum), but product characteristics or strict steel purity demands might dictate a two-stage system. In these installations two magnetic drums are installed in tandem with the second drum generally smaller than the first. Because most of the magnetics will be removed by the first drum, the second drum will receive a more consistent product and less volume.

To suit installation space limitations, the feed can be introduced to the drum at other than the preferred 8 to 9 o'clock position. However, the separated steel has to be carried through a longer arc on the drum surface with these alternate arrangements. Consequently, a special magnetic element is usually required inside the drum.

While scrap processing may not offer the public identification and appeal that other pollution combating areas do, shredding, separating, and reusing scrap steel items can make a favorable impact. Design and application engineering from concept to installation and startup can make possible profitable and efficient operations.

MAGNETS SALVAGE SCRAP—A CASE STUDY[193]

Intercoastal Steel Corp. in Chesapeake, Va., is helping clean up the Virginia-North Carolina landscape. Every working day Intercoastal disposes of 250 to 300 junked cars, remelting the ferrous metal they contain and forming it into angle iron and reinforcing bar for the construction industry. So effective has the recycling program been that old cars, truck bodies, discarded refrigerators and washers are becoming scarce in the area within a 100-mile radius of Chesapeake. The company is now looking farther away for the raw material needed to meet the demand for their finished products.

Junk cars arrive at Intercoastal as either flat-pressed hulks hauled in quantities on trailers or as "whole" autos towed by wreckers. The cars that come in intact are stripped of their gas tanks (to prevent explosions), tires, batteries and radiators.

A crane equipped with an orange-peel grab bucket then picks up the car and lowers it into a shredder. In less than a minute the car is pounded and sheared into a mixture of iron, steel, brass, copper, glass, plastics and fabric. Separating iron and steel from this mass of material is the job of a magnetic removal system. Intercoastal uses a basic two-stage separating process, with two scavenging operations to recover additional ferrous material. Magnetic pulleys on conveyors handle the initial separation, while the final stage of removal is accomplished by an electro-permanent magnetic drum.

Figure 24-13.

Shredded autos are discharged from the masticater into a long-stroke vibrating feeder. The feeder supplies the magnetic tail pulley of a belt conveyor at about the 9 o'clock position. It is placed far enough away from the pulley to allow non-magnetic material to fall freely onto a second belt conveyor which is immediately below and running at right angles to the one above.

Much of the magnetic material is attracted to the magnetic tail pulley and is carried away to the next step. Some falls, however, to the "dirt" conveyor below and is conveyed to a discard pile. Just before reaching the discard pile, the partially separated material passes closely under the magnetic tail pulley of a third belt conveyor. This pulley picks up additional ferrous material mixed in with the nonmagnetic discards, and returns it to the metal-conveying belt leading to the second stage of separation.

The second stage is much like the first, except for a 4-ft-diameter magnetic drum which performs the final separation. Here, there is another "dirt" conveyor beneath the gap between the vibrating feeder and the drum. At the head of the conveyor, there is another conveyor with a magnetic tail pulley to pick out additional magnetic material shaken loose from entrapping nonmagnetics and return it to the feeder for another pass at the drum.

The drum combines both electro and permanent magnets to produce a strong, deep magnetic field, as well as equal holding power across its entire surface. The model used at Intercoastal is 4 ft in diameter, 5 ft wide overall with a 4-ft-wide magnetic area. The drum shell rotates at 250 fpm. Stationary magnets inside the shell produce a magnetic field, which reaches out and grabs the ferrous scrap. Held tightly, the scrap is carried over the top of the drum and away from the magnetic field for discharge on the side of the drum opposite the feeder.

The all-ferrous material recovered by the drum falls to a metal stacking conveyor, which carries it to stockpile areas. From here the scrap goes to one of Intercoastal's two electric furnaces, where it is melted into ingots. The ingots are then transferred to their rolling mill for forming into reinforcing bars and angle iron. The average junk car yields 1600 lb of reusable iron and steel.

GETTING RID OF ABANDONED CARS[184]

Four million junk automobiles have been abandoned behind vacant buildings in small towns and along roadsides of rural areas throughout the United States. It is estimated that 170,000 more will be abandoned each year—unless something is done to check this irresponsible dumping. Because these vehicles are so scattered, it scarcely pays for scrap processors to collect them.

A test project organized and run by General Motors on a community-wide basis has shown that officials of small towns and cities and leaders of civic organizations in rural communities can collect abandoned cars themselves. The roundup and disposal will cost money—but not a prohibitive amount.

Key to success of the project is planning, organization, and a basic knowledge of what must be done, when and how. During GM's two-month program in Traverse City, Mich., 2500 abandoned cars were collected and processed for scrap metal. Based on this test run, GM has compiled the following guidelines to help other communities organize and finance their own cleanup campaign.

Planning and Organizing

The success of an abandoned car cleanup campaign requires thorough planning and organization. Cars must be located, counted and reported to a central clearing point. Legal permission must be obtained to pick them up. Arrangements must be made to have them collected, stripped, crushed and shipped to a shredder or baler where they will be recycled into scrap metals that can be processed for reuse by steel mills and foundries.

All of the work must be done at a reasonable cost. The job requires the wholehearted cooperation of an entire community—governmental, business and civic groups.

The community part of the campaign probably will end when the junk cars have been collected. Stripping, crushing and shipping are jobs best performed by local auto wreckers or scrap processors.

The following groups should be represented on the committee that will run the campaign, or they should be consultants to the committee:

1. Local, state and federal government officials—they can provide legal advice and technical assistance through their agencies. City attorneys, law enforcement agencies and the state highway department should be included.

2. Business and industrial leaders—they can supply certain types of equipment that will be needed.

3. Members of civic groups and service clubs—they can do the locating and counting and, just as important, can spark the enthusiasm and provide the spirit the community needs to do the job.

4. Auto wreckers, scrap dealers and processors—these businessmen can serve as the focal point for reporting and collecting activities.

5. A public relations and communications specialist—he will be needed to keep all the groups—and the public—informed and involved.

6. A legal advisor—he is absolutely essential. He must interpret and explain how the state and municipality regulate the disposal of abandoned cars.

Setting Priorities

Once the campaign committee has been created, it should set priorities and list them in order of importance. Two questions that must be answered at the outset are: what legal steps are involved and how much will the campaign cost. One of the first major jobs of the committee might be to seek changes in local or state laws so that "collectors" will not be accused of car theft when the roundup begins. Other legal questions that might have to be answered include: How is the value of an abandoned car determined and who makes this decision? How long must a car be held before it can be disposed of legally? Who maintains custody of the vehicle during this holding period?

When figuring costs, auto wreckers and scrap processors should be consulted. If these businessmen cooperate, and if the abandoned cars can be located and collected easily using volunteer help and equipment, it may be possible to complete the campaign at little cost.

Federal, state and local government agencies frequently have funds that can be allocated to abandoned car cleanup campaigns. A few states have initiated funding programs for that specific purpose:

- In Columbia County, N.Y., the County Health Department decreed that abandoned cars were breeding grounds for rats. It conducted a highly successful cleanup campaign using funds from the state's rodent control program. The campaign collected and got rid of 12,000 abandoned cars at a cost of $20,000.

- Vermont's Motor Vehicle Department contracts with private companies to remove abandoned cars after they have been picked up and delivered to central locations by municipal workers or volunteer civic organizations. The program began in 1968 and, since then, Vermont has removed 13,151 cars under 32 contracts. Cost averages $10 per car.

- Maryland pays licensed scrap processors $10 for each car turned into scrap. The fund comes from charging an extra $1 on annual title fees for each car. The state has imposed a $200 fine for vehicle abandonment.

Location-Condition-Permission

The exact location of each car is needed to help arrange pickup routes and schedules. Local law enforcement agencies can provide information on the location of abandoned cars, but residents in the community also should be encouraged to participate. They can report abandoned cars and ask the committee to pick up cars they want to discard. Winning citizen assistance and cooperation requires a good publicity campaign. In Traverse City, GM worked with its dealers and the local press to win public cooperation.

The condition of the car dictates the method of pickup. If it has already been thoroughly stripped, it can't be hauled in by tow truck. A flat bed truck or trailer will be needed.

Permission to move the car is definitely necessary. Some people don't want to give up old cars abandoned on their property even though they are virtually valueless. Permission from the property owner is essential even though the car has no registered owner.

Collecting the cars should be contracted out to local garages or auto wrecker firms that have the equipment and know-how. Select a drop point that has sufficient space to store cars as they come in and to store parts and components after the cars have been stripped.

A local auto wrecking yard might be the most suitable drop point. But if volunteer pickup crews will be used, or if contracts have been negotiated with several auto wreckers or scrap processors, it might be preferable to establish a central drop point somewhere else.

Other details requiring consideration before collection begins are:

1. Equipment—Find out what kind of hauling equipment is available in the area, who owns it, and how to obtain its use.

2. Pickup routes—Good planning and a thorough knowledge of the street and highway system are essential. Before each pickup crew makes a run, doublecheck to make sure the cars scheduled for collection have been legally cleared and marked. In the Traverse City campaign, a special deputy spray-painted a large X on cars cleared for processing.

3. Release forms—The legal advisor can determine the types of forms required; usually a form for the vehicle owner and the owner of property from which a car is being removed. To save time, releases should be signed before the actual pickup. Also, simple pickup receipts should be prepared for hauling crews.

Once the abandoned cars have been collected, professionals can handle the jobs of stripping, flattening and shipping the processed hulks to market for recycling. However, the campaign committee must work out the details in advance. For example: Will stripping be done by the wrecker or the processor? When will it be done? Is a portable flattener available? All these points influence the cost of the entire campaign.

The flattener is the equipment that makes it economically feasible for auto wreckers and scrap processors to handle cars abandoned several hundred miles away from scrap processing centers (Figure 24-14). Crushing cars at the collection center makes it possible to ship hulks at significant savings in space and, therefore, cost (Figure 24-15). But to pay for its upkeep, the flattener must process at least 15 cars an hour. This means that stripping crews must work fast or well in advance to assure an adequate supply of hulks for flattening.

The final step in an abandoned car cleanup campaign is shipping them out of the area to the shredder or baler for recycling. This should be handled by the scrap processor. Here again, transportation costs are an important part of the total economics, and the campaign committee must get involved in decisions on how shipments will be made. Costs will vary considerably depending on whether rail, truck or barge transportation is used—and it usually pays to shop around for the best price.

DESTRUCTIVE DISTILLATION OF USED TIRES[5]

Disposal of scrap rubber, particularly scrap tires, is becoming an ever more pressing national problem. Several billion pounds of new rubber are consumed annually in the fabrication of tires. After the treads are worn off, some of the tire carcasses are recapped and some of the rubber is made into reclaim, but most of it ends up on scrap piles.

Figure 24-14. Car flattener in operation.

Figure 24-15. Flattened cars ready for transport.

This type of solid waste makes a particularly difficult disposal problem. Because of its high heat-producing capability, a tire cannot be burned in conventional incinerators. Even if it could, large quantities of particulates, hydrocarbons, sulfur and nitrogen oxides would be generated, thus creating an air pollution problem.

Because of low bulk density and resistance to biodegradation, tires do not lend themselves well to disposal in landfill operations. Areas for landfill are becoming critically short in the areas of high population density—which are also the areas where the most scrap tires are to be found.

The most logical approach to finding uses for this huge quantity of hydrocarbon-rich waste is to recover the hydrocarbons for reuse by pyrolytic techniques.

The Firestone Tire and Rubber Company and the U.S. Bureau of Mines Coal Research Center entered into a cooperative research program (wholly funded by Firestone) to study destructive distillation (carbonization) as a means of disposing of scrap rubber and reclaiming valuable chemicals from the product. Bench scale experimentation has conclusively demonstrated the feasibility of this approach.

At the U.S. Bureau of Mines Coal Research Center in Pittsburgh, Pa., equipment was located as shown schematically in Figure 24-16. This unit, known generally in the coal industry as the Bureau of Mines—American Gas Association (BM-AGA) Pilot Plant, is usually used for studying the coking properties of coals.

The apparatus consists essentially of a hermetically sealed retort heated to the carbonization temperature in an electric furnace. Volatile effluent is passed from the retort into a series of condensers which separate the heavy oils into crude fractions. The lighter components then pass through an electrostatic precipitator to remove entrained particulate matter. Following the precipitator are acid and caustic scrubbers for removal of basic and acidic components from the hydrocarbons and gases. The gases are then dried and light oils removed by condensation in a "dry ice"-acetone trap. The noncondensible gases are passed through a metering system to a gas holding tank where a portion of the gas is retained for analysis. The remaining gas is flared.

A variety of raw materials was tested (Table 24-1). Shredded whole passenger tires, containing both rubber and fabric, as well as ground rubber and containing no fabric were tested. Samples prepared from truck tires were tested also to evaluate the effect of the higher percentages of natural rubber.

Yields of solid, liquid and gaseous products from carbonization of scrap tires under a variety of conditions are shown in Table 24-2. Each test was conducted on approximately 100 lb of raw material. A typical test required 7 to 14 hr to completely devolatilize the charge.

Tests PT-1 and PT-3 give a direct comparison of the effect of carbonization temperature on the product distribution. At the lower temperature (500 C), less residue and gas is produced with nearly 50 percent of the

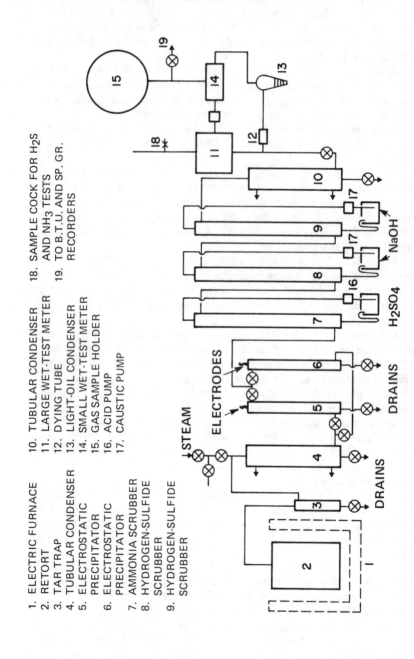

1. ELECTRIC FURNACE
2. RETORT
3. TAR TRAP
4. TUBULAR CONDENSER
5. ELECTROSTATIC PRECIPITATOR
6. ELECTROSTATIC PRECIPITATOR
7. AMMONIA SCRUBBER
8. HYDROGEN-SULFIDE SCRUBBER
9. HYDROGEN-SULFIDE SCRUBBER
10. TUBULAR CONDENSER
11. LARGE WET-TEST METER
12. DYING TUBE
13. LIGHT-OIL CONDENSER
14. SMALL WET-TEST METER
15. GAS SAMPLE HOLDER
16. ACID PUMP
17. CAUSTIC PUMP
18. SAMPLE COCK FOR H_2S AND NH_3 TESTS
19. TO B.T.U. AND SP. GR. RECORDERS

Figure 24-16. Elements of a three-section electrostatic precipitator.

Table 24-1

Description of Raw Materials Tested

Test No.	Source	Description
PT-1	Whole passenger tires	Shredded, bead wire removed
PT-3	Whole passenger tires	Shredded, bead wire removed
PT-4	Whole passenger tires	Shredded, bead wire added
PT-2	Passenger tire tread	35 Mesh (No fabric or bead wire)
PT-5	Whole passenger tires	3-inch pieces, no bead wire
PT-6	Whole passenger tires	Quartered
TT-3	Whole truck tires	Shredded, bead wire removed

Yields of Carbonization Products
Yields, Weight Percent

Test	Temp., C	Residue	Heavy Oil	Light Oil	Gas	Total
PT-1	500	42.0	45.2	4.2	5.0	96.7
PT-3	900	52.3	14.0	6.5	20.8	97.3
PT-4	500-564	41.8	41.3	4.3	5.5	97.4
PT-2	500	44.8	45.6	1.7	3.2	95.6
PT-5	500	40.3	44.6	3.3	4.8	98.0
PT-6	500-900	42.1	42.7	5.5	7.2	100.0
TT-3	500	36.5	48.7	4.3	5.0	95.7

charge being converted to oils. At 900 C a substantially larger portion of the charge is converted to solid residue and gas with much less oil being produced.

Two points of interest and significance are the very high aromatic content of the oil produced at 900 C relative to that produced at 500 C, and the fact that the tire rubber produced significantly more olefins and less paraffins and naphthenes than the passenger tire rubber. The presence or absence of fabric in the rubber had little effect on the nature of the oils.

Properties of the gaseous products were markedly influenced by the carbonization temperature. In general, the heating values of the gases produced at higher temperatures were lower because of higher percentages of hydrogen and low molecular weight hydrocarbons. Truck tire rubber produced a gas with a relatively high content of C_3 to C_5 hydrocarbons and thus had a very high heat content.

All of the volatile products were analyzed by gas chromatography. The oils, both heavy and light, were shown to be extremely complex mixtures. More than 50 components were identified in the various oil fractions. Many additional unidentified components were separated.

That destructive distillation of scrap tires on a large scale is technically feasible has been clearly demonstrated. The products produced can find

Table 24-2

Properties of the Solid Residue				Neutral Oil, Volume Percent				Gas Properties		
Percent Fixed Carbon	Percent Ash	Percent Sulfur	Heating Value, Btu/lb	Percent Residue	Olefins	Aromatics	Paraffins and Naphthenes	Specific Gravity	Btu per cu ft	Percent Hydrogen
86.3	9.6	2.0	13,470	39.6	15.5	51.5	33.0	0.628	922	54.97
90.5	8.3	1.7	13,500	30.1	13.0	84.8	2.2	0.460	765	54.29
87.7	9.2	1.9	13,490	35.5	17.5	51.6	30.9	0.658	835	47.83
74.8	10.2	2.3	13,720	39.8	16.0	54.2	29.8	0.599	890	49.80
83.1	12.7	2.0	13,020	30.4	19.0	50.6	30.4	0.739	921	54.52
85.4	13.6	2.5	12,640	35.2	12.0	56.8	31.2	0.479	762	58.93
80.1	16.5	3.2	12,250	27.6	25.0	51.4	23.6	0.781	1145	49.48

application in a number of areas. Roughly 45 percent of the products are solid carbonaceous residues. These materials could be used in applications such as filter char, as particulate in concrete or asphalt or as smokeless fuels. The liquid and gaseous products can be used as fuels. Although these products are extremely complex mixtures, and as such are not an economical source of a particular hydrocarbon, it is conceivable that a class of compounds for a specific application could be separated economically.

There are two major questions concerning pyrolytic conversion of scrap tires which remain unanswered. The first is whether this type of waste is suitable for continuous processing rather than batch processing. The second is whether product applications can be found which will make this method of scrap rubber disposal economically attractive.

SOLIDS RECOVERY BY SPRAY DRYING[86]

Modern spray drying techniques, now being used to convert many liquid products into marketable powders, are a valuable ally of industry in the battle against pollution. They are also an important method of materials reclamation.

Spray drying is a means of converting a solution or suspension into a dry product. The fluid is reduced to a fine spray and mixed with a stream of heated air to produce a dried powder (Figure 24-17). The powder is then separated from the air.

As the liquid portion of the pollutant is evaporated in the spray dryer, a residual powder is collected and packaged. It may prove to be a commercially valuable product instead of a waste material.

Atomization

The atomization method is the most important single step in the spray drying process. Methods used to atomize the fluid product into the drying air are:

1. **Two-fluid nozzles**—This system uses air and the liquid for atomization. It involves large volumes of relatively high-pressure air. Original applications were in laboratory and pilot-plant dryers. An abrasion resistant nozzle is used in drying specialty ceramics.

2. **Spinning disc or centrifugal atomization wheels**—These are used in many spray dryers. Rpm of the spinning discs may be anywhere from 8000 to 18,000 rpm. The degree of atomization is varied by increasing or decreasing the speed of the wheel and by increasing and decreasing the rate of product feed onto the wheel. The spray angle of a centrifugal atomizing wheel is always 180 degrees as the centrifugal force of the wheel throws the product out in a straight line.

3. **High pressure nozzles**—The nozzles operate at up to 8000 psi. High pressure nozzles usually consist of two basic pieces—the orifice through which the product is swirled, and the spinner, whirl chamber or swirl chamber which gives the product a rotational motion. These

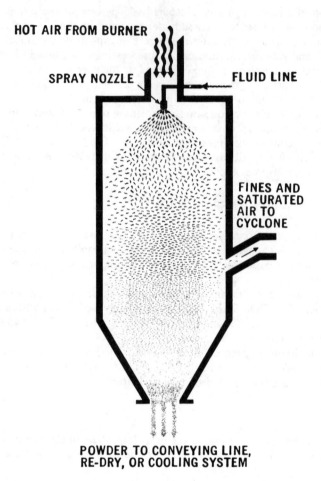

HOT AIR FROM BURNER

SPRAY NOZZLE　　　　　　　**FLUID LINE**

**FINES AND
SATURATED
AIR TO
CYCLONE**

**POWDER TO CONVEYING LINE,
RE-DRY, OR COOLING SYSTEM**

Figure 24-17. In a main dryer using a single-nozzle spray, water is evaporated as the product passes through the chamber. The solid effluent in powder form is removed at the bottom of the unit.

are matched in sets, and at a given pressure thousands of combinations can be made to meter different flow rates and give spray angles of 25 to 120 degrees.

The viscosity of the fluid to be sprayed is of great importance. As viscosity increases, atomization becomes more difficult. Maximum upper limits for the high pressure atomizing nozzle are 10,000 to 12,000 centipoises. Maximum upper limits for atomizing on the centrifugal atomizer are 4000 to 5000 centipoises.

Drying Medium

With non-heat-sensitive products the drying medium is usually heated by direct-fired gas burning. Inlet air temperature can approach 1100 F, with normal exit air temperature between 200 and 300 F.

The drying medium can also be heated by indirect steam or indirect fire. This will keep the inlet air temperature in the 350 to 450 F range and is the method normally used to dry products that contain a flammable, volatile liquid.

Once the product is dry, it is separated from the drying air. Cyclone collectors are usually used as primary collectors. However, as particle size decreases, the effective collective efficiency of a cyclone rapidly decreases. Bag collectors are then used to remove fine particles.

Reclaimed Products

Spray drying is becoming economically feasible in an increasing number of instances because the dry product it produces often can be sold at a profit.

Cheese

Manufacture of cheese provides an example of profitable reclamation. In this industry millions of pounds of whey used to be disposed of in sewers. Spray dryers being used on whey from cheddar, Italian and other cheeses make possible a 3 to 5 cents a pound profit on the product. Reclaimed powder is sold to bakeries and food manufacturers as a food ingredient. A cheese plant handling 400,000 lb of milk a day will produce 24,000 lb of whey powder, which could net an additional daily profit of $1200.

Wood Processing

By using spray drying techniques, one manufacturer of hardboard has turned a pollution problem into a profitable animal feed business. The hemicellulose foodstuff, used primarily in the cattle industry, is a by-product of hardboard production. In this process, hardwood and softwood chips are charged in pressure chambers called "guns." Live steam is then injected until a predetermined pressure is obtained. The chips are then "cooked."

When the contents of the gun are released suddenly through a restricted port, the chips are converted into coarse wood fibers. The pressure and temperature to which the chips are subjected produces soluble products that are removed in a washing process. A large percentage of these solubles are polysaccharide precursors of simple sugar and other five-carbon sugars— all carbohydrates.

The solution from the washing process, which contains 2 - 4 percent solids, is an effluent and possible pollutant. When it is spray dried the resulting powder contains a nourishing mixture of protein, fat, fiber ash and carbohydrates. This product is actually a form of dried molasses.

Sales of the dried product are primarily to feed manufacturers producing pelleted feeds. It is also competitive with fluid molasses which is used as a source of carbohydrate and as an appetite stimulant added to roughages.

Fish Industry

In fish processing, dryers condense and dry soluble proteins, known as "fish solubles," into a powdered product that contains at least 65 percent protein. These solubles are now more profitably marketed as powder, rather than the 50 percent solid slurry previously used because shipping weight is reduced by 50 percent. Also, the powder is more easily blended into other dry food ingredients. The product is used in pet and animal food.

Cost of Spray Drying

A rough estimate of dryer efficiency is: inlet air temperature (Ti) minus the exhaust temperature (To) over the inlet temperature (Ti) minus the ambient air temperature (T2). This gives the cost of removing a pound of water.

$$\frac{Ti - To}{Ti - T2}$$

From the equation it can be seen that the greater the temperature difference between the inlet and exhaust, the greater the efficiency of the spray dryer.

ON-SITE WASTE DISPOSAL AND ENERGY RECOVERY[92]

Many industrial firms are using their production wastes as a recovered resource; often as a fuel resource for steam generation or other forms of energy conversion. Table 24-3 lists some of the common wastes which have been used as fuels.

In recent years there has been a proliferation of shop-assembled package incineration systems. These are usually available in increments of capacity from 700 lb/hr to approximately 7000 lb/hr depending on the physical and chemical nature of the waste to be consumed. The largest standard unit offered by the industry, represented by the Incinerator Institute of America, is rated at an input of 13,000,000 Btu/hr and is classed as starved air (controlled atmosphere) design having primary chamber-modified pyrolysis and secondary chamber burnout, with clear combustion gases discharging at 1600 to 1800 F. A schematic of this unit with a package fire-tube boiler is shown in Figure 24-18.

For larger installations, field erected systems are available. These custom engineered systems are available in capacities up through municipal size,

Table 24-3. Industrial Wastes Used as Fuels for Steam Generation

Bagasse	Pine bark and shavings
Bark	Hardwood bark and shavings
Coconut shells	Rice hulls
Coffee grounds	Roofing plant broke
Corn cobs	Rubber product waste
Cotton seed hulls	Sludges
Fibre cake	Spent solvents and lubricants
Flakeboard broke and dust	Spent sulfite liquor
Furfural residue	Tan bark
Municipal refuse	Tars
Peanut shells	Waste oils
Flooring (linoleum-vinyl) rejects	Wood flour

with 50 to 150 tons/day the usual range. They can be arranged to consume waste oils and solvents, and use a gas-to-air heat exchanger or a boiler unit for heat recovery. The arrangement illustrated has alternate provisions for exhausting products of combustion through a gas washer as well as a fail-safe stack for emergency bypass.

Steam Generating Systems

Many existing steam generators have been adapted to accommodate refuse fuels, separately or in combination with one or more purchased fuels. Coal-fired units are generally more easily adapted for this purpose since their furnaces are usually larger. Insofar as solid waste is concerned, these boiler plants already incorporate material handling equipment and modifications or additional equipment is more easily accommodated.

A small shop-assembled package boiler (Figure 24-19), can be arranged for gas/oil firing, and modified to receive hot gases from an incinerator if the units are in close proximity and space is available for the ducts. Depending on the type of incinerator and wastes to be consumed, boiler flue gas clean-up equipment may be required. Firing light oils or spent solvents into units of this type is also common.

A schematic representation of semi-suspension (thin bed) spreader firing is shown in Figure 24-20. Note that the arrangement provides for spreader firing of coal or refuse fuel separately or in combination. This type of firing system requires that the waste materials be macerated to a maximum size of 4 in. The degree of burning which takes place above the grate bed is contingent upon particle sizing, moisture content, volatility, furnace temperature and turbulence in the ignition-combustion zone.

A full suspension fuel firing system for multi-fuels is shown in Figure 24-21. Suspension firing is also available with fuel burners located all in one wall, divided between two opposite walls or down through the furnace roof. Another version of suspension firing is with a horizontal cyclone

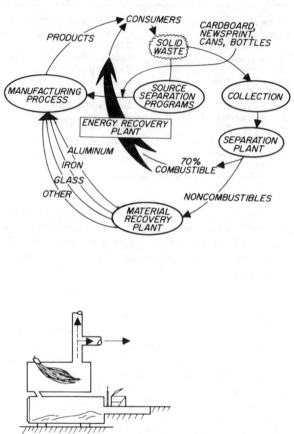

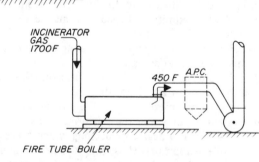

Figure 24-18. Package incinerator with packaged boiler.

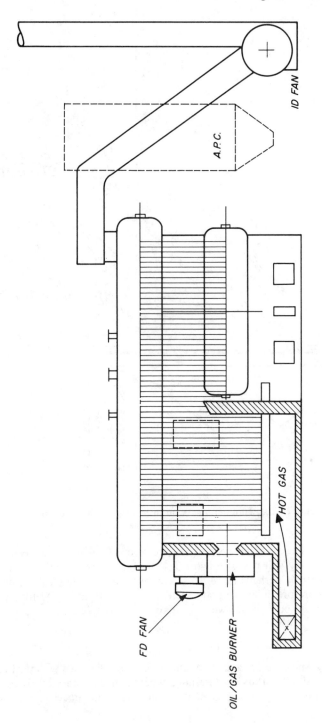

Figure 24-19. Heat recovery of incinerator gas in water wall furnace boiler.

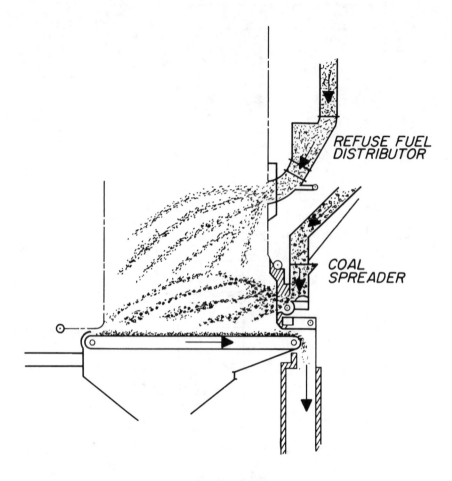

Figure 24-20. Thin burning (semi-suspension)spreader stoker.

furnace. Suspension firing systems require refuse fuels to be reduced in size, so that they will pass through a 1¼-in.-square (maximum) screen. Even so, larger or perhaps more dense combustible materials will enter the furnace. Therefore, a simple grate system at the bottom of the furnace is employed to support this slower burning fuel to completion.

Case History

Spring Mills, Inc. recently conducted a comprehensive technical and economic feasibility study to determine whether it was practical to convert solid waste into steam to supplement the normal plant process generation.

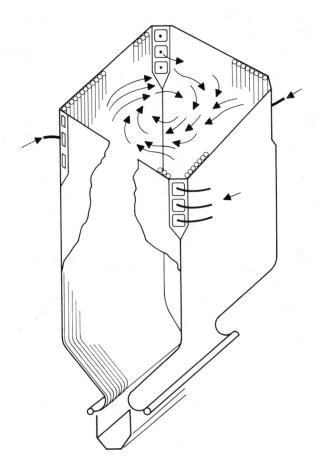

Figure 24-21. Corner suspension fired furnace.

Five of Spring Mills' plants are located close to one another near Lancaster, S.C. Each plant hauled waste to a landfill in 40-cu-yd compacted containers. Recent laws enacted by the state had materially increased disposal costs due to longer hauling distances to an approved sanitary landfill and higher fees for its use.

A solid waste inventory was made of the number, type and size of containers hauled to the landfill for each plant. The filled containers were not sampled to accurately determine the composition of wastes being discarded. The composite trash is, however, essentially corrugated boxboard, paperboard, paper, wood scrap and significant quantities of lint. The greatest waste generation was at the Grace Bleachery—approximately 44 percent of the total generated by all five plants. The total waste projection was

900,000 lb/wk. It was therefore decided that the logical location for a central receiving-processing plant would be at the Grace Bleachery. Based on 24 hr/day, 6 day/week operation, the hourly waste disposal-use rate would be 6250 lb/hr. At 8500 Btu/lb, the waste would release 53,000,000 Btu/hr.

For this study, it was decided to maintain the waste receiving-processing-burning and steam generation facility separate from the existing plant. The components of the refuse processing plant were to be modular and wherever possible, shop-assembled. An unloading-receiving station with a conveyor to a single shredder would serve two identical lines of equipment, each having a fuel surge bin, incinerator, waste heat boiler and air pollution control equipment.

The modular furnace—fire tube boiler systems—selected would each have an input thermal rating of 30,000,000 Btu/hr. The shredder selected was capable of reducing material to a nominal 6 in. top size and processing up to 21,000 lb/hr. The coarse product size was wholly adequate for the burning system and would improve uniformity of the waste fuel, simplify handling and storage and require only a 25-hp shredder. Table 24-4 is a summation of the investment and operating cost data compiled for this 75 tons/day industrial waste heat recovery facility.

Actual return on investment was calculated to be somewhat less than 5 percent. However, as fuels become more scarce, fuel costs continue to rise, additional process steam is required and solid waste disposal costs increase, use of this system should become more appealing.

**Table 24-4. Capital and Operating Costs for a Proposed
75 ton/day Waste Heat Recovery System**

Investment

1. Incinerator system including feed converyors, shredder storage
   hoppers, incinerators, waste heat boilers, and flue gas cleaning $ 792,000

2. Necessary building, erection, steam and condensate system,
   boiler,feed water and treatment, sprinklers, lighting, power,
   and site work, etc.   __400,000__
                                                                        $1,192,000

Annual Savings

1. **Fuel**
   a. 28,750,000 lb waste/yr at 8500 Btu/lb yields 244,375
      MM Btu/yr
   b. 244,375 x 80 percent efficiency yields 195,500 MM
      Btu/yr net
   c. 195,500 MM Btu/yr produces 195,500 M lb steam/yr
   d. Cost of coal and gas to produce 195.5 MM lb steam at
      $0.50/M lb $ 97,500

2. **Hauling and Disposal Fees**
   $483 x 50 $ 24,150

3. **Reduced Operating Costs** at Grace Power Plant
   a. $0.89/M lb transfer rate
      -0.50 Fuel Cost
       0.39
      -0.07 Fixed Cost
       0.32 Cost for other variable
      Save 5 percent of other variable
      5 percent x $0.32 x 2,250,000 M lb $ 36,000
   b. Reduce coal inventory investment of $950,000 by
      5 percent at 10 percent cost of investment $ 4,750
   c. Other savings at Grace Power during weekends $ 13,500

4. **Operating Costs of Incinerator Plant**
   a. Power
      8600 kWh daily x 300 days x $0.0066/kWh $ 17,700
   b. Personnel $ 20,200
   c. Maintenance $ 30,000

                   Net Savings $ 108,000

CHAPTER 25

OCEAN DUMPING OF WASTES[28]

Theoretically, the ocean, engulfing 71 percent of the earth's surface, can assimilate enormous volumes of wastes, but only under direct supervision of international, governmental and environmental agencies. By dumping waste in barren areas devoid of benthic life, ocean dumping can be very successful. These areas can be found along the continental shelf or in the abysmal depths—not in the shallow estuarine waters that are used today, for they are abundant in marine biota and are extremely vital in the life cycle of the ocean.

The United States has been using the oceans adjacent to its coastline for disposal areas for many years. Until recently these actions have gone unchecked by the United States government. But in 1971 Congress investigated several reports concerning environmental pollution, and regulation of ocean dumping was among them. In discussing ocean dumping effectively, one must keep in mind that there are a multitude of contaminating pollutants involved.

OCEANIC POLLUTANTS

These pollutants are grouped into six broad categories, which will be discussed independently. A breakdown of their total wet tonnage can be seen in Table 25-1.

Dredging Spoils

Dredging spoils comprise about 80 percent of the total tonnage dropped in the sea, and also carry the majority of the pollutants. These spoils are transported to specified areas of the ocean for dumping.

Industrial Wastes

These are transported to the dumping sites in towed or self-propelled, double-hull barges ranging in capacity from 1000 to 5000 tons. This is

Table 25-1. Wet Tonnage of Waste Dropped Off United States Coasts

	Atlantic	Gulf	Pacific	Total
Dredging Spoils	15,808,000	15,300,000	7,320,000	38,428,000
Industrial Waste	3,013,200	696,000	981,300	4,690,500
Sewage Sludge	4,477,000	0	0	4,477,000
Construction & Demolition Materials	574,000	0	0	574,000
Solid Wastes	0	0	26,000	26,000
Explosives	15,200	0	0	15,200
Total	23,887,400	15,966,000	8,327,300	48,210,700
Total Polluted	7,120,000	4,740,000	1,390,000	13,250,000

necessary because the wastes contain heavy metals, cyanides, arsenical and mercuric compounds, chlorinated hydrocarbons and various other toxic chemical wastes.

Sewage Sludge

Sewage sludge contains oxygen-demanding pathogenic bacteria and a high concentration of minute solids resulting from municipal sewage treatment. But some sites, such as the New York Blight, are being plagued with raw sewage which carry fat, grease and other floatables, along with the aforementioned products.

Construction and Demolition Debris

This consists of steel, concrete, stone, wood and other inanimate objects. The wastes are usually deposited in the shallow waters and estuaries or in specific dumping areas off the coast, providing their accumulation does not endanger commercial shipping.

Refuse and Solid Waste

Refuse and solid waste is being dumped into the ocean in increasing quantities because land disposal sites are rapidly filling. There are two methods of refuse dumping now in use today—loose dumping and baled dumping. Baling of refuse is more efficient, neat and ecologically sound than loose dumping—and it is naturally more expensive. It keeps the refuse in one place and, because of its increased density, sinks and stays submerged indefinitely. Loose refuse, on the other hand, is harder to handle and requires more storage space, both in transit and at the dumping site. It also floats much more easily and therefore is aesthetically distasteful. Both types have been known to decay at the same rate.

Explosives

Explosives are handled by the United States Navy and are dumped in clearly marked spots on United States Coastal and Geodetic Survey maps. The explosives are obsolete but may be accompanied by toxic gases. Being highly dangerous, these elements are usually towed to their dumping sites aboard old World War II liberty ships. Once there, they are scuttled in water ranging in depth from 500 to 1000 fathoms.

* * *

Pollution in the ocean can be defined as a change in water quality that has an adverse effect on its beneficial use. As a result, certain elements must remain constant in the ocean.

Salinity is as vital to pelagic life as air is to mammalian life. The degree of saltiness in the ocean affects the fish's ability to retain moisture in its tissues.

Temperature in the ocean must remain constant also, for aquatic life cannot adapt to rapid change. Larval forms of life are especially sensitive to this type of change. In tropical waters, some species live in an environment that is within a few degrees of their lethal temperature.

Dissolved oxygen is important as an air supply to the marine world. Fish often suffocate in water where the dissolved oxygen is low. Oxygen is used in many ways in the ocean, not only by aquatic life, but by degeneration of bacteria and by large pollutants such as toxic chemicals and biodegradable sewage.

pH concentration in sea water is buffered by its carbon dioxide content. This must stay near normal for the normal life cycle to proceed uninterrupted.

Turbidity is the occurrence of small particles in the water that interfere with the transmission of light. If turbidity is changed drastically, there will be a marked effect on photosynthesis, the process by which plankton replenish themselves.

Likewise, certain substances must not be allowed to enter the ocean in any great quantity.

Toxic substances directly affect marine life although often not immediately fatally. Their effects may be noticeable only through a lessening of other abilities, such as that of defending against a predator or just performing required body chemistry.

Tainting substances, such as phenols, aromatics and petroleum products, render marine life unfit for human consumption. They absorb the substances and acquire an unwanton taste, if they survive the exposure.

Settleable and floating solids can affect marine life in a variety of ways. Floating objects offer obstacles for pelagic life, while settleable solids may hit or totally cover benthic creatures.

Nuisance organisms such as pathogenic bacteria cause unpleasant odors and use up varied amounts of oxygen. Their growth is usually enhanced by a drastic change in temperature. Equally dangerous is the rapid growth of rich nutrients in plant life. An overabundance of either of these factors could upset the delicate equibalance of the ocean.

Estuarine waters are vital in the development of marine biota for it is here that life itself begins. As a result, ocean dumping should occur in water at depths greater than 500 fathoms. At this depth and pressure, baled and loose refuse is compressed and rendered buoyantless by the hydrostatic pressure of the ocean. Toxic chemicals and metals also dissolve more readily at this depth, for there is a greater quantity of water and dilution becomes easier. In order for this type of ocean dumping to work properly, however, all floatable material must be removed, such as plastics, rubber, wood, grease and oil. If this is done properly, ocean dumping can alleviate the problems encountered by landfill operations in recent years.

Although this method of waste disposal is feasible, it is not readily known what long-range effects it will have on the ocean and the marine life within it. That is why certain precautions and research must be carried out while or before this method is used.

1. Base-line surveys must be taken so that there will be sufficient data with which to compare changes.
2. Ecological research in tracing the pollutants in the food chain must be done so that an understanding of how the pollutant concentrates can be ascertained.
3. Study sewage sludge and determine how it can be assimilated on the continental shelf.
4. Select toxic substances must be studied—how they dissolve and how they affect marine life in the area.
5. Study and relate waste disposal as a total system, integrating incineration, recycling, landfill operations and ocean dumping. Only by complete cooperation from all concerned can this theory become a reality.

OCEAN DUMPING PERMITS

General Permits

General permits may authorize the dumping of materials disposed of in small quantities, such as galley wastes and other nontoxic materials. These permits must be published in the *Federal Register* specifying types and amounts of the material, and the sites for dumping.

Special Permits

Special permits cover the materials not covered by general permits with the exception of toxic metals, oils, inorganic wastes, and oxygen-consuming matter. These permits have a three-year expiration date and are renewable upon reapplication.

Emergency Permits

Emergency permits allow for the dumping of prohibited materials listed under special permits, where there is a risk to human life and there are no other feasible solutions.

Interim Permits

Interim permits are issued for the prohibited materials listed under special permits in excess of the permissible levels under the following conditions:

1. An assessment of the environmental impact is required as a part of the application, along with alternatives to ocean dumping.
2. The development and implementation of a plan to eliminate or bring within agreeable limitations set by the act. The permit's expiration date will be set according to the proposed plan, but shall not exceed one year. However, it is not renewable. A new permit may be obtained upon application, provided certain phases of the proposed plan have been satisfactorily completed.
3. No interim permits may be granted for new facilities, or for expansions of existing facilities.

Research Permits

Research permits may be issued if the scientific merit of the project outweighs the potential damage that may occur and only under the following conditions:

1. A detailed statement of the proposed project and its impact must be provided.
2. There must be a public hearing.
3. Research permits shall expire in 18 months, but are renewable.

Applications for Permits

All applications for ocean dumping permits must include the following:

1. Name and address of the applicant.
2. Name and location of the firm to be used for transportation and dumping.
3. Physical and chemical descriptions of the material and any containers to be dumped.
4. Quantity of material.
5. Means of conveyance and proposed dates and times of disposal.
6. Proposed dump site.
7. Method of disposal at site.
8. Identification of the process which produces the material.
9. Descriptions of alternative methods and why these are not used.

Permit holders are liable for transporting and dumping material not conforming with the permit application.

Permit Fees

All permits for dumping at existing sites shall have a fee of $1000. An additional $3000 shall be charged for dumping at a site other than the designated one. A fee of $700 shall be charged for the renewal of a permit. No agencies of the federal, state or local government shall be charged for dumping.

Materials for Which No Permits Will Be Issued

Under no circumstances will the EPA allow the dumping of the following materials:

1. High-level radioactive materials, such as irradiated reactor fuels.
2. Materials of any form produced for radiological, biological or chemical warfare.
3. Any materials which are not sufficiently described to permit their impact on marine ecosystems.
4. Persistent inert materials, synthetic or natural, which will remain in suspension, unless they are processed to sink and remain in place.

Other Prohibited Materials

The EPA will not permit the dumping of the following materials as other than trace contaminants:

1. Organohalogen compounds or materials which will form these compounds in the marine environment.
2. Mercury and its compounds.
3. Cadmium and its compounds.
4. Crude oil, fuel oil, heavy diesel oil, lubricating oil and hydraulic fluids.

Permissible Concentrations of Prohibited Materials

1. Organohalogen compounds: 0.01 of a concentration shown to be detrimental to the marine environment.
2. Mercury: Solid Phase—not greater than 0.75 mg/kg.
          Liquid Phase—not greater than 1.5 mg/kg.
3. Cadmium: Solid Phase—not greater than 0.6 mg/kg.
          Liquid Phase—not greater than 3.0 mg/kg.
4. Oils: There must not be a visible surface sheen in undisturbed water when it is added at 1 part oil to 100 parts water.

Materials Which Require Special Care

1. The following elements and any of their compounds: arsenic, lead, zinc, copper, nickel, selenium, vanadium, beryllium and chromium.
2. Organosilicon compounds.

3. Inorganic processing wastes such as: cyanides, fluorides, chlorine and titanium dioxide.
4. Petrochemicals, organic chemicals, and organic processing wastes.
5. Biocides.
6. Oxygen-consuming and/or biodegradable organics.
7. Other radioactive wastes.

CHAPTER 26

NOISE COMPLIANCE AND ENFORCEMENT

NOISE, LAWS, CONTROL AND THE ENGINEER[27]

In relating noise to hearing loss six factors must be considered:
(a) frequency of the noise, (b) overall level of noise, (c) time distribution
of the noise exposure during a work day, (d) duration of noise exposure
during a day, (e) total exposure time during an estimated work life, and
(f) individual's age and susceptibility.

Time distribution and duration of noise are most significant. Less
hearing impairment occurs for noncontinuous noise than for continuous,
because the ear is given time to rest or recover. In fact, an ear that has
been given the opportunity to rest between exposures becomes more resistant to permanent hearing loss than does the ear that is exposed continuously.

Sound generates energy which has a distinct relationship to the exposure time. For continuous sound, the amount of time an individual should
be exposed to a particular sound is dependent upon the energy generated
to the ear as well as its frequency and pitch. However, intermittent sound
has quite a different effect. As an example, damage to the ear can occur
after 5 hr of continuous noise exposure of 100 decibels, whereas 10 hr of
exposure (at same level), with varied rest periods, may not be harmful even
though the total energy generated to the ear is equal in both situations.
There is no direct mathematical proportion between continuous exposure
and energy. Tripling noise energy does not generate three times the hearing
loss. However, as energy increases, less exposure time is required to create
the same degree of hearing loss.

Noise effects can be divided into two classifications—nonauditory and
auditory. Nonauditory effects do not produce hearing disability but do
interfere with speech communication. The average level at which speech
is coherent is 50 decibels. If machinery operation requires workers to shout
commands, the noise level should not exceed 68 or 70 decibels. However,
nonauditory effects can hinder the worker's efficiency and produce nervousness. Figure 26-1 shows average machine noise levels, the range at which
probable hearing loss can occur.

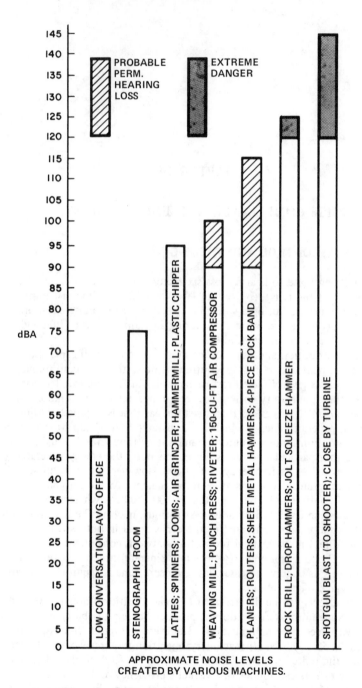

Figure 26-1. Approximate noise levels created by various machines.

Auditory effects can produce temporary hearing disability. Loud unfamiliar noises can cause new workers to have temporary hearing loss, and be the start of permanent damage. It may take several days, a week, or longer, for the worker to recover hearing after exposure has ended, depending upon the individual's age and capacity.

Noises, even moderate ones, have negative side effects upon individuals. The body reacts to sound by a vasoconstrictive reflex which causes small blood vessels to constrict, thus impeding blood flow. The pupils of the eye dilate. Even sounds not detectable to the human ear are harmful. Infrasonic range-frequency noise is too low for the ear to detect. Currently, scientists are working with this range in the hopes of developing a new weapon—an acoustic laser. This range of frequencies can alter the rhythm of brain waves.

Approximately 20 percent of the population not exposed to industrial noise have some type of hearing impairment between the ages of 50 to 59. Twenty-eight percent of all industrial workers in the same age group exposed to industrial levels of 90 dBA for their working lifetime have hearing impairment. Figure 26-2 shows the risk of developing a hearing loss at various ages and levels of noise.

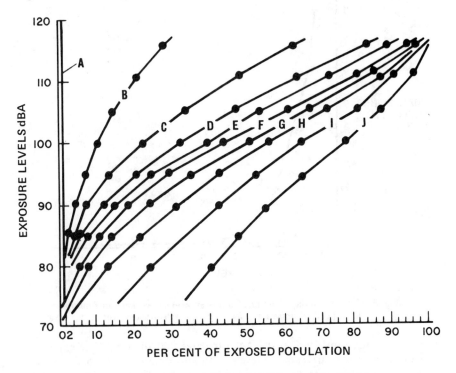

Figure 26-2. Percentage of population risking hearing loss through age and exposure to noise.

On June 30, 1936, President Franklin D. Roosevelt signed a bill which set minimum wages and maximum hours of employment. Safety provisions were also included. The Secretary of Labor was given authority to amend the law as he deemed necessary. In 1969, Secretary of Labor George Shultz implemented a noise control amendment.

The regulation establishes the maximum amount of time employees may be exposed to continuous noises before it is necessary to provide protection for their hearing. Maximum permissible level for continuous sound is 90 dBA for an 8-hr work day. Figure 26-3 shows values designated by the Walsh-Healey Act as a guide to preventing noise-induced hearing loss. The graph shows that as the sound level is increased, the exposure time must decrease in order to deter the danger of deafness.

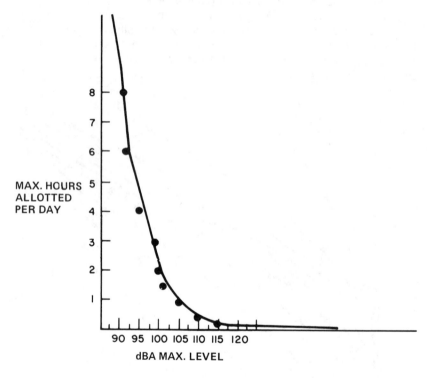

Figure 26-3. Maximum number of hours an average worker can function at a specified noise level according to Walsh-Healey Act.

For interrupted noise, the Act stipulates that the total time exposure is the sum of the various noise levels at various durations. Each duration and level must be recorded and added together to obtain the daily exposure and level. For impact noises (pneumatic hammers, explosions, etc.) the maximum level is 140 decibels as detected on the C network.

Walsh-Healey is a milestone in protective laws for the worker, but it is not the first time attention has been given to noise problems. Presently, 39 states and the District of Columbia provide payment of Workmen's Compensation for noise-induced hearing impairments. Compensation ranges from $765 to $11,000 for one damaged ear, and from $4500 to as high as $33,000 for both.

Noise Control

Laws and regulations are not solutions, merely quantitatively acceptable parameters. Control of the problem may be difficult. A major industrial noisemaker can be found in the forging industry. In plant areas where hammer forges are in operation, levels of 123 to 140 decibels are reached and maintained for extended periods of time. Background noise is continuously around 85 dBA. Around furnaces noise levels range from 98 to 115 dBA. Figure 26-4 shows levels reached by auxiliary operations.

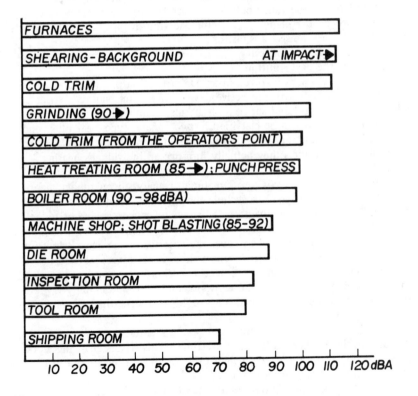

Figure 26-4. Noise levels reached by various operations in the forging industry.

One method to solve this industry's problem is through design and engineering of the equipment. For example, take the hammer forging operation where there is high-speed metal-to-metal contact. Forging presses and upsetters should be considered as substitutes for hammers. The use of personal ear protectors may be required. Suspended acoustical panel on walls and ceilings will greatly reduce noise levels, often to a permissible level.

The chemical process industry has a great variety of noise sources. Unwanted noises are created by the many pumps and motors used in production. Motor noise can be caused by bearings or the motor cooling system. Pumps create hydraulic or mechanical noises. Small pumps may not produce harmful noise levels in the immediate vicinity, but can cause vibration in adjacent rooms. To combat pump noise, vibration isolation devices can be used. Noise within the pump can be due to the sudden collapse of vapor bubbles as they reach higher pressure zones within the pump. This can be corrected by adding small amounts of air to the pump volumes that undergo negative pressure. This air acts as a cushion that suppresses the violent collapse of the vapor bubbles. Noise from pump systems can be managed with proper planning. Careful selection of pipe size and supports will greatly reduce noise transmission.

In some cases, engineering methods may be too costly, and ear protection offers the best solution.

More, however, is involved in an ear protection program than merely fitting the individual with protective devices. Program planning, strict administrative control and supervision are also required.

Ear protectors can be classified in two general types—ear plugs and ear muffs. Ear muffs and plugs do not hinder communication. In noise fields above 85 decibels they actually improve an individual's hearing capabilities.

Ear plugs are made of a soft material, usually silicon rubber, neoprene or plastic, and are designed to seclude the ear canal. Both material and shape influence their effectiveness. Most ear plugs are capable of reducing sound from 25 to 35 decibels.

Ear muffs are designed to cover the exterior of the ear. Below frequencies of 1000 Hz their effectiveness exceeds ear plugs; above 1000 Hz the two devices provide similar protection. Choice between ear plugs and muffs is usually dependent upon working conditions and employee preference. If the working area provides no room for muffs, then ear plugs are more practical.

Probably the least economic and practical solution to a noise problem is reducing the worker's total exposure time. This means changing work periods. However, one way to achieve this result is to give more rest periods. Noise exposure is then intermittent.

The last method of noise control is isolation. This can be accomplished by sealing off different work areas from each other to reduce the number of workers exposed to excessive noise. Or, machinery can be partially or totally enclosed so that the operator controls it from a panel outside.

Materials which are good sound insulators have high density. Acoustical quality lead is an excellent material, but can be difficult to handle in some instances. Lead impregnated cloth is a good acoustical barrier and overcomes the problems of handling pure lead.

A noise insulator material's main function is to prevent the passage of sound waves. It is different from an isolator, which absorbs the vibrations that cause sound. Polymers used to reduce noise from plastic grinders are considered isolators.

The pollution engineer should consider the following when creating a noise control program:

- Evaluate noise exposure and determine equipment causing harm;
- Create a noise monitoring program; and
- Institute a hearing conservation and noise abatement program.

When establishing a hearing conservation program, periodic audiometric testing must be established to evaluate the hearing ability of workers. Audiometric testing can detect hearing losses at an early stage. Workers with hearing impairments are a potential inherent liability in some operations. Monitoring noise levels in manufacturing areas is a means of recognizing a potential hearing problem among employees.

However, to make a noise abatement program succeed the pollution engineer must make sure that employees wear their ear protectors when required. An educational program should be created with incentives or disciplines, so that employees and employers realize that ear protection is just as important as eye protection.

ENVIRONMENTAL LAW— LEGAL ACTIONS TO CONTROL NOISE[7]

Throughout history human activities have produced noises which some individuals have found objectionable. During this same period, legal systems have been attempting to reconcile the rights of some to carry on noise-producing activities with the rights of others to be free from this annoyance. Reviewing some recent court decisions provides a glimpse of some of the legal principles in the battle to control noise.

In the cases described, **P** indicates Plaintiff, the person who starts the action and who claims a law was violated or he was injured. **D** indicates Defendant, the party who is alleged to have violated a law or caused the injury.

A nuisance has been defined as everything that endangers life or health, gives offense to the senses, violates the laws of decency or obstructs reasonable and comfortable use of property. Noise may be a nuisance. If a single individual or small group of persons are disturbed by the noise, a private nuisance action might be brought against the offender. If a large number of individuals are disturbed, a public nuisance action might be brought by a public official or public body empowered to bring such an action. These actions are based on statute, if this exists in a particular state, otherwise under the general principles of law.

Case I

The following is a recent typical private nuisance action.

P brought the action to stop the operation of D's ready-mix concrete plant and for damages based on P's claim that the plant was a nuisance.

P acquired his property first and thereafter D purchased adjacent property and built a ready-mix concrete plant. The primary purpose of the plant was to deliver ingredients for ready-mix concrete into trucks for delivery to customers.

P testified in detail concerning the loud noises made in the operation of the plant including loud noises early in the morning and late in the evening—metal tailgates of trucks were banged and motors made loud whining noises. Conversation could not be carried on in P's home without waiting for the noise to stop. There were no other noise-producing plants in the area. This action was brought under a statute which read:

> "Whatever is . . . offensive to the senses, or an obstruction to the free use of property, so as essentially to interfere with the comfortable enjoyment of life or property is a nuisance and a civil action by ordinary proceedings may be brought to enjoin and abate the same and to recover damages sustained on account thereof.
> "The following are nuisances . . . using any building or other place for the exercise of any trade . . . which by occasioning . . . annoyances becomes injurious and dangerous to the health, comfort, or property of individuals. . . ."

The Court said:

> "Noises may be of such character and intensity as to unreasonably interfere with the comfort and enjoyment of private property as to constitute a nuisance and in such cases injury to the health of the complaining party need not be shown.
> "A fair test of whether the opearation of a lawful trade or industry constitutes a nuisance has been said to be the reasonableness of conducting it in the manner, at the place and under the circumstances in question. . . . Thus the question whether a nuisance has been created and maintained is ordinarily one of fact, and not of law, depending on all the attending or surrounding circumstances. Each case of this nature must depend on its own facts.
> "It is uncontradicted P '[was] there first.' Priority of occupation is a circumstance of considerable weight and it [strongly favors P] here."

The Court found that D's plant constituted a nuisance and went on to say:

> ". . . the Court has reached the conclusion that there is no appropriate order which this Court could enter which would abate the nuisance created by the noise of the operation of this plant as set

forth in this decision and yet permit the plant to continue in operation. Accordingly the Court has concluded that the only course which it may adopt which will properly safeguard the rights of **P** is to [stop] the operation of the concrete plant in its present location." **P** was also awarded damages.

From the above case it can be seen that a plaintiff bringing an action of this type may ask for damages or an injunction (court order) to stop the noise-producing activity, or both.

Courts have been reluctant to grant injunctions where to stop an activity would put many people out of work and cause serious economic consequences. However, with the efforts being made to improve the environment, as noise levels increase and more is learned of the harmful effects of noise, courts may grant injunctions more readily for this purpose.

It was important in this particular case that **P** occupied his property before **D**. In another case, however, property owners purchased their home after a dry cleaning plant was in operation on adjoining land. Noise from the plant greatly disturbed the homeowners who were persons of ordinary sensibilities and rendered them on occasion nervous and ill, and vibrations damaged their home. The laundry operation was stopped by court order and damages were awarded the homeowners.

In the concrete plant case, another important point in **P**'s favor was that the area was quiet before the concrete plant began operations. One of the older cases still stating a correct principle said: "No one can move into a quarter given over to foundries and boiler shops and demand the quiet of a farm."

Case II

In this 1970 case, a homeowner, **P**, constructed an expensive home in a fine residential section near a rriver. At that time the opposite side of the river, 150 to 200 ft from **P**'s home, was used for commercial and light industrial purposes. A shell plant, **D**, was located there for unloading shells from barges. Operations were usually carried on during daylight hours. Cranes, barges, and other equipment used by the shell plant were in satisfactory condition and the kind usually found in this type industry. **P** brought action, claiming **D**'s plant was a nuisance due to the noise and disturbance it caused.

The case was decided in **D**'s favor, the Court reciting the following principle:

"The locality, the occupation of the inhabitants of the neighborhood, the environment, is what determines whether an establishment which uses necessarily noisy machinery, is a nuisance. A manufacturing enterprise that would be a nuisance in one locality might not be so in another. The inhabitants of large cities that are sustained by manufacturing and commercial enterprises must bear the unavoidable discomforts and annoyances thereof. Noises and vibrations from the operation of machinery cannot be a nuisance, subject to injunction, unless they are excessive and unreasonable, depending upon the location of the establishment, its relation to other property, and particularly to other sources of noise or vibration."

Whether or not the noise-producing machinery is old or new, well maintained or poorly maintained, or properly or improperly used is an important factor in cases such as the above. Courts will more quickly decide against those who use obsolete or poorly maintained equipment, or use their equipment improperly.

The nuisance action is an important means for controlling noise. However, it has serious limitations. One is the wide range of opinions which courts and juries have as to how much noise constitutes a nuisance. Some people can withstand noise better than others. Noise is a subjective matter depending on individual experience. Another limitation is that noise is often a combination of sounds from many sources. Here governmental action is the only method of dealing with the problem effectively.

Local Laws Restricting Noise

Local laws of many types attempt to restrict noise

Case III

The ordinance of a major city was challenged in a court case. The ordinance made it a crime to "make, aid, countenance, or assist in making any improper noise . . . [or] disturbance. . . ."

The ordinance was struck down as being vague and indefinite, the Court saying:

"The number of sounds which are constitutionally permitted and protected and which would fall within the proscription of 'improper noise' is infinite. Political campaigns, athletic events, public meetings, and a host of other activities produce loud, confused, or senseless shouting not in accord with fact, truth, or right procedures to say nothing of not in accord with propriety, modesty, good taste, or good manners. The happy cacophony of democracy would be stilled if all 'improper noises' in the normal meaning of the term were suppressed.

"The Supreme Court has held on several occasions that statutes or ordinances prohibiting noise are valid only to the extent that they are necessary to the protection of other important public interests, *e.g.,* the prohibition of horn blowing or other noise in a hospital zone. Prohibition of noise *per se* is unconstitutional and the addition of the vague and indefinite adjective 'improper' does not cure the defect."

One of the real problems in controlling noise by local laws is properly defining the noise which is illegal and restricting that noise without infringing on other lawful activities. The ordinance in the above case was too broad although the matters it dealt with were clearly within the policy powers of the city. An ordinance to control noise must be precise so that a person of ordinary intelligence would clearly understand what noise was prohibited.

Case IV

D's were charged with the offense of disturbing the peace by creating loud and unnecessary noises with motorcycles in violation of a city ordinance. Prosecution offered evidence that D's, in groups of five motorcycles, rode up and down a residential street until 11:30 p.m. Persons in these homes could not hear conversations. D's said they did not make loud noises and the ordinance was vague and indefinite.

The ordinance provided:

"(a)   Subject to the provision of this section, the creation of any un-reasonably loud, disturbing, and unnecessary noise in the city is prohibited. Noise of such character, intensity, and duration as to be detrimental to the life or health of any individual is prohibited.

"(b)   The following acts, among others, are declared to be loud, dis-turbing, and unnecessary noises in violation of this section but said enumeration shall not be deemed to be exclusive, namely:

"(c.4) Use of vehicle. The use of any automobiles, motorcycles, or vehicle so out of repair, so loaded, or in such manner as to create loud or unnecessary grating, grinding, rattling, or other noise."

The D's were found guilty, the Court saying:

"The protection of the well being and tranquility of a community by the reasonable prevention of disturbing noises is within the city's power. . . . The ordinance in question does not define in decibels the intensity of the noise to be prohibited thereby, but such exactness is not required.

"The words loud or unnecessary have a commonly accepted meaning and they give sufficient warning to anyone who has the desire to obey the ordinance.

"D's are in no position to complain that the intensity of the noise from the group was allowed in evidence. It was competent for the prosecution to show the intensity of the group noise without having to show the decibels contributed by each D. Having joined in the violation of the ordinance as a group, D's cannot now be heard to complain that their conduct standing alone would not have constituted a violation. If the contribution of each made the total into an offense condemned by the ordinance, then each would be guilty of the offense."

Actions Against Governmental Bodies

The location of streets, highways, and airports and the noise associated with their use has produced a variety of legal actions. Many of these are actions against a governmental body for "inverse condemnation." The idea here is that the noise and disturbance caused by the public facility has taken away certain private property rights and the private parties involved should be compensated.

Case V

P's, fifty-seven residents of lands near an airport, brought action for inverse condemnation against **D**, the city which owned and operated the airport. The airport had been expanded over the years and also began to handle jets. P's alleged that planes operated over their properties at altitudes of 100 to 150 ft and while landing, taking off, and warming up, the "aircraft engines cause to be heard and felt terrific and overwhelming vibration, concussions and sound waves directed and transmitted from said engines against and upon the P's properties."

The Court cited the following principle:

"If a landowner has a right to be free from unreasonable interference caused by noise, as we hold here he has, then when does the noise burden become so unreasonable that the government must pay for the privilege of being permitted to continue to make the noise? Logically the answer has to be given by [the facts of the particular case]."

The Court said here that P's had a right to compensation based on inverse condemnation.

Results in many cases have been the opposite of the decisions reached by the court in the above cases. This is a very complex and technical area of the law, the outcome of cases depending on the wording of the state constitutions and other factors.

It is generally held that citizens must bear the noise of roadways and activities which are based on governmental policies or sanctions.

The principles brought out in all the cases are frequently encountered, and will achieve even greater importance as efforts to control noise continue.

ASSESSING NOISE IMPACT ON THE ENVIRONMENT[58]

Because of the additive properties of noise, and the decay of sound intensity with distance, it is possible to estimate the sound from a new plant site using mathematical modeling to make an environmental assessment.

This technique, used to estimate community noise, is based upon a Sound Level Computer Model, which relates the noise level between the source and receptor. The model is based upon a straight-line logarithmic decrease of sound level with distance.

The model is based upon the decay formula, where:

$$dB = dB_0 - 10 \log_{10} (d/d_0)^2$$

dB   = sound level at some distance in decibels
dB_0 = sound level at source
d   = distance between source and receiver
d_0 = distance from source that source sound level (dB_0) is measured.

    Certain assumptions have been made in preparing the basic model: (1) micrometeorology of the area is uniform and wind speeds are less than 10 mph, (2) the terrain is flat and topography is not a problem, and (3) row crops, vegetation, and trees in the area do not muffle or interfere with the transmission of sound waves.

    Each of these factors may be included into the basic model to refine the predictions. These factors could each account for a 5 to 10 percent difference in the estimates obtained using the basic model.

    In order to determine the impact of the source on the receptor, the contribution from the plant needs to be added to the background noise levels. Sound readings can only be added as *intensity* levels where:

$$dB_B = 10 \log_{10} (I_B / I_0)$$

dB_B =   background sound level
I_B =   intensity of background
I_0 =   reference level intensity

    The inputs to a predictive noise are: (1) a map of the area in question, (2) location of sound sources and receptor locations, (3) an emission inventory of sound sources, and (4) background sound levels at each grid intersection on the location map.

    The model, in turn, will provide data at each receptor point indicating:

    1. Background sound levels, dB
    2. Contribution of the source(s), dB
    3. Total sound level, dB

    The impact of a source on the community can then be plotted as lines of each noise level (dB), much like contour lines on a topographic map. As an example, consider the noise impact on a community from a proposed 250-ton-per-day (tpd) capacity municipal incinerator (Figure 26-5).

    The computer output is shown in Table 26-1. Background community noise levels can then be plotted as shown in Figure 26-6.

    The impact from the plant on the area (without any background noise levels) and the total noise level isobel curves are illustrated in Figures 26-7 and 26-8. The high noise levels shown for the incinerator and the impact on the community are because of the uncontrolled equipment used in the example. It would be necessary, in this case, to use silencers, mufflers or enclosures to dampen the noise levels and thereby reduce the community noise.

    The minimum noise levels in the community, with an uncontrolled incinerator in operation, are in the range of 40 dBA. Two high noise level areas, one of 80 dBA in the center of the community and one of 70 dBA

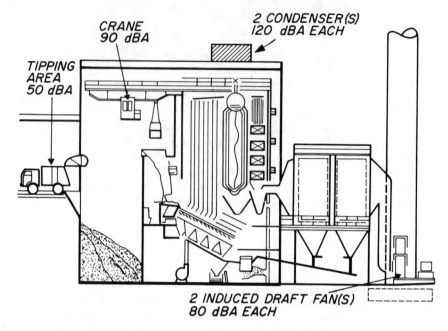

Figure 26-5. Proposed municipal incinerator without sound controls.

Table 26-1. Computer Output of Noise Data for Plotting on Grid Map

Computer Output Sheet

Receptor	Receptor Coordinates		Impact at Receptor Point (dBA)	Background Noise Level (dBA)	Combined Noise Level (dBA)
	X	Y			
1	0.0	0.0	44.33	30.0	44.49
2	1000.0	0.0	45.27	30.0	45.40
3	2000.0	0.0	46.27	30.0	46.38
4	3000.0	0.0	47.33	30.0	47.41
5 (up to 121)	4000.0	0.0	48.42	30.0	48.48

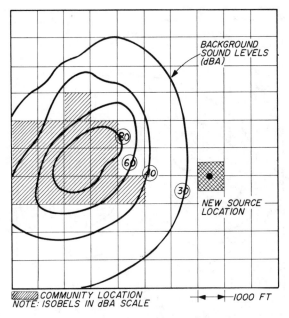

Figure 26-6. Contour of background noise levels within example community.

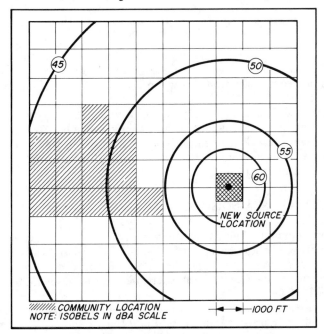

Figure 26-7. Contour of noise impact from proposed incinerator.
(Background noise has not been included.)

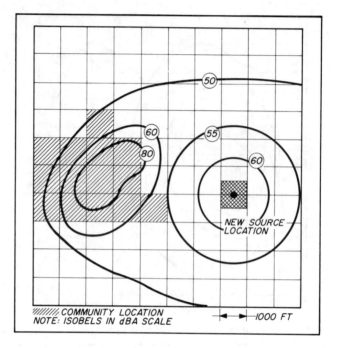

NOTE: ISOBELS IN dBA SCALE
COMMUNITY LOCATION
1000 FT

Figure 26-8. Contour of total community noise levels
including the impact of the incinerator.

at the plant site are indicated by the assessment. The reaction that may be expected in a community from a new intrusive noise source is shown in Figure 26-9.

Because widespread complaints occur between 60 to 70 dB, a considerable portion of the community in this example would be affected by the proposed plant site. Undoubtedly the pollution engineers would consider reducing the community noise levels to below 55 dB where only sporadic complaints might occur.

CALCULATING OSHA NOISE COMPLIANCE[67]

Most pollution engineers are intimately acquainted with the occupational noise criteria set forth in paragraph 50-204.10 of the Walsh-Healey Public Contracts Act as incorporated May 20, 1969, and amended January 24, 1970, and as adopted in the Occupational Safety and Health Act of 1970 in paragraph 1910.95.

The essence of this criteria for occupational noise exposure of employees is nine discrete pairs of sound levels in dBA and associated permissible durations per day in hours, as listed in Table 26-2.

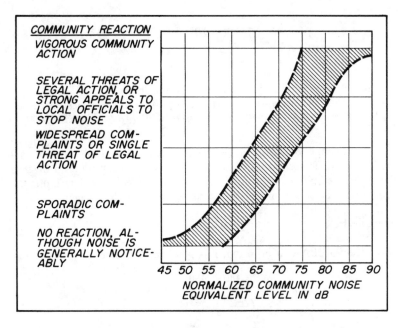

Figure 26-9. Chart of common community reaction to intrusive noise.

Table 26-2

Sound Levels in dBA (slow response)	Duration Per day in Hours
90	8
92	6
95	4
97	3
100	2
102	1½
105	1
110	½
115	¼ or less

Most sound level meters used in industry measure and analyze noise on continuous sampling and integrating electronics schemes. Since the sound pressure levels read with these instruments are not generally limited to the sound levels in dBA of Table 26-2, it is proposed that the following continuous, closed-form function be accepted as a standard formula for employee exposure criteria.

$$T = \frac{480}{\left(\dfrac{L_{P2}-90}{2}\right)}, \text{ min of permissible exposure}$$

where T = permissible exposure time limit in minutes

L_{P2} = the measured "A" scale sound pressure level to which the employee's ears are exposed as referred to the 90 dBA for 8 hr (480 min) criterion.

This simple equation can be used to interpolate Table 26-2 for determining permissible exposure time at a given sound pressure level or given the time of employee exposure. As one can immediately determine, the equation satisfies the nine discrete criteria of Table 26-2 exactly.

Example I:

The 95 dBA-4 hr criterion, L_{P2} = 95 for this case

$$T = \frac{480}{\left(2^{\frac{95-90}{5}}\right)} = \frac{480}{2^1} = 240 \text{ min} = 4 \text{ hr}$$

Example II

The 100 dBA-2 hr criterion

$$T = \frac{480}{\left(2^{\frac{100-90}{5}}\right)} = \frac{480}{2^2} = 120 \text{ min} = 2 \text{ hr}$$

This expression also yields an apparent permissible exposure time even when L_{P2}, the measured sound pressure level, is less than the present 90 dBA level basis of the Walsh-Healey Act and OSHA. This prompts one to ask if there is any physical damage risk data that would substantiate the mathematical result. For example, 960 min of exposure to a sound pressure level of 85 dBA is permissible, but employee exposure at this level for a greater time period would have an adverse effect on human hearing.

There are obviously upper and lower limits to the logical extent of this expression that are represented by practical boundaries of time. For example, 75 dBA would have a time exposure limit of 24 hr (1440 min) which is at least a physical limit on the day, if not the individual employee. At the upper limit, the exposure times diminish very rapidly above sound pressure levels of 135 dBA which gives approximately 56 sec of permissible exposure.

Figure 26-10.

It is interesting to note that the substitution of any future change in the 8-hr based criteria for the 90 (90 dBA) in the exponent of 2 in the denominator of the equation will automatically adjust the equation to the proper form.

It remains for those engaged in establishing the relationship of damage risk criteria and hearing loss to exposure level and time criteria to consider the validity of the proposed equation between levels of 75 to 90 dBA and confirm the accuracy above the 115 dBA level. The fact that the mathematical result seems to have some logical physical significance below the 90 dBA-8 hr threshold should not be ignored.

HEARING CONSERVATION COMPLIANCE GUIDE[70]

Amendments to the Walsh-Healey (Federal Supply) Contracts Act, May 20, 1969 set noise standards and prescribed hearing conservation programs. The standards were extended to almost every business when the provisions of this legislation were adopted as part of the Occupational Safety and Health Act of 1970. The standards state, "In all cases where the sound levels exceed 90 decibels as measured on the 'A' scale for an eight-hour day average, a continuing effective hearing conservation program shall be administered."

What does a noise level of 90 decibels sound like? It is the sound inside a car being passed by a tractor-trailer truck at about 60 miles per hour when the driver of the car has the window open. At a noise level of 80 decibels it is impossible to converse on the telephone. In short, in any area where you have to raise your voice to indulge in conversation at a normal speaking distance you have a noise problem.

To determine which employees must have their hearing measured and recorded, it is necessary to make a noise survey of the manufacturing facilities. The survey should be made using a general purpose sound level meter which meets American National Standards Institute standard S1.4-1961 (or latest) and should be set on the "A" scale slow response. The "A" scale is the setting on the sound level meter which best reflects the range of frequency perception available to the human ear.

Where noise levels exceed those prescribed by the act, employees must be equipped with hearing protection equipment. Qualifying equipment may be in the form of ear plugs, muffs or fine fiber glass wool. However, OSHA says, "Hearing protection equipment the attenuation of which has been determined to be in accordance with ANSI standard Z24.22-1957 is considered acceptable, provided that it affords enough attenuation to provide protection from existing noise."

It is essential that hearing protection be worn in all areas where noise levels exceed the limitations set and that employees be informed of the locations where this equipment is to be worn.

"All employees exposed to noise levels exceeding those set by the standards shall have their hearing acuity determined and recorded. Determination of hearing acuity shall be by pure tone, air conduction and audiometry." Audiometric measurement may be made as frequently as specified by the company's regular physician, but only under very special conditions less than once per year.

According to the OSHA guidelines, hearing measurements must be made in a test room, or booth, which meets background noise level requirements of ANSI S3.1-1960. This ANSI standard states, "Hearing test rooms should be located in as quiet a place as possible, preferably within practical access, but away from outside walls, elevators, heating and plumbing noises, waiting rooms and busy hallways."

If the highest noise levels in a test room do not exceed the levels allowable under the standard, the background noise will not affect test results. To obtain these internal test room ambients, determine the noise level at the test room site with an octave band analyzer. Once this is done, select an audiometric testing room or booth which will provide ample noise reduction to bring the internal noise level down to the prescribed levels (Table 26-3).

There are two types of prefabricated audiometric rooms which are adaptable to industrial hearing conservation programs. They are the compact, pre-assembled booths and the component modular panel rooms. In most instances, the smaller pre-assembled units will be more than adequate. If there is considerable background noise in the test area, then the heavy-duty modular room should be used.

Table 26-3. Maximum Allowable Background Noise for Hearing Measurement Inside Audiometric Rooms

It is assumed that (1) no frequency below 500 Hz will be measured and (2) well-fitted binaural earphones will be worn.

Octave Band Center Frequency	500	1000	2000	4000	8000
Level in dB re 0.0002 dyne/cm^2	40	40	47	57	62

The Audiogram

An employee should be prepared for hearing measurement by a nurse or technician. The attendant should familiarize the employee with the measuring tones and explain the procedure to be used.

The subject is seated with his back to the attendant and told to raise a finger as soon as the sound is heard and to lower it when the sound stops. Responses for each ear are recorded on the audiogram (Figure 26-11).

In order to ensure accurate results of the measurement, the subject should be away from high noise exposure for at least 16 hr, if it is at all possible. High noise exposure within the 16-hr period before the test can cause

AUDIOGRAM

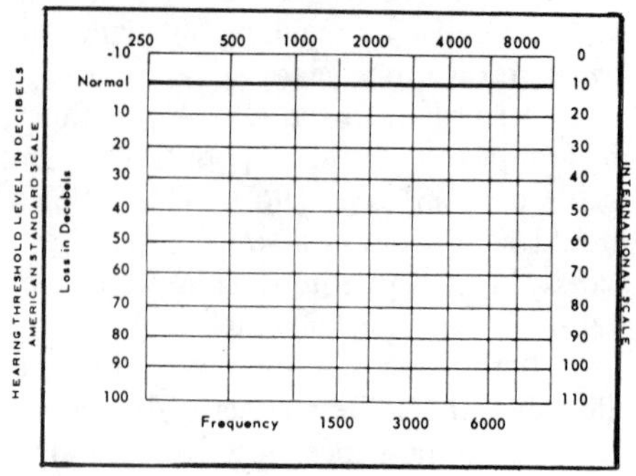

	Right	Left
	Red	Blue
	0	X

This audiogram is plotted on the basis of:
☐ 1951 ASA reference thresholds
☐ 1964 ISO reference thresholds
(Check one of these squares)

(Check Frequency Column Used)

Freq.	Decibel Loss		Freq.
	Right	Left	
250			256
500			512
1000			1024
1500			
2000			2048
3000			2896
4000			4096
6000			5792
8000			8192

HISTORY   No. _____

Prospective Employee ( ) Rehire ( ) Recheck ( )	Address
Name	
Company	Address
Date of Birth	Sex	Occupation	Military Service: Dates Assignment

LAST THREE EMPLOYERS

Company	Type of Work	From	To
Company	Type of Work	From	To
Company	Type of Work	From	To

Ever had:	Head Noises?	Deafness in Family?	Running Ear?

How often do you use firearms?	Ever been to a doctor for ear trouble? Yes( ) No( )
Ever had hearing checked? Yes( ) No( ) When? Where?	When? Why?
Name of Present Ear Doctor	Name of Doctor
Address	Address

Advised to Consult Ear Doctor? Yes( ) No( )   Remarks:

Technician	Station	Make of Audiometer	Date	A.M.	P.M.	Day of Week

Figure 26-11. The audiogram is a written record maintained by an employer containing statistical information about an employee and his hearing.

a slight, temporary threshold shift in the subject's hearing, and the measurement will not accurately depict the person's hearing acuity.

The audiograms must consist of air conduction octave band analysis (air in the ear passage is the medium of conduction, not the bone structure) and shall contain at least the following frequencies: 500, 1000, 2000, 3000 and 4000 cycles per second (Hz). According to the federal guidelines, "the operator of the audiometer should be positioned outside the room or booth, but able to see the interior through the window. The person whose hearing is being measured must face away from the operator and the audiometer to ensure that all responses are based on sound signals alone. It is preferable that the subject be instructed to keep his eyes closed during the examination."

Audiometers

An audiometer which meets the ANSI specification (S3.6-1969) must be used to make the hearing measurements. The audiometer must have a certificate of calibration and must be recalibrated each year. In addition, the audiometer must be checked once per month by an individual with a predetermined hearing curve, and a log of these checks filed with the certificate of calibration. If monthly deviations exceed five decibels, the machine should be electronically recalibrated. There also must be a statement on file indicating whether the audiometer is calibrated to ASA 1951 or ISO 1964 values.

Two basic types of units are available: the manually operated and the automatic recording audiometer. With a manual unit, the operator administers the test tones at each frequency and the subject responds by raising a finger when the tone is heard. The operator then records responses on a chart called an audiogram. A plot of the minimum intensity heard at each given frequency makes up the individual's audiogram. Automatic measurement is different in that the subject responds to test tones by depressing a button in a hand-held control. This activates a pen which records responses for each testing frequency.

No matter which type of instrument is used, testing must be done under the direction of trained personnel. Records of each test must be kept on file and be available for inspection.

Trained Personnel

Dr. Roger Maas, director of hearing conservation, Employers Insurance of Wausau, has the following to say about the personnel who do the testing: "It is no longer possible under the provisions of the Occupational Safety and Health Act to conduct audiometric examination designed for safety, medical and legal purposes without the use of personnel with minimal training of 20 hours as proposed by the Inter-Society committee which studied the problem of preparation requirements for more than 6 years. Federal Bulletin #334, Guidelines to the Department of Labor's Occupational Noise Standards for Federal Supply Contracts, says, 'Audiometric measurements shall be made by a person trained or skilled in audiometry examinations.' I am told, and believe firmly, that skill and competency are obtained by formal instruction as outlined in the Guide for Training of Industrial Audiometric Technicians for Instructors as presented by the Inter-Society Committee."

This guide describes the standard method of administering audiometric examinations as recorded by representatives of the American Speech and Hearing Association, American Industrial Hygiene Association, Industrial Medical Association, and the American Association of Industrial Nurses, and is available from any of these groups. Accredited training courses are available through educational institutions with audiological centers, through individual instruction by otologists and 20-hr courses sponsored by one of the associations mentioned above.

The Guide for Hearing Conservation in Noise says, "The skills necessary to administer air conduction audiometric measurements can be learned by a person of normal intelligence." In many firms the industrial nurse receives this training; however, a medical background is not a requirement. For information on courses in a specific area, contact the president of your State Chapter of the American Association of Industrial Nurses, or the speech and hearing clinic at your nearest college or university.

Economic Sense

Not only is hearing conservation required by law, but it makes good economic sense to establish an effective and continuing program. The average cost of a hearing loss compensation claim is said to be approximately $2000. It doesn't take many claims to develop a figure that could deplete profits significantly when considering the number of American workers exposed to noise levels that exceed the legal limits.

The cost of equipment to establish a program (audiometer, booth, and sound level meter) can be as little as $1600. The 20-hr accredited training course will require another $50-100. So the initial fixed cost is less than the average cost of one claim.

In addition to the monitoring aspect of the hearing conservation programs which will detect any changes is the hearing acuity of employees. A pre-employment audiogram will detect hearing impairments in new employees and also will establish a base line to which future audiograms will be compared.

In the future, it will be management's responsibility to eliminate the problem at its source. Dispensing ear protection alone will not be considered a satisfactory solution as the technology becomes available to solve the majority of industrial noise problems.

BUYING GUIDELINES FOR NOISE CONTROL[87]

"Let the buyer beware" has never been more relevant that it is in specifying and purchasing equipment noise levels, and acoustical materials for the control of noise pollution. This is not to infer that the manufacturers or sellers of materials or equipment are less than ethical. But, the simple fact is that insufficient standardization exists for rating noise levels of equipment. The ratings of acoustical equipment and materials, and how to apply these ratings to obtain the desired effect on the total noise environment are confusing to the layman.

Purchasing of equipment, with respect to acceptable noise levels, must begin by understanding five basic acoustic terms:

1. Sound power level (PWL)
2. Sound pressure level (SPL)
3. Transmission loss (TL)
4. Noise Reduction (NR)
5. Absorption Coefficient ($\propto$)

Sound power level (PWL) of a noise source is a number, expressed in decibels, to describe the power of the sound which is radiating from the source. The PWL number if dependent upon the number of acoustic watts radiating from the source and upon the number of watts to which the source is referenced. In the English units system, the reference level is usually 10^{-13} watts. The PWL is an indication of the acoustic radiation of the noise source without regard to the surrounding environment (Figure 26-12). The source PWL in both cases shown does not change.

Sound pressure level (SPL) of a noise source is a measurement (expressed in decibels) of the existing noise at a specified location. However, PWL and SPL decibels are two different quantities and should not be confused with one another. The SPL is dependent upon the noise source PWL, upon the reference level, distance from the source and the environment around the source. The reference level is usually 0.0002 dynes/cm^2. Figure 26-12 demonstrates the environmental effect. For the same PWL source and same size rooms, SPL measured in a hard-surfaced room will be higher, at some specified distance from the source, than the SPL measured in the soft-surfaced room. This is due to reflections of sound energy off the hard surfaces. Sound pressure levels resulting from a noise source in a large room without reflective surfaces, or outdoors, vary according to distance from the source. The PWL of the source remains the same, but the SPL decreases approximately 6 decibels for every doubling of distance from the source, Figure 26-13.

Transmission loss (TL) rating, in decibels, of an acoustical material is an indication of the effectiveness of the material to stop the transmission of noise. The TL number which has been assigned to the material is a number which has been acquired under laboratory testing conditions. Since TL is a laboratory condition, it does not take into account the environmental surroundings of the noise source, or the noise-receiving area (Figure 26-14).

Noise reduction (NR) of an acoustical material, in decibels, is the measured or calculated result of placing an acoustical barrier between the source and the receiver. The NR is a function of the TL of the barrier and such environmental factors as the reflective sound energy (Figure 26-15).

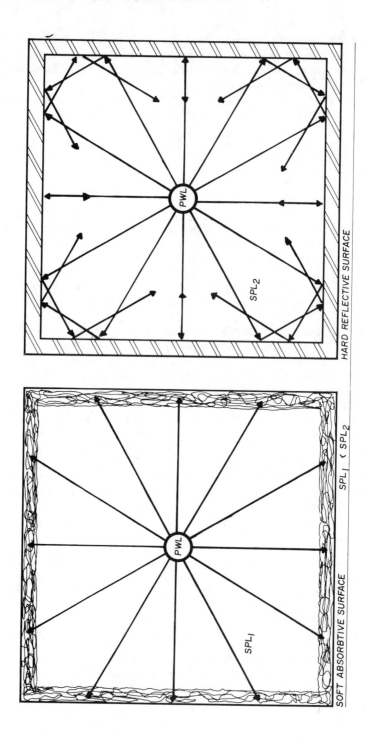

Figure 26-12. Effect of room conditions on a noise source of equal sound power levels.

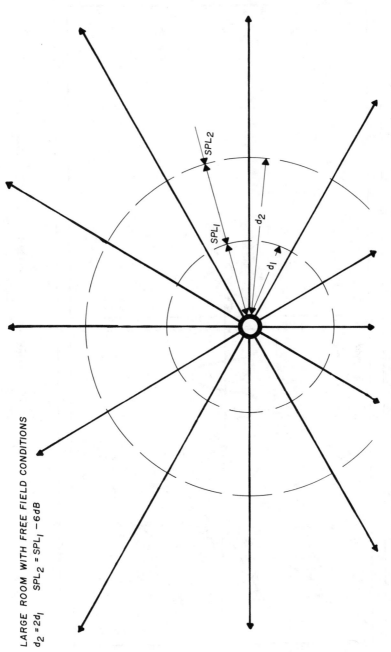

LARGE ROOM WITH FREE FIELD CONDITIONS

$d_2 = 2d_1$ $SPL_2 = SPL_1 - 6dB$

Figure 26-13. Sound pressure level varies with the distance from the source in absence of reflective surfaces.

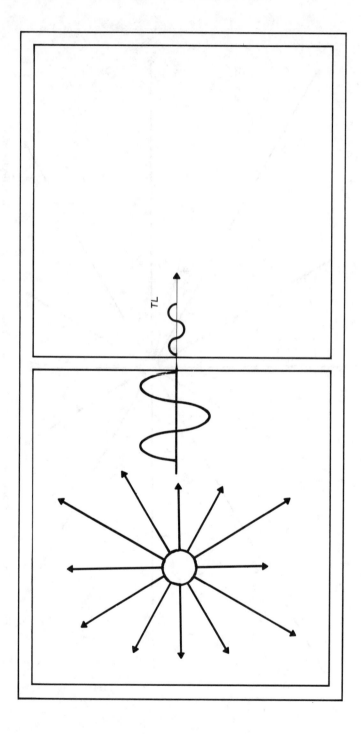

Figure 26-14. Transmission loss (TL) = number of decibel reduction in passing through a material. Environmental factors are not taken into account.

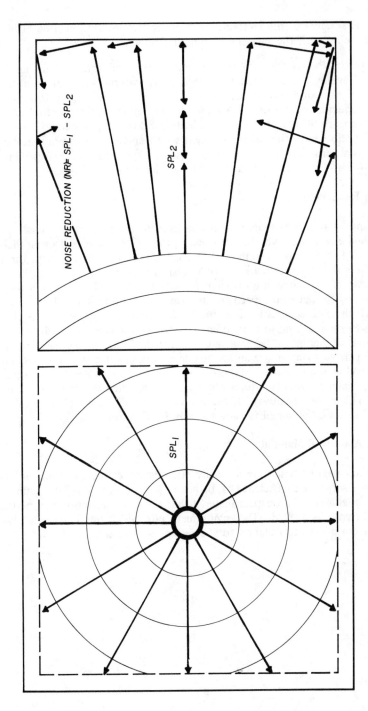

Figure 26-15. Noise reduction (NR) is a function of Transmission Loss (TL) and room environmental effects.

Since the surfaces of the receiving room are reflective, SPL_2 will be higher than it would be if the surfaces were nonreflective.

Absorption coefficient ($\propto$) of a material is an indication of the amount of sound energy that a material can absorb. A material with a rated absorption coefficient of 1.0 theoretically has a capability of absorbing 100 percent of the incident sound energy. A material with an absorption coefficient of 0.9 will absorb 90 percent of the incident sound energy, and could result in a 10-decibel reduction of the incident sound energy. Since only incident energy is absorbed, consideration must also be given to sound energy which could bypass the absorptive surface through defraction or leakage.

Buying Equipment

Special attention should always be given to the purchase of noise-controlled equipment, whether it is plant production equipment or an office machine. It is far more economical to control the noise environment through controlling the noise source. The two most popular means of rating the noise level of a machine are through the octave band PWL or through the octave band SPL at some specified distance from the machine. The distance at which the SPL is specified is normally the operator's station, and is close enough to the machine so that room acoustics or the effect of multiple reflections of energy at the point are minimized. These octave band decibels at the operator's station are an acceptable measurement in rating a machine, if the level at the operator's station is the only area of concern. When noise from a machine seriously affects other work areas, then the machine octave band PWL's must be known. If the octave band PWL's are known, a calculated SPL can be obtained for any location in the room.

Buying Acoustical Material

Acoustical materials are normally rated by either the transmission loss (TL) or absorption coefficient ($\propto$), or both. It is often mistakenly assumed by the purchasers of acoustical material that the noise reduction will be the same as transmission loss. The environment into which the material is placed must be considered to obtain a total effect of the material on the noise level.

CHAPTER 27

NOISE MEASUREMENT AND PERSONAL PROTECTION

SOUND TRANSMISSION AND ABSORPTION[88]

Transmission of sound from one area to another is a phenomenon often misunderstood. Detrimental and unpleasant sound energy transmitted from one area to another is classified as noise. Transmission of this energy should be reduced to an acceptable level.

Noise is transmitted by various means. The most direct means of transmission is normal propagation of sound through the atmosphere. However, transmission is altered by the sound reflecting off hard surfaces, refraction around edges and by placing surfaces into mechanical vibration which in turn radiate sound energy.

Sound pressure is oscillating air pressure, which oscillates above and below atmospheric pressure. At the instant a sound is generated, these pressures oscillate sinusoidally above and below atmospheric pressure. A sound can be thought of as a particle of air which is displaced from its equilibrium position and which bumps surrounding particles. These surrounding particles are put in motion by the bumping and in turn bump adjacent particles. In this manner, sound is transmitted through the atmosphere.

Sound energy radiates from a source in a spherical manner (Figure 27-1). Radiating noise surfaces grow like the surface area of an expanding balloon. As this spherical area increases, the sound energy over a unit area decreases. A sound pressure level will decrease at the rate of 6 dB for every doubling of distance from the source. This decrease is due primarily to the spherical spreading of energy and, to a lesser degree, air absorption of the energy. The 6-dB reduction for each doubling of distance, however, is only true when the sound is radiating from its source into a free field (an area free of reflective surfaces).

Industrial noise environments are not usually free field conditions. The environment is usually affected by a number of obstacles between the noise generating source and the observer, which reflect, refract, absorb or amplify the noise. Figure 27-2 illustrates the complex transmission effect of a sound when it is reflected and refracted. At point **A** the noise source

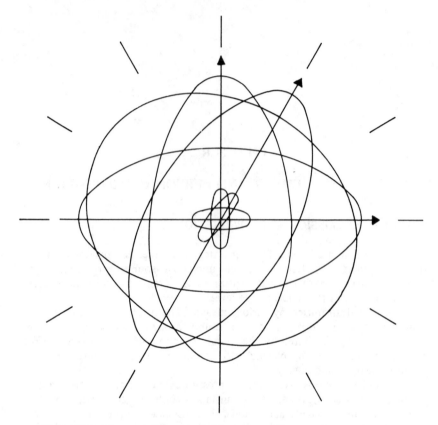

Figure 27-1. Sounds radiate spherically from their source of generation.

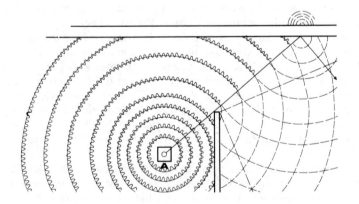

Figure 27-2. Sound transmission in a non-free field.

is radiating energy spherically with the lines indicating equal sound pressure level contours. Assuming perfect spherical radiation the pattern remains spherical until the energy reflects off an obstacle such as a hard-surfaced barrier wall or floor. Some of this energy is reflected into the pattern where it reinforces some of the wave fronts (Figure 27-3). The lobe in the pattern shows the higher sound pressure level that would normally be expected at that distance from the source.

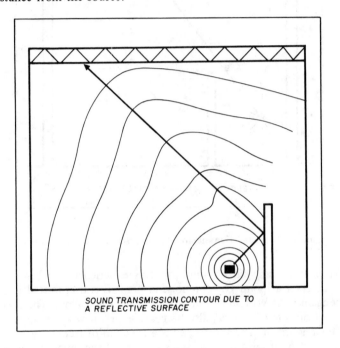

SOUND TRANSMISSION CONTOUR DUE TO A REFLECTIVE SURFACE

Figure 27-3. Sound transmission contour due to a reflecting surface.

This lobe could be eliminated and the pattern returned to its original shape by constructing a barrier wall of acoustical absorption material on the side facing the noise source. In order for the curve to be returned to a spherical front, the material of the barrier wall on the side facing the noise source would have an acoustical absorption coefficient of approximately 1.0. The absorption coefficient of a material is the ratio of the energy absorbed by a surface to the energy incident upon the surface. A ratio of 1.0 means that all incident energy is absorbed and none is reflected back into the wave pattern.

As a direct wave propagation from the source clears the top of a barrier wall, diffraction of the wave front might occur (Figure 27-4). That is, the direction of the wave front can be changed or bent. At this point it can be considered as a new noise source radiating in a spherical manner.

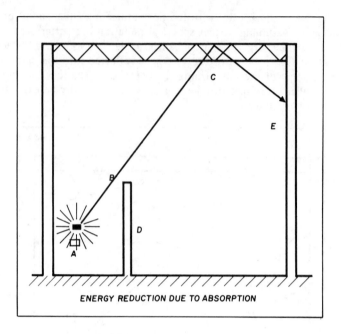

Figure 27-4. Sound transmission due to refraction and reflection.

An observer at point **E** would probably experience a higher sound level than an observer at point **D**. The noise at point **E** is transmitted from diffraction against the wall and from reflection off the ceiling. The point of reflection from the ceiling can also be considered a new noise source.

The amount of energy reflected is a function of frequency, incident angle, amount of absorption, stiffness and mass of the reflective surface. If a reflective surface is massive, stiff and hard, such as a concrete ceiling, the majority of the incident energy will be reflected, and very little lost to vibration of the ceiling. If the surface at the point of the incident energy consists of an acoustical absorption material, then the amount of reflected energy could be very small (Figure 27-5). A ceiling material with an acoustical absorption coefficient of 0.5 has the ability to absorb 50 percent of the incident energy. This absorption would result in a reflected wave front 3 dB lower than the incoming wave front. A ceiling with an absorption coefficient of 0.9 will absorb 90 percent of the incident energy and result in a 10 dB reduction of the reflected wave front. This does not mean, however, that the total noise level has been lowered 10 decibels, since acoustic energy from the diffraction wave is also propagating through that point.

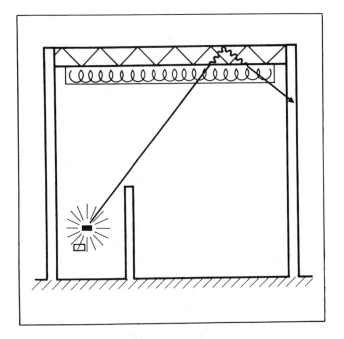

Figure 27-5. Energy reduction due to absorption.

VELOCITY OF SOUND IN GASES[161]

Often it is necessary to know the velocity of sound in gases for solution of formulas relating to pulsation attenuation and silencing devices. Selection of silencer type and size is based on such consideration as type of application and service, noise intensity, flow rate and type of gas being vented, allowable pressure drop, temperature and cost.

Included in some of the calculation is knowledge of the speed of sound in gases which can be found with this nomograph (Figure 27-6) when the molecular weight or specific gravity, "K" value and temperature of the gases are known. It is based on the formula, $C = \sqrt{gKRT}$, where C is the velocity of sound in feet per second, g is equal to 32.2, K is the ratio of specific heats (cp/cv), R is equal to 1544/mol wt and T is temperature, deg R (460 + deg F).

Example: Assume it is necessary to select a silencer to suppress noise from a turbocharged engine exhaust. One requirement is to determine the fundamental frequency of the engine noise. The formula to find the fundamental frequency requires knowledge of the speed of sound. The sound velocity within the exhaust system may be obtained from the nomograph as follows:

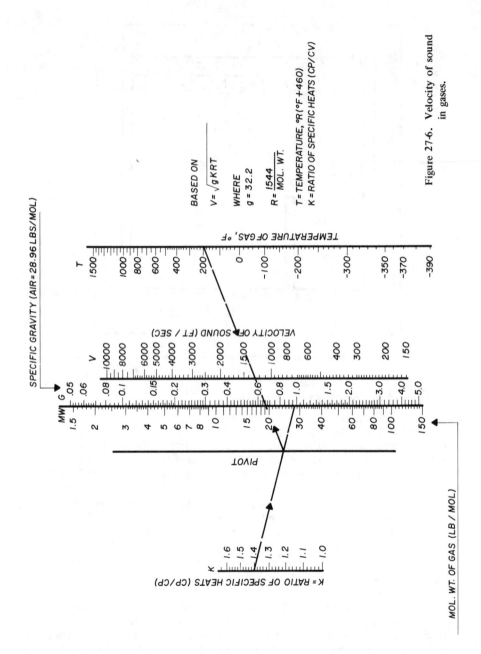

Figure 27-6.   Velocity of sound in gases.

Assume: K value of exhaust gas–1.4, molecular weight–28.0 and temperature–200 deg F.

Solution: Align 1.4 on "K" scale with 28 on MW scale and mark point where line crosses pivot. Next align marked point with 200 deg F on T scale and read 1280 fps where line crosses Velocity of Sound scale.

COMBINING SOUND LEVELS AND
CORRECTING FOR BACKGROUND NOISE[157]

Frequently, sound levels must be combined. Since decibels are logarithms, they cannot be added arithmetically but must be converted to relative powers, combined, then converted back to decibels. It is often necessary to add the effect of one sound to another, or to subtract the effect of one sound from a combination of sounds. The effects of both sound power and sound pressure are important.

The total sound power of a combination of sounds is equal to the sum of the individual sound powers ($W_1 + W_2 + \ldots + W_n$). Due to the logarithmetic character of sound power levels the total sound power level is not equal to the sum of individual sound powers.

Both the ear and the sound level meter respond to the root mean square value of sound pressure rather than to the instantaneous value. Because response is proportional to RMS value, the total sound pressure at a point ($p_1 + {}_2 + \ldots$ n) is equal to the square root of the sum of the squared individual sound pressures ($\sqrt{p_1{}^2 + p_2 + \ldots + p_n{}^2}$) at that point.

Due to the logarithmic character of the sound pressure levels the total sound pressure level is not equal to the sum of the individual sound pressure levels.

Figure 27-7 is constructed for rapid calculation of these values. In using the nomograph the difference between the two known values establishes the difference between the unknown value and one of the known values.

Example 1: Assume a plant with a sound pressure level (SPL) of 74 dB. New equipment is then installed that has a level of 75 dB. What is the resultant noise level?

Solution: First, find the difference between the two dB values, or (75 - 74 = 1). Now, align 1 on Scale **A** with the greater of the two dB values, or 75 on Higher Level Scale **D** and read resultant noise level as 77.5 dB where line crosses the **C** scale.

Example 2: Given $PWL_1 = 90$ dB and $PWL_2 = 100$ dB. Find $PWL_1 + {}_2$.

Solution: $PWL_2 - PWL_1 = 100 - 90 = 10$. Align on Scale **A** with 100 on Scale **D** and read 100.41 dB where line crosses Scale **C**.

Example 3: Effect on background noise. Given $SPL_1 + {}_2 = 60$ dB (over-all including background) $SPL_2 = 55$ dB (background). Find SPL_1.

Solution: $SPL_1 + {}_2 - SPL_2 = 60 - 55 = 5$. Align 5 on Scale **B** with 60 on Scale **D**, extend to Scale **E** and read 58.3 dB.

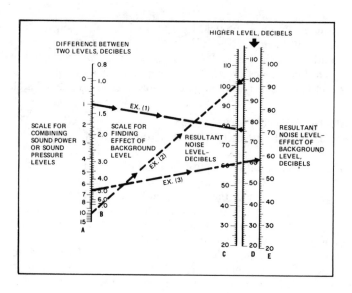

Figure 27-7

PREDICTING PEAK NOISE FREQUENCY
OF VENTS TO ATMOSPHERE[160]

Low frequency noise is produced by low pressure, large valve, mass flow to vents. High frequency noise is generally considered to be produced with high pressure, medium to small size, high velocity vents.

The selection of a silencer type is a function of both the frequency content of the noise and the specific application requirements. However, the basic type is generally set by the fundamental or peak frequency of the noise as given in the following table:

Maximum Performance Range for Silencers

Silencer Type	Peak Frequency, cps
Reactive, multi-chamber	Up to 150
Reactive/dissipative combination	150 to 1000
Dissipative, Absorptive	Above 1000

Peak frequency refers to the fundamental frequency at which maximum noise occurs. The nomograph, based upon the Strouhal formula modified to account for vent gas properties, predicts the peak frequency of unsilenced noise.

$$f_1 Max = 0.2 \ VD^{-1} \ \text{(Strouhal)}$$

$$f_1 Max = 48.2 \ (T/MW)^{\frac{1}{2}} \ D^{-1}$$

where:

$f_1 Max$	=	Peak Frequency, cycles/sec, Hertz
T	=	Temperature, R(F + 460)
MW	=	Mol wt of gas
D	=	Valve throat diameter, ft
V	=	Velocity, ft/sec

The nomograph (Figure 27-8) is based on Gas K (CP/CV) value of 1.4.

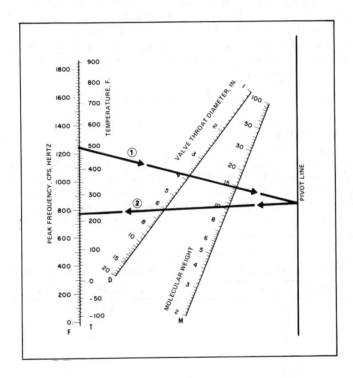

Figure 27-8

Example: What is the peak frequency (f_1 Max) of the unsilenced noise of a gas venting to atmosphere through a valve with a 6-in. throat diameter at 500 F if the Mol. Wt. of the gas is 15?

Solution: Line 500 F on **T** scale with 15 on **M** scale and extend to Pivot Line. From point on Pivot Line connect with 6-in. on **D** scale, extend to **F** scale and read peak frequency as 772 cps.

NOISE AND ITS MEASUREMENT[93]

Noise can be described as an undesirable airborne sound. Sound, in turn, can be described as a repeated pressure fluctuation. It is characterized by its amplitude or sound pressure, and its frequency in time. A third characteristic of sound is its spatial variation or wavelength. For sounds in any medium, the frequency and wavelength can be expressed by the equation: frequency x wavelength = velocity. In air, the velocity of sound is constant, and at standard temperature and pressure is approximately 1130 ft/sec.

Sound propagates in air as a longitudinal wave, that is, a wave where the motion of small regions of the medium is parallel to the direction of propagation. These pressure variations will occur at a given position with the frequency (f) equal to the frequency of the source disturbance. The distance between pressure peaks at any instant in time will be the velocity divided by the frequency:

$$\text{wavelength} = \frac{c}{f} = \frac{1130}{100} = 11.3 \text{ ft}$$

Frequency Spectrum

An octave band is a band in frequency where the lower frequency is related to the upper frequency by the ratio of 2:1. The standard octave bands are named by their center of frequency, and their normal range is shown in Table 27-1. This method of depicting the frequency as a dimension of a sound is often used in noise measurement because the ear seems to sense frequency on such a logarithmic scale.

Table 27-1. Octave Band Center Frequencies (Hz) and Nominal Bandwidths

Center Frequency	Band Edge Frequencies
31.5	22.3-44.6
63	44.6-88.5
125	88.5-177
250	177-354
500	354-707
1000	707-1414
2000	1414-2830
4000	2830-5650
8000	5650-11,300
16,000	11,300-22,600

Sound Pressure

Sound is a pressure phenomenon. The human ear can hear, without damage, a sound pressure 10 million times greater than the sound pressure of the softest sound it can sense. To express this on a linear scale would be extremely difficult, so a system based again on logarithmic ratios has been derived.

The ear also seems to respond to pressure in a logarithmic manner. Therefore, measurements of sound pressure levels are given in *decibels.*

Arbitrarily, zero decibels has been designated as the sound pressure level at the human ear's threshold of hearing, for a 1000 Hertz (cps) pure tone. In real life, variation of sounds that occur in normal living and working makes it difficult for the average person to detect differences of much less than three units or decibels.

Mathematically, this logarithmic range is expressed as the level in decibels, being equal to:

$$10 \text{ Log } \frac{p^2}{10 \ p^2 \ ref}$$

Some problems can arise when working with decibels, since they do not combine in a linear manner. This is due to the fact that we are trying to combine the square of the pressure whereas the units expressing it are in terms of the logarithm of this quantity. When combining two equal sound pressure levels, the sum is not twice the original pressure, but the original pressure plus 3 dB.

Psychoacoustics

The field of psychoacoustics arose from attempts to describe the ear and the mind as a receiver of sound. The ear, coupled with the mind, represents a receiver with radically different properties than electronic instrumentation. The ear itself is basically a nonlinear transducer, and the subjective interpretation of response to sound is widely different. Figure 27-9 shows a set of equal loudness level curves for an average listener. Each curve gives the physical sound pressure as a function of frequency which is judged to sound equally loud as the level of a pure tone at 1000 Hz at the base level in decibels, *e.g.,* a 1000 Hz sound of 50 dB has a loudness of 50 phons. To sound equally loud at 100 Hz, the SPL (Sound Pressure Level) must be 68 dB.

To take this effect precisely into account in instrumentation might require a set of 100 different filters, one for each level. These have been approximated with three different filters, called the A, B and C Scales. The A scale curve corresponds roughly to the ear's response in the range from 0 to 65 dB overall. The B scale corresponds to the response in the range of nominally 65 to 85 dB. The C scale, which is roughly linear from 63 out to 8000 Hz, corresponds fairly well the ear's performance above 85 dB. Data taken with weighting networks are always reported as sound levels to distinguish from sound pressure levels and are always reported with the network name. For example, a measurement of 85 dB on the A scale is reported as a sound level of 85 dBA.

A second factor in determining the suitability of an acoustic environment is speech interference. Noise, because it is a sound, competes with speech for listeners' attention. Sound levels have been rated with respect

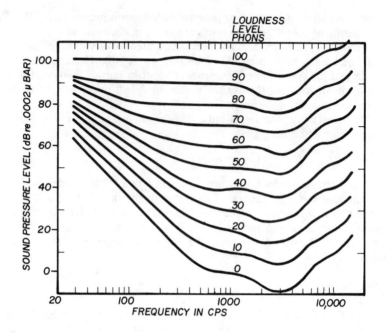

Figure 27-9

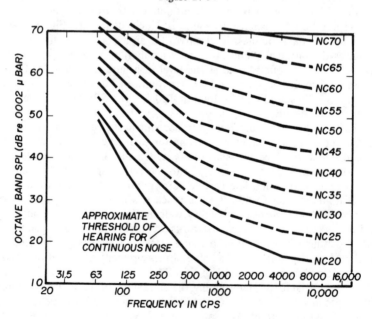

Figure 27-10

to the type of tasks that were required in that environment. The result of investigations is shown in Figure 27-10. These curves are called noise criterion or N.C. curves.

Architectural Acoustics

It is important when making measurements to have some feeling for what the sound field "looks" like. An environment where there are no reflecting walls or objects is referred to as a free field, to indicate the absence of reflections. When hard walls or objects cause the sound to reflect and echo, we find two different kinds of fields. The first, called the direct field, is only discernible very close to the source. This field is similar to what we would see if there were no reflecting walls.

The second type of sound field is the reverberant field, and represents the sum of all reflected sound. One of the reverberant field's characteristics is that its sound level does not change as a function of position. It is, on the average, constant throughout the room. Figure 27-11 shows the approximate profiles of the sound field as a function of position for a source in each of these two environments.

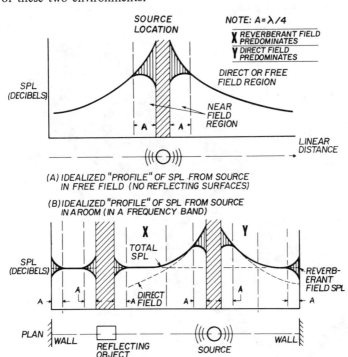

Figure 27-11. Profiles of sound pressure levels will vary depending upon whether measurements are made in a "free" or "reverberant" field, and where the instrument is located.

In the vicinity of a vibrating object with a complex vibrating surface the sound field is not simple or easily described. This region is referred to as the near field, because it is only discernible in a region within about a quarter of a wavelength from the source surfaces.

Instrumentation

Two different types of equipment are available for making measurements in the field. These are the survey-meter class of instrumentation and the precision-sound-level-meter class.

The sound survey meter is a relatively simple, hand-held instrument. It usually provides only the A, B, C weighted sound levels, and perhaps a linear or overall output. For simple requirements, this meter is very useful in indicating the gross or overall characteristics of the sound field. When properly calibrated, it should be capable of providing approximately ±2dB accuracy for measuring the sound pressure level of a plane wave in a free field over the frequency range of 45 to 11,200 Hz.

The precision sound level meter (Figure 27-12) provides more detailed information about a sound field. It is accurate ±1dB for free field, single frequency sounds over a wide frequency range.

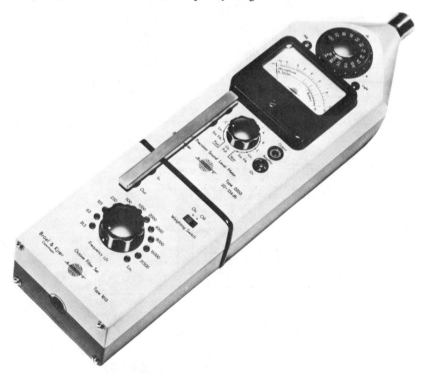

Figure 27-12. Precision sound level meters can measure A, B, C scale weighted sound levels as well as octave band analysis.

For many noise measurements, the sound survey meter should be sufficient, if the end use requires only weighted sound levels. On the other hand, if two noises give similar meter indications, but sound different to the ear, then some type of octave band analysis is required to reasonably compare them.

When an environment or a particular noise source is measured, the data should be as representative of the source as possible. To achieve this, the following precautions are required:

A. evaluating background noise and its effect on the data.
B. determining the sound field type in which measurements are taken (*e.g.,* direct or reverberant field.)
C. avoiding measurements in the near field whenever possible, unless it corresponds to an operator position.
D. using measurements in the direct field, unless the characteristics of the room are well understood, or a power output measurement is being made by a source substitution method.

Corrections for background noise are important. Figure 27-13 provides a plot to make this correction easier. In general, background noise corrections are necessary when the sum measurement (source plus background) is less than 15 dB above background, but are not accurately possible when this difference falls below 4 dB.

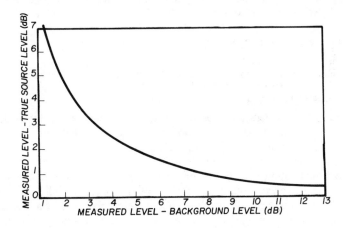

Figure 27-13. Background noise correction chart.

An operator will find the use of a data sheet most valuable. This sheet should contain space for pertinent information including:

1. A sketch of the equipment setup including model and serial numbers and calibration data.
2. A description of the source(s), environment, and ambient factors such as temperature, relative humidity and pressure.
3. A sketch of the measurement position and notes on the observed sound field.

4. The data, including space for noting all control settings, etc.
5. A listing of correction factors (cable corrections, etc.).
6. A checklist for equipment operation and calibration.

Finally, verification of equipment operation is important in any measurement, and one step in this verification should include an on-site system calibration using a portable noise source, which provides a calibrated sound level directly to the microphone. This permits a direct overall calibration of the measuring system at least for a single frequency.

PLANT NOISE SURVEY TECHNIQUES[30]

Legal guidelines, such as the Occupational Safety and Health Act, stipulate acceptable noise levels which may be tolerated in plant operations. Yet it is difficult to institute an exact set of rules simply because of the very nature of sound.

Evaluating Noises

Some attention will have to be spent on mathematics, since existing methods are rather limited in directly measuring quantities necessary for analyzing noises. The only quantity that can be measured directly is sound pressure level. Fortunately, formulas have been derived to relate necessary data.

There are essentially four quantities of importance:

1. sound pressure level,
2. distance from the source at which a change in sound field occurs,
3. intensity of the sound, and
4. sound energy density.

Sound is recorded in decibels. Sound level meters are point-reading instruments to measure sound pressure level directly so that immediate comparison with regulations can be made (Figure 27-14).

Under plant operating conditions sound waves are reflected from walls. The directional sound field (plane or spherical) is found close to the source. At points farther away, sound approaches areas uniformly and randomly from all angles, and thus is diffuse. Since different fields exist at various points, it is not enough merely to take random measurements. As sound generates a new field, the sound pressure becomes distorted. It so happens that the type of field will also affect the accuracy of the equipment. Therefore, one must determine approximately where each field exists, and which field contributes the most problem to the worker.

The distance from the sound source which describes the transition from a directional sound field to a diffuse field can be calculated by $r_G = 0.14\sqrt{\bar{a}A}$ where r_G is the critical radius, or the distance from the sound sources at which the plane field changes to a diffuse field; $\bar{a}$ = absorption coefficient of walls ($\bar{a}$ averages from 0.05 to 0.20 for a factory,

Figure 27-14. Type 1565-A sound
level meter for measuring
in-plant noise levels.
(Photo courtesy of
General Radio Company).

and values can be obtained from available references for specific construction materials); A = total surface area of the room.

Once the position of each field is known, alterations can be made in the equipment, and sound level measurements can be taken in each. The most important data will be obtained in the field which affects the greatest number of workers, which can be called the *critical field*. It is important to realize that the worker actually exists in two fields simultaneously; he is present in the plane field produced by the machine he is working with, and in a diffuse field generated by surrounding equipment.

Sound intensity is a term that can help compare various sources. It is analogous to electrical power and is defined as the rate at which acoustic energy passes through a unit area normal to the direction of sound propagation. It is characterized by the formula, $I = pv$; where I is the sound intensity, p = measured sound pressure, and v = particle velocity. Particle velocity of a sound wave is the speed of an infinitesimal part of the medium, with respect to the medium as a whole, caused by the passage of a sound wave. The units are meters per second. There is a relationship between particle velocity and sound pressure, but values for v at various sound pressures can be obtained from handbooks. Another term for intensity might be acoustical strength. It is a factor telling which sound is stronger in terms of loudness and energy.

Sound generates energy which produces a strain on the human ear. If the sound energy density or the energy per unit volume is known, one can measure this strain. Sound energy density can be obtained by equation: $E = P^2/XC^2$; where E = ergs/cu cm; P = sound pressure, dynes per square centimeter; X = density of air, gm per cu cm; and C = speed of sound cm/sec.

Masking

In plant operations, noises usually originate from many sources at the same time. The noises from different machines have a distinct influence on each other (Figure 27-15). Because sound has wave properties, cancellation generally occurs between several sounds.

Figure 27-15. Checking the noise level of an L. B. Foster Vibro Driver/Extractor on a highly-sensitive decibel meter.

When one sound partially drowns out another, this is known as masking. Total masking occurs when a strong noise renders the lower level completely inaudible. This presents a problem in recording data. The true levels of each source are most important. This may require shutting down several machines and recording data for just one source at a time. The true level is necessary in order to take steps to reduce noise on one particular machine. All measurements can be made under plant operations, but if after analysis the overall levels prove to be dangerous, then the true sound level of each machine will be required to determine its contribution.

Each of the formulas given above are important for obtaining values that cannot be measured directly with instruments. Critical radius, sound pressure level, intensity, and energy each describe something about the noise; its strength, range, frequency, and main area of trouble.

Instrumentation

The most readily measurable aspect of sound by commercially available instrumentation is pressure levels, dB. From sound pressure level the source directivity, sound power, and most effective techniques for control are determined. Sound pressure is recorded by a microphone and amplified by a sound level meter. Careful calibration of the entire setup from microphone to output data is a prime concern.

There are many sound measuring meters on the market today which do the job (Figure 27-16). However, many times a portion of the apparatus

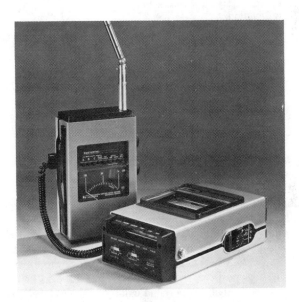

Figure 27-16. Portable precision sound level meter and recorder.

is neglected in specifying and purchasing. This is the microphone. A microphone, or transducer, is subject to change depending upon the testing conditions. Careful consideration should be given to selection in order to obtain accurate readings. Sound pressure and sound pressure level are analogous to voltage and voltage level in electricity which makes it possible to convert audio signal waves to electric. A microphone can convert audio sound pressure into an electrical signal. Its accuracy and that of the meter are dependent upon: (1) the type of field and microphone orientation, (2) the environment, (3) human error and (4) outside interference.

All microphones fall under the following classifications: pressure, velocity, or a combination of pressure and velocity. To convert acoustical fluctuations into the corresponding electrical variations, one can use any of the following transducers: carbon, magnetic, dynamic, condenser, crystal, electronic, hotwire, or ceramic. The three major types are the piezoelectric, the condenser, and the dynamic. Table 27-2 gives a brief comparison of each type. It is worthwhile to compare their accuracies.

Table 27-2. Advantages and Limitations of Major Types of Microphones

Type	Advantages	Limitations
Piezoelectric (or ceramic)	Maintains a flat frequency response down to low frequencies. Can be calibrated with little difficulty. Has a high resistance to moisture and maintains a low sensitivity to vibration.	Has a limited upper frequency range. Much less sensitive than the condensor microphone.
Condenser	Simplest type design. Offers the best frequency response and is insensitive to vibration, although not as resistant as the piezoelectric.	Has a high source of impedance which makes it vulnerable to atmospheric conditions.
Dynamic (moving coil)	Very rugged and moisture resistant. Requires no preamplifiers when using long cables, mainly because its source impedance is a low value.	Very difficult to calibrate. Frequency response is poor compared to the previous models. Quite sensitive to vibration.

The classification of field, diffuse or plane may determine which type of microphone to use and the orientation. Most transducers have directional characteristics. The angle at which the sound field comes in contact with the transducer plays an important role in its accuracy, and adjustments must be made in the meter and microphone. Also, fields help distinguish between which type to utilize, omnidirectional (pressure) or gradient (velocity) microphones. The omnidirectional seems to offer the best results in diffuse fields, whereas a gradient could prove more beneficial in a plane field. In a plane field, the microphone will have to be oriented so that the sound waves strike it perpendicularly. When corrections are required, the best microphones offer accuracy of ± 2 dB.

Humidity has a profound effect on microphones. Moisture may condense within and on the outside of the device. This can create unwanted conduction paths in the amplifier which lower the input impedance. It is therefore desirable to choose a microphone that has its critical circuitry properly encapsulated.

Human error can produce improper data. Wrong setup of equipment and improper reading of meters can destroy the value of the entire survey. It is advisable to be thoroughly familiar with the equipment before starting a survey test.

Outside interference also can influence measurements. The observer or anything producing sounds that are not part of the norm affect data. It may even be desirable for the observer to be absent from the critical field when recording data.

Equipment Arrangement

In measuring sounds, it is vital that the results of the measurements correlate to the worker's sense of hearing. Therefore, the microphone must be mounted on a tripod at the same height as the worker's ears. Several microphones may be used at once, situated at the source and various points within the critical area. Orientation of the microphone in the diffuse field (the distance from the source estimated by r_G) must be neglected, since sound approaches that point randomly. In a directional field it must be situated in such a way that the directional part of the field is frontally incident.

Many times a tape recorder can be set up some distance away with microphones attached by transmission cables. All equipment aiding in measuring—the meter, recorder, correction apparatus—should be outside the testing area. The only device that should be in the area is the microphone.

Survey Guidelines

When conducting a plant noise survey, the following outline may prove helpful:

1. Select sound measuring devices carefully, paying particular attention to the types of microphones necessary. Review the working area situation thoroughly; the environment, the type of sound fields, the number of people affected and their locations.
2. Be familiar with the equipment before testing. Most important, make sure it is correctly calibrated.
3. Determine which machine generates the most sound and find its true sound level.
4. Run the survey under normal plant operations. Remember that a change in humidity or outside interference can alter results.
5. Set up measuring devices properly and have no interference to testing conditions, if possible.
6. Be familiar with the basic formulas for calculating the necessary information.

Once all criteria have been examined and the noise pressure levels compared to OSHA regulations, steps can be taken to eliminate the problem.

INDUSTRIAL HEARING TEST PROGRAMS[25]

Equally important to noise surveys and reduction is an audiometric testing program for conservation of employee hearing. The results of such a program provide a company with data to judge the effectiveness of its overall noise program. Success of a noise control program is inversely proportional to the percentage of noise-induced hearing loss cases uncovered as a result of hearing tests.

The Occupational Safety and Health Act states that an effective, continuing hearing conservation program shall be administered in all cases where the sound levels exceed the following:

Duration—Hours Per Day	Sound Level-dBA
9	90
6	92
4	95
3	97
2	100
1½	102
1	105
½	110
¼ or less	115

Tests must be given to all individuals working regularly or infrequently in areas exceeding 90 dBA. This pertains not only to full time machine operators, but also to maintenance men, shop foremen, clerks, engineers, secretaries, etc., that may happen to enter an area, however briefly, in which the noise level exceeds 90 dBA.

Testing should be performed at least once a year, more often if directed by the company physician. It is also wise to give all job applicants a pre-employment hearing test for several reasons:

1. Their existing hearing may be below the requirements necessary for safe functioning on the job.
2. It will serve as a baseline and permanent record of their hearing before employment. Effectiveness of the noise control program can be measured by deviation from this baseline.
3. It may be used in the future to protect a company from fallacious noise-induced hearing loss claims.

Before attempting to make a decision on whether your company should conduct its own testing or to hire the services of others, the pollution engineer should become familiar with the equipment and time required to conduct a hearing conservation program.

Audiometric Testing

Audiometry is the testing of the hearing acuity of individuals. The testing may be performed at the plant, at a physician's office, or elsewhere, so long as the equipment, facilities, and techniques meet the minimum requirements of OSHA.

The audiometer is the instrument used to measure the hearing acuity of individuals. To comply with OSHA, the audiometer must meet the specifications for limited range, pure tone audiometers as set forth in ANSI Standard S3.6-1969, "Specifications for Audiometers." These specifications may be obtained from the American National Standards Institute, Inc., 1430 Broadway, New York, NY 10018. The limited-range type of audiometer serves quite well in measuring threshold hearing levels for industry-employed persons. The ANSI specification states that the limited range audiometer must produce at least tones of the following frequencies: 500, 1000, 2000, 3000, 4000 and 6000 Hz from 10 dB to at least 70 dB (re: Standard reference threshold level). This must be done for air conduction tests. Bone conduction examinations may then be omitted.

The standard does not differentiate between manual or automatic audiometers; both types are permitted. The manual type requires an operator to be in attendance for the entire test. The operator must manually present the different intensities at specified frequencies. By means of subject response, or lack of response, the operator establishes the threshold levels at each frequency and manually records them.

With automatic audiometers, the subject responds to the intensity presented by the audiometer, usually by pressing or releasing an electric switch. Absence of response causes the instrument signal to increase in intensity until such time as the subject responds, which causes the intensity to decrease again. In this manner, the subject's threshold levels are determined and recorded automatically. An operator is needed only to briefly explain the test, position the earphones, and push a start button.

Table 27-3 is a comparison of the two types of audiometers. Keep in mind that the purpose of testing is to establish thresholds at several frequencies and not for detailed diagnosis of any hearing disabilities.

Audiometric Test Booths

The location selected for audiometric testing is very important, since test tones can be masked easily by ambient noise. Unless the facility has an extremely quiet room, well away from traffic, machines, people, conversations, etc., an audiometric testing booth will be required.

In conformance with OSHA, the ambient noise level for conductance of tests must meet the criteria as set forth in ANSI Standard S3.1-1960 (with latest revision) "Standard for Background Noise in Audiometric Rooms," This standard is available from ANSI. The standard lists maximum

Figure 27-3. Comparison of Manual and Automatic Audiometers

	Manual Audiometer	Automatic Audiometer
Time of test	Dependent somewhat upon operator skill.	Usually 30 seconds per frequency for each ear automatically controlled (6-7 minutes per subject).
Operator's time required	Operator must explain test, fit earphones, present tones at various levels of intensity at each frequency, and record thresholds. Trial period may be given before test to be sure subject understands.	After explanation and fitting of earphones, operator pushes start button. Instrument automatically presents tones, varies intensity and frequency and records thresholds. Trial period may also be given.
Threshold judgments	Made by operator based on subject response.	Recorded by instrument based on subject response.
Simultaneous testing of more than one person	Only one person can be tested at a time.	Limited only by number of audiometers and/or testing booths; however, four or five subject limit is practical.
Retest	Operator can retest a particular frequency immediately, should malingering or inattentiveness be suspected.	Retest can be given at a particular frequency after complete test, or by overriding the automatic if operator has test under surveillance.
Other	Results affected by operator fatigue and possible recording errors.	Instrument is not biased. Tests are uniform. Test-retest reliability is higher, and less operator training required.

allowable sound pressure levels of the ambient noise at each frequency of testing. Table 27-4 lists these allowable levels as given in the referenced standard.

Each location must be first checked for "quietness" by a noise level meter utilizing either 1/3, 1/2 or full octave band settings. For example, if utilizing the 1/2 octave band setting for the range setting including 1500 Hz, the allowable maximum ambient level cannot exceed 39 dB for audiometric testing. A sound level meter with a frequency analyzer having 1/3, 1/2 or full octave band settings must be used. If any table value(s) is exceeded, the site cannot be used as is for audiometric testing. The noise level must be reduced, an audiometric booth used, or another site selected.

Audiometric rooms or booths do not eliminate noise, they only attenuate it. The more attenuation required, the more material required for fabrication, and hence the greater the cost. Therefore, audiometric room design and fabrication are best left to professionals.

Audiometric rooms are available in all sizes, from smaller than telephone booths to 6 ft by 10 ft rooms and larger for multiple-person testing. Some booths are available with wheels or casters, while most are of the permanent type which must be disassembled should relocation be desired. The permanent type are those too big to fit through doorways and do not come pre-assembled. Assembly can be done by plant maintenance men who are well supervised. Or, suppliers of such units can provide installation crews at some cost.

Table 27-4. Maximum Allowable Sound Pressure Levels (dB Ref. 0.0002 Microbar) for No Masking Above Zero Hearing Loss Setting of Audiometers

Test Frequency of Audiometer (Hz)	Octave Band	1/2 Octave Band	1/3 Octave Band
125	40	37	35
250	40	37	35
500	40	37	35
750	40	37	35
1000	40	37	35
1500	42	39	37
2000	47	44	42
3000	52	49	47
4000	57	54	52
6000	62	59	57
8000	67	64	62

Once the size of an audiometric room is established, it is necessary to determine if the attenuation is adequate. This can be done before purchase by requesting attenuation data from the vendor. Data should include the attenuation in dB of the booth or room at each frequency at which testing will be done.

A simple calculation can then be made for each test frequency as follows:

$$\frac{\text{Ambient Noise Level (in dB)} + 10\ \text{dB}}{\text{Safety Factor - Booth Attenuation}} = \frac{\text{Sound level inside}}{\text{Audiometric room}}$$

Repeat calculation for each frequency.

A 10-dB safety factor is recommended as margin for any increase in ambient noise. The values calculated for sound pressure level at each frequency should then be compared against the maximum allowable per Table 27-4. If no values are exceeded, the audiometric room is satisfactory. If one or more values are exceeded, then the room under consideration is unsatisfactory at that location, and either the location must be changed, the noise abated, or another room selected with better attenuation.

Rooms or booths can be furnished with many options; standard features are usually one window, a light, and connections for the audiometer in addition to a door, walls, top and bottom. Costs vary in proportion to attenuation and size. Extras range from an outside shelf for the audiometer, to fluorescent lighting and even carpeting. Two options that should be considered are vibration isolator rails to attenuate vibration transmitted through the ground, and a blower-ventilation system.

Maintenance and Calibration

Before being placed in service, the audiometer must be certified accurate as to frequencies, purity of tones, sound pressure levels, etc. If purchased new, the manufacturer normally provides this service free of charge. However, an audiometer must be recalibrated at least once a year. Such a calibration must be certified and is best left to experts. Prior to purchasing an audiometer, check to see if local servicing or calibration service is available, or else the instrument must be sent to the manufacturer for service. A certified calibration normally costs in the range of $75 to $125. In some cases a service contract covering calibration and repair can be obtained from a manufacturer or his representative.

Further, still in accordance with OSHA, the audiometer must be biologically checked at least once a month, or before each use if used less than once a month. This is performed by having a person with known hearing undergo the actual test, the results of which are then compared to previous tests for stability. This testing can be done by company personnel. Results of these tests must be maintained for possible inspection.

The audiometric booth or room should be inspected on a regular basis, checking seals, gasketing, etc. Additionally, sound level surveys should be made periodically to be certain that the ambient noise level has not increased excessively, and that the booth attenuation has not decreased.

To comply with OSHA, the operator(s) giving the audiometric tests (even if using the automatic type) must be trained and skilled. Courses qualifying operators are offered by many clinics, universities and safety institutes throughout the country. Fees range from $100 to $150. These courses instruct the operator on technique, equipment, interrelation of results and equipment care. The course usually takes 2 to 3 days to complete.

Alternate Programs

An alternative to instituting a plant program is to work out a program with a hospital, a clinic or a mobile laboratory company. The cost of any hospital or clinic arrangement will probably be dependent upon the number of persons requiring testing. Fees could run as high as $15 per person for small groups. This may be the best way for companies to proceed which have a small number of persons to test. However, keep in mind that employees must leave the plant and travel to and from the hospital or clinic, and perhaps wait to be treated.

Several companies offer in-plant testing of employees by means of mobile vans equipped with audiometers and booths. The vans pull onto the plant site, take sound level measurements to qualify the site and then run employees through the test, often at a rate of 20 per hour, utilizing multiple instruments. Prices range from $10 per person tested for small groups, to around $6 per person for several hundred employees tested. Some testing companies require a minimum of 15 to 50 people before they will send out a van.

Mobile van companies do all the work, qualify the site, provide the technicians, maintain records, submit reports, and analyze test results. The employer is responsible only for scheduling people to be tested. However, there are also certain drawbacks. For instance, some people will be absent on the days when the van is on-site. Additionally, there is the problem of testing shift workers and people joining the company after the van has moved on. Of course, all those missing the test can be referred to a clinic or hospital.

Added to this is the problem of "minimum time away from noise" **prior** to taking an audiometric test. A hearing test taken after exposure to noise can result in a temporary or apparent threshold shift and may appear to be a hearing loss or a permanent threshold shift. Authorities differ on the exact amount of rest time from noise before testing, but most agree on at least 16 hours. If adhered to, this represents a very serious scheduling problem, particularly if utilizing a mobile van and not the plant's own equipment.

SELECTING AN AUDIOMETRIC ROOM[71]

The basic purpose of a sound room in audiometric testing is to provide the proper acoustical environment so that tests can be conducted without interference from outside noise. A room of this type should afford adequate ventilation and lighting so that the subject will be comfortable while his hearing is being evaluated. Not only does the room supply environmental control but it eliminates distraction from changes on the visual horizon which may invalidate an audiogram as readily as acoustical interference.

Hearing test rooms should be located in as quiet a place as possible. Preferably they should be within practical access but away from outside walls, elevators, heating and plumbing noises, waiting rooms and busy hallways. If the highest noise levels in a test room do not exceed the levels listed in Table 27-5 test room noises will not affect test results.

Table 27-5. Maximum Allowable Sound Pressure Levels for No Masking Above the Zero Hearing Loss Setting of a Standard Audiometer*

Audiometric Test Frequency (cps)	500	1000	2000	3000	4000	6000
Octave Band Cut-off Frequencies (cps)	300 to 600	600 to 1200	1200 to 2400	2400 to 4800	2400 to 4800	4800 to 10000
Sound Pressure Level (dB) (dB) re 0.0002 Microbar)	40	40	47	52	57	62

*NIOSH recommendations would include background at 250 as well as those listed above.

To obtain these internal room ambients one must know what the outside ambient is going to be. This can be obtained with a noise survey by octave bands of the area where the room is to be located. Once this is done, a room should be selected which will provide ample noise reduction to bring the internal noise level down to those prescribed by the standard (Table 27-5).

Background Noise Levels

The American Standards Association, in its pamphlet on criteria for background noise levels in audiometer rooms, has printed a very useful chart for depicting the outside background noise levels allowable for the use of different types of sound rooms, *i.e.,* regular duty single-wall rooms and double-walled rooms. This chart shows the relationship between outside ambient levels and the amount of performance required to bring down the internal noise levels to acceptable levels for testing without interference.

The following procedure gives a rough indication of the kind of construction that may be necessary. In the octave bands that will contain test tones, measure the sound pressure level at the site of proposed test room and plot those on Figure 27-17. The highest range in which the measured levels fall determines which of the three general classes of rooms is probably necessary.

Specifically the octave band background noise at the room location should be measured at each test tone, *i.e.,* 500, 1000, 2000 and 4000 Hz. From these measured levels the published noise reduction of the proposed audiometric room should be subtracted. To be acceptable the room must bring the noise down below that prescribed in the standard listed in Table 27-5.

Types of Rooms

Basically there are three types of prefabricated audiometric rooms. The first type, which is most familiar to those in audiometric testing, is the single-wall modular panel type. It is fabricated from a 4-in.-thick steel panel which weighs approximately 10 lb per sq ft. This panel has a solid outer surface and a perforated inner surface and is filled with high density acoustical fill and damping material.

These panels require assembly at the location where they are to be used. The components of such rooms include sound doors, window panels and a panel with a jack plate for connection of the audiometer. Separate floor and ceiling panels are necessary.

The second type of room is a hybrid heavy-duty version of the first. It incorporates double-wall panel construction. These are rooms within rooms and are usually constructed using the same 4-in. panels with a 4-in. air space between the inner and outer rooms. Both of these types of modular panel rooms can be assembled to construct a suite of rooms with capacity for many subjects.

The third type of room is a new development in the field of audiometric testing: compact or mini-room. It is a smaller, single-walled, one-person occupancy unit which incorporates the features of the larger room in a smaller pre-assembled package (Figure 27-18). It provides somewhat less noise reduction than the larger rooms but is more than adequate in most industrial testing applications.

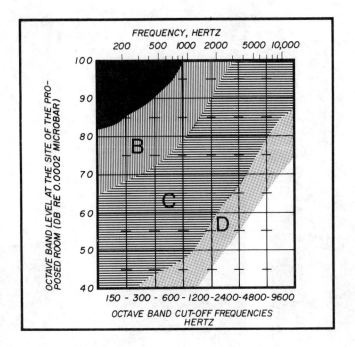

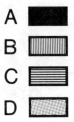

Figure 27-17. Plot the sound pressure level on the chart in the octave
    bands containing test tones. Observe the highest region into
    which these data fall and note the type of probable construction
    required.

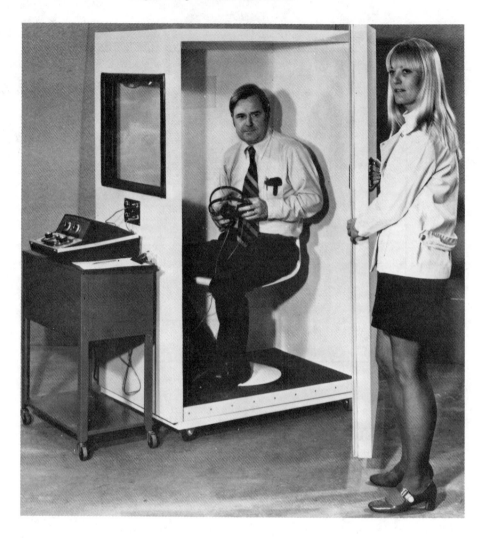

Figure 27-18. Portable audiometric room for industrial areas.

All three types of rooms should include adequate lighting (incandescent or fluorescent with remote ballasts to avoid ballast hum), adequate ventilation (15 air changes per hr) which does not distribute background noise levels, as well as vibration isolation mounts to isolate from structural vibration.

For precise medical and clinical applications, where extremely low ambients are required and radio frequency shielding may be a requirement for diagnostic work, the single- and double-wall modular panel audiometric rooms will probably be required. These rooms are most practical in this type of application and in speech and hearing research.

In industrial testing and some clinical testing the single-wall modular panel and small pre-assembled compact rooms will be adequate depending on background noise. They afford sufficient noise reduction to satisfy these testing requirements.

PERSONAL HEARING PROTECTION DEVICES[29]

One approach to protecting the hearing of employees is to provide individuals with personal ear protection. It is a simple solution, both economically and to the machine operator; however, there are certain drawbacks. Some individuals find personal ear protection devices uncomfortable, irritating, and in many cases, annoying and cumbersome.

There are two general classifications of ear protection: (1) over-the-ear protection, and (2) in-the-ear protection.

Over-the-Ear Devices

Ear muffs provide over-the-ear protection against harmful high frequency noises. However, they should also allow the wearer to hear low frequency sound such as spoken voices and warning signals. This is a problem in personal hearing protection equipment. Workers must be able to communicate with each other conveniently and normally, or they may not bother using the device.

In selecting proper ear muffs, the pollution engineer must determine the sound levels generated in the vicinity of the worker. This can be done with sound measuring meters. Then, attenuation levels of the ear muffs must be carefully examined. This data is available from the manufacturer.

Attenuation is a measure, in decibels, of how effective the device is in reducing damaging high-frequency noises. The engineer can compare hearing protection devices by comparing the attenuation data from various manufacturers. Figure 27-19 shows a typical attenuation graph. The ideal muff should be most effective in high frequency ranges, because it is at these frequencies that most hearing damage occurs.

Attenuation provided by ear muffs varies greatly due to differences in size, shape, seal material, shell mass, and method of suspension. Another influencing factor of attenuation is the wearer's head size and shape. The cushion between the shell and head also has a great deal to do with the attenuation frequency.

There are many ear muffs and cups on the market. Most are made of plastics. There are basically four parts to an ear muff device (Figure 27-20).

1. **Ear Cups**—The outermost covering designed to deflect as much noise as possible. They should be made of a rugged material to withstand mechanical shocks. Although most manufacturers indicate that ear cups are shaped to fit all ear sizes comfortably, this is not always the case. Each worker will probably have to be fitted.

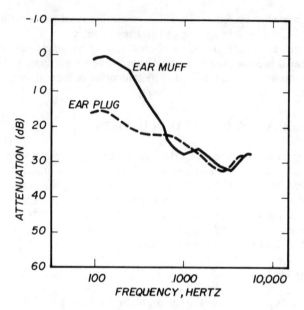

Figure 27-19. Typical attenuation graph for ear plugs and ear muffs. Above 1000 Hertz, both ear plugs and muffs provide about the same protection.

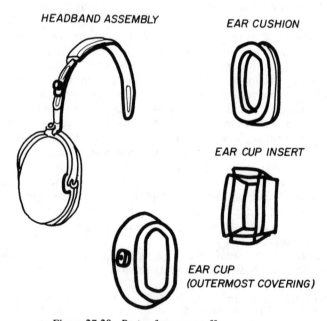

Figure 27-20. Parts of an ear muff.

2. **Sound-Absorbent Earcup Insert**—The outer earcups are lined with soft foam, such as a vinyl or polyurethane foam. This lining insert plays a dual role. First, it provides additional noise reduction, and second, it eliminates the "sea shell" effect of roaring noise generated when one holds an unlined cup over the ear.

3. **Sound-Tight Ear Cushion**—This is optional on some ear muffs. It further reduces sound levels and prevents sound leaks by acting somewhat like a washer. It is generally made of tough vinyl cushion.

4. **Headframe**—This part connects two ear muffs and suspends them from the head. The frame should be easily adjusted with something like a slide lock mechanism. It must be light and have sufficient tensile strength. One of the most important criteria for a headframe is that it provides a comfortable fit. The major criticism of ear muffs is that they are too bulky and uncomfortable. It is therefore necessary to provide a headframe that offers swivel-action around the cups. In this way the wearer can adjust the headframe to provide the most comfortable personal position.

Another problem encountered with ear muffs is that they may interfere with head protection gear. The swivel-action yoke headframe permits the ear cups to be worn with hard hats, bump caps, and welding helmets. There are also models available that can be fastened to safety caps. A spring tension band holds the entire unit on the cap by slight pressure. The earcups grasp the underside of the cap brim.

The important point to remember about personal protection devices is that they depend entirely on a good seal between the skin and ear protector's surface. If a small sound leak occurs, the purpose of the ear muff is destroyed. A typical annoyance encountered by wearers of earmuffs is that these protectors tend to work loose due to talking, chewing or movement. There is nothing that can be done to combat this problem except to reseat the cups from time to time.

Sanitation is another factor in using ear muffs, although this problem is more prominent for inner ear protectors. It is important that the unit can be easily disassembled for cleaning and sanitizing.

Most of the available ear muffs on the market provide about the same degree of protection. The best device is one that is accepted by the employee and worn properly. Individual fittings should be made to ensure comfort and effectiveness. One way of telling if the ear cups are doing the job is to have hearing tests made of the workers before and after use. When properly worn and cared for, ear muffs can provide adequate protection against most industrial noise exposure. However, this cannot be obtained without some initial discomfort.

In-the-Ear Devices

Ear plugs are inner ear protection devices, designed to occlude the ear canal. They are made of a soft material, usually rubber, neoprene, wax, cotton, fiberglass or plastic.

The choice between plugs and muffs largely depends on the work situation and employee's opinion. Both offer similar protection at frequencies above 1000 Hz. If an employee's head is confined in a helmet, he may feel more comfortable with plugs.

There are several problems encountered by individuals using ear plugs. In order to get a good acoustical seal against the sensitive inner lining of the ear canals, the inserts have to apply a certain degree of pressure. This can cause discomfort and is the primary reason why workers refuse to use them. Most workers would prefer to utilize cotton as a sound suppressor; however, this is a poor approach. Figure 27-21 shows the attenuation level of pure cotton. Most ear plugs offer more than double the protection. An employer should discourage employees from improvising any type of inserts.

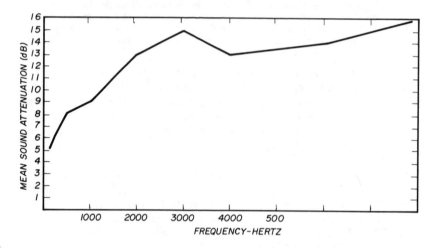

Figure 27-21. Sound attenuation graph for plain cotton.

The second major problem is that they tend to pick up a great deal of dirt. Wearing ear plugs for an extended period of time may cause a "plugged" feeling, dizziness, or vertigo. When this happens, the wearer generally places the plugs on the work area where they pick up dirt, metal filings, germs, etc. The only way to combat this is to fully educate the employee. It must be impressed upon him that it is necessary to clean the plugs every time they are to be used, unless taken directly from a container. Because of the simplicity of design, ear plugs can be easily sanitized.

Still another problem is that plugs tend to lose their effectiveness during the day, due to a break in the acoustic seal between the ear and insert. This occurs because of jaw movements which change the shape of the ear canal. This problem has been arrested to a certain extent by providing a headband connecting the ear plugs. A lightweight headband maintains a slight pressure on the canal caps to hold them in place. When the band is properly adjusted to fit head size and plugs correctly inserted, the caps require less reseating after short periods of use. Women, however, may find headbands troublesome because they muss hair styles. In such cases the headband could be adjusted to fit under the chin.

Ear canals vary in size. Most manufacturers provide inserts in several sizes, but they are still standardized. This, of course, provides a difficult problem in fitting. Often the best approach to effective control is to obtain personally molded ear plugs. There are kits on the market with a soft putty-like material, which can be molded in the individual's ear canal. This method eliminates excessive pressure against sensitive canal walls and provides maximum comfort.

There are certain problems, however, which can occur even when using molded types. Air can become trapped in the ear when first molding the inserts. This causes voids in the plugs which prevent a good acoustical seal. It is important that the plugs fit properly. Even the slightest leakage will lower the amount of attenuation as much as 15 dB in some frequencies.

Some materials tend to shrink upon hardening, thus destroying the seal. Therefore, it is wise to evaluate all literature on such devices. Custom-made inserts should be fitted by trained technicians, nurses or doctors.

The use of personal ear protectors encompasses more than just purchasing and fitting. The employee's ears should be examined and tested at the time he is fitted with ear protectors. Plugs should be fitted individually for each ear. Employees should be educated to the necessity of using the devices and in their upkeep. There may be a degree of discomfort on the part of the worker when first using ear protectors, and he should be made aware of this prior to issue. To generate acceptance an employee should be given the final choice of several different styles of protectors. The most important factor is to purchase the type which will be worn effectively.

POINTERS ON SELECTING HEARING PROTECTORS[143]

There are four basic factors that must be considered when choosing hearing protection: attenuation, comfort, hygiene, and cost. There are two general classifications of hearing protectors—ear plugs and ear muffs.

Ear Plugs

Ear plugs of soft, pliable materials have an average attenuation of 25 dB and can be used up to 105-110 dBA. Since ear plugs attenuate noise by occluding the ear canal, a prerequisite of good attenuation is a proper fit. This does not mean that the ear plug must be inserted to a depth that is uncomfortable to the wearer, but should have a certain amount of pressure to insure an adequate acoustic seal (Figure 27-22). To ensure proper and safe use of ear plugs made of rubber, neoprene or plastic, fitting should be performed by medical or trained personnel. If foam, wax, impregnated cotton, or glassdown is used, directions furnished by the manufacturer should be followed to get the maximum attenuation possible.

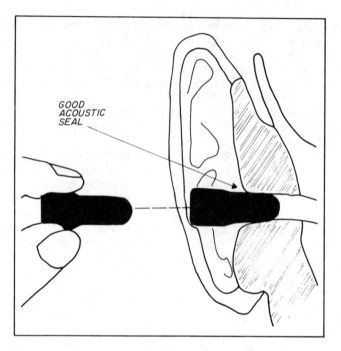

Figure 27-22. Ear plugs must be inserted to a depth to ensure an acoustical seal.

Because of the sensitive nature of the ear, most ear plugs are not comfortable when first worn. The major complaint of ear plug users is that they have the feeling of being "stopped up." This sensation is caused by the initial buildup of pressure, which should begin to dissipate in 5 to 10 minutes. Some plugs have flanges that can irritate the inner ear canal and cause infection. Certain workers may experience allergic reactions from the material used in the plug's manufacture. Infections, irritation, or allergic conditions should be reported immediately to the medical department and the worker provided with another form of protection. In cases where the worker must wear head or face protection, ear plugs afford the most comfortable protection and do not interfere with his normal working operations.

Proper hygiene is probably the most difficult part of any hearing protection program. Because of their size, ear plugs are thrown on work benches, in pockets, tool boxes, and other areas where they pick up dirt and germs. It is difficult to orient the user on hygiene practices and why they must be followed. Instructions on daily cleaning must be provided with ear plugs designed for repeated use. The cleanest form of ear plugs is the disposable type which are thrown away after a day's use.

Ear plugs are the most inexpensive of the hearing protectors but their use must be carefully supervised to combat habitual loss. By providing individual cases for each worker's ear plugs, loss is greatly reduced.

Ear Muffs

Ear muffs afford the greatest amount of protection and have an average attenuation of 35 dB. They can be used up to 115-120 dBA, but offer less attenuation than ear plugs at certain frequencies. Attenuation charts can be obtained from the manufacturer to assist in selecting an ear muff for a particular problem.

In cases where the noise level exceeds 120 dBA, a combination of ear plugs and muffs can ensure maximum protection. Again, as with ear plugs, performance depends entirely on the fitting and the acoustic seal attained. Leakage between the ear cushion and the ear itself will drastically reduce the attenuation of the muff. Safety glasses, hard hats, welding helmets, and goggles increase the chances of improper seal (Figure 27-23).

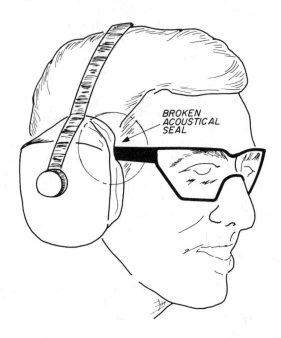

Figure 27-23. Placing ear muffs over eye glass frames can break a proper acoustical seal.

There are two basic fills for ear cushions—liquids and foam. Liquid filled ear cushions are more apt to mold around safety glasses and afford the greatest seal.

Ear muffs that weigh too much or have high tension in the headset spring will soon be discarded or destroyed. Comfort is of the utmost importance in selecting an ear muff, and numerous styles should be investigated. If hard hats and ear muffs are required to be worn together, there are models available that fasten directly to the side of the hat itself.

One complaint commonly heard from wearers of ear muffs is that they are hot and create perspiration, which in turn irritates the skin. Ear muffs with liquid-filled ear cushions stay relatively cool and do not irritate the skin. Although most muffs available on the market today have essentially the same degree of attenuation, caution should be taken to choose a protector that is comfortable, and will be worn by the employee.

Ear muffs are easily sanitized. Liquid germicides are available which can be used in a plastic spray bottle and placed near the lens cleaning station. Signs can be placed above the station to remind the employee of good hygiene practices. The foam lining of the ear cup should be changed periodically, and the ear cushion itself replaced when signs of wear appear.

Although ear muffs are initially the most expensive of the protectors, the rate of loss and replacement is much lower. There is added expense when parts of the muff are lost or damaged, and each part must be purchased as a separate item.

Education and Attitude of the User

Most employees do not understand that a noise-induced hearing loss cannot be corrected. Lack of education concerning the ear and its protection may cause psychological problems when the employee is required to use personal protection. The employee feels that a personal freedom has been taken from him and the psychological resistance occurs. One way to alleviate this problem is to leave the final choice of protection to the worker. Offer him three or four choices so that personal comfort and effectiveness can be determined. When allergies, infections, or discomfort occur, a different type of protection must be issued.

A successful hearing conservation program can depend on the education and instruction that has been given. The employee must realize the noise exposure and risk he is under if he does not comply with the regulations. Hazardous noise areas should be properly posted with signs, ideally showing the type of personal protection that should be worn. Of course, supervision must also be educated to explain why enforcement of the program is of the utmost importance.

Audiometric testing can also help lessen resistance, because of the worker's curiosity. He is usually quite receptive and responds to having his audiogram explained to him. The outcome of the test could suggest measures such as changing or increasing protection, or even changing the employee's working area. The tests also can alert the medical department to a situation that should be referred to a specialist.

A successful hearing protection program is involved and sometimes difficult to initiate. Growing accustomed to using any form of personal protection takes time. With comfortable protectors, worker education, and attitude consideration, a successful hearing protection program can be maintained. It will offer the employee a safer, more pleasant working environment.

POLLUTION ENGINEERING NOISE GLOSSARY[32]

Most pollution engineers are familiar with some acoustical terms. However, much of the terminology in this field can be confusing, and is often not readily available. The primary aim in compiling this glossary has been to provide a quick reference for the most common terms associated with noise work in industry. When possible, definitions have been generalized so as not to preclude the different specific interpretations which may be attached to the term in particular applications; the greatest emphasis is given to engineering application. Also, it is agreed by the authors that the preferred definition is a simple one and the tendency has been toward a simple statement of function rather than the explicit description of all properties included or excluded.

The glossary is divided into three sections:

1. **Introduction to Acoustics**—general terms and definitions related to the science of acoustics. Many of the physical properties of sound are defined.
2. **Designing and Measuring**—terminology pertaining to noise-reducing techniques and sound-measuring devices are listed.
3. **Acoustical Units**—mathematical definitions and acoustical measuring units and related measuring techniques.

It is recognized that the glossary in its present state may not include all terms which might be pertinent to the noise field. This statement applies particularly to fundamental and very specialized terms.

Introduction to Acoustics

Absorption—ability of material to absorb sound energy.

Absorption Coefficients—(1) Normal incidence—method of testing which provides low values of coefficients of absorption. Normal sound incidence on surfaces is rare. Measure of a single, pure tone frequency. (2) Random incidence absorption coefficients—most common values used for various materials. These values indicate the performance of a sound-deadening material.

Acceleration—a vector (→) that denotes the time-rate-of-change velocity. (1) Modifiers such as peak, average, root mean square are often used. The time interval is designated over which the average was taken. (2) Two types of acceleration exist: (a) oscillatory, which is defined by the acceleration amplitude if the motion of waves is simple harmonic, or the rms acceleration if motion is random; (b) nonoscillatory acceleration, in which case it is denoted as transient or sometimes sustained acceleration.

Acoustics—(1) The science of sound including its generation, transmission, and effects on environment. (2) Sum of the qualities that determine the value of a room as to distinct hearing.

Acoustic Output—a measure of the noise source in terms of sound power level or sometimes in sound pressure level (SPL), units expressed in decibels. Usually, data for specific industrial equipment are provided by manufacturers.

Airborne Sound—sound that reaches the point of measurement through the air.

Alias—sampled data which is equally spaced; *i.e.,* two frequencies are aliases of each other if sinusoids of those frequencies cannot be distinguished by the sampled values.

Ambient Noise—conglomeration of different airborne sounds from several sources near and at a distance, from the point of measurement. No individual sound is considered in the environment.

Anechoic—an acoustical environment which approximates a free sound field; free from echoes and reverberation.

Anechoic Room—the more common term is Free Field Room. The boundaries of such a room absorb all sound incident thereon. This generates essentially free-field conditions.

Articulation Index—an arbitrary index numbered from 0 to 1 which expresses the ability to correctly classify certain specified vowel and consonant sounds. This index is the fundamental method of identifying human speech sounds but is rarely computed directly and is most often inferred from other standard tests.

Aural—pertaining to the ear or to the sense of hearing.

Background Noise—noise from various sources in the environment that is unrelated to the particular sound of interest.

Coherence—a measure of a transfer function estimate; its value is zero when the transfer function is of no statistical importance and unit when the estimate is not contaminated by background noise.

Critical Speed—a rotating system's speed that corresponds to a resonance frequency of the system.

Dead Room—a room whose boundaries absorb a great amount of sound (see Anechoic Room).

Diffraction—alteration in the direction of propagation of sound energy in the neighborhood of a boundary discontinuity on the edge of a reflecting or absorbing surface.

Diffuse Sound Field—sound in an area in which sound intensity is independent of direction; an area over which the average rate of sound energy flow is equal in all directions.

Direct Sound Field—area in which most of the sound arrives directly from the source without any contributions from reflection.

Directivity Index—symbolized by DI: (1) of a source in a given frequency band—a description of directionality of a sound from a specific source, units in decibels; (2) difference either plus or minus between sound pressure level measured at a given point in the free field and the sound pressure level averaged over all points in every direction all at the same distance from the source.

Flanking Transmission—transfer of sound from source to receiving point by way of paths other than directly through the acoustical material being tested (generally a partition).

Free-Sound Field—a field free from boundaries. A field in which the effects of the boundaries are negligible over the area of interest.

Hearing Threshold Level—quantity in decibels of the threshold of audibility of the ear exceeding the standard audiometric threshold.

Impact—collision of one mass in motion with another mass which is either at rest or in motion.

Isolation—reduction in the capacity of a unit responding to an excitation that is attained by a resilent support. Isolation is expressed quantitatively as the complement of transmissibility, in a steady-state forced vibration.

Just Noticeable Difference—differential threshold—incremental difference required to be classified as a change in any attribute sound. The median value of the difference indicating where the difference can be noticed in 50 percent of the samplings.

Line Component—a simple tone that may or may not be part of a complex signal.

Live Room—a room that has very little sound absorption.

Loudness—the intensive attribute of an auditory sensation; a term in which sounds are described on a qualitative scale from soft to loud. Loudness depends on a sound pressure of the stimulus and also the frequency and wave form of the stimulus. (See Loudness Contour and Loudness Level.)

Mechanical Shock—when the position of a system is changed significantly in a short amount of time and in a nonperiodic way, characterized by suddenness and large displacement, developing significant inertial forces on the system.

Noise—(1) Unwanted sound: unwanted sound within a useful frequency band. (2) Erratic, intermittent oscillation.

Noise Level—(1) The physical quantity measured (*e.g.,* voltage, decibels); the reference quantity. The instrument utilized and the **bandwidth** are indicated when measurements are made. (2) For airborne sound: the weighted **sound** pressure level is called the sound level; weighting must be noted.

Pitch–characteristic of auditory sensation in which noises are ordered on a scale which extends from low to high. Pitch depends on the frequency of the sound stimulus, sound pressure, and wave form of the stimulus. Pitch may be described by the frequency level of the tone, having a specified sound pressure level, which, when judged by listeners, has equal pitch.

Point Source–(also Simple Sound Source)–a source that generates sound in a uniform fashion in all directions under free field conditions.

Presbycusis–a condition of hearing loss due primarily to aging effects.

Pure Tone (also Simple Tone)–(1) A sound wave in which the instantaneous sound pressure is a simple sinusoidal function of time. (2) Sound sensation that is characterized by singleness of pitch.

Random Noise–an oscillation whose instantaneous magnitude cannot be specified for a given point in time. Only probability distribution functions giving the fraction of the total time that the magnitude lies within a specified range can specify the instantaneous magnitudes. (See Gaussian Random Noise.)

Rate of Decay–rate of decrease of sound pressure level after source has terminated.

Reflective Environment–an environment having large sound absorptive surfaces. PWL (power levels) cannot be measured directly under such conditions but the SPL (sound-pressure level) existing at a point is readily measured with a sound level meter.

Resonance–exists when forced oscillation of a system prevails. This occurs when any change in the excitation frequency causes a decrease in the response of the system.

Resonance Frequency (Resonant Frequency)–frequency where resonance prevails; *e.g.,* velocity resonance frequency.

Response–of a device in the motion that has resulted from a stimulus under specified conditions. (1) Modifying phrases are prefixed to the term response to indicate different inputs and outputs that are being used. (2) Response characteristic, usually expressed graphically, gives the response as a function of some independent variable such as frequency or direction. Other characteristics of the input (*i.e.,* voltage) are held constant.

Reverberation–(1) The continuance of sound in an enclosed area resulting from multiple reflections after the sound source has ended. (2) Persistent sound in an enclosed space, as a result of repeated reflection or scattering after the source has ended.

Reverberation Room–a room whose acoustical design closely approximates the reverberant sound field as a diffuse sound field in both steady state when there is just one sound source and during decay or when sound source has terminated.

Reverberant Sound Field—enclosed space where all or part of the sound is repeatedly reflected from the boundaries.

Reverberation Time—the time required for the mean squared sound pressure level therein originally in a steady state to decrease 60 dB after the source has terminated.

Simple Tone (See Pure Tone.)

Sociocusis—a condition that increases the threshold hearing level as a result of noise exposure that is unrelated to the social environment and exclusive of occupational-noise exposure, physiologic changes with age, and otologic diseases.

Sonics—technology of sound in analysis and processing including the use of sound in any noncommunication system.

Sound—auditory sensation produced by the oscillations in stress, pressure, particle displacement, particle velocity, etc., in a medium with internal forces, or the superposition of such propogated alterations. Note that not all sound waves can generate an auditory sensation; *e.g.,* ultrasound.

Sound Absorption—process of removing sound energy [the property of materials to absorb sound energy (Symbol: A; dimensions: (L^2) units in sabin or metric sabin]. (See Sound Absorption Coefficients.)

Sound Energy—energy added to the medium in which sound travels; consists of potential energy in the form of deviations from static pressure and kinetic energy in the form of particle velocity. (See Sound Energy Density.)

Sound Intensity (Sound Power Density or Sound Energy Flux Density)—in a given direction, at a point, there exists an average rate of sound energy that is transmitted in a specified direction through a unit area normal to this direction at the point being considered.

Sound Power—the rate at which acoustic energy is radiated; rate of flow of sound energy; symbol: W; dimensions: $(ML^2 T^{-3})$ units in watts.

Sound Pressure—in the presence of sound, a fluctuating pressure that is superimposed on the static atmospheric pressure [symbol: P; dimensions: $(ML^{-1} T^2)$ units in dyne/sq cm or newton/sq m.]

Speech Interference Level (SIL or PSIL, Three-Band Preferred Octave Speech Interference Level)—the average of the sound-pressure levels of a sound expressed in dB in three octave bands which have center frequency of 500, 1000, and 2000 Hz.

Standing Wave—a periodic wave that has a fixed distribution in space. It is the result of the interference of progressive waves of identical frequencies. It is characterized by the existence of nodes or partial nodes and antinodes stationed at fixed points in space.

Static Pressure—pressure of the medium on which the alternating sound pressure is superimposed.

Stationary—characteristic of a noise whose spectrum and amplitude direction do not change with time.

Structureborne Sound—sound that reaches the point of measurement by way of vibrations of solid structures.

Threshold of Audibility (Threshold of Detectability)—for a given signal it is the minimum effective-sound pressure level of the signal capable of generating an auditory sensation in a specified fraction of the trials. Note that the ambient noise reaching the ears is considered negligible unless otherwise specified. Threshold is generally given as a sound-pressure level in dB relative to 0.0002 microbar.

Threshold of Feeling (Tickle)—for a given signal it is the minimum sound-pressure level at the entrance of the external auditory ear canal which will stimulate a sensation of feeling different from the sensation of hearing.

Timbre—characteristic of hearing in which the listener is capable of distinguishing between two sounds even though the two sounds are of equal sound pressure and pitch.

Time Series—a succession of discrete trials made at points in time. Spacing of trials is generally uniform on a time scale.

Tone—a sound sensation having pitch, and capable of causing an auditory sensation.

Transient Vibration—a temporary condition in which sustained vibration of a mechanical system exists consisting of both free and/or forced vibration.

Ultrasonics—science of sound at frequencies above the audio range. Confusion sometimes exists between the terms supersonic and ultrasonic. Supersonic pertains to speeds higher than the speed of sound and is thus unrelated to this term.

Vibration—oscillation where the quantity is a parameter (not scaler) that defines the motion of a mechanical system.

Vibration Isolator—resilent support that isolates a system from steady-state excitation.

White Noise—power per unit—frequency independent of frequency over a specified range.

Design and Measure Terms

Acoustical Material—material designed to absorb sound.

Analyzer—combination of filter system and device for indicating relative energy passing through the filter system. The filter is generally adjustable so that the signal measured can be analyzed in terms of relative energy passed through the filter as a function of the adjustment of the filter response-vs.-frequency characteristic.

Audiogram (Threshold Audiogram)—a plot depicting hearing-threshold level (HTL) as a function of frequency.

Audiometer—device for measuring hearing-threshold levels.

Baffle—partition utilized to increase the effective length of the external transmission path between two points in an acoustic system.

Damp—to generate loss of oscillatory or vibrational energy of an electrical or mechanical setup.

Damping—loss of oscillatory or vibrational energy of an electrical or mechanical setup.

Directivity Factor—(1) of a transducer is used for sound emission. It is the ratio of the sound pressure squared to the mean-square pressure, at some fixed distance and specified direction. Distance must be large so that the sound appears to diverge spherically from the acoustic center of the sources. Maximum response is considered for the reference direction. (2) of a transducer is utilized for sound reception, the ratio of the square of the open circuit voltage produced in response to sound waves arriving in a specified direction to the mean-square voltage that is generated in a perfectly diffused sound field of some frequency and mean-square sound pressure.

Directional Gain (Directivity Index)—of a transducer; 10 times the logarithm to the base 10 of the Directivity Factor expressed in decibels.

Earphone (Receiver)—electroacoustic transducer that is closely coupled acoustically to the ear.

Filter—device that separates components of a signal on the basis of their frequencies.

Instrument Noise—electrical sound generated in the measuring devices of airborne sound.

Loudspeaker (Speaker)—electroacoustic transducer that radiates acoustic power into the air, acoustic waveform equivalent to the electrical input.

Masking—amount by which the threshold of audibility of a noise is raised by the existence of another masking sound in the environment. Units in decibels.

Microphone—electroacoustic transducer stimulated by sound waves. It transmits essentially equivalent electric waves.

Noise Reduction—difference in decibels between the space-time average sound pressure levels generated in two rooms by one or more sound sources in them.

Partition—any building unit which divides space; *i.e.,* wall, door, window, roof, floor, ceiling, etc.

Transducer—system or device which is actuated by waves from one or several transmission systems. It supplies related waves to one or more other transmission systems.

Transmission Loss—a measure in decibels of a material's ability to prevent sound from passing through it. (1) For nonporous materials—it is related to mass and stiffness of material in addition to being a function of frequency. Generally, transmission loss rises with increase in mass or frequency (Mass Law). (2) For porous materials—low transmission loss values are associated with porous material even though they are effective absorbers. Examples are fiberglass batts, ceiling tiles.

Vibration Meter—a device that measures displacement, velocity, or the acceleration of a vibrating body.

Weighting—specified frequency response given in a sound-level meter.

Acoustical Units

Amplitude Density Distribution (Frequency Distribution)—a function that expresses the fraction of time that pressure, voltage, or other variable dwells in a narrow range.

Amplitude Distribution Function (Cumulative Frequency Function—a function expressing the fraction of time that instantaneous pressure, voltage, or other variable lie below a specified level.

Autospectrum (Power Spectrum)—a spectrum whose coefficients of components are expressed as the square of the magnitudes.

Confidence Limits—the upper and lower limits of a range over which a given percent probability applies.

Crosscorrelation—measure of similarities of two functions with the displacement between the two being an independent variable. The displacement is generally time.

Cross Spectrum—measure within the frequency domain of the similarity of two functions.

Data Window—(1) Interval including all sampled values in a calculation. (2) Form of a weighting function that is considered as multiplying the data that enters into a calculation.

Decibel—unit of sound level when the base of the logarithm is the tenth root of ten; one-tenth of a bel. Decibel is a unit of sound pressure-squared level.

Degrees of Freedom—a statistical term that is a measure of stability related to the number of independent equivalent terms entering into a distribution.

Displacement—vector quantity that specifies the change of position of a body or particle; usually measured from the mean position or position of rest.

Effective Sound Pressure (Sound Pressure)—the root-mean-square of the instantaneous sound pressures over a time interval at the point of measurement.

Fast Fourier Transfer (FFT)—any of several calculation procedures that gives a set of Fourier coefficients or component amplitudes from a time-series frame with less computational work for large frame sizes than is possible by the classical method of successive calculation of each coefficient.

Folding Frequency—inverse of two times the time interval between sampled values.

Frame—set of values processed or analyzed as a group.

Frame Size—number of sampled points in a frame.

Frequency—time rate of repetition of periodic motion; units are cycles per second or Hertz (Hz).

Frequency Distribution (See Amplitude Density Distribution.)

Gaussian Distribution (Normal Distribution)—a specified amplitude distribution of importance in probability theory; its histogram is bell-shaped. In relation to acoustics, it describes stationary acoustic noise that is not periodic.

Hanning—a smooth data window is used here in the form of a time domain of a raised cosine arch. Weighting is zero at beginning and termination of frame, and unity at middle of frame.

Histogram—plot of amplitude density distribution.

Jerk—third derivative of the displacement with respect to time, vector representing time rate of change of acceleration.

Level—logarithm of the ratio of a quantity to a reference quantity of the same kind. Examples of types of levels are electric power level, sound pressure-squared level, voltage-squared level. In symbols:

$$L = \log_r(q/q_o)$$

where L = level of quantity (determined by the type of quantity being considered—units measured in $\log_r r$),

r = reference ratio and the base of the logarithm,

q = quantity being considered, and

q_o = reference quantity of same kind.

Level Distribution—a collection of numbers that characterize noise exposure giving the length of time that the sound-pressure level dwelled within each of a set of level intervals.

Loudness Contour—a plot which depicts related values of sound pressure levels and frequency that is necessary to generate a given loudness sensation.

Loudness Level—sound equal to the median sound pressure level (dB) relative to 0.0002 microbar of several trials judged by listeners to be of equal loudness at 1000 Hz frequency. Units in phons.

Mechanical Impedance—impedance derived from the ratio of force to velocity during simple harmonic motion.

Mel—unit of pitch; the pitch of a sound that is judged to be n-times that of one-mel tone is n mels.

Microbar—dyne per square centimeter; unit of pressure used in acoustics. One bar equals a pressure of 10^6 dynes/sq cm.

Neper—division of a logarithmic scale. It expresses the ratio of two like quantities that are proportional to energy or power. The ratio is converted to nepers by multiplying the logarithm to the base e by ½.

Noise Reduction Coefficient (NRC)—average of sound absorption coefficients at 250, 500, 1000, and 2000 Hz, expressed to the nearest integral multiple of 0.05.

Noys—dimension utilized to calculate perceived noise level.

Nyquist Interval—period equivalent to the inverse of twice the frequency of that component of the signal having highest frequency. Maximum sampling-time interval that allows reconstruction of a band-limited signal.

Octave—(1) interval between two sounds with a frequency ratio of two; (2) pitch interval that lies between two tones. One of the tones is considered to duplicate the basic musical import of the other tone at the nearest possible higher pitch. Interval in octaves between two frequencies is the logarithm to base 2 of the frequency ratio.

Octave Band—frequency band that has an upper band edge frequency is twice the lower band-edge frequency.

One-Third Octave Band—frequency band that has an upper band-edge frequency of 1/3 times the lower band frequency.

Oscillation—the variation with time of the magnitude of a quantity with respect to a designated reference, when the magnitude is alternately greater and smaller than the reference.

Particle Velocity—root-mean-square velocity superimposed on the other particle velocities of the medium. The velocities of the particles are probably due to thermal agitation, wind, or air currents.

Peak-to-Peak Value—the peak-to-peak value of an oscillating parameter is the algebraic difference between the two extremes of the quantity.

Perceived Noise Level—level of a noise determined by a calculation procedure based on an approximation to subjective evaluations of noisiness; units in decibels.

Periodic Quantity—oscillating quantity that repeats for certain increments of the independent variable.

Phon—unit of loudness level.

Pink Noise—a sound whose noise-power-per-unit-frequency has an inverse relation to frequency over a specified range.

Power Level—units in decibels, 10 times the logarithm to the base 10 of the ratio of a given power to a reference power, reference power is indicated when level is recorded.

Pressure Spectrum Level—the effective sound-pressure level of that unit of a signal held within a one cycle/second wide band and centered at a specified frequency. Generally pressure spectrum level is important only for sound having a continuous distribution of energy over a particular frequency range. Reference pressure must be stated.

Primitive Period (Period)—the smallest increment of an independent variable for which the function repeats itself. If there is no ambiguity, the primitive period is called the period of the function.

Quantization—changing of a value into one of a limited set of values; these values are a discrete series of total numbers equal to the number two raised to an integer power (binary set).

Rate of Decay—time rate of change of sound pressure level, rate at which sound pressure level decreases at a specified point and a given time. Units in dB/sec.

Root Mean Square—square root of the arithmetic mean of the squares of a set of instantaneous amplitudes.

Sabin—unit of measure of sound absorption.

Sone—unit of loudness. A simple tone of frequency 1000 Hz 40 dB above a listener's threshold generates a loudness of one sone. 1 millisone = 0.001 sone. The smallest increment on a psychophysical scale expressing the intensive attribute of complex sounds.

Sound Energy Density—sound energy per unit volume. Sound energy density at a point is the limit of the sound energy in a region divided by the volume of the region, as the volume containing the point goes to zero.

Sound Intensity—average rate of sound energy flow in a given direction divided by the area, perpendicular to that direction, through or toward which it travels. Intensity at a point is the limit of the quotient as the area which holds the point, goes to zero. (Symbol: I; units in watts/sq m.).

Sound Level (Noise Level)—weighted sound-pressure level that is measured by using a metering characteristic. The weighting is A, B, or C; specified in American National Standard Specification for Sound-Level Meters, S1.4-1971. Reference pressure is 20 micronewtons per square meter (2×10^{-4} microbar).

Sound-Pressure Level—20 times the base 10 logarithm of the ratio of the pressure of the sound to the reference pressure. Reference pressure is usually equal to 2×10^{-4} microbar or one microbar. The first value is used for measurements concerning hearing with sound, air, or liquids. The second value is used for calibrations of transducers and various sound measurements in liquids medium.

Sound Transmission Class(STC)—single figure rating system designed to yield a preliminary estimate of the sound insulation properties of a partition or preliminary rank ordering of a series of partitions. It is utilized when speech and office noise present a serious noise dilemma.

Sound Transmission Loss (TL)—of a partition—the ratio of the airborne sound power incident on the partition expressed in decibels to the sound power transmitted by the partition to its other side. A specified frequency band must be noted.

Spectrum—(1) A description of a function of time, of its resolution into components, each having a different frequency, amplitude, and phase. (2) A continuous range of components which have several specified characteristics.

Spectrum Level—ten times the logarithm to the base 10 of the ratio of the squared sound-pressure-per-unit bandwidth to the corresponding reference quantity. Unit of bandwidth is Hertz; reference quantity = 20 (micro-newtons/meter)^2 ÷ Hz.

Standard Deviation (Sigma, σ)—linear measurement of the variability [equal to the square root of variance (T^2)].

Telephone Influence Factor (TIF)—index of the potential interfering effect of a specified power circuit which is on a telephone circuit.

Transfer Function—a measure of the relationship between output signal and input signal of a system. (Ratio of output to input signal.)

Variance—quadratic measure of variability. It is the average of the mean squares of the deviations from the arithmetic mean of a collection of values of some variable.

Velocity—a vector specifying the time-rate-of-change of displacement with respect to some reference plane. Relative velocity refers to a velocity whose reference frame is not inertial.

Waveform—instantaneous waveform expressed as a function of time.

Waveform Averaging (Summation Analysis)—the addition of corresponding ordinates of selected frames of a wave. Summed values may be divided by the number of frames summed to convert to an average.

CHAPTER 28

NOISE SOURCES AND CONTROL APPLICATIONS

TECHNIQUES FOR REDUCING MACHINERY NOISE[1]

Total elimination of noise can be very difficult, but machinery noise can be reduced to safe, livable limits. Sheet lead, lead-impregnated vinyl, sheet lead sandwiched between two layers of polyurethane foam, lead-loaded epoxy, sheet lead laminated to a variety of substrates and other specialty lead products offer excellent sound isolation properties and can be matched perfectly for specific tasks of controlling unwanted noise.

In approaching a noise control problem, the type of machine must be considered. Fully automatic machines may be completely enclosed in a sound-insulating structure. Doors may be built into the enclosure to permit necessary manual adjustments of the machine or tooling changes. Manually fed equipment, however, must provide for operator access so it is not possible to cover the machine completely. In such cases, partial enclosures, efficiently constructed, are capable of reducing noise levels to safe limits.

What are prerequisites of a good sound barrier? Essentially, a good sound barrier should be dense, limp, and impermeable. Additionally, such a product should be unaffected by coolants, cutting oils, drawing compounds and similar industrial materials.

Lead for Noise Reduction

Sheet lead for use in sound insulation applications is normally specified in pounds per square foot. Sheets can be obtained in weights ranging from 1/2 lb (1/128 in. thick) to 8 lb (1/8 in. thick). In most cases, weights normally used for noise reduction applications are 1/2 lb and 1 lb. Sheet lead can be cut with ordinary scissors, formed by hand and applied to most any surface with elastic type adhesives. Lead can be readily laminated to many substrates including steel and aluminum sheet. These lead laminates can be formed, drawn or otherwise shaped using the same tools as employed for the substrates themselves.

Table 28-1. Lead Materials for Noise Insulation

Material	Description	Uses
Sheet lead	Usual weight 1/4 lb to 4 lb	Used alone or laminated to substrates
Lead/foam	1/2 lb and 1 lb sheet lead sandwiched between layers of polyurethane foam	Laminated to enclosures
Leaded plastic sheets	Lead-loaded sheet vinyl or neoprene with or without fabric reinforcement	As a curtain or to line enclosures
Damping tile	Lead-loaded epoxy or urethane tiles	Damping heavy machinery
Casting compounds	Lead-loaded epoxy	Potting, filling complex voids
Trowelling and damping compounds	Lead-loaded epoxy, neoprene and urethane	Damping enclosures, surfaces, resonate members and rattling panels

Leaded vinyl sheet is another useful material for sound attenuation. One manufacturer's adaptation of this product comprises two sheets of lead-loaded vinyl laminated to a core of glass fiber cloth to give the material more supporting strength. Leaded vinyl sheet is often specified in the same manner as lead sheet—by lb/sq ft. It is made in 1/2-, 3/4-, 1-, 1-1/2, and 3-lb weights. If it is not applied in the same manner as sheet lead, it is often hung as curtains around noise–producing sources. This material is often used in conjunction with acoustical wool for reducing noise from machinery. In addition to reducing noise, leaded vinyl sheet is also used to damp vibration in resonating structures.

Sheet lead and polyurethane sandwich material can be used for sound conditioning areas such as the inside of existing machine shrouds or guards. This lead/foam material can be obtained with either 1/2- or 1-lb sheet lead sandwiched between thicknesses of polyurethane foam. Foam can be varied in thickness from 1/4 to 2 in. in increments of 1/4 in. Lead/foam material can also be supplied with either two or three layers of sheet lead in the sandwich. The material can be cut with scissors or simple steel rule dies when repetitive parts are needed. It can be shaped readily to contours and held in position with adhesives. Types are available with a sealed surface that will not absorb oils or coolants.

For structure-borne noise, lead-loaded epoxies make ideal damping compounds. The compounds are made by mixing powdered lead into epoxies. The material is usually applied with a trowel and can be used on resonating structures such as machinery guards to reduce vibration and the resultant noise.

Quieting Machinery

Reducing noise from machinery is still an emerging technique. In this "cut and try" period, much must be borrowed and adapted from the manner in which noise problems have been solved by others.

Mechanical Power Presses

The obvious solution to the problem of noise emanating from an automatic machine is to enclose the unit in a sound-attenuating structure. A large maker of die sets for mechanical power presses took this approach to quiet the presses in its shop (Figure 28-1). One of the presses, a straight-side, 20-ton model fitted with a slide feed, normally operated at 600 rpm.

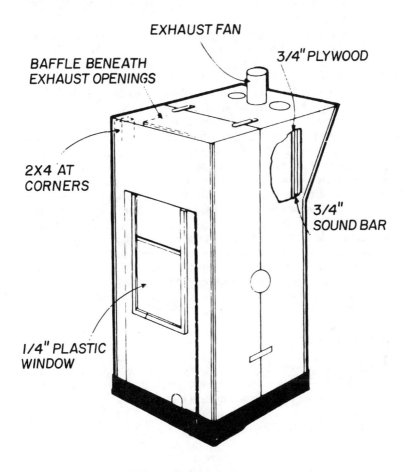

Figure 28-1. Total enclosure for punch press.

Six feet in front of the press, this unit had an "A" scale noise level of 98 dB prior to treatment. The noise level 6 ft from the rear was 98 dB. The first step in the noise reduction program was to build a frame of 2 x 4's around the press as close to it as possible. The frame was then covered with 3/4-in. plywood. The covering provided cutouts for slide feed, finished parts and scrap. The enclosure was lined with a sheet lead/poly-urethane laminate material consisting of a 1/4-in. layer of foam, a septum of 1-lb sheet lead and a layer of 1/2-in. foam. The enclosure could also have been lined with 1-lb lead (1/64-in. thick) or leaded vinyl sheet.

To prevent excessive heat build-up within, louvers were cut into the top of the press enclosure. Flat panels were located beneath the louvers to act as baffles. Exact locations of the baffles were determined by trial and error. Slight shifting of the baffles made a great difference in the reduction of noise. A small exhaust fan was mounted in one of the louvers to further reduce heat build-up.

This press, running under high-speed conditions, now has a noise level in front of the press of 86 dB—A scale. From the rear, the reading is now 85 dBA. This reduction brings the press noise well within Occupational Health and Safety Act requirements.

Power presses with automatic feed devices can also be enclosed in noise barriers made from leaded vinyl (Figure 28-2). These floor-to-ceiling curtains can be used to contain either single presses or banks of units. A press attendant can go in and out of the area to load additional stock in feed devices and to adjust press malfunctions. Total exposure time within the area can be adjusted to the time permitted by the noise level. If fail-safe electronic sensors are used on the presses, they can be adapted to activate alarms when a press needs attention for a malfunction or loading.

Most noise produced by a mechanical power press is from the impact produced by the die set closing on a piece part. In such cases, a bellows constructed of leaded vinyl attached to the press slide can be timed to en-close the impact area when the press is stroked (Figure 28-3). The balance of the upper part of the press with attendant gears or belts and motor can be enclosed with a plywood or lead or lead/foam structure. If such an arrangement does not reduce the noise level to permissible limits, a second structure can be built around the lower section of the press.

Pneumatic Tools

Pneumatic shop tools produce impact noise. By enclosing the barrel of the tool in leaded vinyl sheet and rock wool some quieting can be achieved by muffling the exhaust noises (Figure 28-4).

High Impact Machines

High tonnage presses, drop hammers and other high impact machines create vibrations in addition to airborne noise. These vibrations, unless

Figure 28-2. Lead-loaded curtain enclosure.

short circuited, can be transmitted throughout an entire plant even into office areas where they may be translated into audible noise. To prevent this from occurring, machines must be isolated from their foundations. One highly efficient method of accomplishing this isolation is through the use of lead anti-vibration pads (Figure 28-5).

An excellent example of how lead anti-vibration pads have been used is at a metropolitan New York area newspaper. The editorial offices and the printing plant are housed within a single building. The power and impact of the big press was such that its foundation was separate from the building foundation. However, 44 lead anti-vibration pads, each measuring 24 x 30 in., were used under the press posts. As both foundations go down to bedrock, the pads were the sole effective break in the vibration path.

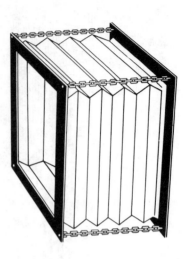

Figure 28-3. Noise control bellows for punch press.

Figure 28-4. Cuff for pneumatic tools.

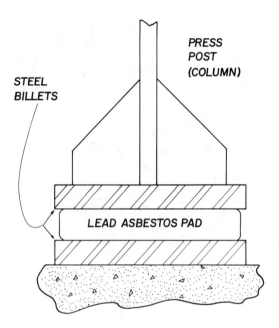

Figure 28-5. Vibration pad.

Lathes and Automatics

Most noise generated by a lathe or automatic screw machine comes from the gears and drive train. Noise emanating from the tool-chip interface should not be of concern in most cases. The solution to a gear noise problem is in the use of partial enclosures. One-pound sheet lead or lead/foam material laminated to any material used for drywall construction can be successful as a sound barrier built around gear cases. Hinged doors or ports should be incorporated in the design of the enclosure to provide access to speed and feed controls. The access doors should be reasonably air tight to be an effective noise seal. The hinge side of the door should be gasketed with leaded vinyl sheet attached to the door and the jamb. The top, bottom and side opposite the hinge should also be gasketed with a lead-loaded material. Smaller gear boxes can generally be quieted through the use of pillows made of lead-loaded vinyl and acoustical wool attached to the top of the gear box (Figure 28-6).

On a larger scale, heavy gears such as those on hoists and crane equipment generate high noise levels. The resonance of these huge gears can be substantially reduced by applying to the gear face damping materials such as lead-loaded epoxies or lead-loaded vinyl sheets.

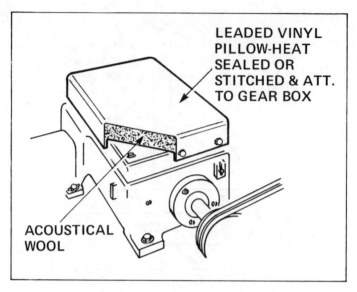

Figure 28-6. Gear box damping pillow.

Machinery Guards

In some instances, large machinery guards of sheet metal have a tendency to set up a high-pitched ringing noise if stiffening ribs have not been added. This problem may be solved by laminating pieces of sheet lead to the guard by trial and error method until the ringing is reduced or eliminated (Figure 28-7). Self-adhesive leaded vinyl sheet may also be used to line the guard with the same result.

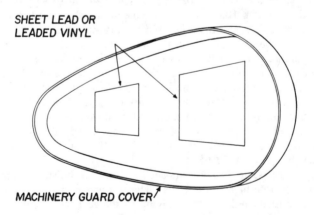

Figure 28-7. Noise control pads for machinery guards.

High Pressure Pipes and Valves

High pressure transmission lines and valves for liquids and gases may be silenced through the application of sheet lead or leaded plastics. By first wrapping the pipe with absorbent material such as glass fiber or mineral wool to act as a thermal insulator, and then covering the material with sheet lead or leaded vinyl sheet, noise levels will be significantly reduced. The final installation step should be the sealing of all the joints on the line or valve with a self-adhesive lead tape (Figure 28-8).

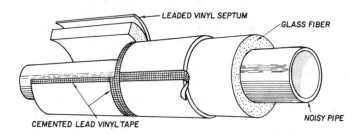

LEADED VINYL SEPTUM

GLASS FIBER

CEMENTED LEAD VINYL TAPE

NOISY PIPE

Figure 28-8. Pipe and valve covering.

In actual practice, noise levels of 110 to 130 dB at a gas pressure reducing station on a pipeline were successfully muffled to 100 dB by wrapping the line with a 2-in. thickness of fine yarn glass fiber and covering it with a single blanket of reinforced leaded vinyl sheet. Greater attenuation could have been attained if a second layer of glass fiber and leaded vinyl sheet had been used.

Air Intakes and Exhausts

Noise generated by the movement of great volumes of air can reach painful and dangerous levels. A leaded vinyl bellows installed between the intake of a large air compressor and the outside of a building will reduce the noise of the air movement within the enclosure. Air ducts which serve as pathways for the intakes and exhausts may be enclosed in glass fiber and sheet lead to further decrease transmission noise.

Air Compressors

Air compressors, a source of irritating noise within an industrial plant, can be silenced by enclosing them in a shroud consisting of sheet lead or leaded vinyl laminated to any material intended for drywall construction. Louvered openings with baffles should be provided for the dissipation of heat build-up within the enclosure. Lead/foam material may also be used as a lining for the enclosure.

Vibratory Feeds

A source of highly distracting noise on production machines devoted to automatic assembly of spark plugs was created by air-vibrated feeds on each of the units. The noise, caused when the steel spark plug shells were vibrated into position on the metal feed track, was like machine gun fire. This unavoidable, metal-on-metal noise was a nerve-wracking distraction to workers. Because of the construction of the feed devices, it was impossible to completely enclose them. To solve the problem, the 2 ft x 4 ft area around each vibrating feeder was surrounded with curtains made from reinforced, lead-loaded vinyl. Even though the curtained area was left open at top and bottom, the noise at the individual feeds was reduced from 93 dB to 86 dB.

Transformers

The hum of transformers located at a substation can be a troublesome noise. When faced with such a noise problem, a West Coast utility used lead adhesive-bonded to steel as the noise barrier because of its weight and because it did not tend to resonate. The barrier was made by laminating 4-lb (1/16-in.-thick) sheet lead to 10-gage steel with epoxy adhesive. The enclosure and its frame were built to be freestanding and isolated from the transformer. Within this sheet, a lining of a 2-in.-thick blanket of glass fiber was spaced about 1 in. from the inner face. Pipes and fittings piercing the barrier were isolated from the walls by neoprene gaskets. After the transformer radiators were remounted, the units showed a noise reduction of 17 dB (Figure 28-9).

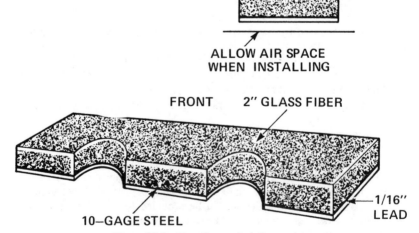

SIDE

**ALLOW AIR SPACE
WHEN INSTALLING**

FRONT   2" GLASS FIBER

1/16" LEAD

10–GAGE STEEL

Figure 28-9. Transformer isolation pad.

GROUND-BORNE VIBRATIONS[12]

While some vibrational phenomena are essential to man—such as the transmission of acoustic and electromagnetic energy—other vibrations can have adverse effects. In addition to a possible degradation of the physical environment, vibrations can cause human discomfort, structural fatigue or failure, increased machinery maintenance and a lowering of the safe or accurate opening limits for certain types of equipment.

Obviously, which vibrations constitute a problem depends on the particular situation or application being considered. In this discussion we will not be concerned with the vibrations *of* machinery or equipment but rather with ground transmitted vibrations *from* industrial, construction or traffic sources.

Earth media such as soils and rocks are efficient transmitters of vibrations in certain frequency bands. Both theory and observation show that the earth acts as a low-pass filter; *i.e.,* the higher the vibrational frequency, the shorter the distance it will travel before being completely damped by the various frictional characteristics of the earth. Ground frequencies in the range of 1 to 500 Hertz are routinely observed emanating from quarry and construction blasting in the distance range from a few hundred feet up to several miles. Large earthquakes send out vibrations in the 0.003 to 1 Hz range that are routinely recorded at seismographic observation stations throughout the world. Thus, the range of frequencies that the earth will transmit span at least some five orders of magnitude. The earth, however, is most efficient at frequencies that are considerably lower than those commonly associated with machinery.

Another important aspect of ground vibration transmission is the existence of two types of waves: *body waves* and *surface waves.* Body waves, having both compressional and shear characteristics, travel the interior of a solid; surface waves travel along and are guided by the free surface of a solid. Body waves travel at higher velocity than surface waves do. Therefore, from a transient disturbance, the arrival sequence at a distant point is first the body waves, then the surface waves. A continuous operating source, such as machinery, generates both body and surface waves, and the resulting motion at a distant point will be some combination of both types. Surface waves have, in general, larger displacements and lower frequencies than body waves and thus have lower particle velocities (the velocity of a particular point rather than the velocity of transmission between two points) than the body waves.

Vibration Zones

Regardless of the specifics in a given case and the types of waves carrying the vibrations, certain essential features of ground vibrational problems can be abstracted that will aid in their solution.

Source of Vibrations or Generation Zone

The problem vibrations may be caused by machinery, pumping stations (pipeline water system), construction activities (blasting, pile-driving) or traffic (rail, highway). The important factors are the *energy level, frequency* and *time history* (transient or continuous) of the source. Also relevant is whether or not we can exercise any *control* over these source characteristics.

Transmission Zone or Path Between
Source and Site

Rock is an efficient vibration transmitter and thus a zone can be quite long. The physical properties (density, Young's modulus, shear modulus, etc.), the geometry and the geologic structure are the principal items of importance in specifying the transmission zone.

Reception Zone or Site

This is the location where unwanted vibrations occur. What is important here is dictated by the specifics of the individual problem. For example, if we were concerned with adverse vibrational effects on delicate instrumentation located inside a building, then the building would be the receptor of the ground-transmitted vibrations. We would have to consider first the building's response to the ground vibrations and then the instrument's response to the building vibrations. This case would be a multiple reception zone problem. If, on the other hand, we were concerned with the integrity of the building then we would have a single or simple reception zone. The obvious point is to carefully consider what happens to the ground vibrations before they actually constitute a problem.

Vibration Measurement Instrumentation

The fundamental problem in ground vibration measurement is that the observer and the instruments partake of the same motion as the ground. This is solved by instruments, called seismographs, that establish an internal point that tends to remain steady while the housing frame vibrates with the earth. Measurement of the motion is made between the frame and the internal steady-point. Inertial elements (masses, magnets, coils) suspended by various spring and/or pendulum configurations are employed to provide the steady reference point.

The selection of a seismograph, as well as the ground motion parameter (displacement, velocity, acceleration) most convenient to measure, are dictated primarily by the ground frequencies being encountered. To understand this consider first simple harmonic vibration:

$$\text{Displacement} = x = A \sin (2\pi ft)$$

where A = maximum displacement amplitude, f = frequency in Hz, and t = time. In this case the Maximum Velocity = v = $2_\pi fA$, and Maximum Acceleration = a = $4_\pi^2 f^2 A$. Thus, the relationship between velocity and displacement is proportional to the frequency while the acceleration-displacement relationship is proportional to the frequency squared.

Next, consider the fact that seismographs are band-pass instruments; *i.e.,* they do not measure all frequencies of ground motion with equal fidelity. This is true because their spring-mass-pendulum configuration has its own resonant (or natural) frequency and the inertial element only becomes steady at frequencies that are much higher than the particular resonant value. We have, therefore, frequency-dependent ground motion parameters and frequency-dependent seismograph response characteristics. These two factors must both be considered to provide a successful measurement of ground vibrations.

For the earth, vibrational frequencies most often encountered in practice (1 to 50-100 Hz) displacement or velocity seismographs having a natural frequency of 1 or 2 Hz are appropriate. The most modern of these portable seismographs are designed to measure ground particle velocity rather than ground displacement. Studies have shown that building damage, caused by vibrations from blasting, is most directly correlatable with ground particle velocity. However, for measurements of the type being considered in this analysis, there is no particular advantage in measuring one parameter over the other.

Frequencies greater than 50-100 Hz are sometimes encountered on hard rock sites near the vibration source. For such situations it is advantageous to employ high-frequency measuring devices such as piezoelectric accelerometers.

Vibration Elimination

There is, of course, no general solution to the problem of vibration elimination because of the many variables involved and the wide range of possible situations that could exist. But again, as in the case of the vibration zones, certain broad classes of solutions can be presented. The approach is to employ methods that are commonly applied only in the reduction of machinery vibration. That particular technology is well developed and has produced three general schemes of vibration reduction:

1. **Decoupling** relates to both the source of the vibrations and the items affected at the site. The approach is to change the amplitudes and/or phases of the outgoing (or received) vibrations by decreasing the number of vibrating components that are physically linked together.

2. **Detuning** attempts to eliminate resonances by changing the natural frequency of the source or the site. Resonance indicates that some physical structure or object is being excited at its own natural frequency of vibration. To change its natural frequency the structure or object must be changed physically; *i.e.,* made heavier or lighter, more or less rigid, etc. If possible, the easiest solution is to change the frequency of source so as to put its emitted vibrations outside of the natural frequency of the affected structure or object.

3. **Isolation and/or Damping**—if the vibrations cannot be eliminated or controlled by any other means, then the affected item should be isolated. Materials and/or structures (isolation springs, shock mounts, etc.) are interjected which are non-responsive and do not transmit the vibrations. Similarly, materials and/or structures (rubber, styrofoam, plastic, etc.) can be used which are absorptive and have a damping effect on the vibrations.

LEAD-LOADED FABRICS KEEP OUTSIDE NOISE OUT[192]

When production facilities are expanded in a plant, additional office space is often necessary for the increased clerical work for the new operation. To keep office workers convenient to the production operation, yet keep noise problems to a minimum, offices are usually constructed at one end of the production area.

Since production area ceilings are high, office walls are often constructed of brick or acoustical partitions. The ceiling is then suspended from the underside of the plant roof. The office walls will usually prove to be good sound barriers. However, the sound-deadening capacity of the ceiling will be far below what is required for conversation, even when acoustical ceiling tile is used in the T-bar suspension systems. The ceiling will control sound originating within the office, but will not serve as a barrier to external noise.

Noise from production equipment can reflect off the underside of the plant roof and focus on the acoustical tile ceiling of the office. As a result of this unique focusing problem, noise levels within the office can be more distracting than in the plant.

Constructing full height insulated office walls to the underside of the plant roof is usually uneconomical in high-ceiling areas. High cost of materials and labor, interruption to production, and modifications to fire control sprinkler systems usually make high-wall office systems prohibitive in cost.

However, using a lead-powder-loaded, vinyl-coated fabric can help reduce outside noise substantially. The material is limp, dense and non-porous and is easily cut with a knife or scissors. It weighs about 0.9 lb per sq ft and is approximately 1/16 in. thick. Most important, the material can be installed easily over one office in an entire group, and its effectiveness evaluated before making a commitment for a total installation.

The lead-loaded fabric can be installed by plant personnel, quickly and easily, usually with only oral instructions to explain the installation technique. Rolls of vinyl are unrolled on the floor, cut to length, and laid directly over the T-bar suspension of the ceiling. No additional support is necessary (see Figure 28-10).

Material is cut to go around suspension wires, and edges of the fabric are overlapped to prevent sound waves from passing through the joints. Edges can be sealed with adhesive tape; access to the sprinkler system is maintained by not sealing them.

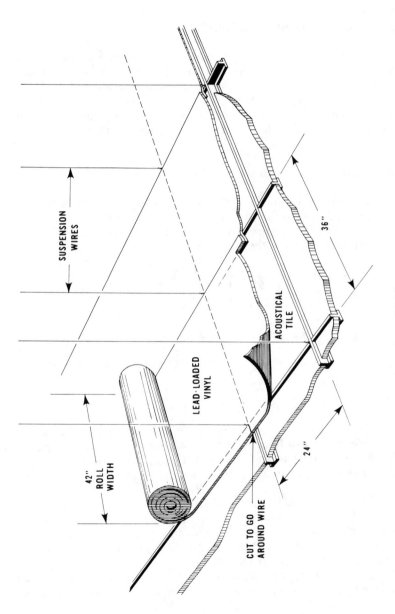

SUSPENSION WIRES

ACOUSTICAL TILE

LEAD-LOADED VINYL

42" ROLL WIDTH

36"

24"

CUT TO GO AROUND WIRE

Figure 28-10.

Noise levels ranging from 95-100 dB in offices have been reduced after installation of lead-loaded fabrics to 75-80 dB.

USING GLASS FOR NOISE REDUCTION[127]

A glass barrier offers the potential for controlling noise while providing visual communications. For the same thickness, it is a better sound barrier than most brick, tile or plaster. An average noise insulation of 35 dB can be obtained by using a single-leaf wall weighing about 6 lb/ft^2. However single-leaf wall construction is not efficient, if an average insulation of 40 dB or more is required. It is better, then, to use double-leaf type construction.

Sound Transmission Mechanism

A single-leaf partition is one in which points on opposite sides of the structure both move in the same direction at the same time at all frequencies (Figure 28-11). The simplest sound insulation treatment of this type of partitition is known as the mass law.

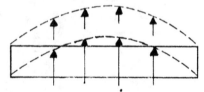

Figure 28-11. Reaction of single-leaf partition.

Monolithic Plates

The ratio of the sound energy incident on one surface of a partition to the energy radiated from the opposite surface is called the "sound transmission loss" of the partition. The sound transmission loss is an inherent characteristic of a barrier and is essentially independent of the barrier's location.

We do not actually "hear" the sound transmission loss, nor can it be measured directly. Instead, we hear and measure the difference in sound pressure level between the spaces separated by a barrier, and we call this difference "noise reduction." It includes the effect of the absorption present in the receiving room and the source room.

If a wall has negligible stiffness, the inertia effects of its mass control the sound transmission loss through the wall. In general, the theoretical "mass law" for sound waves at oblique incidences is:

$$T_L = 10 \, Log_{10} \left[1 + \left(\frac{\omega W}{2 pC} \cos \Phi \right)^2 \right]$$

where
$$\begin{aligned}
T_L &= \text{transmission loss in dB} \\
W &= \text{wall weight per unit area (lb/ft}^2) \\
p &= \text{density of air (lb/ft}^3) \\
C &= \text{velocity of sound in air (ft/sec)} \\
\omega &= \text{angular frequency of excitation (rad/sec)} \\
\Phi &= \text{angle of incidence of sound wave}
\end{aligned}$$

If the incidence sound wave impinges on the panel at normal incidence, when $\Phi = 0$, then the equation becomes

$$T_L = 10\ Log_{10}\ \left[1 + \left(\frac{\omega W}{2pC}\right)^2\right]$$

where T_L is the sound transmission loss for normal incidence.

The quantity $(\omega W/2pC)^2$ is generally large and in practice the sound waves impinge upon the barrier at a wide range of incident angles, in which case the equation becomes

$$(T_L)\ \text{field incidence} = 20\ Log_{10}\left(\frac{\omega W}{2pC}\right) - 5$$

The incidence field equation is the maximum that can be theoretically obtained with a single leaf barrier. When transmission loss is plotted against the frequency spectrum as a logarithmic scale, a straight line having a slope of 6 dB per octave is obtained (Figure 28-12). According to the mass law,

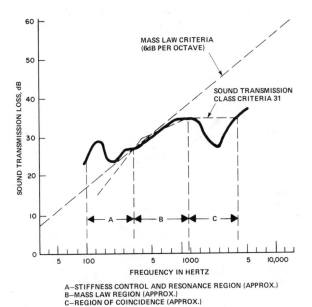

A—STIFFNESS CONTROL AND RESONANCE REGION (APPROX.)
B—MASS LAW REGION (APPROX.)
C—REGION OF COINCIDENCE (APPROX.)

Figure 28-12. Transmission loss characteristics for 7/32-in. thick glass (3.0 lb/ft²).

a 7/32-in. thick glass wall has a potential average of 38 dB, but it achieves only 31 dB. This discrepancy is caused by the influence of stiffness, which becomes significant at certain frequencies.

The bending sound wave speed of a wall varies as the square root of the excitation frequency. At some combinations of excitation, frequency and angle of incidence, the trace wavelength of the incident sound wave will coincide with the wavelength of the bending wave of the wall. The sound wave reinforces the bending wave, and a maximum amount of acoustical energy is transferred to the wall. At this coincidence frequency, a wall becomes quite transparent to sound and the transmission loss drops markedly.

A simple relationship for monolithic plates is:

$$f\Theta = \frac{c^2}{1 \cdot 8h \; C_L \; \sin^2\Theta}$$

The lowest possible frequency at which coincidence can occur is called the *critical frequency* which is obtained when $\sin^2\Theta = 1$

$$f_c = \frac{c^2}{1 \cdot 8h \; C_L}$$

where C_L = longitudinal wave velocity (ft/sec)
    h = plate thickness (inches)
    Θ = angle of incidence of the sound wave (degrees)

Experimental results for coincidence frequency are shown in Figure 28-13. The results indicate that as the thickness of glass plate increases, the

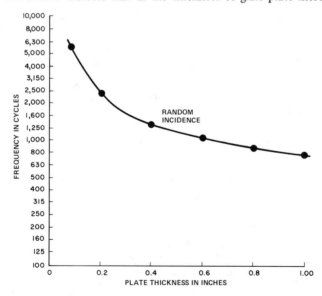

Figure 28-13. Coincidence frequency versus glass plate thickness.

coincidence frequency decreases. In the case of multi-glazed units, the effect of coincidence frequency is the same as that of single glazed units. There is very little effect of the air space on the coincidence frequency. The experiments show that the coincidence dip for multi-glazed units occurs at the same frequency as it does for monolithic plate of the same thickness.

Figure 28-14 shows some experimental results for monolithic glass products for noise control. Plates were tested at the Riverbank Acoustical Laboratory in accordance with the ASTM Standard E90-66T and the sound transmission class (STC) determined by the ASTM Standard RM 14-2.

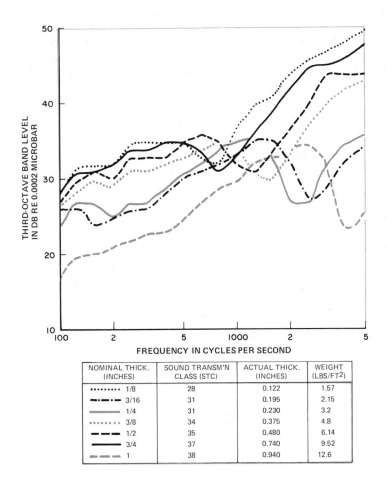

NOMINAL THICK. (INCHES)		SOUND TRANSM'N CLASS (STC)	ACTUAL THICK. (INCHES)	WEIGHT (LBS/FT^2)
....	1/8	28	0.122	1.57
—·—·—	3/16	31	0.195	2.15
————	1/4	31	0.230	3.2
·········	3/8	34	0.375	4.8
— — —	1/2	35	0.480	6.14
————	3/4	37	0.740	9.52
— — —	1	38	0.940	12.6

Figure 28-14. Experimental results of monolithic glass products for noise control.

While the STC rating for 3/16-in. and 1/4-in. plate glass are identical, the performance curves differ markedly. The results show that the coincidence frequency decreases as the thickness of the glass plate increases. The resonance frequency for all these monolithic products falls in the region of 100 to 225 Hz.

NOISE CONTROL EFFECTS OF ROCK WOOL [74]

The following are the results of a continuing study of wool and its effect on acoustical transmission loss. The data presented here show the definite relationship between sound transmission class (STC) and the weight of wool in the stud cavity, lb/sq ft (PSF).

Using a basic metal stud system with 5/8 in. Sheetrock faces, the stud cavity contents were varied between single and multiple layers of 1½ in. 3 lb/cu ft (PCF), and the 1 in. 4 lb/cu ft SAB blankets.

Test	Material Thickness	PCF	PSF	STC
A	no wool in cavity		0	39
B	1 layer - 1 in.	4	0.33	43
C	1 layer - 1½ in.	3	0.38	45
D	2 layers - 1 in.	4	0.67	46
E	2 layers - 1½ in.	3	0.75	46
F	3 layers - 1 in.	4	1.0	48
G	4 layers - 1 in.	4	1.33	48

Note that as the surface density of the blanket increases, the resulting STC also improves (see Figure 28-15). This improvement continues until 4 layers of 1-in. rock wool are used. At this point the assembly begins to "short out" and, consequently, is no longer able to improve the rating. It may be that if we tried to compress 5 in. of material, the STC would drop.

Carrying this analysis one step further, an algebraic equation for this relationship was developed:

$$STC = 40.1 + 8.4 \ (PSF)$$

Using this equation, the effects of various other wool or glass materials within this type of partition assembly can be approximated.

For example:

USG Wool - 2 in., 2.5 PCF
  surface density = 0.417 PSF
STC = 40.1 + 8.4 (0.417)
  = 43.6 or approximately 44.

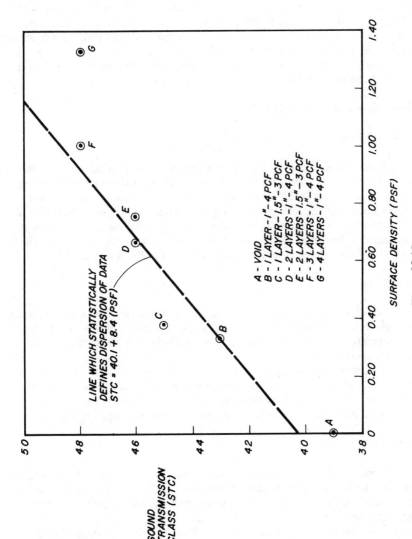

Figure 28-15.

WORKABLE SOLUTIONS TO COMMON
MACHINERY NOISE PROBLEMS[116]

The first inclination that a pollution engineer may have is to build a rigid closure or install a flexible sound curtain or a permanent wall around the equipment. In many cases these approaches constitute the only possible solution and are the most efficient, expedient and economical. However, when the machine can be treated without dramatically altering the production cycle, this is generally the most desirable approach.

The following case histories show successful solutions that can be considered in solving similar equipment noise problems.

Plastic Grinders or Granulators

Existing Situation

Noise levels of 105-115 dBA were encountered in grinding polycarbonates and other tough plastics. Normal frequency was 600 Hz, with intermittent impact peaks of 2000 Hz. Noise sources were: the grinding action itself, hopper impact in the loading and whipping action of the part, electric motor, belt gear drive, and impact at the chipper and material chute. No attempt was made to determine the individual levels of each source, but it was obvious that hopper impact and grinding action were the major contributors.

Considerations

Solutions considered were a flexible curtain closure and a supplemental rigid plywood noise-treated closure. This method could have worked; however, operator accessibility, material flow and light and ventilation would have been sacrificed.

The design of the equipment made it a closure in itself. Also, by treating the machine, the original equipment manufacturer could have reduced the noise levels substantially at time of manufacture, providing a guaranteed noise-treated grinder at lower cost than remedial retrofit.

Solutions

While the following solutions were developed for the OEM supplier, the same techniques can be used by pollution engineers. Mount the machine on vibration isolation mounts (fiber glass precompressed pads). Damp the impact noise areas with a viscoelastic spray-on damping compound. Make the existing steel housing a sound barrier by adding a noise control composite (½-in. polyurethane foam—½ lb/sq ft).

The outside of the metal hopper, pellet tray and chute should be sprayed with a viscoelastic damping compound to a thickness of 1½ times the thickness of the metal. Composite material should be applied to all other metal surfaces using a brush-on, solvent type thermosetting compound with room-temperature cure properties.

Metal components should be treated before the machine is assembled. After assembly, ½-lb/sq ft 2-in. tape of loaded vinyl film with the same thermosetting adhesive should be applied to seal corners and mating joints except at access doors and other removable panels.

Results

There are three areas for potential leaks—(1) the opening at the base of the machine for free-draft motor cooling, (2) the opening in the hopper lid and opening at pellet tray door and (3) engine and other access panels.

Total result was 85 dBA under the most severe operating conditions. Units were also tested without the hopper and pellet tray area treated with damping compound. The compound was, however, found to reduce sound levels 5-7 dBA. There was no reduction in productivity, efficiency or maintenance accessibility. The operator and surrounding workers were protected without the need for ear protectors or helmets. The total noise treatment package added less than 20 percent to the cost of the equipment.

Industrial Wood Saw

Existing Situation

Noise levels of 100-105 dBA were encountered, depending on type of wood being cut. Peak frequency was 1000 to 2000 Hz, indicating that the cutting action and exhaust system were the major noise sources.

Considerations

The machine was already isolated on rubber shear mounts. The saw had a heavy ¼-in. plate closure over the cutting area with small opening at front and back of the saw and a scrap duct on the bottom of the sawing area. The manufacturer used a flexible, non-metallic duct to connect the machine to the main exhaust system.

Total closure was ruled out because the operator had to remain at the machine to preserve production flow. In this case, the OEM supplier wanted to provide saws at an 85 dBA level.

A composite of ½-in. urethane foam and 1 lb/sq ft loaded vinyl barrier had been applied to the outside top, side and bottom of the cutting area closure with only a 4-5 dBA reduction. Two large openings at the inlet and outlet of the saw produced excessive noise levels. Flaps of loaded vinyl were considered for covering these areas; however, possible interference and maintenance replacement ruled them out.

Solutions

A combination of barrier and sound absorption control principles was utilized. A composite of 1-in. urethane foam, 1-lb loaded vinyl film and ½-in. urethane foam was adhered to all outside surfaces of the closure. A 10 to 12 dBA reduction was achieved because air leaks were minimized by the foam (an absorber) and loaded vinyl film (a sound barrier).

Results

The combined treatments resulted in an 85 dBA operating level.

Observations

When there are material passage openings in existing closures, these housings must be treated with absorbing as well as barrier materials. Further noise reductions can be achieved by absorbing sound traps.

Upset Forge Presses

Existing Situation

Noise levels of 100 dB were generated from 36 presses arranged in two rows. The peak frequency was 300 to 600 Hz. The primary noise was generated by high-pressure, water-cooled hydraulic pumps and the hydraulic lines. The noise generated by the forging operation itself was less than 85 dB, low frequency and intermittent.

Considerations

The machines were already isolated on fiber glass pre-compressed pads to protect against structure-borne noise. A complete curtain closure was considered around each machine, but was ruled out when the problem was pinpointed to the pumps and piping.

Solutions

The most practical and economical approach was to cover the pumps with 0.87-lb loaded vinyl with ¼-in. foam. For ease of installation, the composite was adhered to the pump housing with pressure-sensitive adhesive on the foam side of the composite. Joints were taped with a 2-in. wide tape of 0.87 lb/sq ft loaded vinyl. All hydraulic piping on the machine, from the pump to the hydraulic cylinder, was also treated with the same composite and adhesive.

Results

The sound level of the hydraulic system was reduced to below the 85 dB being generated by the press operation.

Observations

Pressure sensitive adhesives are more expensive than brush-on or spray-on rubber-based, contact adhesives with thermosetting properties. However, they are ideal for prototype work, save labor and make installation easier for remedial treatment. Pressure-sensitive adhesives should not be used directly on lead vinyl or loaded vinyl, because the plasticizer in the vinyl will soften and weaken or destroy the adhesive bond. Also, these adhesives will not provide good adhesion to vinyl surfaces.

Lead-loaded and loaded vinyls are available with a barrier coat when pressure sensitive adhesives are required. In all other cases, the pressure sensitive adhesive must be applied on the foam side.

The sound control material in this installation must be removed for pump maintenance. An alternative would have been to vacuum-form clamshell halves of unreinforced loaded vinyl and foam, and join the halves with zippers or catches. This system would provide easier access and replacement. However, in this particular case, when a pump requires maintenance, the unit is removed and replaced by a rebuilt pump. This factor makes the prescribed solution more economical than the vacuum-formed halves.

Gear Housings

Existing Situation

Noise levels generated were approximately 92 to 105 dBA, with peak frequencies of 250 to 4000 Hz. In all cases the equipment or machine performing its operation was below 90 dBA. The excessive noise was caused by structure-borne noise, causing the gear housing or guards to resonate and vibrate.

Considerations

The machines were damped to ensure that structure-borne noise would not reach the floor or building. However, reducing the resonance vibration requires isolation within the machine. Generally, this entails a whole new design of the equipment, in that soft isolators inside the machine can change the work function.

Gear noise can be reduced by having closer tolerances on gears, using different materials for gears and incorporating less backlash in the design.

However, on a remedial problem, complete gear redesign can be costly and time-consuming. If the problem is housing resonance, even redesign of the gears may not help.

Solutions

If the noise is housing resonance, spray a viscoelastic damping compound on the inside or outside—either way is effective. If the problem is gear noise being transmitted through the housing as airborne noise, use a composite barrier on the exterior of the housing. If both housing resonance and gear noise are problems, then use a combination of the damping compound and the composite.

Results

Levels of 90 dBA and below have been achieved.

Observations

If wear and tear on the exterior of the housing is a problem, apply the viscoelastic compound to the interior surface. (Although these compounds are designed to provide good adhesion to metals, some may require the use of a primer. Be sure to check manufacturer's application recommendations.)

Vinyl copolymer viscoelastic compounds provide better resistance to oil and grease than asphalt-based materials and function well in service conditions up to 250 F. For cases of extreme velocity and temperature (200 F or better) of the gear lubricant, epoxy-based damping compounds are recommended.

Turret Lathes

Existing Situation

Noise levels of 100 to 110 dBA, with peak frequencies of 2000 to 4000 Hz are common with turret lathe operation. The primary noise is impact-created at automatic stock feed tubes.

Considerations

Total curtain enclosures and fiber glass sound baffles close to the tool holder have been tried to solve this noise problem. The curtain system has worked well. Fiber glass has had marginal results, because it becomes oil-soaked and loses efficiency, as well as becoming a potential fire hazard.

Solutions

The recommended solution involves a cut-and-try approach, depending on the design of the turret lathe and severity of the noise problem. The methods show some of the measures which can be taken:

1. Install isolation mounts on the lathe.
2. Apply a viscoelastic damping compound on the outside of the stock feed tubes to reduce ringing impact noise.[
3. Check the gear housing and machine panels for vibration. If vibration can be felt, spray on damping materials.
4. If noise is still above acceptable levels, apply composite panels (noise barrier and absorber) around the stock feed tubes. (This is not a total enclosure which will cause material handling and productivity problems.)
5. If excessive noise persists, use noise control composite panels around the tool holder case, in a manner that will not impair operator movement. Make sure the absorbing surface of the noise composite is supplied with an impervious film to eliminate oil absorption.

Results

Levels can be reduced to a range of 80 to 85 dBA.

Observations

Steps 4 and 5 represent a cut-and-try approach to determine the proper location of the composite panels. Place the panels in various locations and take sound level readings to determine the optimum location.

Probably 80 to 90 percent of screw machine noise problems can be solved with one of these five approaches. If not, a movable flexible sound control curtain system can be used. The approaches are also recommended for other types of machinery that have no specific solutions to their noise problems.

HYDRODYNAMIC CONTROL OF VALVE NOISE [19]

Liquid flow through a control valve (called hydrodynamic flow) can and often does create noise. There are three categories of hydrodynamic noise: noise from noncavitating liquid flow, cavitating liquid flow, and flashing liquid flow. Of the three, cavitating flow is the major noise problem. Laboratory testing and field experience show that noncavitating and flashing liquid flow noise levels are quite low and generally not a problem.

Cavitation is a two-stage phenomenon. The first stage involves the formation of vapor bubbles in the fluid stream. As liquid flow passes through the orifice of a control valve, its velocity causes pressure at the vena contracta to drop below the vapor pressure of the liquid, and vapor bubbles are formed.

The vena contracta is the point beyond an orifice where the flow cross section is smallest, pressure is lowest and velocity is highest. However, since clearance between the valve plug and seat ring is the primary restriction in most conventional valves, the vena contracta is formed near the valve seat line.

The second stage is the implosion of these vapor bubbles. As the fluid moves downstream from the vena contracta into a large flow area, velocity decreases with a resulting pressure recovery. When static pressure exceeds the vapor pressure of the liquid, the vapor bubbles implode, generating extremely high pressure shock waves that hammer against the valve outlet and piping. (Pressures in the collapsing cavities reportedly can approach 100,000 psi in magnitude.) Noise and damage result.

Controlling Cavitation

Solving cavitation problems begins first with either controlling the cavitation process, or ideally, eliminating cavitation altogether. In controlling cavitation, the techniques employed are often defensive in nature. For example, valve parts subject to damage are furnished in hardened materials in an attempt to extend valve life against the erosion and shock generated by the imploding vapor bubbles. Another technique simply lets cavitation exhaust itself by destroying some sacrificial part of the piping system (*e.g.,* an elbow downstream of the valve, and orifice plate, etc.).

The third technique, developed within the past few years, involves special cage-type valve trim for globe valves that moves the primary fluid restriction away from the valve plug seat line. In this new trim design, a number of pairs of small, diametrically-opposed flow holes are located in the wall of the cage (Figure 28-16). As the valve plug moves away from the seat, increasing numbers of these holds (always in pairs) are opened to the inside of the cage. Each hole admits a jet of cavitating liquid which ensures substantial pressure recovery in the center of the cage. Collision of this jet with that of the opposing hole creates a continuous fluid cushion (Figure 28-17). The cushion in turn prevents cavitating liquid from contacting the valve plug and seat line of the valve. Under certain conditions, this trim can reduce valve noise as much as 6 dB.

Just developed is an anti-cavitation device for ball valves. This device consists of a bundle of parallel flow tubes which, under certain conditions, create a back pressure within the ball valve to keep static pressure above the vapor pressure of the fluid. However, when cavitation does occur, these flow tubes restrict the size and number of vapor bubbles. Damage tests with soft aluminum rods indicated that damage downstream of the flow tubes was insignificant as compared to damage without them.

When cavitation control methods are used, noise reduction is gained by applying acoustical insulation on the valve and associated piping, but using heavy-walled piping, by installing the valve in an enclosure, or by burying the pipeline. These techniques, commonly referred to as path

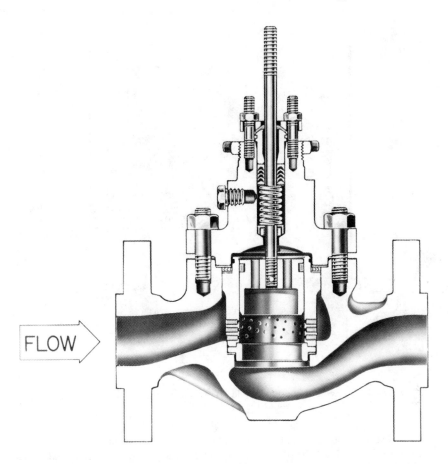

Figure 28-16. Globe style valve body with special cage trim that controls cavitation.

treatment, do not reduce the level of noise carried in the fluid stream—they only shroud it. Therefore, it is important to note that where path treatment stops, fluid noise may annoyingly reappear.

Eliminating Cavitation Noise

If the cavitation is eliminated, cavitation-created noise is also eliminated. Several techniques can be applied to eliminate cavitation.

One involves placing the control valve within the piping system at a point where pressure drop and fluid temperature conditions will not create cavitation. If this proves impossible, then two or more valves in series, each taking a portion of the total desired pressure reduction (called staging), can be used to prevent dropping pressure within the valve below the vapor pressure of the fluid.

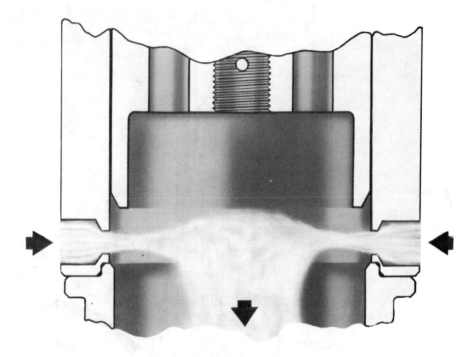

Figure 28-17. Collision of opposing liquid jet streams forms protective fluid cushion about the cavitating liquid.

The technique of staging has been designed into special valve trim for globe bodies. This trim (cage-style) utilizes multiple-flow passages that divide and then recombine the flow stream repeatedly to reduce pressure in a series of relatively small stages. The number of stages required to prevent cavitation depends upon the pressure reduction which must be taken across the cage. As with multiple valves in series, cavitation and its associated noise are eliminated by preventing the static pressure from dropping below the vapor pressure of the fluid.

Problem Definition

To optimize the noise control effort, it first is essential to determine existing or potential operating noise levels. Through research, a hydrodynamic noise prediction technique has been developed which involves the factors of valve style, size and type of trim; size and schedule of adjacent piping; inlet pressure, pressure drop, and liquid vapor pressure; plus valve capacity.

$$SPL = SPL_{\Delta P} + \Delta SPL_{C_v} + \Delta SPL_{\Delta P/P_1-P_v} + \Delta SPL_k$$

where:

SPL = Overall noise level in A-weighted decibels (dBA) at a predetermined point (48 in. downstream of the valve outlet and 29 in. from the pipe surface). This particular point was selected because of the physical dimensions of a "soft room" (a sound absorbing chamber) which was placed around test valves in the laboratory. The point of measurement inside this room was 48 in. downstream of the valve and 29 in. from the pipe surface. There are no control valve industry measurement standards in regard to distance from the noise source; however, since noise attenuates with distance, some reference point always should be selected and recorded.

$SPL_{\Delta P}$ = Base SPL in dBA, determined as a function of pressure drop (ΔP).

ΔSPL_{C_v} = Correction in dBA for required liquid sizing coefficient (C_v).

$\Delta SPL_{\Delta P/P_1 - P_v}$ = Correction in dBA for valve style and pressure drop ratio ($\Delta P/P_1 - P_v$). The pressure drop ratio must be calculated using inlet pressure P_1 in psia) minus the vapor pressure of the liquid (P_v in psia).

ΔSPL_k = Correction in dBA for acoustical treatment; *i.e.*, heavy-wall pipe, insulation, etc.

Use of this prediction technique provides a valid and ready determination of a valve's noise level. The noise level then becomes the basis for evaluating available noise treatment methods. The following example illustrates use of this technique.

Given: Valve style, size and type of trim—a 2-in., 300-lb ANSI rated Fisher Design ED with standard trim.
Adjacent piping—2 in., Schedule 40 pipe
Inlet pressure (P_1)—250 psi
Pressure drop (ΔP)—175 psi
Calculated required C_v—70
Flowing medium—Water at 200 F
Vapor pressure (P_v) of water at 200 F—11.5 psia

Solution: Calculate the pressure drop ratio ($\Delta P/P_1 - P_v$)

$$\frac{\Delta P}{P_1 - P_v} = \frac{175}{250 - 11.5} = 0.734$$

From Figure 28-18 $SPL_{\Delta p} = 59$ dBA
From Figure 28-19 $\Delta SPL_{C_v} = 37$ dBA
From Figure 28-20 $\Delta SPL_{\Delta P_1 - P_v} = -8$ dBA
From Table 28-2 $\Delta SPL_k = 0$ dBA

$$SPL = SPL_{\Delta P} + \Delta SPL_{C_v} + \Delta SPL_{\Delta P/P_1 - P_v} + \Delta SPL_k$$

$$= [59 + 37 + (-8) + 0]_\Delta dBA$$

$$= 88 \text{ dBA}$$

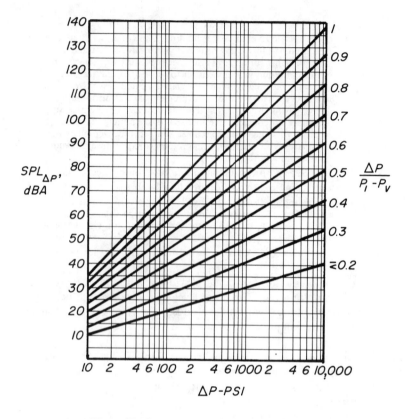

Figure 28-18. Base SPL$_{\Delta P}$ for all valve styles.

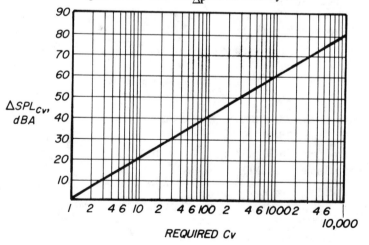

Figure 28-19. ΔSPL$_{C_v}$ correction for all valve styles.

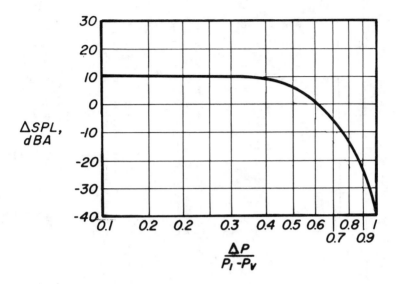

Figure 28-20. $\Delta SPL_{\Delta P/P_1 - P_v}$ correction for standard cage trim in a glove valve, flow down.

Table 28-2. ΔSPL_k Correction for Pipe Wall Attenuation

Nominal Pipe Size, In.	30	40	80	120	160	STD	XS	XXS
2	—	0	-6	—	-12	0	-6	-16
4	—	0	-7	-10	-13	0	-7	-16
6	—	0	-8	-12	-15	0	-8	-18
8	+3	0	-9	-14	-18	0	-9	-16
10	+3	0	-9	-14	-19	0	-6	—
12	+4	0	-10	-16	-20	0	-5	—

Should the predicted noise level exceed established noise standards, federal or other, a choice must be made whether to: (1) control cavitation and use path treatment methods, or (2) prevent cavitation. At first glance, the simple answer is to prevent cavitation, but in actuality, the decision is based on economic trade-offs. Table 28-3 provides a general comparison of noise control effectiveness and economics for the various techniques.

Table 28-3. Hydrodynamic Valve Noise Control Techniques

Technique	Effectiveness	Economic Aspects
Hardened trim parts, sacrificial piping, etc.	Provides no noise attenuation. Must rely on path treatment.	Frequent need for replacement valve parts, new piping, etc. proves very expensive.
Valves in series	Dependent upon service conditions, this technique can eliminate cavitation and noise without requiring additional treatment methods.	Large capital investment in valving plus the cost of designing and installing complex piping configurations may make this method impractical.
Acoustical insulation, heavy walled piping	Insulation can provide up to 10 dB attenuation per inch of thickness. See Table 28-2 for pipe wall attenuation.	Very economical in comparison to special valves, special piping configurations, etc., but cavitation damage will still result unless steps are taken to prevent it.
Cage-style trim for cavitational control	Highly effective in controlling cavitation up to 1440 psi pressure drop. Provides some noise attenuation.	Within its limitations, this technique generally proves to be more economical than valves in series or valve trim designed to prevent cavitation.
Valve trim to prevent cavitation by staging pressure drop.	Effective at most pressure drops. Limited in capacity and restricted to use on clean liquids.	High initial cost, but usually the most economical approach when pressure drops exceed 1440 psi and capacity requirements are not extreme.

AERODYNAMIC CONTROL OF VALVE NOISE

Control valves can prove to be major contributors to industrial noise and, without question, valves handling gas or vapor flow (compressible fluids) are prime offenders. Noise generation occurs in the pressure recovery region immediately downstream from the vena contracta of the valve. There, sudden deceleration due to expansion of the fluid stream creates extreme turbulence and mixing of the flow stream. Noise is the annoying result.

Noise Prediction

The noise potential of a valve must be known before it can be controlled. This points to the use of a proven noise prediction technique. Control valve noise is usually measured as Sound Pressure Level, where the unit of measurement is decibels. It is dependent upon: (1) valve style, size and type of trim; (2) size and schedule of adjacent piping; (3) inlet pressure and pressure drop; and (4) calculated C_g (gas sizing coefficient) or C_s (steam sizing coefficient). Determining the SPL involves a simple additive relationship which includes these variables. The following formula has proven accurate in actual valve installations.

$$SPL = SPL_{\Delta P} + \Delta SPL_{C_g} \text{ (or } C_s) + \Delta SPL_{\Delta P/P_1} + \Delta SPL_k$$

where:

SPL $\quad$ = $\quad$ Overall noise level in dB at a predetermined point (48 in. downstream of the valve outlet and 29 in. from the pipe surface).

$SPL_{\Delta P}$ = Base SPL in dB, determined as a function of pressure drop.

ΔSPL_{C_g} (or ΔSPL_{C_s}) = Correction in dB for required gas (or steam) sizing coefficient.

$\Delta SPL_{\Delta P/P_1}$ = Correction in dB for valve style and pressure drop ratio ($\Delta P/P_1$, P_1 in psia).

ΔSPL_k = Correction in dB for acoustical treatment; *i.e.,* heavy-walled pipe, insulation, in-line silencers, etc.

Methods of Control

Two methods are employed to minimize the operating noise level of a control valve. One is to reduce noise after it has been generated by the control valve; the other is to minimize noise generated within the valve. Generally, the more preferred method is to reduce noise at the source. Several valve manufacturers offer special quiet valves or valve trims.

Three approaches are used in the design of these noise abatement items. One approach employs tortuous flow paths that increase friction pressure losses as flow passes through the valve trim. This results in a gradual pressure reduction and avoids the noise generated by a sudden expansion of the flow.

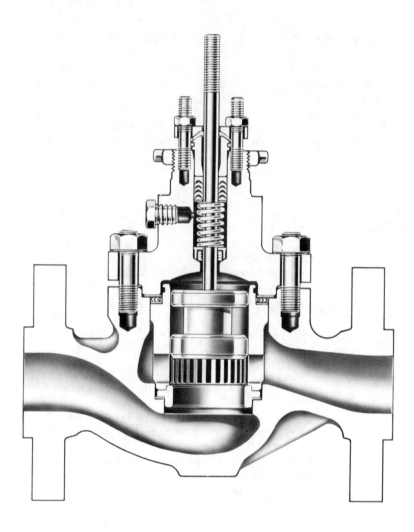

Figure 28-21. Globe style valve with noise reducing trim
that features specially sized and shaped orifices.
Total noise reduction gained with this technique can approach 20 dB.

A second approach utilizes multiple restrictions in series. Each restriction takes a portion of the total pressure drop across the valve. Due to the smaller fluid jet velocities and smaller recovery areas, each small pressure drop generates proportionately less acoustic energy. Quieter valve operation results.

The third approach utilizes specially sized and shaped orifices to minimize the turbulence level of the fluid and provide a favorable velocity distribution in the expansion area.

Figure 28-22. A diffuser installed on the outlet of the valve provides two-stage pressure reduction.

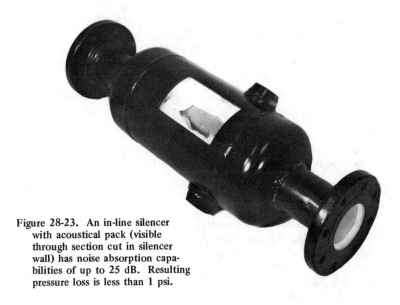

Figure 28-23. An in-line silencer with acoustical pack (visible through section cut in silencer wall) has noise absorption capabilities of up to 25 dB. Resulting pressure loss is less than 1 psi.

To broaden the noise attenuation capability of this last approach, the total pressure drop can be split between the control valve and a fixed restriction downstream of the valve. This two-stage pressure reduction (applicable to high $\Delta P/P_1$ ratios) is best achieved through use of a diffuser. The diffuser, carefully designed for the specific installations, can be very effective in minimizing noise.

Special quiet valve trims prove highly effective, yet there are some drawbacks to their use. One appears in high capacity applications. Quiet trim designs offer low capacity per unit of valve plug travel. For example, where a standard 4- or 6-in. valve might be used, the required valve size for quiet trim might approach 16 to 20 in. on very high pressure ratio applications. Also, pressure staging within some quiet trims occurs in small, individual orifices that are susceptible to clogging if the fluid stream is not relatively clean.

When source treatment does become impractical, economically or otherwise, the alternative is the path treatment method. Path treatment involves using one or more items, including: in-line and vent silencers, heavy-walled piping, or external acoustical insulation.

Of the different types of in-line silencers available, those utilizing internal "acoustical packing" prove to be most effective on the higher frequency ranges characteristic of aerodynamic control valve noise. An in-line silencer can provide up to 25 dB insertion loss when properly applied. Packed silencers, however, are subject to velocity limits and, if not properly designed, can lose the pack material under high pressure conditions.

Heavy-walled pipe and/or acoustical insulation effectively mask noise. The thicker pipe will reduce ambient noise levels only as far as it is used since noise is carried within the flow stream for long distances. Externally applied acoustical insulation will also mask noise. The disadvantage here is that insulation is highly susceptible to damage. Yet another approach is to bury the pipeline. Again, as with heavy-walled pipe and acoustical insulation, noise in the flow stream may reappear where treatment ends.

Prediction Examples

To illustrate use of this technique, the following examples determine first the noise level of an untreated valve, and then show how both source and path treatment can reduce this noise level.

Example

Given: 4-in., 300-lb ANSI rated cage-style globe valve with standard trim
               Adjacent piping = 8 in., Schedule 40
               Inlet pressure (P_1) = 615 psia
               Pressure drop (ΔP) = 348 psi
               Pressure drop ratio ($\Delta P/P_1$) = 0.56
               Calculated required C_g = 4000

Solution: From Figure 28-24, $SPL_{\Delta P} = 103$ dB
From Figure 28-25, $\Delta SPL_{C_g} = -2$ dB
From Figure 28-26, $\Delta SPL_{\Delta P/P_1} = 9.0$ dB
From Table 28-2 (p. 959), $\Delta SPL_k = 0$

Referring back to the formula, $SPL = 103 + (-2) + 9.0 + 0 = 110$ dB.

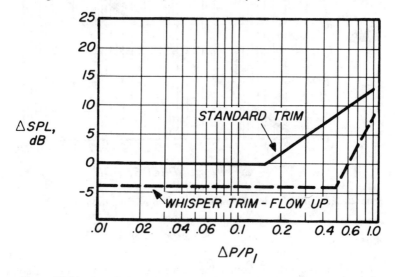

Figure 28-24. Base SPL in dB determined as a function of pressure drop, ΔP.

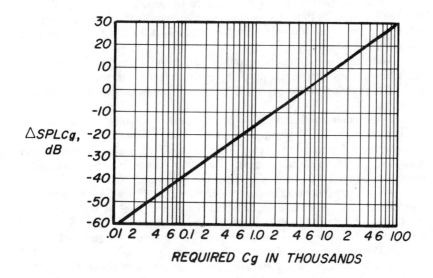

Figure 28-25. Correction in dB for required gas or steam sizing coefficient.

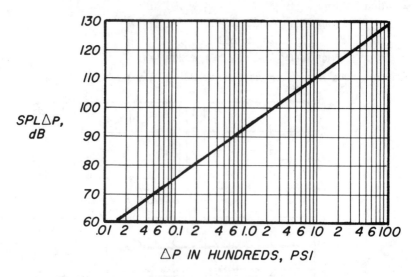

Figure 28-26. Correction in dB for valve style and pressure drop ratio $\Delta P/P_1$.

Example

To reduce the 110-dB Sound Pressure Level, the 4-in. valve was equipped with a special acoustical cage. In addition, the pipe schedule was changed to 80. Other operating conditions remained the same.

$SPL_{\Delta P}$ and ΔSPL_{C_g} are unchanged; however, with acoustical trim, $\Delta SPL_{\Delta P/P_1}$ = -2 dB, and with Schedule 80 pipe, ΔSPL_k = -9 dB. The resulting SPL = 90 dB.

Additional noise reduction could be anticipated through use of an in-line silencer. For example, a properly applied silencer can provide as much as 25 dB reduction in the overall SPL. Acoustical insulation could also add to the reduction.

Determining the best item or items of noise abatement equipment involves far more than "which provides the greatest noise attenuation." Consideration also must be given to ease of installation, average maintenance costs, and cost of replacement parts. Ease of accuracy of noise prediction also enter into overall engineering expense.

When analyzing noise control, remember that decibels are treated on a logarithmic base. For example, reducing the decibel level by 6 dB halves the sound pressure (or audible noise) at a given frequency. Therefore, to spend a relatively large amount for a small dB reduction becomes valid when the actual reduction in audible noise is considered.

The technology of control valve noise has advanced significantly within the past several years. Of particular note is the ability to predict and design out noise for a given application. The pollution engineer should recognize that noise technology and the products that have resulted can be used to advantage now. All that is required is to study the application thoroughly from both the technical and economical aspects, select the product or products most likely to solve a noise problem, and put that product into service with confidence.

SILENCERS FOR RECIPROCATING ENGINE EXHAUST[10]

Noise superimposed on gas flow in a duct can be reduced by silencers. Many types and models of silencers are manufactured. The type needed for a particular application depends on the required noise reduction as a function of frequency, and on the operating conditions.

Silencers are commonly classified as *dissipative* or *reactive*, depending upon the predominant process utilized in reducing the intensity of the noise. Dissipative silencers depend primarily upon porous linings to absorb sound energy as it travels through the flow passages. They are generally used in high frequency applications. Reactive silencers reduce noise primarily by reflecting sound energy back toward the source. These are most effective in low frequency applications. Reactive silencers are used on the exhaust of reciprocating engines because of the low-frequency character of the noise.

The exhaust noise spectrum consists of a number of discrete frequencies superimposed on a background of broad-band noise. These discrete frequencies occur at harmonics of the rotational speed of the crankshaft. The particular level of each is dependent upon the engine setting and the exhaust system.

Dissipative silencers are seldom used alone for exhaust silencing. They may be used in combination with reactive silencers for added high-frequency performance. They are not recommended for liquid fuel engines, however, since an accumulation of fuel in the lining material could present a hazard.

Exhaust silencers are commercially available in models and sizes to satisfy many installation and operational requirements. The units are generally of all-metal construction and consist of multiple chambers connected in series by internal tubes. The internal configuration is designed to provide non-tuned, relatively broad-band acoustical performance. The most common type is illustrated in Figure 28-27. It consists of two unequal chambers

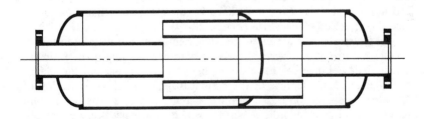

Figure 28-27. Standard two-chamber exhaust silencer.

with intruding inlet and outlet tubes and offset internal connecting tubes. In critical applications, three chambers may be used for increased performance. These silencers are applied primarily to the exhaust of naturally aspirated engines for which the relatively high pressure drop associated with the internal flow reversals is acceptable. They can also be used for turbocharged engines, but larger sizes must be used because of the lower exhaust pressure drop allowed for this type of engine.

For turbocharged engines, a low pressure drop modification is available with straight-through flow passages (Figure 28-28). However, the straight-through configuration sacrifices some acoustical performance, particularly in the high-frequency range. Within back-pressure limitations, silencers may be placed in series for added acoustical performance.

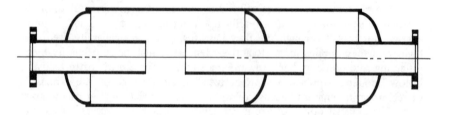

Figure 28-28. Low pressure drop silencer for turbocharged engines.

Selection of Silencers

Silencer selection is usually based on exhaust connection size and on back-pressure limitations established by the engine manufacturer. For most engines, the proper size has already been established and selection is a matter of routine. For each silencer size, manufacturers usually offer three or four models with differing grades of acoustical performance. The acoustical performance of a particular model is dependent upon the operating conditions of temperatures and exhaust gas velocity. Manufacturers publish "Typical Attenuation" curves which illustrate general performance characteristics of their models. These are smoothed curves drawn from test data and are based on conditions typical of the intended service. The most reliable curves include the effect of silencer size (Figure 28-29).

Exhaust silencers are insensitive to orientation and may be installed in either a horizontal or vertical position without affecting their silencing characteristics. Long lengths of piping between the engine and the silencer should be avoided. Because of the nonlinear nature of the propagation of high-intensity sound waves, a shock wave could develop in the piping and produce a loud cracking noise. When this occurs it is necessary either to revise the exhaust piping or to install a small supplementary silencer near the engine.

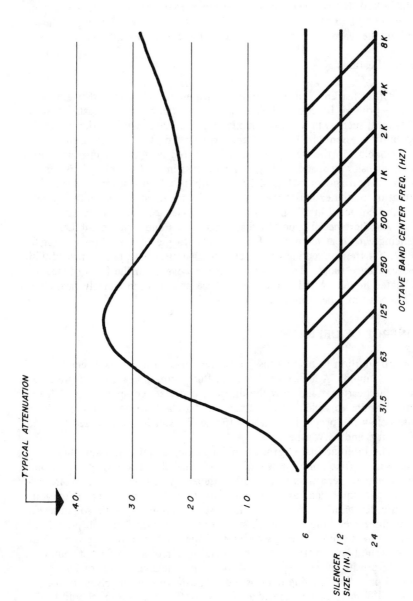

Figure 28-29. Attenuation vs. silencer size and sound frequency.

Tailpipes should also be selected carefully, with consideration given to their effect on emitted noise. At certain frequencies, the interaction of the tailpipe with the silencer can result in more noise than without the tailpipe. A rule of thumb often used in selecting a tailpipe is to avoid lengths that are half the wavelength corresponding to the engine firing frequency.

Special Features

Other features can be combined with engine silencing either as a secondary or primary function. Where sparks from the engine exhaust would be hazardous, such as in a marine application, a silencer with spark-arresting capability is essential. Both wet and dry types of spark-arresting silencers are available. In the wet type, water is injected into the exhaust gas stream to extinguish any sparks and cool the exhaust line. The dry type is similar to the silencer shown in Figure 28-27, with an added feature that imparts a rotation to the exhaust gases. Any sparks are removed by centrifugal action and collected in a built-in trap which is cleaned out at intervals.

Another special type of exhaust silencer is one which combines silencing with heat recovery. This is similar to the silencer shown in Figure 28-27, but the connecting tubes between chambers have been replaced with heat exchange tubes. The exhaust heat is recovered in the form of steam or heated liquid. Whether the use of these units is economically favorable will depend on operating and fuel costs.

CONTROL OF OFFICE NOISE[144]

In addition to hearing loss, noise can have an insidious effect on man's body and mind. A high noise level can flush skin, increase blood pressure, speed the heartbeat and slow the digestive processes compounding all the other tensions to which we are subjected in today's world. Failure to keep noise below the proper level in an office will result in noise-caused fatigue, irritability, and nervousness.

The International Organization for Standardization has guidelines establishing office noise levels of 20 to 70 decibels. The ratings vary from department to department in an office depending on the type of work performed and equipment used. If you want to make a quick approximation of sound levels in an office, take the first measurement with a sound level meter three feet from each work station in the office. This will provide a level of direct sound. Next, walk around the office and note spot readings on the sound level meter that give a blend of direct and reflected sound.

In private offices people work best with the sound level of 28 to 38 dBA. Controlling outside noise is the greatest problem in private offices, so it will be necessary to create a sonic seal. The office door should be solid and tightly fitted in its frame. If it is not, the door frame should be fitted with a gasket of foam weather stripping encased in vinyl. Next, install an automatic door bottom at the threshold. When the door is closed, the seal rises and pushes against the door.

Carpeting and padding will substantially reduce the noise from the floor below and reflected sound within the room. The sound absorbing ability of carpeting depends upon its thickness, density and backing. Tests show that carpeting has a noise reduction coefficient of 0.35 to 0.60. A suspended ceiling of acoustic tile will separate the sound transmitter from the floor above and the sound receiver in the ceiling. Windows should be sealed and weatherstripped. Draperies can be installed to absorb reflected noise.

If the noise is severe, or if major remodeling is planned, consider noiseproofing walls between the offices. Gypsum wall construction is a good sound insulator. However, noise can still be transmitted from one room to the next as long as walls are coupled back to back by studding. To cure this problem, remove the old gypsum board and install additional new studs to support only the new gypsum board. In this manner, the old studding will support the wall in one office, while new studding supports the wall in the other office.   Plaster-over-lathe wall construction is one of the most objectionable carriers. Back-to-back, these walls have a tendency to ring. Again, to correct this problem it is necessary to separate them. Separation of walls involves building a new wall ½ in. from the old. Insulation should be installed between the new studs to further reduce sound transmission.

Some office walls extend only as high as the suspended ceiling, allowing sound to travel over the top to the next office. To cure this, build a vertical baffle above the partition extending from the partitioned ceiling to the constructed ceiling.

Holes or cracks in walls bigger than 1 sq in. will destroy the privacy of an office provided by a 40 dBA partition of 100 sq ft. Electrical outlets also transmit noise when installed back-to-back from one wall to another. The outlets form a tunnel for sound.

Electrical boxes should be at least 2 ft from the electrical box on the other side of the wall.

Blower noise from an air conditioning system can be reduced by installing a fiber glass duct or a fiber glass runner from metal ducts to the office outlets. Sound attenuators can also be installed to reduce cross talk from other offices.

The sound level in secretarial and general office areas should not exceed 48 to 58 dBA, and in typing pools 58 to 68 dBA. Again, doors, ceilings, floors and walls should be examined to be sure that all possible sound is kept out of the office. Next, place acoustical pads under typewriters, and foot pads plus rubber mountings under large pieces of equipment. Phones can be silenced and signal lights installed. High noise level office machines can be quieted with movable wall partitions. Fillers of honeycombed paper, gypsum board, fiber glass, cork or foam rubber sandwiched in the partitions keep sounds from passing through.

Partitions of the type used in landscaping offices are effective acoustical barriers and should hold sound levels in the 48 to 58 dBA range.

Their surfaces are sound-absorbing, not sound-reflecting, so reflected sound dies quickly. To mask speech and office machine noise in landscaped offices it is possible to electronically generate noise or music tailored to the installation. The new noise level is high enough to provide privacy yet low enough not to disturb an individual's concentration.

Computer rooms require extra consideration. The recommended noise level should not exceed 68 to 70 dBA. High-speed printers and card punchers, however, typically produce 80 dBA. Air conditioning and custom door systems needed for cooling the equipment all compound the noise problem. Contain the sound within the room as much as possible. Create a corridor around the outside of the computer room so that a buffer separates peripheral office and equipment spaces. Within the computer room try to locate printing machines, card readers, punches, tape and typewriters and cooling blowers away from the operators.

Openings in the machine cabinets should be directed away from the operator. Noisy equipment should not be placed in corners where it will radiate noise. Computer air conditioning equipment should be located in an adjoining room separated from the computer room by full-height sound-isolating walls.

Sound levels in rooms with equipment such as duplicating machines, keypunch, and accounting machines, and mail handling rooms should range from 68 to 78 dBA. Reflected noise is the villain in machine rooms, and the best way to control it is to put a wall between the machines and the next work station. Wall coverings can be acoustic tile, porous paneling, carpeting, carpet tile, fabric or other commercially available noise-absorbent panels or blocks.

Acoustic hoods for noisy machines are usually constructed of a heavy metal or plastic with a damping material applied to the inner surface, covered with an application of acoustic foam or absorbent material. Noise generated within the enclosure is absorbed or dissipated from reaching the operator.

PROPER CONSTRUCTION
CONTROLS OFFICE NOISE[65]

Careful attention should be given to the layout and design of office areas to create an environment that will attract and keep employees. Sound control for ceiling-high offices must be planned from the beginning. It is very expensive and time-consuming to soundproof an office after it is built.

In the R & D industry, whose needs are constantly changing and whose employees are largely white collar, the office environment must be flexible as well as functional. Facilities should have flexibility to handle diverse projects without future large expenditures for rearrangements and alterations.

Varying degrees of sound control can be considered for office areas: (a) in flexible areas a movable demountable vinyl-covered partition system can be used with a sound transmission class of 42 (using 2-in. sound attenuation blankets between studs increases the STC to 49); (b) in permanent areas a metal stud drywall system should be used with a 3-in. sound attenuation blanket between studs and paneling facing with an STC of 54.

The STC rates a structural assembly's ability to reduce airborne noise. Most authorities agree that dividing walls should have an STC rating of at least 45, while 50 is considered premium construction. Below 40, privacy and comfort may be impaired as loud speech can be audible as a murmur.

Laboratory test data are used to compare, specify and select materials and construction. The STC rating has become the most widely accepted means of indicating a partition's sound resistance—the result of comparing the sound transmission loss of a tested assembly with a "standard contour" of known sound loss performance. Tests conducted using 11 frequencies (ASTM E90-6IT) or 16 frequencies (E90-66T) are matched to standard contours known as STC curves. This procedure brings tests made with either designation into a comparative ratio. Where STC ratings are not available, dB (decibel) transmission loss averages are used resulting from sound tests conducted at 9 or 11 frequencies within a range of 125 to 4000 Hz.

One of the most common mistakes made in drywall construction is erection of starter panel sections. A 2-ft-wide starter panel should be cut from a 4-ft-wide wallboard and installed on one side of the office partition. Starting both sides of the partitions with the standard 4-ft width destroys the STC rating of the partitioning system.

Partitions, floors and ceilings of a building provide resistance to transmission of airborne and structure-borne noise. Their efficiency as sound barriers is dependent upon several factors:

> **Mass**—the heavier the construction, the greater its resistance to sound transmission.
>
> **Isolation**—separation of opposite surfaces of a construction will improve sound isolation. Surfaces may be separately supported with no structural connections between them or may be resiliently mounted to common supports.
>
> **Damping**—addition of sound attenuation wool within the construction effectively dissipates sound energy by converting it to heat.
>
> **Leaks**—cracks, penetrations or any openings, however small, readily conduct airborne sound.
>
> **Flanking paths**—sound is transmitted along the path of least resistance which is usually through the structure and around sound barrier walls and floor-ceilings.

Effective Sound Control

Wall and ceiling construction should provide approximately the same degree of sound control through each assembly. Where a drywall partition is used for sound isolation, the construction should extend from slab to slab. Sound can travel through a suspended acoustical ceiling, up and over a partition attached to a suspended ceiling and into other areas.

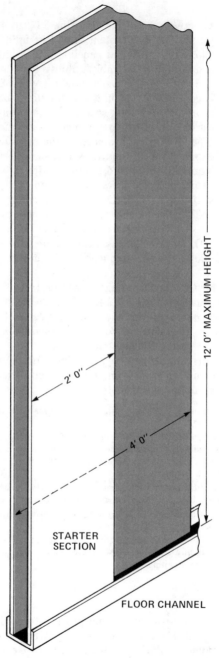

Figure 28-30. For effective noise control, the starter wall panel and the first full panel should be staggered.

Optimum sound isolation requires that the integrity of drywall partitions and ceilings (mass) never be violated by cutting holes for vents or grilles or by recessing cabinets, light fixtures, etc. Instead they should be surface mounted. Door and borrowed light openings are not recommended in party walls. The septum layer in the triple solid drywall partition should always remain whole.

Where holes are necessary, avoid placing them back to back and immediately next to each other. Electrical boxes should be staggered, preferably at least one stud space. A non-hardening, non-skinning resilient caulking material should be used to seal all cutouts, such as around electrical and telephone outlets, plumbing escutcheons, and wall cabinets. The backs and sides of electrical boxes should also be caulked to prevent sound leaks.

Caulking should be used to seal all intersections with the adjoining structure, such as under-floor and ceiling runner tracks, around the perimeter where the assembly meets the floor, ceiling, partition or exterior wall and at vertical intersections occurring at columns and window mullions.

Avoid construction such as ducts, rigid conduits or corridors, which act as speaking tubes to transmit sound from one area to another. Common supply and return ducts should have sound attenuation liners. Conduits should be sealed. Doors leading to a common hall should be gasketed around the perimeter and should not contain return air grilles.

To isolate structure-borne vibrations and sound, resilient ceiling systems and floor coverings are recommended. Vibrating or noisy equipment should have resilient mountings to minimize sound transfer to structural materials. Ducts, pipes and conduits should be broken with resilient, non-rigid boots or flexible couplings where they leave vibrating equipment; and they should be isolated from the structure with resilient gasketing and caulking where they pass through walls, floors or other building surfaces.

COMPUTERS–A WHITE COLLAR HAZARD[53]

Noise levels generated by modern computers are high enough in some circumstances to induce hearing loss. Attention was recently focused on noise problems in computer laboratories because of the relatively simple and recurrent errors made by programmers possibly as a result of distraction induced by the very high noise levels under which they were working. It turned out, however, that the first order of business proved to be the protection of hearing itself.

When anyone calls and reports a noise level problem, a rough rule used for separating the hazardous from the merely annoying is the question: "Do you need to raise your voice?" When we recently investigated one computer laboratory, we could tell even without measuring instruments that the problem there was severe. It was found that a conversation couldn't be carried on over a distance of much more than six feet with the laboratory in its "quiescent" condition (no printing, reading or punching). The filtering action of the ear is good for penetrating about 20 dB into the noise: it is possible to interpret speech when its power is only about one percent of the ambient noise field.

Since normal speech is at a level of approximately 60 dB above the threshold of hearing, the inability to converse implied a noise level of about 80 dB arising from the "quiet" operation of the computer electronics. In fact, "flat" scale readings on the sound level meter proved to be about 79 dB and the A scale reading was 75 dB. This noise seemed to be contributed largely by the blower fans cooling the equipment racks. These noises are essentially steady-state, and can be monitored effectively with a sound-level meter.

However, when the computer and its accessories are in operation, a sound-level meter is not capable of displaying the whole problem. One of the dominant characteristics of the operation of the computer is the impulsiveness of its operation, from jerking cards into the reader, printing out results and generating new punched cards, to further reading, etc. Each operation is so brief that the indicator of the sound-level meter, even on its fastest speed, has scarcely left the stop before the sound is over. To a certain extent, then, the sound-level meter tends to average out the peaks.

But what about the ears of the people working in the room? Other research on the ear indicates that the perception of sounds corresponds to an integration time on the order of 60 milliseconds or less, whereas, because of the ballistics of its meter movement, the sound-level meter integrates over about one second. Thus, the sound-level meter will smooth out features that may be significant in their relation to the ear.

Measuring the Problem

In addition to making direct readings with the sound-level meter at its fastest speed for both octave-band analyses and overall sound-pressure levels, sounds were recorded on magnetic tape for further study in the laboratory. A microphone with good high-frequency response and small size (½-in. dia.) was used in order to mitigate diffraction effects.

For sounds as impulsive as those measured, it is important to monitor the recording process with an oscilloscope because of the importance of retaining the information about the dynamic structure of the signals. A tape recorder was used that permitted monitoring the signal going on to the recording head and the output of the playback head, and compared the two signals by means of a dual-track oscilloscope. Thus, it was possible to ascertain during the recording process that a minimum of peak clipping was taking place, and that the recording stored on the tape was a reasonably accurate representation of the signals coming from the microphone. Ordinary steady-state instrumentation is not adequate for this work. A significant degree of peak clipping does not influence the readings of ordinary equipment for monitoring recording levels.

Using a sound spectrograph of very low inertia, it was possible to study the distribution of the computer noises in frequency and time. These provided considerably more information about the computer noise than did the more conventional octave-band analysis.

Parts A and B of Figure 28-31 show the results of the octave-band analysis of the noise in the computer laboratory under quiescent and operating conditions. These show a generally uniform distribution across the audio

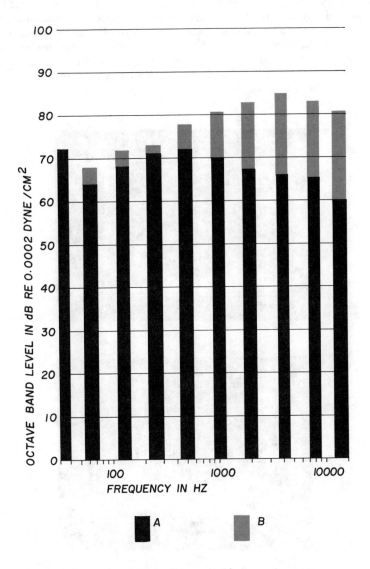

Figure 28-31. Octave band-pressure levels measured in computer room.
(A) "Quiescent"—equipment turned on, but no operations.
(B) Equipment running.

spectrum; the level is higher in the bands centered at 1000 Hz and above when the computer is operating. Nothing in the band analysis could lead one to suspect that the operating noise of the computer had dominant sinusoidal components.

Inspection of the recorded signal with an ordinary oscilloscope showed that peaks in sound level having a duration of about 0.1 second exceeded the general level by about 6 dB, but sinusoidal components were not obvious. However, when the scanning analyzer was used to observe the spectral distribution of the sounds, it could be seen that the printer and reader had essentially sinusoidal peaks of sound output in the region between 1 kHz and 6 kHz (Figure 28-32). These spectral displays are approximately 1-sec samples of the respective sounds.

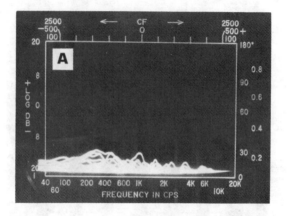

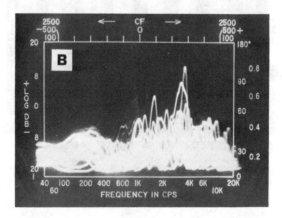

Figure 28-32

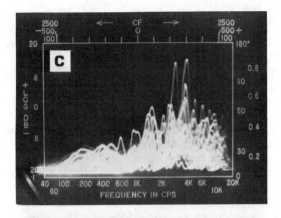

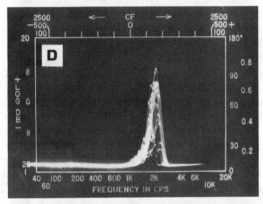

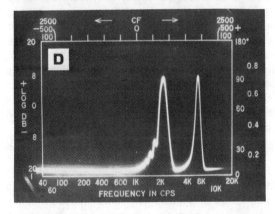

Figure 28-32. Special distribution of computer noise measured at 20 sweeps per second on sweeping wave analyzer. Ordinates are logarithmic. (A) "Quiescent." (B) Printer and reader. (C) Printer, reader and punch. (D) Top: Traces on instantaneous analyzer for two sinusoidal components at 2 kHz and 6 kHz. Bottom: Trace on instantaneous analyzer for 1/3 octave band of white noise centered at 2 kHz. Note top of display peak is broader than for sinusoidal.

Because of the high speed at which the scanning analyzer is swept, a pure sinusoidal signal appears not as a single line in the spectrum, but as a narrow pie-shaped wedge. In Figure 28-32D the top half of the picture shows the characteristic wedge shapes produced by the analyzer for sinusoidal signals of 2 and 6 kHz. The bottom half of the picture shows the pattern for a band filtered "white noise" centered at 2 kHz. Vertical deflections on the x-y patterns are in decibels. In parts A, B, and C, note the shape of the analyzer pattern during the intensity peaks. Their sharpness is greater than that for white noise and indicates the presence of a significant degree of sinusoidal character in the noise. In fact, the spacing of dominant peaks at harmonically related frequencies leads one to suspect that the output has the form of sinusoidal bursts.

All by itself, the punch generated a vast peak of sound centered about 5 kHz. When the punch was operated along with the printer and the reader, it dominated the sounds put out by the rest of the computer. From the sound-level meter indications alone (which were markedly smoothed by its filtering and metering functions) it was evident that even on an average basis the punch radiated just about as much acoustic power as the rest of the equipment combined. Judging by the peaks observed on the analyzer (Figure 28-32C), its instantaneous power at a single frequency was perhaps four times as much as that put out by all of the rest of the equipment.

With the equipment working, the long-term rms noise levels ranged from 89 to 94 dB, depending upon the particular operation taking place. Corresponding A-level weightings were about 4 dB lower. The computer laboratory was quite reverberant, and although these figures were obtained for a microphone location centered between the printers and the punch, the overall level at the counter where the "customers" waited for their printouts was only about 5 dB less. Moreover, the peaks add at least 6 dB to the observed long-term level insofar as the ear is concerned.

The condition here is not simply one of distraction. The noise level is so high that persons in the room are in danger of losing their hearing. It has been proposed upon the basis of medical histories of noise-induced hearing loss that a significant risk of damage to hearing exists if a half-day exposure to levels in excess of 90 dB occurs with any regularity. An exposure of a full 8 hours of a working day should not exceed a level of 85 dB. Figure 28-33 shows a damage-risk criterion that has been proposed. Further, it has been proposed that the criterion level be lowered by another 5 dB if the sound sources are narrowly sinusoidal, as the printer and punch prove to be.

At the time of our measurements, the situation had not yet produced any obvious loss of hearing among the computer operators. Fortunately, computing is a new field and biases the working population toward the young and strong. Even among the age group, however, Jansen (1969) has found that certain changes in the autonomous nervous system correlate instantaneously with the onset of "white" noise signals having a level of 70 dB (flat) and above: the primary physiological effects resemble the responses

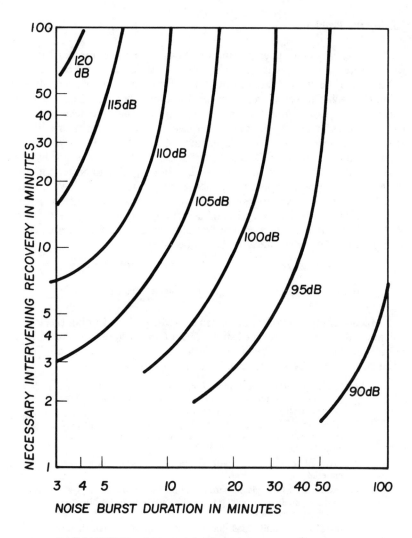

PARAMETER: SPL 600 TO 1200 HZ OCTAVE BAND

Figure 28-33. Recovery times and maximum exposure allowed to remain below proposed damage-risk limits.

to stress or anxiety. The pupils of the eyes dilate, the pulse becomes shallow and rapid, capillary blood vessels contract, and the blood pressure rises. When a person is sleeping, these responses are induced by a still lower sound level, at 55 dB (flat weighting) and above.

Solving the Problem

The ceiling and upper portion of the walls of the computer laboratory were covered with acoustic tile. However, the superficial impression upon walking into the treated laboratory was that nothing had changed. It was still not possible to carry on a conversation even when the machines were "quiet." The basic sound level meter readings for the "quiescent" condition had changed scarcely at all. This is because acoustic treatment has little effect at frequencies below 2 kHz; above that frequency it can absorb enough sound to lower the sound levels in the upper octave bands by about 5 dB. In this instance, because of the distribution of frequencies for the operation of the printers and the punch, it may just possibly have had the effect of reducing an immediate hazard to a potential long-range menace.

The only effective way to eliminate the hazard is to reduce the noise at the source. If noise in a computer facility interferes with normal conversation, the following precautions can be taken:

(a) Test employee hearing at frequent intervals, and wear ear protection.
(b) Do not stay in the vicinity of the noisy equipment unless required.

In the computer room, noise can be reduced by adopting the following recommendations in the order listed. This order is chosen on the basis that noise from the major source must be attenuated before reduction of the noise from subsidiary sources will have any significant effect.

1. Cover all ports or openings around card punches, readers and printers with ¼-in. clear plastic shields edged with soft rubber gaskets to effect an airtight seal.
2. Wherever possible, vibration-mount all drive motors and cooling fans on rubber pads, or use rubber grommet-type sleeves on mounting bolts.
3. Treat all console panels with viscoelastic-type vibration damping material.
4. Install ¾-in. fiber glass board liners on the inside surfaces of console paneling to reduce the noise build-up in the hollow reverberant cavities.
5. Install acoustically lined ducts at intake and discharge sides of cooling fans.
6. Install resilient pads or vibration isolators under all equipment to reduce low frequency vibratory transmission to the supporting false floor.
7. Install acoustically-lined U-shaped partial enclosures around individual card punches, readers and printers.
8. If a choice is available, give preference to many-bladed slow speed fans and wide ducts in providing for the cooling of equipment.

The above measures should be attempted a step at a time in the order listed until the noise level in the computer room is reduced to acceptable limits. However, the most effective way of quieting computers and accessory equipment is to incorporate good noise control techniques early in the design stage.

If the above measures cannot be adopted, locate all computers and associated equipment in a separate room away from office personnel. Operating personnel exposed to the near noise field of the equipment should wear ear protectors.

CHAPTER 29

TRAFFIC NOISE[47]

AUTOMOBILE AND TRAFFIC NOISE

Highway vehicles include automobiles, trucks, buses and maintenance vehicles. Traffic studies of highway vehicles usage in typical urban areas show that about 1600 to 2300 trips are made by automobile drivers and passengers every day for every 1000 people. Approximately 40 to 45 percent of the latter terminate in residential areas. This urban travel represents about 52 percent of the estimated 3 billion highway vehicle miles traveled in 1970.

Automobiles are the primary mode of transportation in the United States and constitute the largest number of highway vehicles. From 1950 to 1970, the number of automobiles in use has increased from 36 million to 87 million. Passenger cars traveled 1000 billion miles in 1970. Automobile sales, including vehicles, equipment and service, reached $92 billion in 1970. Approximately 5 billion people were employed by the automotive industry.

The total number of trucks in use increased from 8.2 million in 1950 to almost 19 million in 1970. Total truck miles increased to 206.7 billion in 1969 from 90.5 billion in 1950. The average annual mileage for trucks is over 11,000 miles. A majority of the total truck operating hours (194 billion) was in population centers, 86 percent of the time in pickup and delivery service and the remainder in long haul service. Thirty-nine percent of all truck miles were on urban streets.

Highway and city buses account for about 27 billion passenger miles in 1970. Mileage has been on a slight decline for a number of years, and bus passengers now constitute 4.2 percent of the commercial total. Around 74 percent of the total of 4000 buses are school buses and account for about one-half of the total mileage. The combination of local and intercity bus lines carried 5.8 billion passengers in 1970, for a passenger revenue of $2 billion, and employed 150,000 people. Due to the increase of all vehicles, an increase in noise pollution is inevitable.

While the federal government will play a very important role in the achievement of a suitable acoustical environment, many noise problems are of such a nature that they must be attacked on a local level. Accordingly, local authorities and interests should be apprised of the nature of particular noise problems, the various means of solving these problems, and the consequences of alternative solutions.

SOURCES OF HIGHWAY NOISE

Trucks

Trucks contribute to our noise problems by their exhausts, cooling fans, engines and tires. The noise levels generated by truck exhaust systems are dependent on factors such as engine type, timing and valve duration, induction system, muffler type, muffler size and location in the exhaust system, pipe diameter, and engine back-pressure. The actual noise-generating mechanism is created by vibrating columns of gas at high pressure and amplitudes which are produced by the opening of the exhaust valve. This noise is communicated directly to the atmosphere. Additional exhaust noise is created by the direct impingement of these released gases on the pipes and muffler shell.

In nearly all applications involving water-cooled engines, an axial flow-type fan is used to draw cooling air through a forward mounted radiator. In many designs, fan noise approaches the level of exhaust system noise and is generally considered an important factor in reducing overall vehicle levels. Generally, fan noise is directly related to fan speed. It has been shown that fan noise increases at a rate of 2 dB per 100 rpm at speeds between 1500 and 2000 rpm. The noise output is also dependent upon tip speed and configuration, blade design and spacing, and proximity of accessories and other objects which affect air flow.

Engine-associated noise in internal combustion engines is produced by the compression and subsequent combustion process which gives rise to severe gas forces on the pistons and to forces of mechanical origin, such as those produced by piston-crank operation, the valve-gear mechanism, and various auxiliaries and their drives. Diesel engines are about 10 dB noisier than gasoline engines. This difference results mainly from their different mechanism of ignition. Gasoline engines initiate combustion with a spark from which the flame front gradually spreads throughout the combustion chamber until the entire fuel/air charge is burned. This yields a smooth blending with the compression. The diesel engine, however, relies on a much higher compression ratio to produce spontaneous combustion which burns a large volume of fuel/air mixture rapidly. This yields a much more severe pressure rise in the cylinder, causing more engine vibration for the diesel engine in comparison with the gasoline engine. At constant speed, diesel engines show only slight reduction in noise, with reduction in load due to high compression pressure even under no load. Gasoline engines, however, show a substantial decrease in noise output with decreasing load. Therefore, the change in noise level between no-load and full-load conditions is rarely more than 3 dB for a diesel engine, but can be as high as 10 dB for gasoline engines.

Truck tire noise presents the major obstacle in limiting overall vehicle noise at speeds over 50 mph. At this speed, tire noise often becomes the dominant noise-producing source on heavy-duty trucks. Typical noise levels from truck tires at 50 mph range from 75 dBA for low-noise trend designs to over 90 dBA for high-noise level tires. The major offender is the cross-bar design used by the vast majority of trucks on their drive wheels. These tires may produce levels in the 80 to 85 dBA range when new, but their noise increases with wear as much as 10 dBA in the half worn condition. This increase is attributable to a change in the tread curvature resulting from wear. Cross-bar retreads pose an even greater problem as their noise level can be as much as 95 dBA at 50 ft when operated at 55 mph in the half-worn condition. Despite their noise, cross-bar retreads are very popular for economical reasons and each tire is recapped an average of two to three times.

Axle loading is also a major factor in the amount of noise generated by tires. Retread tires exhibit the most predominant dependence upon load. One example indicated a decrease of 15 dB resulting when load per tire was reduced from 4500 to 1240 lb. With the tire unloaded, the sides of the retread do not contact the road surface; hence the cups in the tread cannot seal against the road surface and compress small pockets of air. The rib-type tire designs are generally independent of loading due to their uniform tread design across the entire cross section. Variations in road surface also significantly affect tire noise generation.

Another method which has been proposed is that of close fitting noise shields for engines. Noise shields can be placed around specific engine subsystems which contribute significantly to overall engine noises. The shields are rubber mounted to isolate them from the adjoining engine surface. Under test conditions, up to 12 dB of reduction have been obtained over much of the spectrum with the exception of 250 Hz, where no change was found.

In summary, a quieter truck can be achieved by close cooperation between the vehicle and engine designer through:

1. Design of a vehicle giving adequate attenuation of engine noise,
2. Appropriate choice of engine design parameters,
3. Design of quieter engine structure, and
4. Design of quieter tires.

Automobiles

While not as noisy as trucks, the total contribution to the noise environment is significant due to the very large number in operation. Approximately 70 percent of automobiles on the road in 1970 were over 3 years old, the average car being about 5½ years old. Vehicles over 2 years old tend to produce higher noise levels (2 to 3 dB) under most operating conditions. This is due to deterioration of exhaust silencer performance and the response of the vehicle to pavement roughness.

For most automobiles, exhaust noise constitutes the predominant noise source for normal operation at speeds below about 35 to 45 mph, depending on the condition and design of the exhaust system. Above this speed range, in many cases, tire noise becomes equally significant. While exhaust noise does not create a significant interior problem, certain objectionable tones may be audible inside the car. To relate back to the actual exhaust, these gases are produced by the explosions that make automobile internal combustion engines. Mufflers are designed so that the energy of these gases is spent safely and quietly in a set of baffles.

In some cases, the intensity of fan noise is almost equal to that of exhaust. The parameters which govern fan noise generation are essentially the same as those related to trucks. There may not only be rotational imbalances, but the fan is a propeller that sucks air in from the front to flow through the radiator. More work has been done in the field of automobile fans because the objective is to reduce the passenger compartment noise.

Tire noise in passenger cars presents much less of a problem than in trucks. The principal reason for this is that standard automobile tires do not employ the cross-bar tread design. Automobile levels at 50 to 60 mph can be as much as 25 dB less than the worst truck tires.

Much of an automobile's noises come from its engine where there are imbalances due to acceleration of the reciprocating parts of the engine. There are also imbalances in the various rotating parts such as the crankshaft, flywheel, cooling fan, electrical generator, and alternator. The generator rotates at exactly twice the engine speed, producing a very audible and disturbing beat. There are also various gas noises caused by the explosive nature of the cylinder gas pressures.

Buses

Although trucks and buses share many basic design characteristics and some common components, buses are generally quieter due to their increased packaging space. This allows larger mufflers and an enclosed engine compartment. At highway speeds, passenger buses exhibit noise levels primarily in the range of 80 to 87 dBA at 50 ft. This noise is principally due to tire noise. The fact that buses that have been in service for long periods of time become noisier is due to damage to engine compartment seals.

CONTROL OF HIGHWAY NOISES

Noise reduction at the source may be achieved through legislation and enforcement. A combination of two kinds of regulation, usually requiring staged reductions in permissible sound levels in future years, is used. Sound levels emitted by new vehicles are regulated at the point of manufacturer or sale under a standard testing procedure. Although noise control at the source is the most basic approach, it is likely to be a long-term solution. Noise control at the receiver is most effective in reducing in-house noise.

Noise control of the transmission path is the third method of control. One means of controlling the transmission path is to increase the distance between noise source and receiver. This may be accomplished by re-routing of major routes and diverting traffic, particularly truck traffic at night, away from roads passing near or through residential areas and by preserving open space between major routes and adjacent residences. Zoning land adjacent to major highways to less noise-sensitive uses such as industrial or commercial use is proper abatement.

Experimentation with noise barriers has been going on for an extensive length of time. It has been found that different materials are more effective than others in deflecting noise. Some of the materials used are: pre-cast concrete, earth berms, aluminum, wood and porous concrete. Experiments show that noise barriers are fairly effective. These barriers can usually be incorporated in the existing right-of-way. Through application of architectural or landscaping techniques, these barriers can be made pleasing to the highway users. Barrier heights which are greater than the highest sound source must be a prominent design feature.

Foliage barriers are the least effective type although they will result in a 5-dBA sound level reduction for every 100 ft depth of foliage—provided the trees or shrubs are at least 15 ft tall and sufficiently dense throughout the year. Due to the fact that little or no additional right-of-way is generally required, man-made barriers are the most promising means that the highway engineer has at his disposal to abate traffic sounds.

AIRPORT NOISE MONITORING— A CASE STUDY[94]

For many years the city of Inglewood, Calif. proudly called itself the "Harbor of the Air" because of its proximity to Los Angeles International Airport (LAX). The advent of jet planes in 1959 replaced the friendly whirr of propellers with the maddening shriek of jet engines, and Inglewood became immersed in a sea of noise pollution.

Early reaction was angry and fierce. Many wanted to close the airport. Others wanted to sell their homes and move. When the anger subsided, the community took a long hard look at the noise problem and a few basic facts soon became obvious. The airport could not be ordered out, nor was it desirable to do so. The airport, in spite of the noise, was a major economic benefit for Inglewood, a city of 90,000 residents lying two miles east of LAX and directly under the landing patterns of all four runways.

But something needed to be done about the noise and the city took on the responsibility of protecting its citizens from excessive noise as much as possible. The city government established a Noise Abatement Division and charged it with carrying out a 10-point program aimed at alleviating the noise problem. This program is comprehensive in scope, covering both action by the city in such areas as land use, residential soundproofing, and local noise regulation, and actions available to the airline industry such as engine modifications and steeper approach angles.

From the start it was obvious that a noise monitoring program must be included. The monitoring system had to be capable of (1) measuring and documenting in complete detail any particular noise event, be it an aircraft flyover or a loud industrial machine, and (2) following the long-term noise pollution trends in Inglewood. In addition, it was important that the equipment be relatively inexpensive and simple to operate.

The above requirements resulted in a decision to procure two monitoring systems which would complement each other. One is a fixed network of four remote microphones and a central recording station. This measures aircraft noise on a 24-hour-a-day, year-long basis. The second system is a mobile acoustics laboratory that can go anywhere to record and document all aspects of individual noise events. It is designed to be operated by one person although for certain operations it can be set to operate without an attendant. Total investment in both monitoring systems is approximately $50,000.

Fixed Network

The remote microphones for the fixed monitoring network are bolted to telephone crossarms approximately 25 ft above the ground (Figure 29-1). Each microphone is a rain and wind protected type with an integral calibrating system. The noise signal output of the microphone/amplifier feeds equilized telephone lease lines terminated in the Noise Abatement (now Environmental Standards) Office at City Hall.

After a signal boost to recover from line loss, the signal processor displays the information on a noise exposure monitor (Figure 29-2). This monitor presents the total time, in minutes, that the area around the microphone experiences noise exceeding selectable thresholds. The presentation can be considered the cumulative lost time to the resident living nearby due to sound levels exceeding, for example, 89, 90 and 100 dBA.

The noise monitor also provides a mechanism for activating the microphone calibration necessary to assure validity of the measurements. Upon command, a 1000 Hertz signal magnetically drives the microphone diaphragm to an equivalent sound pressure level of 90 dB. Amplifier gain at the central recorder can then be adjusted so the monitor reads 90 dBA.

In addition to activating the noise exposure timers, the monitor supplies an unweighted signal to a tape recorder. In this manner a record of the noise itself can be maintained; the noise can be reprocessed through octave band filters to determine the spectral content; or the noise can be reprocessed to express its level in other common units such as SENEL, PNdB or EPNdB. An unweighted signal from the monitor drives a

Figure 29-1. Pole-mounted microphone.

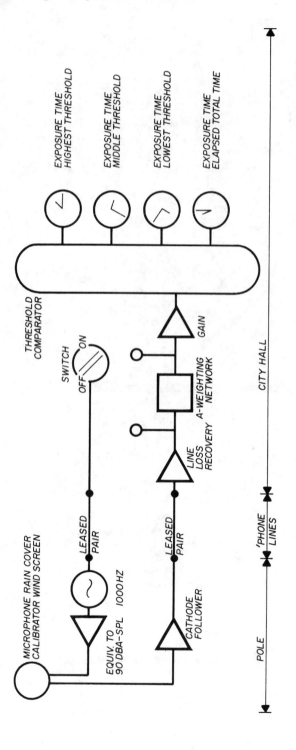

Figure 29-2. Functional diagram of fixed monitoring system.

loud-speaker which permits listening tests. And, finally, an A-weighted signal drives a graphic recorder to obtain sound level distribution plotted against time.

Data from the fixed monitors show some interesting results. Most dramatic are the data from the southeast station. These data seem to clearly indicate a trend toward decreased noise exposure. This decrease is probably the result of increased pilot awareness of the noise problem and an effort to remain at a higher altitude on approach. It is logical that this phenomenon would be observed first at the southeast station. That location is under the flight path to the primary landing runway and is far enough from the runway threshold to allow pilots some degree of latitude in their procedures. With approximately 700 aircraft per day flying over this area creating noise levels near 100 dBA, the trend toward reduced noise is most welcome.

Mobile Monitoring

The mobile acoustics laboratory is housed in a van with a raised roof (Figure 24-3). A microphone rests on a pole 11 ft above ground level. The

Figure 29-3. Mobile acoustical laboratory.

signal enters a precision sound level meter which amplifies, weights, and divides the signal (Figure 29-4). The weighted portion of the signal activates the recording function of the graphic level recorder; the unweighted portion is stored on tape by the audio tape recorder.

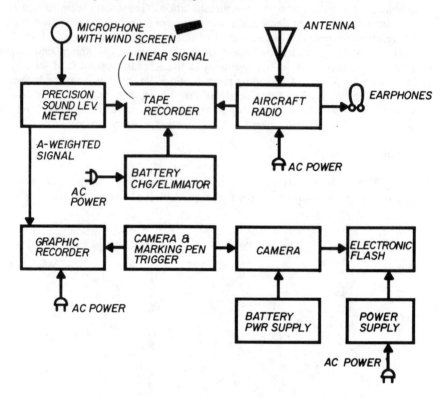

Figure 29-4. Functional diagram of mobile monitoring system.

When monitoring aircraft, the operator numbers each trace serially on the strip chart. This correlates the event with tabled entries which include the time, date, aircraft identification, weather parameters, and photograph number. The operator monitors the control tower radio frequency to ascertain aircraft identification. All communications are recorded on the second channel of the tape recorder. Atmospheric data (humidity, temperature, barometric pressure, wind direction and speed, ceiling and visibility) are noted since they may affect the noise transmission. At the moment each aircraft passes directly overhead the operator snaps a picture with the camera for a positive record of the aircraft type. The size of the image is subsequently used to determine the aircraft's altitude.

The main purpose of the mobile acoustic laboratory is to monitor aircraft flyovers to apprehend any aircraft which violates both Inglewood's noise ordinance and federal altitude restrictions. Inglewood's ordinance specifically exempts any aircraft which is operated in conformity with Federal Aviation Regulations. This exemption is necessary because of Supreme Court rulings which indicate that only the federal government can prescribe the operating rules for aircraft in flight. Inglewood's ordinance prescribes a fine of $500 for each aircraft which, after violating any federal law, creates a noise level in excess of 90 dBA in Inglewood.

The first measurements using this equipment were made in September 1970. After about a month of development and exploratory tests, a statistical survey was made to determine just what constituted a "violation of federal law." The Federal Aviation Regulations are ambiguous with respect to the minimum altitude at which an aircraft may fly when approaching to land. "Normal maneuvers" are permitted below the glide slope for the purpose of remaining on the glide slope. Therefore, Inglewood had to accurately determine by measurement what constituted an "abnormal maneuver." This was done, and indicated that any aircraft more than 100 ft below the glide slope at a point approximately 2.5 miles from the runway was "abnormal."

For a test case the city felt it should cite an aircraft under its ordinance only if it was louder than 99 percent of all other flights and lower than 99 percent of all other flights. This meant that only about one aircraft in 10,000 would violate the criteria.

Nevertheless, such a violation did occur on March 9, 1971, when Continental Airlines Flight 25 passed over the monitoring station at an altitude 118 ft below the glide slope and registered noise of 102 dBA. Inglewood has initiated court action against Continental Airlines and the pilots of Flight 25. The case has now gone to the federal court in a jurisdictional dispute over Inglewood's legal right to cite any commercial aircraft for anything. The Federal Aviation Administration claims complete jurisdiction.

In August of 1971 another statistical altitude survey was conducted and showed that the average altitude of flyovers had increased from the data of November, 1970. The greatest increase shown was for three engine aircraft and amounted to almost a 10 percent increase.

The mobile monitoring unit is now being used for a community-wide general noise survey. The 24-hr composite noise is being calculated in Community Noise Equivalent (CNEL) units. This scale was devised for airport noise monitoring by the California Department of Aeronautics and takes into account noise level, frequency, duration, number of occurrences, and time of day or night. It is similar to the Noise Exposure Forecast (NEF) scale suggested by the FAA, except that it uses the dBA as the basic unit. Preliminary results show a wide variation of composite noise values in Inglewood, from approximately CNEL 55 (NEF 22) far from the flight paths to CNEL 85 (NEF 52) under the flight paths. The California Standard considers CNEL 65 (NEF 32) to be a reasonable level of noise in the vicinity of an airport.

Inglewood feels that solutions to the aircraft noise problem are available. Some will take a long time, others not so long. Some require action on the part of the city, such as land use and soundproofing, and some require action on the part of the flying industry, such as steeper approach profiles and quieter engines. Solutions must be found. Inglewood has already begun gathering hard data upon which sound decisions may be based.

CHAPTER 30

AIRBORNE RADIOACTIVITY RELEASES FROM NUCLEAR POWER PLANTS [72]

The normal operation of electric power plants, both nuclear and fossil-fueled, results in the release of pollutants in the form of heat, radioactivity, and chemicals. Coal and nuclear fuel will be the primary sources of fuel for electric power plants to be built in the next 30 years. The large quantity of airborne pollutants released from coal-fired power plants is well documented, and the magnitude of this environmental pollution problem is recognized by a large portion of the general public. It is interesting to note that in addition to SO_2, NO_x and suspended particulates released, trace amounts of radioactivity in the form of thorium, uranium and radium have also been found in the flyash released from coal combustion. The magnitude of the environmental and health effects of airborne pollution in the form of radioactivity releases, from the normal operation of nuclear power plants, is not very well understood by people outside the nuclear industry.

RADIATION RELEASES

In order to put the level of radioactivity releases from nuclear power plants in proper perspective, it is necessary to understand the levels of radiation to which we are all exposed from natural and man-made sources. The unit used for the measurement of radiation exposure to the human body is the "roentgen-equivalent-man," or rem, which is a measure of energy deposition in the body. Due to the low levels of radiation involved, a more commonly used unit is the millirem (mrem) or one-thousandth of a rem.

People in the United States receive a radiation exposure dose of about 100-250 mrem per year from natural radioactive sources. Some sources are within their bodies, others are in the environment, the amount depending on where they live. In addition, people receive an average dose of about 75 mrem from man-made sources for a total of about 175-325 mrem per year. The national average dose is about 205 mrem per year. The principal sources of radiation are shown in Table 30-1.

Table 30-1. Average Annual Radiation Exposure

Source	mrem
Natural Background	
Cosmic Radiation	50 to 150
Radiation from Ground and Air	15 to 130
Internal Body Radiation	25
Average Radiation from Natural Background	130
Man-Made Radiation	
Global Fallout from Weapons Testing	4
Medical X-rays (diagnostic)	40 to 300
Miscellaneous (TV, Consumer Products, Air Transport)	3
Nuclear Power Facilities	0.003
Average Radiation from Man-Made Sources	75
Specific Examples of Radiation Exposures	
Annual Chest X-Ray	50 to 300
Round Trip Cross-Country Flight	3 per flight
Brick House instead of Wood	30 additional
Medical-Gastrointestinal Tract Exam	22,000
Medical-Fluoroscopic Exam	5,000 to 400,000

The principal sources of natural radiation are cosmic radiation from outer space, radiation from naturally-occurring radioactive materials in the ground and air, and radiation from radioactive materials in our bodies. The internal radiation dosages come chiefly from the potassium in our food and water, of which about 0.01 percent is radioactive potassium—40. The principal man-made sources of radiation are those received from medical and dental X-rays, and from global fallout resulting from past nuclear weapons testing.

The magnitude of both natural and man-made radiation exposure varies with geographical location and with the activities of the individual. For example, people who live in brick houses receive about 30 mrem of radiation more each year than those who live in wooden houses because brick contains more natural radioactive material than wood. Also, a cross-country round-trip flight results in a radiation exposure of about 3 mrem due to the increased exposure to cosmic radiation at high altitudes.

The quantities of radioactive noble gases released from the major light-water-cooled nuclear reactors operated during 1971 have been compiled by the Atomic Energy Commission (AEC) and are given in Table 30-2. This table also identifies the calculated annual whole-body doses that could have resulted at the site boundary, and the average individual dose for the population within 50 miles of the plant. Except for Humboldt Bay, which operated in 1971 with several defective fuel elements, the site boundary

Table 30-2. Noble Gases Released, Site Boundary and
Average Individual Doses for 1971

Facility	Noble gases released, curies	Site boundary dose, mrems	Average individual dose, mrems
	Pressurized—Water Reactors		
Indian Point	360	0.035	0.00005
Yankee Rowe	13	0.3	0.0003
San Onofre	7,670	2.2	0.002
Connecticut Yankee	3,250	5.6	0.003
Ginna	31,800	5.0	0.004
H. B. Robinson	18	0.05	0.00002
Point Beach	838	0.2	0.0008
	Boiling—Water Reactors		
Oyster Creek	516,000	31	0.013
Nine Mile Point	253,000	4.8	0.009
Dresden 1, 2 and 3	1,330,000	32	0.057
Humboldt Bay	514,000	160	0.54
Big Rock Point	284,000	4.6	0.026
Millstone	276,000	5.5	0.0056
Monticello	76,000	4.4	0.0036

doses range from less than 1 to about 15 percent of the average individual radiation exposure from natural and man-made sources. However, individuals do not reside 24 hours a day at the site boundary. The average individual dose is much less than 1 percent of normal man-made and natural radiation exposures.

Even though experience with operating reactors indicates the radiation exposures to the population are insignificant when compared to natural background radiation, steps are being taken by the AEC to reduce further the airborne radioactivity releases from nuclear power plants. The basis for this action is that little information is presently known about the long-term health effects of exposures to low-level radiation, and nuclear power plants will be built in increasing numbers and larger capacities in the future.

Atomic Energy Commission Regulations

In June 1971, the AEC proposed revising Title 10, Part 50 of the U.S. Code of Federal Regulations to provide numerical guides for nuclear power plant design and operation with respect to keeping radioactivity in gaseous effluents as low as practicable. The "as low as practicable" concept was originally added to the AEC regulations in December 1970 and means "as low as is practically achievable taking into account the state of technology and the economics of improvement in relation to benefits to the public health and safety and in relation to the utilization of atomic energy in the public interest." Following issuance of the proposed changes, lengthy public hearings were held and a draft environmental statement was prepared by the AEC in accordance with the National Environmental Policy Act. Comments on the draft statement were reflected in the final environmental statement issued in July 1973. At the present time, the AEC Regulatory Staff has recommended to the licensing hearing board that radiation limits for the release of gaseous radioactive effluents be established at 5 mrem per year exposure to the whole body of an individual located at the site boundary. Separate limits for certain body organs were recommended to be 15 mrem per year to the thyroid of the nearest infant, and 15 mrem per year to the skin of an individual located at the site boundary. The final regulations are expected to be issued within the year.

Although it may appear based on operating experience that very little, if anything, needs to be done to meet the new AEC regulations, this is not the case because very conservative assumptions are being used by the AEC to assess the power plant's design capability to meet these new regulations. Conservative assumptions for anticipated fuel defect levels, meteorology, and living habits of the plant's nearest neighbors will require that augmented gaseous radioactive waste treatment systems be installed on all boiling-water reactor plants and possibly some pressurized-water reactor plants.

Nuclear Power Plant Operations

The commercial nuclear power plants currently in use in the majority of nuclear installations in the U.S. utilize a boiling-water reactor (BWR) or a pressurized-water reactor (PWR) as a heat source for the production of steam. Radioactive gases are produced in both types of reactors by neutron activation of such materials as nitrogen, oxygen and argon, and by the release of some of the gaseous products of the nuclear fission process, such as xenon, krypton and iodine, from the fuel element into the primary reactor coolant stream. The nuclear fission products enter the primary coolant as a result of pinhole defects in the fuel tubing or cladding used to enclose the fuel elements, and as a result of fissioning of uranium existing as a tramp impurity on the surface of the cladding. The nuclear plants are designed to operate with fuel element leakage of up to about 1 percent. This design, in turn, sets the criteria for the design of systems to handle the radioactive gaseous effluents that will maintain releases to the environment within prescribed limits.

In a BWR, the gases released into the primary coolant are carried to the turbine and the condenser along with the steam produced by boiling. Steam is condensed back to water in the condenser, but the noncondensable gases, including the very small volume of radioactive gases, are removed by the condenser steam-jet air ejector and sent to a gaseous radioactive waste treatment system. In a PWR, most of the gases remain in the reactor coolant in a system that is sealed during normal operation. The water in this system is not permitted to boil and is separated from the steam system by a steam generator. As long as this steam generator remains tight, radioactive materials circulate within the closed primary coolant system for a considerable portion of the plant's life. Small quantities of radioactive gases are collected as a result of water chemistry control operations performed on the reactor coolant system and collected during the annual refueling operations when the system is opened up. In currently operating reactors the gaseous radioactivity released by a BWR is significantly greater than that of a PWR of comparable size.

GASEOUS RADIOACTIVE WASTE TREATMENT SYSTEMS

The gaseous fission products are a mixture of a great variety of isotopes of varying half-lives. The half-life of a radioactive isotope is the average time it takes for one-half of the radioactive atoms to decay. Until recently, gaseous radioactive wastes were treated by delaying their release in the form of large pipes buried underground which required a transit time of about 30 min before the gas reached the exit of a tall stack. This holdup time allowed the short-lived gaseous isotopes to decay.

Table 30-3 is a summary of the data collected by the AEC on airborne radioactive effluents from operating BWR's in the United States for the years 1970-1972. These plants are all operating well within the maximum permissible concentrations of radionuclides in effluents prescribed by AEC regulations. Airborne effluents from PWR's are much less than those shown for the plants described in Table 30-3. The unit used to describe the magnitude of the radioactivity is the curie (Ci). One curie of any radioactive isotope is the quantity of the isotope which is decaying at the rate of 3.70 x 10^{10} disintegrations per second. The unit microcurie (μCi), equal to 3.70 x 10^4 disintegrations per second, is used for small values of activity.

Table 30-3. Airborne Effluent Comparison by Year (in curies)

Facility	Noble Gases			Halogens and Particulates (half-life greater than 8 days)		
	1970	1971	1972	1970	1971	1972
Boiling Water Reactors	(X1000)	(X1000)	(X1000)			
Oyster Creek	110	516	866	0.32	2.14	6.48
Nine Mile Point	9.5	253	517	⟨0.001	⟨0.06	0.969
Millstone 1	–	276	726	–	4.0	1.32
Dresden 1	900	753	877	3.3	⟨0.67	2.75
Dresden 2,3	–	580	429	1.6	8.68	5.89
LaCrosse	0.95	0.53	30.6	⟨0.06	⟨0.001	⟨0.712
Monticello	–	75.8	751	–	0.052	0.589
Big Rock Point	280	284	258	0.13	0.61	0.148
Humboldt Bay	540	514	430	0.35	0.3	0.479
*Pilgrim	–	–	18.1	–	–	0.0319
Quad Cities 1, 2	–	–	132	–	–	0.747
*Vermont Yankee	–	–	55.2	–	–	0.171

*Operated less than 1 year.

The gaseous radioactive wastes, primarily the noble gases xenon and krypton, to be treated by the plant's treatment system are mixed in with a larger quantity of nonradioactive gases. The condenser off-gas stream from a BWR contains all gases generated in the reactor, in addition to any air that leaks into portions of the turbine under vacuum. This air in-leakage contains nonradioactive xenon and krypton. Also, due to the intense neutron gamma radiation in the reactor core, some of the water passing through the core is disassociated into hydrogen and oxygen gas. The amount of hydrogen and oxygen in the off-gas stream is much greater than the amount of air. A typical breakdown of the condenser off-gas stream is shown in Table 30-4.

Table 30-4. Typical Condenser Off-Gas Stream for a Large BWR Plant

Air In-leakage	20 scfm
Hydrogen	110 scfm
Oxygen	55 scfm
Water Vapor	33 scfm
Radioactive Fission Product Gases	Negligible
Approx. Total	218 scfm

The constituents of the condenser off-gas stream of a BWR plant vary from time to time. Also, the amount of radioactive fission gases in this off-gas stream vary from time to time. In order to design systems for the treatment of radioactive fission gases, a standardized source term has been defined by the boiling-water reactor manufacturer (General Electric) for large nuclear power stations in the range of 1000 megawatts net electrical output.

This standard source term represents a conservative design criteria. It corresponds to a release rate of approximately 100,000 μci/sec of radioactive fission product gases after a holdup time of 30 min which provides for decay of short-lived radionuclides. This corresponds to a release of over 3 million Ci/yr if the plant operates continuously. For comparison, the actual release for Dresden Unit 1, which had the highest release reported for a BWR, is less than 1 million Ci/yr.

The processing techniques being used for newer BWR plants all rely on the treatment system's ability to delay the release of the radioactive gases until a specified amount of decay has taken place. Table 30-5 identifies the radioactive gases which predominate in the gaseous effluents from BWR's. They are all isotopes of the inert noble gases, xenon and krypton. Very small amounts of solid radioactive particulates and iodine are also part of the off-gas stream and are removed by the treatment system. Any desired decontamination factor can be achieved if suitable means are provided to hold up the release of the radioactive gases for the required amount of time. All krypton isotopes except krypton-85 are eliminated by a holdup of about one day, while there is no practical holdup time that will have any effect

Table 30-5. Effect of Holdup of Radioactive Fission Product Gases
(half-life greater than 1 min) Microcuries Per Second

Isotope	Half Life	30 min	10 hr	24 hr	3 day	9 day	40 day
Kr-89	3.2 min	360					
Xe-137	3.8 min	670					
Xe-138	14.0 min	21,000					
Xe-135(m)	15.6 min	6,900					
Kr-87	1.3 hr	15,000	85				
Kr-83(m)	1.86 hr	2,900	84				
Kr-88	2.8 hr	18,000	1,700	51			
Kr-85(m)	4.4 hr	5,600	1,300	140			
Xe-135	9.2 hr	22,000	10,000	3,600	100		
Xe-133(m)	2.3 days	280	250	200	110	180	
Xe-133	5.27 days	8,200	8,000	7,800	6,900	5,000	850
Xe-131(m)	11.9 days	150	145	140	125	90	15
Kr-85	10.76 years	8	8	8	8	8	8
Approximate totals		100,000	22,000	12,000	7,000	5,000	900

on krypton-85, which has a half-life of about 11 years. Due to its very long
half-life, krypton-85 is not removed quickly from the environment by decay
and could become significant many years in the future. On the other hand,
as an inert gas emitting only low-energy radiation, krypton-85 is not chemically
reactive in ordinary chemical processes, cannot be concentrated by any known
biological mechanism, and does not present a biological hazard at today's
concentrations.

The gaseous radioactive waste treatment systems currently being proposed for new BWR plants can be divided into three basic types—charcoal
and molecular sieve adsorption, selective absorption, and cryogenic distillation. Each of the systems uses processing techniques which are well-known
in principle and use conventional processing technology, modified as necessary for this particular application. Each system uses the following basic
processing steps:

1. Recombination of the hydrogen and oxygen gas which was formed
   by the radiolytic dissociation of water in the reactor core with the
   aid of a catalytic recombiner.
2. Protective measures against a hydrogen explosion by dilution to a
   maximum hydrogen concentration of 4%.
3. Delay of the radioactive gases by various mechanical and physical
   means in order to reduce by decay the level of radioactivity.
4. The retention of solid radioactive particulates by mechanical
   filtration.

A schematic diagram of the gaseous radioactive waste system is shown in Figure 30-1. The basic components of this system are described separately below.

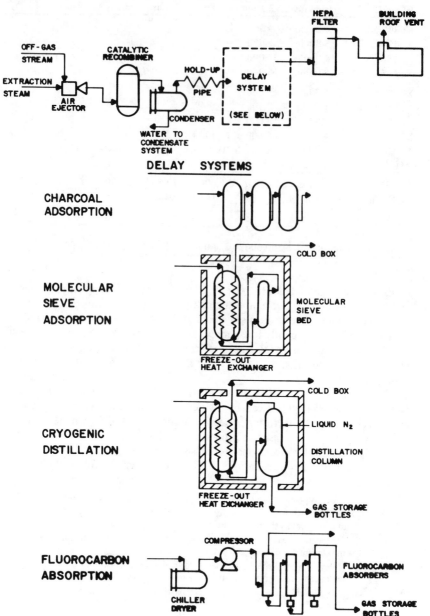

Figure 30-1. Gaseous radioactive waste treatment system.

Recombiner

Catalytic recombination of oxygen and hydrogen in the BWR off-gas stream is performed to reduce the total volume of gas which must be handled and to lessen the possibility of a hydrogen explosion. While recombination of oxygen and hydrogen does not reduce the quantity of radioactive gases, it does reduce the total gas volume by about 80 percent for BWR's leaving a more concentrated smaller volume of gas to handle. A typical recombiner consists of a replaceable cartridge or bed containing a catalyst in a steel tank. The catalyst is generally an array of metal strips or a ceramic material, such as alumina pellets, that have been precoated with finely divided particles of platinum or palladium. Provisions are made for heating and cooling of the catalyst bed to maintain its temperature between approximately 250 and 900 F. At lower temperatures the recombiner is less effective and liquid water may condense in the bed and inhibit diffusion of gas to the catalyst. At higher temperatures, the catalyst has a shorter life.

As generated, the off-gas stream for a BWR is an explosive mixture of oxygen and hydrogen. The concentration of hydrogen is reduced below the flammable limit of 4 percent by diluting the off-gas stream with another gas, usually steam. Since hydrogen makes up about 50 percent of the total volume of the off-gas, the volume of dilution gas is large thus requiring a large recombiner. When steam is used as a dilutant, it can be condensed out downstream of the recombiner while at the same time any water vapor produced in the recombiner, and initially removed from the condenser, is also condensed out.

Charcoal Adsorption

Charcoal adsorption is currently the most widely used method for delaying the release of radioactive gases. The charcoal delay system operates on the adsorptive delay theory. The noble gases, krypton and xenon, migrate through the charcoal beds at a rate which depends upon both the temperature and characteristics of the charcoal, but at a much slower rate than the carrier gas which is air. Each noble gas has a characteristic delay time compared with the nitrogen and oxygen in air. The efficiency of the charcoal beds is a function of temperature, with increased efficiency being achieved as the charcoal temperature is reduced. Fewer charcoal beds are required for a system operating at 0 F than for a system operating at ambient (70 F) temperatures. The reduced number of beds is achieved at the expense of refrigeration; however, reducing the number of beds can also lead to a reduction of the physical size of the building used to house the radioactive waste treatment systems.

The charcoal beds are typically 4 to 6 ft in diameter and have a packed bed height of about 16 ft. Each packed bed will contain about 5 to 7 tons of activated charcoal. At a normal air in-leakage rate of 20 scfm an ambient temperature charcoal system using about 100 tons of charcoal will provide a delay of about 26 hours for krypton and nearly 20 days for xenon. This additional delay results in a decontamination factor of about 100 over that achieved with simply a 30-min holdup line.

The charcoal beds also adsorb iodine and solid particulates and will essentially remove all radioactive iodine and particulates from the gas stream.

Charcoal absorbers can be operated at cryogenic temperatures, which requires only a small volume of charcoal but expensive refrigeration equipment. A significant problem with this system is that any oxygen remaining in the carrier gas will liquefy at the cryogenic temperatures used. Liquid oxygen, ozone and charcoal can form an explosive mixture and this potential problem will hinder acceptance of a cryogenic charcoal system.

Molecular Sieve Adsorption

Molecular sieve adsorption is similar to a charcoal adsorption system operated at cryogenic temperatures. The molecular sieve material is an inorganic substance with predetermined pore sizes that will selectively adsorb the xenon and krypton molecules. Use of an inorganic material eliminates any possibility of an explosive mixture being formed. Preliminary treatment of the off-gas stream is required to remove all water vapor and any carbon dioxide. This treatment is done using cryogenic freeze-out heat exchangers which are operated in pairs so that one can be warmed and vented while the other is freezing out the water vapor and carbon dioxide. After the gas stream is cooled to about -300 F, it goes to a cryogenic delay bed which contains the molecular sieve material which adsorbs the noble gases. At a temperature of -300 F the adsorbent will provide about 2 days holdup for the krypton, and about 200 days holdup for the xenon. This corresponds to a decontamination factor of about 1000 greater than that achieved with a 30-min holdup line.

It is possible to recover and store the radioactive noble gases for longer decay by providing redundant sets of cryogenic delay beds. While noble gases are adsorbed on one set of beds, they would be vented into storage bottles from a second set of beds by warming the beds and reducing their adsorptive capacity. The physical space requirements for this system are small; however, expensive refrigeration systems must be provided.

Cryogenic Distillation

Cryogenic distillation is a system for which much operating experience exists based on commercial use for the distillation of krypton and xenon from air. As in the case for molecular sieve adsorption, it is necessary to remove all of the water vapor and any carbon dioxide from the off-gas stream to prevent freeze-up of the distillation columns. The cryogenic distillation equipment generally consists of distillation column supplied with liquid nitrogen operating at about -300 F and a regenerative heat exchanger. Most of the nitrogen in the off-gas stream passes out from the top of the distillation column. The radioactive xenon and krypton along with the non-radioactive xenon, krypton, argon and oxygen are concentrated in the liquid at the bottom of the distillation column. Due to the small volume of the liquid, it is possible to provide storage bottles for the liquefied gases to permit an indefinite storage time. The liquid can be periodically processed on a batch basis to remove the remaining oxygen and any ozone which may have formed at the cryogenic temperatures. The physical space requirements for the cryogenic distillation equipment are similar to those for molecular sieve adsorption.

Selective Absorption

This process is based on the greater solubility of noble gases than nitrogen and oxygen in fluorocarbon liquids at low temperature and high pressure. As this system operates at low temperatures, a chiller-dryer is used to remove water vapor and carbon dioxide. After drying, the gases are contacted with a fluorocarbon liquid, such as freon, at about -25 F and 400 psi in an absorption column where the krypton and xenon are selectively absorbed by the liquid. The krypton and xenon are stripped from the fluorocarbon in a second column and the gases transferred to pressurized cylinders for storage.

Fluorocarbon absorption systems can achieve a removal efficiency for noble gases well over 99 percent where an overall decontamination factor of 1200 is obtained.

Mechanical Filtration

High-efficiency particulate air (HEPA) filters are provided downstream of the delay beds to remove any particulate matter, charcoal fines, and solid products formed from the decay of the radioactive noble gases.

With the use of sophisticated off-gas treatment systems, the newer BWR plants will not require the costly tall stacks common to older BWR plants. Typically the off-gas stream will be dispersed to the atmosphere through a roof vent without a stack.

MONITORING GASEOUS EFFLUENTS

All normal and potential paths for release of airborne radioactive material are monitored. Specifically, monitors are provided at the vents of the turbine building, reactor building and radioactive waste treatment building. These monitors are normally three-channel monitors with one channel monitoring the gross radioactivity level for noble gases, iodine, and particulates. For a BWR plant, this monitoring is performed continuously, and periodically a detailed isotopic analysis is performed. For noble gases, a sample is analyzed monthly to determine the specific isotopic composition of the noble gases released. For iodine, the level of I-131 is determined weekly, and once a quarter an iodine sample is analyzed for the radionuclides I-133 and I-135. For releases of radioactive material in particulate form, a sample is drawn continuously through a HEPA filter. These filters are replaced and analyzed weekly for gross radioactivity (beta and gamma activity) and an analysis for specified radionuclides such as barium-140, iodine-131 and strontium-90.

The AEC has published a Regulatory Guide 1.21, "Measuring and Reporting Effluents from Nuclear Power Plants," which defines the preferred method for effluent measuring and reporting programs. This Regulatory Guide identifies the minimum sensitivities that should be used in the analysis of the radioactive material samples.

The radioactive material released from the plant, in the form of aerosols, vapors or gases is subject to the same atmospheric dispersion factors as the chemical pollutants from the coal-burning electric power plants. The average concentration of radioactive material in air in the vicinity of the plant release point will depend upon the quantity released and the wind speed, wind direction and atmospheric stability for that site at the time of release.

Airborne aerosols and vapors in the atmosphere eventually are reduced by deposition or by the scavanging action of precipitation. Noble gases are adsorbed to some extent by precipitation, but most simply continue to mix with the atmosphere. In all cases, the amount of radioactive material released will decrease by radioactive decay. Following dilution in the atmosphere, the concentrations of the radionuclides are extremely small. Radiation measurements at the site boundary and beyond may be feasible in some instances, while in others the very low values requires that doses be calculated from radionuclide concentrations at other locations. Radiological monitoring during normal reactor operation thus consists of three phases:

1. Calculations of individual and population radiation exposures based on measured values such as radionuclide concentrations in effluents at the platn vents,
2. Validation studies to confirm these calculations, and
3. Measurements at points of potential population exposure, even if many or all measurements result in "less than" values, to assure that the applicable dose limits have not been exceeded.

Several continuously-recording low-level gamma ray detectors have been found to be highly effective for measuring radiation in the environment. These detectors can quantify an increase of 0.0005 mrem per hour above the natural background. These detectors have been used to determine radiation exposure values in validation studies and for assuring the absence of significant radiation exposure.

Passive integrating dose measuring instruments are considerably less expensive and require less attention than the continuously-recording gamma ray detectors. The passive gamma ray detector normally used is the thermoluminescent dosimeter (TLD). These dosimeters are generally exposed for periods of weeks or months and then collected for reading. At the present time, TLD's can detect increments of 0.5 mrem per month above the natural radiation background levels. This, therefore, corresponds to a yearly level of 6 mrem which is slightly above the whole body exposure limits of 5 mrem which are proposed by the AEC. In view of the relatively sensitive measurements required, to confirm that the radiation limits proposed by the AEC are being complied with, further efforts are required to improve the sensitivity and accuracy of the environmental monitoring instruments.

ENVIRONMENTAL SURVEILLANCE PROGRAM

A typical environmental surveillance program for a large BWR plant is identified in Table 30-6. In addition to the continuous gamma ray detectors and TLD's, which measure atmospheric radiation levels, the effects of the deposition of the airborne radioactive particles on the ground and the effects of the biological concentration of radioactive iodine are determined.

Samples of the leafy portions of natural vegetation, and the edible portions of food and feed crops, are collected during the growing season and analyzed for their radionuclide content. Particular attention is paid to the sampling of milk since radioactive iodine-131 in milk could be a significant source of radiation exposure to the thyroid. The limit established for the release of iodine-131 is the currently specified maximum permissible concentration in air, divided by 100,000. This takes into account the possible concentration of iodine via the plume-grass-cow-milk chain to human ingestion.

The AEC has published Regulatory Guides which describe acceptable environmental surveillance programs. Guides which relate to airborne radioactivity releases are 4.1, "Measuring and Reporting of Radioactivity in the Environs of Nuclear Power Plants," and 4.3, "Measurements of Radionuclides in the Environment—Analysis of I-131 in Milk."

Cost-Benefit Analysis of Treatment Systems

Environmental Impact Statements, prepared in accordance with the National Environmental Policy Act, are required to discuss the amount of environmental benefit to be realized and the costs associated with achieving this benefit. This area has been a source of conflict between the AEC and the nuclear industry during the protracted rulemaking hearings on the proposed AEC regulations limiting radiation releases to as-low-as-practicable. Although the costs to the electric utilities are well known, the benefits to be achieved are not easily quantified nor can a firm monetary value be placed on reduced radiation exposure. The benefits, of course, are a reduction in the total population radiation exposure measured in person-rem (the size of population group X average individual radiation dose reduction). Industry comments on the AEC's draft environment statement indicated that values of $12 to $120 per person-rem reduction, as suggested by the

Table 30-6. Environmental Surveillance Program

	Sampling Summary		
Sample Type	Stations	Sampling Frequency	Analysis
1. Background			
a) Gamma Sensitive Detector	3	Continuous Recording	Background Gamma
b) TLD's	10	Monthly – Annually	Readout and Record at Noted Frequency
2. Air (Particulates and Gas)	10	Weekly	Gross Alpha Gross Beta Gamma Scan & Radioiodine
3. Vegetation and Livestock			
a) Natural Vegetation	10	3 Samples Annually (During Growing Season)	Gross Beta Sr-90 Cs-137 I-131
b) Food and Feed Crops	10	"	Gamma Scan
c) Food Animals	5	"	
4. Soil	5	Quarterly	Gross Alpha Gross Beta Sr-90 Cs-137 Gamma Scan
5. Sediment	5	Quarterly	Gross Alpha Gross Beta Sr-90 Gamma Scan
6. Milk	3	Monthly	I-131 Sr-90 Cs-137 Elemental Calcium
7. Wildlife			
a) Rabbits	5	Annually	Thyroid I-131 Femur Sr-90
b) Waterfowl	5	Annually	Gamma Scan Muscle -p-32, Zr-65

National Academy of Sciences, should be used. The AEC, citing other sources, used values in the range of $100 to $1000 per person-rem to analyze the benefits achieved.

The costs for the BWR gaseous radioactive waste treatment systems described are approximately $4,000,000 in direct costs (equipment, structures) and an additional $2,500,000 in indirect costs (facilities, engineering, interest during construction). The annual costs, including fixed charges, operating and maintenance costs are on the order of $1,500,000.

The magnitude of the reduction in population radiation exposure achieved by expenditure of these costs depends entirely upon the population exposure selected as the base case. Here, the nuclear industry again felt the AEC's assumed base-case was unrealistically high, which resulted in large reductions in population radiation exposures.

It is reasonably certain that the AEC and nuclear industry will never agree on what constitutes a valid cost-benefit analysis for the proposed AEC regulations. In any event, it appears that regardless of the actual benefits the proposed AEC regulations will become law.

HEALTH EFFECTS FROM NUCLEAR FACILITY EFFLUENTS

In 1971 the AEC undertook a detailed study to determine the amounts of radioactive materials that would reach a large segment of the U.S. population by the year 2000. The study concluded, based on the rapidly increasing number of nuclear power plants expected through the year 2000, that the average population dose due to nuclear power will increase from 0.003 mrem per person in 1970 to as much as 0.2 mrem in 2000. Little significance can be attached to this change since this small dose is added to a base of over 200 mrem from other sources, and, after all, a single cross-country round trip flight results in 3 mrem exposure.

The actual health effects of low-level radiation exposures were discussed in a report prepared by the Advisory Committee on the Biological Effects of Ionizing Radiation (BEIR) of the National Academy of Sciences. This report entitled "The Effects on Populations of Exposure to Low Levels of Ionizing Radiation" acknowledged that little is known about the health effects of low-level radiation and that it would be very conservative to assume linear relationship with the effects observed from high-level radiation. Using the assumptions of the BEIR Report the total cancer deaths and those attributable to radiation exposure in the year 2000 are as shown in Table 30-7.

Table 30-7. Cancer Deaths in U.S. Population—Year 2000

Total Cancer Deaths Expected	486,000/yr
Deaths Attributable to Background Radiation	5,310/yr
Deaths Attributable to Medical X-rays	3,720/yr
Deaths Attributable to Nuclear Effluents	10/yr

In reality it will be impossible to attribute individual cancer deaths to the operation of nuclear power plants, since the radiation exposures are such a small fraction of natural background radiation.

SUMMARY AND CONCLUSIONS

The operating experience of nuclear power plants with regard to release of airborne radioactivity has been excellent. Radiation exposure to the general population has been less than 1 percent of the limits established by the Federal Radiation Council and contained in current AEC regulations. In spite of this, the nuclear power industry will be providing new gaseous radioactive waste treatment systems which will reduce even further the quantities of radioactivity released to the atmosphere.

The electric utilities will be expending funds for treatment of airborne releases of nuclear power plants at approximately the same level as for sophisticated SO_2 and NO_x treatment systems for large coal-burning power plants. Although the accompanying reduction in environmental pollution from the coal-burning power plants is significant, the health benefits realized from taking this action for nuclear power plants is considered negligible since the risks already were extremely low. However, this action by the AEC and the nuclear industry should eliminate any fears the general public may have concerning airborne radioactivity released from nuclear power plants.

GLOSSARY

Curie—the unit of radioactivity which expresses the rate at which a substance decays. One curie (Ci) of any radioactive isotope is the quantity of the isotope which is decaying at the rate of 3.70×10^{10} disintegration per second.

Gamma Ray—electromagnetic radiation of very short wavelength and high energy. Its properties are identical to all other electromagnetic radiation. The term gamma is used to indicate the source is the nucleus of an atom, whereas X-ray is used to describe radiation produced by electrons external to the nucleus.

Half-life—the length of time required for one-half of the atoms in a radioactive sample to decay.

Ionizing Radiation—X-rays, alpha, beta and gamma radiation which are energetic enough to disrupt some of the atoms or molecules in their path as they travel through matter. The affected atoms and molecules are converted to negatively and positively charged fragments (ion pairs).

Isotope—atoms of the same element which differ in weight because of a different number of neutrons. For example, C-12, C-13 and C-14 are isotopes of carbon.

Neutron Activation—neutrons released in the nuclear fuel will travel and can react with elements in the cooling water and plant structures creating radioactive isotopes of normally stable elements. For example, stable O-16 + neutron = radioactivie O-17.

Nuclear Fission—the splitting apart of the uranium nucleus upon absorption of a neutron yielding two or three lighter elements and free neutrons. The elements formed are referred to as fission products.

Radionuclide—any radioactive isotope of an element.

Rem—from roentgen-equivalent-man. A unit of radiation used to express different types of radiation (X-ray, alpha, beta, gamma, neutron) on the basis of equivalent biological effects. One rem corresponds to the absorption of 100 ergs of energy per gram.

CHAPTER 31

CONVERSION FACTORS

MULTIPLY	BY	TO OBTAIN
Acres	43,560	Sq Ft
Acres	4,047	Sq Meters
Acres	1.562×10^{-3}	Sq Miles
Acres	4840	Sq Yards
Acre Feet	43,560	Cu Ft
Acre Feet	3.259×10^5	Gallons
Angstrom Units	3.937×10^{-9}	Inches
Atmospheres	76.0	Cm of Mercury
Atmospheres	29.92	In. of Mercury
Atmospheres	33.90	Ft of Water
Atmospheres	10,333	Kg/Sq Meter
Atmospheres	14.70	Lb/Sq In.
Atmospheres	1.058	Tons/Sq Ft
Barrels (British, Dry)	5.780	Cu Ft
Barrels (British, Dry)	0.1637	Cu Meters
Barrels (British, Dry)	36	Gallons (British)
Barrels, Cement	170.6	Kilograms
Barrels, Cement	376	Pounds of Cement
Barrels, Oil	42	Gallons (U.S.)
Barrels (U.S., Liquid)	4.211	Cu Ft
Barrels (U.S., Liquid)	0.1192	Cu Meters
Barrels (U.S., Liquid)	31.5	Gallons (U.S.)
Bars	0.9869	Atmospheres
Bars	1×10^6	Dynes/Sq Cm
Bars	1.020×10^4	Kg/Sq Meter
Bars	2.089×10^3	Lb/Sq Ft
Bars	14.50	Lb/Sq In.
Board-Feet	144 Sq In. x 1 In.	Cu In.
British Thermal Units	0.2520	Kilogram Calories
British Thermal Units	777.5	Foot Pounds

British Thermal Units	3.927×10^{-4}	Horsepower Hours
British Thermal Units	1054	Joules
British Thermal Units	107.5	Kilogram Meters
British Thermal Units	2.928×10^{-4}	Kilowatt Hours
Btu (mean)	251.98	Calories, Gram (mean)
Btu (mean)	0.55556	Centigrade heat units (chu)
Btu (mean)	6.876×10^{-5}	Pounds of Carbon to CO_2
Btu/Min	12.96	Foot Pounds/Sec
Btu/Min	0.02356	Horsepower
Btu/Min	0.01757	Kilowatts
Btu/Min	17.57	Watts
Btu/Sq Ft/Min	0.1220	Watts/Sq In.
Btu (mean)/Hr $(ft^2)\,°F$	4.882	Kilogram-Calorie/Hr $(m^2)(°C)$
Btu (mean)/Hr $(ft^2)\,°F$	1.3562×10^{-4}	Gram-Calorie/Sec $(cm^2)(°C)$
Btu (mean)/Hr $(ft^2)\,°F$	3.94×10^{-4}	Horsepower/$(ft^2)\,°F$
Btu (mean)/Hr $(ft^2)\,°F$	5.682×10^{-4}	Watts/$(cm^2)°C$
Btu (mean)/Hr $(ft^2)\,°F$	2.035×10^{-3}	Watts $(in.^2)\,°C$
Btu (mean)/lb/°F	1	Calories, gram/gram/°C
Bushels	1.244	Cu Ft
Bushels	2150	Cu In.
Bushels	0.03524	Cu Meters
Bushels	4	Pecks
Bushels	64	Pints (Dry)
Bushels	32	Quarts (Dry)
Calories, Gram (mean)	3.9685×10^{-3}	Btu (mean)
Calories, Gram (mean)	0.001469	Cu Ft - Atmospheres
Calories, Gram (mean)	3.0874	Foot Pounds
Calories, Gram (mean)	0.0011628	Watt Hours
Calories, (themochem.)	0.999346	Calories (Int. Steam Tables)
Calories, Gram (mean)/Gram	1.8	Btu (mean)/Lb
Centigrams	0.01	Grams
Centiliters	0.01	Liters
Centimeters	0.0328083	Feet (U.S.)
Centimeters	0.3937	Inches
Centimeters	0.01	Meters
Centimeters	393.7	Mils
Centimeters	10	Millimeters
Centimeter - Dynes	1.020×10^{-3}	Centimeter - Grams
Centimeter - Dynes	1.020×10^{-8}	Meter - Kilograms
Centimeter - Dynes	7.376×10^{-8}	Pound - Feet
Centimeter - Grams	980.7	Centimeter - Dynes
Centimeter - Grams	10^{-5}	Meter - Kilograms
Centimeter - Grams	7.233×10^{-5}	Pound - Feet
Centimeters of Mercury	0.01316	Atmospheres
Centimeters of Mercury	0.4461	Feet of Water
Centimeters of Mercury	136.0	Kg/Sq Meter

Centimeters of Mercury	27.85	Lb/Sq Ft
Centimeters of Mercury	0.1934	Lb/Sq In.
Cm/Sec	1.969	Ft/Min
Cm/Sec	0.03281	Ft/Sec
Cm/Sec	0.036	Kilometers/Hr
Cm/Sec	0.6	Meters/Min
Cm/Sec	0.02237	Miles/Hr
Cm/Sec	3.728×10^{-4}	Miles/Min
Cm/Sec/Sec	0.03281	Ft/Sec/Sec
Cm/Sec/Sec	0.036	Km/Hr/Sec
Cm/Sec/Sec	0.02237	Miles/Hr/Sec
Circular Mils	5.067×10^{-6}	Sq Cm
Circular Mils	7.854×10^{-7}	Sq In.
Circular Mils	0.7854	Sq. Mils
Cord - Feet	4' x 4' x 1'	Cu Ft
Cords	8' x 4' x 4'	Cu Ft
Cu Cm	3.531×10^{-5}	Cu Ft
Cu Cm	6.102×10^{-2}	Cu In.
Cu Cm	10^{-6}	Cu Meters
Cu Cm	1.308×10^{-6}	Cu Yards
Cu Cm	2.642×10^{-4}	Gallons
Cu Cm	10^{-3}	Liters
Cu Cm	2.113×10^{-3}	Pints (Liq.)
Cu Cm	1.057×10^{-3}	Quarts (Liq.)
Cu Cm	0.033814	Ounces (U.S. Fluid)
Cu Ft	2.832×10^{4}	Cu Cm
Cu Ft	1728	Cu In.
Cu Ft	0.02832	Cu Meters
Cu Ft	0.03704	Cu Yards
Cu Ft	7.481	Gallons
Cu Ft	28.32	Liters
Cu Ft	59.84	Pints (Liq.)
Cu Ft	29.92	Quarts (Liq.)
Cu Ft of Water (60 F)	62.37	Pounds
Cu Ft/Min	472.0	Cu Cm/Sec
Cu Ft/Min	0.1247	Gal/Sec
Cu Ft/Min	0.4720	Liters/Sec
Cu Ft/Sec	62.4	Lb of Water/Min
Cu Ft/Sec	1.9834	Acre Ft/Day
Cu Ft/Sec	448.83	Gal/Min
Cu Ft/Sec	0.64632	Million Gal/Day
Cu Ft - atmospheres	2.7203	Btu (mean)
Cu Ft - atmospheres	680.74	Calories, Gram (mean)
Cu Ft - atmospheres	2116.3	Foot Pounds
Cu Ft - atmospheres	292.6	Kilogram - Meters
Cu Ft - atmospheres	7.968×10^{-4}	Kilowatt - Hours

Cu In.	16.39	Cu Cm
Cu In.	5.787×10^{-4}	Cu Ft
Cu In.	1.639×10^{-5}	Cu Meters
Cu In.	2.143×10^{-5}	Cu Yards
Cu In.	4.329×10^{-3}	Gallons
Cu In.	1.639×10^{-2}	Liters
Cu In.	0.03463	Pints (Liq.)
Cu In.	0.01732	Quarts (Liq.)
Cu In. (U.S.)	0.55411	Ounces (U.S. Fluid)
Cu Meters	10^6	Cu Cm.
Cu Meters	35.31	Cu Ft
Cu Meters	61,023	Cu In.
Cu Meters	1.308	Cu Yards
Cu Meters	264.2	Gallons
Cu Meters	10^3	Liters
Cu Meters	2113	Pints (Liq.)
Cu Meters	1057	Quarts (Liq.)
Cu Meters	8.1074×10^{-4}	Acre Feet
Cu Meters	8.387	Barrels (U.S. Liq.)
Cu Yards (British)	0.9999916	Cu Yards (U.S.)
Cu Yards	7.646×10^5	Cu Cm
Cu Yards	27	Cu Ft
Cu Yards	46.656	Cu In.
Cu Yards	0.7646	Cu Meters
Cu Yards	202.0	Gallons
Cu Yards	764.6	Liters
Cu Yards	1616	Pints (Liq.)
Cu Yards	807.9	Quarts (Liq.)
Cu Yards/Min	0.45	Cu Ft/Sec
Cu Yards/Min	3.367	Gal/Sec
Cu Yards/Min	12.74	Liters/Sec
Days	1440	Minutes
Days	86,400	Seconds
Decigrams	0.1	Grams
Deciliters	0.1	Liters
Decimeters	0.1	Meters
Degrees (Angle)	60	Minutes
Degrees (Angle)	0.01745	Radians
Degrees (Angle)	3600	Seconds
Degrees/Sec	0.01745	Radians/Sec
Degrees/Sec	0.1667	Revolutions/Min
Degrees/Sec	0.002778	Revolutions/Sec
Dekagrams	10	Grams
Dekaliters	10	Liters
Dekameters	10	Meters
Drams	1.772	Grams

Drams	0.0625	Ounces
Dynes	1.020×10^{-3}	Grams
Dynes	7.233×10^{-5}	Poundals
Dynes	2.248×10^{-6}	Pounds
Dynes/Sq Cm	1	Bars
Ergs	9.486×10^{-11}	British Thermal Units
Ergs	1	Dyne-Centimeters
Ergs	7.376×10^{-8}	Foot Pounds
Ergs	1.020×10^{-3}	Gram Centimeters
Ergs	10^{-7}	Joules
Ergs	2.390×10^{-11}	Kilogram - Calories
Ergs	1.020×10^{-8}	Kilogram - Meters
Ergs/Sec	5.692×10^{-9}	Btu/Min
Ergs/Sec	4.426×10^{-6}	Foot Pounds/Min
Ergs/Sec	7.376×10^{-8}	Foot Pounds/Sec
Ergs/Sec	1.341×10^{-10}	Horsepower
Ergs/Sec	1.434×10^{-9}	Kg Calories/Min
Ergs/Sec	10^{-10}	Kilowatts
Fathoms	6	Feet
Feet	30.48	Centimeters
Feet	12	Inches
Feet	0.3048	Meters
Feet	1/3	Yards
Feet (U.S.)	1.893939×10^{-4}	Miles (statute)
Feet of Air (1 atmosphere 60 F)	5.30×10^{-4}	Lb/Sq In.
Feet of Water	0.02950	Atmospheres
Feet of Water	0.8826	In. of Mercury
Feet of Water	304.8	Kg/Sq
Feet of Water	62.43	Lb/Sq Ft
Feet of Water	0.4335	Lb/Sq In.
Ft/Min	0.5080	Cm/Sec
Ft/Min	0.01667	Ft/Sec
Ft/Min	0.01829	Km/Hr
Ft/Min	0.3048	Meters/Min
Ft/Min	0.01136	Miles/Hr
Ft/Sec	30.48	Cm/Sec
Ft/Sec	1.097	Km/Hr
Ft/Sec	0.5921	Knots/Hr
Ft/Sec	18.29	Meters/Min
Ft/Sec	0.6818	Miles/Hr
Ft/Sec	0.01136	Miles/Min
Ft/100 Ft	1	Percent Grade
Ft/Sec/Sec	30.48	Cm/Sec/Sec
Ft/Sec/Sec	1.097	Km/Hr/Sec
Ft/Sec/Sec	0.3048	Meters/Sec/Sec

Ft/Sec/Sec	0.6818	Miles/Hr/Sec
Foot Poundals	3.9951×10^{-5}	Btu (mean)
Foot Poundals	0.0421420	Joules (abs)
Foot Pounds	0.013381	Liter - Atmospheres
Foot Pounds	3.7662×10^{-4}	Watt Hours (abs)
Foot Pounds	1.286×10^{-3}	Btu
Foot Pounds	1.356×10^{7}	Ergs
Foot Pounds	5.050×10^{-7}	Horsepower - Hours
Foot Pounds	1.356	Joules
Foot Pounds	3.241×10^{-4}	Kilogram Calories
Foot Pounds	0.1383	Kilogram Meters
Foot Pounds	3.766×10^{-7}	Kilowatt Hours
Foot Pounds/Min	1.286×10^{-3}	Btu/Min
Foot Pounds/Min	0.01667	Foot Pounds/Sec
Foot Pounds/Min	3.030×10^{-5}	Horsepower
Foot Pounds/Min	3.241×10^{-4}	Kg Calories/Min
Foot Pounds/Min	2.260×10^{-5}	Kilowatts
Foot Pounds/Sec	7.717×10^{-2}	Btu/Min
Foot Pounds/Sec	1.818×10^{-3}	Horsepower
Foot Pounds/Sec	1.945×10^{-2}	Kg Calories/Min
Foot Pounds/Sec	1.356×10^{-3}	Kilowatts
Foot Pounds/Sec	4.6275	Btu (mean)/Hr
Foot Pounds/Sec	1.35582	Watts (abs)
Gallons (British)	4516.086	Cu Cm
Gallons (British)	1.20094	Gallons (U.S.)
Gallons (British)	10	Pounds (avordupois) of water at 62 F
Gallons (U.S.)	128	Ounces (U.S. Fluid)
Gallons	3785	Cu Cm
Gallons	0.1337	Cu Ft
Gallons	231	Cu In.
Gallons	3.785×10^{-3}	Cu Meters
Gallons	4.951×10^{-3}	Cu Yards
Gallons	3.785	Liters
Gallons	8	Pints (Liq.)
Gallons	4	Quarts (Liq.)
Gal/Min	2.228×10^{-3}	Cu Ft/Sec
Gal/Min	0.06308	Liters/Sec
Grains (Troy)	1	Grains (Av.)
Grains (Troy)	0.06480	Grams
Grains (Troy)	0.04167	Pennyweights (Troy)
Grains (Troy)	2.0833×10^{-3}	Ounces (Troy)
Grains/U.S. Gal	17.118	Parts/Million
Grains/U.S. Gal	142.86	Lb/Million Gal
Grains/Imp. Gal	14.286	Parts/Million
Grams	980.7	Dynes

Grams	15.43	Grains (Troy)
Grams	10^{-3}	Kilograms
Grams	10^{3}	Milligrams
Grams	0.03527	Ounces
Grams	0.03215	Ounces (Troy)
Grams	0.07093	Poundals
Grams	2.205×10^{-3}	Pounds
Gram Calories	3.968×10^{-3}	Btu
Gram Centimeters	9.302×10^{-8}	Btu
Gram Centimeters	980.7	Ergs
Gram Centimeters	7.233×10^{-5}	Foot Pounds
Gram Centimeters	9.807×10^{-5}	Joules
Gram Centimeters	2.344×10^{-8}	Kilogram Calories
Gram Centimeters	10^{-5}	Kilogram Meters
Gram Centimeters	2.7241×10^{-8}	Watt Hours
Gram Centimeters/Sec	9.80665×10^{-5}	Watts (abs)
Grams Centimeters^{2} (moment of inertia)	3.4172×10^{-4}	Pounds In.^{2}
Grams Centimeters^{2} (moment of inertia)	2.37305×10^{-6}	Pounds Ft^{2}
Grams/Cu Meter	0.43700	Grains/Cu Ft
Grams/Cm	5.600×10^{-3}	Lb/In.
Grams/Cu Cm	62.43	Lb/Cu Ft
Grams/Cu Cm	0.03613	Lb/Cu In.
Grams/Cu Cm	3.405×10^{-7}	Lb/Mil Ft
Grams/Cu Cm	8.34	Lb/Gal
Grams/Liter	58.417	Grains/Gal (U.S.)
Grams/Liter	9.99973×10^{-4}	Grams/Cu Cm
Grams/Liter	1000	Parts/Million (ppm)
Grams/Liter	0.06243	Lb/Cu Ft
Grams/Sq Cm	0.0142234	Lb/Sq In.
Hectograms	100	Grams
Hectoliters	100	Liters
Hectometers	100	Meters
Hectowatts	100	Watts
Hemispheres (Sol. Angle)	0.5	Sphere
Hemispheres (Sol. Angle)	4	Spherical Right Angles
Hemispheres (Sol. Angle)	6.283	Steradians
Horsepower	42.44	Btu/Min
Horsepower	33,000	Ft Pounds/Min
Horsepower	550	Ft Pounds/Sec
Horsepower	1.014	Horsepower (metric)
Horsepower	10.70	Kg Calories/Min
Horsepower	0.7458	Kilowatts
Horsepower	745.7	Watts
Horsepower (Boiler)	33,520	Btu/Hr

Horsepower (Boiler)	9.804	Kilowatts
Horsepower, Electrical	1.0004	Horsepower
Horsepower (Metric)	0.98632	Horsepower
Horsepower Hours	2547	Btu
Horsepower Hours	1.98×10^6	Foot Pounds
Horsepower Hours	2.684×10^6	Joules
Horsepower Hours	641.7	Kilogram Calories
Horsepower Hours	2.737×10^5	Kilogram Meters
Horsepower Hours	0.7457	Kilowatt Hours
Hours	60	Minutes
Hours	3600	Seconds
Inches	2.540	Centimeters
Inches	10^3	Mils
Inches of Mercury	0.03342	Atmospheres
Inches of Mercury	1.133	Feet of Water
Inches of Mercury	0.0345	Kg/Sq Cm
Inches of Mercury	345.3	Kg/Sq Meter
Inches of Mercury	25.40	Mm of Mercury
Inches of Mercury	70.73	Lb/Sq Ft
Inches of Mercury	0.4912	Lb/Sq In.
Inches of Water	0.002458	Atmospheres
Inches of Water	0.07355	Inches of Mercury
Inches of Water	25.40	Kg/Sq Meter
Inches of Water	0.5781	Oz/Sq In.
Inches of Water	5.204	Lb/Sq Ft
Inches of Water	0.03613	Lb/Sq In.
Kilograms	980,665	Dynes
Kilograms	10^3	Grams
Kilograms	70.93	Poundals
Kilograms	2.2046	Pounds
Kilograms	1.102×10^{-3}	Tons (Short)
Kilogram Calories	3.968	Btu
Kilogram Calories	3086	Foot Pounds
Kilogram Calories	1.558×10^{-3}	Horsepower Hours
Kilogram Calories	426.6	Kilogram Meters
Kilogram Calories	1.162×10^{-3}	Kilowatt Hours
Kg Cal/Min	51.43	Foot Pounds/Sec
Kg Cal/Min	0.09351	Horsepower
Kg Cal/Min	0.06972	Kilowatts
Kg Cm^2	2.373×10^{-3}	Lb Ft^2
Kg Cm^2	0.3417	Lb In.2
Kilogram Meters	9.302×10^{-3}	Btu
Kilogram Meters	9.807×10^7	Ergs
Kilogram Meters	7.233	Foot Pounds
Kilogram Meters	3.6529×10^{-6}	Horsepower Hours
Kilogram Meters	9.579×10^{-6}	Lb Water Evap. at 212 F

Kilogram Meters	9.807	Joules
Kilogram Meters	2.344×10^{-3}	Kilogram Calories
Kilogram Meters	2.724×10^{-6}	Kilowatt Hours
Kg/Cu Meter	10^{-3}	Grams/Cu Cm
Kg/Cu Meter	0.06243	Lb/Cu Ft
Kg/Cu Meter	3.613×10^{-5}	Lb/Cu In.
Kg/Cu Meter	3.405×10^{-10}	Lb/Mil Ft
Kg/Meter	0.6720	Lb/Ft
Kg/Sq Cm	28.96	In. of Mercury
Kg/Sq Cm	735.56	Mm of Mercury
Kg/Sq Cm	14.22	Lb/Sq In.
Kg/Sq Meter	9.678×10^{-5}	Atmospheres
Kg/Sq Meter	3.281×10^{-3}	Feet of Water
Kg/Sq Meter	2.896×10^{-3}	Inches of Mercury
Kg/Sq Meter	0.07356	Mm of Mercury at 0 C
Kg/Sq Meter	0.2048	Lb/Sq Ft
Kg/Sq Meter	1.422×10^{-3}	Lb/Sq In.
Kg/Sq Mm	10^{6}	Kg/Sq Meter
Kiloliters	10^{3}	Liters
Kilometers	10^{5}	Centimeters
Kilometers	3281	Feet
Kilometers	10^{3}	Meters
Kilometers	0.6214	Miles
Kilometers	1093.6	Yards
Kilometers/Hr	27.78	Cm/Sec
Kilometers/Hr	54.68	Ft/Min
Kilometers/Hr	0.9113	Ft/Sec
Kilometers/Hr	0.5396	Knots/Hr
Kilometers/Hr	16.67	Meters/Min
Kilometers/Hr	0.6214	Miles/Hr
Km/Hr/Sec	27.78	Cm/Sec/Sec
Km/Hr/Sec	0.9113	Ft/Sec/Sec
Km/Hr/Sec	0.2778	Meters/Sec/Sec
Km/Hr/Sec	0.6214	Miles/Hr/Sec
Kilometers/Min	60	Kilometers/Hr
Kilowatts	56.92	Btu/Min
Kilowatts	4.425×10^{4}	Foot Pounds/Min
Kilowatts	737.6	Foot Pounds/Sec
Kilowatts	1.341	Horsepower
Kilowatts	14.34	Kg Calories/Min
Kilowatts	10^{3}	Watts
Kilowatt Hours	3415	Btu
Kilowatt Hours	2.655×10^{6}	Foot Pounds
Kilowatt Hours	1.341	Horsepower Hours
Liters	10^{3}	Cu Cm
Liters	0.03531	Cu Ft

Liters	61.02	Cu In.
Liters	10^{-3}	Cu Meters
Liters	1.308×10^{-3}	Cu Yards
Liters	0.2642	Gallons
Liters	2.113	Pints (Liq.)
Liters	1.057	Quarts (Liq.)
Liters/Min	5.885×10^{-4}	Cu Ft/Sec
Liters/Min	4.403×10^{-3}	Gal/Sec
$Log_{10}N$	2.303	$Log_E N$ or Ln N
Log N or Ln N	0.4343	$Log_{10} N$
Meters	100	Centimeters
Meters	3.2808	Feet
Meters	39.37	Inches
Meters	10^{-3}	Kilometers
Meters	10^3	Millimeters
Meters	1.0936	Yards
Meters	10^{10}	Angstrom Units
Meters	6.2137×10^4	Miles
Meter Kilograms	9.807×10^7	Centimeter Dynes
Meter Kilograms	10^5	Centimeter Grams
Meter Kilograms	7.233	Pound Feet
Meters/Min	1.667	Cm/Sec
Meters/Min	3.281	Ft/Min
Meters/Min	0.05468	Ft/Sec
Meters/Min	0.06	Kilometers/Hr
Meters/Min	0.03728	Miles/Hr
Meters/Sec	196.8	Ft/Min
Meters/Sec	3.281	Ft/Sec
Meters/Sec	3.6	Km/Hr
Meters/Sec	0.06	Km/Min
Meters/Sec	2.237	Miles/Hr
Meters/Sec	0.03728	Miles/Min
Meters/Sec/Sec	3.281	Ft/Sec/Sec
Meters/Sec/Sec	3.6	Km/Hr/Sec
Meters/Sec/Sec	2.237	Miles/Hr/Sec
Micrograms	10^{-6}	Grams
Microliters	10^{-6}	Liters
Microns	10^{-6}	Meters
Miles	1.609×10^5	Centimeters
Miles	5280	Feet
Miles	1.6093	Kilometers
Miles	1760	Yards
Miles (Int. Nautical)	1.852	Kilometers
Miles/Hr	44.70	Cm/Sec
Miles/Hr	88	Ft/Min
Miles/Hr	1.467	Ft/Sec

Miles/Hr	1.6093	Km/Hr
Miles/Hr	26.82	Meters/Min
Miles/Hr/Sec	44.70	Cm/Sec/Sec
Miles/Hr/Sec	1.467	Ft/Sec/Sec
Miles/Hr/Sec	1.6093	Km/Hr/Sec
Miles/Hr/Sec	0.4470	Meters/Sec/Sec
Miles/Min	2682	Cm/Sec
Miles/Min	88	Ft/Sec
Miles/Min	1.6093	Km/Min
Miles/Min	60	Miles/Hr
Milliers	10^3	Kilograms
Milligrams	10^{-3}	Grams
Milliliters	10^{-3}	Liters
Millimeters	0.1	Centimeters
Millimeters	0.03937	Inches
Millimeters	39.37	Mils
Mm of Mercury	0.0394	In. of Mercury
Mm of Mercury	1.3595^{-3}	Kg/Sq Cm
Mm of Mercury	0.01934	Lb/Sq In.
Mils	0.002540	Centimeters
Mils	10^{-3}	Inches
Mils	25.40	Microns
Minutes (Angle)	2.909×10^{-4}	Radians
Minutes (Angle)	60	Seconds (Angle)
Months	30.42	Days
Months	730	Hours
Months	43,800	Minutes
Months	2.628×10^6	Seconds
Myriagrams	10	Kilograms
Myriameters	10	Kilometers
Myriawatts	10	Kilowatts
Ounces	16	Drams
Ounces	437.5	Grains
Ounces	28.35	Grams
Ounces	0.0625	Pounds
Ounces (Fluid)	1.805	Cu In.
Ounces (Fluid)	0.02957	Liters
Ounces (U.S. Fluid)	29.5737	Cu Cm
Ounces (U.S. Fluid)	1/128	Gallons (U.S.)
Ounces (Troy)	480	Grains (Troy)
Ounces (Troy)	31.10	Grams
Ounces (Troy)	20	Pennyweights (Troy)
Ounces (Troy)	0.08333	Pounds (Troy)
Ounces/Sq In.	0.0625	Lb/Sq In.
Parts/Million	0.0584	Grains/U.S. Gal
Parts/Million	0.7016	Grains/Imp. Gal

Parts/Million	8.345	Lb/Million Gal
Pennyweights (Troy)	24	Grains (Troy)
Pennyweights (Troy)	1.555	Grams
Pennyweights (Troy)	0.05	Ounces (Troy)
Pints (Dry)	33.60	Cu In.
Pints (Liq.)	28.87	Cu In.
Pints (U.S. Liquid)	473.179	Cu Cm
Pints (U.S. Liquid)	16	Ounces (U.S. Fluid)
Poundals	13,826	Dynes
Poundals	14.10	Grams
Poundals	0.03108	Pounds
Pounds	444,823	Dynes
Pounds	7000	Grains
Pounds	453.6	Grams
Pounds	16	Ounces
Pounds	32.17	Poundals
Pounds (Troy)	0.8229	Pounds (Av.)
Pounds (Troy)	373.2418	Grams
Pounds of Carbon to CO^2	14,544	Btu (mean)
Pound Feet (Torque)	1.3558×10^7	Dyne Centimeters
Pound Feet	1.356×10^7	Centimeters Dynes
Pound Feet	13,825	Centimeter Grams
Pound Feet	0.1383	Meter Kilograms
Pounds Feet Squared	421.3	$Kg\ Cm^2$
Pounds Feet Squared	144	$Lb\ In.^2$
Pounds Inches Squared	2,926	$Kg\ Cm^2$
Pounds Inches Squared	6.945×10^{-3}	$Lb\ Ft^2$
Pounds of Water	0.01602	Cu Ft
Pounds of Water	27.68	Cu In.
Pounds of Water	0.1198	Gallons
Pounds of Water Evaporated at 212 F	970.3	Btu
Pounds of Water per Min	2.699×10^{-4}	Cu Ft/Sec
Pounds/Cu Ft	0.01602	Grams/Cu Cm
Pounds/Cu Ft	16.02	Kg/Cu Meter
Pounds/Cu Ft	5.787×10^{-4}	Lb/Cu In.
Pounds/Cu In.	5.456×10^{-9}	Lb/Mil Ft
Pounds/Cu In.	27.68	Grams/Cu Cm
Pounds/Cu In.	2.768×10^4	Kg/Cu Meter
Pounds/Cu In.	1728	Lb/Cu Ft
Pounds/Cu In.	9.425×10^{-6}	Lb/Mil Ft
Pounds/Ft	1.488	Kg/Meter
Pounds/In.	178.6	Grams/Cm
Pounds/Sq Ft	0.01602	Feet of Water
Pounds/Sq Ft	4.882	Kg/Sq Meter
Pounds/Sq Ft	6.944×10^{-3}	Lb/Sq In.

Pounds/Sq In.	0.06804	Atmospheres
Pounds/Sq In.	2.307	Feet of Water
Pounds/Sq In.	2.036	Inches of Mercury
Pounds/Sq In.	0.0703	Kg/Sq Cm
Pounds/Sq In.	703.1	Kg/Sq Meter
Pounds/Sq In.	144	Lb/Sq Ft
Pounds/Sq In.	70.307	Grams/Sq Cm
Pounds/Sq In.	51.715	Mm of Mercury at 0 C
Quadrants (Angle)	90	Degrees
Quadrants (Angle)	5400	Minutes
Quadrants (Angle)	1.571	Radians
Quarts (Dry)	67.20	Cu In.
Quarts (Liq.)	57.75	Cu In.
Quarts (U.S. Liquid)	0.033420	Cu Ft
Quarts (U.S. Liquid)	32	Ounces (U.S. Fluid)
Quarts (U.S. Liquid)	0.832674	Quarts (British)
Radians	57.30	Degrees
Radians	3438	Minutes
Radians	0.637	Quadrants
Radians/Sec	57.30	Degrees/Sec
Radians/Sec	0.1592	Revolutions/Sec
Radians/Sec	9.549	Revolutions/Min
Radians/Sec/Sec	573.0	Rev/Min/Min
Radians/Sec/Sec	9.549	Rev/Min/Sec
Radians/Sec/Sec	0.1592	Rev/Sec/Sec
Revolutions	360	Degrees
Revolutions	4	Quadrants
Revolutions	6.283	Radians
Revolutions/Min	6	Degrees/Sec
Revolutions/Min	0.1047	Radians/Sec
Revolutions/Min	0.01667	Revolutions/Sec
Rev/Min/Min	1.745×10^{-3}	Rad/Sec/Sec
Rev/Min/Min	0.01667	Rev/Min/Sec
Rev/Min/Min	2.778×10^{-4}	Rev/Sec/Sec
Revolutions/Sec	360	Degrees/Sec
Revolutions/Sec	6.283	Radians/Sec
Revolutions/Sec	60	Rev/Min
Rev/Sec/Sec	6.283	Rad/Sec/Sec
Rev/Sec/Sec	3600	Rev/Min/Min
Rev/Sec/Sec	60	Rev/Min/Sec
Seconds (Angle)	4.848×10^{-6}	Radians
Spheres (Solid Angle)	12.57	Steradians
Spherical Right Angles	0.25	Hemispheres
Spherical Right Angles	0.125	Spheres
Spherical Right Angles	1.571	Steradians
Sq Cm	1.973×10^{5}	Circular Mils

Sq Cm	1.076×10^{-3}	Sq Ft
Sq Cm	0.1550	Sq In.
Sq Cm	10^{-6}	Sq Meters
Sq Cm	100	Sq Mm
Sq Cm - Cm Squared	0.02420	Sq In. - In. Squared
Sq Ft	2.296×10^{-5}	Acres
Sq Ft	929.0	Sq Cm
Sq Ft	144	Sq In.
Sq Ft	0.09290	Sq Meters
Sq Ft	3.587×10^{-8}	Sq Miles
Sq Ft	1/9	Sq Yards
Sq Ft - Ft Squared	2.074×10^{4}	Sq In. - In. Squared
Sq In.	1.273×10^{6}	Circular Mils
Sq In.	6.452	Sq Cm
Sq In.	6.944×10^{-3}	Sq Ft
Sq In.	10^{6}	Sq Mils
Sq In.	645.2	Sq Mm
Sq In. (U.S.)	7.71605×10^{-4}	Sq Yards
Sq In. - In. Squared	41.62	Sq Cm - Cm Squared
Sq Kilometers	247.1	Acres
Sq Kilometers	10.76×10^{6}	Sq Ft
Sq Kilometers	10^{6}	Sq Meters
Sq Kilometers	0.3861	Sq Miles
Sq Kilometers	1.196×10^{6}	Sq Yards
Sq Meters	2.471×10^{-4}	Acres
Sq Meters	10.764	Sq Ft
Sq Meters	3.861×10^{-7}	Sq Miles
Sq Meters	1.196	Sq Yards
Sq Miles	640	Acres
Sq Miles	27.88×10^{6}	Sq Ft
Sq Miles	2.590	Sq Kilometers
Sq Miles	3.098×10^{6}	Sq Yards
Sq Millimeters	1.973×10^{3}	Circular Mils
Sq Millimeters	0.01	Sq Cm
Sq Millimeters	1.550×10^{-3}	Sq In.
Sq Mils	1.273	Circular Mils
Sq Mils	6.452×10^{-6}	Sq Cm
Sq Mils	10^{-6}	Sq In.
Sq Yards	2.066×10^{-4}	Acres
Sq Yards	9	Sq Ft
Sq Yards	0.8361	Sq Meters
Sq Yards	3.228×10^{-7}	Sq Miles
Temp (Deg C) + 273	1	Abs. Temp (Deg C)
Temp (Deg C) + 17.8	1.8	Temp (Deg F)
Temp (Deg F) + 460	1	Abs. Temp (Deg F)
Temp (Deg F) - 32	5/9	Temp (Deg C)

Tons (Long)	1016	Kilograms
Tons (Long)	2240	Pounds
Tons (Metric)	10^3	Kilograms
Tons (Metric)	2205	Pounds
Tons (Short)	907.2	Kilograms
Tons (Short)	2000	Pounds
Tons (Short)/Sq Ft	9765	Kg/Sq Meter
Tons (Short)/Sq Ft	13.89	Lb/Sq In.
Tons (Short)/Sq In.	1.406×10^6	Kg/Sq Meter
Tons (Short)/Sq In.	2000	Lb/Sq In.
Watts	0.05692	Btu/Min
Watts	10^7	Ergs/Sec
Watts	44.26	Foot Pounds/Min
Watts	0.7376	Foot Pounds/Sec
Watts	1.341×10^{-3}	Horsepower
Watts	0.01434	Kg Calories/Min
Watts	10^{-3}	Kilowatts
Watt Hours	3.415	Btu
Watt Hours	2655	Foot Pounds
Watt Hours	1.341×10^{-3}	Horsepower Hours
Watt Hours	0.8605	Kilogram Calories
Watt Hours	367.1	Kilogram Meters
Watt Hours	10^{-3}	Kilowatt Hours
Weeks	168	Hours
Weeks	10,080	Minutes
Weeks	604,800	Seconds
Yards	91.44	Centimeters
Yards	3	Feet
Yards	36	Inches
Yards	0.9144	Meters
Years (Common)	365	Days
Years (Common)	8760	Hours

BIBLIOGRAPHY

1. Agne, T. D. "Techniques for Reducing Machinery Noise," *Pollution Engineering,* Vol. 5, No. 4, April 1973.
2. Allen, L. L. "Aerodynamic Control of Valve Noise," *Pollution Engineering,* Vol. 5, No. 10, October 1973.
3. Anderson, C. E. "Chemical Control of Odors," *Pollution Engineering,* Vol. 4, No. 5, August 1972.
4. Beavon, D. K. "Abating Sulfur Plant Tail Gases," *Pollution Engineering,* Vol. 4, No. 1, January/February 1972.
5. Beckman, J. A. "Destructive Distillation of Used Tires," *Pollution Engineering,* Vol. 2, No. 5, November/December 1970.
6. Benkovich, J. F. "Dewatering Screens in Pollution Control," *Pollution Engineering,* Vol. 6, No. 5, May 1974.
7. Betz, G. M. "Environmental Law-Legal Actions to Noise Control," *Pollution Engineering,* Vol. 3, No. 4, July/August 1971.
8. Betz, G. M. "Legal Aspects of Solid Waste Management," *Pollution Engineering,* Vol. 4, No. 3, May/June 1972.
9. Billings, R. W. "FRP for Cooling Towers," *Pollution Engineering,* Vol. 5, No. 10, July 1972.
10. Bishop, C. "Silencers for Reciprocating Engine Exhaust," *Pollution Engineering,* Vol. 6, No. 2, February 1974.
11. Bochinski, J. "Measuring Oxygen Demand? Take Your Pick—TC, BOD, COD, TOC, and TOD," *Pollution Engineering,* Vol. 5, No. 1, January 1973.
12. Bollinger, G. A. "Ground Borne Vibrations," *Pollution Engineering,* Vol. 5, No. 4, April 1973.
13. Bonsard, J. A. "Odor Control of Paper Recycling," *Pollution Engineering,* Vol. 4, No. 5, August 1972.
14. Bowes, D. H. "Magnets Control Many Problems," *Pollution Engineering,* Vol. 4, No. 1, January/February 1972.
15. Brink, J. A. and C. N. Dougald. "Vinyl Plastic Mist Removal," *Pollution Engineering,* Vol. 5, No. 11, November 1973.
16. Busch, J. S. "Design & Cost of High Energy Scrubbers, Part I—The Basic Scrubber," *Pollution Engineering,* Vol. 5, No. 1, January 1973.

17. Busch, J. S. "Design & Cost of High Energy Scrubbers, Part II— Ancillary Equipment," *Pollution Engineering,* Vol. 5, No. 2, February 1973.
18. Busch, J. S. "Design & Cost of High Energy Scrubbers, Part III—The Internal Gas Cooler," *Pollution Engineering,* Vol. 5, No. 3, March 1973.
19. Casciato, A. C. "Hydromatic Control of Valve Noise," *Pollution Engineering,* Vol. 5, No. 9, September 1973.
20. Celenza, G. J. "Controlling Air Pollution From Foundries," *Pollution Engineering,* Vol. 2, No. 1, March/April 1970.
21. Chapman, R. L. "Instrumentation for Stack Monitoring," *Pollution Engineering,* Vol. 4, No. 6, September 1972.
22. Cheremisinoff, P. N. "Broadening the Structure of Plant Safety Through Pollution Control," International Pollution Engineering Congress, October 1973.
23. Cheremisinoff, P. N. "Creating a Corporate Pollution Engineering Operation," *Pollution Engineering,* Vol. 5, No. 3, March 1973.
24. Cheremisinoff, P. N. "Establishing a Central Corporate Department for Company Wide Pollution Control," First International Pollution Engineering Conference, December 4-6, 1972.
25. Cheremisinoff, P. N. "Industrial Hearing Test Programs," *Pollution Engineering,* Vol. 6, No. 5, May 1974.
26. Cheremisinoff, P. N. "Measurement of Pollution in Air and Water," Design Engineering Conference, April 10, 1973.
27. Cheremisinoff, P. N. "Noise, Laws, Control and the Engineer," *Pollution Engineering,* Vol. 3, No. 3, May/June 1971.
28. Cheremisinoff, P. N. "Ocean Dumping of Wastes."
29. Cheremisinoff, P. N. "Personal Hearing Protection Devices," *Pollution Engineering,* Vol. 4, No. 9, December 1972.
30. Cheremisinoff, P. N. "Plant Noise Survey Techniques," *Pollution Engineering,* Vol. 4, No. 3, May/June 1972.
31. Cheremisinoff, P. N. "Pollution Control Interrelationships," 24th National Plant Engineering & Maintenance Conference, March 12-15, 1973.
32. Cheremisinoff, P. N. "Pollution Engineering Noise Glossary," *Pollution Engineering,* Vol. 5, No. 5, May 1973.
33. Cheremisinoff, P. N. "Respiratory Protective Equipment," *Pollution Engineering,* Vol. 5, No. 1, January 1973.
34. Cheremisinoff, P. N. "Sizing Roof Ventilators," *Plant Engineering,* Vol. 27, No. 19, September 20, 1973.
35. Cheremisinoff, P. N. and N. P. Cheremisinoff. "Cyclones," *Pollution Engineering,* accepted for publication.
36. Cheremisinoff, P. N. and N. P. Cheremisinoff. "Electrostatic Precipitators, Part I—Basic Design and Components," *Plant Engineering,* Vol. 28, No. 9, May 2, 1974.

37. Cheremisinoff, P. N. and N. P. Cheremisinoff. "Electrostatic Precipitators, Part II–Properties of Ducts, Fumes and Mists," *Plant Engineering,* Vol. 28, No. 11, May 30, 1974.

38. Cheremisinoff, P. N. and N. P. Cheremisinoff. "Fabric Filters for Dust Collection, Part I–What They Are and How They Work," *Plant Engineering,* Vol. 27, No. 11, May 31, 1973.

39. Cheremisinoff, P. N. and N. P. Cheremisinoff. "Fabric Filters for Dust Collection, Part II–Types of Fabrics," *Plant Engineering,* Vol. 27, No. 12, June 14, 1973.

40. Cheremisinoff, P. N. and N. P. Cheremisinoff. "Fabric Filters for Dust Collection, Part III–Variables Affecting Efficiency," *Plant Engineering,* Vol. 27, No. 13, June 28, 1973.

41. Cheremisinoff, P. N. and N. P. Cheremisinoff. "Fabric Filters for Dust Collection, Part IV–Design and Operation," *Plant Engineering,* Vol. 27, No. 15, July 26, 1973.

42. Cheremisinoff, P. N., N. P. Cheremisinoff, and K. B. Rao. "Understanding Packed Tower Wet Scrubbers," *Plant Engineering,* Vol. 27, No. 23, November 15, 1973.

43. Cheremisinoff, P. N., I. Fedeli, and N. P. Cheremisinoff. "Corrosion Resistance of Piping and Construction Materials," *Pollution Engineering,* Vol. 5, No. 8, August 1973.

44. Cheremisinoff, P. N. and Y. H. Habib. "Converting Contamination Concentration in Waste Discharge to Total Pounds," *Pollution Engineering,* Vol. 6, No. 2, February 1974.

45. Cheremisinoff, P. N. and Y. H. Habib. "The Pert Way of Planning Pollution Control Projects," *Plant Engineering,* Vol. 27, No. 5, March 8, 1973.

46. Cheremisinoff, P. N. and K. Rao. "Fluidized Bed Reactors," *Pollution Engineering,* Vol. 4, No. 6, September 1972.

47. Cheremisinoff, P. N. and R. A. Young. "Automobile and Traffic Noise."

48. Cheremisinoff, P. N. and R. A. Young. "Fans and Blowers," *Pollution Engineering,* Vol. 6, No. 7, July 1974.

49. Cheremisinoff, P. N. and R. A. Young. "Industrial Solid Waste Handling and Disposal," *Pollution Engineering,* Vol. 6, No. 6, June 1974.

50. Chiagouris, G. L. "Analyzing the Cost of Solid Waste Disposal," *Plant Engineering,* Vol. 26, No. 6, March 23, 1972.

51. Chiagouris, G. L. "Common Industrial Practices for Solid Waste," *Plant Engineering,* Vol. 25, No. 1, January 13, 1972.

52. Constance, J. D. "Simplified Method for Determining Inhalable Contaminants," *Pollution Engineering,* Vol. 4, No. 4, July 1972.

53. Corliss, E. L. R. "Computers–A White Collar Hazard," *Pollution Engineering,* Vol. 4, No. 1, January/February 1972.

54. Crawford, G. N. "Solid Waste Incinerators Operation and Design," *Plant Engineering,* Vol. 24, No. 7, April 2, 1970.

55. Crocker, B. B. "Minimizing Air Pollution Control Costs in Older Plants," *Pollution Engineering,* Vol. 1, No. 1, October/November 1969.
56. Crocker, J. D. "Designing A Mechanical Aeration System," *Pollution Engineering,* Vol. 4, No. 6, September 1972.
57. Crocker, J. D. "Economical Dynamic Aeration," *Pollution Engineering,* Vol. 5, No. 6, June 1973.
58. Cross, F. L. "Assessing Noise Impact on the Environment," *Pollution Engineering,* Vol. 5, No. 11, November 1973.
59. Cross, F. L. "Baghouse Filtration of Air Pollutants," *Pollution Engineering,* Vol. 6, No. 2, February 1974.
60. Cross, F. L. "Compactor Sizing and Selection," *Pollution Engineering,* Vol. 5, No. 7, July 1973.
61. Cross, F. L. "Environmental Factors in Plant Site Selection," *Plant Engineering,* Vol. 26, No. 14, July 13, 1972.
62. Cross, F. L. "How to Make Pollution Control Decisions," *Pollution Engineering,* Vol. 3, No. 3, May/June 1971.
63. Cross, F. L. "Introduction to Preparation of Environmental Impact Statements," *Pollution Engineering,* Vol. 5, No. 3, March 1973.
64. Cross, F. L. "Planning Incineration Without Air Pollution," *Pollution Engineering,* Vol. 4, No. 4, July 1972.
65. Damiani, A. S. "Proper Construction Controls Office Noise," *Pollution Engineering,* Vol. 2, No. 5, November/December 1970.
66. Day, R. V. "Cyanide Planting Waste Treatment," *Pollution Engineering,* Vol. 1, No. 1, October/November 1969.
67. Dear, T. A. "Calculating OSHA Noise Compliance," *Pollution Engineering,* Vol. 5, No. 1, January 1973.
68. Dloughy, P. E. "Pretreatment and Filtration of Sludge," *Pollution Engineering,* Vol. 2, No. 4, September/October 1970.
69. Duszynski, E. J. "A Case for Milling Refuse," *Pollution Engineering,* Vol. 3, No. 3, May/June 1971.
70. Eckel, A. "Hearing Conservation Compliance Guide," *Pollution Engineering,* Vol. 4, No. 4, July 1972.
71. Eckel, A. "Selecting an Audiometric Room," *Pollution Engineering,* Vol. 5, No. 2, February 1973.
72. Edelman, R. A. "Air Borne Radioactivity Releases from Nuclear Power Plants," Private Correspondence.
73. Elliot, R. K. "Economics of Solid Waste Compaction," *Pollution Engineering,* Vol. 6, No. 2, February 1974.
74. Ellwood, E. E. "Noise Control Effects of Rock Wool," *Pollution Engineering,* Vol. 5, No. 8, August 1973.
75. Ely, R. B. "Wet Air Oxidation," *Pollution Engineering,* Vol. 5, No. 5, May 1973.
76. Erickson, P. R. "Flocculation of Water and Waste," *Pollution Engineering,* Vol. 5, No. 3, March 1973.
77. Estes, J. E. and B. Golomb. "Oil Spills Can Be Measured," *Pollution Engineering,* Vol. 2, No. 5, November/December 1970.

78. Eweson, E. "Making Compositing Work," *Pollution Engineering,* Vol. 5, No. 6, June 1973.
79. Ferrel, J. F. "Sludge Incineration," *Pollution Engineering,* Vol. 5, No. 3, March 1973.
80. Floros, J. "Designing for Scrap Steel Magnetic Removal," *Pollution Engineering,* Vol. 4, No. 7, October 1972.
81. Floyd, J. D. "Home, Sweet Treatment Plant," *Pollution Engineering,* Vol. 4, No. 2, March/April 1972.
82. Franklin, D. M. "Those Terrible Teepees Can Be Effective," *Pollution Engineering,* Vol. 3, No. 1, January/February 1971.
83. Geiver, R. M. "All About Tape Samplers," *Pollution Engineering,* Vol. 5, No. 2, February 1973.
84. Gilbert, W. "Increasing Venturi Scrubber Efficiency with Steam Jet Ejectors," *Plant Engineering,* Vol. 26, No. 18, September 7, 1972.
85. Gilbert, W. "Selecting Materials for Wet Scrubbing Systems," *Pollution Engineering,* Vol. 5, No. 8, August 1973.
86. Haak, M. P. "Recover Solids by Spray Drying," *Pollution Engineering,* Vol. 3, No. 4, July/August 1971.
87. Hankel, K. M. "Buying Guidelines for Noise Control," *Pollution Engineering,* Vol. 3, No. 1, January/February 1971.
88. Hankel, K. M. "Sound Transmission and Absorption," *Pollution Engineering,* Vol. 2, No. 1, March/April 1970.
89. Hemsath, K. H. and A. C. Thekdi. "Rich Fume Incineration," *Pollution Engineering,* Vol. 5, No. 7, July 1973.
90. Herb, J. H. "Solid Waste Shredders—Applications and Limitations," *Pollution Engineering,* Vol. 3, No. 1, January/February 1971.
91. Hobbs, J. J., G. Medina, and A. Dillon. "Comparing Quality of Our Waters," *Pollution Engineering,* Vol. 5, No. 10, October 1973.
92. Hollander, H. I. and J. D. Lessile. "On-Site Waste Disposal and Energy Recovery," *Pollution Engineering,* Vol. 6, No. 1, January 1974.
93. Holmer, C. "Noise and Its Measurement," *Pollution Engineering,* Vol. 1, No. 1, October/November 1969.
94. Hurlburt, R. L. "Small Town Monitors Big Airport Noise," *Pollution Engineering,* Vol. 4, No. 5, August 1972.
95. Jackson, E. V. "Deep Well Disposal of Liquid Waste," *Pollution Engineering,* Vol. 1, No. 1, October/November 1969.
96. Jones, R. H. "Analyzer Monitors Treatment Efficiency Detects Plant Upsets," *Pollution Engineering,* Vol. 4, No. 8, November 1972.
97. Jones, R. H. "pH Monitoring of Phosphate Removal," *Pollution Engineering,* Vol. 4, No. 5, August 1972.
98. Johnson, D. S. "Mobile Mulcher Chews," *Pollution Engineering,* Vol. 2, No. 1, March/April 1971.
99. Johnston, D. A. "High Density Landfilling with a Mole," *Pollution Engineering,* Vol. 2, No. 3, July/August 1970.
100. Karlson, E. L. "Ozone—Friend or Foe?" *Pollution Engineering,* Vol. 4, No. 3, May/June 1972.

101. Kormanik, R. A. "Aerated Lagoons An Economical Solution," *Pollution Engineering,* Vol. 4, No. 1, January/February 1972.
102. Kormanik, R. A. "Dissolved Air Flotation for Treating Industrial Wastes," *Pollution Engineering,* Vol. 3, No. 4, July/August 1971.
103. Kulwiec, R. and W. Arrot. "PR and the Pollution Engineer," *Pollution Engineering,* Vol. 3, No. 5, September/October 1971.
104. Laberis, S. "Compressed Air Sub-Surface Aeration," *Pollution Engineering,* Vol. 5, No. 4, April 1973.
105. Lauber, J. D. "Air Pollution Control of Aluminum and Copper Recycling Processes," *Pollution Engineering,* Vol. 5, No. 12, December 1973.
106. Lauren, O. B. "Odor Modification," *Pollution Engineering,* Vol. 5, No. 7, July 1973.
107. Lawson, R. "Disposal of Oily Wastes," *Pollution Engineering,* Vol. 1, No. 5, January/February 1970.
108. Lee, J. "Selecting Membrane Pond Liners," *Pollution Engineering,* Vol. 6, No. 1, January 1974.
109. Lin, Y. H. and J. R. Lawson. "Treatment of Oily and Metal-Containing Wastewater," *Pollution Engineering,* Vol. 5, No. 11, November 1973.
110. Lopata, J. R. "Modern Protective Coating Technology," *Pollution Engineering,* Vol. 5, No. 8, August 1973.
111. Maskell, R. "Plastic Pipe—Solutions to Old and New Sewer Problems," *Pollution Engineering,* Vol. 5, No. 11, November 1973.
112. Mazone, R. R. and D. W. Oakes. "Profitably Recycling Solvents From Process Systems," *Pollution Engineering,* Vol. 5, No. 10, October 1973.
113. McIndoe, R. W. "Diatomite Filter Aids," *Pollution Engineering,* Vol. 4, No. 2, March/April 1972.
114. McNeil, W. "Selecting and Sizing Cooling Towers," *Pollution Engineering,* Vol. 3, No. 4, July/August 1971.
115. Meffert, D. P., M. M. McEuen, and R. H. Gilbreath. "Stack Testing and Monitoring," *Pollution Engineering,* Vol. 6, No. 5, June 1973.
116. Meteer, C. L. "Workable Solutions to Common Machinery Noise Problems," *Pollution Engineering,* Vol. 6, No. 1, January 1974.
117. Mikovich, J. J. "What's Happening to Pyrolysis?" *Pollution Engineering,* Vol. 4, No. 2, March/April 1972.
118. Mirra, M. J. "Planning and Automating for Pollution Control," *Pollution Engineering,* Vol. 4, No. 6, September 1972.
119. Mohler, J. B. "Analyzing Atmospheric Corrosion," *Pollution Engineering,* Vol. 4, No. 7, October 1972.
120. Molnar, J. "Duct Weight Calculator," *Pollution Engineering,* Vol. 2, No. 5, November/December 1970.
121. Morne, J. G. "Ultraviolet Water Purification," *Pollution Engineering,* Vol. 5, No. 12, December 1973.

122. Mueller, J. "Cost Comparison for Burning Fumes and Odors," *Pollution Engineering*, Vol. 3, No. 6, November/December 1971.
123. Nolan, M. and A. Marshalla. "Incinerator Emissions: Units, Correction and Conversion," *Pollution Engineering*, Vol. 5, No. 5, May 1973.
124. Ovard, J. "Industrial Cooling Towers—A Use Profile," *Pollution Engineering*, Vol. 3, No. 3, May/June 1971.
125. Parlante, R. "Filtration of Suspended Solids," *Pollution Engineering*, Vol. 1, No. 2, January/February 1970.
126. Parlante, R. "Management of Ion Exchange Systems," *Pollution Engineering*, Vol. 2, No. 4, September/October 1970.
127. Patil, P. G. "Using Glass for Noise Reduction," *Pollution Engineering*, Vol. 3, No. 2, March/April 1971.
128. Pauletta, C. "Fume Incineration," *Pollution Engineering*, Vol. 2, No. 1, March/April 1970.
129. Peters, J. M. "Relating Flares to Air Quality," *Pollution Engineering*, Vol. 5, No. 10, October 1973.
130. Porter, J. J. "Stability and Removal of Commercial Dyes from Process Waste Water," *Pollution Engineering*, Vol. 5, No. 10, October 1973.
131. Potts, C. W. "Cooling Tower Repair and Maintenance," *Pollution Engineering*, Vol. 3, No. 5, September/October 1971.
132. Rabosky, J. G. "Disinfection of Water and Wastewater," *Pollution Engineering*, Vol. 4, No. 9, December 1972.
133. Rabosky, J. G. "Neutralizing Industrial Wastes," *Plant Engineering*, Vol. 26, No. 12, June 15, 1972.
134. Rawa, R. T. "SO_2 Control for Small Boilers," *Pollution Engineering*, Vol. 4, No. 1, January/February 1972.
135. Read, G. W. and F. M. Veater. "Corrosion Resistant Linings for Stacks and Chimneys," *Pollution Engineering*, Vol. 5, No. 8, August 1973.
136. Reigel, S. A. and C. D. Doyle. "Using the Psychrometric Chart," *Pollution Engineering*, Vol. 4, No. 2, March/April 1972.
137. Reilly, P. B. "Wastewater Treatment for Removal of Suspended Solids," *Plant Engineering*, Vol. 26, No. 10, May 18, 1972.
138. Rittmiller, L. A. and I. L. Wadehra. "Handling and Disposal of Chemicals," *Pollution Engineering*, Vol. 3, No. 1, January/February 1971.
139. Roberts, E. G. "Sprays Provide Instantaneous Cooling," *Pollution Engineering*, Vol. 4, No. 4, July 1972.
140. Rohr, F. W. "Suppressing Scrubber Steam Plume," *Pollution Engineering*, Vol. 1, No. 1, October/November 1969.
141. Rosen, K. M. "Practical Ducting Design, Part I—Basic Principles and Terminology," *Plant Engineering*, Vol. 26, No. 20, October 5, 1972.

142. Rosen, K. M. "Practical Ducting Design, Part II—Calculating Duct Losses," *Plant Engineering,* Vol. 26, No. 22, November 2, 1972.
143. Rouston, B. M. "Pointers on Selecting Hearing Protectors," *Pollution Engineering,* Vol. 5, No. 12, December 1973.
144. Rouston, H. "Control of Office Noise," *Pollution Engineering,* Vol. 5, No. 6, June 1973.
145. Schellinger, R. R. "Air Pollution Health Effects," *Plant Engineering,* Vol. 24, No. 4, February 19, 1970.
146. Schiller, B. P. "Removing Coatings with Fluid Bed Reactors," *Pollution Engineering,* Vol. 5, No. 2, February 1973.
147. Schmidt, F. W. "Economics of Incineration," *Pollution Engineering,* Vol. 5, No. 9, September 1973.
148. Schneider, C. "Trickling Filters—From Rocks to Plastic," *Pollution Engineering,* Vol. 4, No. 2, March/April 1972.
149. Schroering, J. B. "Reclaiming Paper Wastes," *Pollution Engineering,* Vol. 1, No. 2, January/February 1970.
150. Schroering, J. B. "Sanitary Landfilling—Applications and Limitations," *Pollution Engineering,* Vol. 1, No. 1, October/November 1969.
151. Schroering, J. B. "Solid Wastes Processing and Reclaiming," *Pollution Engineering,* Vol. 2, No. 1, March/April 1970.
152. Schultz, G. A. "Solid Waste Handling Conveyors," *Pollution Engineering,* Vol. 5, No. 4, May 1973.
153. Schwartz, S. M. "Emission Control of an Aluminum Bonding Oven," *Pollution Engineering,* Vol. 4, No. 4, July 1972.
154. Selldorff, J. T. "Ultrafiltration," *Pollution Engineering,* Vol. 2, No. 1, March/April 1970.
155. Simons, E. L. and W. F. Marx. "E.P.O. at G.E.—A Corporate Example," *Pollution Engineering,* Vol. 5, No. 2, February 1973.
156. Simpson, M. E. "Vacuum Sewage Transport and Collection," *Pollution Engineering,* Vol. 5, No. 9, September 1973.
157. Sisson, W. "Combining Sound Levels and Correcting for Background Noise," *Pollution Engineering,* Vol. 4, No. 2, March/April 1972.
158. Sisson, W. "Determining Wind Chill Factor," *Pollution Engineering,* Vol. 4, No. 9, December 1972.
159. Sisson, W. "Estimating Fly Ash Emissions," *Pollution Engineering,* Vol. 4, No. 1, January/February 1972.
160. Sisson, W. "Predicting Peak Noise Frequency of Vents to Atmosphere," *Pollution Engineering,* Vol. 5, No. 7, July 1973.
161. Sisson, W. "Velocity of Sound in Gases," *Pollution Engineering,* Vol. 5, No. 7, July 1973.
162. Sklarew, R. C. "NEXUS—A New Way to Measure," *Pollution Engineering,* Vol. 4, No. 9, December 1972.
163. Smith, M. F. "Controlling Oil Spills," *Pollution Engineering,* Vol. 2, No. 5, November/December 1970.
164. Smith, S. K. "Economics of Selecting Particulate Control Equipment," *Pollution Engineering,* Vol. 3, No. 1, January/February 1971.

165. Spatz, D. D. "Reclaiming Valuable Metal Wastes," *Pollution Engineering,* Vol. 4, No. 1, January/February 1972.
166. Steinberger, W. "Graphical Determination of Pollution Problems," *Pollution Engineering,* Vol. 4, No. 2, March/April 1972.
167. Stewart, J. "Pushbutton Air Pollution Control," *Pollution Engineering,* Vol. 2, No. 4, September/October 1970.
168. Stott, W. J. "Chemicals for Water Treatment," *Pollution Engineering,* Vol. 4, No. 7, October 1972.
169. Tate, R. W. "Spray Nozzles for Pollution Control," *Pollution Engineering,* Vol. 5, No. 4, April 1973.
170. Taylor, R. W. "Dispersed Air Flotation," *Pollution Engineering,* Vol. 5, No. 1, January 1973.
171. Thomaides, L. "Why Catalytic Incineration?" *Pollution Engineering,* Vol. 3, No. 3, May/June 1971.
172. Thomas, C. D. "Airborne Sensing for Landfill Site Evaluation," *Pollution Engineering,* Vol. 3, No. 6, November/December 1971.
173. Thompson, H. "Ozone, Antidote for Water Pollution," *Pollution Engineering,* Vol. 4, No. 3, May/June 1972.
174. Thompson, H. "Selecting Pumps for Minimum Scrubber Maintenance," *Pollution Engineering,* Vol. 4, No. 3, May/June 1972.
175. Thompson, P. J. "In-Plant Air Cleaning," *Pollution Engineering,* Vol. 5, No. 12, December 1973.
176. Tomany, J. P. "Wet Scrubbers—Applications and Limitations," *Pollution Engineering,* Vol. 1, No. 2, January/February 1970.
177. Train, R. E. "Technology and Pollution Engineering," Environmental Protection Agency.
178. Turk, A. "Odor Source Inventories," *Pollution Engineering,* Vol. 5, No. 10, August 1972.
179. Twitchell, S. B. "Understanding Industrial Water Treatment," *Pollution Engineering,* Vol. 3, No. 2, March/April 1971.
180. Viraraghavan, T. "Tube Settlers for Improved Sedimentation," *Pollution Engineering,* Vol. 5, No. 1, January 1973.
181. Vitez, B. "Improve Your Sampling Techniques," *Pollution Engineering,* Vol. 4, No. 6, September 1972.
182. Wales, R. O. "Simplified Duct Sizing," *Pollution Engineering,* Vol. 3, No. 1, January/February 1971.
183. Walker, W. H. "Soil Erosion—The Unmentioned Polluter," *Pollution Engineering,* Vol. 4, No. 2, March/April 1972.
184. West, P. "Guide to Getting Rid of Abandoned Cars," *Pollution Engineering,* Vol. 5, No. 1, January 1973.
185. Westrom, L. A. "OSHA Checklist," *Pollution Engineering,* Vol. 4, No. 6, September 1972.
186. Williams, E. T. "Freeforward for pH Control," *Pollution Engineering,* Vol. 3, No. 2, March/April 1971.
187. Wiremius, J. D. and S. L. Sloan. "Safely Handling Solid Waste," *Pollution Engineering,* Vol. 4, No. 9, December 1972.

188. Young, R. A. "Airlifting—A Positive Installation Alternative," *Pollution Engineering,* Vol. 6, No. 1, January 1974.
189. Young, R. A. "Combined Treatment Answers Two Problems," *Pollution Engineering,* Vol. 4, No. 4, July 1972.
190. Young, R. A. "Electric Incineration," *Pollution Engineering,* Vol. 5, No. 8, August 1973.
191. Young, R. A. "High Alloy Cost Incinerator Components," *Pollution Engineering,* Vol. 5, No. 8, August 1973.
192. Young, R. A. "Lead Loaded Fabrics Keep Outside Noise Out," *Pollution Engineering,* Vol. 2, No. 5, November/December 1970.
193. Young, R. A. "Magnets Salvage Scrap for Processing," *Pollution Engineering,* Vol. 5, No. 10, October 1973.
194. Young, R. A. "Pulping Process Prevents Pollution," *Pollution Engineering,* Vol. 5, No. 4, April 1973.
195. Zomkowski, B. M. and L. A. Rittmiller. "Comparison of Air and Water Pollution," *Pollution Engineering,* Vol. 3, No. 6, November/December 1971.
196. Carlson, E. L. "Ozone, Friend or Foe," *Pollution Engineering,* Vol. 4, No. 3, May/June 1972.
197. Cheremisinoff, P. N. "Aerators for Wastewater Treatment," *Pollution Engineering,* Vol. 6, No. 3, March 1974.
198. Walker, W. H. "Salt Piling—A Source of Water Supply Pollution," *Pollution Engineering,* Vol. 2, No. 3, July/August 1970.
199. Young, R. A. "Radio System Monitors Pump Stations," *Pollution Engineering,* Vol. 3, No. 6, November/December 1971.

INDEX

INDEX